BIOLOGY OF THE INVERTEBRATES

CLEVELAND P. HICKMAN

Department of Zoology, DePauw University,
Greencastle, Indiana

SECOND EDITION

With 1105 illustrations

THE C. V. MOSBY COMPANY

Saint Louis, 1973

SECOND EDITION

Previous edition copyrighted 1967

Printed in the United States of America

International Standard Book Number 0-8016-2170-4

Library of Congress Catalog Card Number 72-83970

Distributed in Great Britain by Henry Kimpton, London

TS/CB/B   9   8   7   6   5   4   3   2

TO FRANCES

who worked so faithfully with this book

Preface

This revised second edition has some major changes from the first edition. With the restrictions of space demanded by a one-semester invertebrate course and the immense amount of material to be covered, it became necessary to reorganize the material. In the first place, the various invertebrates have been grouped into parts according to common properties and characteristics. Each part has an introductory statement of the major characteristics, together with the phylogeny and the adaptive radiation of the animals within a group. Since invertebrates often present a heterogeneous collection, it may be necessary to explain why certain minor phyla are thrown together without any coherent basis for doing so except as a general convenience. The diversity of invertebrates is so tremendous and our lack of knowledge of relationships is so great that grouping according to functional homologies at various levels of biologic organization may be the only logical plan to follow.

The plan has been to suggest certain unifying principles so that the student may see the basis for such grouping that is here presented. Both structure and function must be presented to have meaningful significance, and so an attempt has been made to present a functional morphology of the animal groups. It is thus possible to comprehend how animals are fitted for the different ecologic niches in which they are found. All animals have certain adaptive restrictions and capacities, both morphologically and physiologically, that determine their position in the biosphere. An appreciation of evolutionary trends is necessary for an understanding of basic adaptive features.

To conserve space, classification is first given in skeleton form, except for some of the larger taxa. Details of the lower skeleton taxa are given in the general context of the work. Stress is also given to recent experimental evidence of vital and puzzling aspects of invertebrate morphology and function. Tables of fundamental properties, comparisons, and characteristics are presented freely in this edition so that the student can quickly grasp the significance of the fact presented.

References to investigations on invertebrates never cease to accumulate. Comprehensive reviews of the literature are becoming more available, as well as symposia on various groups of invertebrates. The great problem has been to select the most significant of these studies. However, since there are limits to what the mind of the student can comprehend in a few months' time, one of our main objectives is to stimulate the interest of the student. For a more thorough study of most of the groups of animals, specialized monographs and reviews can be explored.

Many new illustrations have been added to this edition. Several illustrations from the older edition have been replaced by better ones, and others have been borrowed from active investigators in the field of invertebrate study.

The illustrations used should give the student a better understanding than he could derive from the context description. Most of the drawings in this book were made by Mrs. Frances M. Hickman, but some were contributed by Mrs. Barbara Poor, Mrs. Barbara L. Pickard, and Clifford M. Hickman.

I am grateful to many individuals for suggestions and comments on this and the previous edition. First of all, I wish to thank the scores of teachers who made helpful criticisms of the various sections of the first edition and by whose suggestions I have been guided in this revision. Among those who have been most helpful in some way or other are Dr. W. Andrew, Dr. G. J. Bakus, Dr. G. O. Mackie, Dr. B. J. Kaston, Dr. P. E. Dehnel, Dr. J. D. Costlow, Jr., Dr. W. E. Martin, Dr. W. L. Shapeero, Dr. W. H. Johnson, and Dr. S. Crowell. The following kindly loaned me the original illustrations and drawings from their recently published investigations: Dr. J. H. Bushnell, Dr. A. L. Edgar, Dr. R. P. Higgins, Dr. A. M. McClary, Dr. G. R. Mullen, Dr. J. L. Oschman, Dr. J. J. Poluhowich, Dr. T. W. Porter, Dr. F. A. Pray, Dr. R. M. Sayre, and Dr. D. D. H. Stern. Credit for these and other borrowed material is given under the specific figures.

I also wish to state that I alone am responsible for all errors that may appear in the revision.

Cleveland P. Hickman

Contents

CONTENTS

Biology of the invertebrates

Part I □ Introduction to the invertebrates

Introduction to the invertebrates

An invertebrate (nonvertebrate) refers specifically to an animal without a backbone in contrast to the vertebrates, represented by fish, amphibians, reptiles, birds, and mammals. At present about 1.5 million species of animals have been named and some 8,000 to 10,000 new species are added to the list each year. With the exception of the protozoans (some 30,000 species) and the vertebrates (some 43,000 species) all other animals are invertebrates, comprising altogether about 95% of the animals in the animal kingdom. (Strictly speaking, the protozoans are also invertebrates, but a distinction is sometimes made.) If one were to consider the extinct species, the total number of species that have lived would be many times that of existing forms.

The invertebrates are grouped into about thirty phyla, the number of which varies among different zoologists. A phylum is somewhat difficult to define, but it may be considered a basic structural pattern of evolutionary descent from common ancestry, with its members more less homogeneous and bearing a unique combination of distinct characteristics. Although the members of a phylum may show many variations of external features, they are all constructed on the same general plan. Invertebrates include all the phyla, even some members of the phylum Chordata, which is made up mostly of the vertebrates. Although evolution of new species is a continuous process, the names of new species added annually to the list may result from the discovery of forms that have long been in existence but have only recently come to light. Estimates of the numbers of species are only approximations and vary enormously among zoologists.

The many types of animal life have had their origin in the variations to be found in any population unit. In every creature some minor differences exist that are not shared with others in its group. But whatever their differences, all animals fall into certain natural groups of similar organisms that are called species. All other taxa (genera, order, class, etc.) are ideal, arbitrary categories rather than natural ones. Localities naturally vary in the number of species and their abundance of life, but most regions when carefully investigated, will yield an amazing variety of forms. If the zoologist were to study the fossil record, he would find that a vast change has occurred in most animal forms from geologic age to geologic age. There are excellent reasons for believing, therefore, that variants have been arising from the beginning of life. Many phylogenetic stocks developed into highly specialized groups and then became extinct. Other groups have continued and flourished from the earliest times (Table 1). Many stocks have given rise to numerous species within a relatively short time; others that show few species have remained virtually unchanged for millions of years. An infinite number of species have waxed and waned in response to striking geologic and biologic alterations. The fossil record gives some insight into the structures of some of the early types, but for

Table 1. Geologic record of the invertebrates

Era	Period	Epoch	Time beginning in millions of years	Invertebrate history
	Quaternary	Recent	0.025	
		Pleistocene	0.6-1	
	Tertiary	Pliocene	12	Invertebrates similar to modern kinds
		Miocene	25	First records of uropygids
		Oligocene	34	First records of pseudoscorpions and Siphonaptera
		Eocene	55	First records of Testacea, isopterans, and Lepidoptera
		Paleocene	75	
	Cretaceous		130	First dinoflagellates; first records of trachyline jellyfish; cheilostome ectoprocts appear; first records of pterobranchs; ammonoids become extinct
	Jurassic		180	New foraminiferans appear; palpigrades appear; first records of Dermaptera, Diptera, and Hymenoptera; ammonites evolve rapidly
	Triassic		230	Most modern orders of insects; new hexacoralline corals; brachiopods declining; new articulate crinoids. Many types of invertebrates have died out by end of paleozoic; millepores and sea pens appear; fusulinids and foraminiferans disappear

Table 1. Geologic record of the invertebrates—cont'd

Era	Period	Epoch	Time beginning in millions of years	Invertebrate history
Paleozoic	Permian		260	Rise of modern orders of insects; trilobites become extinct; eurypterids become extinct
	Pennsylvanian*		310	Atremate brachiopods disappearing; first pulmonate snails
	Mississippian*		350	Some winged and some large insects appear; calcareous foraminiferans appear; first brittle star
	Devonian		400	First insects (apterygotes); cystoid echinoderms disappear; freshwater clams appear; spiders may have appeared; first isopods and decapods Invasion of land
	Silurian		425-430	Scorpions and millipedes appear; trilobites decline; branchiopods peak; first barnacles; coral reefs active; eurypterids peak Invasion of land by arthropods
	Ordovician		475	Early sponge pleospongia now extinct; coral reefs and certain corals appear; first crinoids; first scaphopods and pelecypods; first ectoprocts; eurypterids appear; ammonoids appear; gastropods appear
	Cambrian		550	Fossils found indicate considerable evolutionary diversification; invertebrate phyla and many classes established; ostracods; first arthropods (trilobites, xiphosurans, and branchiopods); some radiolarians and foraminiferans; nautiloids appear; Onychophora first appear
Proterozoic (Precambrian)			2,000	Impressions of some soft-bodied forms have been recovered on a few occasions (Burgess Shale, 1911; Ediacara Hills, 1947); sponge spicules, worm burrows, jellyfish, branchiopods Autotrophism established and oxidizing atmosphere
Archeozoic (Precambrian)			4,000-5,000	Reducing atmosphere; origin of life; heterotrophism established

*The Pennsylvania (Upper) and Mississippi (Lower) are often referred to as the Carboniferous period.

most forms, zoologists have only speculations to offer.

There seems to be no doubt that certain broad relationships exist among the various phyla. But the differentiation of the phyla, which must have occurred far earlier than the Cambrian geologic period of some 550 million years ago, has left few or no fossils to indicate primitive relationships. Certain homologies of structure and function appear to be evident, but tracings of exact phylogenies are largely nonexistent. Phylogeny of the different animal groups has been complicated by parallel evolution so that it has been difficult to tell whether animals that appear alike in structure and function do so because they have descended from common ancestors or because they have developed under similar circumstances. Certain criteria of homology have been advanced to distinguish between the two possibilities, such as the position and composition of homologous parts, and historic and embryonic record of the units being considered, but absolute criteria are obscure in numerous cases.

In a general survey of the structure and function of invertebrate groups, one finds the same basic systems appearing, although they may be associated with increasingly complex superstructures. A unifying principle of great importance is gradually emerging as we learn more about the biochemistry of life.

Of the thirty-odd phyla that occur in the animal kingdom, some have been far more successful than others in number of species, or importance in the evolutionary pattern of life. In classifying the phyla, it has been customary to arbitrarily assign major and minor rank to them. Some justification for such a designation will be evident when one considers the part the various phyla play in making up the various animal community populations.

THE BIOSPHERE OF INVERTEBRATES

The biosphere may be defined as that region of the earth where living matter is sustained by a unique structural and functional interrelationship with the environment. It includes that part of the earth's crust inhabited by plants and animals and embraces fresh and salt water, the terrestrial soil, the earth's surface, and the atmosphere. Three basic requirements seem to be present in the biosphere—liquid water, cyclic changes of energy, and interfaces between the states of matter where life can occur (G. E. Hutchinson). Within the environment the invertebrate derives those cycles of energy and materials necessary for its existence. Most species are adapted for a particular kind of environment, or a habitat, that imposes restrictions of size, food resources, unique niches, and other conditions of life. In some way or other every animal is affected by the physical and biotic factors of its constantly changing environment. The environment determines the conditions for life existence, and the organism likewise influences these conditions. All organisms are open systems that exchange materials and energy with their surroundings; they may be regarded as transient organizations that are built out of materials from their environment and return such materials in the form of excretory and decay products.

The biosphere may be divided into major subdivisions of hydrosphere, lithosphere, and atmosphere. The hydrosphere is represented by streams, rivers, ponds, oceans and wherever water is present; the lithosphere includes the crust of the earth or its rocky surface; and the atmosphere is the gaseous envelope surrounding the other two subdivisions. From each of these subdivisions the animal derives certain inorganic metabolites necessary for its existence. Not every part of the potential biosphere is suitable for life, for certain arid regions, high frozen mountain peaks, restricted toxic sea basins, and some other regions are not able to support life that undergoes active metabolism. Spores of bacteria and fungi may be found high in the atmosphere and at low terrestrial depths, but they have little significance for the invertebrate life we are studying.

All the thirty or more phyla probably originated in the sea, but not all phyla have successfully invaded the land surface or its freshwater habitats. However, perhaps 80% of all species of animals are found in terrestrial habitats because certain large groups such as insects have chiefly adapted to a land habitat. The major habitats are the marine (sea), fresh water, terrestrial, and other organisms (hosts for parasites). All these major habitats are subdivided into minor divisions and ecologic niches, for each of which the various invertebrates are more or less specifically adapted. The ecologic relationships of animals to their environment are very complex but are of the utmost importance in understanding animals.

The influence of environmental conditions is well shown by a comparison of freshwater life with that of marine invertebrates. Many phyla (echinoderms, brachiopods, ctenophores, kinorhynchs, priapulids, and some others) have never successfully colonized fresh water. Some other phyla (sponges, coelenterates, and entoprocts) are represented in fresh water, but by fewer species than in marine water. Gastrotrichs have about the same number of species in both fresh and marine waters. Only a few groups (e.g., rotifers and of course insects) are more plentiful both in number of species and individuals in fresh water. Oceans cover 71% of the earth's surface and form an almost continuous body of water from the earliest geologic periods; fresh water is restricted to isolated patches that were invaded by ancestors of those now living there.

R. W. Pennak (1963) has pointed out some of the ways freshwater forms differ from marine ones. Most freshwater invertebrates have lost the motile larval stages common to marine forms, have adopted new osmoregulatory mechanisms for hypotonic fresh water, and have produced structures resistant to desiccation. He concludes that the transition from marine to fresh water has been a slow, arduous task involving mutation, selection, and adaptation and that the whole process of transition is an evolutionary rarity.

The geotectonic factor

The geotectonic factor refers to the structure of the earth's crust, its composition, its distribution, and its changes. Changes in the earth's crust have had profound effects on the evolutionary pattern of animals. The earth's crust is characterized by two great, somewhat antagonistic forces—the elevating and leveling processes. These forces of elevation and leveling are continuous but do not always occur at the same intensity. Cycles of activity alternate with periods of minimum crustal disturbance. Life has had to adapt itself to the ecologic conditions brought about by the continual moving and contraction of the earth's crust.

One influence of the geotectonic factor is the breaking up of the earth's surface into extensive ecologic clines of widely variable physical conditions of climate, temperature, water, etc. This has been responsible for striking morphologic variations and has also afforded a logical explanation for speciation. Most of the existing species owe their origin to geographic isolation. This isolation is a direct outcome of topographic shifting and changes. Isolated biotic communities evolved, at first separately; then they often were reunited because physical barriers were broken down. Continual physical changes produced new biomes and therefore new opportunities for the evolution of new types. Diversity of topography means more ecologic niches for a more diverse life.

Extensive populations are continually being subjected to geographic variations. Spatially segregated populations invariably build up different gene pools from those of other populations of the same species with which they do not interbreed. The degree of differences between discrete populations depends on the degree of geographic variations. Geographic variations necessitate the adaptation of each population to the area it occupies. There are isolating mechanisms other than geographic, but none have contributed more to our understanding of the evolutionary process than the geotectonic factor.

Species and speciation

Diversity of biologic systems is brought about by the slow processes of evolution under the guidance of the chemical and physical properties of the earth's crust. According to the scheme of the origin of life commonly held at present, a chemical preevolution involving the segregation of chemical constituents from many sources occurred. From simple compounds large macromolecules emerged, and under the influence of natural selection, certain aggregations of these molecules developed properties of life. First, a mixture of compounds had the power to absorb selected molecules from the environment. Second, some important compounds acquired the power of duplicating themselves. Third, some compounds served as catalysts or enzymes to facilitate the process.

EVOLUTIONARY AND ECOLOGIC PATTERNS OF INVERTEBRATES

Organic evolution is the result of the reaction between a progressive change in dynamic life and a progressively changing dynamic environment. Life shows an evolution from a less balanced relationship between the external and internal environment to a more closely adjusted relationship. Life as we now know it has been the outcome of a particular kind of biota within a particular kind of environment. It is impossible to understand the major evolutionary factors without at the same time considering the dynamics of the continually changing earth's crust and all the conditions of the external environment. Ecologic evolution is therefore a necessary corollary of biotic evolution. The environment has a broad meaning, because it includes the nonliving (physical) factors as well as all living things that in any way influence the organism (biotic factors).

All living things derive their energy, directly or indirectly, from the sun. Green plants use the energy directly in their photosynthesis to synthesize food products. Animals derive it indirectly by feeding on plants or other animals. The evolutionary direction of life has resulted in three great categories or functional groups of organisms—plants, specialized for photosynthesis; animals, specialized for ingesting plant or animal tissues; and saprobes, or organisms specialized for absorbing organic or inorganic molecules from their surroundings.

Ecologically there are also three large categories of organisms—the producers (plants), which use solar or chemical energy to synthesize organic compounds from the inorganic; the consumers (animals), which live by eating plants or other animals; and the reducers (bacteria, fungi, and others) that decompose protoplasm into simpler constituents for recirculation in the life process. To some extent these ecologic categories are the same as the functional ones; however, some organisms may fit into more than one category; for example, some consumers are also decomposers, and absorption is not limited to decomposers but may be shared by microconsumers, animal parasites, and others.

Evolution and speciation. Organisms are divided into units that are called species. Speciation refers to that process whereby one species is split into two. The process of speciation may be thought of as either sudden or gradual. Some evolutionists believe that new species arose by sudden mutations, or saltation, that produced striking phenotypic effects in a single generation. With the exception of chromosome mutations involving a change in chromosome number (polyploidy), it is doubtful that this method of diversification is common. There is no convincing evidence that new species have arisen in biparental organisms by gene mutations. Speciation is now regarded by most biologists as a gradual process, which was the original darwinian view.

In the cline concept (J. S. Huxley) of the evolutionary process, the subdivisions of the cline, called demes, develop unique genetic characteristics. A deme is an interbreeding population within a species. Adjoining demes that spread over a wide geographic area (cline) show a gradual continuous change in the members of the population because of adjustments to local conditions. As long as the member of one deme can interbreed with that of another when the opportunity offers, there is no valid reason for considering the two demes separate species. However, genetic differences may occur to the point at which fertile hybrids from two demes cannot be produced; then the two groups may be considered separate species. A species must be thought of as a group or a population, and speciation involves the members of a whole population, not merely a new type of organism or mutant form.

Evolutionary change does not always involve an increase in the number of species or a splitting process. It is possible, in time, for one species to undergo a direct transformation into another species without the intervention of any splitting whatsoever. In such a case a new species has been formed without the multiplication of species. Evolutionary change or *phyletic evolution* should be distinguished from branching evolutionary lines, or true speciation.

All major groups of animals may not have followed the same plan of speciation. This is especially the case with the extensive division of invertebrates. In this text only one or two prominent mechanisms of the speciation process can be pointed out.

Isolating mechanisms and speciation. An isolating mechanism is any factor th.... .ends to prevent interbreeding between groups of animals. Whatever method is involved to produce isolation must be considered the chief factor of speciation. The criterion of sterility, however, cannot be used as an absolute difference between species. Two different species may live side by side and have the power to interbreed and produce fertile offspring but rarely do so because of certain isolating mechanisms. Many isolation mechanisms other than geographic prevent interbreeding of species populations.

Biologic isolating mechanisms may be classified in various ways. Interbreeding between two species in the same region may be prevented by differences in the specific habitats of the two populations. This preventive method may be very effective among species that have limited mobility or among sedentary animals. Snails, for instance, may have soil preferences that hold them more or less to particular habitats, although different species lie fairly close together spatially. Even if such different species are completely fertile with each other, their habitat preferences normally prevent interbreeding. Seasonal differences in breeding can also be an effective barrier to the interbreeding of two species. One species may breed early and another late in the season, resulting in reproductive isolation. One of the most effective isolating mechanisms is based on differences in behavior patterns (ethologic barriers). Much mating depends on courtship displays, and appropriate stimuli between male and female must be initiated to ensure specific mating. The courtship display of the male of one species may not arouse the mating response of the female of another species. Another important isolating factor is genetic isolation. Even if mates of different species copulate, they cannot produce fertile offspring. Each species seems to have a delicately integrated genetic system that fits it for a definite ecologic niche. This factor nearly always operates in time, because the longer two species are separated from each other by isolating mechanisms, the more incompatible their gametes become in functional relationships.

In any form of speciation, whenever genetic isolation is fully established between species, each will have its own evolutionary development, and hybridization will be subsequently impossible.

Evolution of communities and biomes. The many hundreds of thousands of animal species do not live in random mixtures but are assembled into groups with greater or lesser organizations and degrees of adjustments between the members and their environments. Two important aspects of group populations—communities and biomes—should be considered in the overall picture of evolutionary and ecologic patterns. All organisms meet the impact of their environments, not as single organisms but as members of populations that behave more or less as units. There are various degrees of population units, but the most important are the community and the biome. A biotic community is commonly defined as a natural assemblage of organisms or populations living together in a relatively uniform biologic area, self-sustaining and relatively independent of adjacent population assemblages. Certain modifications of this definition may be made for some communities because it is a broad one and includes assemblages that do not always meet arbitrary distinctions. The community may be considered the natural unit of organization in ecology. A *major community* such as a forest is one that is self-sufficient as long as it receives sun energy from the outside; a *minor community* such as a cave depends to some extent on the energy supplied by adjacent communities. The evolution of the community may be considered to be a result of ecologic selection, by which organisms have been able to exist only when they fitted into some scheme of interdependence among the members and produced a self-sustaining unit. Communities vary greatly, but in general they contain only those species that fit into the ecologic conditions of the community and form a definite part of the web of interrelationships. Such an organization may be considered obligatory. The evolutionary history of any community must involve an extensive background and must have consisted of a slow process of gradually accumulated relationships. No doubt most communities started out with a few simple relationships, and as the community increased in complexity, more and more of these unit relationships were involved. The dynamics of the earth's crust, referred to earlier, have been a responsible factor in forming congregations of species. Such changes could disrupt some existing communities and merge others. Communities have thus been in a flux from the beginning of life, and continuous change in their structure has been the general plan of their development. Various kinds of relationships have been evolved along with the

evolution of communities, such as the predator-prey relationship (e.g., food chains), symbiotic relationships (e.g., mutualism and commensalism), parasitic relationships, and competitive relationships for the same food or other requirements. Each community develops a definite functional unity and characteristic pattern of energy flow. Also, each community passes through orderly processes of change in which communities replace one another in a particular area in sequence—a general process commonly called ecologic succession.

The largest land community is the biome, which is characterized by a uniform climax vegetation and is produced by the interaction of regional climates with regional biota. Although the names of the various biomes are associated with more or less specific plant formations (deciduous forest, coniferous forest, etc.), they are total community units comprising both plants and animals. The term *biome* has a broad meaning and includes the whole series of relatively transitory communities or seral stages as well as mature climax communities. A biome usually has a characteristic climatic condition involving temperature, humidity, and rainfall. Biomes may be of great areas such as grasslands and deserts, or they may be relatively small, such as in mountainous countries where many biomes are often found close together. In the evolution of biomes each may have changed its dominant vegetation many times, even though the same ecologic conditions may have prevailed throughout. The origin and evolution of biomes are intimately associated with the great geologic changes in the physical environment of the earth's surface. Whenever the seral succession and development of communities have occurred through geologic periods, the climatically controlled regions with their communities become biomes. Terrestrial biomes have undergone slower evolutionary changes than have biotic communities, but in the long geologic history of the planet many characteristic biomes have appeared and disappeared. New biomes appear under the impact of great climatic changes, the colonization by plants and animals of regions where no life existed previously, and the succession of dominant vegetative forms that have changed the physical features of a given region. The fossil record plainly indicates that many regions of the earth which now have cold climatic conditions were once tropical or subtropical in nature. Tropical biomes have always existed to some extent, whereas many biomes in temperate and arctic regions have become extinct during geologic periods. Biomes continually undergo changes, although at a slow rate. Many changes are brought about by man himself. In certain regions he has enlarged some biomes and restricted others. Aside from man's influence, other factors produce additions and deletions in community structures and in the biome of which they are a part.

Paedomorphosis. The general term *paedomorphosis* refers to those conditions in which larval or immature features of ancestors become adult characters of descendants. During the evolution of an organism, it is possible for an adaptation that fits an early stage of a life history to its particular habitat to move into a later ontogenetic stage. Paedomorphosis is the idea that the adaptations to a larval or young life may profoundly influence the evolutionary course of the organism involved. Many terms (paedogenesis, neoteny, fetalization) have been used for this phenomenon, but they all refer broadly to sexual maturity in a larval or juvenile stage or to a change in relative rates of development of characters during phylogeny. The concept first introduced by W. Garstand years ago has been brought into sharper focus in recent years. The process has been fruitful in the production of evolutionary novelties. Man appears to be the result of such an evolutionary tendency. As an adult he shows many of the fetal characteristics of apes. Selection for a particular larval adaptation may be strong enough to overcome a contrary selection in the adult life. Behavior as well as morphologic characters are influenced by paedomorphosis.

Among invertebrates, paedomorphosis has undoubtedly played an important role in evolution. A larval millipede with few segments is supposed to have served as a structural ancestor of insects. The burrowing acorn worm (Hemichordata) may represent the larval condition of an echinoderm. The wingless female glowworm is actually the larval form of a firefly. Comparable cases of paedomorphosis have also been found among the ammonoids and brachiopods.

The concept of polymorphism. Many species, including both plants and animals, are represented in nature by two or more clearly distinguishable kinds of individuals. Such a condition is called *polymorphism*. This polymorphic variation may involve not only color but many other characters as well—physiologic and structural. The term has been used with many meanings, but the definition of the English geneticist E. B. Ford is generally accepted at present. His definition emphasizes the occurrence together in the same habitat of two or more distinct forms of a species in such proportions that the rarest of them could not have been maintained by recurrent mutation. This definition rules out seasonal and some other variations, such as the winter and summer pelage and plumage of certain mammals and birds. Polymorphism always refers to variability within a population and is restricted to genetic polymorphism. It results when there are several alleles or gene arrangements with discontinuous phenotypic effects. It may be expressed by just two alternative types such as male and female dimorphism, or there may be many morphologic types within the species. The genetic pattern is well shown by the example of the ladybird beetle *Adalia bipunctata,* studied by N. W. Timofeeff-Ressovsky. In this genus some of the individuals are red with black spots while others are black with red spots. The black color behaves as a mendelian dominant gene and red as a recessive gene. The black and red forms live side by side and interbreed freely. Studies show that the black form is predominant from spring to autumn, and the red form is predominant from autumn to spring. It is thought that the changes are produced by natural selection, which favors the black form during summer and the red form during winter. For instance, more black forms, produced by dominant genes, die out during winter while the recessive red form survives. It is thus possible for the recessive gene, at least during part of the seasonal cycle, to occur more commonly than its dominant allele.

Sometimes an advantageous gene may spread through a population and tend to replace its allele. Such an example is industrial melanism in moths in certain industrial regions of England. In this case a mutant gene (melanism) of rather low frequency has a selective advantage over the normal form when the environment becomes black from pollution. Two polymorphic forms may exist side by side because of a balance of selective forces (balanced polymorphism) or because one may have a selective advantage and gradually replace the other (transient polymorphism).

Polymorphism has adaptive value in that it enables the adaptation of the species to different environmental conditions. Polymorphic populations are thus better able to adjust themselves to environmental changes and to exploit more niches and habitats. Without a doubt, the potentialities of polymorphism are much greater than its realization because every population must have many allelic series that are never expressed in the visible phenotype. Up to the present time there is no satisfactory explanation for the phenotypic uniformity of some species in contrast to the polymorphism of others.

Polymorphism is widespread throughout the animal kingdom, and often different forms have been mistaken for separate species. Some classic examples of polymorphism are the right-handed and left-handed coils of snails of the same species, the blood types of man and other animals (a biochemical distinction), sickle cell anemia in man, albinism in many animals, silver foxes in litters of gray foxes, and rufous and gray phases of screech owls in the same brood.

Regressive evolution. Regressive evolution is found to some extent among all groups of animals and may be defined as a loss in the internal stability of an organism. Organisms that show it have lost to some extent functional adaptations characteristic of their ancestors. Such loss of structural adaptations is usually manifested by vestigial organs that persist long after their functions have disappeared. Organisms are known that possess vestigial legs, eyes, wings, mouths, or other organs. The most logical explanation for the presence of such nonfunctional structures is that they have come from ancestors in which they were functional. Regressive evolution is probably caused by decreased or vanished selection pressure and by a convergent degeneration of homologous or analogous functional structure in different animals because of similar habitats.

Regressive evolution may not always involve the complete loss of function of an organ; it may be converted into another function. Some vestigial organs may not take on another function, but being harmless, they persist because they are correlated with a useful structure. Cave animals (troglobionts) represent one of the best examples of regressive evolution. Eyes and photoreceptor organs are reduced or absent in most cave beetles, millipedes, crayfish, isopods, and spiders, as well as in cave vertebrates. Some of the vestigial organs found in this group may be caused by mutations affecting the relative rate of growth or by the selective value of low metabolism associated with growth retardation.

Regressive evolution may occur at any level of biologic integration. The ameba may have regressed from more complex flagellates. Examples of regression among invertebrates are numerous. Stages in life histories may have dropped out, examples being the coelenterates that may lack either the polyp or the medusa stage, wingless insects, and extreme degeneration of body organs among many insects.

Extinction has also been a major factor in evolution. In all phylogenetic lines there has been an enormous turnover of organic patterns so that the diversity of animals has been restricted. Some extensive groups (ammonites, trilobites, graptolites, etc.) are wholly extinct, and many present groups are mere shadows of their former diversity.

Invasion of land

It is generally agreed by zoologists that life originated in the sea, probably in the littoral regions. It is also thought that all the great groups, or phyla, arose in the sea. There are many divergent opinions on these matters, but the fossil record, fragmentary as it is, indicates that early life existed mostly in ocean waters. The types of body fluids in most animals is often advanced as further evidence for this view because of the similarity between animal blood and seawater; however, some authorities do not consider this to be valid evidence. Although life may have originated in the sea, much evolution has occurred in fresh water and on land. Animals that are preadapted for such habitats spread to wherever there are vacant adaptive zones. It is known that various groups of aquatic arthropods invaded the land independently at least eight times. Invasion of the land therefore did not take place at any one time, and doubtless it is still occurring today.

The most successful invertebrate colonizers of the land are the arthropods and the mollusks (snails). Certain basic adaptive features of these groups made successful land invasion possible. The arthropods with chitinous exoskeletons and the snails with slime and shells were admirably fitted for resisting the desiccation of land conditions. Other useful adaptations of these two groups for a land existence were highly efficient breathing systems such as tracheae, book lungs, and lung devices. Burrowing invertebrates could also meet the challenge of a land existence. Land invasion no doubt occurred on a large scale as soon as a flourishing land flora was developed, such as that of the Devonian and Carboniferous periods (Table 1). Early invaders of the land became dominant because of the lack of competition. By succes-

sion and spreading, animals explored and occupied new habitats.

Most migrations onto the land have occurred in places where there were gradual environmental changes, such as on ocean beaches, in estuaries, tidewater pools, etc. Conditions found in such places permitted experimental trials without fatal consequences. Abrupt changes from sea water to fresh water must have occurred rarely. A. S. Pearse, 1950, in *The Emigrations of Animals From the Sea* has summarized the various problems animals had to meet in their invasion of fresh water and land.

Influence of estuaries. An estuary is that zone where a stream or river meets the sea. It represents a mixture of fresh water from the river and salt water from the sea. Estuaries usually consist of an arm of the sea or the wide mouth of a river. This zone has unique environmental characteristics such as brackish water, oscillating tidal currents, water mixing, and sedimentary deposits. It is a region of constant variations of temperature, salt content of the water, and bottom deposit. The instability of stream waters, as in volume, speed, and levels, is reflected in the estuaries at the mouths of rivers. Stratification of fresh water over the denser salt water also occurs there. Estuaries show many individual variations, depending on the volume of water discharged by the stream, its swiftness, the topography of the surrounding land, wind currents, tides, etc.

In spite of their unstable conditions, estuaries represent a type of threshold by which some animals pass from the sea to fresh water and land and vice versa. Most estuarine animals, however, are marine. All degrees of structural and physiologic adjustments to varying degrees of salinity are found in the organisms that inhabit estuaries. Some are able to adapt themselves without difficulty to salty, brackish, or fresh water. One great problem is their ability to meet adequately the varying processes of osmosis in such environments. Most organisms established in estuaries have mechanisms for stabilizing their internal salt concentration by increasing their hydrostatic pressure through body contractions, by getting rid of excess water through excretory devices, by taking in excess salts, and by having an impermeable skin. Other devices of osmotic regulation are also known. A varied assortment of animals is found in many estuaries, often in associations not found elsewhere.

An estuary is a difficult passageway by which an organism may move to fresh water and land, and some authorities think that relatively few marine species became established in fresh water by migrating through estuaries. Some major phyla such as echinoderms have never been able to pass the barrier to a freshwater existence, although some are found in estuaries, because they have never been able to regulate their osmotic pressure sufficiently. Mollusks, on the other hand, have been more successful, and two of their six orders are found in fresh water or on land. It does seem, however, to be easier to go from marine to fresh water than in the reverse direction.

Psammolittoral-phreatic pathway. Besides the pathway of estuaries for invading the land, there is another potential pathway—the psammolittoral-phreatic pathway. Psammolittoral refers to sandy beaches particularly the intertidal areas; phreatic refers to the groundwater habitat. To move from the marine psammolittoral region into phreatic groundwater and thence into freshwater localities would not be difficult for small invertebrate forms. Both the psammolittoral and phreatic habitats have dense and similar populations.

In the past few years much study and investigation has centered on the interstitial fauna that lives in the sandy beach (psammolittoral) habitats of fresh, brackish, and marine waters. Such a biotope is a mixture of sand grains, water, air, and detritus. The microfauna and micro flora inhabiting the interstices between the grains of sand are called psammon (Gr. *psammos*, sand) or mesopsammon. Psammon includes bacteria, diatoms, protozoans, and micrometazoans. Probably most of the animal phyla are represented among these minute forms, and the populations are sometimes very large. In the labyrinths between the sand grains the water is often deficient in oxygen and may contain a great deal of decaying organic substance. The type and grain size of the sand may influence the distribution of the fauna. Much of the fauna tends to avoid very fine sand, as the micropassages are too confining. Most forms live near the sand surface but many live beneath the surface, sometimes several feet down where the water tends to be more acidic and low in oxygen. In the surf zone considerable disturbance (temperature, rain, and water displacement) occurs, but the fauna gliding in their passageways or clinging to the sand grains can usually adjust to such disturbances.

PALEOECOLOGY AND INVERTEBRATES

The study of the relationship of ancient animals to their environment and to each other is called paleoecology. The biota of the past had to face environmental conditions just as their descendants do today. Paleoecology tries to reconstruct the nature of those environments in which fossil forms lived in the remote past. Interpretations of past conditions are based on data obtained from fossils and from rock deposits in which fossils are found. All possible types of environments and ecologic relationships are involved, but marine environments are emphasized more than others because the paleontologic record is better known in marine deposits.

Many of the present ecologic concepts, such as communities, ecologic succession, and food chains, apply also to the ecology of the geologic periods. Other concepts may be vague because many organisms either do not fossilize or their composition undergoes radical alterations by the addition of substances foreign to them while they were living. Some fossil assemblages are also mixed and represent more than a single environment. Much information, however, can be obtained about them, such as the methods of locomotion, temperature of their environments, chemical makeup of their surroundings, kinds of existence for which they were adapted (swimming, burrowing, creeping, etc.), nature of catastrophic death, relative abundance of certain forms, extinction of others, and other data.

The most reliable source of information is that of the fossil-bearing rocks. The nature of the sediment and its manner of deposition often reveal a great deal of information about environmental conditions of the past. Specific types of sediment formation, such as shales, limestones, and sandstones, may yield data about turbidity, currents, landslides, etc., as well as information about the successions of fossils, the existence of geographic barriers that prevent mingling of contemporaneous animals, the correlation of individual fossil assemblages from widely different regions, the restricted occurrence of some species, and the wide distribution of others.

Hypotheses to explain the conditions of ancient life from the study of living creatures are not always trustworthy. Some animal lines have vastly changed their habits of living over long periods of time. Some are now found only in deep ocean water, whereas their ancient ancestors were largely shallow-water dwellers. Others were formerly pelagic, whereas their present descendants are relatively sessile. In most cases such hypotheses lack adequate means for testing. However, it seems to be true that most ecologic adaptations of animal groups were acquired early in their history and were retained with minor changes.

FORM AND ORGANIZATION OF ANIMALS

Most animals have characteristic shapes and forms. This is especially true of multicellular organisms that, although they may undergo considerable changes in body form during their development, usually conform in the adult stage to a specific pattern. There are also some individual variations among the members of a species, but these are usually minor, such as size. Specific forms of the members of a species should be compared at the same stage of the life history. Among the protistans, however, variations of body form are more widespread because they are unicellular and reflect to a greater extent the continually changing system of protoplasm during the life of the organism. Few protistans have a development that corresponds to that of metazoans. Usually among unicellular forms the life history consists of interdivisional stages that vary in form with a particular stage. The more complicated the life history of the protistan organism, the greater the body variation.

Body variations found among animals are usually grouped around a mean that is common for the species as a whole. The causes for these variations are either the hereditary makeup of the animals or environmental factors during their life histories.

What determines the form of animals?

Animals have a bewildering variety of forms. This diversity is expressed in shape, size, and general structural organization. Size varies from that of the smallest protistans to that of the largest whales. Variety in organic structures ranges all the way from the inconstant ameboid unicellular pattern to the bizarre structures of the Sargasso sea prawns or the echinoderms. As all animals have a unifying basic plan of structure and function, why do they have so many specific body plans? In our present stage of knowledge we can only speculate here and there in regard to why the animal assumes and maintains its own characteristic body plan.

Whatever the structural plan of an organism, it must be considered in the light of the unique properties of life itself. This uniqueness of the life process imposes certain requirements on the physical and functional makeup of the animal. Life is a dynamic process that can take place only in a certain complex organization. The animal, first of all, is a separate entity (colonies and monsters excepted) and has the power to maintain its characteristic form throughout life, although its body substance is continually undergoing change. It is an open system for energy flow of materials from the outside. It is able to incorporate some of these materials into its own unique organization, and it uses others for energy purposes. In this metabolic process it forms waste substances that are discarded from the body.

This relationship between living organisms and their environment is expressed in patterns of adaptations. The animal has a characteristic life history and passes through stages of development and growth from the zygote to maturity. It undergoes many structural changes during its development and continues to do so throughout its life history. It has some power to regenerate lost parts with structures similar to those lost so that animals of a particular species tend to retain the same mean size. The general activities and responses of the organism to its environment are coordinated into characteristic behavior patterns that make the animal a completely functional living unit. The dynamic quality of animals is displayed by their power of autonomous movement, which is performed by a variety of effector organs but chiefly by muscles.

Organisms have many ways of performing the same function, and this has resulted in many structural forms and patterns. The interaction of the organism and its environment with all the implications of natural selection has been a primary cause. Cumulative hereditary variability under the impact of natural selection, which works on the restricted potentiality of the basic body plans of animals, has mainly determined organic form. The way that this operates to produce such a variety of forms is another matter. Physical and chemical principles of which we are still largely ignorant have operated throughout the evolutionary process. Most groups of animals have many structures whose presence cannot satisfactorily be explained; or, if a logical reason can be advanced, the method by which they evolved is unknown.

Physiologic and morphologic adaptations to varied habitats are generalizations of wide application in the evolution of animals. By adaptive radiation a primitive group can move into many ecologic niches. Whenever there is a shift into a new niche, there is a basic adaptive feature of a group that is modified for specialized purposes, e.g., the feathers or wings of birds that originated from reptiles. Adaptations to various forms of life usually involve morphologic changes, although some appear to be purely physiologic. Thus the radially symmetric forms are best adapted for a sedentary or sessile life because the animals tend to grow equally in those directions in which the incidental forces of feeding are equal. Cephalocaudal differentiation and streamlining are characteristic of bilateral, actively moving animals. Along with this tendency the elongated serpentine form is popular for those animals that creep. Modifications of this plan are found in those that have flat or round bodies and those that have feeble locomotor powers. Segmentation of the body made possible a wide range of adaptive habits because it created a more flexible construction and aided greatly in wriggling movements. The addition of locomotor appendages brought about much adaptive divergence in body forms so that animals could swim, dig, or run. Changes in jaw and feeding methods have been responsible for a variety of modifications of the head end and may be considered a primary factor in the varied patterns of body forms. Other factors that have played an important role in determining body form are the effects of the parasitic habits, the differential growth rates of body parts for specialization, the erect position, and the possession of shells.

Experimental morphogenesis throws some light on how these body plans are produced in the development of a specific form, but the overall picture can be obtained only by the integration of evolutionary, ecologic, and populational genetic factors. So far, little has been accomplished in this line of investigation. Throughout such a study the three universal dimensions of living matter—development, function, and structure—must also be effectively integrated. Basically the relatively new science of molecular biology may afford the key to many of the problems of morphology.

Among invertebrate metazoans there is one type of body plan around which many others are built. This is the saclike, radially symmetric form in which the cavity is surrounded by a double wall with an outer layer of cells for protection from the outside and an inner layer for digestion of food. A single opening to the outside serves as both mouth and anus. With the addition of tentacles for food gathering and protection, such a plan is found among the coelenterates, which are considered to be close to the ancestral base of metazoans. Digestion in such forms may be partly or wholly intracellular at first, but a better start is made toward the extracellular digestion of larger forms than is possible among unicellular organisms. Specialization of the cells of the inner and outer layers leads to (1) the formation of muscular and nervous elements for coordination and movement, (2) sensory elements for responses to the varied external and internal stimuli, (3) secretory cells that form protective noncellular layers of cuticle or calcareous shells, and (4) mesenchymal elements between the two layers.

Such a body plan has great possibilities for modification and specialization. The digestive tube may be much folded, branched, elongated, or shortened. The body form may be elongated and narrow or short and broad. This body plan is ideal for division of labor among communities of cells and is capable, with certain modifications, of attaining a variety of sizes. Although many such forms are sessile or restricted in movement, others are active swimmers.

The various body plans of the invertebrate Metazoa that have been built around, or derived from, a saclike or similar body plan include the flatworm type with bilateral symmetry and differentiated cephalocaudal axis, the roundworm type with tube-within-a-tube organization, the annelid type with segmented body, the arthropod type with specialized somites and jointed appendages, and the molluscan type with mantle and specialized foot.

Axial gradients and regeneration. Axial gradients refer to an axial organization in which differential regions of dominance affect quantitatively the metabolic processes of the animal's body. Axial organization may be expressed in physiologic and developmental gradients of various kinds and also in the regeneration of parts. Axial gradients are usually most pronounced in the organization of lower types of animals such as planarians, in which the existence of gradients was first established. In higher animals they appear to play a lesser part, except in very early developmental stages, because of the less plastic condition of the animals' differentiated tissues and body structures. The axial organization is built into the general plan of the body and is a property of protoplasm. In general, a dominant center of influence is in the anterior part of the body, and lesser fields of influence are distributed over the body.

Axial gradients may express themselves by gradients of suceptibility and by processes of regeneration. In susceptibility, cells of different regions and types may be affected in different ways and rates by the same poison. This seems to be correlated with gradients of oxidative metabolism. Some cells are far more susceptible to poisons than others; the more active the metabolism of a cell, the more susceptible it is to the action of a poison. The power of regeneration varies greatly, depending mainly on the level of animal organization, being found far less frequently in higher than in lower forms. Only very small parts of the body are regenerated in the adult condition of higher vertebrates. The simpler multicellular organisms may regenerate body organs or lost appendages, but most insects will not. A regenerated part may be a different type from the original lost part. The eye of a crustacean may be replaced by an antenna. Usually the larger the part lost, the more rapid is the growth rate of the regenerating tissue. The process of regeneration commonly follows a definite sequence, although there are many variations. First, a regeneration blastema is formed from a mass of undifferentiated cells (neoblasts) that have come from the loss of differentiation in tissues near the lost part. During the growth of the blastema its cells undergo differentiation into the tissues that are to form the regenerated structures. Then the tissues are organized into a pattern that duplicates the lost structure by differential growth. Regenerated tissue is usually derived from the same kind of tissue in the parent body, but there are exceptions to this. Although much is known about the patterns of regeneration in many groups of animals, no clear explanation can be given about the underlying reasons for this phenomenon.

Growth and body form. Differential growth is characteristic of all structures in the animal body. Under most conditions each tissue has a definite maximum rate of growth. Change in form and size are found throughout the life history of all multicellular animals. This not only occurs during the development of the animal but also during the adult condition. Growth may cease in mammals when they reach maturity; however, in most invertebrates (except insects) growth continues in the adult, but at a slower tempo. Growth tends to cease at a certain level in many terrestrial animals because land forms have more size restrictions than have aquatic organisms. The buoyancy of a watery medium permits large size, as in the whale. Some land forms such as reptiles continue to grow during adulthood. In a strict sense, body size never remains constant in most adult organisms because of loss of water and the relative increase of nonliving connective tissue. In many arthropods (insects) with chitinous exoskeletons, however, body size does not change in the adult.

Growth may take place via one or all of three types of growth changes. There may be an increase in *number* of structures, in *size* of structure, or in *spaces* between structures. Growth is mainly a molecular phenomenon; molecules are manipulated to form body structures. Molecules ready-made from the outside may be incorporated into the body, or may be synthesized within the cells. They may increase in size by combining with other molecules to form macromolecules, and they may be separated from each other by spaces filled with fluid or cement substances. In this way cells increase in size, and by cell division they increase in number.

Most changes in body form are produced by differential growth of body structures. Growth involves both quantitative and qualitative changes. One organ or its part may grow faster or slower than another. It may also grow slower or faster than the body as a whole. When an organ is growing at the same rate as the mean growth rate of the body, it is called *isometric* growth. If the organ's growth rate differs from the mean growth rate of the rest of the body, it is called *allometric* growth. Allometric growth is positive when the organ grows faster than the body's rate, negative when it grows more slowly. Allometry may be expressed on a logarithmic scale in which the line gradient represents a growth coefficient, which is a relation of the relative growth rate of the organ to that of the body. Allometry factors may be an intrinsic part of the animal's pattern of body proportions, but they may be modified by ecogeographic principles. Body proportions seem to be very sensitive to natural selection factors that may be expressed by differences in temperature and climate. Ecogeographic rules apply to both vertebrates and invertebrates. Insects especially show adaptive patterns to geographic variations. Butterfly pupae from coastal southern California are larger and heavier than are those of the same species from the Mojave Desert (W. Hovanitz). Beetles of the same species are larger and have wider abdomens in temperate central Europe than do members along the warmer Mediterranean region (B. Rensch).

Growth gradients appear to be under the control of the axial organization, or the morphogenetic fields of development, but many factors are unknown. It is the differential growth patterns that in the long run determine body form and proportions, although the exact mechanism is unknown.

BASIC ADAPTIVE PATTERNS OF INVERTEBRATES

Each higher group, or taxon, of animals has some basic adaptation around which the evolution of the group has occurred (G. G. Simpson). This basic adaptation, which is usually laid down early, determines to a great extent the amount and direction of evolutionary divergence and thus the number of ecologic niches for which the members of the group are fitted. Such adaptations may consist of a single outstanding characteristic, or each may involve several structures. In some cases there may be a certain basic adaptation for a class and another for a phylum. Many different groups may have similar adaptations, but one group may exploit the possibilities of the adaptation in question, whereas others may make little use of it. When animals change from one kind of environment to another, it is almost always made possible by some basic adaptive pattern. Within the new zone the nature of the adaptation will determine the degree of specialization for new ecologic niches. The adaptation may not be retained by all the taxa of a group. It may be lost entirely, and the nature of the original adaptation may be obscure. The same basic adaptation may undergo an entirely different evolution in different closely related taxa, or it may evolve at a different rate in different groups. In general, the broader and less specialized the adaptation, the more likely it will be retained in the evolution of the group. Some distant and not closely related taxa have the same basic adaptation (convergent evolution). If two or more groups have many adaptations in common, such adaptations can usually be traced to similar basic adaptation or adaptations of a common ancestor.

The concept of basic adaptive patterns of invertebrates is of the utmost importance in understanding the evolutionary history of any group. It shows the trends and possibilities of the group and points the way for future development. It helps to explain the reasons why certain animals are found in some regions and why others cannot exist there, to explain the presence of some of the coadaptations found in the lower taxa by the restrictions imposed from the basic adaptations of the higher taxa, and to explain why the phyla are basically different and how these differences arose as adaptations by the early members of each phylum. The concept of basic adaptive patterns is also important in comparing the varied adjustments or adaptations of lower taxa of a group with the basic adaptation of the large taxonomic category. In the study of every organism this concept thus shows the importance of determining the nature of the basic adaptive feature. Such a determination is not always easy because the successive changes of the environment are so involved in the long evolution of a group that the fundamental adaptive plan may be obscured by coadaptations of lower taxa to restricted ecologic niches. It is not enough to know the present adjustments of animals; one must determine the ecologic relationships of their ancestors. The fossil record, when sufficiently complete, may throw some light upon such paleoecologic information. In dealing with specific groups later in this text, this important concept is emphasized throughout.

Adaptive patterns and adaptive radiation

Adaptive radiation is an evolutionary process by which a single relatively unspecialized phyletic line evolves into several descendant lines that are specialized for different niches or adaptive zones. Adaptation is a gradual change in the gene pool of the phyletic line involved so that adaptive genotypes are formed.

Radiate evolution is a broad term and refers to both physiologic and morphologic characters. Forms that are totally indistinguishable by morphologic criteria may have entirely different reactions to physiologic gradients. It is also common for two groups not closely related to have similar ecologic behavior patterns (convergent evolution) because a particular pattern of ecologic factors will select similar characteristics in the two groups by natural selection. Primitive characteristics also are likely to be best fitted for a good display of adaptive radiation patterns because primitive characteristics are more general, more labile, and more capable of a variety of specializations. Some characters are far more labile than others and do not fit into the primitive category at all. Specific color patterns vary enormously throughout the animal kingdom, and it is impossible to state what the primitive pattern was like even if there should happen to be one.

Among invertebrates the insects represent the best illustration of adaptive radiation because no other group has equaled them in the patterns of evolutionary diversity. In a work by Lord Rothschild in 1961 (*A Classification of Living Animals*) some 29 orders and 700,000 species of insects are listed. The orders of insects represent a striking case of radiation because the fossil record, incomplete as are all fossil records, partly indicates the way in which these orders emerged and became specialized for varied ecologic niches. The fossil record shows the progressive changes in structure that can be confirmed by morphologic and embryologic studies of existing forms. Probably originating as far back as the Devonian period (400 million years ago), this amazing group has passed through three major evolutionary steps (F. M. Carpenter): the development of simple, functional, nonfolded wings from the thoracic wall; the development of wing articulations for folding; and the development of complicated metamorphoses. Insects have a very high potentiality for specialization, which has caused them to evolve species that are fitted for restricted ecologic niches such as different parts of the same plant. A. E. Emerson has pointed out the variety of adaptive patterns among the Coleoptera (beetles), the largest order of insects. Their chief basic adaptive features are the thick, chitinous exoskeleton (as in the elytra, for example) and their complete metamorphosis. More than fifty ecologically adapted patterns can be listed among them, although some apply to the same species. Their varied adaptations involve behavior patterns food habits, luminescent organs, locomotor defensive mechanisms, warning and concealing coloration, divided eye vision, reproductive patterns, aquatic adaptations, and many others. Some of these adaptive patterns may be widespread, and others may be very rare. Insect adaptations are especially common to meet drastic changes in the environment, for example, the resistance of flies and lice to DDT. Many phylogenetic branches may have evolved similar adaptations by convergence.

Invasion of new habitats. Wherever an animal enters a new habitat or niche, it faces a whole array of selection pressures and must be equipped with suitable characteristics if it is to meet successfully the conditions of the new environment. This means that the organism is in some way preadapted to the new niche in which it is now going to live. Any major shift of an animal into a new region usually involves a basic divergence of a basic adaptive character. Preadaptations need not be, and seldom are, complete for a major shift. Persistent habit over a long period within a particular habitat fits an animal to live better and increase its degree of successful adaptation. The successful invasion of the land by arthropods is mainly caused by such preadaptations as the chitinous exoskeleton, segmented body, and jointed appendages. This type of morphology made possible specializations for efficient methods of locomotion and support. Terrestrial conquest is possible only to forms with strong skeletons or some sort of armor. The arthropodan type of integument, legs, and efficient striated muscle has produced one of the best forms of locomotion among invertebrates. Most successful evolutionary groups are explained by their ability to exploit such preadaptations in situations in which they can have full deployment of their potentialities. Arthropods could not have undergone their amazing diversity in aquatic environments only.

Evolutionary novelties (E. Mayr) are those newly acquired morphologic or physiologic characters that enable an animal to fit into a new adaptive zone. C. Darwin and others have emphasized that the most important cause of the appearance of new structures is a change of function. Change of function alone is insufficient; there must also be inherent in the structural pattern of the animal the potentiality to perform the new function (preadaptation). The adaptive radiation of the arthropods, mentioned earlier, could take place only by the combination of their basic adaptive or preexisting structures into the adaptive pattern of the new habitat. Most evolutionary novelties are considered to be modifications of structures already present. It is possible, of course, that the right combination of genes or mutations could produce something new or a phenotype expression so novel that nothing like it existed before. The belief that novelties could thus suddenly emerge by a lucky mutation or combination of mutations (saltation) is not considered a valid premise in light of the theory of gradual evolution that is held by most biologists at present. Most mutations, moreover, have only a slight effect on the phenotype and simply replenish the gene pool. Preadaptations must be thought of as functional rather than morphologic shifts. Most preadapted structures can perform their functions before the need of the new functions arises, which is probably what L. Cuénot had in mind when he gave us the concept of preadaptation. Preadaptation may involve many traits, called generalized preadaptation, or a specific structure if the new niche is restrictively specialized. Whenever an old structure evolves toward the preadapted condition, it is modified by old selective forces until the new preadapted structure assumes its function; new selective forces now control its development in the new environment.

Basic adaptations and classification of invertebrates

Common characters in invertebrates are usually a result of basic adaptations. Throughout the phylogenetic tree the main branches are represented by a common basic structural plan within which has occurred a great variety of adaptive modifications. According to Mayr and others, higher categories (orders, classes, phyla) are divided into lower categories such as families, which have a great variety of adaptations and usually occupy distinctive ecologic niches. According to this scheme, the higher categories have few obvious adaptive characters because basic adaptations have been overlaid by the specific adaptations of the lower groups. Only general adaptive features are evident in higher taxa. For instance, it is easier to describe the adaptive patterns of each of the orders of insects than to explain those of arthropods in general.

The old typologic method of taxonomy is being replaced by methods emphasizing the population structure of species. Although type forms are convenient units for classification and are still used, the newer method tries to avoid forcing natural populations into artificial schemes. The clinal pattern for explaining evolutionary diversity shows that all populations of a species display one or more of three marked characteristics: (1) gradual changes in a population brought about by clinal variations, (2) geographically isolated populations, and (3) so-called hybrid zones of variable populations between two stable populations.

The character, or phenotypic, gradient of the cline (J. S. Huxley) is an outcome of differential selective factors of environmental variations to be found in the extensive range of a cline. These environmental factors vary according to gradients themselves and rarely show abrupt changes. Gene flow among adjacent populations of a cline tends to produce uniformity in the populations that share in the gene flow. Geographic isolates refers to a population that is prevented by barriers from sharing gene flow with other populations of the species. Almost any cline of extensive range will show such "pockets" of population. Such populations are powerful forces for incipient speciation. Contact belts or zones between phenotypically different populations offer opportunities for members of different populations to interbreed. Mayr calls this mixing of gene flow *allopatric hybridization*. Such populations of secondary intergradation may show any degree of differences in characters. Although most hybrid zones have been described for vertebrates, hybrid belts also involve invertebrates, such as those of insects (T. H. Hubbell) and snails (E. Mayr and C. B. Rosen).

In view of the previous concepts it is difficult to select a specimen of a population and designate it as a type. Species structure shows many variations that are determined by many factors. The taxonomist must understand these factors if a representative classification of a form is to be made.

From this standpoint of basic adaptations in relation to the preceding classification principles, it may be stated that a few key characters, which usually distinguish a major group, may show such variations under the impact of phylogenetic divergence that they tend to lose their analytic fitness for classification purposes. The classification of the larger groups is often based on a single or a few major characters, but the failure of the systematist to make a wide and comparative analysis of all variable characters that bear on the phylogeny of the forms being classified will usually result in an unnatural and misleading classification. A striking example of this principle is demonstrated by the classic case of the tunicate. The dramatic transformation during its individual history from a free-tailed tadpolelike larva to the sessile, soft, baglike adult that has lost many of its chordate characters made it impossible to correctly classify this form until its complete life history was known. Before that time, tunicates were classified with the mollusks because of the superficial resemblance between the two groups.

HISTOLOGIC STRUCTURE OF INVERTEBRATES

In both invertebrates and vertebrates, tissues are made up of cells and intercellular material. The various types of cells appear early in all metazoans, and the principal types of tissues (epithelium, connective tissue, muscle, nerve, and reproduction) occur in invertebrates the same as they do in vertebrates. Are tissues alike in invertebrates and vertebrates? It is known, of course, that the tissue cells are arranged in similar ways to form the various types of tissues, with the exception of greater complexity in vertebrates. Only a chemical analysis of the characteristics of the tissues can give an answer to the above question. Histochemistry, however, had not yet advanced far enough to give precise answers. The functional significance of the substances that make up tissues is still obscure. For instance, it has long been believed by most biologists that true cartilage is not found in invertebrates, but in recent years evidence has been accumulating to indicate that true cartilage does occur in invertebrates and that there is considerable overlap between invertebrate and vertebrate types of cartilage (Person and Philpott, 1969). How the cells in tissues are manipulated depends on how they respond to similar needs. From a common pool of analogous structures these needs may evoke similarities in tissue at cell and molecular levels in both invertebrates and vertebrates, even though the chemical makeup of the tissues may vary greatly.

PHYSIOLOGIC MECHANISMS OF INVERTEBRATES

All living systems must react to changes in their environment. Reactions may be quick, direct responses to stimuli, or they may be long-delayed responses to changes in the environment. Many such reactions are the outcome of a complex syndrome of causes and responses that partake of the nature of an adaptation. Organisms do not all respond in the same way to environmental changes. Each type of organ-by other layers of materials of contrasting re-fractile indices. A soap bubble, the color of which changes as the distance between layers change, is a good example of interference of light. Among invertebrates the transparent, thin wings of flies and dragonflies, scales of the wings of Lepidoptera and Coleoptera, carapace of many crustaceans, and others represent this type of coloration. *Scattering* is produced by the dispersal of the shorter wavelengths by very small particles, such as the blue color of the sky. In the invertebrates the blue color of some dragonflies and wood lice appears to be caused by interference, although this type of coloration is far more common among vertebrates.

Pigments, or biochromes, are definite chemical compounds that have the power to impart color to animals. Biologic pigments may be suspensions, colloids, or even true solutions. A pigment transmits some but not all the wavelengths of visible light. If a pigment absorbs all the incident light, it is black; if it absorbs all the shorter wavelengths of visible light except the red, the color of the pigment is red, and so on with the other wavelengths of light. The most common biologic pigments are the carotenoids, melanin, and guanine, but there are many others (quinones, flavins, anthocyanins, hemoglobin, chlorocruorin, etc.). We are mainly interested in the first three.

Carotenoids or lipochromes are common among invertebrates and produce red, orange, and yellow colors. They are found in beetles, coelenterates, sponges, crustaceans, and echinoderms. These pigments are commonly found in branched chromatophores, known as xanthophores. Melanin is fairly common among invertebrates and is responsible for black or brown coloration wherever it is found. This pigment is usually in the form of small granules that are mainly found in branched cells, known as chromatophores or melanophores, but it is sometimes found in unbranched ectodermal cells. Melanogenesis is influenced by many factors such as movement, hormones, humidity, and low temperatures. A group of pigments closely related to melanin are the ommochromes, which isolate the ommatidia of the compound eye in their optical activity. The ommochromes are usually brown but may be other colors. These pigments are also found in invertebrates other than the arthropods. Guanine is the purine base of one of the nucleotides of the nucleic acids (DNA and RNA). It is found therefore in all cells, but only when it accumulates abundantly in granular or crystal form does it enter into the coloration of animals. Such animals give a white or silvery light because they reflect all incident light. This pigment is usually found in special cells (guanophores). It has a restricted distribution among invertebrates, being found in cephalopod skin, which displays iridescence by light interference in its cells (iridocytes).

Pigment cells are often unicellular, and changes in coloration are produced when the pigment is dispersed in the branched cells (colored effect) or concentrated in the center of the cell (blanching). In some crustaceans the pigment unit is multicellular and may contain many different pigments, but only one color is found in a cell. In some mollusks, especially in cephalopods, the pigment cell is enlarged or constricted by radially disposed muscles. This is the only type of pigment cell controlled directly by nervous action in the invertebrates; in the other types, color changes are affected by neurohumors through the internal medium. The direct nervous control is probably responsible for the rapidity of color changes in cephalopods.

Color changes in invertebrates are caused by many kinds of stimuli and are usually adaptive in nature. Many crustaceans become pale at night and dark during the day, although there are one or two exceptions. Color of background is a common stimulus; most crustaceans will become dark on a black background and pale on a white background. Light, temperature, and humidity also influence color changes.

Respiratory pigments as oxygen carriers represent another important group of pigments among invertebrates. These pigments are spe-

ism has its own characteristic adaptive patterns, which means that it has its own particular physiologic mechanisms. But all functions of an organism are based on the methodical arrangements of structures, and the way these systems are put together in the final analysis determines the life process. The science of physiology tries to describe the processes that go on in the living organism and cannot be divorced from the structural components. However, physiologic mechanisms are more subject to variations with environmental change than are morphologic components. The way in which an animal adapts itself to a given environment depends on its functional capacity. Its functional responses, however, are always determined by its genetic constitution and by its environment, but these two factors are not always equally effective. The same genotype may produce different phenotypes under varied environments because the living system has the ability to modify its properties in accordance with changes in its environment. The vital functions of all animals must be adapted to environmental changes. Such adaptations are the result of the selection of the best adapted individuals, and favorable mutations must occur for natural selection to choose those individuals favorable for adaptations.

It is possible to classify organisms in various ways such as morphologic, ecologic, and physiologic, depending on the properties one is emphasizing. As C. L. Prosser has often pointed out, biologists have stressed the evolutionary trends of morphology more than they have those of physiology. Functional similarities and dissimilarities in the phylogeny of organisms are just as important in understanding evolution as are structural features. In the long and important chemical evolution that preceded the advent of living organisms, the groundwork for the basic plan of physiologic mechanisms was laid down. The whole scheme of our present concept of the origin of life involves physiologic aspects at nearly every step. Throughout the origin of complex carbon compounds under the

influence of radiation energy, synthesis of protein molecules, formation of nucleoproteins and high-energy phosphate bonds, electron transfer system of metabolism, selective diffusion of organic molecules, enzymatic action at all steps of the metabolic process, and other functional aspects have evolved during the preliminary biochemical evolutionary process. During the so-called organic evolution that gave rise to the great diversity of life as we now know it, very few biochemicals have been added to organic life. Physiologic mechanisms have evolved early and were necessary prerequisites for living systems.

Physiologic regulation

Physiologic regulation is the adjustment an organism makes in controlling the output of its body processes. By such means every activity operates under influences that determine its rate and amount. The mechanisms by which quantitative responses by each property and component of the organism's body are made are triggered, in most cases, by specific detectors (E. F. Adolph). This indicates that detectors relay information to effectors, which in turn respond in a more or less specific way. Most regulations tend to restore the organism to normal conditions whenever its bodily states are disturbed by external stresses. Often delicate mechanisms of feedback are involved in these adjustments. To what extent an animal can restore itself depends on its genetically determined limits of regulation. All organisms have large numbers of specific response mechanisms that are compressed into a very limited number of body structures or organs. A problem of all organisms is to regulate a maximal number of components with a minimal number of detectors and effectors.

Many regulative mechanisms maintain the constancy of internal states (homeostasis). Organisms show many variations in this respect. In some animals internal conditions change with the altered environment. This is illustrated by the body temperature of poikilothermic animals, which increases or decreases in conformity with the external environment. In others there is a relatively precise regulation of internal conditions in the presence of environmental change so that such organisms can tolerate a wider range of environmental fluctuation but have a narrow range of internal conditions. Some animals are able to regulate certain functions and conform to the environment with respect to other functions. Physiologic division of labor operates in all regulative control of physiologic functions. One organ may be specialized for one aspect of a regulating mechanism, and another organ may be specialized for another role. Responses may be specific in some regulations and highly nonspecific in others. Regulative responses also vary with stages in the life history of an organism. Those of a caterpillar may be quite different from those of a butterfly. Some responses are quick acting, and others are delayed for long periods of time.

The numerous investigations concerning the physiologic differences of species and animal populations have centered upon two chief questions: (1) Are physiologic variations genotypic or phenotypic? (2) What is the mechanism involved in these variations? Organisms subjected to persistent environmental changes usually undergo compensatory changes. When a single environmental factor is involved, the compensatory change of the organism is called *acclimation;* when multiple environmental factors are considered, the compensatory changes are referred to as *acclimatization.* Acclimation is commonly done in the laboratory under controlled conditions. Acclimatization deals with a complex of environmental stresses such as climatic and seasonal changes. Many biologists consider acclimatization to have a wider meaning—that of genetic adaptation. Acclimation has been more widely studied because it is much easier to study the influences of one factor than of many. Whenever an organism adjusts in a new environmental condition at the same rate as in the old one, it is considered to have perfect acclimation.

Behavior patterns

Every living organism has built into its organization certain patterns of behavior. Behavior may be defined as externally directed activity originating from the stimulation of internal mechanisms. It often but not always brings about some change between the organism and its environment. Behavior directly or indirectly is determined or influenced by mechanisms involving a sequence of stimulation and response. A reaction may be initiated by a stimulus, but its nature depends on the type of organism involved. Animals have different mechanisms of response, depending on their effectors (muscles, glands, sol-gel relations, stinging organelles, etc.). Reactions may be initiated by external stimulation, or they may also arise in the organism spontaneously or automatically. Sometimes the same reaction is produced by stimuli of many kinds. Some behavior patterns are simple and involve only local reactions of certain parts of the body; others are concerned with centralized reactions of the whole organism. All behavior patterns vary with the general physiologic condition of the organism. A hungry hydra, for instance, will react toward food stimuli differently than will a satiated specimen.

Behavior patterns vary all the way from the simple taxes or trial-and-error movements of Protista to the highly complicated patterns in those organisms with well-coordinated nervous systems. Many of the simpler metazoan forms have behavior similar to that of the Protista. Some different types of stimulation may produce the same reaction in both protozoans and metazoans, but the latter have more types of reaction. Perhaps most behavior has some adaptive significance, but such adaptation is not necessarily perfect. Everything the animal does—food getting, defense and offense tactics, courtship display, shelter seeking, and scores of other activities—involves behavior in some form. Invertebrates display a wide range of behavior adaptations that are correlated with their structural organization. As we go up the scale of organization levels, we find that uncentralized behavior tends to be replaced by centralized control in which there is a definite pattern of receptor organs, conductors, association centers, and effector systems. In coelenterates there is little centralized behavior, and most of their reactions are localized because their nerve net acts mainly as a conducting system. As H. S. Jennings pointed out long ago, some of their reactions appear to be random and spontaneous, whereas others such as the swimming of the medusae may be regarded as a simple form of central control. Flatworms have advanced in central control with their so-called "brain" and ladder type of nervous system. When stimulated, reactions of the whole organism usually take place. Planarians have some capacity for learning—perhaps the first group that may be said to have this faculty. Both flatworms and the more highly organized annelid worms can be trained in simple ways, such as being conditioned to master the right turns in elementary mazes. Echinoderms with a highly modified nerve net, including associative neurons and reflex arcs, have better coordinated reactions than do coelenterates and may display unified action in some behavior patterns. However, many actions of echinoderms seem to be aimless or of the trial-and-error variety.

Among the invertebrates, however, it is in the arthropods that we find the most complicated patterns of behavior. This is particularly true of the large group of insects. These organisms have adapted themselves to so many ways of life and new environments largely closed to other invertebrates, that they have developed a central and coordinated control far beyond that of most others in the invertebrate kingdom. Their receptor system and nervous connections are specialized for quicker conduction and response, and the addition of the power of flight has made possible the development of greater speed and distance of travel. Their compound eyes and antennary sense organs are admirably adapted for perception of the environment. Their brain controls the tone of their muscles

and the sensitivity of the whole body. But the common form of insect behavior is innate with combinations of chain behavior, simple reflexes, and some degrees of learning. Chain behavior, in which each action acts as a stimulus to the next step, is common among them. However, they do not operate by instinct alone because their behavior can be modified. It may be said therefore that innate structural behavior patterns and reflexes of great complexity, with learning of the simplest conditional reflexes, have been built into the insect's nervous system in an evolution of hundreds of millions of years.

Pigment

Pigmentation of some kind is found in almost all invertebrates. It may be permanent or it may be changeable. Pigmentation has many functions such as those for protection, concealment, regulation of light, formation of vitamins, mating reactions, humidity adjustments, and many others. In some cases, pigmentation may be purely incidental to physiologic processes. It may be produced by the deposition of organic pigments (biochromes), or it may result from structural coloration or schemochromes (color effects produced by diffraction, interference, or scattering of light rays by physical surfaces). Structural colors may be altered or destroyed mechanically, but no chemical substance can be extracted from surfaces that display such coloration. Diffraction and interference colors are also iridescent, that is, the visible colors of the spectrum will vary according to the angle with which the object is seen.

Diffraction is chiefly produced by a surface that is broken up into a form of grating. When a beam of white light strikes the units of such a grating, spectra that have an iridescent effect appear. The moving ciliary combs of ctenophores provide such a surface and are responsible for the beautiful colors found in these creatures. *Interference* is far more common than diffraction in the animal kingdom and is produced whenever light rays are reflected from surfaces of very thin multiple layers separated by other layers of materials of contrasting refractile indices. A soap bubble, the color of which changes as the distance between layers change, is a good example of interference of light. Among invertebrates the transparent, thin wings of flies and dragonflies, scales of the wings of Lepidoptera and Coleoptera, carapace of many crustaceans, and others represent this type of coloration. *Scattering* is produced by the dispersal of the shorter wavelengths by very small particles, such as the blue color of the sky. In the invertebrates the blue color of some dragonflies and wood lice appears to be caused by interference, although this type of coloration is far more common among vertebrates.

Pigments, or biochromes, are definite chemical compounds that have the power to impart color to animals. Biologic pigments may be suspensions, colloids, or even true solutions. A pigment transmits some but not all the wavelengths of visible light. If a pigment absorbs all the incident light, it is black; if it absorbs all the shorter wavelengths of visible light except the red, the color of the pigment is red, and so on with the other wavelengths of light. The most common biologic pigments are the carotenoids, melanin, and guanine, but there are many others (quinones, flavins, anthocyanins, hemoglobin, chlorocruorin, etc.). We are mainly interested in the first three.

Carotenoids or lipochromes are common among invertebrates and produce red, orange, and yellow colors. They are found in beetles, coelenterates, sponges, crustaceans, and echinoderms. These pigments are commonly found in branched chromatophores, known as xanthophores. Melanin is fairly common among invertebrates and is responsible for black or brown coloration wherever it is found. This pigment is usually in the form of small granules that are mainly found in branched cells, known as chromatophores or melanophores, but it is sometimes found in unbranched ectodermal cells. Melanogenesis is influenced by many factors such as movement, hormones, humidity, and low temperatures. A group of pigments closely related to melanin are the ommochromes, which isolate the ommatidia of the compound eye in their optical activity. The ommochromes are usually brown but may be other colors. These pigments are also found in invertebrates other than the arthropods. Guanine is the purine base of one of the nucleotides of the nucleic acids (DNA and RNA). It is found therefore in all cells, but only when it accumulates abundantly in granular or crystal form does it enter into the coloration of animals. Such animals give a white or silvery light because they reflect all incident light. This pigment is usually found in special cells (guanophores). It has a restricted distribution among invertebrates, being found in cephalopod skin, which displays iridescence by light interference in its cells (iridocytes).

Pigment cells are often unicellular, and changes in coloration are produced when the pigment is dispersed in the branched cells (colored effect) or concentrated in the center of the cell (blanching). In some crustaceans the pigment unit is multicellular and may contain many different pigments, but only one color is found in a cell. In some mollusks, especially in cephalopods, the pigment cell is enlarged or constricted by radially disposed muscles. This is the only type of pigment cell controlled directly by nervous action in the invertebrates; in the other types, color changes are affected by neurohumors through the internal medium. The direct nervous control is probably responsible for the rapidity of color changes in cephalopods.

Color changes in invertebrates are caused by many kinds of stimuli and are usually adaptive in nature. Many crustaceans become pale at night and dark during the day, although there are one or two exceptions. Color of background is a common stimulus; most crustaceans will become dark on a black background and pale on a white background. Light, temperature, and humidity also influence color changes.

Respiratory pigments as oxygen carriers represent another important group of pigments among invertebrates. These pigments are specialized for the storage and transport of oxygen from a respiratory surface to the body tissues, usually by a circulatory system. Oxygen can form loose combinations with these pigments when exposed at high tensions. These combinations are capable of releasing the gas at the lower tensions in the tissues. The pigments vary biochemically in the various phyla and also in some cases in the same phylum. As a group they may be described as chromoproteins containing a metallic atom in their makeup and having the power to combine with oxygen and to a certain extent with carbon dioxide. Respiratory pigments are derived chiefly from the porphyrins (organic compounds of tetrapyrrole that have great biologic importance in many metabolic activities of the body).

Although all vertebrate animals have one type of respiratory pigment (hemoglobin), invertebrates, on the other hand, make use of several types. Hemoglobin is the most widespread of all the invertebrate respiratory pigments. It is the dominant pigment of annelids, although the polychaetes have a variety of respiratory pigments. Among the invertebrates the respiratory pigments are usually carried in the plasma, but there are a few exceptions. Besides the hemoglobin, three other respiratory blood pigments are found in the invertebrates. Chlorocruorin (closely related to hemoglobin) is greenish in dilute solutions and occurs in certain families of polychaetes, in one species of which (*Serpula*) are found both hemoglobin and chlorocruorin. Hemerythrin is another iron-containing pigment found in some polychaetes (*Magelona*), in some sipunculids, and a few others. It is violet in color and is present in the coelomic or blood corpuscles. Hemocyanin contains copper instead of iron, has no porphyrin group, and is always found in the plasma. In the presence of oxygen it gives a blue color to the blood. It is fairly widespread, being found in many mollusks and some arthropods. The various pigments have originated independently many times and thus have no real phylogenetic significance. Their molecular weights vary enormously (from a few thousand to millions).

The oxygen-carrying capacities of pigments also vary greatly among different invertebrates and are related to the ecologic conditions of the animals. The oxygen resources of the environment may be such that certain animals have adjusted to small quantities of oxygen, and their evolutionary development has restricted them to such habitats. It is known that the values and limiting factors of an animal species have become established in its phylogeny in response to the available oxygen. In emergencies, some animals may become anaerobic when they have insufficient oxygen to meet their demands.

Biologic clocks

A physiologic mechanism that has been studied extensively in recent years is the biologic clock. A biologic clock refers to an automatic timing device that is built into the behavior pattern of animals in such a way that they can utilize the natural cycles of their physical environment to the best advantage. Such timing devices have been the outcome of the organism's adaptation to the rhythmic nature of its external environment. These internal clocks are now known to play important roles in many aspects of animal behavior and physiology. Biologic clocks have been studied and demonstrated in most major groups of organisms, with the possible exception of some primitive forms such as bacteria and blue-green algae. Timing devices are known to be involved in migratory activities, bird migration, orientation problems, diurnal and nocturnal activities, breeding cycles, seasonal cycles, color transformations, etc.

Although biologic clocks have evolved and are synchronized under the impact of external cycles, they are not altogether dependent on the persistence of external rhythms. Most biologic clocks are built around circadian rhythms, that is, the 24-hour cycle of the day. Organisms usually establish their rhythmic activities both externally and internally in a given order or sequence and do not function at the same rate over the daily cycle. By such means they can adapt their activities to the regularly recurring changes in the environment. Other rhythms or cycles are also correlated with the lunar phases, tidal phases, and even animal phases. Carefully performed experiments have demonstrated that biologic clocks are based on intrinsic processes and are a definite part of the animal's organization. This is strikingly shown by experiments in which animals that normally have rhythms correlated to daily changes in their environment will still have these same rhythms even in the absense of external stimuli. For instance, fiddler crabs normally become dark in color by the dispersal of their pigment in pigment cells at daybreak and become pale in the evening by concentration of their pigment to the center of pigment cells. They thus have a dark color during the day in contrast to the lighter one at night. When kept continuously in a darkroom for long periods of time, the crabs retained their dark-light rhythm synchronized with the outside day-night cycle without noticeable variation. However, it is possible to reset the clock by subjecting the crabs to certain external conditions. It was found by F. A. Brown, Jr., that if the crabs were left for 6 hours in the darkroom in sea water near the freezing point and then warmed to room temperature, the clock stopped or slowed so that when the crabs resumed their 24-hour cycle, the rhythmic change was 6 hours slow. Furthermore, the crabs showed no tendency to return to the original rhythm and remained at the reset time. The same investigator also was able to reset the clocks by illumination effects at critical times in the daily cycle.

Among invertebrates, timing devices are probably most highly evolved in honeybees, although these organisms have been more extensively studied than other forms. Bees have the capacity to determine their direction by the sun and can make suitable corrections for the shifting of the sun's position at various times of the day. They thus share with birds a compass mechanism for determining correct orientations.

An explanation of the precise way organisms adjust themselves to the rhythmic cycles of their environment is impossible from our present state of knowledge. Biologic clocks are inherited and are relatively independent of temperature, chemicals, drugs, and other inhibitors. It is not known whether the clock is localized or involves the makeup of the whole organism. It appears to be a metabolic mechanism under enzyme-controlled metabolic reactions and thus requires a certain amount of energy for its operation. However, the usefulness of the clock to the living organism can be demonstrated in many ways.

Bioluminescence

Interesting effector organs among invertebrates are those capable of producing light. The power to produce living light is found in most major groups of organisms. It is found in the Protozoa (Sarcodina and Dinoflagellata) and in virtually all metazoan phyla, both marine and terrestrial, such as Coelenterata, Annelida, Arthropoda (Crustacea, Insecta), Mollusca, and some others. Although the power to produce light has evolved several times in organic forms, the mechanism of light generation is much the same in all forms. The functional significance of light production is not apparent in all cases, but it appears to serve a useful function for the attraction and recognition between sexes of the same species, for protection against predators, and for the attraction of prey. Emission of light appears to provide no survival value for the organism in some species.

The mechanism of light production involves the coordinated action of water, inorganic ions, oxygen, two organic substances called *luciferin* and *luciferase,* and the high-energy bond, adenosine triphosphate (ATP). Light production is a chemical reaction in which luciferin is oxidized in the presence of the enzyme luciferase (Fig. 1-1). Both luciferin and luciferase differ in composition in different species, although they are probably similar in chemical composition, both being proteins. Extracts of

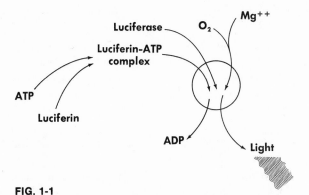

FIG. 1-1

Bioluminescence in animals. The energy compound (ATP) reacts with luciferin to form an active luciferin-ATP complex, which, in the presence of oxygen, magnesium ions, and the enzyme luciferase, emits light. ATP becomes ADP when ATP reacts with luciferin.

luciferin will not luminesce by themselves, but if ATP is added to luciferin, ATP complex is formed. In the presence of water, oxygen, and magnesium ions this complex emits light when luciferase is added. In this process oxygen is used up and ATP is reduced to adenosine diphosphate (ADP). Some forms (ctenophores) will luminesce in the complete absence of dissolved oxygen but may derive their oxygen from some bound form. It is evident that all animals do not produce light in the same way, because a multiplicity of processes are involved.

In most animals the production of light is not a continuous process. Light is usually produced intermittently by nervous stimulation of specialized cells in light-producing organs. However, the control or stimulation is not the same in all organisms. The energy emitted in light production falls within the visible spectrum but may range across a narrow or broad band. The light rays may be of any length, but yellow and blue are more common. More than one color may be given out by different luminous organs in some animals. The intensity of the light emitted varies with different animals, being brightest in the ctenophore *Mnemiopsis.*

So far as is known, no heat is associated with the light, and therefore it is often called "cold light."

Feeding mechanisms

Ciliary, or filter, feeders are common among invertebrates and are found to some extent in larval forms of some vertebrates. This is a primitive feeding method commonly used by many animals for collecting food and getting it into the mouth. That so many animals have met this need in much the same manner indicates that a great deal of convergent evolution has taken place in the mechanisms of ciliary feeding. Forms not closely related often use similar methods. The method, which varies somewhat with different groups, consists essentially of ciliary currents induced by cilia, hairs, or setae that draw food particles into the mouth, often with the assistance of mucus that collects food particles into stringy masses. It is the typical feeding method of most sessile organisms, but is also found in those with active locomotion.

In some protozoans and rotifers, currents of water are produced by the beat of the cilia on the disk, and food particles are carried into the mouth or gullet. In ciliates such as *Paramecium* the mechanism of food collecting is assisted by the currents of water induced by the strong cilia in the oral groove. The disk of rotifers consists of two incomplete rings of cilia that create vortices, and accepted food particles pass between the two rings of cilia into the mouth while rejected food is cast away in places where the rings are incomplete.

Most metazoan invertebrates make some use of ciliated surfaces and mucus in food collecting. Particles of food are collected by the sticky mucus and conveyed to the mouth and inside the body by cilia. The lophophore of tentacles in bryozoans, ectoprocts, and brachiopods utilizes this type of filter feeding. Water currents are driven into the crown of the lophophore between the tentacles, which are provided with cilia for carrying the food into the mouth. Ciliary currents and mucous secretions are used for carrying food along the tentacles of tube-dwelling annelids and the ambulacral groves of crinoids. Many mollusks such as the freshwater clam have elaborated an efficient method of filter feeding. Protochordates, which may be considered the only invertebrates with gill slits, have a variant form of the same method. In the classic *Amphioxus* the wheel organ and other ciliary bodies draw currents of water into the mouth and pharynx, where mucus secreted by the endostyle collects the food particles into stringy masses or nets that are carried by ciliary action through the hyperpharyngeal groove into the intestine. In the intestine the food is digested and absorbed with the mucus. In most filter-feeding devices the suspended particles that are not food or those that are too large are sorted out and discarded.

The transport of the gut food cord has been reviewed by J. E. Morton. Transportation through the intestine of most filter feeders involves some peristalsis, but chiefly it depends on ciliary movements. The gut cord is usually rotated by a rod or ciliated esophagus that aids in circulating the food particles in the stomach and shedding them from the cord for extracellular enzymatic digestion. Many ciliary feeders also retain a great deal of intracellular digestion that requires a sorting of food particles suitable for phagocytosis.

Filter-feeding forms are generally omnivorous because they must depend mainly upon planktonic organisms that contain both plant and animal forms, although some do make a selection.

There are other feeding mechanisms among invertebrates besides filter feeding. In particulate feeders small particles may be ingested by pseudopodia or moved along by ciliary action, ciliary mucus fields, or other ciliary mechanisms. But there are also many ways of

REFERENCES

Ager, D. V. 1963. Principles of paleoecology, New York, McGraw-Hill Book Co. An excellent introduction to the ecologic conditions under which animals and plants lived in the past geologic ages.

Carter, G. S. 1961. A general zoology of the invertebrates, ed. 4. London, Sidgwick & Jackson, Ltd., Publishers. This work is not intended as a text of invertebrates but represents a fine appraisal of the many aspects of invertebrate morphology and physiology.

Handler, P. 1970. Biology and the future of man. New York, Oxford University Press.

Hedgpeth, J. W. 1957. Treatise on marine ecology and paleoecology, vol. 1. New York, Geological Society of America. A pretentious treatise on various ecologic aspects of animal life.

Hutchinson, G. E. (editor). 1970. The biosphere. Sci. Amer. **223**:45-208 (September).

Marcus, E. 1958. On the evolution of the animal phyla. Quart. Rev. Biol. **33**:24-58. An evaluation of the present status of animal phylogeny.

Mayr, E. 1963. Animal species and evolution. Cambridge, Harvard University Press. A masterly analysis of the present basic concepts of evolutionary biology; perhaps the most outstanding treatise of its kind published in the last decade.

feeding on large particles. Some invertebrates fragment large particles by crushing them with mandibles (insects and crustaceans), by other means such as the radula (mollusks), or by the peculiar arrangement of teeth called Aristotle's lantern (sea urchins). Some have internal trituration to assist in the fragmentation or other mechanical devices for preparing food for the final chemical processes of enzymatic action. Earthworms and others simply ingest large quantities of earth from which organic parts are drawn for food. Sucking mechanisms involving piercing mouthparts for the ingestion of fluids or of soft tissues are found in arachnids, some insects, and others. Many internal body parasites simply absorb food that has been predigested by their hosts. Almost any form of natural organic material may be used as food by some invertebrates. However, as common as cellulose is, few invertebrates secrete enzymes (cellulases) for its digestion.

Morton, J. E. 1960. The functions of the gut in ciliary feeders. Biol. Rev. **35**:92-140

Nursall, J. R. 1959. Oxygen as a prerequisite to the origin of the Metazoa. Nature (London) **183**:1170-1172.

Pearse, A. S. 1950. The emigrations of animals from the sea. Washington, D. C., The Sherwood Press. An attempt to explain how animals met and overcame problems in the invasion of the land.

Person, P., and D. E. Philpott, 1969. The nature and significance of invertebrate cartilages. Biol. Rev. **44**:1-16.

Reid, G. K. 1961. Ecology of inland waters and estuaries. New York, Reinhold Publishing Corp. An introduction to the chemical and physical factors that operate in fresh water and estuaries and their bearing on biota.

Rensch, B. 1960. Evolution above the species level. New York, Columbia University Press. A modern, synthetic theory of evolution, based on systematics, comparative morphology, and paleontology.

Rothschild, Lord N.M.V. 1961. A classification of living animals. New York, John Wiley & Sons, Inc. Includes an index of orders, suborders, and many genera.

Simpson, G. G. 1961. Principles of animal taxonomy. New York, Columbia University Press. A useful, up-to-date account of the principles underlying classification. No specific treatment is given of animal taxa.

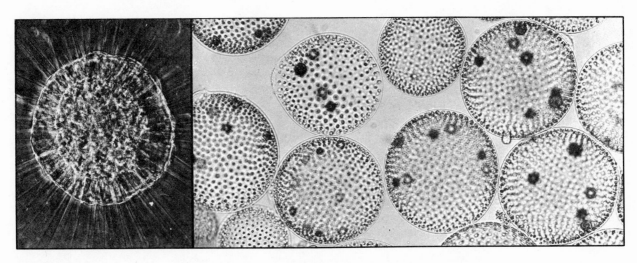

Part II □ The acellular animals

PHYLUM PROTOZOA

Phylum Protozoa

GENERAL CHARACTERISTICS

1. Early single-celled organisms evolved into Monera (bacteria and blue-green algae) and into Protista (algae, fungi, slime molds, and protozoans). Monera have dispersed genetic material without nuclear membranes; the Protista have nuclear membranes. Both Monera and Protista may have arisen from the same kind of early cell (monophyletic origin), or they may have come from different types of cells (polyphyletic origin). The Protista as a group evolved four methods of nutrition—photosynthesis, saprotrophism, parasitism, and holotrophism. Primitive protists may have had both animal and plant characteristics together. From such a plant-animal organism a purely photosynthetic branch, or plant, and a purely heterotrophic branch, or animal (Protozoa), is supposed to have evolved.

2. A protozoan is therefore a protistan that is exclusively heterotrophic (acquires food from preexisting organisms). This heterotrophism may be expressed as holotrophism (bulk feeding), parasitism, or saprotrophism. However, by tradition some photosynthetic flagellates (often considered to be algae) are also included in the general scheme of protozoans.

3. Since protozoans are complete organisms and correspond to an entire multicellular animal, they are often called acellular, in which case they could be considered masses of protoplasm not divisible into cells. It is a matter of small importance whether they are called acellular or unicellular, but all the basic functions of the higher organisms are carried out by a single cell in the Protozoa. On the other hand, the cell concept implies that the functional units of the cell are chiefly differentiated into nuclear, or mostly genetic, structures and the cytoplasm, or mostly nongenetic, structures. In this sense, a protozoan cell is homologous to a metazoan cell.

4. Protozoan evolution is similar to that of multicellular animals, since protozoans have specializations within the single cell that correspond with the differentiation of cells and tissues of the metazoan. Within the single cell of the protozoan the specialized unit is called an organelle.

5. Within the restrictions of a single cell the Protozoa or their ancestors have furnished the roots of all the functional units that are complicated and specialized in the metazoan form, such as the differentiation of sex, stimulus-receptor response, skeletal structures, and contractile structures.

PHYLOGENY AND ADAPTIVE RADIATION

1. Protozoans probably have had a polyphyletic origin rather than a monophyletic one. Since the members of a phylum should have a monophyletic origin, protozoans probably should have a higher rank than that of phylum, e.g., superphylum. The four groups we shall consider are the flagellate, ameboid, spore-forming, and ciliate protozoans. They have established certain interrelations, but many aspects of these relationships are obscure.

2. The protozoan body type has been successful in its adaptations to many varied habitats

and ecologic niches. They have been guided in their evolution by the basic adaptive features of their cortex or ectoplasm. Within a single cell plan they have undergone division of labor and the formation of organelles that have enabled them to meet the requirements of particular habitats. They are found in all trophic levels of the communities they inhabit. They are fitted to a great extent to occupy most of the ecologic niches in the ecosystem, and perhaps because of their small size, some niches that are denied to multicellular forms. All modes of existence, whether free living or symbiotic, are represented among them.

CLASSIFICATION

The following classification of Protozoa is based mainly on that proposed by B. M. Honigberg and others in 1964. Any scheme for protozoans at the present time is more or less tentative. Some taxa have been omitted and only the larger ones are defined. Common examples are given in most cases, although the restrictions of space prevent others from being used as examples.

Phylum Protozoa. Unicellular (acellular) or colonial; mostly microscopic; 30,000 to 40,000 named species.
Subphylum Sarcomastigophora. Monomorphic nuclei; locomotor organs of flagella or pseudopodia, or absent.
  Superclass Mastigophora. Having flagella for locomotion; autotrophic, heterotrophic, or both.
  Class Phytomastigophorea. Plantlike flagellates with chromoplasts.
    Order Chrysomonadida. *Synura.*
    Order Cryptomonadida. *Chilomonas.*
    Order Dinoflagellida. *Noctiluca.*
    Order Volvocida. *Volvox.*
    Order Euglenida. *Euglena.*
    Order Chloromonadida. *Gonyostomum.*
    Order Heterochlorida. *Heterochloris.*
  Class Zoomastigophorea. Flagellates without chromoplasts.
    Order Rhizomastigida. *Mastigamoeba, Tetradimorpha.*
    Order Kinetoplastida. *Trypanosoma.*
    Order Choanoflagellida. *Proterospongia.*
    Order Diplomonadida. *Giardia.*
    Order Retortamonadida. *Chilomastix.*

    Order Hypermastigida. *Trichonympha.*
    Order Trichomonadida. *Trichomonas.*
  Superclass Opalinata. Body with longitudinal, oblique rows of flagella.
    Order Opalinida. *Opalina.*
  Superclass Sarcodina (Rhizopoda). Pseudopodia as feeding and locomotor.
  Class Actinopodea. Pseudopodia radiating from spherical body.
    Subclass Heliozoia. *Actinosphaerium.*
    Subclass Radiolaria. *Thalassicolla.*
    Subclass Acantharia. *Acanthometra.*
    Subclass Proteomyxidia. *Vampyrella.*
  Class Rhizopodea. Having lobopodia, filopodia, or reticulopodia.
    Subclass Lobosia
      Order Amoebida. *Amoeba.*
      Order Arcellinida (Testacida). *Arcella.*
    Subclass Filosia. With tapering and branching filopodia. *Allogromia.*
    Subclass Granuloreticulosia
      Order Foraminiferida. *Globigerina.*
    Subclass Mycetozoia. Slime molds.
  Class Piroplasmea. Rod-shaped or ameboid parasites. *Babesia.*
Subphylum Sporozoa. Usually with spores but no polar filaments.
  Class Telosporea. Spore formation after growth of sporozoon.
    Subclass Gregarinia. *Monocystis.*
    Subclass Coccidia. *Eimeria.*
  Class Toxoplasmea. Spores lacking and only asexual reproduction. *Toxoplasma.*
  Class Haplosporea. Spores and only asexual reproduction. *Haplosporidium.*
Subphylum Cnidospora. Spores with polar filaments.
  Class Myxosporidea. *Ceratomyxa.*
    Order Actinomyxida. *Synactinomyxon.*
    Order Myxosporida. *Unicapsula.*
  Class Microsporidea. *Nosema.*
Subphylum Ciliophora. Cilia present in adult or young stages; tentacles in some.
  Class Ciliata. With characters of the subphylum.
  Subclass Holotrichia. Cilia usually over entire body surface.
    Order Gymnostomatida. *Coleps.*
    Order Chonotrichida. *Spirochona.*
    Order Trichostomatida. *Colpoda.*
    Order Hymenostomatida. *Paramecium.*
    Order Astomatida. *Anoplophrya.*
    Order Apostomatida. *Foettingeria.*
    Order Thigmotrichida. *Boveria.*
  Subclass Peritrichia. Restricted ciliation; enlarged disk at apical end.
    Order Peritrichida. *Vorticella.*

Subclass Spirotrichia. Reduced body cilia and well-developed buccal ciliature.

    Order Heterotrichida. *Stentor.*

    Order Oligotrichida. *Halteria.*

    Order Tintinnida. *Tintinnus.*

    Order Entodiniomorphida. *Entodinium.*

    Order Odontostomatida. *Saprodinium.*

    Order Hypotrichida. *Euplotes.*

Subclass Suctoria. Stalked with tentacles; no ciliature.

    Order Suctorida. *Podophrya.*

FOSSIL RECORD

Protozoan fossils do not reveal a great deal about the evolution of the phylum. Most protozoan fossils belong to the orders Radiolaria and Foraminiferida because the members of these two orders are provided with shells, or tests, that fossilize fairly easily. Radiolarian fossils have been found in Precambrian rocks, but they show us little about the nature of the earliest protozoans, which were probably without shells and were not preserved. The early foraminiferans, which are the most abundant of all the protozoan fossils, did not produce calcareous shells until the Jurassic period; before this their tests were largely composed of sand or other foreign substances. Most of the largest forams became extinct in the remote past. They were important in the formation of chalk, limestone, and marble, and their tests make up a large part of the sediment covering the ocean floor.

Other groups of Protozoa that were fossilized to some extent were the dinoflagellates, the silicoflagellates, and a few test-bearing ciliates. Many protozoan fossils bear remarkable resemblance to modern species. Most of their fossils are of marine origin.

The study of protozoan fossils has practical importance. Many protozoans were sensitive to temperature changes of the oceans, and their cycles of abundance or decrease indicate climatic changes of past geologic periods and can be used to predict future climatic cycles. But one of their most important utilitarian aspects is the information they yield about oil deposits. This is especially true of foraminiferans that have played a role in the formation of hydrocarbons. By examination of borings and castings from wells and quantitative measurements of fossils, it has been possible to locate geologic traps or pockets of oil.

ECOLOGIC ASPECTS

Protozoans are especially suited for moist conditions, but they may be found in a dormant state of encystment wherever life can exist. Protozoans have many means of dispersal or distribution because they are small and active, and their cysts can be carried by air or water currents, the feet of birds, digestive systems of animals, and in innumerable other ways.

Protozoans possess great adaptibility and acclimation to a wide variety of habitats and have occupied an enormous range of ecologic niches. They are found in water everywhere, from the arctic regions to fairly hot springs with temperatures of more than 60° C. Many have the power to acclimate themselves by gradual change to temperatures and other environmental conditions far beyond their usual range. In dry regions (deserts, arid soils, mountain elevations, etc.) their cysts are viable for years. They are found in both fresh and salt water but are more abundant in standing water than in running water. Some are definitely pelagic and others are benthonic. They may be found at all depths of the sea.

Limiting factors necessarily vary with different species. Moisture seems to be necessary for reproduction in all species, although many live encysted for a long time in extreme dessicating conditions. All species, perhaps, have certain optimum ranges of restricting factors. Many species thrive best within the range of 10° to 25° C. In the cyst state, however, they can withstand much lower temperatures—even freezing in some cases—although freezing conditions are fatal in most instances. Most species favor a pH of neutrality or one that is slightly alkaline. Adaptation to the salt content of water naturally varies with marine and freshwater forms, but many protozoans can be acclimated gradually either way. Some ciliates, such as *Prorodon,* are found in the Great Salt Lake, where the salt content varies from 20% to 27%. Nutritional requirements, such as certain minerals, amino acids, vitamins, and sugars, are also limiting factors, but these requirements vary greatly. They all require molecular oxygen or oxygen from some other source.

All forms of relationships are found in Protozoa. Many types of symbiosis, both useful and harmful, are found in the ecologic interactions of Protozoa. Mutualism, or a state of mutual benefit to both partners, is well illustrated by zoochlorellae and zooxanthellae, algae that live in certain protozoans. The same relationship is also found between termites and their flagellates. Commensalism, in which one partner receives benefit without harm to the other, is well shown by *Kerona* and *Trichodina,* which use the hydra's body surface as a substratum—although these species may be ectoparasitic. Parasitic forms are found in all classes of Protozoa, and the Sporozoa are exclusively parasitic. It is thus seen that protozoans have their ecologic niches just as do higher forms. In nature each type of protozoan inhabits a particular territorial range, and within its ecological niche it uses the raw materials for its existence and gives off byproducts. Many factors, both physical and biotic, may regulate the environment where protozoans may or may not exist.

Within recent years much investigation (as mentioned in Chapter 1) has been done on the interstitial fauna (psammon) in that biotope made up of sand grains of marine, brackish, and freshwater beaches (B. Swedmark, 1964). The minute fauna found here is probably made up of representatives of every animal phylum, including many protozoan forms. Within this habitat of sand grains, air, water, and detritus, the interstitial fauna live in the passages and spaces between the grains of sand or on the surface of the sand grains (A. C. Borror, 1968). Although such an environment may undergo drastic and abrupt changes in oxygen supply, tidal effects, rain, and extremes of temperature and light, most members of the psammon escape these disturbing factors by shifting to more favorable regions. The nature of the sand grains and their location (marine or freshwater) also plays a part in the ecology of the forms found there. The interstitial milieu supports an enormous population of animals. Most types of eating habits—herbivores, carnivores, omnivores, detritus eaters, scavengers, etc.—are represented here. Important food sources are the flora of bacteria, algae, and diatoms. Among the Protozoa, ciliates and foraminiferans seem to be most common in the psammon.

GENERAL FEATURES: STRUCTURE AND FUNCTION

Most protozoans are microscopic, but some may be 2 to 3 mm. long, and are visible to the naked eye. The smallest may not be more than 2 microns long; the largest are certain foraminiferans that may exceed 70 mm.

The body may be very plastic and without a limiting pellicle, as in the ameba, which has only a thin cell membrane; however, many protozoans have a body whose rigidity is determined by surface and internal structures. These factors determine the form, which varies greatly throughout the phylum. All varieties of symmetries are found, but strict bilateral symmetry is uncommon because of spiral torsion. The kinds of adjustment protozoans make to their habitats are often correlated with their shape. Radial symmetry and conical shape is the rule among sessile forms such as the vorticellids and their relatives; spherical symmetry or its modifications are more common among floating types, for example, *Volvox.* Creeping forms are usually flattened. Many protozoans are asymmetric.

Besides the cell membrane, protistans often have other body coverings of nonliving secreted materials. These range from a variety of pellicles to rigid tests or shells. Some secreted coverings consist of organic materials such as cellulose or tectin and may be gelatinous in structure. They often contain embedded materials such as minerals, sand grains, or diatom shells. Dinoflagellates may have a hardened armor of cellulose plates in elaborate designs. Tests and shells contain mostly inorganic substances. Foraminiferans build shells of calcium carbonate. Radiolarians may have complex internal skeletons of silica or strontium sulfate.

The protoplasm of protozoans is usually divided into distinct regions—a relatively clear ectoplasm and a somewhat opaque endoplasm. This differentiation is best seen in some ameboid forms in which the colloidal reversibility of ectoplasm and endoplasm is evident.

The ectoplasm contains many structures and, as already stated, is chiefly responsible for the varied evolutionary trends of the phylum. Among the structures the ectoplasm bears are the trichocysts and toxicysts of ciliates, the nematocysts of some dinoflagellates, and the polar capsules of the Cnidosporidia. Some of these are offensive and defensive organelles, and others may be used for anchoring the organism to the substrate. Trichocysts are minute oval-shaped bodies in the deeper layer of the pellicle and are arranged at right angles to the body surface (M. A. Jakus, 1945). When stimulated, they discharge long filaments through pores in the pellicle (Fig. 2-39, C to E). The polar capsules of cnidosporidians and the nematocysts of dinoflagellates discharge spirally coiled threads that resemble those of coelenterate nematocysts. The peculiar sucking tentacles of the order Suctorida are also ectoplasmic structures, as are the various types of locomotor organelles such as pseudopodia, flagella, and cilia.

Ciliary arrangement varies with the species. In some species the cilia are arranged in rows and spirals originating in the region of the mouth and proceeding to the posterior region. Many cilia may be grouped together to form compound organelles such as membranes, membranelles, and cirri (Fig. 2-39, A and F). Typically each cilium has a basal granule (kinetosome) from which a filament (basal fibril or kinetodesma) extends into the cytoplasm, joining with other filaments to form a longitudinal bundle running parallel to the row of cilia (Fig. 2-38). These ectoplasmic fibrils form the neuromotor apparatus, which conducts stimuli and coordinates locomotor activities. In some complex ciliates, fibrils radiating out from a central organelle, or motorium, control movements of fused cilia called cirri.

Both cilia and flagella show by the electron microscope a similar pattern of structure (P. Satir, 1961). There is no absolute distinction morphologically and physiologically between a cilium and a flagellum (H. P. Brown and A. Cox, 1954). Cilia are usually present in greater numbers, are shorter in length, and have a rhythmic coordination over a ciliated surface. Flagella, on the other hand, are usually longer and are present in fewer numbers (although there are some exceptions, e.g., in the symbiotic flagellates of termites). The beat of a flagellum is often symmetric, with several waves included within a short instance. The beat of a cilium is asymmetric and involves only one wave with two marked phases—an active phase and a re-

Table 2. Dimensions of some cilia and flagella*†

Organism and part of body	Organelle	Length (μ)	Diameter (μ)
Peranema	Flagellum	100	1
Trichonympha	Flagellum	20-150	0.25
Paramecium	Cilium	10-12	0.27
Mnemiopsis (comb plate)	Cilium	2,000	0.3
Rattus (trachea)	Cilium	5	0.24
Mus (retinal rod)	Cilium	0.7	0.16-0.24
Homo			
Entire sperm tail	Sperm tail	55	
Middle piece		4-5	1
Principal piece		45	0.3-0.4
End piece		5-7	0.18

*Modified from many sources.
†A sheath is present in some of those listed.

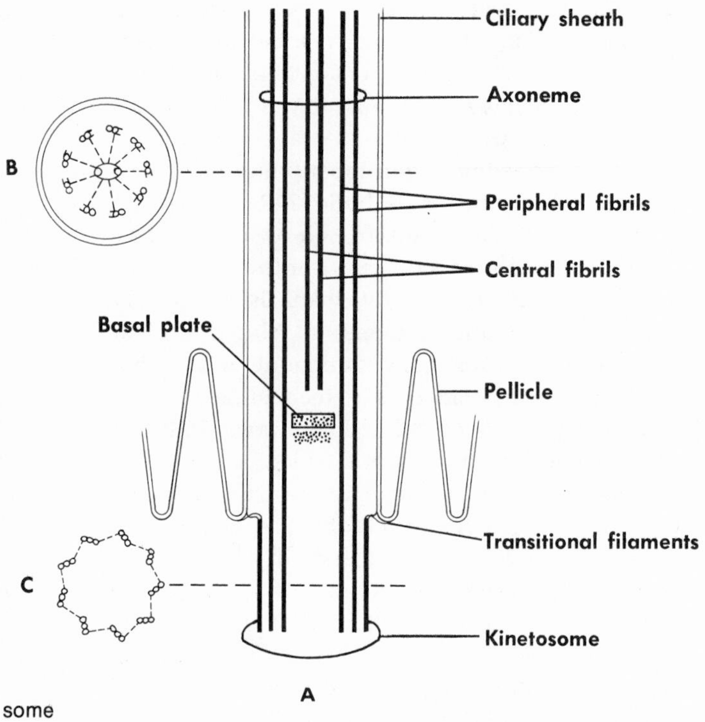

FIG. 2-1

A, Fine structure of a cilium or flagellum. **B,** Transverse section of cilium showing typical arrangement of fibrils. **C,** Transverse section of basal body (kinetosome) showing position of third subfibrils and twisted arrangement of fibril triplets.

Ciliary sheath

Axoneme

Peripheral fibrils

Central fibrils

Basal plate

Pellicle

Transitional filaments

Kinetosome

A

FIG. 2-2

Electron micrograph of cross section through some flagella of *Pseudotrichonympha.* (×140,000.) (Courtesy I. R. Gibbons, Harvard University, Cambridge.)

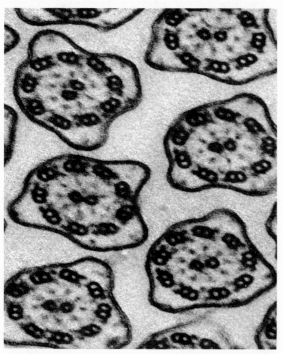

covery phase. Typically there are only one or two flagella per cell. Cilia commonly have a length of 3 to 20 microns, but may be longer when combined into compound structures. Flagella may be as short as 5 microns and as long as as 150 to 200 microns. Sperm tails of metazoans may have flagella much longer. Although cilia and flagella may vary in length, their diameter is more constant, varying usually between 0.1 and 0.3 micron. (See Table 2.)

A flagellum or cilium consists of an outer sheath or membrane enclosing a structureless matrix, within which is the axoneme. The membrane forming the sheath is continuous with the plasma membrane of the cell surface (Fig. 2-1). In the axoneme 9 longitudinal tubular fibrils form a cylinder surrounding 2 longitudinal central fibrils, thus forming the well-known 9 + 2 pattern (Figs. 2-1 and 2-2). This uniformity of pattern may indicate a significant correlation between the makeup of the flagellum or cilium and its function, as well as an evolutionary implication. The 9 peripheral fibrils are double (each with 2 subfibrils), whereas the 2 central ones are single. The 9 peripheral fibrils merge at their base to form a hollow basal granule (the kinetosome or blepharoplast) and are continuous with the 9 triplet microtubules of the basal body (Fig. 2-1). The 2 central fibrils of the axoneme terminate at the cell surface near a transverse plate and do not reach the basal body. One of the 2 subfibrils of each peripheral fibril bears a double row of short arms that point in a clockwise direction. The proximal lumen of the basal body may have an elaborate cartwheel structure made up of delicate fibrils that run from a central ring to the inner subfibril of each triplet fibril. Some flagella may bear rows of delicate filaments (mastigonemes) along their outer surface.

The similarity of the basal body, or kinetosome, to the centriole (the active agent in cell division) indicates a close relationship between the two. It is known that a basal body may act as a centriole during cell division and that a centriole may be connected with the basal body by a fibril called a rhizoplast. Both the basal body and centriole contain an identical arrangement of peripheral fibrils so that the two structures are interchangeable (E. Fauré-Fremiet, 1961). Attached to each basal body (in ciliates) is a kinetodesmal fibril, which passes to the left and joins other similar fibrils from the basal bodies to form a compound fibril, known as a kinetodesma (Fig. 2-38). A kinety is a row of basal bodies and their kinetodesma. All of the kinetia taken collectively (J. N. Grim, 1970; A. V. Grimstone, 1961) form the neuromotor system (infraciliature).

Flagella and cilia perform many functions besides locomotion of protozoans. Flagella in the collar cells of sponges create feeding currents. They also propel sperm in most animals and fluid in the pronephridia of flatworms. Many low forms use the coordinated beat of cilia in filter-feeding devices, and cilia are especially effective in moving fluids over surfaces and through tubes. For producing distant currents, cilia may be longer and larger, or they may be parts of membranelles or other compound structures. One organelle associated with movement is the undulating membrane of the trypanosomes and some others (P. C. C. Garnam, 1966).

Little is known about the role of the various parts that go to make up the cilia and flagella. The infraciliature provides coordination for the rhythmic movement of cilia. Nervelike transmission is not now considered valid, although the beat of cilia and their rhythmic activities can be altered by experimental nervous stimulation, but the real details of coordination at present remain unknown. Both contraction and compression elements are involved. No simple theory can account for the control and coordination observed in ciliary movement. Ciliates are known to have a body, or somatic, ciliature and an oral ciliature around the mouth region (C. B. Metz and associates, 1954). However, the distribution of cilia shows many variations— from primitive forms with the entire body covered to those that have restricted regions of cilia distribution. When a ciliate divides in cell division, new kinetia are formed by the growth and development of new kinetosomes.

Pseudopodia are retractile extensions of the body and exist in several forms, such as the lobopodia with broad and rounded tips; axopodia with axonemes in their slender bodies; and filopodia (rhizopodia), which are also slender, tapering, without axonemes, and tend to branch and fuse (Fig. 2-3). Other types of pseudopodia include the filamentous myxopodia that branch and anastomose into networks. In some foraminiferans a reticulum of fine branching and anastomosing pseudopodia extends out at the test opening, and in each pseudopodium there is a two-way streaming movement of granules passing outward on one side and inward on the other (T. L. Jahn and R. A. Rinaldi, 1959).

The ectoplasmic cortex also includes myonemes, or contractile fibrils, found in certain ciliates and sporozoans.

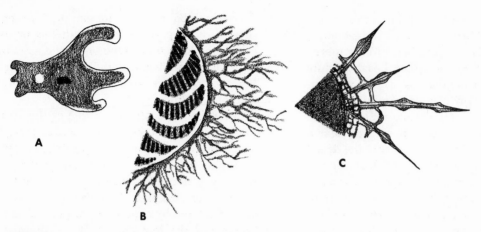

FIG. 2-3

Types of pseudopodia. **A,** Lobopodia, as in *Amoeba.* **B,** Portion of foraminiferan showing rhizopodia. **C,** Portion of heliozoan showing axopodia. Each axopodium contains axoneme surrounded by cytoplasmic sheath.

Endoplasmic inclusions consist of chromatophores, mitochondria, stigmata, nuclei, food vacuoles, water expulsion vesicles (contractile vacuoles), and others. Chromatophores are common in phytoflagellates, which, strictly speaking, do not belong to Protozoa. Chromatophores contain chlorophyll that may be masked by other colored pigments. Pyrenoid bodies are often associated with chromatophores and may function in starch formation. Starch-related carbohydrates known as paramylum bodies are found as food reserves in many phytoflagellates, including *Euglena.* Light-perceiving organelles such as stigmata are also characteristic of many phytoflagellates, whereas more complex photoreceptors (ocelli) are present in some dinoflagellates. Symbiotic bacteria and green flagellates occur in many Protozoa.

Mitochondria, which are of great importance in cellular metabolism, are very small rodlike or spherical bodies that have been described in many species of Protozoa and may be found in all Protozoa. Another common cytoplasmic body, the Golgi apparatus, appears to be present in some protozoans according to studies with the electron microscope (P. P. Grassé and N. Carasso, 1957). Dictyosomes (disk-shaped organelles) and parabasal bodies (long cylindric structures usually attached to a centriole by a filament) exist in some classes of Protozoa. They are probably secretory in function, the same as the Golgi apparatus.

The contractile vacuole, now better known as the water expulsion vesicle, is found in most freshwater protozoans, as well as some marine and parasitic species. In most species the water expulsion vesicle is a reservoir fed by pulsating radial canals, which discharge their contents into the vesicle during the filling stages of the latter (Fig. 2-4). Each radial, or collecting, canal has a swollen ampulla near its medial end and is surrounded with a sponge network of nephridial tubules. During the early part of evacuation of the vesicle, water flows back into the collecting tubules forming the ampullae while the vesicle remains collapsed. The radial canals and their ampullae then empty into the main vesicle as it expands. The vesicle then moves to an outlet pore and is anchored by a system of microtubules. Rupture of the pore membrane is effected by the radial tension of the microtubules (A. M. Elliott and I. J. Bak, 1964) and by the contents of the vesicle. A new pore membrane re-

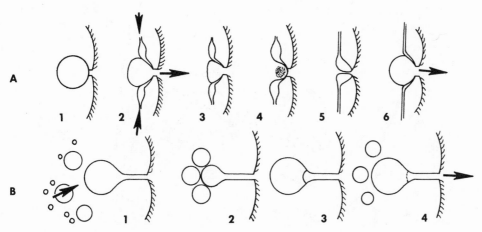

FIG. 2-4

Operation (diagrammatic) of two types of water expulsion vesicles (contractile vacuoles). **A,** In *Paramecium multimicronucleatum,* vesicle in **1** is full, with membrane closing pore; in **2** and **3** pore is open, vesicle emptying, collecting canals filling; **4,** pore is closed; **5** and **6,** vesicles fuse to form water expulsion vesicle (contractile vacuole). **B,** In *Paramecium trichium,* water expulsion vesicle is fed by smaller vesicles, has no collecting canals; vesicle empties by permanent tube and pore (many species lack tube). **1,** Vesicle full; secondary and tertiary vesicles forming; **2** and **3,** vesicle empties, then refills by fusion of secondary vesicles formed by fusion of tertiary vesicles; **4,** primary vesicle full again. (Modified from King.)

generates after systole, or the collapse of the water expulsion vesicle (I. L. Cameron and A. L. Burton, 1969). There is often more than one expulsion vesicle in a particular species, and the outlet pore may or may not be in a specific place. They are mostly osmoregulatory in function, but they may have other functions, including a slight one of excretion of ammonia as the chief nitrogenous waste.

Protozoa have various devices for obtaining their food. In saprozoic feeding, food in solution simply diffuses through the body wall. In holozoic feeding, which seems to be the most common method, solid particles pass through temporary or permanent cytostomes, and the food is digested intracellularly in food vacuoles (A. M. Elliott and G. L. Clemmons, 1965; S. O. Mast, 1947). Some protozoans synthesize their

own food by photosynthesis (autotrophic or holophytic nutrition); others make use of symbiotic algae living within them. Combinations of these methods may be found in the same species. Suctoridans have unique sucking tentacles by which the liquid contents of their prey are drawn into food vacuoles (Fig. 2-41).

Protozoans other than the Ciliophora have only one kind of nucleus, but there may be several nuclei in the same organism. In addition to chromosomes, a nucleus usually contains an endosome (karyosome), which is a central mass of chromatin material (Fig. 2-5, *H* and *I*). Electron microscopic studies indicate that the nuclear membrane consists of two membranes (J. D. Robertson, 1960), each about 70 to 80 Angstroms (Å) thick and separated from each other by a space of 300 to 400 Å. Many pores up to 800 Å in diameter are found in the membrane. The inner and outer membranes meet at the boundaries of the pores. The Ciliophora have two kinds of nuclei—the micronucleus and the macronucleus. The macronucleus, the larger of the two, is concerned mainly with the metabolic activities of the organism, but it also has a regenerative capacity; on the other hand, the micronucleus is concerned with the

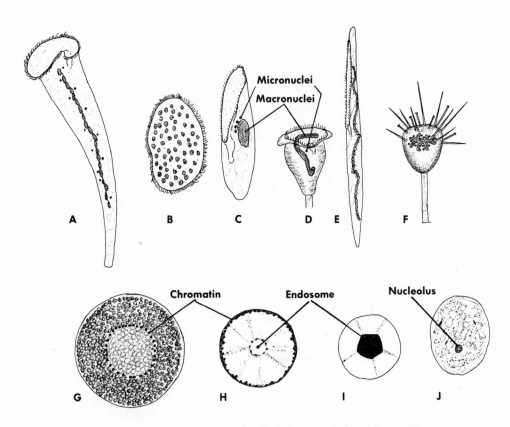

FIG. 2-5

Types of nuclei. **A** to **F,** Ciliophorans with both micronuclei and macronuclei (except **B**). **A,** *Stentor.* **B,** *Opalina* with many vesicular nuclei and no micronuclei; shows more affinity to flagellates than to ciliophorans. **C,** *Paramecium multimicronucleatum.* **D,** *Vorticella.* **E,** *Spirostomum.* **F,** *Ephelota,* a suctoridan. **G** to **J,** Types of vesicular nuclei. **G,** *Endamoeba.* **H,** *Entamoeba.* **I,** *Endolimax.* **J,** Typical vesicular nucleus with nucleolus. Endosomes contain chromatin, nucleoli do not.

genetic functions of the organism. There are usually not more than two of each kind of nuclei, but some protozoans have more. Protozoan nuclei vary a great deal with the group and the species. They may be of almost any shape such as oval, crescent, ribbonlike, or beaded (Fig. 2-5).

Both asexual and sexual methods of reproduction are found in Protozoa. Asexual reproduction is present in all protozoans and is the only method in many. Reproduction involves division (fission) of the body and is usually preceded or accompanied by nuclear division. The most common asexual method is simple binary fission that is usually mitotic (i.e., involving precise chromosome distribution) and may be transverse, longitudinal, or oblique (Fig. 2-6). There are variant forms of fission. Ordinary binary fission results in 2 daughter cells about equal in size; budding produces organisms unequal in size; plasmotomy is division of multinucleate forms into multinucleate progeny without division of nuclei; and schizogony (multiple fission), is the formation of a number of uninucleate buds from a multinuclear parent. Some organelles may be self-reproducing, such as blepharoplasts of flagella, basal granules (kinetosomes) of cilia, and parabasal bodies of trypanosomes.

Sexual reproduction is found in some protozoans and includes such processes as syngamy, conjugation, and autogamy. Syngamy involves the fusion of two gametes that may be similar

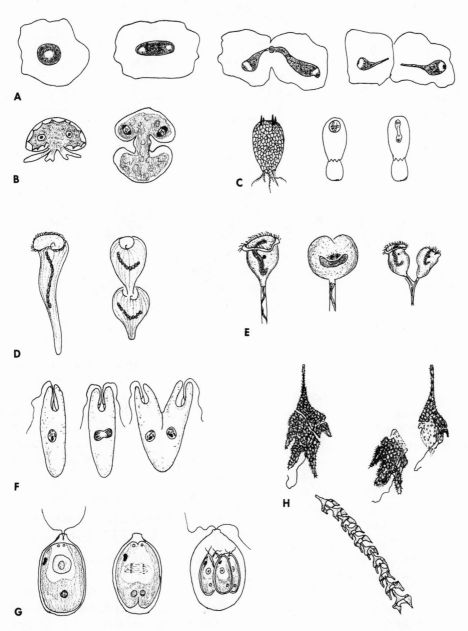

FIG. 2-6

Binary fission in protozoans. **A** to **C,** Sarcodinians. **D** and **E,** Ciliates. **F** to **H,** Flagellates. **A,** *Amoeba proteus,* typical mitotic division. **B,** *Arcella* extrudes cytoplasm and one of its two nuclei; cytoplasm secretes new shell. **C,** *Euglypha.* **D,** *Stentor,* transverse fission. **E,** *Vorticella,* longitudinal fission. In **D** and **E** micronuclei divide mitotically; macronuclei divide amitotically. **F,** *Euglena.* **G,** *Chlamydomonas,* multiple fission. **H,** *Ceratium* divides diagonally, each half regenerating missing parts; apical horn of one may remain attached to girdle of another, forming chains, with original halves forming ends of chains.

FIG. 2-7

Some protozoan cysts.

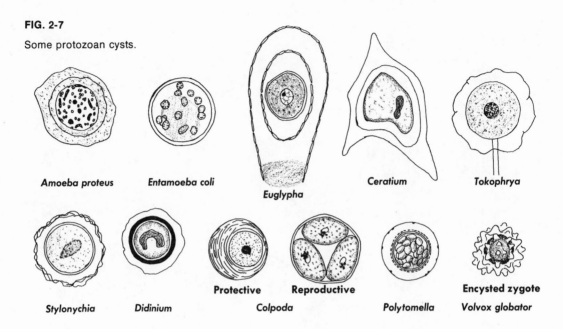

Amoeba proteus   Entamoeba coli   Euglypha   Ceratium   Tokophrya

Stylonychia   Didinium   Protective   Reproductive   Polytomella   Encysted zygote
Colpoda   Volvox globator

(isogametes) or dimorphic (anisogametes). This process usually involves meiosis, which may occur in the formation of the gametes or after their fusion (postzygotic). Conjugation is an exchange of gametic nuclei between two animals that adhere together temporarily, followed by about three divisions of the micronucleus (Fig. 2-42). In conjugation a zygote nucleus, or synkaryon, is formed in each conjugant. Autogamy is more restricted than conjugation and involves similar nuclear behavior except that there is no exchange of gametic nuclei. Instead there is a fusion of gametic nuclei in the same individual that produces the gametes.

The life cycle of protozoans, which may be very complex, often involves encystment at some stage (Fig. 2-7). The process varies greatly among the protozoans and is absent in a few species. In general, encystment involves an accumulation of reserve food, loss of water, a rounding up of the organism, dedifferentiation and loss of organelles, and secretion of an enclosing wall of chitin, mineral salts, keratin, or other materials. Cysts vary greatly in form. Most have a spherical or discoidal shape.

Encystment has many advantages, such as providing the protozoan with protection against desiccation, extreme temperatures, and harsh environmental factors and with increase in longevity and widespread distribution. Some cysts survive for years. Reproductive phases such as fission, budding, and syngamy may occur in the cysts of some species. No encystment has been found in *Paramecium,* and it is rare or absent in marine forms.

Subphylum Sarcomastigophora
SUPERCLASS MASTIGOPHORA

The superclass Mastigophora is made up of members that bear at some stage (usually the adult stage) one or more flagella. They represent a large heterogeneous assemblage of species and are regarded as being closer than any other protozoan group to the primitive ancestral stock of acellular forms. Often they are regarded as a type of connecting link between animals and algae. Of all protozoans, they have the most varied characteristics of size range, life cycles, types of reproduction, modes of living, and types of nutrition. Many are free living, but others are parasitic. They have a widespread distribution, and some can exist in regions devoid of other forms of life if physical conditions are not too harsh. Stagnant pools contain them whereever carbon dioxide is abundant. They are the most abundant members of both freshwater and marine plankton. A few species can live in both fresh water and sea water if the transfer is gradual. Most of them are solitary and free swimming, others are sessile and stalked, and still others form colonies of varied structures and relationships.

The Mastigophora may be divided into two subdivisions or classes, the Phytomastigophorea and the Zoomastigophorea. Members of the former are often referred to as the phytoflagellates, or plantlike forms, and those of the latter as the zooflagellates, or animallike forms. The principal differences between the two groups are their manner of nutrition and reproduction. Phytoflagellates mostly have chromatophores that contain the chlorophyll necessary for photosynthesis, and their nutrition is holophytic. Starch or paramylum is their chief anabolic food product. Zooflagellates lack chromatophores, and they have animal starch or glycogen. Their nutrition is either holozoic or saprozoic. Some phytoflagellates also have sexual stages that are lacking in nearly all zooflagellates. Phytoflagellates, moreover, are usually provided with a pellicle or cellulose cell wall and rarely have more than two flagella. It is thought that zooflagellates have been derived from phytoflagellates by the loss of pigment in the latter.

In the evolution of the flagellates it seems that the loss of chlorophyll was one of the first changes, followed by changes to a saprozoic, holozoic, or symbiotic method of nutrition. Colorless heterotrophs occurred frequently by loss of plastids. Actually, such changes can be produced by artificial methods—for example, the production of colorless euglenas by treatment with streptomycin. Present primitive zooflagellates are usually free living, holozoic, and resemble colorless flagellate algae. The collared flagellate *Protospongia* (order Choanoflagellida), for instance, resembles certain flagellate algae and is of interest because of its possible affinity with the sponges (Fig. 2-16). The evolutionary trend among the zooflagellates has been toward a parasitic life, some possessing complex life histories. Although most of the small primitive members of this group have only one or two flagella, some of the highly specialized symbiotic species have hundreds of flagella attached to ribbon-shaped kinetosomes, e.g., *Trichonympha*, the symbiont of termites (Fig. 2-18).

Colony formation is another trend in mastigophorans. Some colonies are simple aggregates of nonmotile cells in a gelatinous matrix; others are free swimming and are composed of hundreds and even thousands of cells. These larger colonies show some differentiation into somatic and reproductive cells, as in *Pleodorina* (Fig. 2-10, *D*) and *Volvox* (Fig. 2-11). Colonial forms may be spherical, flattened, linear, tubular, dendroid, etc. Some are compact colonies and resemble certain stages in the embryology of the Metazoa. Certain authorities suggest that ancestral protozoans were similar to the simpler flagellates of the present time and gave rise on the one hand to algae and higher plants and on the other hand to colorless flagellates, sarcodinians, and other protozoans. The relationship between flagellates and the Sarcodina must be a close one because some flagellates lose their flagella and become wholly ameboid in their movements.

GENERAL FEATURES

The most common feature of mastigophorans is the presence of one or more flagella, although this organelle is absent in a few of them. The form of these animals may be oval, spherical, elongated, etc., usually with definite anterior and posterior ends. Only a cell membrane or a very thin pellicle is present in those that have free ameboid movement or can change their body shape. Many, however, have firm pellicles that may be ridged in various ways. In addition, calcareous or siliceous shells and armors of considerable variety are common in the group. Differentiation between ectoplasm and endoplasm is usually not distinct. Some species have a pit or groove (cytopharynx) at the anterior end that serves as the insertion of the flagella but is probably not for food intake. Chromatophores or chloroplasts of various shapes are found in the colored species. The chromatophore may consist only of chlorophyll, which may be hidden by red, brown, or other pigments. Hematochrome, a reddish pigment, may be scattered through the cytoplasm. It may shield the chromatophores from certain rays, and it is often concentrated in the cup forming the stigma, or eyespot, of some species. Pyrenoids, the protein bodies concerned with starch formation, are found in association with the chloroplasts.

Water expulsion vesicles are found in many flagellates, especially freshwater forms. The water expulsion vesicle is usually formed by the fusion of small vacuoles and discharges into the cytopharynx, or reservoir, if one is present.

In some dinoflagellates there are nematocysts similar to those in coelenterates and also trichocysts, the organelles so common among ciliates.

Most flagellates have a single nucleus, but a few (e.g., *Giardia*) have two. The nucleus may have an endosome (karyosome) and may be quite large in some species and small in others.

Locomotion. The typical flagellum is made up of a longitudinal axoneme enclosed in a sheath. The electron microscope reveals that the axoneme is made up of a bundle of 11 fine fibrils, 2 of which are central; 9 are present in a ring around the other two (Figs. 2-1 and 2-2). The fibrils are enclosed in a spiral sheath continuous with the cell membrane. The 9 outer fibrils may each be made up of 2 or more finer fibrils. In some flagella, by special preparations, numerous small lateral mastigonemes, which may increase the flagellar surface in movement, can be seen extending from the flagellar sheath. The flagellum originates from the blepharo-plast (kinetosome), a kind of basal granule just inside the pellicle that seems to be essential for the movement of both flagella and cilia. The blepharoplast may function as a centriole in cell division, but in other cases there may be a separate centriole fused to the nuclear membrane. The blepharoplast appears to have genetic continuity and is connected to the centriole by a filament, the rhizoplast (Fig. 2-13). Some flagellates also have one or more parabasal bodies, which vary in shape and are connected by fine fibrils to the blepharoplast. Another body found in trypanosomes and some other flagellates is the kinetoplast, which is connected to the blepharoplast. The functions of parabasal bodies and kinetoplasts are obscure at present.

The mechanism of flagellar movement is not yet fully understood. Its method of use is not the same in all flagellates. An early theory (Bütschli) emphasized a spiral turning of the flagellum like a propeller, serving to pull the organism forward. The undulating motion necessary for this action passes from the base to the top of the flagellum, and spiral rotation of the body ordinarily occurs along with the flagellar action. But some species move with the flagellum directed behind the body. Some flagellates have a rowing stroke in which the flagellum is held rigid and concave in the direction of the stroke. It then bends back like a cilium when it recovers its position so that it offers less resistance to the water. Whatever the movement, adenose triphosphatase (ATPase) activity is involved.

Most flagellates have 1 to 4 flagella, although the symbionts of termites may have hundreds. They may have various functions such as steering, anchoring, pushing, or pulling. The undulating membrane of some parasitic species has a border formed from a flagellum. Flagella may originate from almost any part of the body —anterior end (most common), posterior end, lateral surface, or cytopharynx.

Nutrition. The Mastigophora make use of every type of nutrition except chemosynthesis. The chlorophyll-bearing flagellates are almost entirely holophytic, whereas the colorless forms may be saprozoic, holozoic, or both. Some species may use all three types of nutrition (*Ochromonas*). *Euglena gracilis* is holophytic in the light but saprozoic in the dark. Ameboid phases of flagellates engulf their food much like the Sarcodina. In most cases holozoic flagellates ingest their food at the anterior end near the base of the flagellum by a mouth (cytopharynx) or some kind of depression. Internal skeletal rods (trichites) may reinforce the mouth and gullet, although this type of internal skeleton is far more common in ciliates.

Precise experimental investigation on the nutritional requirements of protozoans seems to show that these requirements are simpler for flagellates than for any other protozoan group. This is especially the case with the chlorophyll-bearing forms. In those that have lost their plastids, food requirements are more exacting (S. H. Hutner and L. Provasoli, 1951). It is apparent that in the evolution of microorganisms the primitive, simple plantlike protozoans can synthesize their proteins with nitrates; some colorless (saprozoic) species require ammonium salts, and other colorless flagellates require amino acids and peptones. Thus there has been a physiologic loss of function along with morphologic evolution (W. H. Johnson, 1941).

Life cycle and reproduction. Most flagellates reproduce asexually by longitudinal division. In dinoflagellates, fission is transverse or oblique. Mitotic processes vary greatly within the group. Sexual reproduction is mainly restricted to some of the phytoflagellates and complex polymastigotes. In their life cycle there is usually a cyst stage. Multiplication may also occur by multiple fission and, in some, reproduction may take place within the cyst. Multiple fission is most common in the trypanosomes and dinoflagellates.

Binary fission usually begins with the division of the basal body or a centriole. The fate of the flagella at binary fission is not the same in all species. In some they may be lost entirely and new ones may be regenerated in the daughter cells. When there is only one or a few flagella, each flagellum may split longitudinally (rare), or, more commonly, a second flagellum grows out from the basal body, which divides in fission. In multiflagellate forms the flagella are distributed in approximately equal numbers to each daughter cell, each of which then regenerates new flagella from basal granules to complete its normal number. In armored species at cell division the theca may be divided between the daughter cells, or one cell may receive the whole theca, and the naked cell regenerates a new one. Sometimes the theca is lost entirely, and the daughter cells have to regenerate new ones. A spindle is formed within the nucleus in some cases, with the halves of the basal body acting as centrioles if the latter are lacking. In other cases in place of a spindle there may be only a strand connecting the centrioles. Usually the dividing mechanism (spindle and centrioles) as well as the chromosomes develop from the endosome. There are some variations of this plan. Chromosomes are usually constant for a particular species and behave as chromosomes do in other forms. The division of the nucleus is followed by that of the cytopharynx (if present), and the organism divides posteriorly from the anterior tip. In some dinoflagellates, fission is incomplete, resulting in a chain of individuals, e.g., *Ceratium* (Fig. 2-6, *H*). Mitosis in termite flagellates has been extensively studied by L. R. Cleveland.

In those flagellates that have sexual reproduction, the gametes may be morphologically alike (isogametes), or they may be unlike (anisogametes). In the anisogametes, microgametes (small and motile) and macrogametes (large and nonmotile) are recognized. The sexual fusion of two gametes is called syngamy. The phytomonac genus *Chlamydomonas* is a good example of isogamy, although some of its species have heterogamy. *Volvox* in its varied reproduction produces both eggs and sperm and is an example of heterogamy. Meiosis among flagellates may be prezygotic or postzygotic.

The life cycle varies among the different flagellates. Cyst formation is common with many
of them, especially among the holozoic species.
It is not found in the chlorophyll-bearing species
to any extent. Stages in the same life cycle may
show polymorphism, in which individuals show
distinct differences from each other. Intracystic
formation, in which secondary cysts are formed
inside primary ones, may also occur, especially
in dinoflagellates. Palmella, or ameboid stages,
in which individuals lose their flagella and become ball-like, are found in the life history of
many flagellates, especially dinoflagellates. By
fission this stage sometimes forms a colony-like
aggregation in a matrix of mucus and may remain in this condition for some time.

■ CLASS PHYTOMASTIGOPHOREA

The class Phytomastigophorea is made up of
the plantlike flagellates, or phytoflagellates.
Most of them are characterized by bearing
chromatophores of various pigments such as
chlorophyll, carotin, and xanthophyll, although
some may lack chloroplasts (*Chilomonas*, Fig.
2-8, *B*). These pigments may be represented in
most colored species but in different proportions
so that chloroplasts (green chromatophores)
contain more of the chlorophyll; and xantho-
plasts (yellow chromatophores) have more of
the xanthophyll than of other kinds (*Dinobryon*,
Fig. 2-8, *A*). However, some chromatophores
may consist entirely of chlorophyll, as in members of the Euglenida and Volvocida. Those that
have chlorophyll can make their food by photosynthesis and are nutritionally independent
except for a few outside requirements (salts
and vitamins).

This class is no doubt made up of the most
primitive members of Protozoa and bears a
close relationship to algae. As a group the members display a great range of structure and
physiology. Many are solitary, but others are
colonial. Although most possess flagella, some
are nonflagellated or ameboid at one stage or
throughout their existence. The green phytoflagellates are almost always autotrophic, but

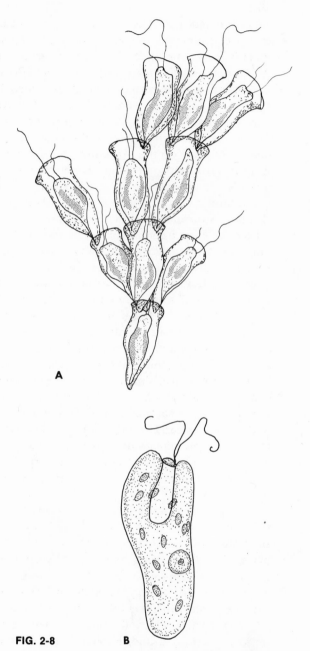

A

FIG. 2-8   **B**

A, *Dinobryon sertularia* (order Chrysomonadida); each
cell is enclosed in vaselike test, bears two unequal flagella, and usually has two lateral yellow-brown chromatophores; cells are 30 to 40 microns long. **B,** *Chilomonas
paramecium* (order Cryptomonadida); lacks chloroplasts;
has trichocysts at base of cytopharynx; oval bodies are
starch grains; widely used in nutritional studies; 20 to
40 microns long.

others have saprozoic and holozoic nutrition. Some members are symbionts in other organisms.

Phytoflagellates have a wide distribution in both fresh and salt water. Green flagellates make up one of the most important members of plankton. They are abundant in some stagnant pools and ponds. Except in the cyst form they require moisture of some kind to exist. They also form important food chains for many other organisms. A few may be parasitic, such as certain ectoparasitic dinoflagellates on the gills of fishes and other forms.

As type forms in this class, we shall select members of three common orders, Dinoflagellida, Euglenida, and Volvocida. A study of these three orders will give a fair conception of the range of organization in this class.

■ ORDER DINOFLAGELLIDA

The dinoflagellates represent an unusual group of protozoans that are found primarily in marine water, although some species occur in fresh water. Some have chromatophores of various colors, and others are colorless. Their chief characteristic is the arrangement of the two flagella, one of which usually runs in an encircling groove transverse to the body axis and the other in a longitudinal groove directed backward for propelling the animal. The transverse flagellum causes the organism to rotate. Dinoflagellates are usually of fixed shapes and bizarre forms (Fig. 2-9). The encircling groove, or girdle, divides the animal into two parts—the epicone and the hypocone. Chain, ameboid, and palmella aggregates are common in the group. Many of them have a stigma, or eyespot.

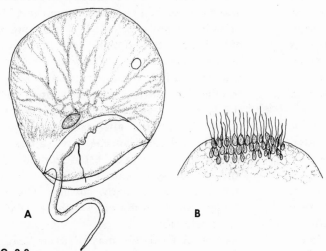

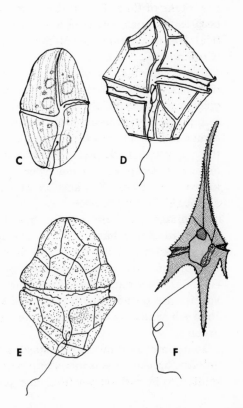

FIG. 2-9

Order Dinoflagellida. **A,** *Noctiluca scintillans;* noted for phosphoresence in the sea; 200 to 2,000 microns long. **B,** *N. scintillans* budding; buds become isogametes that fuse in pairs to form zygotes. **C,** *Gynodinium costatum;* typical unarmored dinoflagellate; related species *G. brevis* causes certain "red tides." **D,** *Gonyaulax;* responsible for "red tides" off California coast; 50 microns long. **E,** *Glenodinium cinctum,* with both epitheca (anterior to girdle) and hypotheca (posterior to girdle) made up of plates; 45 microns long. **F,** *Ceratium hirudinella;* has one apical and two or three basal horns; many varieties, depending on depth of water in which they are found; 95 to 700 microns long.

Pouchetia is noted for having a complex light-perceiving organ with a lens.

Some species are naked or enclosed in a simple cellulose envelope, but most of the marine forms are armored with a cellulose coat of two valves or of one or many plates cemented together. The armor is sometimes elaborately embossed, often with spines. Unarmored species are often brightly colored. Some, too, such as *Noctiluca* (Fig. 2-9, *A*) are phosphorescent. Nematocysts and trichocysts are found as organelles in a number of species.

Most dinoflagellates have holophytic nutrition, but the colorless ones may be ameboid and holozoic. Some may entrap their prey by branched pseudopodia extruding through theca pores. Some are ectoparasitic or endoparasitic in a nonflagellate encysted condition.

One of their characteristic structures is the pusule system organelle, which consists of vacuoles connected by canals to the exterior. The system is noncontractile and may serve for the intake of fluid. Two of these organelles may be present, one connected to each flagellar orifice. The exact function of the pusule system is unknown.

Reproduction is by oblique or transverse binary fission (Fig. 2-6, *H*). Usually each daughter cell receives part of the armor and secretes the remaining part, but in some cases the entire armor is lost. There may be multiple fission in the encysted condition. Encystment is rare among chlorophyll-bearing species. Little is known about sexual stages in dinoflagellates, although they may occur.

Dinoflagellates are mostly free living and have a wide distribution. They make up much of the plankton in both fresh and marine water. One of the most common species is *Ceratium* (Fig. 2-9, *F*), which bears three long spines. Its species are found in both fresh and salt water. In fresh water its chromatophores are usually green.

Sometimes certain marine species of dinoflagellates appear in enormous numbers (up to 25 million to 50 million per liter) and produce the "red tides" off the coasts of many regions, especially Florida and southern California. These organisms produce a toxin highly fatal to fishes and other marine life. The toxin produces its injurious effect on the nervous system. The genus *Gonyaulax* (Fig. 2-9, *D*) is mainly responsible for the California outbreaks and the genus *Gymnodinium* (Fig. 2-9, *C*) for those in Florida.

■ ORDER VOLVOCIDA (PHYTOMONADIDA)

Protozoans of the order Volvocida are mostly grass green, but a few are colorless. This is a large group of about 100 genera, and they closely resemble typical plants. Two flagella are found in most, although some have more and a few have only one. Unicellular species are more common than the colonial forms. Their size is small and their body is usually oval or elongated and enclosed in a cellulose membrane. There is no gullet. Their chlorophyll is the same as that of higher plants, and their pyrenoids have starch as the reserve material. They are provided with a stigma and 2 water expulsion vesicles (contractile vacuoles). Some are stained red with the pigment hematochrome. Their nutrition is chiefly holophytic, although some are mixotrophic. Colorless species are saprozoic. Reproduction is asexual (longitudinal fission) and sexual. Encystment and palmella stages are common in many species. The central nucleus divides mitotically with typical chromosome behavior.

Colonial Volvocida may be irregularly flat or spherical. Most colonies have 4 to 128 cells, but some species of *Volvox* may have several thousand. Individual zooids in a colony typically have two flagella and have most of the other structures found in solitary species. The zooids may be loosely connected, but they are chiefly held together in a common cellulose membrane or embedded in a mucilaginous jelly. Familiar genera of colonial species are *Gonium* (4 to 16 cells, Fig. 2-10, *A* and *B*), *Pandorina* (16 cells, Fig. 2-10, *C*), *Eudorina* (32 cells), and *Pleodorina* (32 to 128 cells, Fig. 2-10, *D*). In all these

FIG. 2-10

Order Volvocida. **A,** *Gonium pectorale;* usually 4-16 zooids; fresh water; colony up to 90 microns in diameter. **B,** *Gonium sociale;* 4 zooids; each may give rise to daughter colony by cell division, or each may serve as isogamete; fresh water; cells 10 to 22 microns long. **C,** *Pandorina morum;* 16 cells; each cell divides four times to form new colony; colony 20 to 50 microns or more in diameter. **D,** *Pleodorina illinoisensis;* 4 small anterior zooids sterile; other 28 zooids capable of both asexual and sexual reproduction; colony up to 450 microns in diameter. **E,** *Eudorina elegans;* 32 cells; each forms new colony; colony 40 to 150 microns in diameter.

except *Pleodorina,* each zooid is capable of both asexual and sexual reproduction. The zooids at the anterior pole of *Pleodorina* are sterile. Asexual reproduction takes place by the repeated fission of each zooid to form a colony, which is set free by the rupture of the parent. Sexual reproduction may occur by the zooids becoming free and serving as gametes, or gametes may be formed by fission of the zooids. The gametes may be isogamous or anisogamous. After encystment the zygote becomes a free flagellate that divides to form a colony, or the zygote may hatch directly into a small colony.

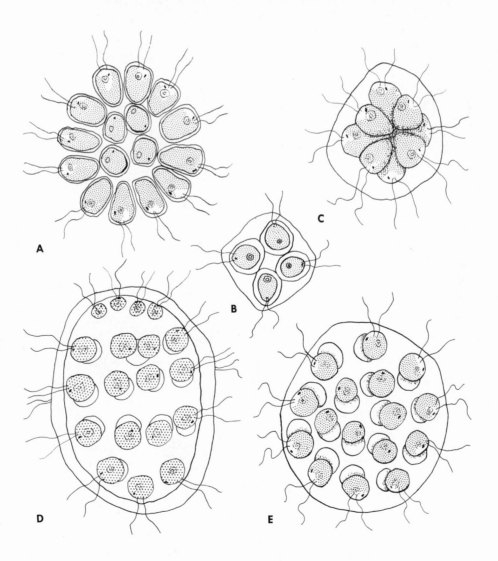

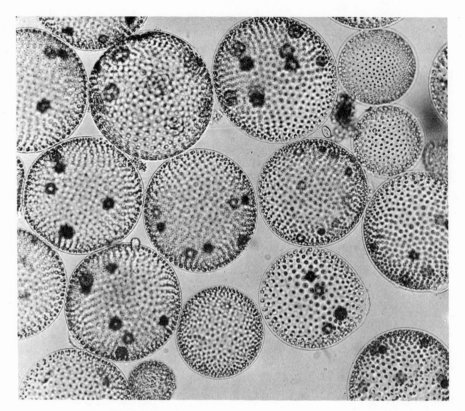

FIG. 2-11

Volvox colonies (order Volvocida); 1,000 to 3,000 cells
in hollow ball; 350 to 500 microns in diameter.

FIG. 2-12

Portion of colony of *Volvox globator* (order Volvocida)
showing protoplasmic connections between zooids. **A,**
Side view. **B,** Surface view.

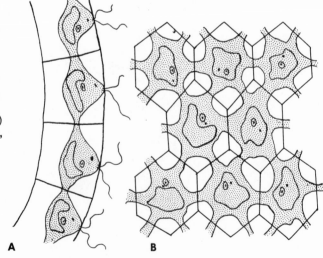

The most striking colonial form is the genus *Volvox*. The most familiar species is *Volvox globator* (Fig. 2-11). It is a hollow sphere about 1.3 mm. in diameter with many thousands of biflagellate zooids. The zooids are embedded in a gelatinous matrix, and adjacent cells have cytoplasmic connections between them (Fig. 2-12). The cells in these colonies are always green, with the exception of the sexually differentiated cells. The majority of the zooids each has a pair of flagella, a stigma, and chromatophores. Some zooids are gonidia that form new colonies asexually by multiple fission. These colonies remain in the hollow of the parent colony until set free. Other zooids enlarge to form macrogametes or eggs, and still others,

by repeated division, produce small biflagellate microgametes or sperm. Eggs are liberated into the hollow colony where they are fertilized by the sperm to form zygotes with thick, spiny shells. The zygotes eventually hatch into new colonies. *V. globator* is monoecious, and male, female, and asexual reproductive zooids are found in the same colony. However, in some species the sexes are separate (dioecious).

■ **ORDER EUGLENIDA**

The order Euglenida is made up of forms that usually show an anteroposterior elongation with a definite shape. The pellicle, often ridged or striated, may be soft enough for contortions or too rigid for change of shape. Euglenida have

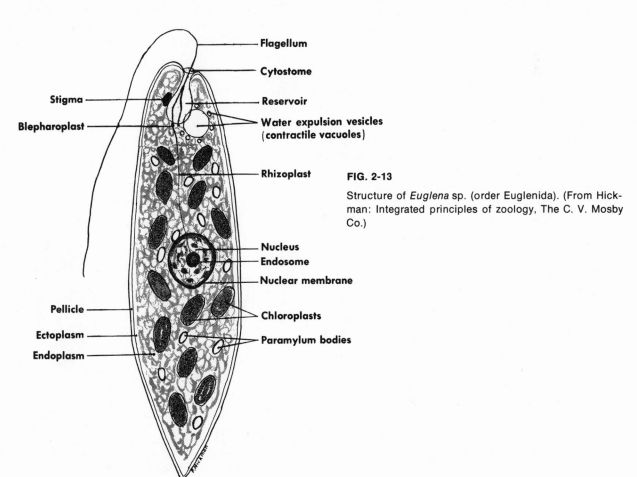

FIG. 2-13

Structure of *Euglena* sp. (order Euglenida). (From Hickman: Integrated principles of zoology, The C. V. Mosby Co.)

Labels (clockwise from top): Flagellum, Cytostome, Reservoir, Water expulsion vesicles (contractile vacuoles), Rhizoplast, Nucleus, Endosome, Nuclear membrane, Chloroplasts, Paramylum bodies, Endoplasm, Ectoplasm, Pellicle, Blepharoplast, Stigma

one or two flagella, which may be equal or unequal in length. A stigma of red hematochrome granules is found at one side of the cytopharynx, but the stigma is absent in colorless species. Many species have green chromatophores (never masked by other pigments), which may be irregular, circular, or stellate in shape. Hematochrome is found in a few. Pyrenoids are sometimes associated with the chromatophores, and their reserved food supply is paramylum (a starchlike polysaccharide), which is stored in definite bodies.

A cytostome and cytopharynx (gullet), from which the flagellum or flagella emerge, are present at the anterior end. In holophytic forms the gullet is not concerned with obtaining food but serves as a passageway for fluid or waste and for the flagella, which are usually attached to its wall by two roots. In holozoic species the cytostome and gullet serve for food intake as well (Fig. 2-13).

The lower part of the gullet widens into a reservoir, into which the large water expulsion vesicle (contractile vacuole) discharges its contents. The water expulsion vesicle is formed each time after discharge by the fusion of many small vacuoles. Green members are almost entirely holophytic, but some absorb soluble organic matter (saprozoic); most of them under certain conditions can become wholly saprozoic. The colorless species are holozoic or heterotrophic and live on bacteria and other small organisms. Reproduction in the order occurs by asexual longitudinal fission in the free or encysted condition. Fission may occur in the cysts of some, producing palmella stages. Sexual reproduction has never been fully established, although a primitive form of it has been reported in one or two species.

Euglenida are widespread in almost all quiet surface waters, although the green species are found mostly in fresh water, where they form typical blooms.

This order takes its name from its most characteristic family, the Euglenidae. There are several species of *Euglena* (Fig. 2-13), and they vary a great deal in structural features. The following are the most common species that one is likely to find in fresh water: *Euglena viridis* (60 microns), with a spindle-shaped but plastic body and stellate arrangement of chromatophores from a central point; *E. spirogyra* (100 microns), with a cylindric but changeable body and spiral arrangement of tiny knobs; *E. oxyuris* (500 microns), with a very large body and spirally twisted pellicle; *E. gracilis* (50 microns), with an elongated oval body and fusiform chromatophores; and *E. rubra* (35 microns), with a cylindric body and numerous red hematochrome granules.

Of the colorless members of the order, one of the most familiar is the genus *Peranema* (Fig. 2-14, *A*), which has an elongated body with a more or less truncated posterior end and long tapering flagellum. Two rods near the reservoir

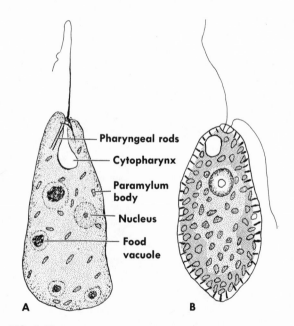

FIG. 2-14

A, *Peranema trichophorum* (order Euglenida); colorless euglenoid; one flagellum (not seen) is attached to pellicle; 20 to 70 microns long. **B,** *Vacuolaria virescens* (order Chloromonadida); nucleus large; many chromatophores; commonly found in acid bog waters but are poorly known; 50 to 150 microns long.

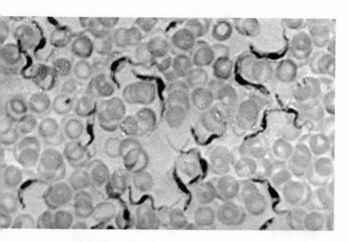

FIG. 2-15

Trypanosoma gambiense (order Kinetoplastida) scattered among human red blood cells; carried by the tsetse fly; it causes African sleeping sickness; 22 microns long.

are used in the ingestion of prey. It occurs very commonly and has been much used in experimental nutrition studies.

The rarely found members of the order Chloromonadida (Fig. 2-14, *B*) bear a superficial resemblance to the Euglenida.

■ CLASS ZOOMASTIGOPHOREA

The Zoomastigophorea lack chromatophores and vary in their organization from simple to complex. They do not possess starch bodies or paramylum, and they often have more than two flagella, but some have none. Some have pseudopodia as well as flagella. A parabasal body of debatable function is often present. Cells are usually naked or have delicate membranes; most are small. Many form colonies that show a variety of forms. Nutrition is holozoic, saprozoic, or parasitic. The Zoomastigophorea are found in both fresh and marine waters in both free-living and parasitic form. Asexual reproduction is by longitudinal fission; sexual reproduction is unknown. Encystment is common among them. Because of these and other characteristics, these flagellates belong to the animal division of the Mastigophora.

This class may have arisen from the Phytomastigophorea by loss of chromatophores and by other changes. Some of the Phytomastigophorea show variations in color and nutrition not far removed from typical members of Zoomastigophorea.

The following orders and some of their members are of the most interest from the standpoint of this text.

■ ORDER KINETOPLASTIDA (PROTOMONADIDA)

The Kinetoplastida have one or two flagella, their body usually has no definite pellicle, and they are often ameboid. Free-living forms are holozoic, whereas the many parasitic species are saprozoic. In addition to longitudinal fission, multiple fission is found in some parasitic forms.

The trypanosome (genus *Trypanosoma*, Fig. 2-15) is a common parasite in the blood and fluids of vertebrates and is carried from host to host by bloodsucking invertebrates, in which it undergoes a series of changes in its life history. *Trypanosoma gambiense* (or the similar *T. rhodesiense*) causes the dreaded African sleeping sickness. Its vector is the tsetse fly (*Glossina*). The flagellate is elongated with pointed ends and a flagellum arising from a basal granule near the parabasal body. The basal portion of the flagellum forms the outer margin of the undulating membrane that runs along one side of the body. The parasites do their chief damage by invading the nervous system, resulting in lethargy, coma, and death. There are many other species of pathogenic or nonpathogenic trypanosomes that infect different species of vertebrates. One of these, *T. lewisi*, is nonpathogenic in the rat and has been much studied. This flagellate undergoes multiple fission in the stomach of a flea and changes to trypanosome form before being discharged in the feces. Rats are infected by licking their fur, where the feces are deposited by fleas. Chagas' disease of man in the tropics is caused by *T. cruzi*, which is transmitted by a bug (*Triatoma*). In this disease

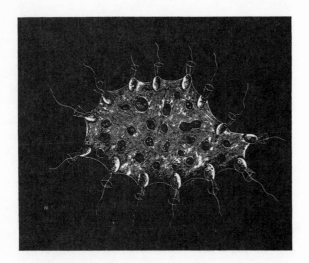

FIG. 2-16

Proterospongia (Protospongia), a colonial choanoflagellate; by repeated division of one cell, spores are formed, each of which may form new colony of 6 to 60 cells (order Choanoflagellida). (From Hickman: Integrated principles of zoology, The C. V. Mosby Co.)

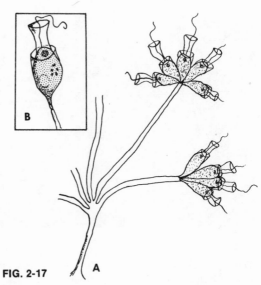

FIG. 2-17

Codosiga botrytis (order Choanoflagellida), a choanoflagellate. **A,** Colony. **B,** Single zooid; cells 10 to 22 microns long.

the heart and skeletal muscles show swellings caused by cystlike bodies of the parasites.

Another dangerous parasite of this order is *Leishmania,* which is probably transmitted by sandflies *(Phlebotamus).* Stages in the life history of the parasite are found in both the vector and man, where it causes kala-azar, a disease of the liver, spleen, and related organs of the reticuloendothelial system.

■ ORDER CHOANOFLAGELLIDA (PROTOMONADIDA)

An interesting nonparasitic group are the choanoflagellates, which have a thin protoplasmic collar around the base of the single flagellum (Figs. 2-16 and 2-17). They include the colonial *Proterospongia (Protospongia,* Fig. 2-16), which forms a gelatinous mass with the flagellated collared membranes on the outside and ameboid individuals embedded in the interior. Particles of food are attracted by the flagellum to the outside of the base of the collar, where they adhere and are transferred to a

vacuole in the protoplasm. This organism has been regarded as a possible link between the unicellular protozoans and the sponges because of the unique collared cells (choanocytes) lining internal cavities of the sponges. However, *Proterospongia* may have no phylogenetic significance (L. H. Hyman).

■ ORDERS DIPLOMONADIDA AND RETORTAMONADIDA

Orders Diplomonadida and Retortamonadida were formerly placed in the old order Polymastigina. *Giardia* (order Diplomonadida) and *Chilomastix* (order Retortamonadida) (Fig. 2-19, *B* and *C*) are both common human parasites in the digestive tract, although the latter are also found in the gut of insects. *Giardia* produces stools with much mucus.

■ ORDER HYPERMASTIGIDA

The members of the order Hypermastigida are the most complex flagellates, both in their structure and modes of division. They live in the

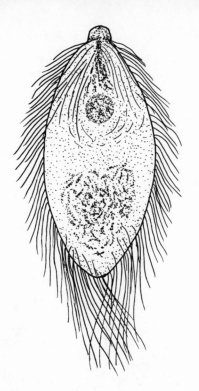

FIG. 2-18

Trichonympha agilis (order Hypermastigida), the classic symbiont of termites; 55 to 115 microns long. (Drawn from photograph of living trophozoite.)

alimentary canals of termites, cockroaches, and wood roaches, where they often form a classic symbiotic relationship with the host termite. They have many flagella that originate in the very complex blepharoplasts and related structures in the anterior part of the body. Their method of nutrition is holozoic or saprozoic. Termites and cockroaches eat wood, which they digest to a soluble sugar for the host. Without their flagellate symbionts, the host is unable to utilize the wood diet and will die. The flagellates benefit by obtaining food and lodging (L. R. Cleveland). At each molt the flagellates are shed by the termites. Reinfection occurs when recently molted individuals lick the anus of infected individuals (termite) or eat the cysts passed in the feces (roaches). These flagellates live under anaerobic conditions, but exist in organic rich environments.

Trichonympha (Fig. 2-18) is the common genus of this order that is found in termites; *Lophomonas* is a genus that lives in the domestic cockroach.

SUPERCLASS OPALINATA

Opalina (Fig. 2-19, *A*) is the most familiar member of the superclass Opalinata and of its one order (Opalinida). It is found in the rectum of frogs and toads and is common. Its nutrition is entirely saprozoic because it has no cytostome. It reproduces by binary fission and has many nuclei. Cysts are passed out of the host and are eaten by tadpoles, in which they hatch to form anisogametes. After fusion of the gametes the zygote gives rise to the adult by nuclear division.

Until recently the taxonomic status of *Opalina* has been in doubt. It had always been classified among the ciliates but is now placed among the flagellates because its several nuclei appear to be all alike and its plane of cell division is like that of flagellates, not ciliates (J. O. Corliss, 1961).

SUPERCLASS SARCODINA (RHIZOPODA)

The extensive protozoan superclass Sarcodina is chiefly characterized by ameboid movement that involves the formation of cytoplasmic extensions called pseudopodia, which are also used for food intake. Some sarcodinians at certain stages develop flagella. Most of them creep or float, although a few are sessile. They have no pellicle or a very thin one so that the body is plastic unless it is enclosed with a skeleton of some kind. They are mostly solitary or free living, but some are parasitic and colonial. In their phylogeny their affinities to the Mastigophora are evident because some flagellates have ameboid stages and some sarcodinians have flagellate stages or flagellate gametes. The flagellate order Rhizomastigida includes *Mastigamoeba*, which resembles amebas but also has a flagellum (Fig. 2-20).

The fossil record indicates that the sarcodinians are probably the oldest of protozoans because foraminiferan and radiolarian shells date back to Precambrian times. The only older fossils that have been found are those of algae and fungi.

Sarcodinians are found in both fresh and ma-

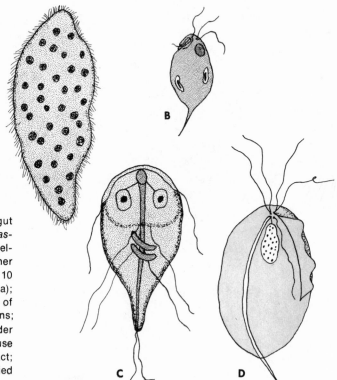

FIG. 2-19

A, *Opalina* sp. (order Opalinida); common parasite in gut of frogs and toads; 100 to 840 microns long. **B,** *Chilomastix mesnili* (order Retortamonadida); commensal flagellate in human colon; related species found in other vertebrates; no clinical significance; trophozoite 10 to 15 microns long. **C,** *Giardia* (order Diplomonadida); stained trophozoite; is common intestinal parasite of man, causing digestive disturbances and irritations; cysts common in feces. **D,** *Trichomonas vaginalis* (order Trichomonadida); common human parasite; may cause severe irritation in vagina and male urogenital tract; related species found in mouth and colon. (**C** modified from Jahn.)

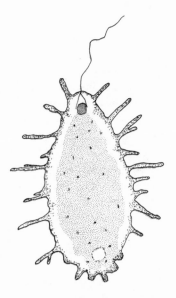

FIG. 2-20

Mastigamoeba aspera (order Rhizomastigida); often considered borderline form between flagellates and Sarcodina because they possess both flagella and pseudopodia; 150 to 200 microns long.

rine water. Many are almost exclusively freshwater forms such as the shelled amebas, and others such as the radiolarians are exclusively marine. A number of species live in both salt and fresh water. The foraminiferans and radiolarians are plankton forms, and as they die their shells sink to the ocean floor to form the bottom ooze. Other members of this group are not restricted to the surface but live at considerable depths. Some species of amebas are able to live in moist surface soils.

The Sarcodina are divided into two classes—the Actinopodea, or those with axopodia (raylike pseudopodia with axial filaments), and Rhizopodea, or those that lack axopodia but have various types of pseudopodia such as lobopodia, filopodia, and rhizopodia.

GENERAL FEATURES

The Sarcodina as a group show less organization in some particulars than do the flagellates.

For example, their shape is less definite and often without symmetry of any kind, although some may have spherical or other symmetry. Naked forms may change shape constantly because of temporary pseudopodia. Although a pellicle is poorly developed or absent altogether, the cytoplasm is usually differentiated into ectoplasm and endoplasm. In place of a specialized cuticle, shells, tests, and skeletons are secreted by the cytoplasm. These may contain silica and chitin strengthened by the addition of inorganic salts and foreign substances. Openings in the shells permit the extension of pseudopodia. Internal skeletons are present in radiolarians. The endoplasm contains the nucleus or nuclei, food vacuoles, granules, and contractile vacuoles. Most have only one nucleus, but some are multinucleated.

The protoplasmic extrusions (pseudopodia) represent one of the most characteristic features of this class. There are several types of pseudopodia, depending to some extent on the order. The lobopodium type is broad and cylindric with rounded tips and is composed of both ectoplasm and endoplasm; the ectoplasmic filopodium is slender with sharp tips; the rhizopodium is often much branched and forms extensive networks; and the axopodium type is usually straight and raylike with a central axial rod (axopodium).

Locomotion. The most characteristic locomotion of the Sarcodina is ameboid movement. Although this form of locomotion typically belongs to the order Amoebida, nearly all sarcodinians use some modification of it. Ameboid movement is not the same in all amebas but varies from species to species. In some species there may be protoplasmic flow without the formation of typical pseudopodia. *Amoeba proteus* is often taken as the type example of ameboid locomotion (S. O. Mast, 1931). In this species the rigid pseudopodia may appear without restriction of size; in active movement the animal appears to flow into a pseudopodium, which appears to direct the locomotion. The theory is based on the view that the protoplasmic flow is dependent on the sol-gel reversibility of the cytoplasm. The real driving force is the pressure exerted by the less fluid plasmagel, the external layer of cytoplasm. When this layer contracts in the formation of a pseudopodium, the more fluid inner plasmasol flows into that region where the pressure of the plasmagel is lessened. This movement causes the beginning pseudopodium at that point to protrude as the plasmasol flows into it. When the plasmasol reaches the tip of the pseudopodium, it tends to "fountain" back peripherally in places where it becomes a part of the plasmagel just under the pellicle. At the same time, the plasmagel at the temporary posterior end is changing into plasmasol, which now flows anteriorly. Recent studies with the electron microscope fail to show any fibrillar structures that could account for protoplasmic streaming in the ameba, although ameboid movement is considered a type of contraction similar to muscle contraction. At the molecular level, however, the endoplasm is thought to contain long protein chains that undergo contraction.

There may be more than one mechanism of ameboid movement. The tail-contraction, or Mast's, theory just described is considered by some authorities to be too simple, and in its place a theory has been proposed that involves the pulling forward of the cytoplasm by a contraction force at the tips of the pseudopodia (front-contraction theory). Evidence indicates that the endoplasmic axial core has about the same viscosity as the ectoplasmic tube, instead of being much less as in Mast's theory. The chief regions of low viscosity and places where there is a rapid flow of cytoplasm are zones around the axial endoplasm and the region where the endoplasm is being formed from the posterior ectoplasm (R. D. Allen, 1962). Strands of freely movable cytoplasm move forward in the axial endoplasm to the anterior fountain zone, where they diverge outward and join the ectoplasmic tube. The endoplasm anchored to the ectoplasmic tube contracts, everts, and pulls the axial endoplasm forward. Continuous streaming movements are kept up by the propagated contraction along the ectoplasm. The hyaline cap is formed from water squeezed from the endoplasm. This water eventually moves backward to the tail in the channel just beneath the plasmalemma.

Neither one of the foregoing mechanisms explains the reticulopodia formation of foraminiferans and radiolarians in which streaming movements always occur in two directions simultaneously. Some workers have proposed an unknown active shearing force or antiparallel force to explain such movement. All streaming movements of the cell are at present poorly understood.

T. L. Jahn and E. C. Bovee (1965) divide the Sarcodina into two classes on the basis of differences in the kind and movements of pseudopodia—the Autotractea (with slender and filamentous pseudopodia and active shearing mechanisms of locomotion) and the Hydraulea (with tubular pseudopodia and contractile hydraulics in a tube of gel). The Autotractea include radiolarians, heliozoans, foraminiferans, etc., and the Hydraulea consist of amebas, lobose arcellinids, mycetozoans, and others.

Nutrition. Nutrition in Sarcodina is holozoic, food consisting of almost any small organisms such as other protozoans, algae, bacteria, and diatoms. There are no permanent organelles for food intake, which may occur at any region of the body. A pseudopodium forms a food cup by flowing around and enclosing the food object. The food vacuole so formed is engulfed or sucked into the cytoplasm. Engulfment of the prey may take place by two methods: (1) circumfluence, in which the encircling pseudopodium remains in close contact with the food, and (2) circumvallation, in which the pseudopodium by an enlarged cup surrounds not only the prey but also some water along with it. The first method is commonly employed by shelled forms with filopod pseudopodia, the second by naked amebas. Those with threadlike pseudopodia (axopodia and rhizopodia), such as foraminiferans and radiolarians, capture their prey in a network arrangement. These pseudopodia produce a sticky and paralyzing substance that sticks to the prey. Intracellular digestion within the food vacuole occurs through enzymatic action.

Life cycle and reproduction. The Sarcodina show many types of life cycles. Asexual reproduction takes place by binary fission, which varies in the different groups. Sexual reproduction occurs in all the orders but is uncommon among naked amebas. Encystment occurs among some but is absent in others. The nucleus may be single or multiple.

Division is accompanied by true mitosis with spindle, chromosomes, and centrioles (usually). The number of chromosomes varies in those studied from a few to hundreds (*Amoeba dubia*). In most cases the nuclear membrane persists during cell division. Division is usually simpler in the naked Sarcodina than in the shelled species. In some shelled amebas the shell divides into two parts, and each daughter cell receives one part and forms a new half. *Arcella vulgaris*, which has two nuclei, at fission extrudes part of the cytoplasm and a nucleus through the shell aperture just before fission. The extruded cytoplasm secretes a new shell, and when the two-shelled animal divides, each daughter has a shell (Fig. 2-6, *B*). A variation of this fission is found in *Euglypha*, which has a test composed of secreted scales and spines. Before fission an individual stores up an extra supply of these platelets, which are used to form a new test by the extruded cytoplasm (Fig. 2-6, *C*).

Life cycles are known for a few of the Foraminiferida. They have two phases, one asexual and the other sexual. They are also dimorphic, having two types of individuals in each species, based mainly upon the size of the initial chamber (proloculum) of the test. The megalospheric type (gamont) has a large proloculum, one nucleus, and is relatively small in size; the microspheric type (schizont) has a small proloculum, many nuclei, and is relatively large. This dimorphism results from a reproductive cycle in which the asexual and sexual generations alternate. In a typical life history (although

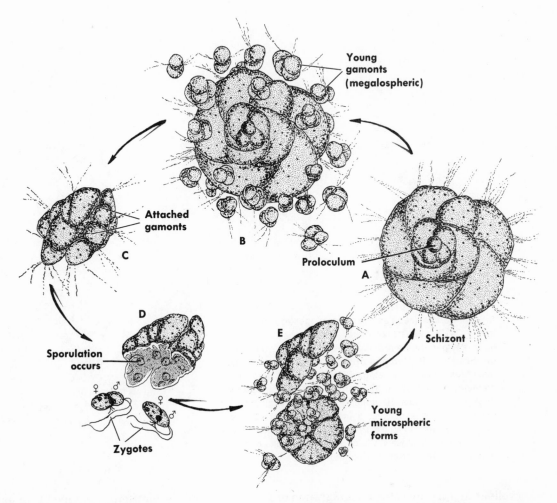

FIG. 2-21

Life cycle of *Discorbis* (order Foraminiferida); see text for description of stages. **A,** Schizont, or microspheric individual, produces by multiple fission (asexual) many amebalike forms that secrete megalospheric shells and become gamonts. **B,** Release of gamonts from parent schizont. **C,** Gamonts mature and pair. **D,** Attached gamonts produce biflagellated anisogametes that fuse in pairs (sexual) to form zygotes. **E,** Development of zygotes into schizonts with microspheric shells, each of which grows into mature schizont, **A.**

there are some variations) a schizont, or microspheric individual (Fig. 2-21, *A*), becomes multinucleated and by multiple fission (asexual phase) produces many small amebalike young, each of which secretes a megalospheric shell (Fig. 2-21, *B*) and becomes a gamont after

escaping from the parent shell. Pairs of gamonts attach (Fig. 2-21, *C*), and in time the gamonts undergo sporulation and release many flagellated anisogametes (Fig. 2-21, *D*). These fuse in pairs (sexual phase) to form zygotes that develop into schizonts with microspheric shells (Fig. 2-21, *E*).

The life cycle of Heliozoia, or sun animalcules, is well known for certain genera, especially *Actinosphaerium* (Fig. 2-24) and *Actinophrys* (Fig. 2-22, *A*). These show some variations from other Sarcodina. Both asexual and sexual reproduction occur. In *Actinosphaerium*, which is multinucleated, the nuclei divide mitotically, and eventually the organism divides by plasmotomy to form two organisms. The sexual

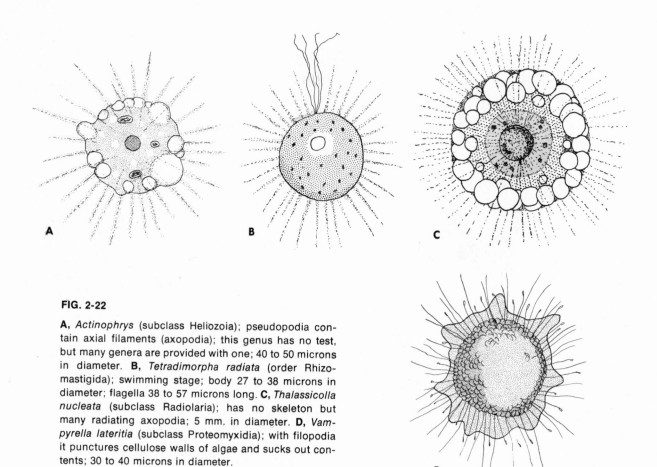

FIG. 2-22

A, *Actinophrys* (subclass Heliozoia); pseudopodia contain axial filaments (axopodia); this genus has no test, but many genera are provided with one; 40 to 50 microns in diameter. **B**, *Tetradimorpha radiata* (order Rhizomastigida); swimming stage; body 27 to 38 microns in diameter; flagella 38 to 57 microns long. **C**, *Thalassicolla nucleata* (subclass Radiolaria); has no skeleton but many radiating axopodia; 5 mm. in diameter. **D**, *Vampyrella lateritia* (subclass Proteomyxidia); with filopodia it punctures cellulose walls of algae and sucks out contents; 30 to 40 microns in diameter.

cycle at first involves the disintegration of most of the nuclei and the encystment of the organism. Within the cyst each nucleus with some cytoplasm forms a secondary cyst. Then each organism with a cyst divides into two daughter cells that undergo two typical maturation divisions to produce gametes; these unite to form the zygote. The latter, after a period of encystment, hatches into a young helizoan that eventually acquires a multinuclear condition.

Life cycles of radiolarians are still obscure. It is known that asexual reproduction involves binary fission in which the skeleton is divided, and the daughter cells each regenerate the missing half; or one daughter cell receives the whole skeleton, and the other daughter cell regenerates a new skeleton. True mitosis with hundreds

of chromosomes is typical. In sexual reproduction, the occurrence of which is not completely known, flagellate microspores and macrospores may be produced, or they may be isospores. It is not known definitely whether or not these unite in sexual union.

■ CLASS ACTINOPODEA
Subclass Heliozoia

The members of the well-known Heliozoia are characterized by having a more or less spherical form with radiating axopodia (Figs. 2-22, *A*, and 2-24, *A*). These pseudopodia are usually characterized by having rows of granules along them. In some the axonemes of the axopodia converge on a central granule. The cytoplasm in some is differentiated into a highly vacuo-

FIG. 2-23

Types of radiolarian skeletons; note great variety of forms representing all symmetries. (Courtesy General Biological Supply House, Inc., Chicago.)

lated ectoplasm with contractile vacuoles and a less transparent endoplasm with few granules. The inner and outer cytoplasm are not separated by a central capsule. Many of them have a siliceous skeleton or test; some have a one-piece test with pores through which the axopodia are extended. This test may have a latticelike arrangement with spicules. Some are enclosed by a gelatinous mantle in which spicules of silica and foreign objects are embedded. The nucleus may be single or multiple.

Heliozoians are holozoic and live on small animals, which they trap with axopodia that are supposed to contain a paralyzing substance. The prey may be pulled back by the shortening of the axopodia, or the axial filaments may disappear and the pseudopodia surround the prey and carry it into the body. Most heliozoians are found in fresh water in which they float, or they may roll in contact with a substratum by successive contractions of axopodia.

The most common genera are *Actinophrys*

(Fig. 2-22, *A*), which has no skeleton and *Clath-rulina*, which has a latticed test of tectin.

Subclass Radiolaria

The subclass Radiolaria is the oldest known group of animals and contains chiefly pelagic marine organisms. They are among the largest of protozoans, some of which may be several millimeters in diameter. The variations in the different species have centered around their skeleton (Fig. 2-23), which has undergone diversified specializations. They have a central capsule enclosed by a central capsule membrane that separates the inner from the outer cytoplasm and is composed of chitin or tectin. It is usually spherical, ovoid, or lobate and is perforated for cytoplasmic continuity. The intra-

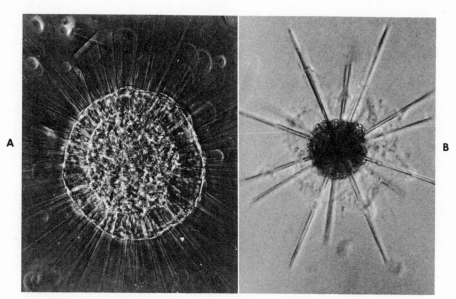

FIG. 2-24

A, *Actinosphaerium,* a heliozoian. **B,** *Tessaraspis,* a radiolarian from marine plankton. (From Encyclopaedia Britannica film, "Protozoa: One-Celled Animals.")

capsular part is concerned with reproduction; the extracapsular part, with nutrition and hydrostatic functions. Water expulsion vesicles (contractile vacuoles) are not found. The extracapsular part, or calymma, consists of a mucilaginous substance that forms the walls of large vacuoles of fluid (Fig. 2-22, *C*). Pseudopodia are of the axopodium or filopodium type, but axonemes are not always present. Skeletal parts are composed of strontium sulfate or siliceous materials. Spines often radiate out from the center of the body through the central capsule (Fig. 2-24, *B*). A latticework test of spherical or other shapes is fused with the spines and lies at the surface of the body. In some species the rods and spines are outside the central capsule. Most types of symmetry are found in the group, as Haeckel long ago discovered.

The central capsule contains the nucleus and often many oil globules. Some species are multinucleated.

Radiolarians are chiefly holozoic, feeding on small organisms, microplankton, etc. The food is caught by the pseudopodia and passed into the deeper regions of the cytoplasm for digestion. Some species have symbiotic zooxanthellae that may furnish the products of holophytic nutrition to the host.

Most species live within the upper 1,500 to 2,000 feet of the sea, although some have been found much deeper. When they die, their shells sink to the floor of the sea to form bottom sediments or ooze. Their shells can withstand greater pressures than those of the foraminiferans so that the ooze of the deepest oceans is largely made up of radiolarian shells.

Some authorities separate the order into two subdivisions. The term "Acantharia" is applied to those few with geometrically arranged spicules of strontium sulfate; the true "Radiolarida" is the term applied to those with an odd assortment of skeletons composed of other substances.

Subclass Proteomyxidia

The members of the subclass Proteomyxidia are poorly defined because they are not well known, and they show many variations. In some species the mature forms are plasmodia; in others they are ameboid uninucleate organisms. Their pseudopodia are filopodia, which usually anastomose to form branches. Their nearest relatives seem to be the Mycetozoia. Most of them are found as parasites on algae or other plants in either marine or fresh water. Some parasitize eelgrass. The formation of a motile pseudopodium by the aggregation of individuals before encystment occurs in some. Asexual reproduction is by binary fission and multiple fission in the encysted condition, resulting in biflagellate swarmers. Gametes and zygotes are produced in the life cycle of some members. Filamentous tracks along which the individuals glide are often secreted.

Some common genera are *Vampyrella* (Fig. 2-22, D), *Pseudospora*, and *Biomyxa*.

■ CLASS RHIZOPODEA
Subclass Lobosia
■ ORDER AMOEBIDA

The members of the order Amoebida are considered the most characteristic and the best known of all sarcodinians. Their size varies from tiny forms of 3 to 4 microns diameter to the large members of 3 mm. diameter. They typically form lobopodia, which display ameboid movement involving a flow of protoplasm. In some there may be protoplasmic movement without the formation of pseudopodia. Their shape is irregular and constantly changing. They may have a delicate pellicle, although there is little cortical differentiation. The ectoplasm is hyaline, and the endoplasm is granular. Both these layers are usually found in the pseudopodia. A flagellate stage may be found in some dimorphic species (soil amebas), but most have a monomorphic life cycle. Binary fission is the common method of reproduction, although multiple fission may also occur. Their nuclear division is mitotic with spindles and chromosomes. The spindle is often intranuclear. Sexual reproduction is doubtful. Encystment is common. Water expulsion vesicles are present in freshwater species but are mainly absent from marine and parasitic forms.

Amebas are found in all kinds of fresh, salt, and brackish waters, in most soil, and in the digestive tracts of invertebrates and vertebrates.

The order is divided by some authorities into three groups or families. Family Dimastigamoebidae includes in their history a dominant ameboid stage, a transitory biflagellate stage, and a cyst. This family includes the soil amebas, fecal amebas, and some others. Under certain conditions the ameboid phase may be transformed into the flagellate phase. *Naegleria (Dimastigamoeba)*, is the best known genus.

Family Amoebidae includes the familiar genus *Amoeba* of several species. They are free living and are monomorphic in their life history. A single water expulsion vesicle (contractile vacuole) is found. They are mostly naked but may have a delicate pellicle. The various species differ with respect to type of pseudopodia. *Amoeba proteus* (Fig. 2-25), which may reach a diameter of 600 microns, has large lobopodia with longitudinal ridges and a disk-shaped biconcave nucleus. In its endoplasm are many truncate or other shaped crystals. *A. dubia* is another familiar species and is somewhat smaller than *A. proteus*. It usually has many lobopodia with smooth surfaces and an oval nucleus. Its few endoplasmic crystals are large (of different types), and it may have more than

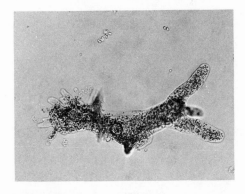

FIG. 2-25

Amoeba proteus (order Amoebida); animal is preparing to extend three new pseudopodia to the left.

one water expulsion vesicle. Other species one may encounter in cultures are *A. discoides* with few smooth pseudopodia, often confused with *A. proteus; Amoeba (Thecamoeba) verrucosa* with short, blunt pseudopodia and wrinkled body surface; and *A. limax* with no definite pseudopodia and rounded body. The very large amebas are represented by the genus *Pelomyxa* (Fig. 2-26), formerly and often still called *Chaos*, which reaches a size of 3 to 5 mm. and can easily be seen with the naked eye. *Pelomyxa carolinensis* moves by means of pseudopodia, as does *A. proteus; P. palustris* moves by protoplasmic waves and with little change in shape.

The third family of amebas is the Endamoebidae, which are parasitic or endocommensal in the digestive tracts of invertebrates and vertebrates. One or two are highly pathogenic. There are two genera often confused because of similarity of spelling: *Endamoeba* is the genus found in cockroaches and termites; *Entamoeba* is the one commonly found in vertebrates. *Entamoeba coli* is found in the intestine of man and monkeys. This species is small with blunt and granular pseudopodia. Mature cysts contain eight nuclei. It may be considered a commensal, but heavy infections result in gastrointestinal disturbances. *Entamoeba gingivalis* is a species living in the mouth. The most harmful member of this genus is *Entamoeba histolytica* (Fig. 2-27), which produces amebiasis or amebic dysentery in man. In its life history the trophozoite, which is about 20 to 40 microns in diameter, lives on the intestinal tissues and blood of man causing ulceration of the colon. By binary fission the trophozoite gives rise to small precystic stages that eventually form a cyst (5 to 20 microns diameter) with 4 nuclei. Dysenteric feces may contain trophozoites, but normal feces may have only the cysts. Man becomes infected by ingesting cysts in drinking water. When the cyst breaks up in man, a single tetranucleate ameba hatches and divides mitotically to form 8 uninucleated trophozoites.

■ ORDER ARCELLINIDA (TESTACIDA)

The ameboid forms of order Arcellinida are enclosed in a one-chambered test that is made of siliceous materials or pseudochitin, plus foreign substances (sand) and secreted plates embedded in a gelatinous matrix. Some of the tests have two layers. Some tests are flexible, and others are firm. They have lobopodia or filopodia that extend, become attached, and pull the body forward. A single aperture is provided

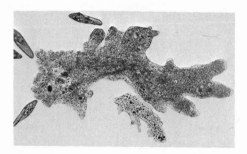

FIG. 2-26

Comparison in size of *Pelomyxa* (1 to 5 mm. long) and *Amoeba proteus* (up to 600 microns long); several paramecia are also shown. (Courtesy Carolina Biological Supply Co., Burlington, N. C.)

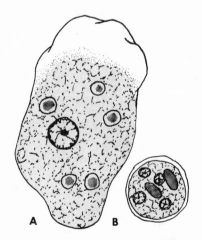

FIG. 2-27

Entamoeba histolytica (order Amoebida). **A,** Mature organism (trophozoite); 15 to 35 microns long. **B,** Cyst; dark, rodlike structures are chromatoid bodies.

for the extrusion of the pseudopodia. The endoplasm is granular or vacuolar and contains food vacuoles, contractile vacuoles, one or more nuclei, and often numerous extra nuclear chromidia. Asexual reproduction is by longitudinal or transverse fission in which usually one daughter cell receives the parental shell and the other secretes a new shell (Fig. 2-6, *B*). In those with hard tests, part of the cytoplasm exudes and forms a new shell before division occurs. Multiple fission and encystment may occur. Sexual reproduction is by ameboid anisogametes that copulate (syngamy). Arcellinidans live mostly in fresh water, although a few are marine. Some are found in moist soil, moss, etc.

Difflugia, a familiar genus with species up to 500 microns long, has a pyriform shell of foreign particles cemented together (Fig. 2-28, *A*). *Arcella*, with species up to 250 microns in di-

ameter, has a transparent, bowlike test of chitin or siliceous plates embedded in tectin (Fig. 2-28, *B*). Another common genus is *Euglypha* (Fig. 2-6, *C*), which has an oval test of siliceous scales in longitudinal rows and is often provided with one or more spines.

Subclass Granuloreticulosia
■ ORDER FORAMINIFERIDA

Some authorities think there are as many as 20,000 species in the order Foraminiferida. They are characterized by having rhizopodia or myxopodia (Fig. 2-3, *B*), and a one-chambered or multichambered test that may be calcareous or siliceous with various substances such as sand or sponge spicules. These substances are cemented together by pseudochitinous or gelatinous materials. Foraminiferans are among the largest of protozoans, some reaching a diameter of more than 100 mm., although most are in the range of 0.5 to 10 mm. The tests are of various shapes—oval, tubular, branched, spiral, etc. Most shells are multichambered or multilocular, consisting of a series of successively larger chambers that are separated internally by calcareous septa, with pores or canals connecting the chambers (Fig. 2-29). The cytoplasm is found in all the chambers and is continuous through the septal pores. One or more nuclei are found in the endoplasm. By passing through the mouth or pores of the test, the cytoplasm forms a layer (ectoplasm) over the test. The ectoplasm and endoplasm are connected by pores in the test. As the animal grows, successive chambers are added to the initial chamber, the proloculum.

Foraminiferans capture their prey by means of their pseudopodial network, which exhibits active streaming movements.

Most foraminiferans are found in marine or brackish water; only a few are freshwater forms. A few are pelagic or sessile, but most are creeping bottom dwellers (benthonic). A third of the ocean bottom is covered with *Globigerina* ooze, which is made up mainly of the accumulation of the tests of this common

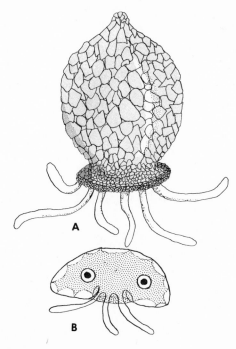

FIG. 2-28

A, *Difflugia urceolata* (order Arcellinida); test covered with sand grains; 200 to 230 microns long. **B,** *Arcella vulgaris* (order Arcellinida); test circular and domelike; 30 to 100 microns in diameter.

FIG. 2-29

Foraminiferan shell, *Elphidium* sp.; 3 to 4 mm. in diameter.

foraminiferan. Their tests are rarely found in ocean water deeper than 4,000 meters, because at greater depths their tests dissolve in the larger concentration of carbon dioxide. Foraminiferans have left an impressive fossil record starting about the close of the Devonian period. They have formed important rock strata such as the limestone used in the great pyramids of Egypt. They are also important in detecting oil-bearing strata.

Subclass Mycetozoia

Organisms belonging to Mycetozoia are commonly called slime molds and show a remarkable resemblance to fungi because the former put forth sporangia with spores. They seem to exist on the borderline between Protozoa and Protophyta, and they have been the subject of

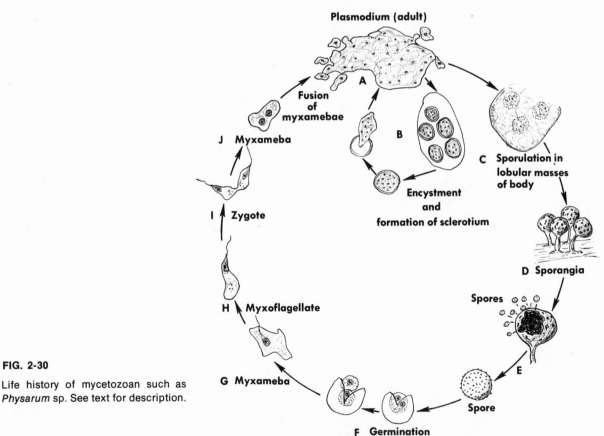

FIG. 2-30

Life history of mycetozoan such as *Physarum* sp. See text for description.

intensive investigation by J. T. Bonner in recent years because of their interesting life history and primitive nature. Three or four subgroups are recognized in this order, each of which apparently has a different ancestry. It has been suggested that they evolved from soil amebas. They resemble plants in having cellulose walls, and they are like amebas in having ameboid stages and similar methods of ingesting food. They are holozoic or saprozoic in their nutrition. They have water expulsion vesicles (contractile vacuoles) and often have pigments of various colors. Most are free living, but some are parasites of plants. Most of them live on decaying vegetable matter, an old rotten log in moist surroundings being a favorite locality. In their mature stages they are either a plasmodium (Fig. 2-30, *A*) or a pseudoplasmo-

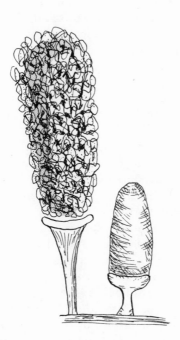

FIG. 2-31

Sporangium of *Arcyria denudata* (subclass Mycetozoia); fibrous network (capillitium) bearing spores is supported by cuplike base on stalk. On left, sporangium wall has broken and capillitium is expanded; sporangia are 2 to 3 mm. high.

dium. Streaming movements to and away from the periphery are found in the plasmodia. Under dry or other unfavorable climatic conditions the plasmodium divides into cysts, each with a dozen or so nuclei and surrounded by a wall (sclerotium, Fig. 2-30, *B*). When revived, the contents of the cyst fuse and produce plasmodia. Lack of food causes the plasmodium to develop sporangia from lobular masses (Fig. 2-30, *C* and *D*). The walls of the sporangia reach down to the substratum, where they may spread out as networks that serve as supporting structures. Spores are formed in the sporangia by mitosis and become covered with a cellulose membrane. When the sporangia disintegrate, the spores are scattered by the wind (Figs. 2-30, *E*, and 2-31). Spores germinate into myxamebas, which undergo ameboid movement and later form myxoflagellates, or swarmers (Fig. 2-30, *F* to *H*). Myxoflagellates may serve as gametes and undergo syngamy to form ameboid zygotes, or they may feed, grow, multiply by binary fission, and encyst. In either case they eventually form myxamebas (Fig. 2-30, *I* and *J*) that by growth and nuclear division produce a small plasmodium. Fusion of these plasmodia may produce the large plasmodia.

In the Acrasina, or cellular slime molds, the small myxamebas may exist as solitary individuals feeding and undergoing binary fission. Sometimes they aggregate (possibly under the influence of acrasin or evocator secreted by certain myxamebas) and form a pseudoplasmodium in which there is no protoplasmic fusion. This pseudoplasmodium can move as a unit and may increase in size by fission of its amebas. Eventually the pseudoplasmodium assumes an upright condition and forms a pseudosporangium with base, stalk, and spores. Sexual reproduction does not occur.

■ CLASS PIROPLASMEA

The Piroplasmea are small intracellular parasites that live in vertebrate erythrocytes. Some are carried by ticks. They have no cilia, flagella, or spores. They reproduce by binary fission or schizogony. *Babesia bigemina,* which causes Texas cattle fever, is the common species. It was with *B. bigemina* that T. Smith worked out the etiology of Texas cattle fever in 1891 — a famous landmark in the study of arthropod vector diseases.

Subphylum Sporozoa

The Sporozoa are made up entirely of endoparasites. Their cosmopolitan nature is shown by the fact that their hosts are distributed among all the animal phyla. They are distinguished from all other protozoan classes by the absence in adults of all organelles of locomotion, such as pseudopodia, cilia, or flagella. Some sporozoans are of great importance to man. The malaria parasite causes more illness and death than any other disease agency. Many serious disorders among animals other than man are caused by these parasites. Hyperparasitism or the infection of another parasite by them is also found. They show an amazing degree of host specificity—a particular sporozoan species may infect only one kind of host. As a group they do not show a close bond of homogeneity but must be considered an assortment of organisms that have few affinities with each other. About the only character they have in common is their parasitic habit and their spore formation in some stage of their life history.

Since the sporozoans are such a heterogeneous group, it is possible that each of the major taxa had a separate origin. The characteristics of certain stages in their life history may give a clue in some cases to their affinities, such as the presence of flagella on the microgametes of Telosporea, which is supposed to indicate flagellate ancestry. Ameboid movement of many sporozoans may indicate relationship with Sarcodina. Spore formation, so characteristic of sporozoans, is also found in both Mastigophora and Sarcodina. Perhaps the subphyla of Protozoa are polyphyletic, as suggested by many authorities (L. H. Hyman, G. A. Kerkut).

The taxonomy of the sporozoans is difficult, and authorities are not in agreement on certain taxa. At the present time the scheme that probably meets with the greatest favor is the one that divides the old group Sporozoa into two subphyla—Sporozoa and Cnidospora. This classification is based mainly on the characteristics of the spore.

GENERAL FEATURES

The Sporozoa and Cnidospora have a great variety of shapes. This variation is not restricted to groups of species but is found in stages of the life history of a given species. Their method of infecting hosts is usually by means of spores enclosed in hard walls, or sometimes the transmission is by naked young. An intermediate host such as mosquitoes, leeches, flies, and other vectors is used in some cases.

Sites of infection may be body spaces, intestine, bladder, coelomic cavity, blood cells, muscle cells, and other organs.

Organelles are uncommon in Sporozoa. The cytoplasm has one nucleus in young stages but becomes multinucleated about the time of multiple fission. Differentiation of ectoplasm and endoplasm is common in the group. Myonemes and adhesive organs are present in gregarines.

Locomotion. All locomotor organelles are absent in the adult stage, but microgametes are often flagellated. Certain stages exhibit ameboid, gliding, and squirming movements. Cytoplasmic fibrils (myonemes) present in some may explain the sinuous motion in those that have such fibrils. Ameboid movement appears to be fairly common in the active stages of some species of *Plasmodium.* Many of their movements are difficult to explain.

Nutrition. There are no organelles for the capture and ingestion of food in Sporozoa. Nutrition is mostly saprozoic, or by absorption of liquefied substance from the host over most of the body surface, except in a few forms in which absorption may occur at the anterior end. However, work by M. A. Rudzinska and W. Trager shows that a certain species of malaria plas-

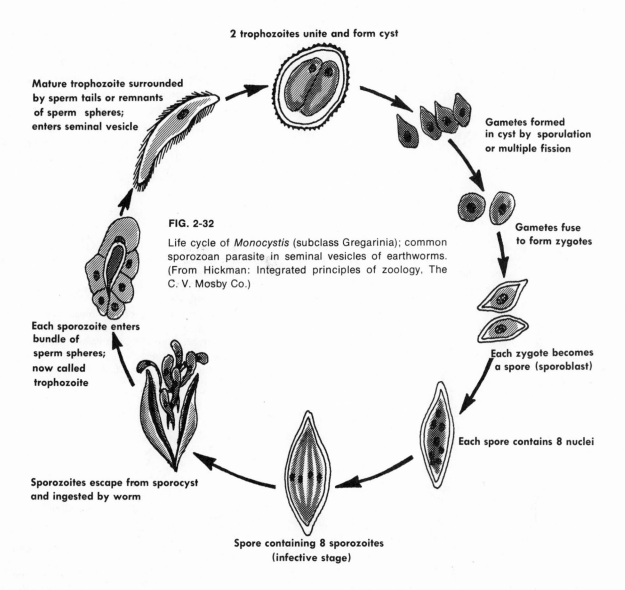

2 trophozoites unite and form cyst

Mature trophozoite surrounded by sperm tails or remnants of sperm spheres; enters seminal vesicle

Gametes formed in cyst by sporulation or multiple fission

FIG. 2-32

Life cycle of *Monocystis* (subclass Gregarinia); common sporozoan parasite in seminal vesicles of earthworms. (From Hickman: Integrated principles of zoology, The C. V. Mosby Co.)

Gametes fuse to form zygotes

Each sporozoite enters bundle of sperm spheres; now called trophozoite

Each zygote becomes a spore (sporoblast)

Each spore contains 8 nuclei

Sporozoites escape from sporocyst and ingested by worm

Spore containing 8 sporozoites (infective stage)

modia can ingest hemoglobin in small solid particles by ameboid movement. It has also been shown that trophozoites of *Nosema* (Fig. 2-37), a microsporidian, can ingest solid bits of their host's tissues.

Life cycle and reproduction. There is an asexual and a sexual phase in the life cycle of the Sporozoa, but there is a great deal of diversity in the life cycles of the different forms. Some groups, such as the gregarines (Fig. 2-32),

may restrict their reproduction to the sexual phase. Typically, in the asexual phase the vegetative trophozoite attains full size and becomes a schizont, which by multiple fission (schizogony) forms agametes, or merozoites. The merozoites either reinfect the host and produce another generation of schizonts, or they may develop directly into microgametes, which can produce a zygote or oocyst by fertilization (sexual reproduction). The oocyst undergoes a series of multiple fissions to form sporoblasts and sporozoites (sporogony, or spore formation), which may be naked young or walled spores.

Only the zygote is diploid; all other phases are haploid. If there are two different hosts, the life cycle usually involves schizogony in one and sporogony in the other.

In some species the life cycle may not involve sporogony at all, and the infective stage is an ameboid sporoplasm, not a sporozoite. Life

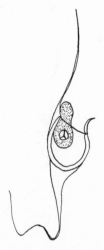

FIG. 2-33

Haplosporidium chitonis (class Haplosporea); spore stage. Spore develops into multinucleated plasmodium in amphineuran host. Plasmodium then produces uninucleated sporoblasts that develop into spores. (Redrawn from Kudo: Protozoology, Charles C Thomas, Publisher.)

FIG. 2-34

Life cycle of *Eimeria tenella* (diagrammatic) (subclass Coccidia) in the chicken. These coccidian intracellular parasites produce coccidiosis in higher invertebrates and vertebrates. In chickens they attack intestinal epithelium, especially lining of cecal glands, where several species are known to produce infestation. In rabbits *E. stiedae* attacks epithelium of bile ducts. Other species of *Eimeria* invade blood or other cells in various species of animals. Much of their life history is characterized by a prolific phase of asexual reproduction, which may account for heavy infestations.

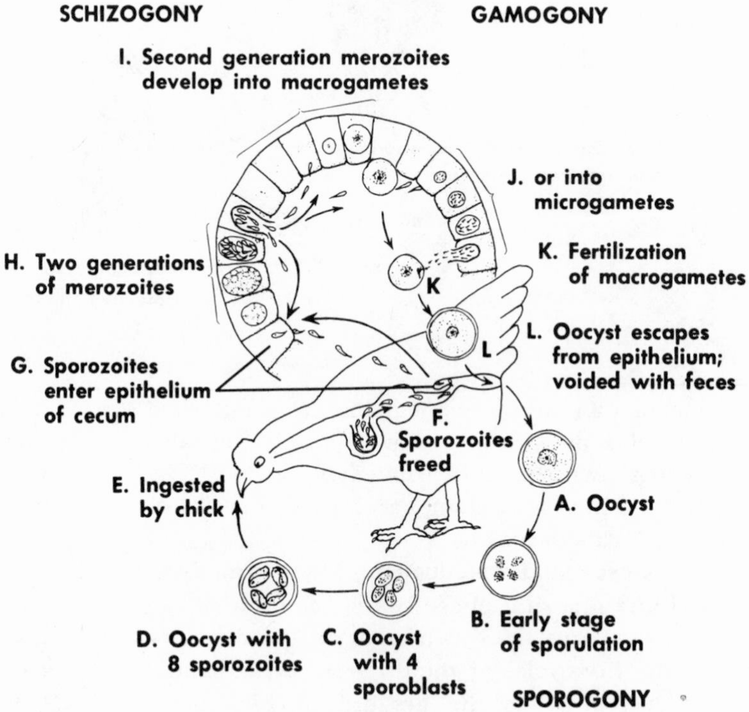

SCHIZOGONY   GAMOGONY

I. Second generation merozoites develop into macrogametes

J. or into microgametes

K. Fertilization of macrogametes

H. Two generations of merozoites

G. Sporozoites enter epithelium of cecum

L. Oocyst escapes from epithelium; voided with feces

F. Sporozoites freed

E. Ingested by chick

A. Oocyst

D. Oocyst with 8 sporozoites

C. Oocyst with 4 sporoblasts

B. Early stage of sporulation

SPOROGONY

histories are incompletely known for many Sporozoa.

■ CLASS TELOSPOREA

The class Telosporea contains some common parasites characterized by having spores that lack polar capsules and produce one to many sporozoites. In some cases they give rise to naked sporozoites instead of spores. Their life histories have both asexual and sexual phases,

FIG. 2-35

Life cycle of *Plasmodium vivax*, coccidial blood parasite that causes malaria in man. **A,** Sexual cycle produces sporozoites in body of mosquito. **B,** Sporozoites infect man and reproduce asexually, first in cells of liver sinusoids and finally in red blood cells. Malaria is spread by the mosquito, which sucks up gametocytes along with human blood and later, when biting another victim, leaves sporozoites in new wound. (From Hickman-Hickman, Jr.: Biology of animals, The C. V. Mosby Co.)

although schizogony is lacking in most of the gregarines, and they increase only by sporogony.

The subclass Gregarinia includes some common parasites of arthropods and annelid worms. Two gregarines often adhere end to end to form what is called syzygy. In such unions the anterior member becomes the female and the posterior, the male. Usually when a spore enters a host, the emerging sporozoites enter the epithelial cells of the digestive system. When they become trophozoites, they are restricted mainly to the gut lumen.

A common gregarine genus is *Monocystis* (Fig. 2-32), which parasitizes the seminal vesicles of oligochaetes. *Gregarina rigida*, common in the gut of certain grasshoppers, is another gregarine.

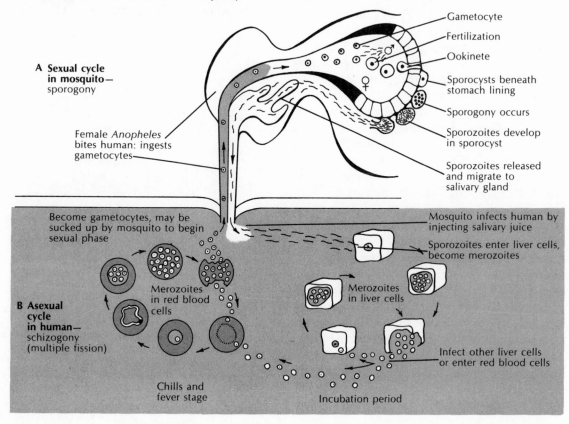

The subclass Coccidia has a wide distribution in both vertebrates and invertebrates. Most of them attack the epithelium of the digestive tract and glands. Both asexual schizogony and sexual sporogony are found. The genus *Eimeria* (Fig. 2-34), which includes many species, is a common parasite of the bile duct and liver of rabbits and of the intestines of chickens. This order has a very wide host range, being found in cattle, pigs, frogs, fishes, etc. These parasites may be very destructive.

The best known members of the subclass Coccidia are the four species that cause malaria in man, but there are many other species, some of which infect other vertebrates. Members of this group are intracellular parasites of vertebrate blood corpuscles. Their life history always involves two hosts—a vertebrate and an invertebrate. Asexual reproduction (schizogony) takes place in the blood of vertebrates, and sexual reproduction, or sporozoite formation, occurs in the alimentary canal of a bloodsucking invertebrate. Since they always remain within the body of one or the other host, the parasites have no cyst wall, and naked sporozoites (the infective forms) hatch directly from the zygote. All the true malarial parasites are placed in the genus *Plasmodium* (Fig. 2-35). The vector that carries malaria for mammalian hosts is the *Anopheles* mosquito; those vectors for birds are mosquitoes of the genera *Aedes* or *Culex*. Certain other genera such as *Haemoproteus* and *Leucocytozoon*, closely related to *Plasmodium*, cause malaria in birds and reptiles. Certain flies serve as vectors for the parasites of these genera. The paroxysm of malaria consisting of chills and high fever corresponds with the liberation of merozoites and their toxins from corpuscles. This schizogonic cycle varies with the different species of *Plasmodium* responsible for human malaria. In *Plasmodium vivax* (benign tertian malaria) the cycle is every 48 hours, in *P. malariae* (quartan malaria) every 72 hours, in *P. ovale* every 48 hours, and in the highly malignant *P. falciparum* (tertian malaria) every 48 hours.

Subphylum Cnidospora

The sporozoans of the subphylum Cnidospora usually have spores with polar capsules, each containing a coiled filament that by inversion can anchor the spore to its host. Each spore may contain 1 to 4 capsules. Spores in this subclass also contain one or more ameboid sporoplasms, which in the host become amebulas and later become trophozoites that feed and grow on the host. The cnidosporan spore (in some orders) is unique in being formed from a number of cells instead of from a single one, a condition that may indicate an affinity with the Mesozoa. The method of multiplication in the group is both schizogonic and sporogonic, but schizogony is very restricted in this subclass. Details of their life cycles vary in the different orders.

The Cnidospora are exclusively parasites of the lower vertebrates and invertebrates; none parasitize man. There is no secondary or intermediate host. However, they have economic importance for man's interests because their in-

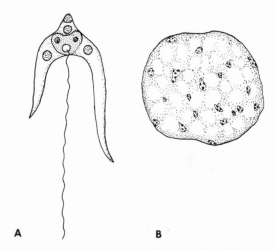

FIG. 2-36

A, *Synactinomyxon tubificis* (order Actinomyxida); spore stage; found as parasites in annelids, especially in gut of family Tubificidae. **B,** *Myxidium serotinum* (order Myxosporida); trophozoite stage; this sporozoan is common in gallbladder of certain toads and frogs. (Modified from Kudo.)

fections may seriously affect silkworms, honey-bees, and fishes. Two classes are recognized — Myxosporidea and Microsporidea (Fig. 2-37). Myxosporidea includes two orders — Myxosporida (Fig. 2-36, *B*) and Actinomyxida (Fig. 2-36, *A*).

The more common members of this subphylum are *Nosema bombycis* (Fig. 2-37), the cause of pébrine disease of the silkworm, and *Unicapsula muscularis,* found in muscles of halibut fish and often producing epidemics. The control of *Nosema* was a result of classic experiments by L. Pasteur.

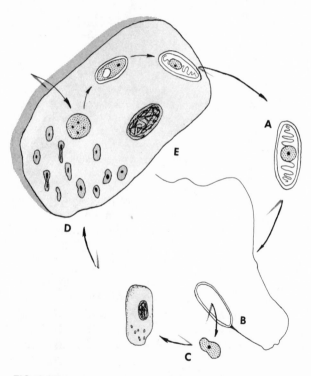

FIG. 2-37

Life cycle of *Nosema bombycis* (class Microsporidea); cause of pébrine, fatal disease of silkworms. When spore, **A,** with coiled filament and sporoplasm, is ingested by worm, filament is extruded, **B,** and sporoplasm is freed, **C,** from other part of spore and invades gut epithelial cell, **C,** where, by binary fission or budding, many uninucleated amebas are formed and, **D,** may enter any tissue in the body. In time, multinucleated cells, **E,** arise and develop into characteristic spore, **A,** made up of sporoplasm, spore envelope, and polar capsule.

Subphylum Ciliophora
■ CLASS CILIATA

The ciliates represent one of the most homogeneous groups among the protozoans. They are also distinguished by their wide distribution and the ease with which they can be found in water that will support any life. More than 6,000 species have been described, and doubtless many more exist. Their distinctive characteristics are their cilia (present in some stage) for locomotion and for food getting and the presence of two kinds of nuclei — a large macronucleus and a smaller micronucleus. The ciliates are larger than most other protozoans, although their range in size is from 10 or 12 microns to 3,000 microns, or 3 mm. long.

A variety of forms is found in ciliates. Free-swimming organisms may be spherical or elliptic, whereas creeping forms tend to have a flattened shape. Some species have a kind of radial symmetry, and others approach bilateral symmetry. Most of them are free living and solitary, but some are sessile, colonial, ectocommensal, or endocommensal. Most are colorless, but a few are brightly colored, such as the different species of *Stentor*. Color may be caused by chromatophores or by symbiotic bacteria. The color of *Blepharisma*, for instance, is in the pellicle. This can be observed when the pellicle is shed in a strychnine sulfate solution. Ciliates are found in both freshwater and saltwater habitats.

Perhaps no group of animals has been more widely used for experimental purposes than the ciliates. This use is probably due to the ease with which they can be obtained and cultured, their rapid multiplication, and the many fundamental biologic problems that they pose. Although many ciliates are simple protozoan organisms, some are the most complex animals to be found among unicellular forms. Some ciliates appear to have exploited nearly all the potentialities to be found within the restrictions of a single membrane or cell.

Classification of ciliates has undergone many changes since their discovery by Leeuwenhoek in 1676. At the present time the scheme of the French investigator E. Fauré-Fremiet and the American investigator J. O. Corliss is widely accepted by many protozoologists. Their scheme is based on the probable evolutionary pattern of the group. Chief specific structures that are used in the taxonomic arrangements of ciliates are the ciliary patterns and the nature of the cytostome. Although there is only one class, there are four subclasses and subordinate orders.

In recent years, Corliss, a world authority on Ciliophora, has proposed a phylogenetic tree of the class showing possible relationships among its numerous orders. The fossil record of ciliates yields little or no evidence to support any scheme of phylogeny, but certain indirect evidences have proved useful, such as the data obtained from comparisons of life histories, cytostomes, infraciliature, etc. Corliss and others have laid great emphasis on changes in the somatic and buccal ciliation in the evolution of the group. As mentioned elsewhere, the groups of Protozoa are probably polyphyletic, and each has originated by convergent evolution from several stocks. Ciliates pose peculiar problems of nuclear dimorphism, sexual phenomena, and complex infraciliature that makes them especially difficult to relate to the other groups. Their kinship to the flagellates is apparent in the similarity of flagella and cilia and the apparent homology of kinetosomes and other structures of the two groups.

GENERAL FEATURES

Most ciliates have a distinct ectoplasm and endoplasm. From the ectoplasm the pellicle, cilia, infraciliature, and trichocysts originate; from the endoplasm the nuclei, contractile vacuoles, food vacuoles, pigment, etc., originate. The endoplasm undergoes a rotating movement, or cyclosis. The pellicle may be very thin and hardly distinguishable. In a few species it may be differentiated into an armor of calcareous grooves. Longitudinal and diagonal grooves in the pellicle indicate places of attachment of

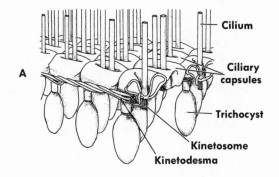

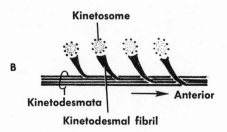

FIG. 2-38

A, Portion of pellicle and infraciliature of *Paramecium.* **B,** Diagram of a ciliate kinety (kinetodesmata and their basal granules, or kinetosomes). All kinetia together form neuromotor system. (**A** after Corliss, from Ehret and Powers, 1959; **B** modified from Grimstone.)

ciliary rows. The pellicle may also have a pattern of various figures bounded by ridges.

The pellicle has been most thoroughly studied in *Paramecium,* in which, according to C. F. Ehret and E. L. Powers (1959), the outer stratum of the pellicle is made up of ciliary corpuscles, each composed of an outer and inner peribasal membrane. One or two cilia arise from a pit in the center of each corpuscle (Fig. 2-38).

Cilia are short, slender, and hairlike. They are usually arranged in longitudinal or diagonal rows. The electron microscope reveals that they, like flagella (Figs. 2-1 and 2-2), have 2 central and 9 peripheral fibrils. Cilia may clothe the entire surface (somatic ciliation), or they may be restricted to certain regions (oral ciliation). Compound ciliary organelles are complex structures formed by the compounding or fusion of cilia into a sheet (undulating membrane), a triangular or rectangular block of cilia 2 to 4 rows wide (membranelle, Fig. 2-39, *F*), or

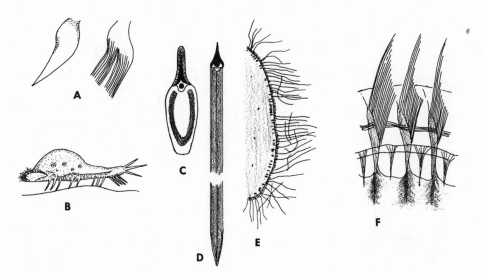

FIG. 2-39

A, Left, intact cirrus; right, cirrus unraveled into cilia by acid treatment. **B,** *Stylonychia,* using cirri for locomotion. **C,** Undischarged trichocyst. **D,** Portion of discharged trichocyst. **E,** Diagrammatic portion of paramecium with both discharged and undischarged trichocysts. **F,** Structure of membranelles of *Stentor* with fibrillar attachments. (Modified from Hyman.)

the fusion of a tuft of cilia (cirrus, Fig. 2-39, *A* and *B*). Undulating membranes and membranelles help propel food into the cytopharynx and may also aid in locomotion. Membranelles are found in a specialized zone leading to the cytostome.

In connection with the ciliation, just underneath the pellicle is the infraciliature, composed of the basal granules (kinetosomes) of the cilia and their associated kinetodesmal fibrils (Fig. 2-38). The cilia and their kinetosomes are arranged in rows. The kinetosomes of a row and the fine fibrils (kinetodesmata) that connect them and run anteriorly in a bundle parallel to the row make up what is known as a kinety. All ciliates seem to have kinety systems, even those that are without cilia at some stage. Kinetosomes appear to be self-duplicating and can give rise to cilia, fibrils, and trichocysts. The silver line system of staining by impregnation with silver (first developed by B. M. Klein in

1926) has been important in the study of the infraciliature and trichocysts. More recently, electron microscopic studies have revealed more of the ultrastructure of these organelles.

The fibrillar system, often known as the neuromotor apparatus, was first described in detail by R. G. Sharp (1914) in his classic experiments on *Epidinium* (*Diplodinium*, Fig. 2-49), a ciliate symbiont of cattle. He described a central motorium near the cytostome from which the fibrils radiated to connect with the basal granules (kinetosomes) of the cilia. The neuromotor system is apparently a motor-coordinating system for ciliary movement.

Trichocysts (Fig. 2-39, *C* to *E*), another distinctive character of ciliates, are restricted mainly to the subclass Holotrichia. They are small rodlike or oval bodies in the ectoplasm at right angles to the body surface. They may be distributed all over the body, or they may be limited to certain regions such as the anterior end. When discharged, they turn inside out as long filaments, which pass through the pellicle. Upon extrusion, the trichocyst hardens except for the tip, which remains sticky for attachment. The stimulus for discharge seems to be mechanical touch, but other modes, e.g., chemical and electric, are known. Their functions are not always clear, but they may be offensive and

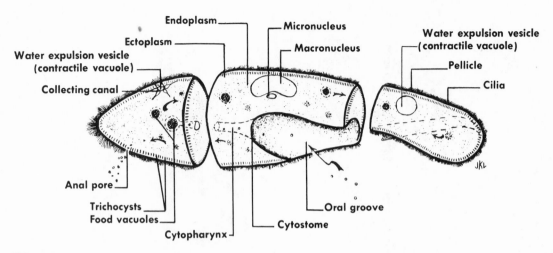

FIG. 2-40

Structure of *Paramecium* (diagrammatic). Arrows show direction of protoplasmic flow (cyclosis), which keeps food vacuoles moving. (From Hickman and Hickman: Laboratory studies in integrated zoology, The C. V. Mosby Co.)

defensive. The trichocysts of some ciliates are known to paralyze some organisms. A number of different types of trichocysts are recognized. The electron microscope reveals them as tack-shaped barbs with long, finely striated shafts.

Some ciliates are capable of contractile movements—either of the whole body, as in *Stentor* (Fig. 2-47, *B*) or *Spirostomum*, or of their stalks, as in *Vorticella* (Figs. 2-46, *A*, and 2-47, *A*) or *Carchesium* (Fig. 2-46, *B*). Myonemes, contractile fibrils in the ectoplasm, bring about these movements. Myonemes are mostly of ectoplasmic origin, although in peritrichs they appear to be derived from both ectoplasm and endoplasm. They usually run lengthwise in the cortex near the grooves where cilia are attached, although in some such as *Stentor* they are also wound spirally around the gullet. In contractile stalks the fibrils are spirally arranged and bring about the peculiar bobbing motion of these attached forms.

Most ciliates have a mouth (cytostome) usually associated with a buccal cavity (cytopharynx (Fig. 2-41) with compound ciliary

organelles. The mouth may be in a lateral or ventral position. The region around the cytostome is the peristome and is specialized for water currents. It may be an oral groove or some other specialization. A funnellike cytopharynx often extends from the peristome (and cytostome) down into the endoplasm. Special cilia and membranelles are found in connection with the peristome for creating water currents. Myonemes encircling the cytopharynx of many ciliates facilitate swallowing.

Water expulsion vesicles (contractile vacuoles) with one to many collecting canals (Figs. 2-4, *A*, and 2-40) are found throughout the Ciliophora. Some ciliates have no collecting canals at all but are fed by smaller secondary and tertiary vesicles (Fig. 2-4, *B*). Most water expulsion vesicles are fixed in position, open by a permanent pore to the exterior, and are of constant number for the species. Some marine as well as freshwater ciliates may have water expulsion vesicles.

Two kinds of nuclei are the rule in the majority of Ciliophora (Fig. 2-5). Usually one of each kind is present, although there may be more. The smaller, or micronucleus, contains chromosomes and has a generative function. It undergoes mitosis like any metazoan cell except that centrioles are often absent. The micronucleus is usually ovoid or rounded. The larger macronucleus has somatic or metabolic functions and does not undergo mitosis but divides by amitosis. It is usually single and may be oval, round, crescent, or other shapes. The macronucleus is known to have genic control over certain hereditary characters, and it arises from the micronucleus.

Two other cytoplasmic organelles of almost universal presence in ciliates are (1) mitochondria, spherical or rodlike structures, and (2) Golgi apparatus, which may appear as networks, spherules, or rings. The former of these two cytoplasmic inclusions furnish important teams of enzymes in cellular metabolism; the latter is apparently involved in secretion.

Locomotion. Ciliates are among the most active of all protozoans, and their locomotory mechanism is very efficient. Cilia produce movement by a rowing action, each cilium producing a lash like an oar, involving both an effective stroke and a recovery stroke. The effective stroke is made with the cilium outstretched to offer as much resistance as possible; in the recovery stroke the cilium is bent and brought back in position like the arm of a swimmer, with the least amount of resistance. The cilia are coordinated in their beat by the neuromotor system because the numerous cilia must beat together. Most ciliates move in a spiral path that is generally characteristic for the species. Various theories try to account for this spiral rotation on its longitudinal axis. H. S. Jennings thought that the body form and the stronger beat of the oral cilia were chiefly responsible; others think that the spiral movement actually results from the oblique strokes of the cilia working together in the same direction. Cilia are also used for crawling and walking (e.g., *Oxytricha*, *Stylonychia* [Fig. 2-39, *B*], and *Euplotes* [Fig. 2-48, *C*]). Many forms have large cirri that have originated from the fusion of tufts of adjacent cilia (Fig. 2-39, *A* and *B*). Some cilia may also be sensory in function, as in the order Thigmotrichida. Contractile movements brought about by myonemes have already been discussed.

Nutrition. Most ciliates are holozoic and are the most expert food catchers among the protozoans. Many of them are provided with a cytostome at the anterior end, and as they move through the water, they come in contact with food, which is either swept into the mouth by special cilia or membranelles or is seized directly. Preoral chambers in the form of grooves, depressions, vestibules, or buccal cavities provided with undulating membranes or membranelles aid in sweeping the food down the gullet into the endoplasm, where food vacuoles are formed. This arrangement is common among the current feeders. Many ciliates are raptorial and ingest prey even larger than themselves. *Didinium*, for instance, has a proboscis for seizing paramecia and then engulfing them (Fig. 2-43, *A*). The sessile suctorians, while adhering to their prey with their knobbed, tubelike tentacles, can paralyze them with a hypnotoxin. The contents of the prey are then sucked through the tentacular tubes and digested in the body (Figs. 2-41 and 2-43, *B*).

Ciliates live on a great variety of foods such as bacteria, other protozoans, algae, diatoms, etc. *Paramecium* has an especial fondness for bacteria and may consume millions in a single day. Some ciliates are more fastidious in their food selection than others. Some, like *Didinium* (Fig. 2-43, *A*) prefer paramecia and rarely seize other kinds. Those that practice food selection may choose their prey by chemical attraction; many are restricted by the limitations of their food-getting mechanisms.

Ciliates have no digestive tube, and when a food vacuole is formed at the base of the cytopharynx or other region, it follows a definite circuit induced by the cyclosis of the endoplasm (Fig. 2-40). A food vacuole consists of the food particle, a film of water, and some endoplasmic secretions; its size is somewhat constant for a given species. In the digestive process the reaction in the vacuole is at first acid (about pH 4) and then gradually becomes alkaline as digestion proceeds. Proteolytic amylases and other enzymes have been detected in certain

FIG. 2-41

A, *Tokophrya* feeding on ciliate. **B,** Tentacles of *Tokophrya;* left, extended; middle, feeding on contents of ciliate; right, contracted. **C,** *Choanophrya* feeding on organic particles in water. (Subclass Suctoria.)

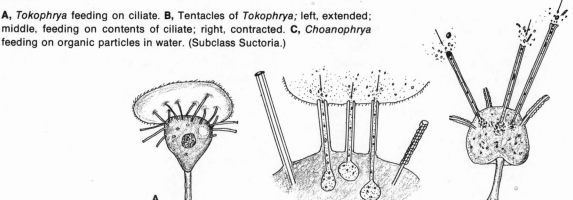

A   B   C

cases. Digestion takes place rapidly and may be completed in 20 to 30 minutes. The indigestible part, or fecal material, may be discharged through a special opening, the cytoproct, or anal pore.

Some parasitic ciliates are saprozoic, but others do not differ from free-living holozoic forms in their nutrition.

Life cycles and reproduction. The reproductive life cycle of Ciliophora involves asexual reproduction, a type of sexual reproduction, and encystment. True sexual reproduction with fusion (syngamy) of free gametes probably does not occur in this extensive group, but conjugation with meiosis and chromosome recombinations does occur, a process that certainly represents sexual phenomena. Some species also have variant forms of conjugation, such as autogamy and cytogamy.

Asexual reproduction. Asexual reproduction is mostly transverse binary fission, except in the Peritrichia, in which it is longitudinal (Fig. 2-6, *D* and *E*). Fission involves both micronuclei and macronuclei. The micronucleus divides first by a mitotic process with spindle and chromosomes as already mentioned. On the other hand, the macronucleus divides amitotically simply by elongation and then constriction in the center; no mitosis or chromosomes are involved in the process. However, the ciliate macronucleus is known to be highly polyploid

and contains many genomes produced by repeated endomitosis. There are many variations in nuclear behavior among the different ciliate species.

Budding occurs in some ciliates in which small daughter cells arise from the outer surface of the parent body; or the budding may be internal (Suctoria) in a brood pouch from which the young ciliated individuals escape at the end of development.

Conjugation. In ciliated Protozoa a process called conjugation occurs at certain times in the life cycle of many ciliates. Not all ciliates undergo conjugation but can live indefinitely without the process. Whenever it occurs, the major details of conjugation follow much the same pattern with the exception of certain minor variations. Conjugation may be defined as a temporary union of two individuals during which there is an exchange of nuclear material, together with a series of macronuclear and micronuclear divisions and reorganization processes. After the two individuals separate, each of the two conjugants undergoes further nuclear and cell divisions before resuming its regular cycle of growth and binary fission (Fig. 2-42).

Conjugation involves the presence of two ciliates of different mating types. In some cases members of the same mating types have been induced to conjugate by the introduction of

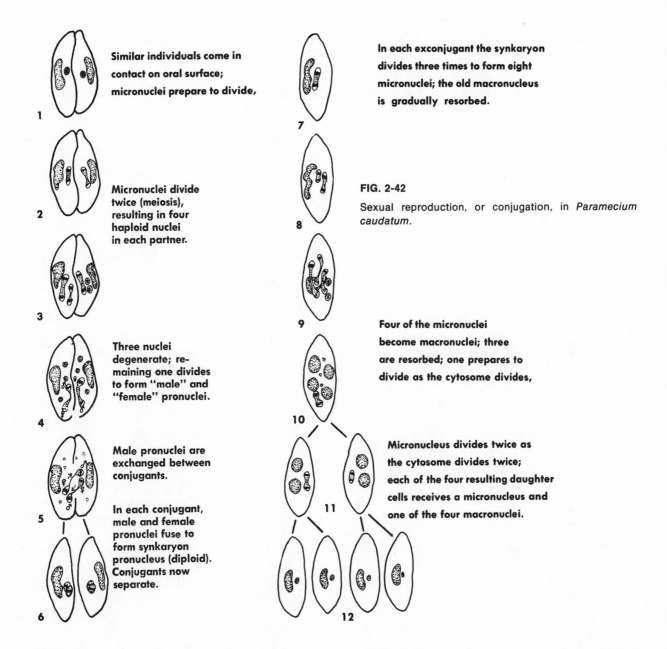

1 Similar individuals come in contact on oral surface; micronuclei prepare to divide,

2 Micronuclei divide twice (meiosis), resulting in four haploid nuclei in each partner.

3

4 Three nuclei degenerate; remaining one divides to form "male" and "female" pronuclei.

5 Male pronuclei are exchanged between conjugants.

6 In each conjugant, male and female pronuclei fuse to form synkaryon pronucleus (diploid). Conjugants now separate.

7 In each exconjugant the synkaryon divides three times to form eight micronuclei; the old macronucleus is gradually resorbed.

FIG. 2-42
Sexual reproduction, or conjugation, in *Paramecium caudatum.*

8

9

10 Four of the micronuclei become macronuclei; three are resorbed; one prepares to divide as the cytosome divides,

11 Micronucleus divides twice as the cytosome divides twice; each of the four resulting daughter cells receives a micronucleus and one of the four macronuclei.

12

cilia from a ciliate of a different mating type. The addition of such cilia seems to serve as a glue for holding similar mating types together. Mating types vary in different species of ciliates. In some there are only two mating types; other species may have a dozen or more types. Visible differences between complementary mates are

found in only a few ciliates; in most cases different mates can be detected only by physiologic reactions. As a general rule, the cilia of the surface of one conjugant will adhere to the surface cilia of a different mating type.

All ciliates are divided into a variable number of genetically isolated varieties, each of which includes two mating types. Conjugation can occur only between individuals of the same variety but of different mating types. The varieties (syngens) are comparable to physiologic species and cannot normally undergo cross fertilization with each other, but only the mating types of a single variety can do so; thus each variety has its own distinctive mating types.

Using *Paramecium caudatum* (which has only one macronucleus and one micronucleus) as an example, the process of conjugation occurs in the following way (Fig. 2-42). When two different mating types of the same variety are mixed together under conditions suitable for conjugation, the individual paramecia usually clump together in large groups. Within an hour or so they begin to separate and conjugate together in pairs by attachment of their ventral (oral) surfaces with a protoplasmic bridge between the conjugants. Now a series of nuclear changes occurs in each conjugant. This change involves the gradual fragmentation of the macronucleus and two meiotic divisions of the diploid micronucleus to produce 4 haploid micronuclei (pronuclei), 3 of which disintegrate. The remaining haploid micronucleus now undergoes a mitotic division to form one haploid stationary pronucleus (female) and one haploid wandering pronucleus (male). The wandering pronucleus in each conjugant now moves through the protoplasmic bridge into the opposite member of the conjugating pair and fuses there with the stationary pronucleus to form a diploid zygote nucleus (synkaryon). The conjugants now separate as exconjugants. In each exconjugant, the synkaryon divides three times to produce 8 nuclei, 4 of which become micronuclei and four macronuclei. Three of the micronuclei then disintegrate. When the exconjugant undergoes two cytosomal divisions, the resulting 4 daughter cells each receives one macronucleus. The single remaining micronucleus in each synkaryon undergoes mitosis at each of the two divisions so that each cell now has the normal complement of one macronucleus and one micronucleus. These small daughter cells now grow in size and undergo binary fission in the normal way. In ciliates of more than one macronucleus and micronucleus, the synkaryon merely divides several times to produce the normal number of macronuclei and micronuclei.

As in regular sexual reproduction, conjugation brings together and fuses reciprocally in two individuals two haploid nuclei similar in many ways to male and female gametes in metazoans.

Various factors may induce conjugation, such as starvation, sudden darkness in light conditions, low temperatures, the presence of certain salts, etc. The process of conjugation may be beneficial in producing nuclear reorganization and heritable variation and may be useful in rejuvenation of species. Whatever value conjugation may have in this way, however, does not apply to all ciliates, for some survive without the process. Conjugation is definitely restricted to certain types.

Autogamy. In autogamy (formerly known as endomixis) there is no conjugation of two individuals and there is no cross fertilization, but nuclear reorganization similar to that in conjugation occurs in the same individual. It is internal fertilization in which two daughter nuclei fuse within the same cell to form a homozygous animal. Autogamy may take place in the absence of conjugation and may have the same rejuvenating effect.

Cytogamy involves a pairing of animals, but there is no cross fertilization between the two individuals, and the fusion nucleus (zygote) is formed from two pronuclei in the same animal, as in autogamy. Cytogamy may be due to the failure of the male nucleus to pass into the mate.

Encystment. Encystment occurs in many protozoans, and Ciliophora are no exception (Fig. 2-7). Environmental conditions, such as crowding, lack of food, accumulation of waste products, lack of essential nutrients, dessication, distilled water, and rich feeding are inducements to encystment.

There are many variations in the process

among the ciliates. Precystic changes in the organism usually occur before a cyst wall is secreted. Gelatinous materials for the ectocyst membrane sometimes accumulate in the peripheral cytosome. Starch and glycogen may accumulate, but food vacuoles are ejected. Locomotor organelles are mainly resorbed. The fluid content of the organism is lessened and the cytosome becomes denser. The cysts display a great variety in form and size and in their methods of formation. Some are spherical and smooth, some are warty or covered with knobs, and some are colored. The ectocyst is composed of a hard substance that in some species seems to be a polysaccharide in composition; in others it appears chitinous or keratin-like.

Excystment, or the emergence from the cyst, involves many changes such as a regeneration of organelles and an internal reorganization. Organelles such as water expulsion vesicles and cilia first appear. Cyclosis starts and the organism grows larger. The membrane becomes thinner and finally disappears. The cyst itself is often ruptured by the absorption of water and organic substances into the cytoplasm, or it may be dissolved by enzymes.

FIG. 2-43

A, *Didinium nasutum* (order Gymnostomatida); proboscis armed with trichocysts is efficient organ for seizing paramecia. **B,** *Podophrya fixa* (order Suctorida). **C,** *Spirochona* (order Chonotrichida); asexual reproduction by lateral budding is common in this group.

Subclass Holotrichia
■ ORDER GYMNOSTOMATIDA

The order Gymnostomatida is considered the most primitive of all the orders of Ciliophora; members are numerous and widespread. They are found in both fresh and salt water, the sands of intertidal zones being a favorite habitat. Many of them are large (2 mm. or longer). The simple cilia are usually distributed all over the body, but there are no cilia in the buccal area. The cytostome may be at or near the anterior end, at the lateral region, or on the ventral surface. Rodlike trichites reinforce the cytopharynx. A few of these organisms are harmless commensals, but most of them are carnivorous or herbivorous. There are many families and genera in this order, the most familiar genera being *Didinium* (Fig. 2-43, *A*), which has an expansible cytostome used in catching paramecia and other ciliates; *Coleps*, which has a barrel-shaped body and ectoplasmic plates; *Loxophyllum*, which is flask shaped with trichocysts and chain nuclei; *Dileptus,* which has a long neck with rows of trichocysts and a round mouth with trichites; and *Prorodon*, which is ovoid or cylindric with a terminal contractile vacuole.

■ ORDER CHONOTRICHIDA

Members of the order Chonotrichida are mostly ectocommensals on marine crustaceans. The group is not large. The body is commonly vase shaped with an apical peristome that is

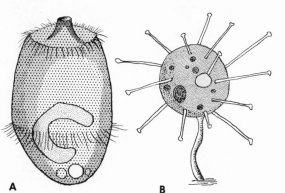
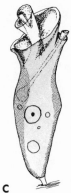

A   B   C

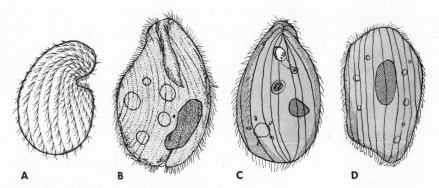

FIG. 2-44

A, *Colpoda cucullus* (order Trichostomatida). Divides only in encysted stage; 40 to 110 microns long. **B,** *Balantidium coli* (order Trichostomatida); only parasitic ciliate in man; note cytopyge, or cell anus, at posterior end; 40 to 80 microns long. **C,** *Tetrahymena geleii* (order Hymenostomatida); genus *Tetrahymena* has been widely used in physiologic experiments on growth and metabolism; 40 to 60 microns long. **D,** *Anoplophrya orchestii* (order Astomatida); often found in gut of sand flea, *Orchestia;* 6 to 68 microns long.

provided with a relatively complicated ectoplasmic funnel. This funnel contains rows of cilia leading to the cytostome and cytopharynx. *Spirochona* (Fig. 2-43, *C*) and *Stylochona* are common genera.

■ ORDER TRICHOSTOMATIDA

Ciliates of the order Trichostomatida, like the gymnostomes, have no buccal ciliation but have a vestibulum that passes from the outside of the body to the cytostome. A few species are endo-commensals in domestic animals, but most are found wherever free-living protozoans live. *Colpoda* (Fig. 2-44, *A*), with its vestibulum in the middle of the flattened ventral side and its convex left border, is one of the most familiar examples of the order. It has been used frequently in nutrition experiments. *Balantidium coli* (Fig. 2-44, *B*), with an oval body, anterior peristome, and poorly developed cytopharynx, is a common parasite in the intestine of pigs and is the only ciliate that is parasitic in man. This species, which man may acquire by in-

gesting cysts from pigs, may cause balantidiasis, a disease of the large intestine characterized by ulceration and severe dysentery; it can be fatal.

■ ORDER HYMENOSTOMATIDA

Ciliates of the order Hymenostomatida usually possess a ciliation of an undulating membrane on the right side of the buccal cavity and an adoral zone of membranelles on the left side. There are variations of this plan among the different species. This order includes some of the best known and most studied genera among the protozoans, such as *Paramecium* (Fig. 2-45) and *Tetrahymena* (Fig. 2-44, *C*). The former is

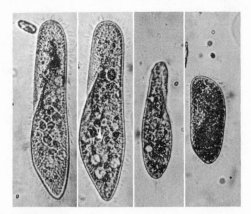

FIG. 2-45

Four common species of *Paramecium* photographed at same magnification. Left to right: *P. multimicronucleatum, P. caudatum, P. aurelia,* and *P. bursaria* (order Hymenostomatida). (Courtesy Carolina Biological Supply Co., Burlington, N. C.)

studied in all zoology laboratory courses as a ciliate type and has been the object of intensive investigations in sexual phenomena, mating types, taxonomy, and genetics. The latter is widely used for nutritional studies because it can be grown in media free from other organisms (axenic cultures). Other common genera of this order include *Frontonia,* with an oval to ellipsoid body, long collecting canals, and uniform distribution of fusiform trichocysts; and *Colpidium,* with an elongated, kidney-shaped body and lateral cytostome.

■ ORDER ASTOMATIDA

All members of the order Astomatida lack a mouth or cytostome. They are parasitic forms found mostly in oligochaetes. Some species are provided with holdfast organelles such as suckers or other devices. Asexual reproduction is by transverse fission and may also involve budding. The group is ill defined and may be secondarily degenerate forms. Some authors suggest making them a suborder of Thigmotrichida. *Anoplophrya* (Fig. 2-44, *D*), a common genus, has an oval or ellipsoid body, dense ciliation, and is parasitic in the gut of earthworms and other forms.

■ ORDER APOSTOMATIDA

The ciliates of the order Apostomatida are characterized by having rosettelike cytostomes, spiral ciliary rows, and complex life histories. Most of them are commensals on marine crustaceans. In their life histories different developmental phases with division of labor among them are correlated with the molting cycles of their hosts. The genera *Foettingeria* and *Spirophrya* are two common examples of the order.

■ ORDER THIGMOTRICHIDA

Members of the order Thigmotrichida may or may not have a cytostome. They take their name from the thigmotactic cilia with which they attach themselves to their hosts (usually mollusks). Their buccal ciliation, when present, is

FIG. 2-46

A, *Vorticella* (order Peritrichida); a solitary stalked form; 35 to 150 microns long, depending on species. **B,** *Carchesium polypinum* (order Peritrichida); a colonial form whose stalks contract individually; 100 to 125 microns long. **C,** *Boveria teredinidi* (order Thigmotrichida); cytostome at posterior end; on gills of *Teredo navalis;* 27 to 173 microns long. **D,** *Stentor coeruleus* (order Heterotrichida); 1 to 2 mm. extended.

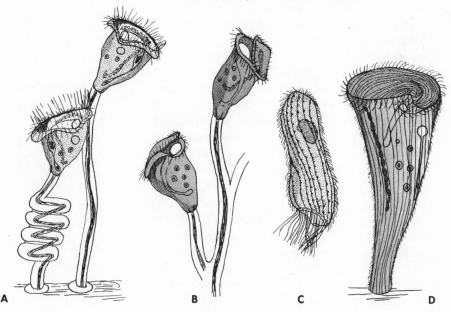

A   B   C   D

often located at or near the posterior end. Some species have larger and longer cilia on one side than the other. Some of their structures are modifications associated with their parasitic habit. Their common habitat is the mantle cavity or gills of clams. Some have tentacles for sucking out the contents of their hosts' epithelial cells. They show affinities to both hymenostomes and peritrichs. *Ancistruma mytili*, with ovoid and crescent-shaped body, lives in the mantle cavity of *Mytilus*. Another common genus is *Boveria* (Fig. 2-46, *C*), with conical body and posterior cytostome, which lives in the shipworm *Teredo*.

Subclass Petritrichia
■ ORDER PERITRICHIDA

Forms of the Peritrichia are characterized by a disklike anterior region with an adoral zone of membranelles and cilia that wind counter-clockwise when viewed from the anterior end. Many species look like an inverted bell attached to a stalk. Most are sessile and attached as ectocommensals to animals and plants. Stalked forms have a free-swimming stage called a telotroch, which makes possible a wider distribution. Asexual reproduction is by binary fission, and conjugation is common. With the exception of the adoral zone, the body is devoid of cilia. These ciliates may be solitary or colonial. The best known examples of peritrichs are the solitary *Vorticella* (Figs. 2-46, *A*, and 2-47, *A*) and the colonial *Epistylis*. Both are stalked. *Vorticella* has an inverted, bell-shaped body and contractile stalk and is common in both fresh and salt water. *Epistylis* has a vase-shaped body and noncontractile stalk. It may form large colonies. Another stalked form is *Carchesium* (Fig. 2-46, *B*), which is similar to *Vorticella* but is colonial. Each stalk, however, can contract independently. A common motile peritrich is *Trichodina*, which is ectozoic on hydras and parasitic in or on fish and amphibians. It has a low, barrel-shaped body with a posterior circlet of cilia by which it glides. It may cause serious infections in fish.

Subclass Spirotrichia
■ ORDER HETEROTRICHIDA

The order Heterotrichida is a large group of the subclass Spirotrichia and contains many ectocommensals and endocommensals of both invertebrates and vertebrates, although they are not found in man. Ciliation is more or less uniform over the body except in the adoral zone of membranelles or undulating membranes. In a few species, ciliation is lacking altogether. Pigments of various colors may be found in their cytoplasms. Bodies are of various shapes, including oval and elongated forms, but are not depressed. Some species are quite large. The order includes many interesting types that are frequently used in cultures. One of these is *Stentor* (Figs. 2-46, *D*, and 2-47, *B*), with a trumpet-shaped or cylindric body, a peristome encircled by a row of membranelles, a spirally coiled cytopharynx, and a beadlike macronucleus. Capable of great contraction because of its myonemes, it can extend 1 to 2 mm. in length. Some of its species are blue. *Stentor* is often used in morphogenetic investigations. *Blepharisma* is pink—a rare color in ciliates. *Spirostomum* is an elongated, cylindric form that may reach a length of 3 mm. It has well-developed myonemes and is highly contractile. *Nyctotherus*, with an oval, compressed body, is a common commensal in the intestines of frogs and cockroaches.

■ ORDER OLIGOTRICHIDA

Members of the order Oligotrichida have scanty cilia, as the name implies. The body is usually spherical or fusiform, and the adoral zone of membranelles at the anterior end is well developed. All members of this order are free living. *Halteria* (Fig. 2-48, *A*) is a familiar form that often can be recognized by the jerky movements that it makes with its long bristles.

■ ORDER TINTINNIDA

Members of the order Tintinnida are conical or trumpet shaped and live in secreted shells or loricas. They are mostly pelagic, but a few are

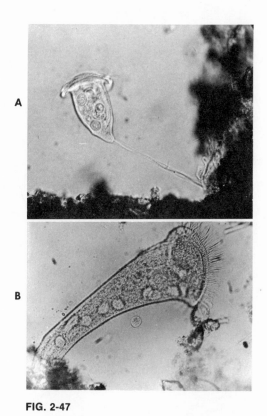

FIG. 2-47

A, *Vorticella,* living (order Peritrichida). **B,** *Stentor* (order Heterotrichida) can extend to 1 to 2 mm. in length. (**A** from Hickman: Integrated principles of zoology, The C. V. Mosby Co.; **B** from Encyclopaedia Britannica film, "Protozoa: One-Celled Animals.")

found in brackish or fresh water. Their adoral zone of membranelles often extends out from the lorica. Other cilia are scanty. They have left the only known ciliate fossil record. *Tintinnus* (Fig. 2-48, *B*) has a lorica that is often covered with foreign bodies. It may be found in fresh or salt water.

■ ORDER ENTODINIOMORPHIDA

The honor of being the most complex of all ciliates belongs to the members of the order Entodiniomorphida. The body is somewhat elongated, oval, and asymmetric. Members have no external ciliation except for one or two zones of membranelles. There may also be one or two tufts of specialized cilia. Along the right ventral side of some species is a row of skeletal plates. Posterior spines occur in some. A tough pellicle encloses the body of the animal, and internal structures are numerous and complex. Besides the cytostome, there is a gullet, rectum, and anus. Myonemes are common and have an extensive neuromotor system with a motorium ("nerve center"), also the regular organelles of the ciliate.

FIG. 2-48

A, *Halteria* (order Oligotrichida); 25 to 50 microns long. **B,** *Tintinnus* (order Tintinnida); diagrammatic, with lorica in optical section; 100 microns long. **C,** *Euplotes* (order Hypotrichida); 70 to 90 microns long. **D,** *Epalxis mirabilis* (order Odontostomatida); 35 to 45 microns long. (**B** redrawn from Jahn and Jahn: How to know the Protozoa, William C. Brown Co., Publishers.)

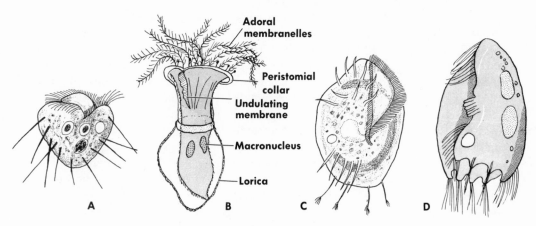

Adoral
membranelles

Peristomial
collar

Undulating
membrane

Macronucleus

Lorica

A   B   C   D

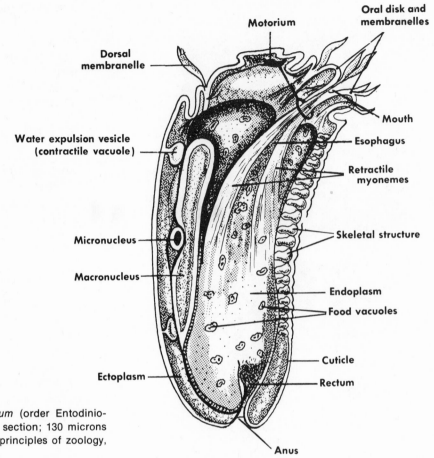

Dorsal
membranelle

Motorium

Oral disk and
membranelles

Mouth

Water expulsion vesicle
(contractile vacuole)

Esophagus

Retractile
myonemes

Skeletal structure

Micronucleus

Macronucleus

Endoplasm

Food vacuoles

Cuticle

Rectum

Ectoplasm

Anus

FIG. 2-49

Epidinium (Diplodinium) ecaudatum (order Entodiniomorphida); shown in longitudinal section; 130 microns
long. (From Hickman: Integrated principles of zoology,
The C. V. Mosby Co.)

These organisms are exclusively endocommensals in the stomach (rumen and paunch) of
herbivorous animals such as cattle, sheep, goats,
and others. They are present in enormous numbers (10 billion to 50 billion in a single animal)
and were once thought to aid in the digestion of
their host's food. However, as E. R. Becker
(1932) showed, herbivores can survive just as
well without them; but the protozoans cannot
survive without their hosts. This finding indicates that the protozoans are strictly harmless commensals. Some are found in the colon
of higher mammals including apes. *Epidinium* (Fig. 2-49) and *Ophryoscolex* are common
genera.

■ ORDER ODONTOSTOMATIDA (CTENOSTOMATIDA)

Members of the order Odontostomatida bear
carapaces and are sparsely ciliated. Their adoral zone has 8 membranelles, and their cilia
are restricted to a few rows on the carapace.
Their bodies are highly compressed. This is an
insignificant group that lives in sapropelic habitats, i.e., the black bottom slime of foul waters
rich in carbon dioxide and decaying nitrogenous
products. Some species are found in sewage
tanks, but there are both freshwater and saltwater forms. The common genera are *Sapro-
dinium* and *Epalxis* (Fig. 2-48, D).

■ ORDER HYPOTRICHIDA

Members of the order Hypotrichida are common wherever protozoans occur. They are oval and strongly flattened dorsoventrally, with the ventral side specialized for creeping. On the ventral surface the cilia are replaced by cirri that are designated according to location—frontals, marginals, etc. The dorsal surface usually bears sensory bristles with only a few cilia. The peristome and its adoral zone of membranelles are well developed. In some species the buccal area may occupy a large part of the ventral surface. This order represents a group that has reached a high degree of specialization in the evolution of ciliates.

The following genera are familiar to all who study protozoans: *Stylonychia* (Fig. 2-39, *B*), with an ovoid shape and three caudal bristles; *Oxytricha*, similar to *Stylonychia* but without caudal bristles; *Euplotes* (Fig. 2-48, *C*), with ovoid body and 4 caudal bristles; and *Kerona*, with a kidney-shaped body and six oblique rows of ventral cirri but no caudal bristles. *Kerona* is ectozoic, or parasitic, on hydra, where it is often seen creeping on the body and tentacles.

Subclass Suctoria
■ ORDER SUCTORIDA

The Suctorida are often considered as a separate class of Protozoa because of their highly specialized tentacles and other structures. Cilia are present only in the young. These ciliates have a pellicle and sometimes a lorica. Their tentacles serve as mouths when they become fastened to prey and suck out the prey's protoplasm (Fig. 2-41). One kind of tentacle is suctorial and bears knobs on the end (Fig. 2-43, *B*); another kind is sharp for piercing. Sucking may be effected by expansion of the body surface.

Most species are stalked, and young are produced by both external and internal budding. Binary fission and encystment also occur. Sexual reproduction involves fusion of conjugants. An infraciliature is found throughout their life history. Common genera are the following: *Podophrya* (Fig. 2-43, *B*), which has tentacles over the entire body surface; *Acineta*, which has tentacles usually arranged in two fascicles and a lorica; and *Ephelota*, which has a stout stalk and both suctorial and prehensile tentacles. Suctoridans are frequently attached to hydroids, bryozoans, and the algae on the backs of turtles.

REFERENCES

Allen, R. D. 1962. Amoeboid movement. Sci. Amer. **206**:112-122 (February). The newer interpretation of ameboid movement.

Borror, A. C. 1968. Ecology of interstitial ciliates. Trans. Amer. Microscop. Soc. **87**:233-243.

Brown, H. P., and A. Cox. 1954. An electron microscopic study of protozoan flagella. Amer. Midl. Nat. **52**:106-117.

Cameron, I. L., and A. L. Burton. 1969. On the cycle of the water expulsion vesicle in the ciliate *Tetrahymena pyriformis*. Trans. Amer. Microscop. Soc. **88**:386-393.

Cleveland, L. R. 1925. Method by which *Trichonympha* ingests solid particles of wood. Biol. Bull. **48**:282-287. In this and other papers Cleveland has described his classic studies on the symbiosis between termites and roaches and their intestinal Protozoa.

Corliss, J. O. 1961. The ciliated Protozoa: characterization, classification, and guide to the literature. New York, Pergamon Press, Inc. An excellent evaluation of the class Ciliata. One of the most important recent studies on the Protozoa.

Edmondson, W. T. (editor). 1959. Ward and Whipple's freshwater biology, ed. 2. New York, John Wiley & Sons, Inc. Many useful keys to freshwater Protozoa are found in this important handbook.

Ehret, C. F., and E. L. Powers. 1959. The cell surface of Paramecium. Int. Rev. Cytol. **8**:97-133.

Elliott, A. M., and G. L. Clemmons. 1965. An electron micro-related structures in *Tetrahymena pyriformis*. J. Protozool. **11**:250-261.

Elliott, A. M., and G. L. Clemmons 1965. An electron microscope study of digestion in *Tetrahymena pyriformis*. Amer. Zool. **5**:734-735.

Fauré-Fremiet, E. 1961. Cils vibratelles et flagelles. Biol. Rev. **36**:464-536.

Garnam, P. C. C. 1966. Locomotion in the parasitic Protozoa. Biol. Rev. **41**:561-586.

Gibbons, I. R., and A. V. Grimstone. 1960. On flagellate structure in certain flagellates. J. Biophys. Biochem. Cytol. **7**:697-718.

Gojdics, M. 1953. The genus *Euglena*. Madison, University of Wisconsin Press. A comprehensive account of the taxonomy and morphology of the genus.

Grassé, P. P. (editor). 1952. Traité de zoologie. Paris, Masson et Cie. Parts 1 and 2 of volume 1 of this comprehensive treatise are devoted to the phylogeny, general features, and other aspects of the Protozoa together with a detailed treatment of the subphylum Plasmodroma (Sarcomastigophora).

Grassé, P. P., and N. Carasso. 1957. Ultrastructure of the

Golgi apparatus in Protozoa and Metazoa (somatic and germinal cells). Nature **179**:31-33.

Grim, J. N. 1970. *Gastrostyla steinii:* Infraciliature. Trans. Amer. Microscop. Soc. **89**:486-496.

Grimstone, A. V. 1961. Fine structure and morphogenesis in Protozoa. Biol. Rev. **36**:97-150. An excellent article on the structure of the protozoans, as revealed by the electron microscope.

Hirshfield, H. I. (editor). 1960. Biology of the *Amoeba.* New York, New York Academy of Sciences. An entire treatise by many specialists is devoted to the many aspects of this classic protozoan.

Hutner, S. H., and A. Lwoff (editors). 1951-1955. Biochemistry and physiology of the Protozoa, vols. 1 and 2. New York, Academic Press Inc. A comprehensive account of the physiology and chemistry of the group.

Hyman, L. H. 1940. The invertebrates; Protozoa through Ctenophora, vol. 1. New York, McGraw-Hill Book Co., Inc. An extensive section of this fine treatise is devoted to the morphology and physiology of the Protozoa. Knowledge about the group is more recent in volume 5 (1959).

Jahn, T. L., and F. F. Jahn. 1949. How to know the Protozoa. Dubuque, Iowa, William C. Brown Co., Publishers. A useful manual for the identification and description of protozoan forms.

Jahn, T. L., and E. C. Bovee. 1965. Mechanisms of movement in taxonomy of Sarcodina. I. As a basis of a new major dichotomy into two classes, Autotractea and Hydraulea. Amer. Midl. Nat. **74**(3):30-40; 293-298.

Jahn, T. L., and R. A. Rinaldi. 1959. Protoplasmic movement in the foraminiferan *Allogromia laticollaris* and a theory of its mechanism. Biol. Bull. **117**:100-118.

Jakus, M. A. 1945. The structure and properties of the trichoceptes of *Paramecium.* J. Exp. Zool. **100**:457-485.

Jennings, H. S. 1906. Behavior of the lower organisms. New York, Columbia University Press. A classic study that always has its appeal.

Johnson, W. H. 1956. Nutrition of Protozoa. Ann. Rev. Microbiol. **10**:193-212.

Kudo, R. R. 1961. Protozoology, ed. 5. Springfield, Ill., Charles C Thomas, Publisher. An authoritative work that stresses taxonomy.

Manwell, R. D. 1961. Introduction to protozoology. New York, St. Martin's Press, Inc. This introductory study stresses the biologic aspects of the protozoans with less emphasis on the taxonomy of the group. An excellent text for learning the ways of these small forms.

Mast, S. O. 1931. Locomotion in *Amoeba proteus.* Protoplasma **14**:321-330.

Mast, S. O. 1947. The food vacuole in *Paramecium.* Biol. Bull. **92**:31-72.

Mayr, E. (editor). 1957. The species problem. Washington, D. C., American Association for the Advancement of Science. One of the papers is by Professor T. M. Sonneborn, who presents a fine analysis of the status of mating types and the concept of the syngens.

Metz, C. B., D. R. Pitelka, and J. A. Westfall. 1953. The fibrillar system of ciliates as revealed by the electron microscope. I. *Paramecium.* Biol. Bull. **104**:408-425.

Metz, C. B., and J. A. Westfall. 1954. The fibrillar system of ciliates as revealed by the electron microscope. Biol. Bull. **107**:106-122.

Pennak, R. W. 1953. Fresh-water invertebrates of the United States. New York, The Ronald Press Co. An excellent section on freshwater Protozoa with keys of the common genera. Introductory accounts of protozoans in general are especially worthwhile.

Pettersson, H. 1954. The ocean floor. New Haven, Conn., Yale University Press. A description of the role the forams and radiolaridans have played in building up the sediment carpet of the ocean floor.

Pitelka, D. R. 1963. Electron-microscopic structure of Protozoa. New York, Pergamon Press, Inc.

Robertson, J. D. 1960. A molecular theory of cell membrane structure. Verh. IV Intern. Kon. Elektronenmikroskopie. Berlin, 1958 **2**:159-171.

Satir, P. 1961. Cilia. Sci. Amer. **204**:108-116 (February).

Sharp, R. G. 1914. *Diplodinium ecaudatum,* with an account of its neuromotor apparatus. Univ. Calif. Pub. Zool. **13**:43-122.

Sleigh, M. A. 1963. The biology of cilia and flagella. New York, Pergamon Press, Inc. An analysis of the movement of those organs from the viewpoint of their comparative anatomy and function. Many revealing diagrams of structural organization and some excellent electron micrographs.

Swedmark, B. 1964. The interstitial fauna of marine sand. Biol. Rev. **39**:1-42.

Tartar, V. 1961. The biology of *Stentor.* New York, Pergamon Press, Inc.

Wichterman, R. 1953. The biology of Paramecium. New York, McGraw-Hill Book Co., Inc. One of the finest and most exhaustive monographs ever published on a single genus of the Protozoa.

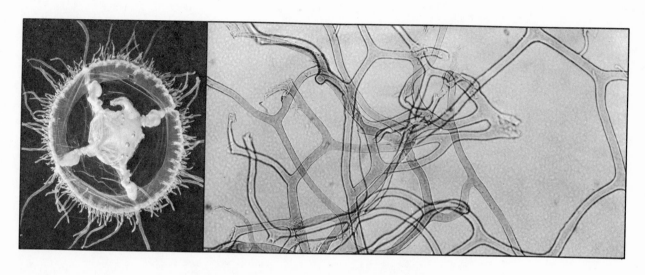

Part III □ The lower metazoans

3

Origin of the Metazoa

ORIGIN AND NATURE OF THE METAZOA

1. Although protozoans have attained great success within the limits of a plasma membrane, the very nature of such an organization imposes certain restrictions, especially that of size. If too large, an undivided mass of protoplasm lacks that mechanical and integrative organization which makes for successful occupation of the many kinds of habitats present in the biosphere. A much greater degree of diversification can be found in a group of different types of cells in which they can follow different specializations.

2. As a result of such limitations, many protozoans have become colonial to overcome such restrictions. At first these colonial forms consisted of irregularly arranged groups of cells that were chiefly independent of each other in a gelatinous matrix. Later development included protoplasmic strands linking the cells with each other. The cells of a colony may have been alike at first, but by division of labor and differentiation, some of the cells were somatic and some were germ cells. Examples of these sequences are found in some present forms. In the life cycle of many protozoans there are stages that consist of many cells.

3. Of the different theories that try to explain the origin of metazoans or multicellular animals, a logical and much favored one considers the origin of the Metazoa as coming from the Protozoa through stages that resemble the colonial protozoans of the present time. The requirements of life are very much the same in both protozoans and metazoans. Life activities are essentially alike, and similar principles must be followed in life's existence. All animals, small or large, poorly integrated or highly organized, must have food, oxygen, osmoregulation, waste disposal, reproduction, and other functional features. The Protozoa are able to carry out all these functions within one cell; the Metazoa distribute these functions among groups of cells by division of labor. It is understandable that a hard and fast line between some colonial forms in which there has been division of labor and differentiation and simple metazoans cannot be made. Some of the colonial protozoans may be considered multicellular animals, especially if they have attained an anteroposterior axis and have cells differentiated with respect to that axis.

4. No one route of metazoan origin is known with certainty, and all theories are highly speculative, with only suggestions of evidences for those proposed. The ingenious way in which zoologists have applied biologic principles in support of their ideas for the origin of metazoans is worthy of some admiration, although flaws can be found in all of them.

Origin of the Metazoa

Although the evolution of the isolated cell has been very creative, as seen in the many types of protozoans, it was not until the transformation of the unicellular to the multicellular state that animal life could realize most of its potentiality. As multicellular forms, animals could increase in size and become adapted to many environmental niches denied to the protistan form of life. As groups of cells, metazoans by specialization could form tissues, organs, and systems as functional assemblages. Many new architectural patterns were now available. At some stage in their evolutionary history the protistans gave origin to the higher plants and animals (Metaphyta and Metazoa). Both groups have reproductive tissues and usually organs such that gametes (sperm and eggs) are differentiated. Metazoans possess larval stages, whereas metaphytes do not. Metazoans have heterotrophic nutrition and are mostly motile, whereas metaphytes are photosynthetic and sessile.

Pressures exerted on unicellular animals. The basic structural pattern of the protistans and the way their patterns can be modified indicate that they must have had enormous evolutionary potentialities. But they encountered certain disadvantages in their adaptive evolution. All their differentiation had to occur within the limits of a plasma membrane. The nucleus can control the chemical activities of only a small amount of cytoplasm. Although many protistans overcame this restriction somewhat by having many nuclei and by having specialized organelles, they still lacked adequate circulatory and conducting systems for specific integration. As the body increased in size, diffusion alone was insufficient to supply needed materials, and more efficient transportation was required, which may be the chief advantage they obtained when some of them formed colonies.

Single-celled forms were thus confronted by two selective pressures. They had to be larger to exploit successfully all the potential environments in which to live, and they needed that complexity of structure necessary to adjust to these possible environments. Size has been very important in the evolution of animals, and size is achieved by cell aggregations that have some selective advantages. Larger size enables organisms to be less dependent on environment because much of the protoplasmic environment of

the organism is remote from the fluctuating changes of the external environment. But even in metazoans there are limitations upon size. As an organism becomes larger, controlling restrictions of support, coordination, nutrition, and communication begin to operate.

Fossil record. The fossil record throws no light on the nature of early metazoans. The early fossils date from the early Cambrian period of the Paleozoic era, such as the trilobites. But trilobites had already attained a high degree of complexity and must have had a background of simpler ancestors. Various theories have been proposed to account for the lack of Precambrian fossils, for example, the Brooks theory stating that early metazoans lacked a heavy calcareous shell; the Daly theory stating that there was insufficient calcium in the sea water of Precambrian times to form calcareous shells; and the Axelrod theory stating that the littoral (shallow shoreline) regions, where metazoans may have evolved, were subject to such drastic changes that fossilization could not occur.

J. R. Nursall suggests that a metazoan fauna probably did not exist for most of the Precambrian era. He postulates that there was a rapid diversification of metazoan forms in the late Precambrian times because of the addition of free oxygen to the environment. Early primitive organisms were heterotrophs that obtained their nutrients dissolved in the sea and lived under anaerobic conditions. The evolution of autotrophic or photosynthetic organisms meant that organisms were no longer dependent on sea nutrients. The addition of free oxygen to the atmosphere by photosynthesis made oxidative metabolism possible. This more efficient method of extracting energy by food breakdown made possible a greater range of organization such as the tissues and organs of the multicellular organism. The fossil record then would be a natural outcome of the greater diversity of life that resulted.

Theories of metazoan origin. It is thought that all many-celled animals arose from flagellate protistans and that the sponges (Parazoa) may have arisen from a different group of flagellates than did the other metazoans or else may have diverged very early from the metazoan stem. Early protists experienced many selective pressures. Some protists apparently changed little, others became photosynthetic and plantlike, and still others became heterotrophic, such as rhizopods and ciliates. Somewhere in their evolution, protists gave rise to terrestrial green plants (Metaphyta) and the multicellular animals (Metazoa) as we now know them both.

There are many theories regarding the exact origin of the metazoans. The polyphyletic origin of different groups has made it difficult to arrive at definite conclusions. This problem is all the more difficult because of the lack of fossil and biochemical evidences. Whereas there is general agreement that metazoans came from the protists, there is no agreement regarding which group of protists may be the actual ancestor. At present speculation is based largely on evidence presented by the study of the conditions of living forms, as in embryology and comparative anatomy. So many phases have to be accounted for in any theory that great ingenuity has to be exercised in explaining how this or that condition arose.

The first serious attempt at explaining metazoan origin was that proposed by Haeckel in the nineteenth century. Many aspects of his theory were so logical that they were long accepted by others. Haeckel attacked the problem of phylogeny from an embryologic standpoint. He came to the conclusion that the succession of embryonic stages of an animal actually represents a succession of its past evolutionary stages. According to his view, the zygote would be the unicellular protistan stage. He postulated a hypothetical ancestral animal, a blastaea, with a resemblance to an embryonic blastula. This type of animal was similar to a colonial protozoan form such as *Volvox*. He believed that the gastrula stage would correspond to a hypothetical adult animal he called a gastraea and that this type of animal arose from blastaeas by embolic invagination. The gastraea type is

represented by living coelenterates today, and the coelenterates, he thought, were diploblastic with no sign of mesoderm (which was later added, along with bilateral symmetry, by the flatworms). He expressed his views in brief by his familiar law of recapitulation ("ontogeny recapitulates phylogeny"). In other words, the embryonic development of the zygote (ontogeny) repeats the evolutionary development of the phylum (phylogeny). This theory has been largely invalidated. Many coelenterates form a stereoblastula, or solid morula, instead of a hollow coeloblastula. Gastrulation generally does not form by embolic invagination but often forms in a different manner. Many other objections have been advanced against his theory, which is considered oversimplified.

Two other theories may be briefly described. One of these is the *colonial theory*, which states that the metazoans evolved from colonial flagellates and may have been in the form of hollow, spherelike bodies similar to *Volvox* or other colonial forms. A common series of colonial organisms occurs in the holotrophic order Volvocida, which by asexual reproduction forms colonies of individuals, such as *Gonium*, with 4 to 16 zooids; *Eudorina*, with 32 zooids; and *Pleodorina*, with up to 128 zooids. Such a progenitor may exhibit polarity, functional differentiation of cells (mostly reproductive), and sexual reproduction. In some the differentiation and division of labor may have extended further: individuality in cells finally became lost, and the whole colony became a single organism. The colonial theory considers that blastula-like colonies took place in the development as Haeckel believed, but that the inner layer was solid instead of hollow as Haeckel maintained.

The *syncytial theory* of metazoan origin states that the progenitor of the metazoans is to be found in a multinuclear ciliate form that already had organelles. Simple division of the protist by developing membranes divided the single cell into multiple cells, each with one nucleus, thus forming a flatworm-like metazoan. According to this compartmentalization theory, coelenterates arose from ancestral pro-flatworms and not the reverse.

Of the two theories, the colonial theory is preferred by most biologists.

Organization of the metazoan

Most animals have two major traits in common, a structural one and a functional one. With the metazoans, the structural organization is usually manifested by organs and organ systems. The functional trait revolves around the type of nutrition, which is usually heterotrophic, that is, its food cannot be made within its body (photosynthesis) but must be obtained from preexisting organisms. These two traits set the pace for the basic nature of the animal.

The structure of each animal has a specific architectural plan or design; all its constituent parts are arranged in some pattern. Such a structural arrangement is fitted for adjustments to its particular habitat. Usually the more generalized features of an animal appear first in its organization and in the more specialized features later. This is why embryos of many forms often tend to be very much alike, but become different as specializations occur.

Basic organization features depend on the organization level of the animal. However, there is a principle of uniformity throughout the animal organization. The life processes are about the same in all animals. They all maintain a continuous exchange of energy and materials with their environment. There is a basic unity in structural and functional patterns no matter how different animals may seem to be.

Differentiation and division of labor. All animals, even the simplest, undergo division of labor that leads to specialization of body parts and increased effectiveness of the living process. A unicellular organism carries on the entire life process of many functions. In a multicellular form these functions may be distributed among several or many cells. No cell is limited to a single function, for every cell must have the materials for metabolic and self-perpetua-tive functions to survive. The fewer additional functions a cell must perform, the more specialized it is.

Whatever the design of an animal, that pattern may be sufficient for its own particular relation to its environment. Environments vary in difficulty, and it may require a greater complexity of design to adjust to some particular niches. The organisms that fit into a more difficult environment usually have a higher division of labor and specialization and reach a complexity of structure not found in lower forms.

Symmetry and polarity. Symmetry has adaptive significance and various types are found in animals. Symmetry refers to the regularity of body parts and the geometric relations that exist among their structures. An animal may have primary symmetry as an embryo and a different (secondary) symmetry as an adult. Symmetry may be modified to meet the conditions of life, and most animals tend to be asymmetric in regard to distribution of organs.

Polarity refers to a differential gradation of materials along an axis. There may be also a gradient of activity in polarity. Motility of an animal has a significant symmetric effect on the architecture of an animal.

In *spherical symmetry* there is a symmetric center and every plane passed through the center divides the body into similar halves. Some colonial protozoans approach this type of symmetry, which is especially fitted for floating or rolling.

In *radial symmetry* an animal can be divided into similar halves by any plane passing through the longitudinal (oral-aboral) axis. Such symmetry is particularly suitable for forms that are adapted to a sessile existence and is characteristic of the coelenterates and, to some extent, the ctenophores. However, in the ctenophores the symmetry is modified and is called *biradial symmetry* because of the presence of some part that is single or paired so that usually only one or two of the longitudinal planes divide the form into truly similar halves.

Bilateral symmetry, in which two halves of

the animal, one on each side of a dorsoventral plane, are more or less mirror images of each other, is by far the commonest type. Such symmetry is usually associated with cephalization and motility, and the animals are usually elongated in the direction of movement.

Asymmetry refers to a lack of symmetry and is demonstrated by most sponges and by the coiled shells of adult snails.

Animals are commonly grouped according to their symmetry. The radiate animals, or those with radial or biradial symmetry, are called the Radiata; the bilateral animals, or those with bilateral symmetry, are called the Bilateria.

Segmentation and tagmatization. Segmentation refers to a series of linearly arranged somites, or metameres, each of which is similar to those in front of and behind it. Such a condition must have appeared at least twice in evolutionary history—in annelids and in chordate animals. Because of the close relationship of the annelids and arthropods, the arthropods are also segmented.

Metamerism is a mesodermal phenomenon and is restricted to the body wall musculature and body cavity. A corresponding metamerism is found in certain other organs such as blood vessels, nerves, and excretory organs. The only organs that are not involved in the metamerism are those derived from endoderm—in other words, the digestive organs. Some other wormlike forms show conditions of pseudometamerism, especially among the flatworms.

Several theories have been advanced to explain the origin of metamerism. Some state that somites arose as incompletely divided chains of zooids, and others maintain that metamerism first arose in the muscular tissue as an aid to locomotion and that this influenced other mesodermal organs to follow suit. In general the problem of its origin is obscure.

Whatever its origin, there is no doubt about the significance of metamerism in the locomotion of animals. With the aid of septa the coelomic cavity is divided into a series of repeated units. The pressure of the coelomic fluid within these units creates a type of hydrostatic skeleton that gives form to soft bodied organisms. Varying the amount of turgor within individual segments increases the capacity for local changes of shape and becomes a valuable aid to locomotion. Even in animals possessing a skeleton the serial arrangement of the body wall musculature ensures more effective and efficient body movements.

Metamerism is modified in various ways. Segments may be united into functional groups, each group being structurally separated from other groups and specialized to perform a certain function for the animal. These groups are called *tagmata,* and tagmosis is especially common in the arthropods. In most cases, metamerism is best shown in the embryonic condition but is frequently masked in the adult.

Embryology of the metazoans

Types of cleavage. In the embryology of the metazoans there is an early period of cleavage that differs somewhat from ordinary cell division (mitosis). In ordinary mitosis in the adult body the individual cells undergo a period of growth between successive divisions, but in cleavage no such interval of growth occurs, and each succeeding division produces smaller blastomeres, or cells. Cleavage is considered as terminating upon the establishment of the blastula.

The pattern of cleavage depends to a great extent on the amount of yolk in the egg. Yolk tends to retard cleavage by presenting a barrier to the orderly rearrangement of nuclear material and the formation of the spindle. The locations of cleavage planes depend largely on the positions of the mitotic spindles, which are controlled to a certain extent by the organization of the cytoplasm and local differences in the egg cortex. Invertebrate eggs differ enormously in size from the tiny microscopic eggs of the bryozoans to the very large ones of the squid.

There are two major types of cleavage—spiral and radial. In radial cleavage the early cleavage planes are either parallel with or at right angles to the polar axis, thus producing blastomeres in tiers or layers of cells. Radial cleavage is *indeterminate* (regulative), and the early blastomeres are equally potent. If the first 2 or 4 cells are separated from each other, each cell is capable of continuing cleavage and forming a small blastula or even a diminutive larva. It is thus possible for twinning to occur. In this type of cleavage the localization of specific formative substance is delayed and decided by interactions with adjacent cells (embryonic induction).

In spiral cleavage, on the other hand, the cleavage planes are diagonal to the polar axis and produce alternate clockwise and counterclockwise quartets of unequal cells around the axis of polarity. Unlike radial cleavage, spiral cleavage is *determinate* (mosaic), and the fate of each cell is fixed early in development. When a 2 celled embryo is separated into 2 blastomeres, each cell is capable of producing only half a larva. The formative substance of the cytoplasm is early localized, and all blastomeres are necessary for the formation of a complete embryo.

Cell lineage is a type of embryologic study that involves tracing blastomeres, or the cells formed during differentiation of the zygote, to their ultimate fate in tissues and organs. To trace the lineage of cells, a method of designating each blastomere and its descendents is employed. This system can be used only in those forms that follow a rigid, stereotyped pattern, as in those eggs that have determinate cleavage.

As will be pointed out later, spiral cleavage occurs chiefly in the Protostomia and radial cleavage in the Deuterostomia. However, there may be specialized modifications of cleavage patterns in both these groups.

In spiral cleavage the egg divides into four equal blastomeres in the first two meridional cleavages. These blastomeres are called A, B, C, and D. The third division is unequal and at right angles to the first two and results in four large macromeres and four small cells forming the first quartet of micromeres. These micro-

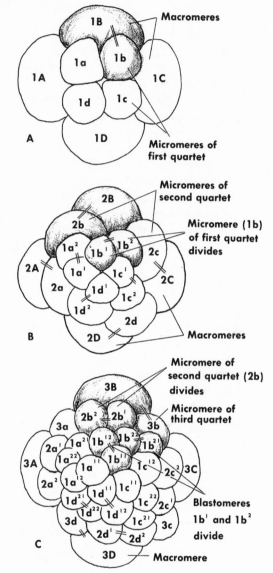

FIG. 3-1

Cell lineage, polar view, as found in spiral determinate cleavage characteristic of protostomes (polyclads, annelids, mollusks, ascidians, etc.). In such cleavage, fate of blastomeres is determined early so that removal of one early blastomere will result in defective larva. **A,** Third cleavage, resulting in micromeres and macromeres. **B,** Second quartet has formed; first quartet has divided. **C,** Third quartet has formed; second quartet divided again. One quadrant of cells (the progeny of one cell of the four-cell stage), composed of blastomere **1B** and its descendants, is shaded. (See text for explanation.)

meres are designated 1a, 1b, 1c, and 1d, and the macromeres are 1A, 1B, 1C, and 1D (Fig. 3-1, *A*). In successive cleavages, second, third, etc. quartets of micromeres are given off by the macromeres. Micromeres of the second quartet are numbered 2a, 2b, 2c, and 2d, whereas the macromeres become 2A, 2B, 2C, and 2D. When a third quartet is given off (3a, 3b, 3c, and 3d), the macromeres become 3A, 3B, 3C, and 3D, etc. (Fig. 3-1, *B* and *C*). In this scheme lower case letters represent micromeres; capital letters represent macromeres.

As the original quartets divide, their subsequent divisions are represented by exponents (Fig. 7-9, *C*); thus the two daughter cells of 1b become $1b^1$ and $1b^2$. When these two daughter cells divide, more exponents are added. In this way the daughter cells of $1b^1$ become $1b^{11}$ and $1b^{12}$, and those of $1b^2$ will be $1b^{21}$ and $1b^{22}$. The division of $1b^{21}$ produces $1b^{211}$ and $1b^{212.}$ The exponent 1 is given to the cell nearer the animal pole. The macromeres are located at the vegetal pole. It will be noted that all the blastomeres named "a" are derived from blastomere A of the original four blastomeres (A, B, C, D), those named "b" come from blastomere B, etc., and each set occupies about one quadrant of the embryo. The number before the letters designates the quartet from which the cell is derived, and the exponent indicates the number of subsequent divisions after the formation of a particular quartet of cells. The lower the number the nearer the micromere is to the animal pole; the macromeres are found at the vegetal pole. Cleavages are oblique and alternate right and left. It is thus possible to follow the origin of any particular cell.

The first three quartets and the cells derived from them form the ectoderm and the ectomesoderm. The fourth quartet produces the endoderm of the enteron. Micromere 4d of the fourth quartet gives rise to the endomesoderm, but there are some variations of this pattern in some species.

Blastula and gastrula formation. The blastula may be hollow and have a blastocoel (coelo-blastula), or it may be solid and have no blastocoel (stereoblastula). Or in meroblastic cleavage, which is common in birds and certain invertebrates, cleavage may be restricted to the upper part of the zygote so that the blastula is only a plate of cells on top of an undivided cell mass. Cleavage in centrolecithal ova, found in some invertebrates, produces a continuous layer of cells surrounding a central undivided yolk mass (a periblastula).

Gastrulation occurs in many ways, giving rise to the germ layers and to the archenteron, or primitive gut, which opens to the outside by the blastopore. The blastocoel may persist, or it may be obliterated completely. The outer layer of the germ layers is the ectoderm and the inner layer is the endoderm. Gastrulation and germ layer formation are chiefly processes of morphogenesis rather than of differentiation, which comes later in the development of organs.

Mesoderm and coelom. Mesoderm is formed in a variety of ways. It may be derived from ectoderm or endoderm, since all cells of a gastrula belong to one or the other of these two primary layers. Mesoderm may be considered the secondary germ layer. In the radiate phyla, mesoderm is derived from ectoderm (ectomesoderm); in other metazoans it comes chiefly from the endoderm (endomesoderm). Most Bilateria derive mesoderm from both ectoderm and endoderm. From the germ layers the various tissues and organs are derived.

The formation of mesoderm is best understood in relation to the formation of the body cavities. In the radiate and lower bilateral animals the digestive tube is the only body cavity of the adult. Such forms are called *acoelomate* because there is no true body cavity, or coelom. In radiate forms a few cells from the ectoderm form between the ectoderm and endoderm a mesenchymal layer that secretes a jellylike matrix, the mesoglea, in which there may be some cellular ingression. In acoelomate Bilateria (flatworms and others) the mesoderm completely fills the region between the ectodermal integument and the endodermal digestive tube. This middle layer, the parenchyma, may contain jelly-secreting mesenchyme and specialized cells of organ systems.

In the *pseudocoelomates* the body cavity is a remnant of the blastocoel; it is lined on the outside by the body wall tissues and on the inside by the tissues of the digestive tube, with some mesoderm aggregated in places.

The *eucoelomates* (animals with a body cavity that is a true coelom) include the remainder of the Bilateria. The coelom may be formed in either of two ways—the schizocoelic or the enterocoelic method.

In the schizocoelic method the adult mesoderm arises in the embryo from two endoderm-derived cells (primordial mesoderm cells), one on each side of the archenteron. At first it is a mass filling the space between ectoderm and endoderm, but by delamination it splits into an outer somatic layer and an inner splanchnic layer on each side of the archenteron. The somatic layer adheres to the body wall and the splanchnic layer to the digestive tube. This results in a coelomic space completely surrounded by mesodermal tissue.

In the enterocoelic method the mesoderm arises as paired lateral pouches from the endoderm. When the pouches lose continuity with the endoderm, their inner splanchnic portions remain against the developing digestive tube while their outer somatic layer is applied against the body wall. Between the two layers is the coelom as in the other method.

Other patterns of coelom formation are also found. For example, in some lophophorates loose mesenchyme may form and become arranged in a regular manner to form coelomic spaces.

The value of the coelom cannot be overestimated. It makes the activity of the digestive system independent of the body wall. Filled with fluid, the coelom can act as a hydrostatic skeleton. It may also aid in transporting food, waste, and other materials. It may form tubular organs connecting the coelom to the outside (coelomoducts) for the transport of gametes.

Also, it forms a space for the location of organ systems that are connected to the body wall by mesenteries.

Protostomia and Deuterostomia

The terms Protostomia and Deuterostomia were introduced arbitrarily by K. Grobben (1908) to describe the two groups into which the bilateral animals of the animal kingdom are divided. According to this concept, after the origin of a bilateral ancestor two evolutionary lines of these animals developed. One line became the phyla in which the blastopore of the gastrula forms the mouth (Protostomia); the other line became the phyla in which the blastopore forms the anus (Deuterostomia) (Fig. 3-2). The radially symmetric coelenterates and ctenophores are naturally excluded.

Other fundamental differences besides the method of formation of the mouth and anus are found in the two groups. All protostomes have determinate cleavage and, with the exception of the arthropods, the prostomes have spiral cleavage. And in protostomes the coelom arises as a schizocoel by a splitting of the mesoderm. Deuterostomes have radial and indeterminate cleavage, and the coelom forms as an outgrowth of the enteron (enterocoelic method). In those protostomes that have a larva, it is usually of a trochophore type; in deuterostomes the larva of indirect development is of a dipleurula type.

Accordingly, the bilateral animals may be said to have a diphyletic origin—the protostomes leading to the annelids, arthropods, mollusks, and others and the deuterostomes leading to the vertebrates and related groups. There are many exceptions to be noted among the animals that make up the two groups. The development of the mesoderm varies greatly among the various animals. The development of the coelom is known to vary; in vertebrates, for example, it arises as a schizocoel instead of an enterocoel. There are many modifications in every phylum, for nature follows no set plan of development for each major group.

In general, the concept of Protostomia and

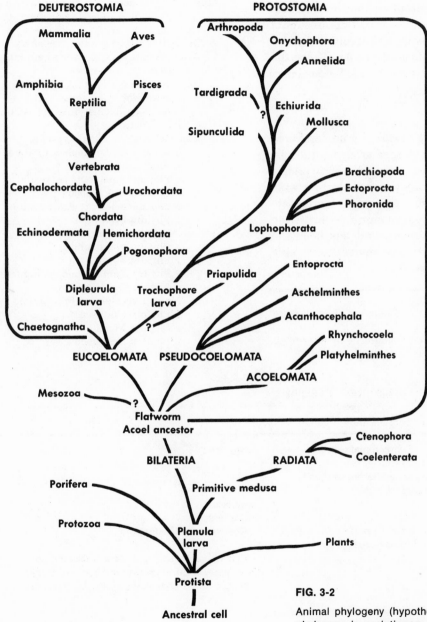

DEUTEROSTOMIA

PROTOSTOMIA

Mammalia

Aves

Amphibia

Pisces

Reptilia

Arthropoda

Onychophora

Annelida

Tardigrada

Echiurida

Sipunculida

Mollusca

Vertebrata

Cephalochordata ———— Urochordata

Chordata

Brachiopoda
Ectoprocta
Phoronida

Echinodermata / Hemichordata

Pogonophora

Lophophorata

Dipleurula
larva

Trochophore
larva

Priapulida

Entoprocta

Aschelminthes

Acanthocephala

Rhynchocoela

Chaetognatha

?

Platyhelminthes

EUCOELOMATA   PSEUDOCOELOMATA

ACOELOMATA

Mesozoa

?

Flatworm
Acoel ancestor

Ctenophora

Coelenterata

BILATERIA

RADIATA

Porifera

Primitive medusa

Protozoa

Planula
larva

Plants

Protista

FIG. 3-2

Animal phylogeny (hypothetical). Basis for such a tree
phylogeny is evolutionary relationship, or common ancestry, but such information is scanty. In general, time is
represented by vertical levels—the higher the branch, the
more recent the taxon—but this is also uncertain because
fossil record is incomplete. Interpretations of relationship
are based on common characters, similarity of embryologic development, basic structural patterns, fossils, etc.

Ancestral cell

Deuterostomia may be considered logical, in spite of many shortcomings, for each group has tended to follow potentialities inherited from a common ancestry. How the Deuterostomia arose from the Protostomia is largely a speculative point; only a fossil record could resolve the problem, and that at present is a nebulous hope.

A BRIEF CLASSIFICATION OF THE INVERTEBRATES

The various animal phyla may be combined arbitrarily into a few larger groups united by common traits of structure and embryology. Such a classification may be of help in a wide ranging comprehension of the invertebrate kingdom. The following scheme represents the opinion of many zoologists and has been the gradual result of the development of taxonomic principles (see also Table 3).

Subkingdom Protozoa (acellular or unicellular)— phylum Protozoa
Subkingdom Metazoa (multicellular)— all other phyla

Branch A (Mesozoa)—phylum Mesozoa
Branch B (Parazoa)—phylum Porifera
Branch C (Eumetazoa)—all other phyla
  Grade I (Radiata)—phyla Coelenterata (Cnidaria), Ctenophora
  Grade II (Bilateria)—all other phyla
    Acoelomate—phyla Platyhelminthes, Rhynchocoela (Nemertina)
    Pseudocoelomate—phyla Rotifera, Gastrotricha, Kinorhyncha, Gnathostomulida, Nematoda, Nematomorpha, Acanthocephala, Entoprocta
    Eucoelomates—all other phyla
      Eucoelomate Lophophorate animals
        Phylum Phoronida
        Phylum Ectoprocta (Bryozoa)
        Phylum Brachiopoda
      Minor Eucoelomate Protostomia
        Phylum Priapulida
        Phylum Echiurida
        Phylum Sipunculida
        Phylum Tardigrada
        Phylum Pentastomida (Linguatulida)
        Phylum Onychophora
      Major Eucoelomate Protostomia
        Phylum Mollusca
        Phylum Annelida
        Phylum Arthropoda

Table 3. Outline résumé of the invertebrate phyla*

Grade of organization	Phyla
I. Unicellular (or acellular), or colonial with undifferentiated somatic cells	Protozoa
II. Multicellular (Metazoa) with differentiated cells and typically larval stages	
A. Stereoblastula grade	Mesozoa
B. Cellular grade (Parazoa)	Porifera
C. Tissue grade (Eumetazoa, in part)	Coelenterata, Ctenophora
D. Organ system grade (Eumetazoa, in part)	
1. Acoelomate	Platyhelminthes, Rhynchocoela (Nemertinea)
2. Pseudocoelomate	Rotifera, Gastrotrichia, Kinorhyncha, Nematoda, Nematomorpha, Acanthocephala, Entoprocta
3. Eucoelomate	
a. Protostomes (blastopore forms mouth; spiral, determinate cleavage)	
1. Lophophorate protostomes	Ectoprocta, Brachiopoda, Phoronida
2. Schizocoelous protostomes	Priapulida, Sipunculida, Mollusca, Annelida, Arthropoda
b. Deuterostomes (blastopore forms anus; radial, indeterminate cleavage)	Chaetognatha, Pogonophora, Hemichordata, Echinodermata, Chordata

*Modified from Hegner-Engemann.

Eucoelomate Deuterostomia
  Phylum Echinodermata
  Phylum Chaetognatha
  Phylum Hemichordata
  Phylum Pogonophora (Brachiata)
  Phylum Chordata
    Subphyla Urochordata (Tunicata),
    Cephalochordata

REFERENCES

Balinsky, B. I. 1970. An introduction to embryology, ed. 3, Philadelphia, W. B. Saunders Co. One of the best general works in this field. The student of invertebrates will find much of interest in it.

Bullock, T. H. 1965. In search of principles in integrative biology, Amer. Zool. **5:**745-755.

Butt, F. H. 1960. Head development in arthropods, Biol. Rev. **35:**43-91.

Dougherty, E. C. (editor). 1963. The lower metazoa. Los Angeles, University of California Press.

Grimstone, A. V. 1959. Cytology, homology and phylogeny— a note on 'organic design,' Amer. Natur. **93:**273-282. All students of invertebrates should read this little article.

Hadzi, J. 1953. An attempt to reconstruct the system of animal classification. System. Zool. **2:**145-157.

Scheer, B. T. (editor). 1957. Recent advances in invertebrate physiology, a symposium held at the University of Oregon, Eugene, Ore., University of Oregon Publications.

CHAPTER 4

Mesozoan and parazoan animals

PHYLA MESOZOA AND PORIFERA

GENERAL CHARACTERISTICS

1. The mesozoan and parazoan groups are often considered primitive multicellular animals.

2. The multicellular animals, or Metazoa, are typically divided into three branches: Mesozoa, Parazoa, and Eumetazoa.

3. Two phyla (Mesozoa and Porifera) have in common the status of being separate branches of the Metazoa in contrast to Eumetazoa, or true metazoans. Both phyla have unconventional morphologic organizations when compared with the Eumetazoa.

4. Some colonial protozoans may actually overlap the diagnostic characteristics of the metazoans by having differentiation of somatic cells into different types.

5. The great branch of Eumetazoa is characterized by having many types of somatic cells that function as an integrated organism together with definite polarity and body axes.

6. The phylum Mesozoa has the simplest pattern of organization of the three branches mentioned in 2 above, being composed of an outer layer of somatic cells or syncytium that encloses one or more reproductive cells. Morphologically, the Mesozoa may have remained at the morula (steroblastula) stage and do not have the ectodermal and endodermal layers. All are parasitic and have marine invertebrate hosts.

7. The more complex branch Parazoa, or phylum Porifera (sponges), is characterized by many negative metazoan features, such as the lack of organs and definite tissue layers. Their incipient tissues are organized at low levels of integration but are sufficiently differentiated in their cellular structure to produce a functional whole. As sessile animals they have developed an efficient pore and canal system for trapping and filtering food from currents of water passing through.

8. The developmental patterns of sponges do not correspond to those patterns found in Eumetazoa. Embryonic layers of sponges are not homologous to those of the true metazoans. An inversion occurs in the embryo in which the original outer layer of flagellated cells becomes the inner layer of flagellated collar cells (choanocytes) associated with food getting, and the inner nonflagellated layer becomes the outer layer of epithelial-type cells (pinacocytes).

9. The body layers of the Mesozoa are not homologous to those of higher metazoans. The mesozoan outside layer has digestive functions; the inside, reproductive.

10. Of the many thousands of cells in sponges, three types have predominated in building the sponge's general morphologic structure: (1) the pinacocytes of the surface epithelium and the lining of some inside passages; (2) the flagellated choanocytes, which are inside and drive the water through the colony and collect food on their collars; and (3) the amoebocytes, which

wander through the sponge to form the skeleton, distribute food, and perform other functions.

PHYLOGENY AND ADAPTIVE RADIATION

1. Some consider the Mesozoa to be degenerate trematodes; others regard them as primitive and as a link between the protozoa and true eumetazoans. If primitive, the first metazoans must have had a solid blastula (stereoblastula) instead of a hollow one. A protozoan derivation of Mesozoa may be indicated by their cilia and the differentiation of only somatic and reproductive cells, as in certain colonial protozoans.

2. The phylum Porifera may have evolved from flagellate protistans or developed early from the metazoan stem and represent an evolutionary dead end (not in direct line of ancestors). Their chief limitations involve the absence of differential muscular tissue, a hydrostatic skeleton, or muscular integration. They also show some protozoan characteristics, such as the phagocytic method of nutrition and the resemblance of their flagellated larvae to colonial protozoans. The absence of a nervous system together with few contractile elements indicates a low level of integration.

3. The Mesozoa are adapted to a parasitic existence, which may account for their complicated life history. In the Porifera the unique water-current system has undergone considerable diversification from simple to the more complex patterns and represents their major evolutionary trends in the production of many species.

■ PHYLUM MESOZOA

The Mesozoa are wormlike animals that are endoparasites in the bodies of certain marine invertebrates. Since their discovery in 1839, they have been found as parasites in cephalopods, flatworms, rhynchocoels, brittle stars, annelids, and clams. The members of this phylum are considered the simplest multicellular animals, and their taxonomic status is a subject of perennial controversy among students of the group. One common interpretation considers their simplicity as a secondary adaptation to a parasitic life and that they should be placed as a subdivision (class) under Platyhelminthes; an alternative view emphasizes their primitive nature and that they are a well-defined group meriting phylum status in their own right. They are small forms, the largest not exceeding 9 mm. in length; most are much smaller than this, and some species are under 1 mm. in length. They number about 50 species.

CLASSIFICATION

Class Moruloidea
Order Dicyemida (Rhombozoa). Elongated symmetric, vermiform body; single layer of flat cells or syncytium (usually ciliated) enclosing a central axial cell or cells giving rise to reproductive cells (agametes); two reproductive phases (nematogen and rhombogen); nematogen phase gives rise to vermiform larva; rhombogen phase produces infusoriform larva; life cycle consists of sexual (primary host) and asexual generations in intermediate host; cell constancy; all members are endoparasites in cephalopods. *Dicyema, Pseudicyema, Dicyemennea.*

Order Orthonectida. Asexual stage is a multinucleated ameboid plasmodium; plasmodium asexually gives rise to males and larger females; hermaphroditic or dioecious; inner cell mass of many cells gives rise to sex cells; internal fertilization; ciliated larva. *Rhopalura, Stoecharthrum.*

GENERAL STRUCTURAL FEATURES AND LIFE HISTORIES

Details of structure as well as life histories vary in the two orders of Mesozoa. In the order Dicyemida the common form of the parasite (primary nematogen) is an elongated cylindric body, with the posterior end somewhat pointed and the anterior end relatively thick and blunt (Fig. 4-1). It is found exclusively in the kidneys of cephalopods and is about 8 or 9 mm. long in some species, shorter in others.

The organism is made up of about 25 cells (but the actual number varies with the species) arranged in a single layer of large, ciliated cells, the somatoderm (somatic cells), enclosing 1 or more elongated axial cells. The somatoderm of the head contains 8 or 9 polar cells that form the cap (calotte). This cap consists of 2 circlets or tiers of cells, with 4 cells in the anterior circlet and 4 or 5 cells in the other circlet. The other somatoderm cells (10 to 17 in number) form the trunk and are constant in number for the species. These cells may be arranged more or less spirally.

The elongated axial cell contains, in addition to its own nucleus, reproductive cells called agametes that have been formed without maturation. In some species there appears to be an earlier stage (stem nematogen) than the primary nematogen. These stem nematogens usually have 3 axial cells arranged in linear series in the long axis, and they may also have a different number and arrangement of somatic cells (B. H. McConnaughey).

Axoblasts, or germ cells, in the stem nematogen give rise to ordinary, or primary, nematogens that replace the stem nematogens as the population of organisms increases. As soon as the primary nematogens have completed their differentiation and become vermiform larvae they leave the mother nematogen and increase the infection in the kidney of the cephalopod host. Primary nematogens keep on increasing until the host attains sexual maturity or until there is a heavy population of parasites; the parasite is then transformed into a new phase called the rhombogen. From infusorigens, or hermaphroditic individuals, which remain in the axial cell of the rhombogen, eggs and sperm are produced. The fertilized egg gives rise to free-swimming ciliated larvae, the infusoriform larvae. The surface of the larva consists of 2 unciliated apical cells and a number of ciliated cells. Within the interior of the larva is the so-called urn, made up of a number of cells that surround the urn cavity, which may represent a remnant of either blastocoel or gastrocoel. These larvae are discharged into the sea with the cephalopod urine, and their fate is unknown. There are variant accounts of this process among the authorities.

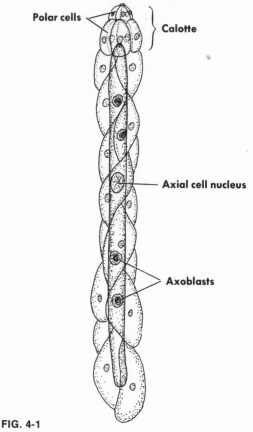

FIG. 4-1

Primary nematogen of *Dicyema* (order Dicyemida).

The order Orthonectida shows several differences from the preceding order. This parasite has been found in a number of invertebrates such as flatworms, mollusks, annelids, and brittle stars. The asexual stage in their life history consists of a multinucleated ameboid plasmodium that causes damage to the host by spreading. After they multiply by fragmentation, the plasmodia give rise asexually to males and females. Usually only one sex comes from the agamete of a plasmodium, but sometimes both sexes may arise from the same plasmodium. Some sexual forms are hermaphrodites. Males are smaller than females.

In structure they consist of a single layer

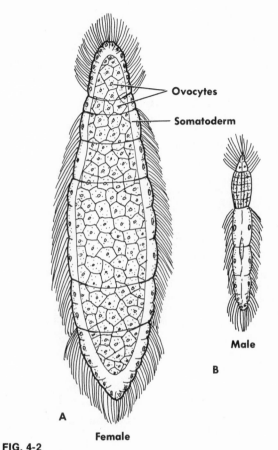

FIG. 4-2

A, Female and, **B,** male orthonectid *(Rhopalura)* (order Orthonectida).

of ciliated epithelial cells surrounding an inner mass of sex cells (Fig. 4-2). The epithelial cells form rings around the body, each ring consisting of rows of cells. The rings give an appearance of segmentation. The cilia on the anterior group of rings (the cone) are directed forward; those on the rest of the body point backward. The male has fewer rings than the female but has the same body regions. The inner cell mass consists of eggs in the female and sperm in the male. Hermaphroditic forms have a testis anterior to the egg mass.

The males and females escape from their hosts and swim in the sea. The male fertilizes the female by discharging his sperm through a genital opening in the female. After insemination, the eggs develop in the female into ciliated larvae. When liberated from the female, the larvae infect new hosts of the same kind. The larvae have superficial layers of ciliated cells and an inner mass of germinal cells. In the host the larvae disintegrate, and each germinal cell gives rise to a plasmodium.

Salinella, often included with the Mesozoa, was a genus (as yet unclassified) found in Argentina salt beds by J. Frenzel in 1892. This wormlike form consists of a single layer of cells enclosing a digestive cavity with a mouth and anus. The digestive cavity and ventral body surface are ciliated. Reproduction is chiefly by transverse fission. There are no internal cells. Its food seems to have been detritus picked up from its substratum. It is less than 1 mm. long. It has been an animal of mystery since its discovery.

■ PHYLUM PORIFERA

The sponges are multicellular organisms of the incipient tissue level of organization. Their whole structural organization is different from that found in other phyla. They consist of two or more types of specialized cells, but they have no organs or mouth, and certain of their cells are capable of independent existence for a few days. The body has numerous pores, canals, and chambers through which water circulates and openings through which water passes to the outside. Certain internal cavities are provided with collar cells (choanocytes) that, by the beating of their flagella, keep currents of water moving through the various canals and chambers. The sponges are almost all sessile, and they are aquatic organisms that have relatively simple responses, little contractility, and little irritability. For supporting their body structure most have a unique skeletal system of spicules or organic fibers or a combination of both.

With the exception of one family, they are all marine forms that exist from the intertidal zone to a depth of about 9,000 meters. More than 5,000 species have been described, but their taxonomy is very much in a state of flux.

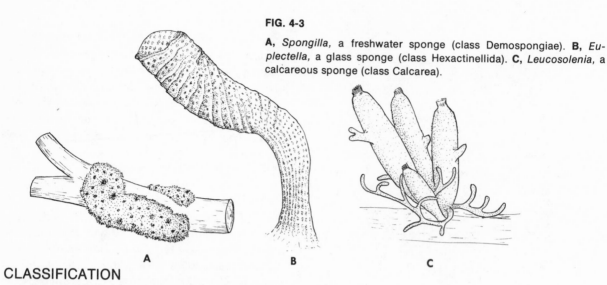

FIG. 4-3

A, *Spongilla,* a freshwater sponge (class Demospongiae). **B,** *Euplectella,* a glass sponge (class Hexactinellida). **C,** *Leucosolenia,* a calcareous sponge (class Calcarea).

A   B   C

CLASSIFICATION

Class Calcarea (Calcispongiae). The calcareous sponges. Skeleton of calcium carbonate spicules. All three grades of sponge structure—asconoid, syconoid, and leuconoid. Spicules formed on a 3- or 4-rayed plan with some monaxons.

Order Homocoela. Asconoid type of structure. *Leucoselenia* (Fig. 4-3, *C*).

Order Heterocoela. Syconoid or leuconoid type of structure. *Scypha* (Fig. 4-8).

Class Hexactinellida (Hyalospongiae). The glass sponges. Siliceous spicules formed on a 6-rayed principle and often united into a network. Syconoid or leuconoid type of structure. *Euplectella* (Fig. 4-3, *B*).

Class Demospongiae. The complex sponges. Siliceous spicules, spongin fibers, or a combination of spongin and spicules may form skeleton. Leuconoid type of structure.

Subclass Tetractinellida. Spicules mostly 4 rayed; no spongin; shallow-water forms. *Halisarca.*

Subclass Monaxonida. With or without spongin; spicules monaxon. *Mycale* (Fig. 4-4), *Cliona* (Fig. 4-6), *Spongilla* (Fig. 4-3, *A*).

Subclass Keratosa. Skeleton of spongin; no spicules; horny sponges. *Spongia* (*Euspongia*).

Class Sclerospongiae. A fourth class of sponges has been proposed for certain coral reef sponges found in the West Indies. Their massive calcareous skeleton of aragonite is laid down on a network of organic fibers, the centers of calcification. They also have siliceous spicules embedded in the fiber matrix. They appear to be related to the class Demospongiae. Because of their location and general structure, they are commonly referred to as the coralline sponges.

FOSSIL RECORD

Although most sponges are not preserved well as fossils because their spicular skeleton tends to separate into fragments, a fossil record of sponges extends from late Precambrian to Recent times.

All three classes of living sponges have fossil records. Sponge fossils have been most abundant in certain Jurassic, Cretaceous, and Terti-

FIG. 4-4

Mycale adherens encrusting scallop shell *(Hinnites)* (class Demospongiae). From Smith Island, Washington. (Courtesy G. J. Bakus.)

ary formations. The Calcarea are recorded from Devonian to Recent times, although a few date back to the Cambrian.

Most sponge fossils are of the pharetrone type in which the spicules are united in a rigid network. The glass sponges (Hexactinellida) have been found in most formations from Cambrian to Recent, although they reached their greatest abundance in Cretaceous and Jurassic periods. Well-preserved specimens have been recovered from the Devonian period. Demospongiae show a fine fossil record from Precambian to Recent times. Purely horny sponges are not preserved as fossils, but most Demospongiae also have siliceous spicules that do fossilize.

ECOLOGIC RELATIONSHIPS

Ordinarily few animals eat sponges because of their spicules and the bad taste of the sponge's body substance. Some sponges are actually toxic. However, coral reef fishes prey on most species of sponges (G. J. Bakus, 1964). A few sponges with horny skeletons may be eaten by turtles. Sponges also have regular predators in nudibranchs, chitons, and gastropods. Freshwater sponges (*Spongilla*, Fig. 4-3, *A*) are parasitized by the larvae of spongeflies (order Neu-

FIG. 4-5

Suberites ficus containing hermit crab. Sponge grows over and dissolves snail shell, leaving crab encased in sponge for shelter (class Demospongiae). From San Juan Island, Washington. (Courtesy G. J. Bakus.)

roptera) that live in them. Myriads of different animals live in the natural cavities or on the surface of sponges in different types of biologic association. These animals include arthropods, annelids, nematodes (which may eat sponge cells), hydroids, zoanthids, and many others. Certain crabs (*Dromia*) fasten sponges or pieces of sponge to their backs for protective purposes. *Euplectella* (a glass sponge) (Fig. 4-3, *B*) harbors and imprisons certain shrimp for life. Some freshwater sponges are symbiotic with zoochlorellae (an algae), which may manufacture sugars that are utilized by the sponge cells. The siliceous sponge *Suberites ficus* (Fig. 4-5) often lives on a snail shell occupied by a hermit crab and may completely enclose both shell and crab except for the shell opening. In regions where this association occurs, such sponges will probably not grow unless their larvae become attached to a shell inhabited by a crab.

Some sponges may do injury to other organisms. The larvae of the boring sponge *Cliona* (Fig. 4-6) settle on the shells of pelecypods, burrow through, and occupy channels in the shell interior. The method of boring is not known. The mollusk is frequently weakened and may become susceptible to infection. Oysters are often victimized. Also, sponges that are unable to burrow may occupy the burrows of the boring sponges.

M. W. deLaubenfels, who made an extensive study of marine sponge distribution, found that one of their favorite habitats is offshore from the mouth of a river. In some cases a particular species is abundant in a habitat where no other sponges are present.

The English investigator G. P. Bidder (1923) has worked out the relation of sponges to currents of water. He found that sponges grow more rapidly if they are elevated 10 to 20 cm. above their substratum because the water currents are more vigorous there or because the sponges can avoid contamination from sediments. Apparently for this reason, many sponges are stalked. In nonstalked sponges he found the angle of supply between the intake and outflow

FIG. 4-6

Cliona celata, sulfur sponge, a boring sponge, on oyster shell (class Demospongiae).

currents to be 90 degrees but somewhat greater in stalked forms.

GENERAL MORPHOLOGIC FEATURES

The structure of a sponge is best understood by studying the asconoid type or the embryologic development of a freshwater sponge. In the asconoid type of sponge (Fig. 4-7, *A*) the vase-shaped and radially symmetric body consists of a wall enclosing a cavity, the spongocoel, which opens at the summit by an osculum. The wall has both an outer and inner epithelium between which is mesenchyme. The outer epithelium, or epidermis, is made up of a single layer of flat cells, the pinacocytes; the inner epithelium, which lines the spongocoel, contains the choanocytes, also arranged in a single layer. Between the epithelial layers is the mesenchyme, containing skeletal spicules, various amebocytes, and a gelatinous matrix. Within the wall are tubular cells (porocytes), each of which

extends from the external surface to the spongocoel. By the beat of the flagella of the choanocytes, water is drawn through these pores into the spongocoel, where food particles are trapped by the choanocyte cells, and then passes out through the osculum. The bore of the ostia and osculum can be regulated by myocytes (modified pinacocytes), which form circlets around these openings. The simple asconoid type of sponge thus described is of restricted occurrence because the inefficient extraction of food particles imposes size limitations.

The syconoid type of canal system overcomes some of these restrictions. This type is produced by the outpushing of the walls of the asconoid type to form radial canals (Fig. 4-7, *B*). These radial canals fuse and are arranged with incurrent (inhalant) canals lying between them (Fig. 4-8). By this arrangement the incurrent canals open to the outside by dermal ostia (pores), and their blind inner ends lie between the radial canals; the radial canals open into the spongocoel by internal ostia, and their blind outer ends lie between the incurrent canals.

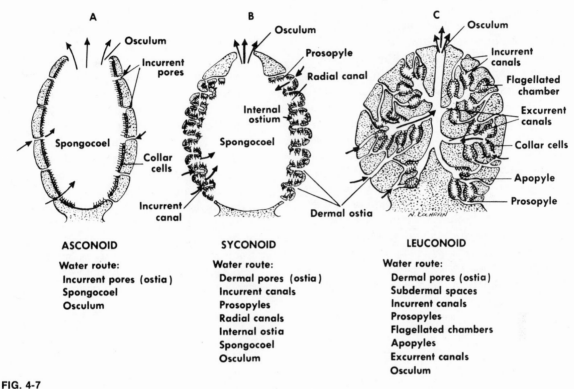

ASCONOID

Water route:
Incurrent pores (ostia)
Spongocoel
Osculum

SYCONOID

Water route:
Dermal pores (ostia)
Incurrent canals
Prosopyles
Radial canals
Internal ostia
Spongocoel
Osculum

LEUCONOID

Water route:
Dermal pores (ostia)
Subdermal spaces
Incurrent canals
Prosopyles
Flagellated chambers
Apopyles
Excurrent canals
Osculum

FIG. 4-7

Three types of sponge structure. (From Hickman: Integrated principles of zoology, The C. V. Mosby Co.)

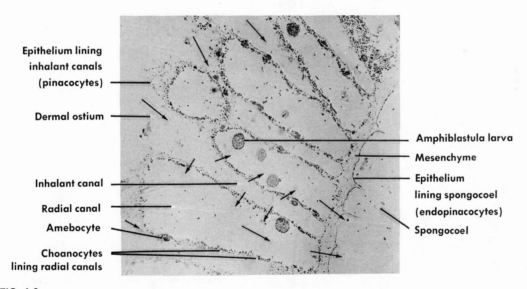

FIG. 4-8

Cross section through wall of *Scypha* showing canal system and cell structure (class Calcarea). (Photomicrograph of stained slide.) (From Hickman: Integrated principles of zoology, The C. V. Mosby Co.)

Between the radial and incurrent canals, the wall has small pores (the prosopyles), which allow the water to pass from the incurrent canals into the radial canals. In this way a wall is produced that contains alternating radial and incurrent canals. The inhalant canals are lined with pinacocytes, the radial canals are lined with choanocytes, and the spongocoel is now lined with an epithelium of endopinacocytes (Fig. 4-8). This increases the area of food-catching surface, as compared to that of the asconoid type. In the syconoid type the water flows through dermal ostia, incurrent canals, prosopyles, radial canals with choanocytes, internal ostia, spongocoel, and osculum.

The leuconoid type of structure, which is found in the majority of sponges of Calcarea and Demospongiae, is the most complex of the three types. Most leuconoids form large, ill-defined masses with many oscula. Here out-pocketings of the choanocyte-lined canals have resulted in many choanocyte-lined chambers surrounding and opening into each excurrent canal (Fig. 4-7, C). Small excurrent canals unite to form larger canals that finally carry the water to the osculum. Water enters the choanocyte chambers by way of a system of dermal pores, incurrent canals, and prosopyles.

An example of the leuconoid type is *Spongilla*, the freshwater sponge (Fig. 4-9). Here the thin outer layer, or cortex, contains many small dermal pores and is supported by columns of large spicules (megascleres). Water flows from the subdermal cavity beneath the cortex into the choanocyte chambers, carrying food and oxygen to flagellated collar cells; it leaves

FIG. 4-9

Section through peripheral portion of freshwater (leuconoid) sponge (diagrammatic) (class Demospongiae).

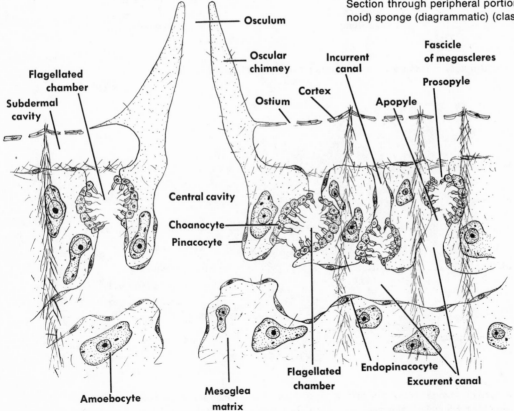

by way of excurrent canals to a small spongocoel and out the osculum. *Spongilla* attaches in
small masses to submerged objects. It often contains symbiotic green zoochlorella, which lend a
green color to the sponge.

Sponge cells. Since sponges have reached the
incipient tissue level of differentiation, many
cell types are to be expected. The chief ones and
their functions will be described.

Pinacocytes (Fig. 4-10) are flat, polygonal
cells forming the epidermis, the lining of the
inhalant and exhalant cavities (syconoid sponges), and also the lining of the larger canals
and spaces of the leuconoid type. They have the
ability to contract and can reduce the surface area of a sponge. They also can be differentiated into spindle-shaped, contractile
myocytes, which form circlets around the ostia

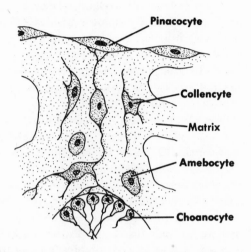

FIG. 4-10

Types of sponge cells. Some authors have suggested a
nervous function for collencytes.

FIG. 4-11

A, Lophocyte (class Demospongiae). **B** to **E,** Secretion of
triradiate spicule. **B,** Group of 6 scleroblasts have started
3 rays. **C** to **E,** Further development of spicule with 3
founder cells at base and 3 thickeners at tips. (**A** modified from Pavans de Cecatty, 1955; **B** to **E** modified
from Woodland, 1905.)

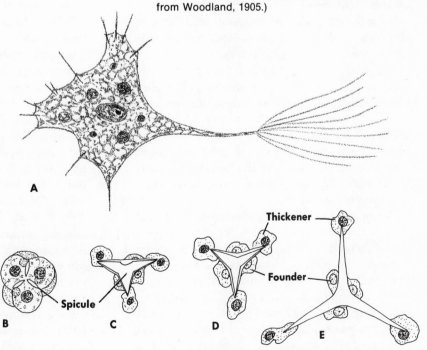

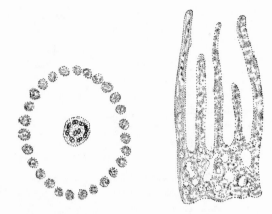

FIG. 4-12

Structure of collar cell based on electron microscope studies. Left, transverse section through collar showing protoplasmic tentacles and flagellum (center). Right, longitudinal section showing portion of cell with its protoplasmic tentacles. (Modified from Rasmont and others, 1957.)

and oscula for regulating water passage. A modification of pinacocytes are the porocytes, tubular cells containing central canals that act as incurrent pores in asconoid sponges. Porocytes may occur also as transient structures in siliceous sponges and have been described in freshwater sponges (Spongillidae).

The mesenchyme contains many types of ameboid cells, such as star-shaped cells called collencytes that are found among the inhalant canals (Fig. 4-10); scleroblasts that secrete the spicules (Fig. 4-11, *B* to *E*); spongioblasts that form spongin; archaeocytes or undifferentiated embryonic cells that give rise to sex cells; and gland cells that are derived from amebocytes and secrete mucus.

The most highly differentiated cells of the sponge are the collar cells, or choanocytes, whose flagella create the currents of water through the sponge. Each is made up of a spherical, oval body surmounted by a contractile, cytoplasmic collar with a long flagellum extending from the base of the collar (Fig. 4-12). Electron microscopy shows the collars to be composed of microvilli for trapping food (Fig.

4-12). Choanocytes vary in size in the different types of sponges, being the largest in the Calcarea.

A new type of cell (lophocyte, Fig. 4-11, *A*) has been described for some siliceous, freshwater sponges. It is an ameboid cell with a tuft of fibrils at one or both ends and occurs in connection with the pinacocyte epithelium of the subdermal cavity. Its function is unknown, but it may secrete fibrillar bundles.

Over the years, sponge investigators have described cells that bear some resemblance to nerve cells. O. Tuzet (1953) and M. Pavans de Ceccatty (1955) revived interest in the concept by describing what they consider to be nerve cells. Most students, however, are skeptical concerning physiologic evidence of a nervous system in sponges. C. L. Prosser (1946), who worked on both marine and freshwater sponges, concluded that there was only local contraction of the myocytes in reaction to stimuli and that there was no evidence of true conduction. Contractile cells may stimulate one another, or a fall in the fluid pressure of the sponge may be responsible for integrative transmission. The whole problem has been reviewed by W. C. Jones (1962).

Sponge skeletons. The skeleton is one of the most typical structures of sponges, being absent in only a few genera. From a taxonomic viewpoint, the skeleton is one of the most useful diagnostic characteristics for classifying sponges. The skeletons consist of spicules (Figs. 4-13 and 4-14), spongin (Fig. 4-15), or both. Spicules, or sclerites, are crystalline in structure and consist of rays or spines radiating from a point. They are secreted by scleroblasts (Fig. 4-11, *B* to *E*), a kind of mesenchyme cell. Both calcareous and siliceous spicules have a core of organic material around which is deposited either calcite ($CaCO_3$), in calcareous sponges, or silica (SiO_2nH_2O). Traces of other minerals (manganese, zinc, copper, etc.) also occur.

Spicule shapes and sizes vary greatly and an extensive terminology has been developed to describe them. According to size, they are com-

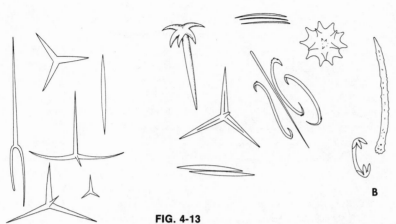

FIG. 4-13

Some types of spicules. **A,** Class Calcarea. **B,** Class Demospongiae.

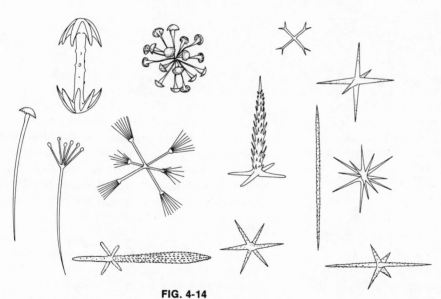

FIG. 4-14

Some spicules from hexactinellids (class Hexactinellida).

monly divided into the larger macroscleres, or megascleres, that make up the supporting framework of the sponge and smaller microscleres, which are scattered through the mesenchyme. This distinction does not apply to all sponges (e.g., calcareous sponges), however. On the basis of their number of axes and rays, spicules are divided into many types. Monaxons are straight or curved spicules that are formed by growth along one axis. They have pointed, knobbed, or hooked ends. Tetraxons (quadriradiates) consist of 4 rays, each pointing in a different direction. Sometimes this type loses rays and becomes 3, 2, or 1 rayed. The triradiate spicule is the most common spicule in Calcarea. The rays may be equal or unequal in length.

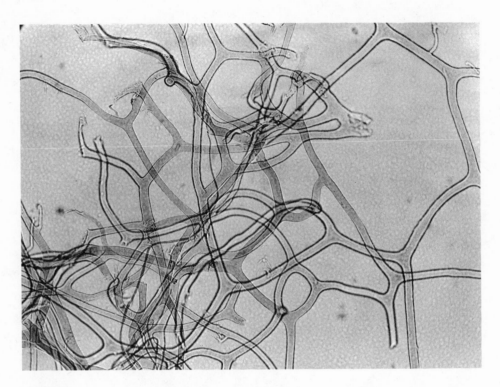

FIG. 4-15

Spongin fibers. The subclass Keratosa have skeletons composed entirely of spongin fibers, but other members of the Demospongiae also have spongin for holding together spicules and providing support for the colony.

Triaxons (hexactinals) have three axes meeting at right angles, producing 6 rays. Many modifications of this type are found, such as a reduction in rays, curving and branching of rays, and development of hooks, spines, and knobs on their ends or elsewhere. Polyaxons, or asters, have several rays radiating from a central point and are found chiefly among the microscleres. Spicules may be detached from each other, or they may form a loosely jointed framework. In some cases they form a very firm and rigid latticework.

Spongin (Fig. 4-15) is the fibrous skeleton found along with siliceous spicules in many sponges, or it may make up the entire skeleton in others (horny or bath sponges). In species that have siliceous spicules, spongin often serves to hold the spicules together. Spongin is a scleroprotein, as is also the familiar keratin of hair and fingernails, and is secreted by spongioblasts in the form of separate rods that fuse together to form long fibers (Fig. 4-15).

PHYSIOLOGY

Most known physiologic phenomena in sponges involve their water-current systems because sponges have practically no locomotion and very restricted contractile powers. They do have some coordination and behave as integrated organisms, although many sponges are regarded as colonies rather than true metazoan animals. Some contraction is found in the majority of sponges, and it is probably limited to the contractile powers of the pinacocytes, myocytes, and collencytes. Most of the activities of sponges, however, center on the regulation of the water currents on which they depend for their nutrition. By means of myocytes around the oscula and dermal pores they are able to narrow or enlarge these openings and thus regulate the water flow. In an undisturbed condition most sponges have their orifices wide open in both quiet and running water, although there are some individual differences. As a rule, the dermal pores are not as sensitive as the oscula to injurious conditions such as unoxygenated and foul water, harmful chemicals, extreme temperatures, and exposure to air. In freshwater sponges, mechanical touch or stroking may not cause closure of the dermal pores or oscula. Their most sensitive region appears to be the edge of the osculum. The exhalent opening (osculum) may close independently without altering the activities of other structures, and induced prolonged contraction may cause the sponge to burst, indicating little coordination.

Contraction of the pinacocytes can produce a decrease in the entire body size of a sponge, especially in the asconoid type. In sponges with oscular tubes (e.g., certain freshwater species) the tube may lengthen under certain conditions and shorten under others. There is little or no evidence of conductivity from one region of a sponge to another; contractions are believed to be direct responses to stimuli. Some transmission of excitatory states may occur, however, through contractile tissue because mechanical stimulation produces contraction of the body. Distant excitation may occur in some cases far from the point of stimulation due to a fall in fluid pressure.

The chief problem of the sponge is to move food particles to the choanocytes. This process is rather simple in the asconoid sponge in which the water current passes directly through the porocyte to the choanocytes lining the spongocoel. In the leuconoid type the problem is more involved because there are more openings, canals, and chambers through which the water must pass. The hexactinellid sponges make use of the steady current of their deep sea habitat and spread their netlike body over it to catch their food supplies. The whole incurrent system acts as a sieve; only the smallest particles can reach the choanocytes because prosopyles

have diameters of only a few microns. The collars of the choanocytes trap the small food particles, which are engulfed by the cell body and digested or else passed on to amebocytes in which digestion occurs (calcareous sponges). In other sponges, particles of food small enough to get through the dermal pores are ingested by the specialized amebocytes (archaeocytes and collencytes) that form a reticulum in the inhalant canals. It is probable that certain particles too large for the dermal pores are ingested by the pinacocytes of the epithelial surface. Digestive enzymes, however, are more common in the choanocytes than they are in the amebocytes (Hartman, 1960).

Egested waste is discharged by the amebocytes either into the water currents of the exhalant canals or to the dermal epithelium for voiding. Gaseous exchange of oxygen and carbon dioxide occurs between the amebocytes and the water flowing through the canal systems or through the thin epithelium.

REPRODUCTION AND REGENERATION

Both asexual and sexual reproduction are found in sponges. Asexual methods involve reduction bodies, gemmules, and budding. Reduction may occur during unfavorable conditions. It involves masses of many kinds of amebocytes with epidermal covering that persist after the disintegration of the parent sponge. These reduction bodies (Fig. 4-16) can develop into complete sponges when conditions become favorable.

Gemmules occur in all freshwater sponges and some marine sponges. Freshwater gemmules originate as clumps of food-laden amebocytes (archaeocytes) around which is secreted a hard layer by other amebocytes (Fig. 4-17). Scleroblasts secrete spicules between the outer and inner spongin of the hard coat. The sponge forms large numbers of these gemmules in the fall, and when the parent sponge disintegrates, the gemmules can withstand adverse conditions. In the spring the cells within the gemmule escape to the outside through an outlet,

119

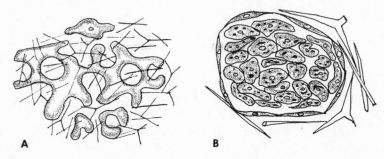

FIG. 4-16

A, Reduction bodies of calcareous sponge. **B,** Section through reduction body showing mass of amebocytes enclosed in epithelium. Gemmules differ from reduction bodies in having an external layer of columnar cells with spicules.

FIG. 4-17

Gemmule formation in freshwater sponge. **A,** Young gemmule of *Heteromyenia* in which trophocytes and cylindric layer are secreting inner membrane. **B,** Portion of cylindric layer of same into which scleroblast is placing its amphidisk. **C,** Completed gemmule of *Heteromyenia*. **D,** Gemmule of *Spongilla* hatching. (**A** to **C** modified from Evans, 1901; **D** modified from Brien, 1932.)

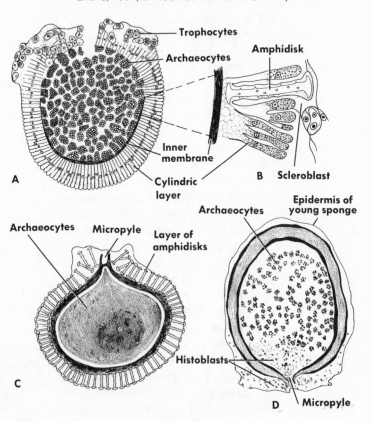

the micropyle (Fig. 4-17, *D*) and form a sponge. Gemmules of marine sponges are also enclosed by spongin with or without spicules. In some species of Demospongiae, gemmules without spongin develop into flagellated larvae resembling those from fertilized eggs.

In budding, masses of archaeocytes migrate to the tips of spicule clumps that project from the surface. Later these buds become detached and form new sponges.

Sexual reproduction, which may not occur in tetractinellid sponges, involves germ cells, ova, and sperm. Germ cells may arise from archaeocytes, other amebocytes, or from choanocytes. The ova grow at the expense of other amebocytes or special nurse cells (trophocytes). After reaching full size the ova undergo maturation. The various stages of spermatogenesis occur during the development of the sperm, which may arise by the transformation of an amebocyte or a choanocyte. The sperm fertilizes an egg with the help of a choanocyte that loses its collar and flagellum and acts as a carrier for the sperm (Fig. 4-18). The sperm loses its tail, enlarges, and is transferred to the egg after the carrier cell fuses with the ovum. In some other sponges, sperm are carried by water currents. Fertilization and early development occur in situ. Both dioecious and hermaphroditic sponges are found, but the majority of sponges are hermaphroditic.

Most sponges have remarkable powers of re-

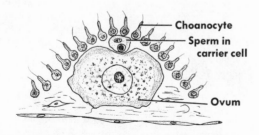

FIG. 4-18

Carrier cell with sperm near egg. A carrier cell develops from a choanocyte after the latter has resorbed its flagellum and collar.

generation. Fragments may grow into complete sponges, but this process is a long one. Suspensions of sponge cells prepared by squeezing pieces of sponge through bolting cloth into seawater can reorganize or reunite into functional sponges (H. V. Wilson). In this reorganization the various types of cells migrate to their appropriate positions in the new organism, or some types of cells are formed anew from amebocytes. Mixed cell suspensions of two species may form a temporary aggregation but later separate so that each cell mass is composed exclusively of one species or the other.

EMBRYOLOGY

The embryology of sponges shows variations in the different classes. These major differences chiefly revolve around the manner of formation of the two types of larvae—the amphiblastula (mainly calcareous sponges) and the parenchymula (Demospongiae). The amphiblastula is hollow; the parenchymula, or stereogastrula, is solid. In calcareous sponges such as *Scypha*, the egg after fertilization remains in the mesenchyme of the parent and there undergoes development (Fig. 4-19). Segmentation, which is holoblastic, produces first a ring of 8 similar cells, each of which divides horizontally into a small and a large cell (Fig. 4-20). The 8 large cells so produced are destined to form the future epidermis; the 8 small ones lying next to the maternal choanocytes will form choanocytes. The future choanocyte cells divide rapidly, whereas the future epidermal cells remain undivided. This produces a blastula with a blastocoel. An opening in the middle of the large cells (macromeres) serves for ingesting maternal cells. Each small cell (micromere) acquires a flagellum on its inner end next to the blastocoel. Then a process called inversion occurs in which the embryo turns inside out through its mouth so that the flagellated cells are now on the outside. Most of the surface of the larva is made up of these flagellated cells, but the posterior part contains the large nonflagellated cells. This amphiblastula larva breaks out of the mesen-

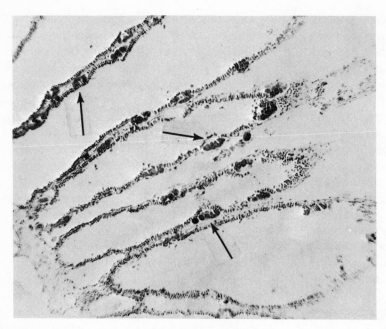

FIG. 4-19

Portion of cross section of *Scypha* showing eggs in early
cleavage stages.

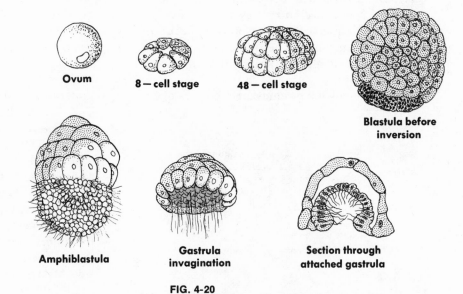

Ovum   8 — cell stage   48 — cell stage   Blastula before
inversion

Amphiblastula   Gastrula
invagination   Section through
attached gastrula

FIG. 4-20

Development of calcareous sponge *(Scypha).* (Modified from F. E. Schulze.)

chyme and passes with the water current through the osculum of the parent. After the larva swims around for a short time, the gastrula is formed either by overgrowth of the larger nonflagellated cells or by invagination of the flagellated cells into the interior. The micromeres are now located inside the gastrula and the macromeres are on the outside, which is just the reverse process of gastrulation in the Metazoa. The nonflagellated cells give rise to pinacocytes, scleroblasts, and porocytes; the flagellated cells form choanocytes, archaeocytes, and amebocytes.

In the early development of sponges a so-called olynthus stage of an asconoid type develops (Fig. 4-21). Depending on the species, this olynthus stage may remain as an asconoid type, or it may differentiate later into a syco-

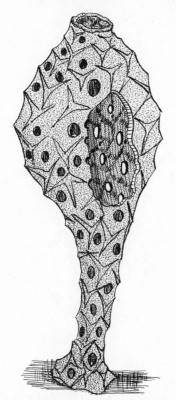

FIG. 4-21

Olynthus stage in development of calcareous sponge. It has ascon organization.

noid or leuconoid structure. In the parenchymula (stereogastrula), or solid type of larva, the outer flagellated cells lose their flagella, migrate into the interior, and transform into choanocytes. In some cases the formerly flagellated cells may degenerate after reaching the interior, and the choanocytes are formed anew by the archaeocytes (H. Meewis, 1939).

The parenchymula (Fig. 4-22, A and B) differs chiefly from the amphiblastula in having most of its surface (except the posterior region) flagellated and in having the interior filled with amebocytes that are formed by a proliferation of cells from the blastula wall. Most of the ameboid cells eventually migrate to the outside and form the dermal epithelium while the flagellated cells move into the interior.

Although the leuconoid type may pass through developmental stages that resemble the asconoid and syconoid structures before attaining the adult leuconoid condition, many Demospongiae as well as other leuconoid sponges are derived directly from a simple rhagon type of canal system. The rhagon (Fig. 4-22, D) consists of a broad base and a narrow summit with a relatively large spongocoel that opens to the exterior by an osculum at the top. Its walls are thick with rounded chambers that open into the spongocoel by large apopyles and open to the outside by prosopyles of inhalant canals. In the wall between the chambers and subdermal spaces is the mesenchyme. The leuconoid type is developed from the rhagon by outfoldings of the flagellated chambers and by the formation of excurrent channels. The rhagon with its flagellated chambers and canals is more advanced in its organization than olynthus, but none of the adult Demospongiae remain in this condition. Many variations of the leuconoid type may have arisen independently several times in the evolution of sponges.

Few developmental stages of the Hexactinellida have been studied. Y. Okada (1928) has studied the development in *Farrea sollasii* in which he describes the cleavage of the fertilized eggs as developing into a stereogastrula (Fig.

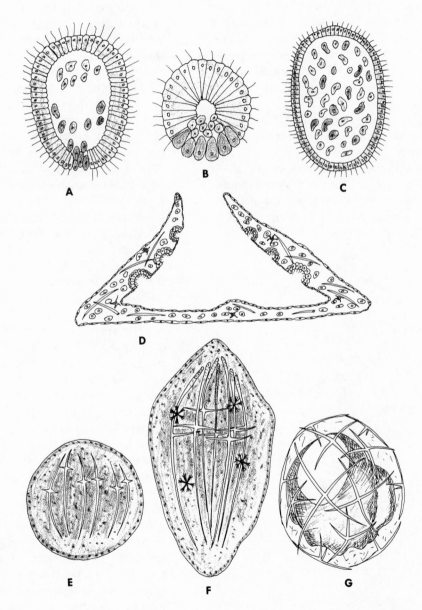

FIG. 4-22

Some types of sponge larvae. **A** to **C**, Parenchymula (stereogastrula) larvae. **A**, Larva of *Clathrina* showing unipolar immigration. **B**, Newly hatched larva of *Leucosolenia*. **C**, Completely ciliated larva of horny sponge. **D**, Vertical section of rhagon stage. **E** to **G**, Larval stages of *Farrea* (class Hexactinellida). **G**, Osculum and flagellated layer are forming. (**E** to **G** modified from Okada, 1928.)

4-22, *E* to *G*). The embryo consists of an outer layer of closely packed cells and a central jelly-like mass of ameboid cells, some of which later become choanocytes arranged around a central cavity. Cross-shaped tetractinal spicules appear at the surface before the larva leaves the adult sponge. After the larva settles down, the sponge assumes a tubular form with an osculum but lacks a definite epidermis.

Both the hollow amphiblastulae and the solid stereogastrulae have a brief free-swimming existence. The free-swimming period varies greatly; in some species it may last for a few days and in others for only a few hours. Usually the larvae swim with the anterior end directed forward, but sometimes they may swim with the posterior end forward. A calcareous larva, such as that of *Leucosolenia* (Fig. 4-22, *B*), which has most of the body surface ciliated when newly hatched, may gradually increase its nonciliated region at the expense of the ciliated portion until both areas are about equal before the larva becomes attached. Bakus (1964) has shown that when the parenchymula of *Tedania gurjanovae* becomes attached to a substratum by its anterior end, its cilia become motionless in a few minutes, and the larva undergoes a radical metamorphosis that involves a change into a platelike, amorphous mass of irregular cells before assuming the adult structural features.

REFERENCES

Phylum Mesozoa

Caullery, M. 1912. Le cycle evolulif des orthonectides. Proceedings of 8th International Congress of Zoology at Graz (1910), pp. 765-774. An active investigation of the orthonectids.
Cheng, T. C. 1964. The biology of animal parasites. Philadelphia, W. B. Saunders Co. A good account of the mesozoans and other parasites.
Frenzel, J. 1892. Untersuchungen über die Mikroskopische Fauna Argentiniens. Arch. Naturgesch. **58**:66-96. An account of the strange and unique animal Salinella which consists of a single layer of cells surrounding a digestive cavity with a mouth at one end and an anus at the other. It has no internal cells and does not fit into any known metazoan phylum. It has not been found since its original discovery.

Hyman, L. H. 1940. The invertebrates; Protozoa through Ctenophora, vol. 1, chap. 4. New York, McGraw-Hill Book Co., Inc. The most comprehensive American account of the morphology and phylogeny of the Mesozoa.
Hyman, L. H. 1959. The invertebrates; smaller coelomate groups, vol. 5. New York, McGraw-Hill Book Co., Inc. The author reaffirms her stand that the Mesozoa is a well-defined phylum and not a degenerate flatworm.
Lameere, A. 1916-1919. Contributions a la connaissance des Dicyemides. Bull. Sci. Fr. Belg. **50**:1-35; **51**:347-390; **53**:234-275. This is one of the best accounts of the order Dicyemida by one of the leading French investigators of the Mesozoa.
McConnaughey, B. H. 1951. The life cycle of the dicyemid Mesozoa. Univ. Calif. Pub. Zool. **55**:295-336.
McConnaughey, B. H. 1963. The Mesozoa. In Dougherty, E. C. (editor). The lower Metazoa; comparative biology and phylogeny. Berkeley, University of California Press. The most recent account by the foremost American student of the group. The author points out the resemblance between the Mesozoa and the parasitic flatworms and Aschelminthes.
Stunkard, H. W. 1954. The life history and systematic relations of the Mesozoa. Quart. Rev. Biol. **29**:230-244. This investigator presents evidence to show that the Mesozoa are degenerate flatworms, being derived from a remote ancestor of all flatworms.
Van Beneden, E. 1876. Recherches sur les Dicyemides. Brussels Acad. Roy. Belg. Bull. Clin. Sci. **41**:1160-1205; **42**:35-97. The famous Belgian cytologist was one of the first to study this group of animals discovered in 1839, and he considered them intermediate between the Protozoa and the Metazoa.

Phylum Porifera

Bakus, G. J. 1964. Morphogenesis of Tedania gurjanovae Koltun (Porifera). Pacific Sci. **18(1)**:58-63. A description of larval metamorphosis is given in addition to adult morphology.
Bakus, G. J. 1964. The effects of fish-grazing on invertebrate evolution in shallow tropical waters. Allan Hancock Foundation Publications (occasional papers, No. 27). Los Angeles, University of Southern California Press. The author presents evidence that the grazing of reef fishes has been a potent force in the survival and distribution of many shallow-water sponges.
Bullock, T. H., and G. A. Harridge. 1965. Structure and function of the nervous system of invertebrates. San Francisco, W. H. Freeman and Co., Publishers.
deLaubenfels, M. W. 1936. Sponge fauna of the Dry Tortugas with material for a revision of the families and orders of the Porifera. Washington, D. C., Carnegie Institution of Washington, Tortugas Laboratories, pub. 30. An important systematic work on American sponges.
Fry, W. G., (editor). 1970. The biology of the Porifera. New York, Academic Press, Inc.
Hartman, W. D. 1960. Porifera. In McGraw-Hill Encyclopedia of Science and Technology, vol. 10, New York, McGraw-Hill Book Co., Inc. A concise account of the basic morphology and physiology of sponges.
Hyman, L. H. 1940. The invertebrates; Protozoa through Ctenophora, vol. 1. New York, McGraw-Hill Book Co., Inc.

The section on Porifera represents one of the best accounts of the group. In volume 5 of this treatise (1959) the section on Porifera is brought up to date.

Jewell, M. Porifera (section 11). In Edmondson, W. T. (editor). 1959. Ward and Whipple's fresh-water biology, ed. 2. New York, John Wiley & Sons, Inc. A taxonomic key to the fresh-water sponges of the United States.

Jones, W. C. 1962. Is there a nervous system in sponges? Biol. Rev. 37:1-150. A notable review of this controversial subject in which the author refutes the idea of sensory and ganglionic cells in sponges but believes that excitatory transmission is restricted to contractile tissue. A comprehensive bibliography is included.

Jorgensen, C. B. 1966. The biology of suspension feeding. New York, Pergamon Press, Inc.

Lackey, J. B. 1959. Morphology and biology of a species of Protospongia. Trans. Amer. Microscop. Soc. 78:202-206. Ever since Kent (1880) described the new genus *Protospongia*, or *Proterospongia*, as a possible transition between the choanoflagellates and sponges, some biologists have cast doubt on the identity of his single specimen, considering it simply as a sponge fragment. But in recent years the genus has been rediscovered and appears to be a typical choanoflagellate.

Levi, C. 1956. Étude des Halisarca de Roscoff; embryologie et systematique des demosponges. Arch. Zool. Exp. Gen. 93:1-184. An important study on the embryology and histology, and a proposed new classification of the Demospongiae.

Meewis, H. 1939. Contribution a l'etude d'embryogenese des Myxospongidae. Arch. Biol. (Liege) 50:3-66.

Minchin, E. A. Porifera. In Lankester, E. R. (editor). 1900. Treatise on zoology, part 2, pp. 1-178. London, Adam & Charles Black, Ltd. A work that will always be useful for the basic structure of sponges.

Okada, Y. 1928. Development of *Farrea sollasii*. Tokyo Univ. Faculty Sci. J., sec. 4, vol. 2, part 1, pp. 1-27. One of the few accounts of the embryology of the Hexactinellida.

Pavans de Ceccatty, M. 1955. Le systeme nerveus des eponges calcaires et siliceuses. Ann. Sci. Natur. Zool. series 11, 17: 203-288. A description of bipolar and multipolar cells with long processes that connect structures has been advanced as evidence for the presence of nerve cells in sponges.

Pennak, R. W. 1953. Fresh-water invertebrates of the United States, chap. 3. New York, The Ronald Press Co. An account of the biology and taxonomy of the freshwater sponge family Spongillidae.

Tuzet, O. The phylogeny of sponges according to embryological, histological, and serological data, and their affinities with the Protozoa and the Cnidaria. In Dougherty, E. C. (editor). 1963. The lower Metazoa; comparative biology and phylogeny, pp. 129-165. Berkeley, University of California Press. The author believes that the sponges are intermediate between the protozoans and the coelenterates and should be regarded as the most primitive living metazoans.

Wilson, H. V. 1907. On some phenomena of coalescence and regeneration in sponges. J. Exp. Zool. 5:245-257.

Wilson, H. V., and J. T. Penney. 1930. The regeneration of sponges (Microciona) from dissociated cells. J. Exp. Zool. 56:73-148. A repetition and extension of the early classic investigations of Wilson in this field.

The radiate animals
PHYLA COELENTERATA AND CTENOPHORA

GENERAL CHARACTERISTICS

1. The Coelenterata and Ctenophora have primary radial or biradial symmetry. This pattern of symmetry consists of an oral-aboral axis with the parts arranged concentrically around it. Arrangements of body parts may be indefinite or may be built around a definite number—four or six or a multiple thereof. Two morphologic types, the polyp (hydroid) and the medusa (jellyfish), are characteristic of the phylum Coelenterata.

2. The body may be cylindric, globular, or spherical, with the body wall of two distinct layers (diploblastic), epidermis and gastrodermis. Between these two layers is the structureless, gelatinous mesoderm, or there may be the beginning of a third germ layer, the mesoderm (triploblastic). Within the mesoglea there may be an extensive system of collagenous fibers arranged to form a lattice network.

3. The gastrovascular cavity is the only internal cavity and is provided with a single opening, the mouth, which also serves as an anus. The cavity may be a single epithelial sac, or it may have many branches and outpocketings. In some there may be a definite pharynx, or gullet. In lieu of a vascular system the gastrovascular cavity serves as a system of transportation in the radiate animals.

4. The two phyla (Coelenterata and Ctenophora) are considered the most primitive eumetazoans and are primarily composed of tissue and incipient tissue. Their embryonic germ layers are homologous to those of the higher metazoans.

5. The radiate phyla have digestive, muscular, nervous (nerve net), and sensory systems but lack the respiratory, excretory, and vascular systems. Some coelenterates may have skeletal systems, but these are lacking among the ctenophores. The central body cavities of both phyla contain fluid and aid in forming a simple kind of hydrostatic skeleton. Their sex cells develop from interstitial cells in either epidermis or gastrodermis. Asexual reproduction may occur in both phyla.

6. Each phylum has distinct features. Coelenterates have stinging organoids (nematocysts) that are lacking (with one exception) among the ctenophores. Ctenophores have colloblasts, or adhesive organoids, on their tentacles. Polymorphism, so common among the coelenterates, is not found in ctenophores. Ctenophores have the characteristic eight meridional rows of comb plates for which coelenterates have no counterpart.

7. The radiate phyla have contributed the germ layers, ectoderm and endoderm, with the beginnings of the third germ layer, mesoderm. They have also contributed the first nervous cells (protoneurons), and some have made advances toward a bilateral metazoan. A gastrovascular cavity represents an important advancement in metazoan evolution because it affords digestive surface or storage capacity.

PHYLOGENY AND ADAPTIVE RADIATION

1. The coelenterates may have come from a ciliated, free-swimming, gastrula type of animal similar to the planula larva found throughout the group. The trachyline medusae (Hydrozoa) are considered the most primitive of modern coelenterates because of their direct development from the planula or actinula to medusae stages and because their ontogeny repeats the phylogenetic history better than does that of other members of the phylum. On this basis the coelenterates may have come from a primitive medusa, as most trachylines lack or have a poorly developed polyp stage. Ctenophores may have arisen from primitive trachyline medusoid lines and developed a greater structural specialization than the coelenterates.

2. Each phylum has adhered to its basic plan of structure—the coelenterates to the polymorphic form of polyp and medusa and the ctenophores to comb plates and biradial symmetry. Adaptive radiation has produced modifications for different environments so that some emphasize the polyp and others the medusa stages. To overcome the restriction imposed by a body structure that provides little space for specialized organs, the coelenterates compensate by the evolutionary development of different individuals specialized for performing different functions in their life cycle.

■ PHYLUM COELENTERATA (CNIDARIA)

The members of the phylum Coelenterata have reached the tissue grade of organization and belong to the radiate phyla of the Eumentazoa. They have primary radial symmetry (biradial in some); one opening (mouth) into the enteron, which may be simple or modified by diverticula or canals; and a body wall composed of two distinct layers—epidermis and gastrodermis (from ectoderm and endoderm), with a structureless mesoglea between them. In some members there may be the beginning of a third germ layer, the mesoderm. The body parts are built concentrically around an oral-aboral axis that extends from mouth to base and that gives the characteristic radial symmetry. Arrangement of bodily parts may be indefinite or built around a definite number—four or six or a multiple thereof. Typically, with few exceptions they all have extensible tentacles that encircle the oral end and possess diagnostic stinging organelles (nematocysts).

Polymorphism (more than one kind of form in one species) is characteristic of the group, although it is absent in one class. The individuals of the polymorphic forms are (1) the cylindric polyp attached at the aboral end, the free end provided with mouth and tentacles, and (2) the free-swimming medusa with a bell-shaped or saucer-shaped gelatinous body provided with marginal tentacles and having a centrally located mouth on a projection of the concave side.

The phylum is a large one; its members are exclusively aquatic and mostly marine. They include jellyfish, sea anemones, corals, hydras, and hydroids. Individual organisms may be single or in colonies.

This phylum holds a unique position in the animal kingdom. In the first place, the coelenterates represent the first of the Eumetazoa and the nearest representatives of the ancestral stock that gave rise to the long line of other metazoans. They represent the basic structural organization of the multicellular animal, and with their modifications they have laid the basis for the animals that have come after them. Because of their great regenerative powers and general plasticity of structure, it has been possible by extensive investigations to formulate many of the basic concepts of animal structure and function. For the past 200 years coelenterates have been one of the favorite subjects for investigating the important problems of animal nature and life in general. It is not an exaggeration to state that they have furnished more classic experiments than have any other group of animals. Some investigators think that a thorough understanding of coelenterates gives fundamental clues to the basic requirements of animal life. They have a tissue grade of construction that all metazoans must pass through in their development.

In general, coelenterates lack organs, but they possess all the important cells, and as an increase in complexity is chiefly a matter of the aggregations of various cells into organs and systems, they have the basic groundwork for the higher types of organization. Their gastrula-like organization has emphasized a diploblastic type of germ layers—ectoderm and endoderm (embryologically) or epidermis and gastrodermis (maturity). In the absence of a typical mesoderm, which is the source of muscle origin in all other eumetazoans, coelenterates have developed their muscles from differentiated epithelial cells. The lack of a mesoderm and a body cavity restricts organ development. There is no brain, no stomach, no excretory system; coelenterates show a general lack of other organs.

Ever since A. Trembley performed his classic experiments on the freshwater hydra (1740 to 1744), investigators of the coelenterates in every generation have come up with new discoveries. Some of the basic concepts that have arisen from such investigation are those on regeneration, nature of nerve cells and nerve systems, nematocysts (cnidae) and their functional varieties, neuromuscular facilitation, non-nervous conduction in siphonophores, spontaneous rhythmic activity, symbiotic commensals, algal symbionts and coral formation, germ plasm theory, interstitial cells in the role of mesenchyme cells, hydranth regression and replacement, polymorphism, and many others.

CLASSIFICATION

The classification of coelenterates varies with different students of the group, with little agreement among them except for some of the major taxa.

Phylum Coelenterata (Cnidaria). Body with radial or biradial symmetry; individual either a sessile polyp or free-swimming medusa; stinging nematocysts; tentacles about the mouth; presence of mouth but no anus; only one cavity (enteron); reproduction usually asexual in polyps, sexual in medusae; freshwater or marine; 9,600 species (N. M. V. Rothschild).

Class Hydrozoa. Both polyp and medusa in life cycle; medusa typically craspedote (possessing a velum); gastrovascular cavity without stomodeum or nematocysts and lacking partitions; sex cells in epidermis.

Order Hydroida

Suborder Gymnoblastea (Anthomedusae). *Hydra, Tubularia, Pennaria, Eudendrium, Hydractinia.*

Suborder Calyptoblastea (Leptomedusae). *Obelia, Campanularia, Sertularia, Plumularia.*

Order Milleporina. *Millepora.*

Order Stylasterina. *Stylaster, Stylantheca.*

Order Trachylina

Suborder Trachymedusae. *Liriope, Aglantha.*

Suborder Narcomedusae. *Cunina.*

Suborder Pteromedusae. *Tetraplatia.*

Suborder Limnomedusae. *Gonionemus, Craspedacusta.*

Order Siphonophora. *Physalia, Stephalia.*

Order Chondrophora. *Velella.*

Class Scyphozoa. Medusa is main stage in life cycle; no true velum; no stomodeum; gastrovascular system with gastric tentacles and pouches; mesoglea extensive and cellular; gonads endodermal in gastric cavity; marginal tentaculocysts with endodermal statoliths; polyp stage absent or reduced, consisting of a scyphistoma that produces medusae directly or by transverse fission; medusae mostly dioecious.

Order Stauromedusae. *Haliclystus, Lucernaria.*

Order Cubomedusae. *Carybdea, Tamoya.*

Order Coronatae. *Periphylla.*

Order Semaeostomae. *Aurelia, Cyanea, Chrysaora.*

Order Rhizostomae. *Cassiopeia, Rhizostoma.*

Class Anthozoa. Exclusively polypoid; medusae absent; mouth with stomodeum (gullet); siphonoglyph usually present; oral end expanded into oral disk; gastrovascular cavity divided into compartments by complete or incomplete septa with nematocysts; mesoglea cellular (connective tissue); endodermal gonads in septa; skeleton present or absent.

Subclass Alcyonaria. With pinnate tentacles.

Order Stolonifera. *Tubipora, Clavularia.*

Order Telestacea. *Telesto.*

Order Coenothecalia. *Heliopora.*

Order Alcyonacea. *Alcyonium, Xenia.*

Order Gorgonacea. *Gorgonia, Corallium.*

Order Pennatulacea. *Renilla; Pennatula.*

Subclass Zoantharia. With simple and unbranched tentacles.

Order Actiniaria. *Metridium, Adamsia, Actinia.*
Order Ptychodactiaria. *Ptychodactis.*
Order Corallimorpharia. *Corynactis.*
Order Scleractinia (Madreporaria). *Astrangia, Fungia, Orbicella.*
Order Zoanthidea. *Zoanthus.*
Order Antipatharia. *Antipathes.*
Order Ceriantharia. *Cerianthus.*
Order Rugosa. Extinct.
Order Tabulata. Extinct.

FOSSIL RECORD

Coelenterates have one of the longest fossil records in the animal kingdom. Representatives of all classes have been found as fossils, although some have left better records than others. Those without skeletons or hard parts naturally have left far fewer fossils than corals, for example. Only under the most fortunate circumstances have jellyfishes been preserved, such as those in the famed Burgess Shale of British Columbia. Some Precambrian fossil medusoids have been found, supplying evidence for the theory that the medusa is more primitive than the polypoid stage which, according to the fossil record, came later in Cambrian times.

Of the hydrozoans, the orders Milleporina and Stylasterina with their calcareous skeletons have left the best fossil records. Fossils of both orders are found in Tertiary and Cretaceous deposits. Most hydrozoans have soft parts only and have left few fossils, although some extinct orders of this class that had skeletons are present in certain Mesozoic deposits.

By far the best fossil record is left by members of the subclass Alcyonaria (Fig. 5-1) of the anthozoans. To this group belong the corals, which are important components of many rock formations from Ordovician to present times. Common types of such fossils are the sea pens, the sea fans, blue corals, and the organ-pipe corals. The other subclass (Zoantharia) of the anthozoans that includes corals as well as sea anemones have also left many coral fossil records, but the sea anemones (order Actiniaria)

FIG. 5-1
Large scleractinian coral fossil; such corals usually date from Mesozoic era.

have left no distinctive fossils because of their lack of hard parts. Two extinct orders of corals, Rugosa and Tabulata, have left extensive fossil records dating back to the Ordovician period.

GENERAL MORPHOLOGY AND PHYSIOLOGY

As mentioned in the introduction, coelenterates have two morphologic types, the polyp (hydroid) and the medusa (jellyfish). Each of these phases can be derived from the other. The presence of two different and distinct individuals in the group is an example of polymorphism, a condition that is widespread throughout the phylum and that is carried in some taxa of the phylum to a far greater extent than anywhere else in the animal kingdom. Polymorphism, however, shows many variations among the coelenterate classes (Table 4). It is most striking in Hydrozoa, in which polyp and medusa stages are fairly well balanced; it is present in Scyphozoa, in which there is a distinct tendency to suppress the polyp phase and em-

Table 4. Life cycles of some representative coelenterates*

Hydrozoa	
Hydra	Fertilized ovum—polyp (sexual)—gametes (sperm or ova)
Obelia	Fertilized ovum—planula larva—polyp (asexual)—medusa (sexual)—gametes
Gonionemus	Fertilized ovum—planula larva—reduced polyp (asexual)—medusa (sexual)—gametes
Tubularia	Fertilized ovum—polyp with gonophores (sexual)—gametes
Liriope	Fertilized ovum—planula larva—actinula larva—medusa (sexual)—gametes
Scyphozoa	
Aurelia	Fertilized ovum—planula larva—scyphistoma (fixed polyp, asexual)—medusa (sexual)—gametes
Pelagia	Fertilized ovum—medusa (sexual)—gametes
Haliclystus	Fertilized egg—planula larva—polyplike attached medusa (sexual)—gametes
Anthozoa	Fertilized egg—planula larva—polyp (sexual)—gametes

*It will be noted that some of the above forms have alternation of asexual (polyp) and sexual (medusa) generations. This differs from alternation of generations of plants in that both sexual and asexual forms are diploid and only the gametes are haploid. It is thought that the ancestral coelenterate form was the medusa and the polyp is a retention of the larval form (neotenous). In the evolutionary sequence the medusa in some is entirely suppressed and in others the polyp stage disappears or is reduced.

phasize the medusa; and it is entirely absent in Anthozoa, in which only the polyp form is found. Polymorphism, as pointed out in a previous section, has adaptive value because it promotes ecologic diversity of habitat distribution. As E. Mayr (1963) has emphasized, the advantages of this morphologic pattern are so great that one wonders why it is not more widespread in the animal kingdom.

The term *polymorphism,* widely used by zoologists in reference to the coelenterates, has a somewhat different meaning here than among other groups because it refers to division of labor among diversified individuals of a species.

The term *alternation of generations,* wherein sexual and nonsexual forms succeed each other, has largely been discarded in describing the life histories of those coelenterates with medusoid and polyp stages. The polyp is considered a persistent larva, and the medusa is considered an adult. But there are also other types of so-called polymorphism in the coelenterates, such as in *Hydractinia,* in which a colonial pattern of sexual and nonsexual individuals is found together; the same is true of the Siphonophora. In many Scyphozoa a single polyp may produce several medusae, and the blastostyle of a colony of *Obelia* buds off many medusae. There are many other variations on this theme in this extensive phylum. In some coelenterates it is very difficult to distinguish between a colonial system and an individual system. The problem of individuality, therefore, enters the picture in fine distinctions.

Polyps are usually sessile, although some, such as hydras and sea anemones, have limited sliding or creeping movements; hydras may also move by a somersaulting method. A polyp is usually radially symmetric but may tend in some cases toward biradial or bilateral symmetry. It has a longitudinal oral-aboral axis through the cylindric trunk, or stem. The aboral end, or base, is specialized in various ways for attachment to a substratum. The base may be in the form of an adhesive pedal disk, rootlike stolons, or simply a pointed end. The oral or free end bears a mouth, usually surrounded by a set of varying numbers of tentacles, and it may be expanded into a vase-shaped hydranth or an oral disk. The mouth may be circular as in hydras, or it may be elongated as in anthozoan polyps and may bear ciliated grooves called siphonoglyphs at one or both ends. The wall of the polyp is made up of an outer epithelial wall (epidermis) and an inner epithelial layer (gastrodermis) with a mesoglea of gelatinous substance or mesenchyme between the two epithelial layers. The central cavity of the cylindric body or trunk is the gastrovascular cavity, or enteron. In hydroid polyps the enteron is a

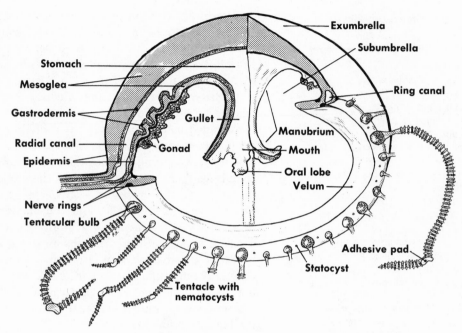

Stomach
Mesoglea
Gastrodermis
Radial canal
Epidermis
Gullet
Gonad
Nerve rings
Tentacular bulb
Tentacle with nematocysts
Exumbrella
Subumbrella
Ring canal
Manubrium
Mouth
Oral lobe
Velum
Adhesive pad
Statocyst

FIG. 5-2

Structure of *Gonionemus* (diagrammatic), a hydrozoan medusa. (From Hickman: Integrated principles of zoology, The C. V. Mosby Co.)

simple tube; in scyphozoan and anthozoan polyps the cavity is partly divided, and its surface is greatly increased by septal ridges that may, in anthozoans, form complete partitions in part of the gastrovascular cavity.

The medusa individual is built on basically the same coelenterate plan as is the polyp but is specialized for a free-swimming or floating existence. It is usually bell shaped or bowl shaped with a mouth at the end of a suspended stalk, or manubrium, on the underside (Fig. 5-2). The aboral surface is the convex side of the umbrella and is called the exumbrella; the oral side is the concave side, or subumbrella. The mouth leads into the gastrovascular stomach that may be a simple chamber, or its periphery may be divided into 4 perradial pouches by 4 interradial septa. From the stomach 4 or more radial canals run through the umbrella to the ring, or circular, canal in the bell margin. Centrifugal canals may lead off from the ring canal and pass throughout the length of the tentacles. Sense organs such as eyespots and statocysts are often found along the ring canal, on the bases of the tentacles, or on marginal specializations of the

bell. The arrangement of the radial canals, tentacles, and sense organs gives a tetramerous radial symmetry to most medusae. Both the exumbrellar and subumbrellar surfaces are covered with epidermis whereas the stomach, manubrium, and various canals are lined with gastrodermis. Between the epidermis and gastrodermis is a thick jelly layer, the mesoglea, which may or may not contain cells. The mesoglea is usually much thicker in medusae than in polyps.

The medusae of class Hydrozoa contain a circular shelf, the velum (Fig. 5-2), which projects inward from the subumbrellar margin of the bell and partly reduces the subumbrellar space. Such medusae are called craspedote medusae in contrast to the acraspedote scyphozoan medusae that lack a velum.

Swimming in medusae is accomplished by a type of jet propulsion in which the subumbrellar muscles contract suddenly and force the water from the subumbrellar space. This expulsion propels the medusa in an aboral direction. The umbrella is returned to its original form by the elasticity of the thick mesogleal layer, and the cycle is repeated. The swimming movement results from coordination of the nerve net and the marginal sense organs. The medusa thus has two main functions—locomotion and sexual reproduction—and its many modifications involve one or the other of these two functions. In some cases it may become a mere float by loss of tentacles and other specialized structures; in other forms it may serve only in a reproductive capacity.

The body form of the medusa is determined to a great extent by the firm, resilient mesoglea that is much more abundant in the jellyfish stage than in the polyp. The body form of the polyp depends partly on the rigidity of its tissues and partly on the water in the gastrovascular cavity. Large polyps are often provided with ciliated grooves in the mouth region for conducting water into the cavity. When the polyp is expanded, water performs the duties of a skeleton by giving rigidity to the body. Water in the body of the polyp is regulated by the action of circular and longitudinal muscles in the body wall. When water in the gastrovascular cavity is trapped by the closure of the mouth, the polyp can become shorter and broader by contracting the longitudinal muscles, or extended and thin by contracting the circular muscles.

Tentacles represent one of the most specialized structures in coelenterate morphology, although they may be absent in both polyp and medusa stages. Throughout the group, tentacles may be divided into two categories: (1) capitate, or those that are short, with a small cap provided with nematocysts, and (2) filiform, or those that are elongated, with nematocysts generally placed along their entire length in diversified arrangements (warts, rings, etc.). Either or both kinds may be present in the same species; tentacles display a large diversity in number, arrangement, and structure. This diversity is not restricted to differences between polyp and medusa phases but also extends to differences within each phase. Tentacles may be hollow or solid; when hollow the tentacular canal may be continuous with the gastrovascular or ring canal, or it may be separated by a partition. Solid tentacles are filled with a gastrodermal core of vacuolated cells. Tentacles are usually unbranched, but in a few species they are branched. They may be arranged in one or more circlets, or they may have an irregular arrangement. They may be definite in number for a particular species, or they may be indefinite in number and increase with age. Especially in the medusa phases, tentacles are usually liberally provided with sensory organs as well as nematocysts. Specialized sense organs are often found in their enlarged bases (tentacular bulbs). The bases may also function to produce nematocysts and aid in digestion. In some cases the tentacular bulb is the only part of the tentacle present.

Stinging organelles. One of the diagnostic features of coelenterates is certain stinging organelles, or capsules, that are used for food gathering and for defense against enemies. These organelles are divided into two kinds, nematocysts and spirocysts. Nematocysts are of many forms (Figs. 5-3, A to E, and 5-4) and are widely distributed throughout the phylum; spirocysts are all alike (Fig. 5-3, F and G) and are restricted to the subclass Zoantharia of Anthozoa.

Nematocysts are organelles produced in special cells called cnidoblasts, which are transformed interstitial cells. Nematocysts may be produced in body regions where they are to be used, or they may develop in other locations, the cnidoblasts migrating by ameboid movement to their definitive sites. The nematocysts on the tentacles of hydroid polyps originate on the hydranth body or stem; in medusae the basal enlargements of the tentacles serve as sites of formation. In Hydrozoa nematocysts are restricted to the outer epidermis; in the other two classes they are found in both epidermis and

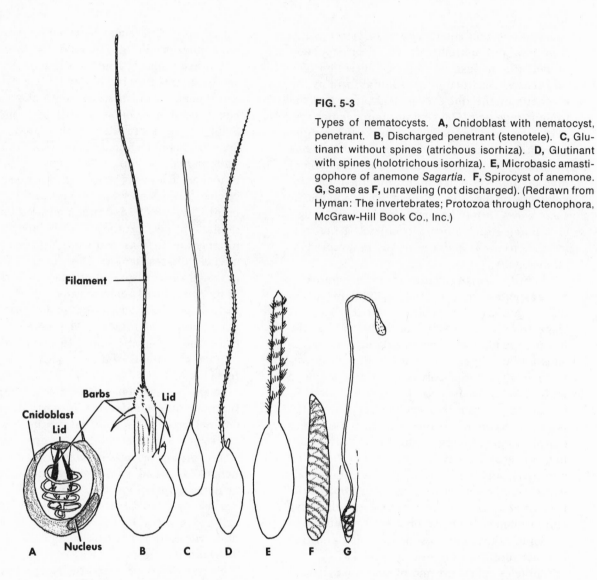

Filament

Barbs   Lid

Cnidoblast
Lid

Nucleus

A   B   C   D   E   F   G

FIG. 5-3

Types of nematocysts. **A,** Cnidoblast with nematocyst, penetrant. **B,** Discharged penetrant (stenotele). **C,** Glutinant without spines (atrichous isorhiza). **D,** Glutinant with spines (holotrichous isorhiza). **E,** Microbasic amastigophore of anemone *Sagartia.* **F,** Spirocyst of anemone. **G,** Same as **F,** unraveling (not discharged). (Redrawn from Hyman: The invertebrates; Protozoa through Ctenophora, McGraw-Hill Book Co., Inc.)

gastrodermis. In both polyps and medusae certain regions of the body are favorite locations for nematocysts. They are most common on tentacles where they are grouped into rings, ridges, and knobs, where cnidoblasts have migrated and are constantly renewed (P. Brien, 1960). In the tentacles of hydroid polyps such as hydras, they form clusters known as batteries. They are far more common in the oral than in the aboral region, and they are totally absent where the body has an exoskeleton. In

Scyphozoa and Anthozoa they are common on the gastric filaments, septal filaments, and acontia. Some nematocysts are only a few microns long after discharge, but others may reach a length of more than 1 mm.

The cnidoblast (Figs. 5-3, *A*, 5-4, and 5-5), which forms the nematocyst, is an oval or pear-shaped cell with a basal nucleus; it is attached to the epidermis by a slender stalk. The double-walled capsule of the nematocyst is first secreted by the cnidoblast, and later the tube or

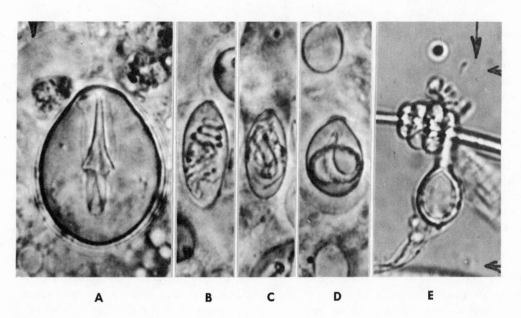

A   B   C   D   E

FIG. 5-4

Nematocysts of *Hydra littoralis*. **A,** Penetrant. **B,** Streptoline glutinant. **C,** Stereoline glutinant. **D,** Volvent. **E,** Discharged volvent attached to copepod bristle. (Courtesy Carolina Biological Supply Co., Burlington, N. C.)

FIG. 5-5

Portion of cross section of body wall of hydra. (From Andrew: Textbook of comparative histology, Oxford University Press, Inc.)

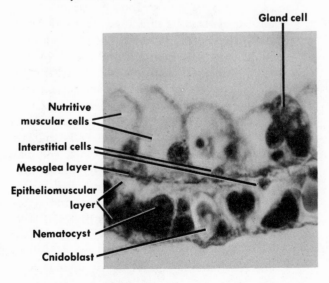

Gland cell

Nutritive muscular cells

Interstitial cells

Mesoglea layer

Epitheliomuscular layer

Nematocyst

Cnidoblast

thread becomes differentiated and coiled inside this capsule. The capsule has a small lid, or cap, directed to the outside. The nematocyst is placed near the free surface of the cnidoblast. A bristle-like structure, the cnidocil or trigger, projects from the tip of the undischarged nematocyst. A cnidocil is probably a modified flagellum, since it has the characteristic nine peripheral elements. In an anemone nematocyst the trigger is actually a flagellum. Although there are seventeen major categories of nematocysts (R. Weill), all are discharged by eversion of the thread tube to the exterior. The various kinds of nematocysts can be grouped into two main divisions; those that have threads with closed ends (astomocnidae), and those that have threads with open ends (stomocnidae). The first type is used for attachment or as a lasso to catch prey; the other type is used for penetration and injection of toxins into prey. The toxins are probably complex proteins in nature.

When a nematocyst is discharged, it cannot be used again. A discharged nematocyst consists of the old capsule, or bulb, and a thread tube that often bears spines and thorns at its base. Nematocysts are useful in determining the taxonomic status of coelenterates. Most coelenterates have three or four types of nema-tocysts, but in the primitive order Trachylina and some others they are all of one type. Hydra has four kinds of nematocysts: desmonemes (volvents), stenoteles (penetrants), holotrichous isorhizas (glutinants with spines), and atrich-ous isorhizas (glutinants without spines) (Figs. 5-3 and 5-4). The first two types are discharged by food stimulation (L. E. R. Picken and R. J. Skaer, 1966); the third type responds to me-chanical stimuli other than food; and the last type responds to contact with the substratum and is used for attachment when the animal is walking on its tentacles. Nematocysts have been called "independent effectors" because their discharge appears to be independent of nervous excitation pathways, at least in many species studied. However, G. O. Mackie and others have shown that nonnervous conduction is possible in nerve-free epithelia such as the membranes of epitheliomuscular cells. Con-tact with the cnidocil, or trigger, is involved in some discharges, but in many cases chemical stimulation (as from food) is necessary to lower a threshold before the trigger operates (C. S. Jones, 1949). The mechanics of the actual process of discharge seem to involve an intake of water because of increased per-meability of the capsule wall and, in some cases, probably involve contraction of the cnidoblast. The electron microscope reveals the presence of supporting rods and fibrous bands around nematocysts, and these may be contractile (G. B. Chapman and L. C. Tilney, 1959). This increased pressure on the contents of the cap-sule forces open the lid (operculum) and causes the thread tube to evert.

The spirocysts are thin-walled capsules one cell layer in thickness (Fig. 5-3, F and G). They contain a long, unarmed tube of uniform dia-meter that everts when discharged. Within the body of the tube are bundles of rodlets running parallel to the long axis of the tube (J. A. West-fall, 1964). Spirocysts stain with acid dyes, whereas nematocysts have an affinity for basic dyes. The chief content of spirocysts is muco-protein or glycoprotein. The exact function of spirocysts and their methods of discharge are unclear. Since they have no flagellum, a chemi-cal stimulus has been suggested for their dis-charge. They are restricted mainly to the oral disk and tentacles of sea anemones.

MORPHOLOGY AND PHYSIOLOGY OF BODY SYSTEMS

Some systems, such as respiratory and ex-cretory systems, are entirely lacking in the coelenterates, and the circulatory and digestive systems are combined into one—the gastro-vascular system. Respiration and excretion are effected by the individual cells without any specialized structures or organelles.

The epithelial system is represented by the epidermis, or outer epithelium, and by the gas-trodermis, the inner epithelium (Fig. 5-5). Epidermis is derived from ectoderm. The epider-mis usually consists of a single layer of cells that vary from flat squamous to tall columnar types. In some cases the epithelium is a syncytium. It may be ciliated or flagellated in most polyps and in some medusae. The bases of the epithelial cells reach to the mesoglea, to which they may be fastened by pseudopodial processes.

Many specialized cells are found in the epi-dermis. In epitheliomuscular cells the bases may be drawn into strands containing myo-nemes or contractile fibers that run in a longi-tudinal direction in or near the mesoglea. In some groups (Trachylina, Scyphozoa, and An-thozoa) the bases lack contractile extensions because the independent myoneme bundles may be embedded in the mesoglea or near it. Other cells found in the epidermis are the cnido-blasts with nematocysts; gland cells that are mucus producing and are especially common on the tentacles and pedal disk; elongated and branched sensory cells that are attached at their bases to the general nerve plexus; and small, rounded, undifferentiated interstitial cells that are general factotums in producing various types of cells (sex, cnidoblasts, etc.) as needed.

The gastrodermis, derived from endoderm, is

constructed on the same plan as the epidermis, and it shares some of the same kinds of cells. It lines the gastrovascular cavity and has two major functions to perform: (1) the movement of food materials about the gastric cavity by means of its flagellate cells and (2) digestion of food both extracellularly and intracellularly. The chief cells found in the gastrodermal layer are the large nutritive-muscular cells whose bases are drawn into contractile extensions containing myonemes that run in a circular direction near the mesoglea (Fig. 5-5). Nutritive-muscular cells are generally columnar, highly vacuolated, and contain food vacuoles. Their free ends usually bear two flagella to each cell. In some coelenterates (Anthozoa) the gastrodermal muscles run in a longitudinal direction as well as in a circular direction. Other types of gastrodermal cells are gland cells (some of which secrete enzymes) with abundant secretion granules, sensory nerve cells similar to those of the epidermis, and small interstitial cells. Cnidoblasts are found in the gastrodermis in such specialized regions as gastric filaments, septal filaments, and acontia but are absent in the gastrodermis of hydrozoans.

The mesoglea between the epidermis and gastrodermis varies greatly in different taxa of the phylum. It also varies in the polyp and medusae stages; in the medusa it makes up the greater part of the animal. In Hydrozoa it is almost entirely acellular, but in the other classes the mesoglea contins fibers and cells (collenchyme) forming something that resembles a third germ layer. In all medusae the extensive mesoglea has a content of almost 95% water. In the medusae of all classes the mesoglea is sharply marked off from epidermis and gastrodermis by a very thin jelly layer. The mesoglea may be thought of as a type of skeleton, and in some anthozoans it may serve as the origin and insertion of muscle cells.

The digestive system, as already mentioned, is a part of the gastrovascular system. Digestion may be described as partly extracellular and partly intracellular. Glandular cells in the gastroderm produce proteolytic enzymes that are discharged into the gastric cavity, where they partially break down the food into small particles. These particles are then engulfed by gastrodermal cells of all types and not merely by the nutritive-muscular cells. Within the food vacuoles so formed, the digestive process is finally completed. The digested products are then diffused to all parts of the organism. Epidermal cells cannot engulf food materials.

All members of the phylum are carnivorous and strictly herbivorous diets are unknown in the group. However, some xeniid corals apparently do not feed and are thought to depend on their algal symbionts for nutritional purposes. Most coelenterates cannot digest starches. The nematocysts are effective weapons in immobilizing prey, and the tentacles manipulate the food to the mouth region where ciliary action and muscular movements force it into the enteron.

Occasionally throughout the phylum symbiotic algae (zooanthellae) are found that, by photosynthesis, produce by-products which are incorporated into the protoplasm of the coelentrate. In some anthozoans, such as reef-forming corals, this accessory food supply may form an important item in their economy. The green hydra (Chlorohydra) gets its color from gastrodermal zoochlorellae (Chlorella). In some symbiotic relationships, the apparent advantages may be largely in favor of the algae, but it may be assumed that the coelenterate also derives some benefit.

Although the gastrovascular cavity is a relatively simple sac in many polyps of Hydrozoa, it becomes more complicated in both polyp and medusa of the other classes. In polyps it may be subdivided by septa, and in medusae it consists of a central stomach, with radiating canals that join the circular, or marginal, canal. The endodermal tentacles, or gastric filaments, in scyphozoans are the chief sites of enzyme secretion and to some extent of intracellular digestion as well. Anthozoans are provided with a sinuous cord, the septal filament, along the free edge of each septum, and these filaments have gland cells that secrete digestive enzymes into the gastrovascular cavity. Both gastric filaments and septal filaments are liberally supplied with nematocysts.

In coelenterates the muscular system is in the simplest condition found among true metazoan animals. The muscular system in simple hydroid polyps consists of an outer sheath of epidermal, longitudinal fibers at the bases of epidermal epitheliomuscular cells and an inner sheath of circular fibers at the bases of gastrodermal cells. These sheaths are not continuous sheets but are rather networks of anastomosing strands. As previously mentioned, these muscle fibers are parts of epithelial cells in both sheaths. Such a muscle arrangement makes possible contraction and extension of the cylindric body of a sessile, radial animal.

In all medusae and in most members of Scyphozoa and Anthozoa the musculature appears to be a subepithelial layer except in the tentacles, where the primitive sheath is largely retained. The muscle fiber (myoneme) is always in a separate part of the cell from that portion containing the nucleus, and the connection between the two portions may be drawn into one or more long strands (depending on the number of myonemes). This arrangement may give the appearance of a separate muscle layer where the epidermis is a thick one, as in many of the larger coelenterates. In some cases the epitheliomuscular cells lie entirely below the surface of the epithelium—for example, in the swimming muscle of large scyphomedusae. The deep-lying tissue may actually be enclosed in mesoglea and may become partially isolated from the superficial layer.

In medusae the swimming muscles of the subumbrella are well-developed circular and radial bands. Muscle cells are elongated and, in the circular fibers of medusae, are striated. In anthozoans and scyphozoans, in which muscle fibers form independent bundles, the mesoglea is closely applied to the muscles and may actually enfold them.

Skeletons are of two types in coelenterates: (1) the mesoglea and (2) secreted epidermal exoskeleton. The mesoglea has already been mentioned as a type of skeleton that may give considerable support and rigidity to the body of coelenterates, especially in the medusae. The secreted exoskeleton of the epidermis exists in a great variety of types ranging all the way from the chitinous tubes (periderm or perisarc) of hydroids to the calcareous skeletons of corals. In hydrozoan polyps especially, the tubular skeleton of chitin gives rigidity against wave action in habitats where colonial hydroids are found. Such a skeleton may form a cup (hydrotheca) around the hydranth (Fig. 5-8), may be restricted to the stem (Fig. 5-11), or may be absent altogether (as in hydras). In stony corals the hard, calcareous exoskeleton lies outside the polyp's body and exists in various patterns of skeletal vertical ridges or septa radiating from the center to the periphery. The axial skeleton, another type, is found in sea pens and gorgonians (Anthozoa). This skeleton has an axial rod of sclerified proteins (Fig. 5-46, B) derived from invaginated ectoderm or from the mesoglea. Such a skeleton is tough and flexible. Sometimes the axis is a rigid structure of calcium carbonate. In the common sea anemone the fluid in the gastrovascular cavity acts as a hydrostatic skeleton.

The nervous system of coelenterates has been the subject of many investigations over the years, and many new concepts about it have originated from time to time. Since coelenterates comprise the lowest metazoan phylum that possesses a nervous system, and since the classic nerve net they possess is considered to be one of the most primitive types, interest in its morphology and physiology has never waned. The system consists of an unpolarized network, or plexus, of bipolar and multipolar neurons and their neurites, together with the neurites of epidermal sensory cells that unite with the plexus. The chief nerve plexus lies just beneath the epidermis outside the muscular sheath. A similar but usually less developed plexus is found in a subgastrodermal position. Both plexuses may be regarded as one system in some forms, because neurite connections between the two nets may be found in the mesoglea. In other cases the two plexuses have little or no contact and behave as separate entities. *Velella*, for instance, has two distinct plexuses. One of these is made up of continuous giant fibers (syncytial or closed system, Fig. 5-6), and the other consists of discontinuous neurons and is known as the open system (G. O. Mackie). Some studies have shown two nerve nets in *Aurelia* and others—the subumbrella net that controls the simultaneous contraction of circular and radial muscle, and a diffuse net in the epidermis that

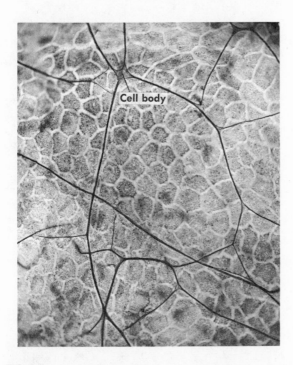

FIG. 5-6

Portion of syncytial giant fiber system of *Velella*. This form is provided with two nervous systems—one of syncytial (continuous) giant fibers and one of nonsyncytial (discontinuous) neurons. (Courtesy G. O. Mackie, University of Alberta, Edmonton.)

controls local reactions. The nerve net in the much-studied hydra is still customarily described as a continuous net with nerve cell bodies at the angles; but synaptic or open systems have been described in other hydrozoans, in sea anemones, and in scyphomedusae. The neurosensory cells in the epithelium may make synaptic junctions with the general nerve net as S. Leghissa has shown in the anthozoan *Actinia*. In colonial forms the nerve net may be continuous from one member to another in the whole colony, but there are many variations of this pattern. Response to stimuli may involve the whole colony, or it may be restricted to a localized region.

In hydra bipolar or multipolar cells are situated above the muscular processes of epithelio-muscular cells. These ganglion cells may send processes to muscle fibers and sensory cells, or they may terminate in the processes of other ganglion cells. Some of the ganglion cells are known to be neurosecretory and certain dense granules are concentrated at the terminations of neurites (T. L. Lentz and R. J. Barnett, 1965).

R. K. Jha and G. O. Mackie (1967) have studied the ultrastructure of hydrozoan nerve elements. They found that the nerve cells are essentially the same in hydroid polyps and hydromedusae, but only the hydromedusa *Sarsia* is known to have structurally polarized synapses that can conduct in a one-way arrangement, as in higher animals. They also found secretory vesicles in the transmission junctions as previously reported in the large jellyfish *Cyanea* (a scyphozoan).

Theoretically a stimulus at any one point of a nerve net can be transmitted to all points of the system. However, local activities, e.g., locomotion, feeding, etc., are possible. Whether synaptic or continuous, the nerve net differs from the majority of nervous systems in being able to conduct equally well in all directions, because polarization is either lacking or very restricted. Independent effectors are common in coelenterates; the action of nematocysts, cilia, and chromatophores, for example, may be largely or entirely independent of nervous elements. In medusae coordination of swimming is controlled by the balance organs, or statocysts, of the bell margin. Light sensitivity may be generalized, as in *Hydra* (L. M. Passano), or localized in ocelli, as in *Sarsia* (Fig. 5-12).

Sensory cells of coelenterates are elongated and may or may not have varicose enlargements at the end. They usually terminate near the surface in a point or a flagella-like bristle. Their sensory function is chiefly at the level of single receptor cells, although some receptors are multicellular. Some may be general receptors for various stimuli, but others are differentiated for mechanical, photo, or other kinds of stimuli. However, their chief receptors are adapted for chemoreception. D. M. Ross (1966) has presented a great deal of evidence for this view, based on the feeding habits of coelenterates and especially on the chemoreceptor responses involved in the symbiotic association between the sea anemone *Calliactis parasitica* and the shells of hermit crabs. Ross has shown that the sense cells and perhaps the nematocysts of the anemone have a specificity to some shell factor, perhaps to organic deposits on the shell.

In hydra, L. M. Passano and C. B. McCullough (1965) have described a pacemaker interaction that helps explain a characteristic and unique behavior pattern found in these animals. In the periodic exploration of their environment at 5- to 10-minute intervals (less frequent at night), a hydra may assume a ball shape by contracting its longitudinal muscles. A little later the hydra extends itself again. This behavior is called a contraction burst and is initiated by a pacemaker located in the subhypostome, with a conducting system running the length of the animal. Another pacemaker system, the rhythmic potential system located near the base of the polyp (also with a conducting system through the length of the hydra), has no obvious behavior activity but produces a rhythmic output of potential, as recorded by electrode contact. Evidence shows that there is a frequency correlation between this rhyth-

mic potential system and the contraction burst, being longest during the burst and shortest just after. Each system has been shown to influence the other in the production of a coordinated response to light or other factors not yet thoroughly understood.

Both asexual and sexual reproduction are common throughout the phylum. Most coelenterates are dioecious, although there are a few exceptions. Sex cells may develop from interstitial cells in either the epidermis or the gastrodermis. With few exceptions the gonads are restricted to medusae or sessile gonophores. Gonophores are medusa buds that do not develop to a free stage. In some coelenterates the gonads are transient organs, and their location

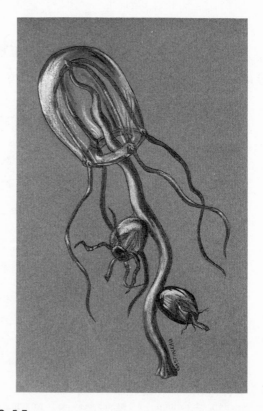

FIG. 5-7

Medusa of *Sarsia gemmifera,* a gymnoblastic hydroid. Asexual reproduction occurs by budding along manubrium. Height of parent bell is 2 to 4 mm.

on the body varies. In hydromedusans the gonads may appear on the manubrium or in the gastrodermis beneath the radial canals; in hydroid polyps they may occur almost anywhere. Scyphozoan gonads are located on the floor of the stomach, and germ cells escape by way of the mouth. In anthozoans the gonads are located on the septa that subdivide the coelenteron. In all cases ducts and accessory organs are absent. Fertilization usually occurs in the surrounding water, but in some species the eggs are fertilized in situ, where they develop to a late stage.

In asexual reproduction, polyps may bud from stems, stolons, or other polyps, and medusae may bud from hydranths, stolons, or other medusae. In hydrozoans, in which both polyp and medusa stages are usually well represented, most medusae bud from hydroid colonies, but they may also bud from other medusae (*Sarsia*, Fig. 5-7). A rare method of asexual reproduction in hydrozoans is by means of frustules—small,

FIG. 5-8

Life cycle of *Obelia*. (From Hickman: Integrated principles of zoology, The C. V. Mosby Co.)

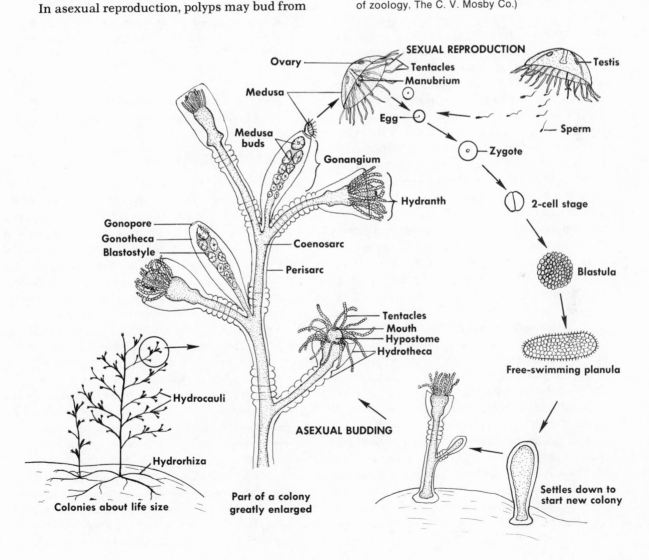

elongated, planula-like but nonciliated bodies that are budded off from a polyp or stolon and creep about on the bottom until they eventually develop into new polyps or colonies. In the Anthozoa, polyps undergo both transverse and longitudinal fission. Transverse fission is also common in the development of the medusa from the larval stage in Scyphozoa, and it may also occur in the polyp of Hydrozoa. The scyphistoma larvae, common in many scyphozoans, may give rise to medusae directly by budding off the side or by strobilation (Fig. 5-40) or may send out stolons that bud off other scyphistomae. A scyphistoma in process of strobilation gives rise by transverse fission to ephyrae (Figs. 5-40 and 5-42). Medusae sometimes arise by direct asexual budding from other medusae, as in the Anthomedusae, but no polyps are derived directly from medusae by budding.

Polyps have great powers of regeneration, and most parts cut from a parent stock will develop into a new polyp. Gradients of regeneration are found in polyps similar to those present in planarian worms. Much of this regenerative capacity is lost in the medusa stage.

Although there are some differences in the developmental stages of coelenterates, a certain basic plan is followed (Fig. 5-8). Cleavage follows fertilization, and a ciliated blastula is formed. The gastrula may be formed by an infolding of one end of the blastula or by the inwandering of surface cells so that a second layer is formed. In all cases a stereogastrula results. The stereogastrula elongates into the planula larva (Fig. 5-9, *A*), which is provided with cilia and is free swimming. In a short time the planula becomes attached to a substratum by its anterior end or by its side and forms mouth and tentacles at its posterior end. From such a polyp a colony may form by growth and budding. Some variations of this pattern occur. The planula in many hydroids (Tubulariidae and others) develops tentacles and a mouth and becomes an actinula, which resembles a short, stalkless polyp (Fig. 5-9, *B*). The actinula creeps around, attaches itself, and becomes either a

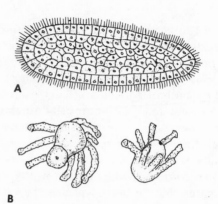

FIG. 5-9

A, Planula larva in longitudinal section, with outer layer of ectodermal cells and inner core of endodermal cells; characteristic of many coelenterates. **B,** Actinula larvae of *Tubularia*. (Sketched from life.)

polyp or a medusa. The larvae of medusae that have no polypoid stage develop directly into new medusae.

■ CLASS HYDROZOA

The members of Hydrozoa may be either polymorphic, wholly polypoid, or wholly medusoid in their life histories. They have a tetramerous or polymerous radial symmetry and differ from the Scyphozoa and Anthozoa by having no subdivisions of the gastrovascular cavity, no nematocysts in the digestive cavity, and no stomodeum. Other differences are the presence of a velum and the presence in the medusa of highly specialized sense organs differing from the rhopalia of scyphozoans.

A large diversity of form exists within the group. A typical hydrozoan polyp consists of an attached base, a stem, and a terminal hydranth bearing the mouth and tentacles. Solitary forms are usually anchored to a soft substratum by short, unbranched roots or by modified stolons. In hydra, the freshwater polyp, the aboral end is an adhesive pedal disk. The base in colonial forms is made up of stolons (rhizomes) that form an anastomosing rootlike structure (hydrorhiza, Fig. 5-8) fastened to a substratum. The stolons give rise to buds that form the new individuals of a colony. The main stalk of a colony is called the hydrocaulus. The stolons and hydrocaulus are made up of a living hollow tube (coenosarc) that is usually covered by a chitinous tube (periderm or perisarc) secreted by the epidermis. The coenosarc has both epidermis and gastrodermis separated by a thin layer of mesoglea and encloses a gastrovascular cavity that extends throughout the colony. The hydranth is usually vaselike and bears an elevated part, the hypostome or manubrium, in which the terminal mouth is situated. Around the base of the hypostome and mouth is usually a circlet of tentacles, but other arrangements of tentacles are also found. In the Anthomedusae (Gymnoblastea) the hydranths lack a periderm (athecate); in the Leptomedusae (Calyptoblastea) each hydranth is enclosed in a secreted cup of periderm, the hydrotheca (thecate).

The hydrozoan medusa has tetramerous radial symmetry and is typically bell shaped or dome shaped (Fig. 5-2). From the center of the subumbrella hangs the quadrangular manubrium that may bear lobes and oral tentacles provided with nematocysts. A stomach, or gastric cavity, is found at the base of the manubrium, but variations of this plan exist. Four radial canals, 90 degrees apart, usually radiate from the stomach to the ring canal in the bell margin. Some forms may contain more radial canals that may also branch. The gastrovascular cavity includes all these canals as well as the stomach and manubrium. A velum is present, although in some cases, e.g., *Obelia*, it may be rudimentary (Fig. 5-26).

The margin of the circular bell is usually unscalloped and bears the sense organs and tentacles. Sense organs are commonly represented by ocelli (Fig. 5-12) and statocysts. Tentacles may be few or numerous and in a few species may be branched. They may be either solid or hollow extensions of the ring canal. The base of each tentacle is enlarged to form the tentacular bulb that may bear an ocellus or other sensory structures. The bulbs also serve as sites of nematocyst formation.

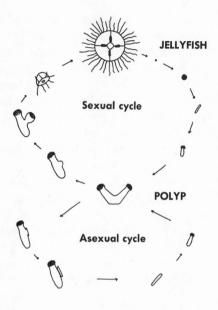

FIG. 5-10

Life cycle of freshwater hydrozoan *Craspedacusta sowerbyi*. (Courtesy Charles F. Lytle.)

Diversity of body form among hydrozoans is chiefly caused by variations between the hydroid and medusoid generations. Some hydrozoans have no medusoid stages, and others have no hydroid stages. Still others may have both generations combined and present in the same specimen. This composite nature is well illustrated in the Siphonophora in which many different types of polypoid and medusoid components are found together in the same colony. The basic adaptive nature of the phylum, as already mentioned, plays a part in this evolutionary pattern. The typical balanced system of hydroid and medusoid stages found in many hydrozoans (Figs. 5-8 and 5-10) permits existence in varied assortments of environments. The polyp fitted for a sessile existence and the medusa with a swimming adaptation promote a wide distribution of the species. But a pure medusa form without a polypoid stage (e.g., *Liriope*, Fig. 5-29) can live far out at sea where a polyp could not exist. On the other hand, a purely sessile life in a favorable situation could lead to suppression of the medusoid stage and the persistence of the polyp form.

Class Hydrozoa is commonly divided into six orders, each of which has certain distinctive characteristics. They are Hydroida, Milleporina, Stylasterina, Trachylina, Siphonophora, and Chondrophora.

Few species of coelenterates are found in fresh water. C. F. Lytle (1964) in his analysis of the freshwater distribution of Hydrozoa lists four distinct and separate invasions of this class from the sea. These include (1) some forty species of hydra, (2) five species of freshwater medusae (one each in America, China, Japan, India, and Africa), (3) one colonial gymnoblast (*Cordylophora*), and (4) one parasitic hydroid (*Polypodium*) that parasitizes the sturgeon egg in the Caspian Sea and southern Russia.

■ ORDER HYDROIDA

The order Hydroida includes most of the hydrozoans. It includes the suborders Gymnoblastea and Calyptoblastea.

The Gymnoblastea (Anthomedusae) include species without a protective cup (athecate) around the hydranths and gonozooids, e.g., *Eudendrium* and *Bougainvillia* (Fig. 5-11). The Calyptoblastea (Leptomedusae) include species that are thecate, or have a hydrotheca cup around the hydranth and a gonotheca cup around the gonozooid, e.g., *Obelia* (Fig. 5-8) and *Campanularia* (Fig. 5-24, *B*). Gymnoblastic

FIG. 5-11

Bougainvillia, a gymnoblastic hydroid. Left, portion of colony. Right, medusa.

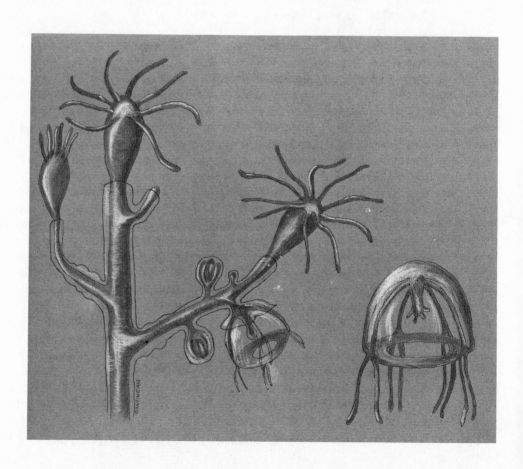

hydroids usually begin as juveniles with few tentacles on their hydranths; as they grow they add more tentacles (Fig. 5-13). On the other hand, the hydranths of calyptoblastic hydroids do not grow after emerging from a bud, do not add tentacles, and are absorbed by the colonies after a few days. The medusae of the Gymnoblastea (Fig. 5-7) possess tall bells, manubrial gonads, tentacular bulbs, usually with ocelli (Fig. 5-12), no statocysts, and few tentacles. The medusae of Calyptoblastea usually have saucer-shaped or bowl-shaped bells, gonads on the radial canals, statocysts (but lack eyespots), and many tentacles (Fig. 5-26). Some calyptoblasteans do not have free medusae. Many morphologic variations exist within these two suborders.

In the life histories of the suborders Gymnoblastea and Calyptoblastea there is typically both a polyp and a medusoid stage, but there is a distinct tendency toward suppression of the medusoid stage in many members of these suborders. Such suppression may involve rudimentary medusae attached to the polyp, sessile gonophores, reproductive undeveloped medusae (sporosacs) on the polyp, and even complete loss of the medusoid generation.

In the typical life history the fertilized egg (zygote) produced by the medusa develops into a planula larva. The ciliated planula larva swims around, attaches itself to a substratum, and develops a mouth and tentacles at its free end. It is now a polyp and by developing stems, stolons, and other polyps, begins a hydroid colony (Fig. 5-13). In some hydroids (*Tubularia,* Fig. 5-14) the planula larva remains in the gonophore, develops into an actinula larva with mouth and tentacles, becomes free to creep around, and finally transforms into a polyp. A colony of polyps gives rise to two types of individuals (dimorphia)—gastrozooids for nutrition and gonophores, or buds, that develop into medusae. Since most Hydroida are dioecious, the medusae produced by a particular colony are all of the same sex. Gonophores may arise from the hydrocaulus, from a blastostyle stalk, from the hydrorhiza, or from the gastrozooid directly. Some of these variations will be described in the following accounts of some common genera of hydroids.

In members of the gymnoblastic type the perisarc does not extend beyond the base of the polyp and does not form a hydrotheca.

Many of the Hydroida display bioluminescence. The light may be produced in the cellular tissue, or it is possible for the enzymes (luciferases) and luciferins to be extruded and light emitted by reactions between the two. Many medusae of different species (e.g., *Aequorea*) may emit light as a blue flash from the tentacular bulbs. In some polyps (*Campanularia, Obelia, Plumularia,* etc.) the luminescence may originate in the hydranths and even in the stalks. Because of its spotty distribution, it is difficult to assign special uses for bioluminescence in those hydroids that possess it. In all hydroids it appears only in response to stimuli.

Patterns of growth and regeneration in hydroids. Many experiments have been made on coelenterates because of their plasticity for growth and regeneration. Many classic experiments indicate that as a group they have great regenerative power, although there are variations in their patterns. Many coelenterates, for

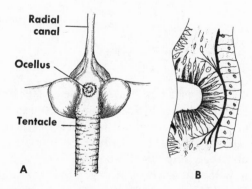

FIG. 5-12

A, Base, or bulb, of marginal tentacle of *Sarsia tubulosa,* bearing ocellus. **B,** Section through ocellus of *Sarsia*. Pit is filled with transparent lens and lined with pigmented epithelial cells and nerve cells. (**A** modified from Russell.)

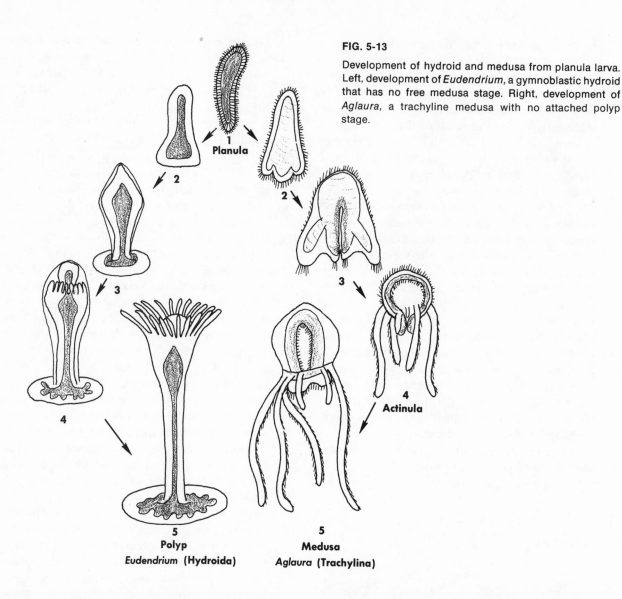

FIG. 5-13

Development of hydroid and medusa from planula larva. Left, development of *Eudendrium,* a gymnoblastic hydroid that has no free medusa stage. Right, development of *Aglaura,* a trachyline medusa with no attached polyp stage.

1
Planula

2

2

3

3

4

4
Actinula

5
Polyp
Eudendrium (Hydroida)

5
Medusa
Aglaura (Trachylina)

instance, can be made to diminish and dedifferentiate under adverse conditions to small gastrula-like bodies but to regenerate to a normal condition when fed. A hydra may be cut into small pieces a tiny fraction of 1% of its mass, and each piece will develop into a complete animal with tentacles and body. By splitting the hypostome region, a 2-headed hydra can be obtained. In all pieces the basic polarity of the animal is usually retained: the anterior end becomes the head and the posterior

end the basal disk. A hypostome grafted onto the side of another hydra will induce the formation of another individual, but tissue grafts from other parts of the hydra are reorganized or absorbed into the tissue characteristic of the new location where it is grafted. For example, E. B. Harvey grafted the hypostome and tentacles of a bleached hydra into a green hydra; the new individual produced some white tentacles from the graft tissue and some green ones from the host.

FIG. 5-14

Tubularia crocea, a gymnoblastic hydranth, with two whorls of tentacles and clusters of gonophores. (Sketched from life.)

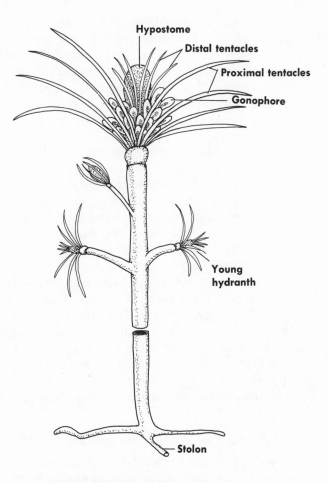

A recent evaluation of regenerative patterns in coelenterates indicates that investigators differ a great deal in their experimental results on the same animal group. It has long been thought, for example, that there is a localized growth zone in hydra located just under the tentacles at the top of the column (N. J. Berrill, 1961; A. L. Burnett, 1961; and many others). This idea of a localized growth zone has been challenged, however, by R. D. Campbell (1967) and G. Webster (1971), who find that dividing (mitotic) cells are spread over the body above the stalk and that there is no sharply mitotic activity in the subtentacular region where the growth zone is supposed to be. Webster in his extensive analysis of pattern formation thinks cell movement and changes in cell contact, rather than cell division, are more involved in patterns of regeneration.

Colonial hydroids have been a favorite subject for coelenterate developmental investigation. Their colonies are made up of repeated patterns of hydranths, and they also grow asexually by the elaboration of stolons arranged on a substratum. Most hydroids are colonial, but a few, including *Hydra* and *Tubularia,* are solitary. Most colonial hydroids are marine littoral forms that become attached to rocks, algae, and often to other animals. Some are also found at considerable depths. Usually they are of small size (from several millimeters to a few inches in height).

S. Crowell (1961) has studied *Campanularia* from several aspects, such as patterns of colonial growth, aging, regression and replacement of hydranths, and others (S. Crowell, 1950, 1953, 1960, 1961). He has made a comparison of the morphogenesis of thecate, or calyptoblastic, species (those with a protective hydrotheca around the hydranth) with the athecate, or gymnoblastic, species (those that have no protective sheath). In the athecate hydroid the oldest hydroid is terminal in position (monopodial growth); in the thecate species the oldest hydranth is the one nearest the base of the colony (sympodial growth). A fully developed thecate hydranth does not grow (S. Crowell, 1960), will not regenerate removed parts, and will regress and become resorbed within a few days. He found that the athecate species, however, will do the reverse of all these actions. In *Cordylophora,* an athecate gymnoblastic species and the only freshwater colonial hydroid, C. Fulton (1961) has studied the development and growth of a colony and found that peristaltic waves begin at the tip of each hydranth and spread rhythmically and synchronically two or three times per hour throughout the colony. This may occur to circulate the nutrients. The rate of peristalsis during feeding is more than this, which may indicate a high degree of integration in a colony, as shown also by the same phenomenon in other groups of coelenterates.

In the solitary gymnoblastic species *Tubularia* (Fig. 5-14), G. O. Mackie (1966) believes that the hydranths can probably live indefinitely under favorable water conditions. He finds that the hydroid emerges from the bud in an immature condition, adds tentacles, grows in size, and is able to regenerate lost parts.

The power of an animal to regenerate decreases with the degree of differentiation and its grade in the evolutionary scale. However, in all phyla there are certain groups that have restricted or poor capacity for regeneration. In general, coelenterates have amazing powers of regeneration as befits their low taxonomic status in the animal kingdom, but distinctive differences are indicated in the few examples given.

Suborder Gymnoblastea (Anthomedusa)

Some familiar gymnoblastic genera follow.

Tubularia. Tubularian colonies are made up of large polyps with long stems. They may consist of single, unbranched individuals, or branching may occur from stolons that do not form a network. The hydranth bears two whorls of solid tentacles, the basal circlet of large tentacles and oral circlet of smaller tentacles around the mouth (Fig. 5-14). The sessile gonophores grow in grapelike clusters from stems attached between the two circlets of tentacles. The medusae always remain attached to the parent polyp, and their zygotes develop into planula larvae in the subumbrella before leaving as actinula larvae (Fig. 5-9) to start a new colony.

Clava. The elongated hydranths of these clublike polyps possess 15 to 30 filiform tentacles scattered over them (Fig. 5-15). The hydranths

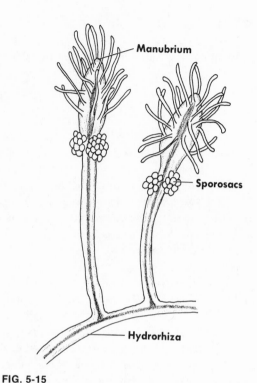

FIG. 5-15

Clava; polyps about ¼ inch tall with filiform tentacles (suborder Gymnoblastea).

spring from a hydrorhiza. The undeveloped medusae or reproductive buds (sporosacs) are attached in clusters just beneath the tentacles and never become free.

Corymorpha. This is a solitary hydroid about 2 inches long (Fig. 5-16). The hydranth has two circlets of filiform tentacles, with branching gonophores attached between them. The upper circlet of small tentacles surrounds the mouth at the top of the tall hydranth. The stem, which is narrow at the top and broader at the base, terminates in a branched, rootlike structure attached to sandy or muddy bottoms. This stem has a number of small canals instead of a single gastrovascular cavity, and it is especially adapted for support by being filled with gastrodermal cells. The medusae that develop from the gonophore buds are free swimming, and each is provided with a single trailing tentacle. *Corymorpha* has no asexual reproduction.

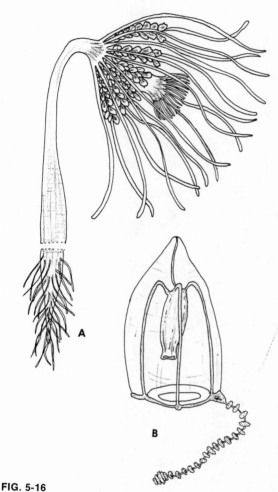

FIG. 5-16

Corymorpha, a gymnoblastic hydroid. **A,** Polyp. Height is about 5 cm. **B,** Medusa. (Redrawn from Hyman: The invertebrates; Protozoa through Ctenophora, McGraw-Hill Book Co., Inc.)

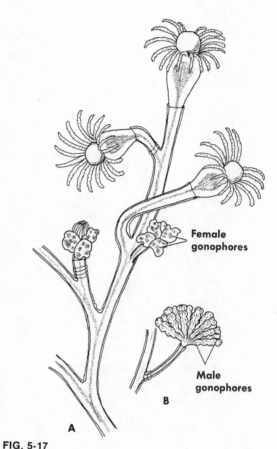

FIG. 5-17

Eudendrium (suborder Gymnoblastea). **A,** Part of colony, with female gonophores. (Sketched from life.) **B,** Male gonophores.

Eudendrium. This is a striking example of a bushy, much-branched colony (Fig. 5-17). The hydranth is provided with a single circlet of filiform tentacles (17 to 28) surrounding a funnel-shaped hypostome. The male and female gonophores arise from the gastrozooids and differ from each other in appearance. No medusa generation occurs.

Hydractinia. This genus is a polymorphic, colonial form found on rocks and especially on shells occupied by hermit crabs. Several types of hydroid individuals are connected at the base by a network of stolons (Fig. 5-18). At least three types of zooids exist—nutritive zooids, reproductive zooids, and defensive zooids. Each of these types has its own distinctive features. Nutritive zooids (gastrozooids) are tube shaped with a terminal mouth surrounded by filiform tentacles. They ingest and digest food for the colony. Reproductive zooids (gonozooids) are

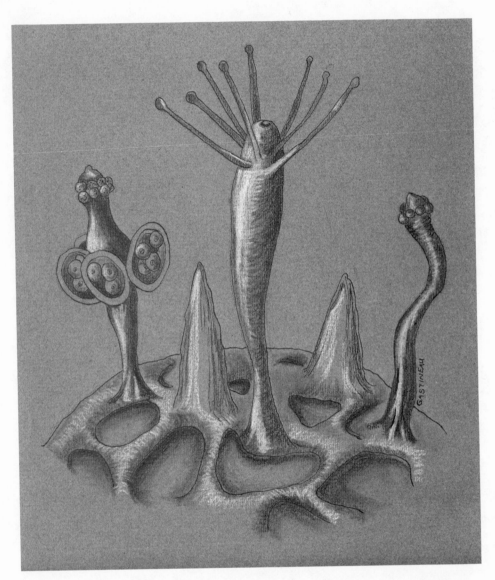

FIG. 5-18

Portion of colony of *Hydractinia*, with three types of individuals connected at base by network of stolons. Left, gonozooid. Center, gastrozooid, shown between 2 spines. Right, dactylozooid. (Suborder Gymnoblastea.)

short with no mouth or tentacles, although they possess nematocysts. They bear gonophores that produce sporosacs, as there is no medusa stage. Some zooids are male and others are female, but not in the same colony. The defensive zooids (dactylozooids) are extensible, bear nematocysts, and are most abundant near the margin of the colony. Some have short capitate tentacles; others are elongated and few in number.

Hydra. These athecate forms belong to the family Hydridae. There are fifteen species of this family in the United States, most of which are members of the genus *Hydra*. Hydras are solitary in habit, possess no medusoid stages, and are distinguished by complete absence of a

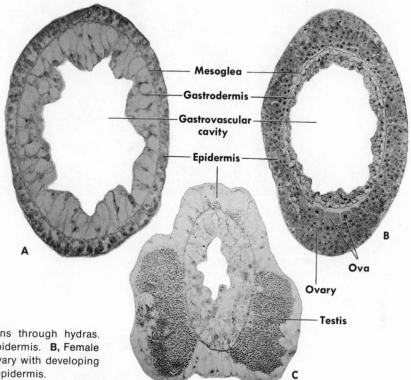

FIG. 5-19

Photomicrographs of cross sections through hydras. **A,** Normal individual with typical epidermis. **B,** Female with epidermis thickened to form ovary with developing ova. **C,** Male with testes formed in epidermis.

perisarc and the presence of hollow tentacles. When extended, the body varies in length from 2 to 25 mm. depending on the species. Hydras are strictly freshwater forms and are usually attached to a substratum of some kind (vegetation, leaves, stones, etc.) by a pedal disk of glandular cells.

The body consists of a long column and a distal region, the hypostome, that bears the mouth at its tip and a circlet of extensible tentacles at its base. Five or 6 tentacles are usually present, but there may be more and sometimes less than this number. The column is usually divided into a stout distal part where the stomach is located and a short proximal stalk that terminates in the pedal disk.

The body wall is composed externally of a layer of epidermis and internally of a layer of gastrodermis lining the gastrovascular cavity (Figs. 5-5 and 5-19, A). Between these two layers is the secreted noncellular mesoglea. Epithe-

liomuscular cells form most of the epidermis. The bases of these cells terminate in contractile fibrils (myonemes) next to the mesoglea and form a longitudinal muscular layer, or sheath. The nutritive-muscular cells of the gastrodermis are also provided with contractile fibrils that form a circular layer of muscle. Sensory cells are present in the epidermis, and interstitial cells are found in both epidermis and gastrodermis. The elongated sensory cells, which are generalized receptors for touch, temperature, etc., are connected with processes of the nervous system that form a nerve net lying at the base of the epidermis or in the mesoglea. The four kinds of nematocysts found in a hydra were described in the section on stinging organelles (Fig. 5-4). They are found throughout the epidermis except on the basal disk. Nematocysts are especially abundant on the distal column and on the tentacles. Each nematocyst organelle is contained in a modified interstitial cell, the

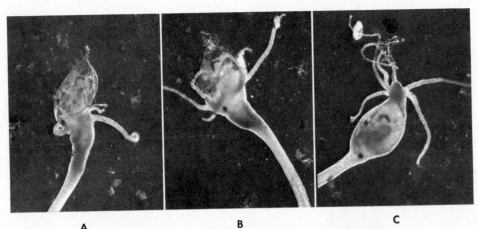

FIG. 5-20

Living hydra in process of eating *Daphnia*. **A** and **B**, The mouth stretches to accommodate the now paralyzed prey. **C**, With *Daphnia* safely swallowed, hydra has already captured a small protozoan on tentacle. (From Hickman: Integrated principles of zoology, The C. V. Mosby Co.)

FIG. 5-21

A, Hydra with testes. (From stained slide.) **B,** Living hydras, mostly males with testes. **C,** Living hydras, mostly females with mature ovaries and developing eggs. (**B** and **C** courtesy General Biological Supply House, Inc., Chicago.)

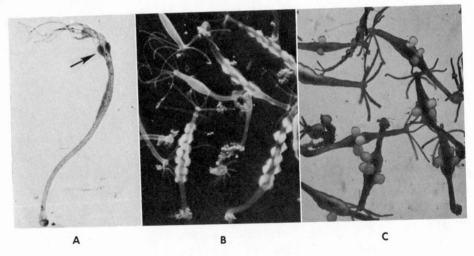

cnidoblast. Nematocysts are used as defensive and food-getting devices (Fig. 5-20).

Budding is the most common method of asexual reproduction. The first buds usually originate near the junction of the gastric and stalk region, and successive buds appear spirally toward the distal region. The bud, or new hydra, at first has a common gastrovascular cavity with the parent since it starts as an outpouching of the wall on the side of the parent. The bud later becomes cylindric, develops tentacles, and eventually pinches off at the base so that the new hydra becomes a separate individual. Hydras may also reproduce asexually by transverse or longitudinal fission, but such fission does not occur in normal individuals; it is found only in those that suffer unusual conditions of injury or lowered metabolic state.

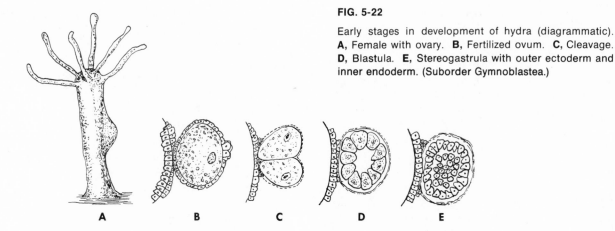

FIG. 5-22

Early stages in development of hydra (diagrammatic). **A,** Female with ovary. **B,** Fertilized ovum. **C,** Cleavage. **D,** Blastula. **E,** Stereogastrula with outer ectoderm and inner endoderm. (Suborder Gymnoblastea.)

A   B   C   D   E

Sexual reproduction is mainly seasonal, occurring in autumn or early winter when water temperature is dropping, or in spring or summer for certain species. Hydras may be either dioecious or hermaphroditic; most American species are dioecious. In hermaphroditic species the ovaries are proximal and the testes are distal to the gastric region. Sometimes the ovaries and testes of hermaphrodites develop simultaneously, or they may develop at different times. The gonads develop from cell aggregations in the epidermis and are temporary. Clumps of interstitial cells merge to form a single ovum in a rounded swelling called the ovary (Figs. 5-19, *B*, and 5-21, *C*); the testes (Figs. 5-19, *C*, and 5-21, *A* and *B*) are conical elevations (formed by the repeated divisions of an interstitial cell) provided in some species with a nipple through which the sperm escape. In dioecious species, testes or ovaries may be located throughout the gastric region. Males far outnumber females. Both eggs and sperm have two maturation divisions in their development.

When mature, the ovary ruptures through the epidermis, exposing the egg (Fig. 5-22), which remains attached throughout fertilization and early development. The egg may be fertilized by sperm from the same or from another hydra. During late cleavage (which is total and equal) the embryo secretes around itself the embryonic theca of chitin or other material. These thecated embryos are sticky, and when they drop off from the parent, they become attached to a substratum and complete their development. Such embryos are highly resistant to adverse conditions of drying and freezing. After a dormant period of several weeks the theca ruptures, and the juvenile hydra emerges equipped with tentacles and gastrovascular cavity. Sexual reproduction in many species may be induced by placing them in a refrigerator for 2 or 3 weeks; feeding should continue during this period.

Besides the fifteen species of hydra, the only freshwater coelenterates in the United States are a colonial hydroid, *Cordylophora lacustris* (Fig. 5-23), with attached medusoid gonophores, and a freshwater jellyfish with its polyp stage, *Craspedacusta sowerbyi* (Fig. 5-28).

Cordylophora. Cordylophora is a branching, colonial hydroid colony (Fig. 5-23) consisting of a basal hydrorhiza fastened to substrata of sticks and rocks. It has been found in certain inland rivers of the Middle West and in many other parts of the world. This form seems to prefer brackish inlets or estuaries, in which it reaches its largest growth. Its appearance is usually sporadic. The colonies attain a height of 10 to 100 mm.; the smaller sizes live in fresh water. The hydranths are naked, with 10 to 20 filiform tentacles arranged irregularly over them. Nervous elements are restricted to the ectodermal layer (R. K. Jha, 1965).

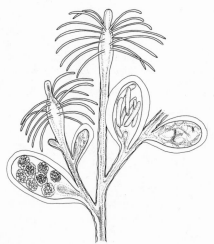

FIG. 5-23

Cordylophora lacustris with attached medusoid gonophores. This freshwater hydroid forms colonies 1 to 2 inches high. (Suborder Gymnoblastea.)

FIG. 5-24

Some calyptoblastic hydrozoans. **A,** *Sertularia cornicina;* left, portion of stem and one gonophore; right, natural size. **B,** *Campanularia verticillata;* left, portion of stem with hydrothecae and gonophores; right, natural size. **C,** *Plumularia corrugata;* left, portion of stem with small hydrothecae and large female gonophores; right, about natural size. (Modified from Fraser, 1937.)

The gonophores or sporosacs are bulbous bodies located below the hydranths. Few gonophores are produced on colonies in fresh water. Mature ova remain on the female gonophores, where they are fertilized by sperm that swim from the male gonophores. The embryo develops in situ on the gonophore to the planula stage and then escapes to form a new colony. There is no true medusa. Most of the hydranths disintegrate in the fall or early winter, leaving only a few basal portions of the colony to start new colonies in the spring.

Suborder Calyptoblastea (Leptomedusae)

The Calyptoblastea include those hydroids that have a protective periderm or perisarc cup around the hydranths (hydrothecae) and around the gonozooids (gonothecae). Most members of this suborder have no free medusae. The Leptomedusae, or jellyfish of this group, usually possess flattened or saucer-shaped bells with gonads on the radial canal, statocysts, and numerous tentacles. Most of them lack ocelli on the tentacular bulb. Some of the more common members of this group follow.

Plumularia. This colonial form originates from a creeping hydrorhiza and is made up of

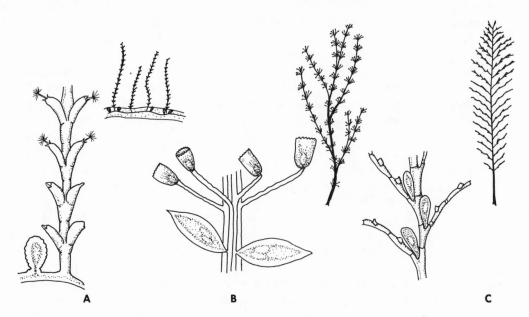

A   B   C

plumelike, feathery branches bearing small sessile, bell-like hydrothecae arranged on one side only (Fig. 5-24, *C*). It also bears nematophores provided with nematocysts and adhesive cells that are used to ensnare prey. The hydrothecae are too small for the polyps to be completely retracted within them.

Sertularia. In this colony the hydrothecae are arranged on opposite sides of the branching stems (one pair to each internode), and they are large enough to enclose the polyps when retracted (Figs. 5-24, *A* and 5-25). The colony may grow to a height of about ½ to 2 inches. The gonothecae are large and vaselike. The lid, or operculum, of the hydrotheca is made up of 2 flaps. These forms are most commonly found on floating seaweeds.

Campanularia. The cup-shaped hydrothecae are borne at the ends of branches in this form (Fig. 5-24, *B*). Gonophores are sessile. Medusae are provided with lithocysts on the margin of the

FIG. 5-25

Sertularia colonies attached to bit of seaweed (preserved) (suborder Calyptoblastea).

umbrella, and gonads are attached to the radial canals. In some species the gonangia are much larger than the hydrothecae. Colonies are usually ½ to 1 inch in length. *Campanularia* differs from *Obelia* in not producing free medusae.

Obelia. This colonial genus contains many species and has a wide distribution throughout the world. Colonies are attached to rocks and shells sometimes as deep as 40 fathoms (1 fathom = 6 feet). This animal is one of the most frequently studied in general zoology courses.

The colony (Fig. 5-8) is usually quite bushy and may grow to a height of from less than an inch to more than a foot. The branched stem bears flowerlike hydranths, with gonangia arising from axils of the branches. The medusae are free swimming and originate from medusoid buds on the blastostyle, the central stalk of the gonangium. The branching stem is attached to the substratum by a rootlike hydrorhiza. The stem consists of an inner, hollow coenosarc with epidermal, mesogleal, and gastrodermal layers. The coenosarc cavity is a part of the gastrovascular cavity, which is continuous throughout the colony. Around the stem is the horny perisarc. The hydranths are the asexual polyps and resemble hydras in structure. Each hydranth is enclosed in a cup-shaped or vaseshaped hydrotheca, which is a continuation of the perisarc. Each hydranth or polyp has about 20 solid tentacles.

The gonangium is the reproductive polyp; it is made up of a central clublike blastostyle enclosed in a perisarc sheath, the gonotheca. The blastostyle bears on its surface many small medusoid buds. When mature, these medusoid buds escape through the constricted opening, the gonopore, at the top of the gonotheca. The sexes of the medusae are separate in *Obelia*.

The bell of the medusa is about 1 mm. in diameter when the medusa escapes from the blastostyle and usually bears 16 to 24 tentacles. When mature, the medusa may attain a diameter of 5 mm. and possess many tentacles, but the velum is rudimentary (Fig. 5-26). Sperm and eggs are shed into seawater when the jellyfish become sexually mature. The zygote develops into a ciliated planula larva that attaches to a substratum, develops a mouth with tentacles, and becomes a polyp. From this asexual hydroid form develop the stolons, stems, and other polyps to start a new colony.

■ ORDER MILLEPORINA

The members of the order Milleporina are the stinging corals of shallow tropical seas. They are rarely found below a depth of 30 meters. Their structure is similar to that of hydroids with the addition of a calcareous skeleton (coenosteum), and they are often brightly colored. They make up part of the fauna of coral reefs and form leaflike calcareous growths that attain a height of 1 to 2 feet. Their skeleton is covered with a thin layer of tissue pitted with pores, through which the bodies of the polyps are extended. These pores open into cavities and interconnecting tubes crossed by horizontal calcareous plates. The minute polyps are said to cover the coenosteum like a layer of white

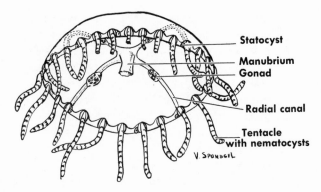

FIG. 5-26

Medusa of *Obelia*. Although hydrozoan, velum is rudimentary (suborder Calyptoblastea).

FIG. 5-27

A, Group of polyps of *Millepora* (order Milleporina). Gastrozooid is surrounded by 5 dactylozooids (2 are shown cut off). **B,** Portion of surface of dry *Millepora,* magnified to show 3 gastropores surrounded by smaller dactylopores. **C,** Portion of corallum of *Astylus* (order Stylasterina). Each cyclosystem on branch has deep central gastropore surrounded by circlet of slitlike dactylopores. This genus, as its name implies, lacks spines.

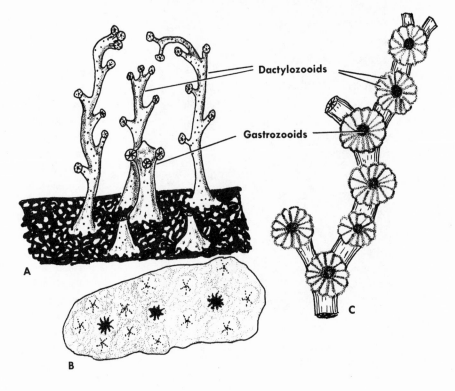

felt. The two types of polyps are the gastrozooids, with knobby tentacles and a mouth, and the dactylozooids, which bear two or three kinds of nematocysts and short, capitate tentacles but no mouth. Gonophores bud from the coenosarc tubes and develop medusae that disintegrate after releasing sex cells. *Millepora* (Fig. 5-27, *A* and *B*) is the only genus in this order.

■ ORDER STYLASTERINA

The Stylasterina resemble the Milleporina and are often included with them in a single order, the Hydrocorallina. They are found mostly in tropical seas and are brightly colored. Like the millepores, their calcareous skeleton is covered by tissue that bears pores opening into ramifying tubes. The gastrozooids are found in deep cups along the edges of the branches or only on one face of the colony (Fig. 5-27, *C*). Gastrozooids also differ from those of millepores by having in most genera a spine, or style, at the base of each cup. The other type of polyp (dactylozooid) is without tentacles or mouth, but is provided with nematocysts. Eggs and sperm are produced by sporosacs on the surface of the colony, from which the planula larvae escape. The order is represented by such genera as *Stylaster, Stylantheca* (California), and *Cryptohelia*.

■ ORDER TRACHYLINA

Members of the order Trachylina are medusae that differ from other hydrozoan jellyfish in possessing gastroderm lithocysts, or balancing organs (Fig. 5-28), and in almost completely lacking the polypoid stage.

The order Trachylina is of great interest in the evolutionary development of coelenterates because their direct mode of development (planula-actinula-medusa) is supposed to indicate a primitive condition. From the primitive actinula larva other larvae may have budded off and, in time, developed into the polyp or hydroid form, which, with many evolutionary modifications, became an established part of their life history.

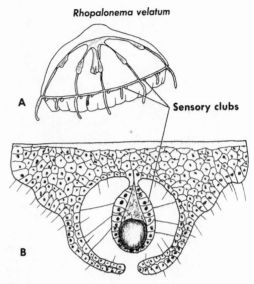

FIG. 5-28

A, *Rhopalonema velatum,* a trachymedusan, showing sense organ at right of each tentacle base. **B**, Sensory club with single concretion nearly enclosed in capsule (*Rhopalonema*). In young medusa, sensory club is free. Later, ectodermal capsule grows from margin and finally completely encloses club.

The four kinds of medusae found in this order are sometimes considered as separate suborders. They are the Trachymedusae, the Narcomedusae, the Pteromedusae, and the Limnomedusae.

Suborder Trachymedusae

Most of the Trachymedusae are found in the open seas and have solid tentacles. *Liriope* is an example (Fig. 5-29). Their gastrovascular system contains 4 to 8 radial canals beneath which are the epidermal gonads. Along the margin of the bell are the statocysts. Development is direct—the egg gives rise to a swimming planula that becomes an actinula larva by the formation of a mouth and tentacles. The actinula transforms directly into a medusa.

Suborder Narcomedusae

Narcomedusae differ from Trachymedusae in having lobed stomachs, tentacles attached

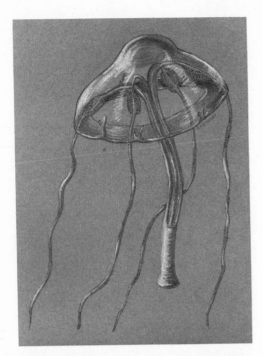

FIG. 5-29

Liriope scutigera (suborder Trachymedusae) lacks hydroid stage. Bell is about ¾ inch in diameter, with pink tentacles and yellow-green manubrium. Common along Atlantic coast.

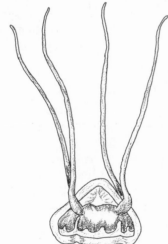

FIG. 5-30

Aegina citrea (suborder Narcomedusae) with tentacles attached above bell margin. Deep-water form (order Trachylina). Hardy reports that in an aquarium it remains almost motionless, with tentacles stretched upward. (×2½.)

above the margin of the bell, and an absence of peripheral canals (Fig. 5-30). They also have lithostyles or sense clubs hanging from the bell margin, and their gonads are located in the gastric pockets.

Suborder Pteromedusae

Pteromedusae are represented by *Tetraplatia*, an unusual medusa found in marine plankton. It is slender and bipyramidal in shape, with a groove about the equator from which extend four swimming lobes, each with two statocysts. These lobes apparently represent the velum. The upper pyramid is the bell; the lower one is the manubrium, with the mouth at the lower pole.

Suborder Limnomedusae

The Limnomedusae are considered by some to be a suborder of the Hydroida and by others to belong to the Trachylina.

The Limnomedusae may have hollow or solid tentacles and a minute polyp stage. The polyps of Limnomedusae give off nonciliated planula-like frustules that creep about and develop into polyps. Medusae also arise from gonophores that are budded from the sides of the polyps. Two common Limnomedusae are *Gonionemus* (Fig. 5-31) and the freshwater form *Craspedacusta* (Fig. 5-32). *Gonionemus* has hollow tentacles that bear adhesive pads for clinging to seaweeds (Fig. 5-2); *Craspedacusta* has solid tentacles.

The histology of the musculature of *Gonionemus* (L. A. Fraser, 1962) may be typical of that found in many types of medusae. He describes a layer of circular fibers (attached to overlying epitheliomuscular cells) in the subumbrella and the upper side of the velum. Some 4 to 6 muscle fibers are attached to each of the epidermal cells. Radial fibers are present in the tentacles, manubrium, and lateral walls of the radial canals. Several radial fibers are attached to each of the epidermal cells. The mesoglea on which these radial muscles rest is found in ridges for increased surface for muscle attachment.

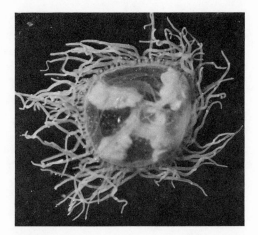

FIG. 5-31

Gonionemus, exumbrella view, preserved specimen (suborder Limnomedusae). Note well-developed gonads beneath radial canals and adhesive pads on tentacles. Polyp stage of *Gonionemus* is small, buds off frustules to form polyps, and has gonophores that give rise to medusae.

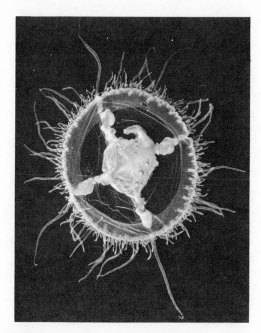

FIG. 5-32

Craspedacusta sowerbyi, a freshwater limnomedusan (preserved specimen); 15 to 20 mm. in diameter. Resembles *Gonionemus,* but tentacles are more varied in length and lack adhesive pads. (Courtesy Carolina Biological Supply Co., Burlington, N. C.)

Craspedacusta sowerbyi is the only freshwater medusa, or jellyfish, in North America. It has been found sporadically in artificial bodies of water located in most of our states. This jellyfish appears to have a wide distribution and may have originated from marine ancestors in certain river basins of China. A mature medusa has the typical bell shape and is about 15 to 20 mm. in diameter (Fig. 5-32). There are from 50 to 400 or 500 tentacles arranged in three (sometimes more) sets. The bell margin is provided with many statocysts, and there is a thick velum. Nematocysts of one kind are abundant around the mouth, at the margin of the bell, and on the tentacles. Swimming movements are produced by rhythmic contractions of the bell. Most of the medusae of a locality are of one sex only (monosexual populations). The 4 gonads are convoluted masses on the subumbrella on or near the radial canals. Eggs and sperm are discharged into the water, where fertilization occurs.

The zygote develops into a colony of just a few individual polyps. The naked polyps, or hydranths, are only about 0.5 mm. long, are devoid of tentacles, and can creep around on a substratum to feed on small organisms. Debris often collects on the polyp because of a sticky secretion. A colony of polyps may reach a size of 2 or 3 mm. Three methods of asexual reproduction occur in the hydroid stage (Fig. 5-10). By budding new polyps are formed from the parent and may remain attached to form a colony; or planula-like buds (frustules) may constrict off and develop into polyps that creep about and form a colony. A third method is the formation of polyp buds that develop into free-swimming medusae. C. F. Lytle has shown that the production of the three kinds of buds is differentially affected by feeding rates. At high feeding rates, for example, more frustules and fewer of the other two kinds of buds are formed. The medusa at first is very small and has only 8 tentacles, but the number of tentacles increases rapidly as the medusa matures.

The feeding habits of the polyp have been

reported by a number of investigators. Their prey appears to be small worms, algae, larvae, and crustaceans. J. H. Bushnell, Jr., and T. W. Porter (1967) have described the cooperative feeding activities of two polyps eating a tendipedid (midge) larva. After immobilizing the larva with the large nematocysts of their capitula, each polyp takes hold by its mouth of one end of the larva. As the process of ingestion and digestion proceeds the two polyp mouths move closer together until the capitula of the two polyps meet (Fig. 5-33). In this "kissing" posture they may remain for some time while digestion proceeds.

■ ORDER SIPHONOPHORA

Members of the order Siphonophora are large floating colonies of modified polypoid and medusoid components; several different types exist. The various components may be connected by a stem or coenosarc, or they may be compactly grouped together. Such colonies represent the highest degree of polymorphism to be found among the coelenterates. It is thought that siphonophores evolved from trachyline ancestors with a swimming actinuloid

stage and an adult medusoid stage, and that the merger of these two stages produced the siphonophoran colony. Both polypoid and medusoid components are found in several modifications (Fig. 5-34), but few are strictly hydroid forms. The commonest polypoid components are (1) gastrozooids, each with a mouth and a hollow tentacle armed with nematocysts for the capture and the digestion of food; (2) palpons, similar to gastrozooids but simpler, serving as accessory digestive organs; and (3) gono-

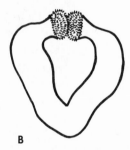

FIG. 5-33

A, Two polyps of a colony of *Craspedacusta sowerbyi* cooperatively devouring a tendipedid larva (midge). It has been immobilized by nematocysts from capitula of polyps and each end seized by one of the polyps. **B,** Polyps' mouths move closer together until capitula meet in the "kissing posture," where they remain until digestion of midge is completed. After separating, portions of larval exoskeleton are discharged through mouths of the polyps. (From Bushnell, J. H., and T. W. Porter. 1967. Trans. Amer. Microscop. Soc. **86:**22-27.)

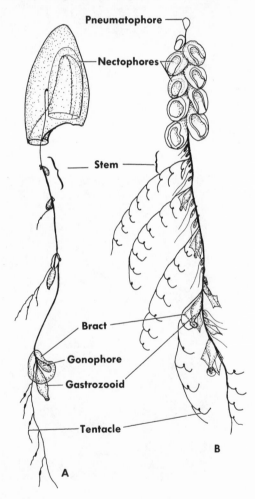

FIG. 5-34

Two types of siphonophore colony. **A,** Calycophora. Colony has swimming bells (nectophores) but no float. **B,** Physonectae. Colony has both bells and float (pneumatophore).

zooids with or without a mouth but with no tentacle and bearing clusters of gonophores for reproduction. The medusoid components are made up of (1) swimming bells (highly modified medusae without mouth and tentacles) for locomotion; (2) bracts (thick, gelatinous, many-shaped structures with a gastrovascular cavity) for streamlining, protection, and flotation; and (3) gonophores, basically medusa-like, for carrying the sex cells. The float or pneumatophore, formerly considered to be an altered medusa, is now regarded as an inverted polyp, and the bracts may be altered tentacles instead of altered medusoids. On the basis of swimming bells and floats, the Siphonophora are often divided into the following groups: Calycophora with swimming bells (nectophores) (Fig. 5-34, *A*), Physonectae with both floats and swimming bells (Fig. 5-35), and Cystonectae

with only floats (Fig. 5-36). *Stephalia* (Fig. 5-35), a classic example of a form with swimming bells, belongs to the Physonectae.

One of the most familiar of siphonophores is the Portuguese man-of-war (*Physalia*, Fig. 5-36). This colonial form (a cystonect) has a float that may reach a diameter of 12 inches or more. The stem is rudimentary beneath the float. A gas gland secretes into the float a gas composed of nitrogen, oxygen, carbon monoxide, and argon but varies greatly in composition from specimen to specimen. This form lacks the swimming bells and bracts characteristic of many siphonophores, but it possesses the other components. Lacking swimming bells, *Physalia* is carried along by wind and wave action. Whenever a fish comes in contact with

FIG. 5-35

Stephalia (family Physonectae of order Siphonophora). Note ring of swimming bells (nectophores) at base of float.

FIG. 5-36

Physalia physalia, the Portuguese man-of-war, is best known of siphonophores. This one has caught fish. (Courtesy New York Zoological Society.)

the dangling, trailing dactylozooids, it is paralyzed by the nematocysts and drawn up by the tentacles to the region of the gastrozooids. The lips of the latter spread over the fish and form a type of bag in which digestion begins. Nematocysts are large, numerous, and very virulent. They are found on most members of the colony, although the float and parts of the gastrozooids are deficient in mature nematocysts. Each of the cnidoblasts is provided with a cnidocil. The nature of the toxin found in the nematocysts has been studied by C. E. Lane, who found it to be a simple protein consisting of a few toxic peptides. He also found that the toxin is synthesized by gastroderm cells and passes to the nematocyst (in the epidermis) through the mesogleal layer. *Physalia* is very dangerous to bathers in southern waters.

The Siphonophora represent true super organisms, for their specialized zooids are interdependent individuals serving as organs of the superorganism.

Regarding the siphonophores, G. O. Mackie (1963) makes the interesting observation that they are the most complex coelenterates to have fully explored the possibilities of colonialism. By doing so, they have escaped some of the limitations and restrictions imposed by a diploblastic body plan. By converting whole individuals into organs, siphonophores have reached the organ grade of construction. Higher animals have reached this higher grade by using mesoderm (lacking in siphonophores) in a triploblastic arrangement for the formation of organs.

■ ORDER CHONDROPHORA

Another group of polymorphic floating colonies is the Chondrophora, which includes *Velella* and *Porpita*. Although this group is usually included under Siphonophora, some taxonomists now consider it to be an order in its own right or a suborder of the Hydroida. These colonies are probably the most modified polymorphic forms of all. *Velella*, for example, has a flat, disklike float with air chambers and

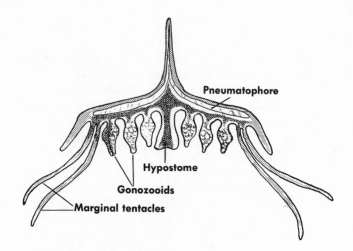

FIG. 5-37

Section of *Velella* (diagrammatic), the purple sail (order Chondrophora).

a vertical extension that acts as a sail (Fig. 5-37). The stem is a flat coenosarc forming part of the float. Beneath the disk is a large central gastrozooid surrounded by gonozooids that bear the medusiform gonophores. On the margin of the disk are the dactylozooids, or fringing tentacles. The gastrovascular cavities of the various zooids communicate with gastrodermal canals above the central gastrozooid, forming a type of excretory system. The air chambers of the disk open to the outside by pores and canals. Gonophores are set free in reproduction, but they quickly disintegrate after the discharge of the sex cells.

■ CLASS SCYPHOZOA

The class Scyphozoa consists of members that have emphasized the medusa stage in the typical coelenterate life history of polypoid and medusoid structures. They include the schyphomedusae and medusalike polypoids with distinct 4-part radial symmetry (tetramerous) and a gastrovascular cavity divided by 4 longitudinal septa. The polypoid stage is usually reduced (scyphistoma) or absent. Diversity in

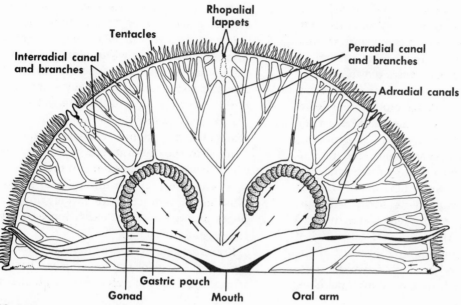

FIG. 5-38

Portion of *Aurelia*, oral view, showing direction of circulation in ciliated canal system of gastrovascular cavity. Food-laden water is drawn through mouth into gastric cavity, then into the 4 gastric pouches, and from there to bell margin and tentacles through straight adradial canals. It returns by way of branched interradials and perradials. Ciliated tracts also move particles along oral arms.

their life histories is shown by some that are polyplike and sessile throughout their life cycle, and by others that are always pelagic and lack the sessile polyp stage. In general the scyphomedusae are of large size and the scyphopolyps are of small size. Scyphomedusae differ from hydromedusae in possessing a mesoglea with cells (a true cellular layer), gastric tentacles or oral lobes about the mouth, division of the gastrovascular cavity into a central stomach and 4 gastric pouches, scalloped or notched bell margin containing sensory organs and no velum. In some members, including *Aurelia*, a velarium that is analogous to the velum exists.

The schyphomedusae vary in shape in the different groups of Scyphozoa; they are dome shaped, cone shaped, or saucer shaped. The

umbrella, as in hydromedusae, has an upper convex surface (exumbrella) and a lower concave surface (subumbrella). Nematocysts are found in both surfaces of the bell in definite regions.

Tentacles may be solid or hollow and are usually definite in number (few or many) for a species. They are provided with numerous nematocysts of two or three types. In some of the orders 4 interradially located subgenital pits, or shallow depressions, located on the subumbrellar surface may function in respiration.

The endoderm-lined manubrium with its terminal mouth is found in the center of the subumbrella. The mouth is four cornered, and each of its angles may be drawn into a frilly oral arm, which is liberally supplied with nematocysts. (The oral arms are fused and the central mouth is closed in the order Rhizostomae; in place of the mouth many suctorial mouths open by canals into the interior.) The mouth leads to the large gastrovascular system. Usually 4 septa composed of gastrodermis and mesoglea divide the gastrovascular cavity into a central stomach and 4 perradial gastric pouches. Gastric filaments of mesogleal and gastrodermal derivation are found on the free inner edge of each septum. The filaments bear glands and nematocysts. In some orders this arrangement of the gastrovascular system is not found in adults but is restricted to the scyphistoma, or larval stage. From the periphery of the stomach many radial canals, simple or branched, run to the bell margin, where they join the ring canal, if present (Fig. 5-38).

Many medusae are suspension feeders. Aurelia, by pulsating movements of the bell, rises toward the surface, and then as it sinks it traps planktonic forms in its surface mucus. Ciliary tracts on the exumbrella move the food to the bell margin. Here the oral arms lick it off and it moves along ciliated grooves of the arms to the mouth and stomach. From the stomach food is drawn into the gastric pouches.

Here gastric filaments secrete digestive enzymes, and nematocysts are probably used to quiet the prey. Flagellar action moves the resulting fluids outward along the adradial canals to the ring canal where it moves in both directions to reach the tentacles and rhopalia. Fluids return by way of the interradial and perradial canals. (See Fig. 5-38.)

Muscles are mainly epidermal and are restricted to the subumbrella and tentacles. The coronal, or ring, muscle band of the subumbrella serves to contract the umbrella in swimming. Muscles of tentacles are largely longitudinal fibers and are used in extension and contraction in catching prey.

Present evidence seems to indicate that there are 2 nerve nets in scyphomedusae. One of these is the main subumbrellar net, or plexus, which

FIG. 5-39

Structure of rhopalium, or sense organ, of *Aurelia* (diagrammatic). **A,** *Aurelia,* with part of body removed, showing location of rhopalia. **B,** Outer or aboral view of rhopalium. **C,** Radial section of rhopalium.

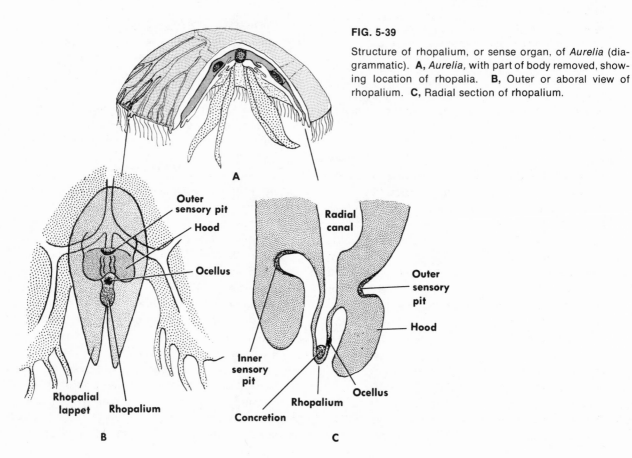

controls pulsation of the bell by activating simultaneously the muscles necessary for locomotion. The other, more diffused nerve net, with cell bodies in both subumbrellar and exumbrellar epidermis, controls local reactions of food ingestion and may also inhibit the bell pulsations. Investigations seem to indicate that the nervous system is synaptic and not continuous. There is no marginal nerve ring such as that found in hydromedusae except in one order, the Cubomedusae.

In most orders the bell margin is notched or scalloped, with a small pair of lappets in each notch (Fig. 5-38). Sense organs, or rhopalia, usually four or a multiple of four, are found between the lappets (Fig. 5-39). Each rhopalium is composed of a statocyst containing a concretion (statolith), a pair of sensory pits, and

sometimes an ocellus that may contain the typical parts of the vertebrate eye (cornea, lens, retina, etc.).

Most Scyphozoa are dioecious. The gonads are formed from endodermal (gastrodermal) cells just below the gastric filaments. In medusae with septa the gonads are arranged on both sides of each septum, forming 8 gonads

FIG. 5-40

Life cycle of *Aurelia*. After fertilization of eggs by sperm from another medusa, zygotes develop through early stages on oral arms of female. A ciliated Planula larva swims away and becomes an attached scyphistoma that, by transverse fission, develops into strobila stage made up of several saucerlike ephyrae. The ephyrae detach, swim around, and develop into mature jellyfish. (Order Semaeostomae.)

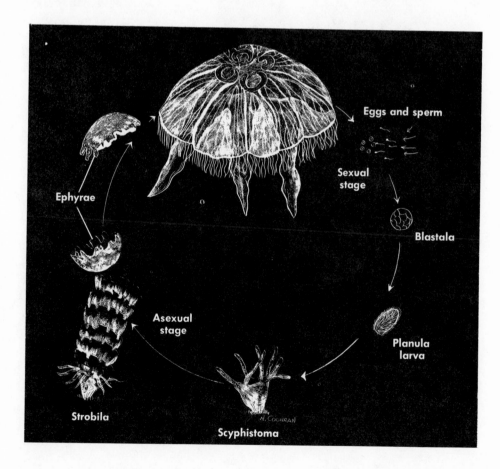

altogether. In most medusae the ripe sex cells rupture into the gastrovascular cavity and reach the outside through the mouth. The zygotes may develop, in seawater or in the folds of the oral arms, into coeloblastulae at first and later into ciliated planulae (Fig. 5-40). After attachment

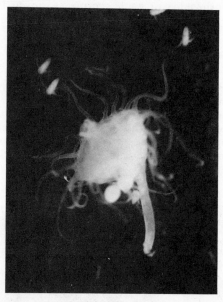

FIG. 5-41

Scyphistoma of *Cassiopeia*. Buds shown at left of the stalk develop into pseudoplanulae that give rise to more scyphistomae. In the spring the scyphistoma will bud off ephyrae one at a time (monodisk strobilation). Ephyra has an evenly scalloped edge with 16 or more rhopalia. (Order Rhizostomae.)

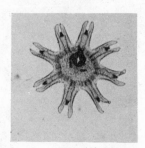

FIG. 5-42

A typical scyphozoan ephyra with 8 bifurcated arms and 8 rhopalia.

the planula develops into a scyphopolyp, or scyphistoma (Fig. 5-41) that undergoes asexual budding and finally matures into a strobila (Fig. 5-40). The strobila asexually produces ephyrae (Fig. 5-42), which are shed as young, free-swimming medusae and grow into adult medusae. Radial cleavage is characteristic of the group.

The five living orders of Scyphozoa are Stauromedusae, Cubomedusae, Coronatae, Semaeostomae, and Rhizostomae. A brief description of these orders and their common genera is herewith presented.

■ ORDER STAUROMEDUSAE

Stauromedusae are sessile scyphozoans that possess a type of combined polyp and medusa form. The medusoid part is made up of a cuplike bell (calyx) that is eight sided and bears eight groups of capitate tentacles and eight sensory bodies (rhopalioids or anchors). The polyp part is usually a stem that terminates in a pedal disk, which is attached to a substratum. The whole often forms a trumpet-shaped or goblet-shaped body. The four-cornered mouth is situated at the center of the calyx, and on each side of the quadrangular manubrium is located

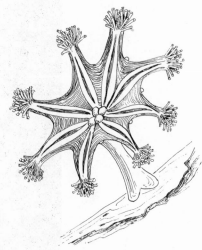

FIG. 5-43

Haliclystus, the "clown" (order Stauromedusae).

a subumbrellar funnel into the interior. The gastrovascular cavity is continued into the arms, tentacles, and rhopalioids. Although sessile, most Stauromedusae can move about by attaching and releasing their pedal disks and anchors.

Haliclystus, or the "clown" (Fig. 5-43), is one of the most familiar genera of this order.

■ ORDER CUBOMEDUSAE

The bell of the Cubomedusae has four flattened sides and a relatively cuboid form. A gelatinous, leaf-shaped blade, or pedalium, is present at the base of each exumbrellar ridge. Each pedalium bears a tentacle or cluster of tentacles provided with nematocysts. Above the margin in the center of each flat side is located a sensory organ (rhopalium) (Fig. 5-44, A). Each rhopalium has a statocyst and 6 ocelli. The short, quadrangular manubrium that connects the mouth with the stomach is present in the center of the subumbrellar cavity. An extension of the subumbrella forms a velarium.

The nervous system features a marginal nerve ring that connects with the sensory organs, an arrangement not found in other Scyphozoa. Reproduction is typically scyphozoan, and there may be a polypoid stage — such a stage has been described in some members. The best known genera are *Carybdea* (Fig. 5-44, A) and the "fire medusae," *Chiropsalmus.* Most members of the group can give dangerous stings. More than fifty fatalities due to stings from these have been reported from the Australian coastline. Most dangerous species prefer shallow water and are especially dangerous to bathers (J. H. Barnes, 1966).

■ ORDER CORONATAE

Medusae of the order Coronatae are dome shaped, conical, or flattened, but the exumbrella

FIG. 5-44

Three scyphomedusans. **A,** *Carybdea* (order Cubomedusae). **B,** *Periphylla* (order Coronatae). **C,** *Chrysaora* (order Semaeostomae).

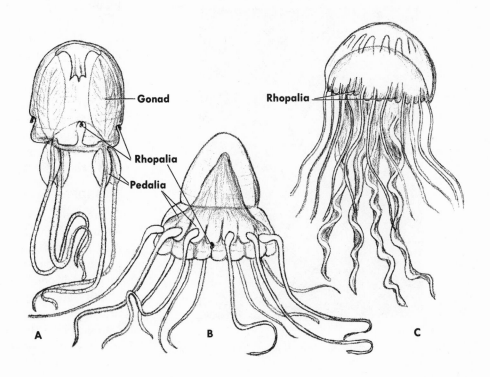

is divided by a horizontal coronal furrow into two parts — an upper domelike disk and a lower coronal part. The coronal part has thick, gelatinous pedalia separated from each other by grooves that point toward the centers of marginal lappets. Usually each of the pedalia is provided with a tentacle. Sensory organs (rhopalia) are found at the ends of those pedalia that have no tentacles. There are 2 stomachs, an upper central stomach and a coronal stomach, which communicate through gastric ostia.

The members of this group are found chiefly in deep marine waters (abyssal) and usually are known only from deep-sea dredgings. Some members, however, such as *Nausithoe,* are found in the surface waters of warm seas. *Periphylla* (Fig. 5-44, *B*), is a deep-sea form.

■ ORDER SEMAEOSTOMAE

Medusae of the order Semaeostomae are the most familiar ones of the Scyphozoa and are

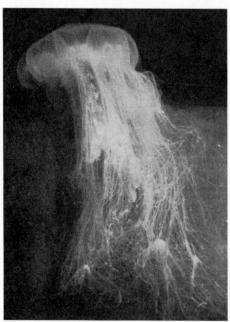

FIG. 5-45

Cyanea, the largest jellyfish (order Semaeostomae). (Courtesy C. P. Hickman, Jr.)

the ones commonly studied in zoology classes. The bell is flat to dome shaped, and its margin is scalloped into many lappets. In the niches between the lappets are the sensory organs (rhopalia). Tentacles are distributed in different places according to species. They are located along the exumbrellar margin in *Aurelia;* in the grooves between the lappets in *Pelagia;* and on the subumbrella in *Cyanea* (Fig. 5-45). Tentacles are short in *Aurelia* (Fig. 5-40), but long in most genera. The oral arms are conspicuous structures in most members, being frilly, leaf-like, and provided with ciliated grooves and nematocysts. The stomach is cross shaped and lacks septa and gastric pouches in the adults, but has numerous gastric filaments. From the stomach numerous simple or branched radial canals run to the bell margin, where they connect with sense organs and tentacles. A ring canal is absent in most members. Gonads are located in the floor of the stomach and discharge their products through the mouth.

The life history is the typical pattern of planula, scyphopolyp, strobila, and ephyra (Fig. 5-40). All are dioecious except the striking and beautiful *Chrysaora* (Fig. 5-44, *C*). In *Pelagia* the planula transforms directly into an ephyra medusa.

Most Semaeostomae are coastal forms, and the order includes the largest coelenterate, *Cyanea* (Fig. 5-45). Specimens of this genus taken from northern waters have bells up to 8 feet in diameter and 800 or more tentacles that may reach 100 to 200 feet in length.

■ ORDER RHIZOSTOMAE

The chief distinctive features of the Rhizostomae are the fusion of the oral arms and the lack of marginal tentacles. The bell varies in shape from a distinct bowl or dome to those that are flat and concave on top. The central mouth is usually obliterated by the fusion of the 8 oral arms. In its place are many suctorial mouths connected to the complicated brachial canals that run to the central stomach. From the cruciform stomach numerous radial canals in the

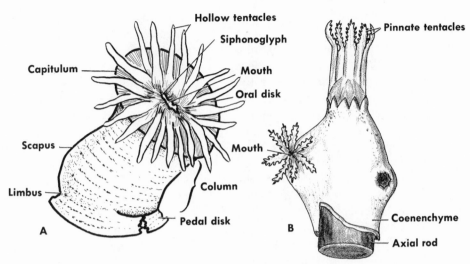

FIG. 5-46

Anthozoan polyps. **A,** Sea anemone *Bunodosoma* (subclass Zoantharia). **B,** Portion of soft coral *Gorgonia* (subclass Alcyonaria, order Gorgonacea).

FIG. 5-47

Cross section (diagrammatic) through pharynx of *Alcyonium* to show octomerous septal arrangement typical of alcyonarian polyps.

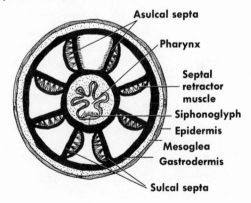

form of networks run to the periphery of the bell, which may or may not have a circular canal. Filamentous appendages loaded with nematocysts are found on the arms. Life histories are similar to those of the Semaeostomae.

Cassiopeia is the most familiar American form and is common around Florida. *Rhizostoma* has a wide distribution in shallow waters of the tropical and subtropical regions.

■ CLASS ANTHOZOA

The Anthozoa are characterized by having an exclusively polypoid, attached form. They may be solitary or colonial and include the sea anemones, sea fans, sea pansies, sea pens, corals, and some others. Externally the polyp exhibits the characteristic radial symmetry of the coelenterate, but internally it presents a distinct bilateral tendency that is definitely primary, as revealed by the embryologic record. According to their symmetry, Anthozoa may be divided into those that are octamerous, or built around a plan of eight (Alcyonaria, the soft corals, Figs. 5-46, *B*, and 5-47) and those that are hexamerous or polymerous, or built on a plan of 6 or multiples of 6 (Zoantharia, the sea anemones and hard corals [Figs. 5-46, *A*, and 5-54]).

The general morphology of the group includes the oral disk with an oval or slitlike mouth in its center. The mouth bears a ciliated groove at one or both ends, the siphonoglyph, and opens into the gastrovascular cavity by the pharynx, the lining of which appears to be ectodermal. The siphonoglyph aids in driving water into the gastrovascular cavity for respiratory purposes. The tentacles vary in the two subclasses. In the Alcyonaria 8 pinnate tentacles arise from the margin of the oral disk (Fig. 5-46, *B*). These rather thick tentacles bear short projections, the pinnules, arranged in a row on each side of the tentacles, giving the whole a feathery appearance. In the zoantherians the simple tubular tentacles are arranged in one or more circles to form a tentacular crown (Fig. 5-46, *A*). A pedal disk is found at the basal end of the body. That part of the body between the pedal and oral disks is called the column.

In anthozoans, the gastrovascular cavity is partitioned by longitudinal mesenteries, or septa, provided on one side with longitudinal retractor muscles. These mesenteries are actually vertical folds of the mesoglea and gastrodermis of the body wall, radiating inward toward the pharynx, and connecting to the oral disk at the top. Septa that reach from the wall to the pharynx are called complete, or primary, septa. *Metridium* (Figs. 5-53 and 5-54) usually has six pairs of primary septa. Pairs of primary septa associated with the siphonoglyphs are called directives. Below the pharynx the inner edges of the septa are free. Between the pairs of primary septa may be successively shorter cycles of paired incomplete septa that do not reach to the pharynx, known as secondary, tertiary, and quarternary septa. The free edge of each incomplete mesentery forms a type of sinuous cord called the septal filament (Figs. 5-53 and 5-55), which bears nematocysts and gland cells for digestion. In some forms 2 septal filaments (asulcal) are ectodermal derivatives from the stomodeum and are provided with flagellate cells for driving water toward the mouth (Fig. 5-47). In many anemones the lower ends of the septal filaments are prolonged as free threads (acontia), which may protrude through the mouth or through the cinclides (pores in the column for the emission of water) when the body contracts.

The mesoglea in Anthozoa is a mesenchyme consisting of a jelly matrix and stellate cells.

The muscular system of epidermal longitudinal fibers and gastrodermal circular fibers varies in the different orders but is usually not well developed. Septa possess both circular gastrodermal fibers and longitudinal gastrodermal fibers (retractors).

Skeletal structures (when present) are external or internal. The external skeleton, such as the calcareous exoskeleton of corals, is formed from the secretions of ectodermal cells. Internal skeletons, such as those of the soft corals, are sclerites or calcareous spicules secreted by ectodermal scleroblasts in the mesoglea.

The nervous system is made up of subepidermal and gastrodermal plexuses, which in most cases form a continuous nervous connection.

Both sexual and asexual reproduction occurs in anthozoans. Gonads develop on all septa except the two asulcal ones. Sex cells are formed by the interstitial cells of the gastrodermis. Both dioecious and monoecious forms are found. Zygotes develop into planulae or polyps. Asexual reproduction may involve fission and budding.

The class Anthozoa is usually divided into two subclasses, Alcyonaria (Octocorallia) and Zoantharia (Hexacorallia), each of which contains a number of orders. The distinction between the two subclasses is based chiefly on the nature of the tentacles, septa, siphonoglyphs, and skeletons. Several evolutionary lines are recognized in the two groups, but the order Antipatharia of the subclass Zoantharia is considered to contain primitive living anthozoans because of their simple organization of septa, tentacles, mesoglea, and siphonoglyph.

A brief résumé of the two subclasses and their their orders is presented here.

Subclass Alcyonaria

Alcyonarian polyps have octomerous radial symmetry with a distinct tendency toward bilateral symmetry (Figs. 5-46, *B*, and 5-47). Only one strongly developed siphonoglyph (ventral) is present. Some forms are dimorphic (e.g., small siphonozooids without tentacles and

FIG. 5-48

Tubipora, the organ-pipe coral (order Stolonifera). (From Hickman: Integrated principles of zoology, The C. V. Mosby Co.)

regular autozooids). They are colonial, and the gastrovascular cavities of the various polyps of a colony are interconnected by complex canal systems, or solenia, of the common mesogleal mass (coenenchyme) in which the polyps are embedded (Fig. 5-49, *E* and *F*). An endoskeleton of horny material or calcareous spicules is secreted by ectodermal cells or by ectodermal scleroblasts.

FIG. 5-49

Some Alcyonarian corals. **A,** *Telesto* (order Telestacea). **B,** Enlarged portion of *Telesto.* **C,** *Heliopora* (order Coenothecalia). Single zooid with adjacent coenenchymal tubules (skeleton removed). **D,** Architecture and mode of growth of colony of *Heliopora* (diagrammatic). Numbers indicate calyx of mother zooid, **1,** and calices of successively formed zooids. **E,** Small colony of *Alcyonium* (order Alcyonacea). **F,** Section through coenenchyme of *Alcyonium* showing expanded and contracted polyps and solenial network. (**A** redrawn from Hyman: The invertebrates; Protozoa through Ctenophora, McGraw-Hill Book Co., Inc., **B** to **D** modified from Bourne, 1895; and **F** modified from Hickson, 1895.)

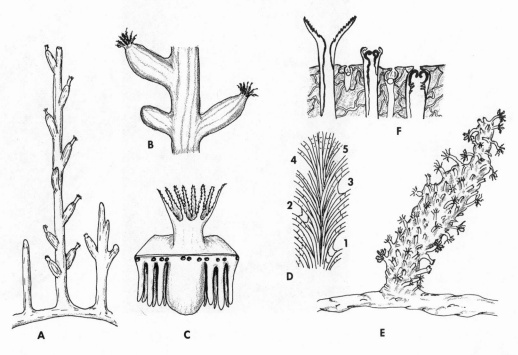

■ ORDER STOLONIFERA

Stoloniferan polyps arise singly from a creeping stolon that forms a thin, flat mat for a colony, and there is no coenenchymal mass. The polyp is cylindric and thin walled, and the oral end, or anthocodium, is retractile into a nonretractile anthostele usually protected by spicules. Polyps are connected by solenial tubes of the stolons and sometimes by crossbars. Their skeleton (when present) is composed of calcareous spicules, which, in the common organ-pipe coral *Tubipora* (Fig. 5-48) found in coral reefs, are fused together into tubes. Polyps never bud from the wall of a primary polyp. Another common genus is *Clavularia*.

■ ORDER TELESTACEA

The members of Telestacea form branching colonies by lateral budding from a primary polyp that arises from a creeping stolon. Their spicules, or sclerites, may be scattered singly or fused into tubes. *Telesto* (Fig. 5-49, *A* and *B*) is the best-known member.

■ ORDER COENOTHECALIA

The blue corals, present on certain coral reefs, have calcareous skeletons composed of crystalline fibers of aragonite that are fused onto lamellae. The polyps occupy numerous wide, cylindric cavities and the erect solenial systems are found in smaller cavities. A flat coenenchyme with a network of solenia connecting polyps and erect solenial tubes is found on the surface. The skeleton grows by the pushing up of polyps and solenia to successive higher levels. *Heliopora* (Fig. 5-49, *C* and *D*) is a common blue coral of coral reefs.

FIG. 5-50

Gorgonian soft corals, dried. **A,** Sea fan *Gorgonia.* **B,** Sea plume. These specimens are about 15 inches tall.

A   B

■ ORDER ALCYONACEA

The Alcyonacea are the fleshy or soft corals that have the polyps embedded in a gelatinous coenenchyme from which retractile anthocodia or oral ends protrude. In some species the anthocodia are nonretractile. Dimorphic forms are present in a few species. The polyp base may be protected by spicules and is called a calyx. The skeleton consists of isolated calcareous spicules, or sclerites, that are scattered over the coenenchyme. The colony may be mushroom shaped or plantlike in form. *Alcyonium* (Fig. 5-49, *E* and *F*) and *Xenia* are common genera; most members are found at considerable depths in warm oceans.

■ ORDER GORGONACEA

The Gorgonacea are horny corals that have a gorgonin axial skeleton and are often found in fanlike or featherlike colonies (Fig. 5-50). The colony may branch in a radial direction or in one plane (sea fan). The axial rod (Fig. 5-46, *B*) consists of a hard outer cortex of gorgonin and an inner medulla of a loose horny material. The axial skeleton in some members (*Paragorgia, Corallium*) contains calcareous spicules, but in the more typical gorgonians (*Gorgonia, Paramuricea*) the skeleton is spiculeless. A shallow layer of some form of coenenchyme gelatinous mesoglea and epidermis with polyp tubes, solenia, and calcareous spicules covers the central axial rod. In some ways the gorgonians are the most striking of all corals. They have a wide distribution, especially in the warmer seas, and add much color to coral reefs.

■ ORDER PENNATULACEA

The members of Pennatulacea corals are the sea pens (Fig. 5-51), sea feathers, and sea pansies (Fig. 5-52). They have one long, primary axial polyp with many secondary lateral polyps that arise as direct buds from the primary polyp. The body consists of a distal rachis bearing many polyps and a proximal peduncle without polyps. No stolons are present, and the lower end of the peduncle (sometimes enlarged) is

FIG. 5-51

Sea pen, preserved specimen (order Pennatulacea).

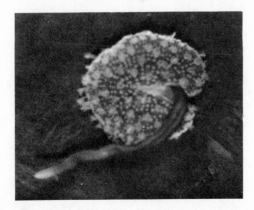

FIG. 5-52

Renilla, the sea pansy (about life size). Polyps grow on only one side of the rachis. Long peduncle attaches to substratum (order Pennatulacea).

embedded in the soft substratum of the sea floor. The rachis may be a radially symmetric, club-shaped body over which the polyps are scattered, or it may be bilaterally symmetric, with the polyps restricted to lateral positions. Modifications of the last plan exist, such as the presence of polyps on the dorsal surface and a polyp-free central surface. All members of the order are dimorphic, with typical autozooids that may or may not be retractile and siphonozooids without tentacles, retractor muscles, or gonads. The gastrovascular cavity of the primary polyp is divided by partitions into 4 longitudinal canals enclosing a horny skeletal axis. A spongy network of solenia is found around the canals. The cavities of the secondary polyps communicate with the longitudinal canals by solenia. Special gastrodermal muscles (both circular and longitudinal) are found in the peduncle and to some extent in the rachis for circulating the water. Most of the colonies are provided with a horny, calcareous, unbranched axial skeleton, and the coenenchyme contains smooth calcareous spicules of many shapes and forms.

The Pennatulacea are found in warm coastal waters on soft substrata. Many have been obtained from fairly deep dredgings. Common genera are *Renilla,* the sea pansy (Fig. 5-52), *Pennatula,* and *Funiculina.*

Subclass Zoantharia

Members of the subclass Zoantharia of Anthozoa differ mainly from those of the subclass Alcyonaria in having simple tubular tentacles (mostly retractile), mesenteries (complete or incomplete) in sixes or multiples of six, usually siphonoglyphs in both edges of the stomedeum, a skeleton (when present) not of mesogleal calcareous spicules, and a solitary as well as a colonial existence. These anthozoans are monomorphic, and the number of tentacles varies from 6 to many hundreds. The septa have a distinct bilateral or biradial arrangement and correspond on each side of the plane of symmetry that bisects the mouth (Figs. 5-53 and

5-54). In some members the septa are always found in pairs close together. Two pairs of complete septa, one at each constricted end of the mouth, are called directives and are characteristic of sea anemones and some others. This subclass contains far more variations in most morphologic features than does the subclass Alcyonaria.

■ ORDER ACTINIARIA

Actiniaria are the most widely distributed of all the anthozoans, being found in coastal waters from warm to frigid regions. They are commonly called sea anemones because of their flowerlike appearance (Figs. 5-46, A, and 5-53). They are solitary and sessile forms that live below the low-tide mark attached to a substratum by their pedal disk. They are not immovably fixed, however, for they can glide on their pedal disks by waves of muscular contraction. Varieties of movement are found in the group, and a few can swim by using their tentacles. Some burrow into the mud or sand, where they live in sandy mucous tubes.

The body is cylindric and divided into oral disk, column, and base. The circular fold between the column and oral disk divides the animal into an upper capitulum and a lower scapus (Fig. 5-46, A). A deep groove, or fossa, may be found between the collar and base of the capitulum. Between the pedal disk and column, the junction is called limbus. In extreme contraction a marginal sphincter muscle at the base of the capitulum may function to close the upper end of the scapus over the retracted oral disk. The oral disk is flat, usually circular, and bears the tentacles and mouth, which is present in a smooth region called the peristome.

Tentacles with nematocysts may be few in number, or they may be numerous and cover most of the oral disk. They may be arranged in cycles or radiating rows. When the number of tentacles is definite, the number is usually some multiple of six and doubles in each successive cycle after the innermost cycle. There

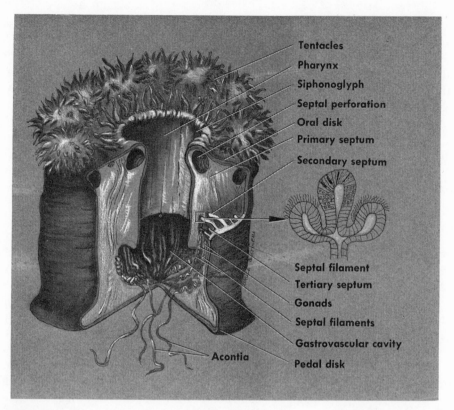

Tentacles
Pharynx
Siphonoglyph
Septal perforation
Oral disk
Primary septum
Secondary septum

Septal filament
Tertiary septum
Gonads
Septal filaments
Gastrovascular cavity
Pedal disk

Acontia

FIG. 5-53

Metridium cut to show pharynx and septal arrangement. Inset, section through septal filament showing its central cnidoglandular band and outer ciliated bands. (Subclass Zoantharia, order Actiniaria.)

FIG. 5-54

Transverse sections (diagrammatic) through actiniarian polyp such as *Metridium*. **A**, Through pharynx. **B**, Below pharynx.

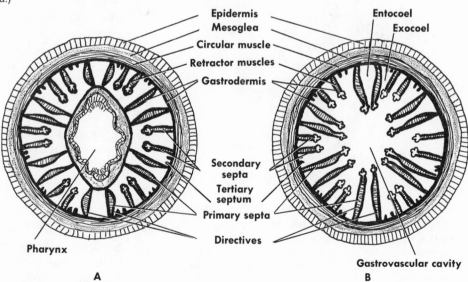

Epidermis
Mesoglea
Circular muscle
Retractor muscles
Gastrodermis

Entocoel
Exocoel

Secondary septa
Tertiary septum
Primary septa
Directives

Pharynx

Gastrovascular cavity

A

B

are thus 6 primary tentacles (innermost and oldest), 6 secondary tentacles, 12 tertiary tentacles, 24 quaternary tentacles, etc. The primary tentacles are usually the largest, tentacle size decreasing toward the marginal region.

In most sea anemones the septa are in pairs (Figs. 5-53 and 5-54). Although many variations exist, the pairs of septa occur in cycles or multiples of six. The first cycle consists of six pairs of primary septa that are complete and reach the stomodeum. Two pairs of these, one at each end of the pharynx, are directives and there are two pairs on each side between the directives (Fig. 5-54). A second cycle of six pairs is made up of incomplete secondary septa. In the third cycle, twelve pairs of incomplete tertiary septa occur. Complete septa are perforated, below the oral disk and near the pharynx, with holes (stoma) for water flow between the interseptal chambers. Marginal stoma near the column may also be present.

The septal, or mesenteric, filaments already

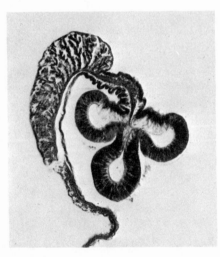

FIG. 5-55

Septal filament of sea anemone, *Metridium,* with trefoil arrangement of free edge. Central cnidoglandular band is flanked on each side by flagellated band. (From Andrew: Textbook of comparative histology, Oxford University Press, Inc.)

described are made up in sea anemones of 3 tracts, or lobes (Fig. 5-55). The middle lobe (cnidoglandular) is provided with nematocysts and enzyme-secreting gland cells and is found in all septa below the stomodeum. The lateral lobes bear ciliated cells for water circulation and alveolar tracts of phagocytic cells for digestion. Since the functions of the septal filaments are performed by different parts of the same filament and not by different filaments as in Alcyonaria, some variations exist in the distribution of the tracts in the various sea anemones.

Muscles in sea anemones are specialized for many different kinds of movement, such as the retractor muscles of the septa for shortening the polyp, the circular muscles of the wall mesoglea for extending the polyp, the transverse muscles of septa near the stomodeum for openin the mouth, the muscles of the tentacles, etc. Special circular muscles beneath the oral disk can form a sphincter for closing the margin over the retracted oral disk when the animal contracts. Most of the muscles are gastrodermal in origin.

The nematocysts of sea anemones include both types, spirocysts and nematocysts proper. The former are restricted to the tentacles and oral disk, whereas the latter are found everywhere. Several kinds of the nematocysts proper are found.

Sea anemones feed on various organisms with which they come in contact. Some are able to master small fish that are paralyzed by their nematocysts. Cilia on the surface of the smaller sea anemones may carry small food particles to the tips of the tentacles on the oral disk. The tentacles then deposit the food on the lips of the mouth. K. J. Lindstedt (1971) has found a biphasic feeding response in the sea anemone *Anthopleura* in which asparagine controlled the contraction and bending of the tentacles while reduced glutathione regulated the ingestion of food reaching the mouth.

The gonads are bands on the septa near the septal filaments (Fig. 5-53). They may be present on all filaments or on just certain filaments. Some species may be dioecious and others monoecious. Asexual reproduction by fission, budding, and pedal laceration occurs in the group. Fission may be either longitudinal or transverse. Fertilization may occur in the gastrovascular cavity, and development may occur in septal chambers, or it may occur in the sea.

Total cleavage may be equal or unequal and gives rise to a coeloblastula. By ingression or invagination the blastula is converted into a ciliated planula larva. All sea anemones in their development pass through a phase called the Edwardsia stage because of its similarity to the primitive *Edwardsia*, a small, burrowing sea anemone. In this stage 8 septa with bilateral symmetry occur similar to the eight mesenteries or macrosepta of *Edwardsia*. In a later Halcampoides stage two couples of septa develop and pair with the 4 lateral septa to form 6 complete pairs, the basic plan of sea anemones. Additional paired septa may form patterns characteristic of the different species.

■ ORDER SCLERACTINIA (MADREPORARIA)

The order Scleractinia is made up of the true or stony corals. They are chiefly responsible for the coral reefs and islands of the warm coral seas. They may be solitary polyps, but most are colonial and are attached to a firm substratum. They possess a hard, calcareous exoskeleton secreted by the epidermis and found outside the polyp body. In general the polyp (usually small) has the same basic plan as the typical sea anemone (Fig. 5-56). A pedal disk and cinclides, however, are absent. The typical oral disk bears tentacles in cycles of six so arranged that there is a tentacle over each interseptal cavity. The tentacles may be found in alternating cycles of the hexamerous pattern similar to that found in Actiniaria. Outer cycles, however, may not have their regular number of tentacles because of irregularities in the forma-

FIG. 5-56

Polyp of stony coral (diagrammatic) (order Scleractinia) cut away to show internal structure.

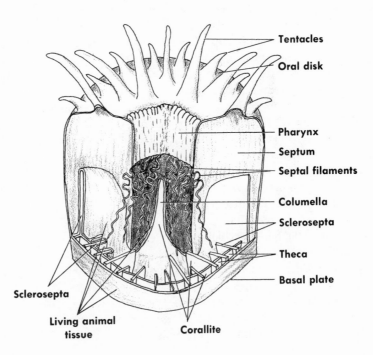

Tentacles
Oral disk
Pharynx
Septum
Septal filaments
Columella
Sclerosepta
Theca
Basal plate
Sclerosepta
Living animal tissue
Corallite

tion of septa. The tentacles, which vary in size and length, usually bear nematocysts in some form of cluster.

The mouth is circular or oval surrounded by a peristome. There are no siphonoglyphs. The septa show many variations of the hexamerous plan. In some members, the primitive arrangement is similar to that of *Edwardsia,* i.e., 4 single complete septa, 4 incomplete septa, and two pairs of directives. But corals often possess alternating cycles of incomplete septa to form the hexamerous pattern of 6, 6, 12, 24, etc., pairs. Usually only three cycles of septa are present. The septal filaments bear only the cnidoglandular tract and may protrude through the mouth during ingestion and digestion of food. Septal muscles are poorly developed, and stoma are lacking.

The skeleton of corals is secreted by the epidermis of the polyp base and the lower part of the column. After a basal plate of skeleton is formed by the polyp base, the epidermis may secrete a colloidal mass in which calcareous crystals are precipitated. In this process the epithelium is thrown into radial folds and a circular fold that are responsible for the formation of the radial vertical walls (sclerosepta) and the skeletal cup (theca). These epithelial folds also project upward into the septal spaces of the gastrovascular chamber into which the radiating sclerosepta fit.

Most true corals are colonial and may form masses of various shapes, such as spherical masses, branches, plates, etc. The thecae of a colony may be widely spaced or close together. Colonies usually arise by budding or division from a single, sexually producing polyp. Superficially the corallum appears to lie within the polyp, but morphologically it is entirely outside the polyp body; the polyp is merely molded upon the skeleton. The surface of the colony between the thecae is occupied by a sheet of living tissue (coenenchyme), which is an extension of the polyp walls above the theca. The coenenchyme thus connects together all the members of a colony and contains a gastrovascular cavity that

is continuous with the gastrovascular cavities of individual polyps. The lower epidermal surface of the coenenchyme, which is made up of both epidermis and gastrodermis, secretes the skeletal parts between the thecae.

Most corals, especially the reef-forming ones, prefer tropic waters not over 25 to 50 meters in depth, but there are exceptions. The symbiotic zooxanthellae that are common in reef corals may impose this restriction of depth. Some of the common colonial stony corals are the branching ivory coral *Oculina* (Fig. 5-57), with widely separated thecae; *Astrangia* (Fig. 5-58), with closely placed thecae and protruding polyps (very common along the North Atlantic coast); *Orbicella,* with massive form, a common reef builder of the West Indies; *Acropora,* with branches and small cylindric cups; and the brain coral (*Meandrina*), with rows of confluent thecae that produce a surface of curved depressions and ridges. Solitary corals are best represented by *Caryophyllia* and *Balanophyllia.* Perhaps the most abundant source of reef-forming corals is the Great Barrier Reef of Australia where more than 200 species have been described.

As already mentioned, reef-forming corals have a symbiotic relationship with unicellular algae or zooxanthellae (Dinophysidae). These plants are found in specialized carrier cells in the gastrodermal epithelium. The revealing work of T. Goreau and others indicates that the zooxanthellae are actively involved in coral skeletogenesis by promoting the mechanism of calcification. Investigations show how the process of calcium deposition is speeded up by photosynthesis of the zooxanthellae, which removes the excess CO_2 formed at the site of mineralization. Excess CO_2 in the form of H_2CO_3 tends to decrease the velocity of the reactions necessary for coral formation. In this process calcium ions (Ca^{++}) in sea water first combine with bicarbonate ions (HCO_3^-) of metabolic origin to form calcium bicarbonate ($Ca[HCO_3]_2$) and then calcium carbonate ($CaCO_3$), which is deposited in the coral frame-

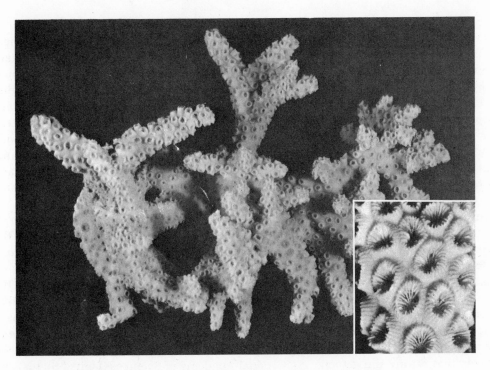

FIG. 5-57

Oculina, the ivory coral; cluster about 4 inches tall. Inset at right is closeup of some of cups (order Scleractinia).

FIG. 5-58

Astrangia danae, a scleractinian coral, polyps extended from their corallites. This beautiful little coral grows in colonies up to 3 inches across, encrusted on rocks and shells in sheltered places. (Courtesy General Biological Supply House, Inc., Chicago.)

work. In addition to assisting in coral skeletogenesis, zooxanthellae also supply oxygen and, according to some authorities, food as well. It is thought that these algae must have played an important role in the growth and evolution of shallow-water corals by promoting the rapid deposition of calcium to offset the constant erosion of coral material under the impact of the severe battering of shore wave action.

Coral reef formation. The Madreporaria have formed most of the coral atolls, coral islands, barrier reefs, and fringing reefs. They may form only a minor part of such existing structures now, but they have built the interlocking framework on which algae and other plant life live. The coral framework also forms diverse habitats for various forms of animal life. To a limited extent the alcyonacean and the gorgonacean corals also contribute to coral formations. The extinct Rugosa order, especially the colonial types that were active from the Ordovician to the Permian periods were responsible for early coral formations. The hydrozoan Milleporina or stinging coral may also be abundant in the surf portions of coral reefs.

The three chief types of coral reefs are (1) the fringing reef near the shore and usually separated from it by a narrow lagoon; (2) the atoll, more or less circular, enclosing a lagoon and resembling fringing reefs without an island; and (3) the barrier reef separated from shore by many miles of deep water. Fringing reefs are common in southern Florida and the South Pacific; atolls occur in southern seas; and the Great Barrier Reef extends for 1250 miles along the northeast coast of Australia, separated from the continent 10 to 100 miles.

A number of theories have been proposed to account for coral formations. Charles Darwin proposed the subsidence theory for atolls, by which they began as fringing reefs around elevated land masses (or volcanic cones) and as the land sank, the coral grew upward and became the only solid substance near the surface. The Daly theory states that during the glacial period large volumes of water were locked up in ice and the level of the ocean very much reduced. As warmer geologic periods occurred, the ice melted and the water so released caused the ocean levels to slowly rise. By wave or other action flat areas were formed near land masses on which coral began to grow, building higher as the ocean level rose. There is some evidence in favor of each theory. Core borings on a number of atolls have been made to a depth of thousands of feet and the core samples showed coral sediments at depths far below the depth at which coral can live, indicating that coral formations occurred when the ocean levels were much lower than they are now.

A unique atoll ecosystem is Aldabra in the Indian ocean some 260 miles northeast of the Malagasy Republic (formerly Madagascar) and 400 miles from the coast of Africa. The atoll is less than 22 miles long and less than 10 miles wide at its widest part. Being an atoll, it has a fringe of high limestone land of about 60 square miles surrounding a central lagoon.

The land mass is divided into four islands, on one of which is a small settlement of fishermen. It is one of the two places where the giant tortoise (*Testudo gigantea*) has a natural population. It is also the habitat of the last surviving members of the flightless rail (*Dryolimnas cuvieri*), and many other birds also nest on the islands. In many other ways it has a unique fauna characteristic of an undisturbed oceanic island, and its ecosystem is simplified and lends itself to profitable study by interested naturalists.

This atoll is of current interest because of an agreement between Great Britain and the United States, for defense purposes, to make the atoll a development project involving an airfield, a dam at the main entrance to the lagoon for a harbor, and other changes that would make it a defense center. Such changes would be sure to disturb the general ecosytem. On this account wild life students of the National Academy of Sciences (United States) and of the Royal Society Council (Great Britain) have made vigorous protests in favor of keeping the atoll as it is, and for the defense base to be located elsewhere. At the present time the atoll is under the control of the Royal Society (J. Walsh, 1967).

■ ORDER ZOANTHIDEA

Although most members of the order Zoanthidea are colonial, some solitary species exist. They are best described as skeletonless, anemonelike anthozoans without a pedal disk. Their body is divided into a capitulum and scapus and is often encrusted with detritus such as sand grains, sponge spicules, and foram shells that may be embedded in the surface and mesoglea. The oral disk bears two circlets of unbranched tentacles. Their most unique feature is the arrangement of the septa. Only the sulcal (ventral) pair of directives is complete. The other septa form pairs of one complete septum and one incomplete septum so that in the entire polyp there is only one pair in which the two members are complete (the sulcal directives). In the colonial forms polyps are united by stolons with solenia. A common colonial form, *Epizoanthus,* is found epizoic on other corals and snail shells inhabited by hermit crabs. Although most abundant in warm tropical waters, this order has a wide distribution.

■ ORDER ANTIPATHARIA

The order Antipatharia consists of the black or thorny corals, most abundant in the deeper waters of the tropic and subtropic zones. They usually form slender, branching, plantlike colonies several feet tall. A basal plate of the main stem attached to a substratum is usually present. The polyps, which are arranged on various parts of the branches, are short, cylindric bodies, each possessing a circlet of 6 nonretractile tentacles provided with clusters of nematocysts. Two slightly differentiated siphonoglyphs are found in the stomodeum. Adjacent polyps of a colony are united by a coenenchyme but have no common gastrovascular cavity. Their body surface is ciliated, and their body musculature is weakly developed. The polyps are mostly dioecious. *Antipathes* is one of the most familiar genera.

■ ORDER CERIANTHARIA

Members of Ceriantharia bear some resemblance to anemones in that their bodies are cylindric, muscular, and smooth. By mucus secreted from ectodermal cells each polyp forms a type of sheath or tube of sand, nematocysts, and detritus in which it is enclosed up to the oral disk. The sheathed animal lies buried in the sandy bottom with only the oral disk and two circlets of tentacles protruding. Other than this sheath, there is no protective skeleton. The polyp is provided with one dorsal siphonoglyph, which also forms a groove (hyposulculus) in the pharynx. A pedal disk is lacking.

Cerianthus, the most familiar form, is common on sandy bottoms in both American and European waters, and is found in both deep and shallow water.

■ PHYLUM CTENOPHORA
GENERAL CHARACTERISTICS

1. The ctenophores are a group of biradial jellyfish known as comb jellies because of the presence of ciliated comb plates, used in locomotion.

2. They are entirely marine and are mostly pelagic and planktonic; some are modified for creeping.

3. The gastrovascular system in the form of canals is the only body cavity.

4. The body wall is composed of an epidermis, a layer of collenchyme, and a gastrodermis. Collenchyme contains amebocytes, connective tissue, and true muscle cells, and so it is more advanced than coelenterate mesoglea.

5. Most ctenophores possess on their tentacles specialized adhesive cells called colloblasts. The tentacles are used for catching prey and as balancing organs. (Nematocysts are found in one species of ctenophore).

6. Skeletal structures as well as excretory and respiratory organs are lacking.

7. The shape is varied, including football, pyriform, globular, bell, and ribbon shapes. They are entirely monomorphic.

8. There is a well-developed statocyst at the aboral pole and a nerve net system in the epidermis.

9. They have determinate cleavage (coelenterates are indeterminate), and a cydippid larval form with bilateral features.

PHYLOGENY AND ADAPTIVE RADIATION

1. That both coelenterates and ctenophores have arisen from the same basic stock seems to be a logical suggestion. The presence of nematocysts in one group of ctenophores (*Euchlora rubra*) may suggest that the ctenophores have arisen from a trachyline medusa ancestor. The biradial symmetry characteristic of ctenophores may easily have been derived from the tetramerous radial symmetry of hydromedusae. Within the Ctenophora the order Cydippida is considered the most primitive and the least

modified of the ctenophoran plan. The other orders supposedly evolved along independent lines at a later time.

2. The basic structures that have guided the evolutionary development of the ctenophores are the arrangement of the comb plates and the general biradial symmetry. Both these features are found with modifications throughout the group.

CLASSIFICATION

Fewer than 100 species of ctenophores have been described to date, although some new species have been added in the last decade or so. Ctenophores are classified on the basis of their tentacles, body form, oral lobes, gastrovascular divisions, and body compressions. On this basis two classes and five orders make up the phylum at the present time.

Phylum Ctenophora. Biradial, modification of radial symmetry; ellipsoid, spherical, or ribbon shaped; 8 rows of ciliated comb plates at some stage; ectoderm, endoderm, and ectomesoderm (mesoglea); gastrovascular cavity with one opening (mouth), a large stomodeum, and many branches; tentacles (in most) with colloblasts in place of nematocysts (except in one species); nervous system of a subepidermal plexus with concentrations under the comb plates; endodermal gonads in walls of gastrovascular canals; hermaphroditic reproduction; mosaic development; cydippid larva.

    Class Tentaculata. With tentacles.
        Order Cydippida. *Pleurobrachia.*
        Order Lobata. *Mnemiopsis.*
        Order Cestida. *Velamen.*
        Order Platyctenea. *Ctenoplana.*
    Class Nuda. Without tentacles.
        Order Beroida. *Beroë.*

GENERAL MORPHOLOGIC FEATURES

As a group, ctenophores are perhaps the most beautiful animals of the sea. Their glasslike

FIG. 5-59

Pleurobrachia showing major structures (diagrammatic). (From Hickman: Integrated principles of zoology, The C. V. Mosby Co.)

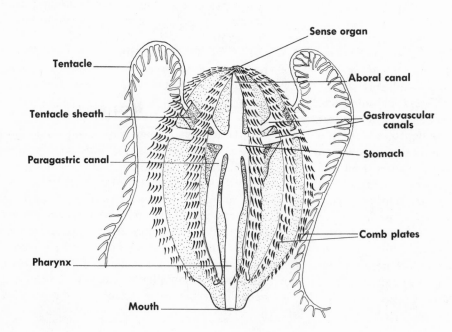

transparency, the delicacy of their structure, and their iridescence by day and luminescence at night have given them a distinct beauty rarely found in the animal kingdom.

Ctenophoran morphology is perhaps best understood in the most typical and least modified members such as those belonging to the order Cydippida. The ctenophores of this order most familiar to American and British zoologists are *Pleurobrachia* (Fig. 5-59) and *Hormiphora,* which resemble each other except in a few details. These genera are oval shaped or egg shaped and possess transparent bodies. The body wall contains an outer epidermis, which may be syncytial or cellular, of cuboid or columnar cells and an inner endodermal epithelium, with collenchyme (mesoglea) between the two layers containing true muscle cells arranged in an anastomosing network. The mouth is situated at one pole and a sense organ at the other pole (Fig. 5-59).

On the surface of the body are 8 rows of comb plates arranged meridionally and equally spaced. These plates start near the sense organ

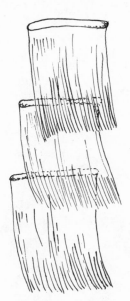

FIG. 5-60

Portion of comb row showing three plates, each composed of transverse rows of long fused cilia.

and extend for about four fifths of the distance to the oral pole. Each plate consists of a transverse row of fused cilia borne on modified ectodermal cells; the plates are arranged in succession and united as a ciliary tract to form a complete comb-plate row (Fig. 5-60). In locomotion the beat in each row starts at the aboral end and, as each plate beats in succession, proceeds in a wave to the oral end causing the animal to move with the mouth in front because the effective movement of each comb is toward the aboral pole. The direction of the wave is reversible. The synchrony and coordination of the beat are controlled by nervous impulses from the statocyst through nervous connections.

The comb plates fit ctenophorans for a pelagic and planktonic existence for which feeble movements are sufficient. However, in highly modified ctenophores such as *Cestum* the swimming action of the plates is supplemented by the sinuous movement of the ribbonlike body, and in the Platyctenea a creeping locomotion is evident.

Two tentacles are found in most ctenophores, situated at opposite ends of the transverse or tentacular axis (Fig. 5-59). The base of each tentacle is fastened to the bottom of a pouch, or sheath, into which the tentacle may be retracted. When fully extended, the highly extensile tentacle may be much longer than the body. The tentacle is solid, composed mainly of mesoglea and longitudinal muscle strands, and covered with epidermis. The sheath is an invaginated epidermal pouch that bears at its bottom the formative tissue for the tentacles and the colloblasts. The tentacle has many lateral branches that are covered with special adhesive cells (colloblasts or lasso cells). Each colloblast (Fig. 5-61) is composed of a sticky, crescent-shaped body that is fastened on its concave side by 2 filaments, one of which is spirally twisted and contracted around a stiff, central filament that is derived from a flagellum. Food (small organisms, ova, etc.) is captured by the sticky colloblasts of the tentacles, which are drawn over the lips of the mouth where ingestion of food occurs. Tentacles are used not only to catch prey but also to serve as balancing organs when the ctenophore is floating. The colloblast may correspond to the nematocyst of coelenterates, but these two structures are not homologous because they have a different origin. A colloblast is considered a cell; a nematocyst is a secretory product (organelle) of a cell.

The sense organ, or statocyst (Fig. 5-62), is located at the aboral pole and is somewhat complicated in structure. Its general structure consists of a cyst containing a centrally located statolith, which is composed of fused calcareous bodies. The floor of the statocyst is composed of ciliated epidermal and secretory cells. The statolith is supported on 4 pillars of fused cilia (the balancers) located interradially on the sensory floor. Two ciliated furrows lead from the base of each balancer to the pair of comb rows in the same quadrant. These tracts may carry impulses to the comb rows from the sense organ. The dome-shaped cover of the statocyst is formed of fused cilia. On each side of the statocyst in the sagittal plane is a ciliated area of unknown function called the polar field.

G. A. Horridge (1965) has described four pathways of coordination in ctenophores, although not all the pathways are equally developed in all species. He finds the four conducting systems to be the following: (1) The comb-plate activation system whose beat starts at the 4 balancer

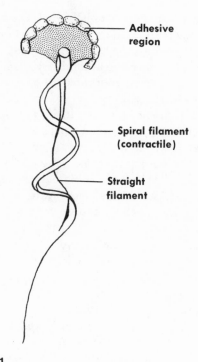

FIG. 5-61

A colloblast. This is a type of adhesive cell located on ctenophore tentacles.

FIG. 5-62

Statocyst enlarged to show details. (Modified from Chun.)

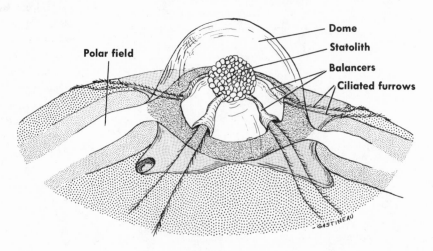

cilia and is controlled by the direction of the statolith is supported by the balancers. Its beat is transmitted cell by cell along a system of ciliary cells in the groove whose depolarization triggers the beat of the cilia. (2) The comb-plate inhibitory system of the subepidermal nerve net is found throughout the surface of the epithelium and consists of a synaptic net of tripolar neurons. Activity of this net after stimulation of body surface will cause all ciliary action to cease, but will cause a propagated muscle twitch. (3) A transmission between muscle cells transmits direct cell-to-cell conduction, as indicated by various experiments involving muscular contraction. (4) A system (possibly endodermal) coordinates the luminescent waves that are provoked by stimulating the ctenophore when dark adapted. Horridge attempts to explain how separate pathways of excitation arise in primitive nervous systems and later in the evolution of nervous transmission converge to one central pathway, as found in the central nervous system of higher forms.

The gastrovascular system consists of an extensive canal system that is closely involved with the biradial symmetry of the group. The slitlike mouth opens into a long tubular pharynx, or stomodeum (ectodermal lining), that is flattened in the tentacular, or transverse, plane and lies directly in the oral-aboral axis. Its walls are much folded, and most digestion occurs here. Digestion in Ctenophora is both extracellular and intracellular. The pharynx extends upward for about two thirds of the way toward the aboral pole and opens through a short esophagus into the stomach (infundibulum). The stomach is lined by endoderm and is flattened in the sagittal plane at right angles to the tentacular plane. Extracellular enzymes are mostly proteases and are responsible for the preliminary digestion. Final digestion and absorption occur intracellularly in the gastroderm cells.

From the stomach the digested food products are distributed to the body by vertical and hori-zontal sets of canals. The vertical canals consist of 2 blind pharyngeal, or paragastric, canals, and a single aboral canal. This aboral canal, just below the statocyst, divides into 4 short excretory canals that end in small sacs. Two of these sacs (anal canals) open by pores to the surface; the other 2 end blindly. A pair of horizontal canals branches to form 8 canals and serves to connect the stomach with the 8 meridional canals located beneath the comb rows. Each of the 8 canals so formed passes to the underside of a comb row to produce, orally and aborally, a long meridional canal that extends the length of each comb row.

Ctenophores are hermaphroditic, and male and female gonads occur close to each other in the meridional canals or their branches. The bandlike gonads are formed from endodermal epithelium lining the meridional canals; the male gonad is on one side, and the female gonad is on the other side of each canal. In the canals the ovaries are always found on the side nearest the sagittal or the tentacular plane (perradial), and the spermaries are found on the interradial side of the wall. In some species (e.g., *Euchlora*) gonads are found only on 4 of the 8 meridional (subtentacular) canals. Eggs may be fertilized in situ, either by sperm carried in by gastrovascular canals or by adjacent sperm (self-fertilization). In most cases, however, the mature germ cells are shed into sea water, where fertilization occurs. Fertilized eggs in the canals are shed directly through the epidermis over the meridional canals and not through the mouth. *Coeloplana* and *Ctenoplana* appear to be the only members with definite genital ducts (for sperm).

In their embryology ctenophores are the only animals that have biradial cleavage, and their development follows a biradially symmetric scheme. At the 8-cell stage 4 blastomeres are large and 4 are smaller. These cells are so arranged that the large blastomeres form a central mass with two smaller ones at each side. The long axis at this stage later becomes the tentacular plane of the adult. Succeeding divisions result in an increase of micromeres (the smaller blastomeres) so that a ring of these blastomeres is formed on the concave aboral surface of the large macromeres, which divide much more slowly. As the micromeres continue to divide they spread increasingly over the surface of the macromeres as a single layered sheet (an epibolic process). During this process the macromeres invaginate into the interior so that gastrulation involves both epiboly and invagination. The micromeres become the epidermis, and the macromeres become the endoderm; thus the cleavage is determinate (mosaic), and all the early blastomeres are required to form a complete organism. In further development the cells of the collenchyme apparently come from the ectoderm; the mesenchyme is thus an ectomesoderm. The musculature comes from cells in the collenchyme that are not epitheliomuscular cells. The characteristic and unique comb rows arise from interradial bands on the ectoderm where cells divide rapidly. From the aboral ectoderm the sense organ and polar bands are also derived. The young larval form is called a cydippid larva and has many features of the adult belonging to the order Cydippida. In other orders the larva undergoes various modifications before becoming an adult.

Most ctenophores display luminescence. The luminous materials are granules secreted by endodermal cells in the walls of the meridional canals often in or near the gonads. Ctenophores emit their light by flashes and control its production by nervous stimulation. External light seems to inhibit luminescent action, for they do not usually flash until they have been in the dark for several minutes.

■ CLASS TENTACULATA
■ ORDER CYDIPPIDA

The chief characteristics of the order Cydippida were described in the previous section. As already mentioned, the members of this order retain the characteristics of the larva until the adult stage is attained. They are often considered to be the most typical and least modified of all ctenophores. They are globular or egg shaped and possess 2 tentacles that have lateral branches covered with colloblasts. Their gastrovascular canals are simple, end blindly, and do not form a peripheral canal system. In

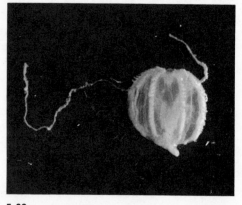

FIG. 5-63

Pleurobrachia (order Cydippida), a common ctenophore; about ¾ inch in diameter. (Preserved specimen.)

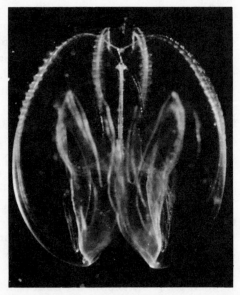

FIG. 5-64

Mnemiopsis (order Lobata). (Photograph by Roman Vishniac, courtesy Encyclopaedia Britannica Films, Inc.)

some members the body is compressed in the sagittal plane. The comb rows are mostly equal in length.

Familiar examples are *Pleurobrachia* (Fig. 5-63), *Hormiphora,* and *Euchlora* (the genus with nematocysts).

■ ORDER LOBATA

The body of ctenophores in the order Lobata is compressed in the tentacular plane and is helmet shaped. A large contractile lobe is found on each side of the oral end in the sagittal areas of the body. These lobes cause the subsagittal comb rows to be longer than the subtentacular ones. From the ends of the latter comb plates 4 auricles or lappets with ciliated edges extend above the mouth, a pair on each side. The tentacle sheaths of the larva disappear during

metamorphosis, and the reduced tentacles of the adult are located on each side of the mouth. In addition, numerous accessory tentacles are found in ciliated auricular grooves that extend from the mouth to the auricles. In the gastrovascular system the 4 interradial canals arise directly from the stomach, and transverse canals are lacking.

Common examples of the order are the genera *Bolinopsis, Mnemiopsis* (Fig. 5-64), and *Ocyropsis.* The first two are common along the Atlantic coast.

■ ORDER CESTIDA

Members of this order of ribbon-shaped ctenophores are compressed in the tentacular plane and elongated in the sagittal plane. They may be more than a meter long. The subsagittal comb rows extend over the entire length of the aboral surface, whereas the subtentacular plates are rudimentary or reduced to only a few. The tentacle sheaths and the filamentous tentacles are shifted orally toward the mouth. Two rows of short tentacles extend in tentacular grooves from the mouth to the extremities of the body.

FIG. 5-65

A, *Cestum veneris* (order Cestida) commonly called "Venus' girdle." Common in European and tropic waters (Florida). **B,** *Beroë* (order Beroida). No tentacles or tentacle sheaths. (**B** redrawn from Hyman: The invertebrates; Protozoa through Ctenophora, McGraw-Hill Book Co., Inc.)

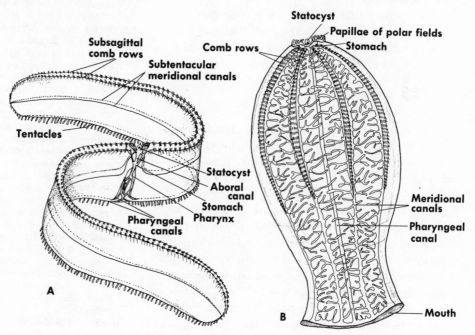

The subtentacular meridional canals run down the middle of the elongated body and unite with the subsagittal and paragastric canals, all of which run horizontally.

Cestum, or "Venus' girdle" (Fig. 5-65, *A*), is the familiar example. *Velamen* is the other genus and occurs off the eastern American coast.

■ ORDER PLATYCTENEA

Members of the order Platyctenea are the most highly modified of all ctenophores. Much of this modification centers around their creeping or sedentary methods of existence. Their body is flattened in the oral-aboral axis and elongated in the tentacular axis so that they actually have dorsal and ventral surfaces. The gastrovascular system is shortened, and part of the oral (ventral) surface is everted pharynx. The canals from the stomach anastomose to form a peripheral network. Comb rows may or may not be present. The gonads are present in the meridional canals, and each testis opens to the surface by a duct and pore. They possess tentacle sheaths and retractile tentacles as well as a statocyst.

The most striking examples are *Coeloplana* and *Ctenoplana* found chiefly off the coasts of Indo-China and Japan; *Coeloplana* has also been found off the coast of Florida.

■ CLASS NUDA
■ ORDER BEROIDA

These bell-shaped or cone-shaped ctenophores have no trace of tentacles or tentacle sheaths at any stage of their development. They have a wide mouth with a greatly enlarged stomodeum (pharynx), and the body is compressed in the tentacular plane. In some members the comb rows may extend the full length of the organism, although in *Beroë* they do not reach this far. The 8 meridional canals unite with the paragastric canals and send out numerous diverticula, which anastomose to form a peripheral network all over the body. The polar fields are fringed with branched papillae.

The widely distributed genus *Beroë* (Fig. 5-65, *B*) has a pinkish cast and may reach a length of several inches in colder waters, although its size is much smaller off the New England coast.

REFERENCES
Coelenterata

Barnes, J. H. 1966. Studies on three venomous Cubomedusae. In Rees, W. J. (editor). The Cnidaria and their evolution. New York, Academic Press, Inc.

Berrill, N. J. 1957. The indestructible hydra. Sci. Amer. **197:** 48-58. A study of the hydra's structure and regenerative capacity.

Berrill, N. J. 1961. Growth, development and pattern. San Francisco, W. H. Freeman & Co., Publishers.

Brien, P. 1960. The fresh-water hydra. Amer. Sci. **48:**461-475.

Buchsbaum, R., and L. J. Milne. 1960. The lower animals; living invertebrates of the world. Garden City, Doubleday & Co., Inc. Superb illustrations of hydroids, jellyfishes, and corals are given in Chapter 3.

Bullock, T. H., and G. A. Horridge. 1965. Structure and function in the nervous system of invertebrates. San Francisco, W. H. Freeman & Co., Publishers. A comprehensive summary of what is known about the form and function of the nervous system of invertebrates.

Burnett, A. L. 1961. The growth process in hydra. J. Exp. Zool. **146:**21-84.

Bushnell, J. H., Jr. and T. W. Porter. 1967. The occurrence and prey of *Craspedacusta sowerbyi* (particularly polyp stage) in Michigan. Trans. Amer. Microscop. Soc. **86:**22-27.

Campbell, R. D. 1967. Growth pattern of hydra: distribution of mitotic cells in *Hydra pseudoligactis*. Trans. Amer. Microscop. Soc. **86**(2):169-173.

Chapman, G. B., and L. G. Tilney. 1959. Cytologic studies of the nematocysts of hydra. I. Desmonemes, isorhizas, cnidocils, and supporting structures. J. Biophys. Biochem. Cytol. **5:**69-78.

Crowell, S. 1953. The regression-replacement cycle of hydranths of *Obelia* and *Campanularia*. Physiol. Zool. **26:**319-327.

Crowell, S. 1960. Non-regulative differentiation in the thecate hydroid, *Campanularia*. Anat. Rec. **138:**341-342.

Crowell, S. (editor). 1965. Behavioral physiology of coelenterates. Amer. Zool. **5:**335-589. An excellent symposium by many specialists on the present status of the physiologic aspects of this group. Some of the papers also discuss the ctenophores and echinoderms.

Darwin, C. 1962. The structure and distribution of coral reefs. Berkeley, University of California Press. A reprint of the classic study by the great naturalist.

Fraser, L. A. 1962. The histology of the muscularis of *Gonionemus*. Trans. Amer. Microscop. Soc. **81:**257-262.

Goreau, T. 1961. Problems of growth and calcium deposition in reef corals. Endeavour **20:**32-39. A summary of the recent concept involving the role of zooxanthellae in coral formation.

Hand, C. 1959. On the origin and phylogeny of the coelenterates. Syst. Zool. **4**:191-202. A summary of the planuloid ancestral theory of coelenterate origin.

Hardy, A. C. 1956. The open sea. Boston, Houghton Mifflin Co. Beautiful color plates of jellyfishes are included.

Hyman, L. H. 1940. The invertebrates; Protozoa through Ctenophora, vol. 1. New York, McGraw-Hill Book Co. Inc. A comprehensive and authoritative treatise with extensive bibliographies and revealing illustrations. Volume 5 (1959) of this work summarizes the investigations on the coelenterates since 1940.

Jha, R. K., and G. O. Mackie. 1967. The recognition, distribution, and ultrastructure of hydrozoan nerve elements, J. Morphol. **123**:43-61.

Jones, C. S. 1949. The control and discharge of nematocysts in hydra. J. Exp. Zool. **105**:25-60.

Lankester, E. R. (editor). 1900. Treatise on zoology; part 2, The Porifera and Coelenterata. London, Adam & Charles Black, Ltd. The account of the coelenterates by G. H. Fowler and G. C. Bourne in this outstanding treatise will always be of use to students of this group, although many investigations and interpretations have been made in the past seventy years.

Lenhoff, H. M., and W. F. Loomis. 1961. The biology of hydra and of some other coelenterates. Coral Gables, Fla., University of Miami Press. The record of a noteworthy symposium discussing the structure of hydra and the results of research on various problems of coelenterate patterns.

Lentz, T. L., and R. J. Barnett. 1965. Fine structure of the nervous system of hydra. Amer. Zool. **5**:341-356.

Lindstedt, K. J. 1971. Biophasic feeding response in a sea anemone: control by asparagine and glutathione. Science **173**:333-334.

Loomis, W. F. 1959. The sex gas of hydra. Sci. Amer. **200**:145-156. An account of the factors that influence sex in hydra.

Lytle, C. F. 1964. Zoogeography of the freshwater Hydrozoa. Amer. Zool. **4**(4):436. (Abstr.)

MacGintie, G. E., and N. MacGintie. 1968. Natural history of marine animals, ed. 2, New York, McGraw-Hill Book Co.

Mackie, G. O. 1960. The structure of the nervous system in *Velella*. Quart. J. Microscop. Sci. **101**:119-131. A description of the two types of nervous systems — the syncytial and the nonsyncytial — found in certain coelenterates.

Mackie, G. O. 1963. Siphonophores, bud colonies, and super-organisms. In Dougherty, E. C. (editor). The lower Metazoa; comparative biology and phylogeny. Berkeley, University of California Press. The author discusses some of the difficulties in distinguishing between individuality and colonial patterns of animals and shows how the members of the siphonophore colony are integrated and controlled by the "colonial will."

Marcus, E. 1958. On the evolution of the animal phyla. Quart. Rev. Biol. **33**:24-58. An appraisal of the status of the origin of the various animal groups.

Payne, F. 1924. A study of the freshwater medusa, *Craspedacusta ryderi*. J. Morphol. **38**:387-430. An excellent account of this interesting freshwater coelenterate.

Pennak, R. W. 1953. Fresh-water invertebrates of the United States. New York, The Ronald Press Co. One chapter of this important reference treatise is devoted to the freshwater coelenterates. Keys, illustrations, and specific descriptions are included.

Picken, L. E. R., and R. J. Skaer. 1966. A review of researchers on nematocysts. In Rees, W. J. (editor). The Cnidaria and their evolution. New York, Academic Press, Inc.

Rees, W. J. (editor). 1966. The Cnidaria and their evolution. New York, Academic Press, Inc.

Ross, D. M. 1966. The receptors of the Cnidaria and their excitation. In Rees, W. J. (editor). The Cnidaria and their evolution, New York, Academic Press, Inc.

Russell, F. S. 1953. The medusae of the British Isles. New York, Cambridge University Press. A magnificent monograph of the British jellyfishes, with full text descriptions and excellent plates.

Smith, F. G. W. 1948. Atlantic reef corals. Coral Gables, Fla., University of Miami Press.

Tardent, P. 1963. Regeneration in the Hydrozoa. Biol. Rev. **38**:293-333.

Totton, A. K., and G. O. Mackie. 1960. Studies on *Physalia physalia* (L). Discovery Rep. **30**:301-407. An exhaustive account of the morphology, behavior, and histology of this species.

Uchida, T. On the interrelationships of the Coelenterata, with remarks on their symmetry. In Dougherty, E. C. (editor). 1963. The lower Metazoa; comparative biology and phylogeny. Berkeley, University of California Press. The author's views fall into line with the planuloid ancestry that Hand and Hyman have stressed. Uchida believes that this planuloid ancestor gave rise to two groups — those with radial symmetry (Hydrozoa and Scyphozoa) and those with bilateral symmetry (Anthozoa).

Walsh, J. 1967. Aldabra: biology may lose a unique island ecosystem. Science **157**:788-790.

Webster, G. 1971. Morphogenesis and pattern formation in hydroids. Biol. Rev. **46**(1):1-46.

Wells, J. W. Coral reefs. In Hedgepeth, J. W. (editor). 1957. Treatise on marine ecology and paleoecology. Washington, D. C., Geological Society of America. Memoir **67**:609-632.

Westfall, J. A. 1965. Nematocysts of the sea anemone *Metridium*. Amer. Zool. **5**:377-393.

Yonge, C. M. 1931. A year on the Great Barrier Reef, London, William Collins Sons & Co., Ltd.

Ctenophora

Bourne, G. C. Ctenophora. 1900. In Lankester, E. R., (editor). Treatise on zoology. London, Adam & Charles Black, Ltd. A good basic description of the morphology of ctenophores.

Horridge, G. A. 1965. Relations between nerves and cilia in ctenophores. Amer. Zool. **5**:357-375.

Hyman, L. H. 1940. The invertebrates; Protozoa through Ctenophora, vol. 1. New York, McGraw-Hill Book Co., Inc. One of the best general accounts of the ctenophores. Discusses the morphology, physiology, phylogeny, and classification of the group. Well illustrated and provided with an extensive bibliography.

Hyman, L. H. 1959. The invertebrates; smaller coelomate groups, vol. 5. New York, McGraw-Hill Book Co., Inc. The chapter entitled "Retrospect" summarizes work on the Ctenophora.

Komai, T. 1951. The nematocysts in the ctenophore *Euchlora rubra*. Amer. Natur. **85**:73-74. The author believes that the nematocysts, which exist in this species instead of colloblasts, have been acquired by eating medusae and are not derived from a natural development within the animal.

Komai, T. 1963. A note on the phylogeny of the Ctenophora. In Dougherty, E. C. (editor). The lower Metazoa; comparative biology and phylogeny. Berkeley, The University of California Press. The author thinks that the Ctenophora should be considered a subphylum under phylum Coelenterata and that they are more closely related to the Hydrozoa than to any other group.

Mayer, A. C. 1912. Ctenophores of the Atlantic coast of North America. Washington, D. C., Carnegie Institution of Washington, Pub. No. 162. A technical account with many plates and figures.

Picard, J. 1955. Les nematocysts du ctenaire *Euchlora rubra*. Recueil Trav. Stat. Marine Endoume, Bull. No. 9, fasc. 15, pp. 99-103. The nematocysts are restricted to the epidermis along longitudinal tracts and are not acquired by eating medusae; they are formed in the epidermis of the tentacle bases.

Rees, W. L. 1966. The Cnidaria and their evolution. New York, Zoological Society of London and Academic Press, Inc.

The acoelomate animals

PHYLA PLATYHELMINTHES AND RHYNCHOCOELA

GENERAL CHARACTERISTICS

1. The acoelomates are bilaterally symmetric animals with three definite layers of cells (triploblastic). The third germ layer (mesoderm) has arisen largely from the endoderm.

2. Platyhelminthes have only one internal cavity, the digestive or gastrovascular cavity, with the region between the ectoderm and endoderm filled with mesoderm in the form of muscle fibers and mesenchyme. A hydrostatic skeleton is present in each phylum with a deformable parenchyma in flatworms and a fluid hydrostatic skeleton in the functioning of the proboscis of the nemertines (phylum Rhynchocoela).

3. Acoelomates have more specialization and division of labor than do the Radiata, resulting in a higher grade of organization. The Platyhelminthes have attained the tissue-organ level of organization; the Rhynchocoela have advanced to the organ-system level by collecting some organs into systems. The development of the mesoderm in the two phyla has meant a great source of tissues, organs, and systems.

4. Contributions to the evolutionary blueprint have been made by both phyla. Among these are the mesoderm, bilateral symmetry, cephalization, subepidermal and mesenchymal musculature, centralization of the nervous system, excretory system, circulation and blood system, and complete digestive system. Their muscular system especially is far better developed and is not restricted to the ectodermal and endodermal epithelia, as it is in most coelenterates.

PHYLOGENY AND ADAPTIVE RADIATION

1. Both Platyhelminthes and Rhynchocoela are closely related. Although opinions differ regarding the origin of the Bilateria, most zoologists think that all Bilateria have originated from the acoelomates. Evidence supports the view that all other Protostomia have the acoelomates as their ancestors. Cleavage, for instance, is of the spiral type in both acoelomates and protostomial coelomates.

2. Order Acoela of the Platyhelminthes has a mouth and no definite gut. One widely accepted theory holds that acoeles may have originated from a planuloid ancestor and in turn gave rise to the other flatworms (Fig. 6-1). The flatworm body plan with its creeping adaptation could facilitate bilateral symmetry, cephalization, caudal differentiation, etc. The Rhynchocoela could have arisen from the Platyhelminthes because the body construction of ciliated epidermis, muscles, mesenchyme-filled spaces, etc. is similar in both groups. But the ribbon worms (Rhynchocoela) have advanced further by having a complete digestive system and a vascular system.

3. Flatworms have evolved into one free-living and two parasitic groups. The free-living Turbellaria best represent the basic archetypic

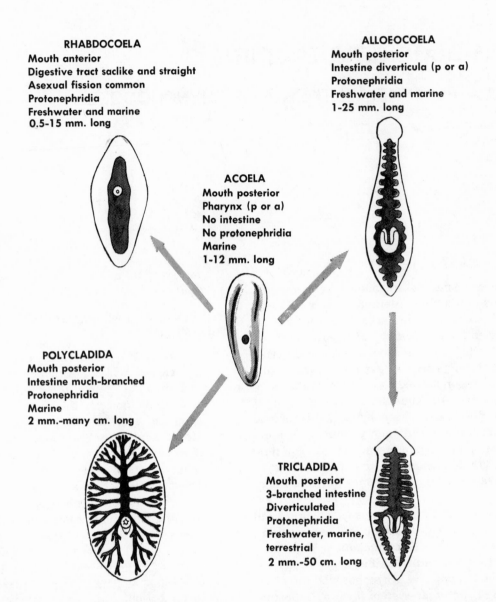

RHABDOCOELA
Mouth anterior
Digestive tract saclike and straight
Asexual fission common
Protonephridia
Freshwater and marine
0.5-15 mm. long

ALLOEOCOELA
Mouth posterior
Intestine diverticula (p or a)
Protonephridia
Freshwater and marine
1-25 mm. long

ACOELA
Mouth posterior
Pharynx (p or a)
No intestine
No protonephridia
Marine
1-12 mm. long

POLYCLADIDA
Mouth posterior
Intestine much-branched
Protonephridia
Marine
2 mm.-many cm. long

TRICLADIDA
Mouth posterior
3-branched intestine
Diverticulated
Protonephridia
Freshwater, marine, terrestrial
2 mm.-50 cm. long

FIG. 6-1

Classic orders of class Turbellaria showing chief diagnostic characters. (**p** means present; **a** means absent.) (From Hickman: Integrated principles of zoology, The C. V. Mosby Co.)

characteristics. All the numerous species are variations on the same theme of body structure and functional behavior. Rhynchocoela have stressed the proboscis apparatus, which at first may have been a sensitive organ for exploring the environment but later became adapted for seizing prey. The Platyhelminthes have been guided in their evolutionary trends by their creeping habit and vermiform body. The group as a whole has set the pattern for the establishment and development of the organ systems of higher metazoan animals.

■ PHYLUM PLATYHELMINTHES

Members of the large phylum Platyhelminthes are known as flatworms. Platyhelminthes represents one of the dozen or so major phyla of the animal kingdom. They are bilaterally symmetric, triploblastic, nonsegmented, dorsoventrally flattened worms that are also characterized by lack of anus (usually), coelom, or hemocoel, and possess no true circulatory or respiratory systems or skeleton. Other characteristics are the presence of an excretory system of flame-bulb protonephridia; a complex, usually hermaphroditic reproductive system; and a solid mesenchyme (parenchyma) filling spaces not occupied by organ systems. Within the group a large diversity of morphologic features as well as of ecologic habits exists, although all have some common features that justify the inclusion of members within the same phylum. Much of this diversity of structure results from the widespread parasitic habits that are found throughout most of the phylum. Members within the phylum illustrate all stages from free existence to partial and complete dependence on other members of the animal kingdom.

In the evolutionary blueprint of the animal kingdom the phylum Platyhelminthes occupies a unique position. They introduce certain morphologic features that, with some refinement and modification, become the adaptive structural characters of all higher groups. Among these important contributions are bilateral symmetry, true triploblastic organization in which mesoglea is replaced by a cellular, endomesodermal layer, a subepidermal musculature suggestive of the muscles of more advanced animals, an excretory system for a more demanding metabolism, and the start of a centralized nervous system of the ladder type in place of the diffused nerve net of radially symmetric animals. It is of physiologic interest that flatworms are considered by animal behaviorists to be the lowest group that can be studied for basic concepts of behavior and learning. Not the least of the flatworm contributions to the blueprint of the animal kingdom is their organ-tissue architecture in which a tendency occurs for an anatomic unit to be made up of several tissues. This tendency definitely points the way toward the complexity of higher forms.

CLASSIFICATION

Some 10,000 to 15,000 species of flatworms have been listed to the present time. Students of this group differ a great deal concerning its taxonomy, and several schemes have been proposed for its classification, including those of L. H. Hyman, B. Dawes, J. G. Baer, G. R. La Rue, and others. This disagreement is not merely restricted to taxa of lower levels but involves those of class rank as well. However, the scheme here followed is one that corresponds with the opinion of many zoologists.

Phylum Platyhelminthes. Body bilaterally symmetric, dorsoventrally flattened, acoelomate; soft and vermiform; body covered by ciliated epithelium or cuticle; some provided with suckers and hooks for parasitic existence; digestive system usually incomplete (gastrovascular) or absent; excretory system of flame-bulb protonephridia; body spaces filled with parenchyma (mesenchyme); no true circulatory or respiratory systems; reproductive system complicated and usually hermaphroditic; development direct or with metamorphosis of larval stages; free living, commensal, or parasitic habits.

    Class Turbellaria. Mainly free living, from 5 mm. to 50 cm. long; cellular or syncytial epidermis usually ciliated and with rhabdoids; body undivided with ventral mouth; gastrovascular cavity (usually); no suckers; simple life cycles.

        Order Acoela. *Convoluta.*

        Order Rhabdocoela. *Microstomum.* (Suborders raised to ordinal rank by some are the Catenulida, Macrostomida, and Neorhabdocoela.)

        Order Alloeocoela. *Plagiostomum.* (Suborders raised to ordinal rank by some are the Archoophora, Lecithoepitheliata, Holocoela, and Seriata.)

        Order Tricladida. *Dugesia.*

        Order Polycladida. *Notoplana.*

    Class Trematoda. Body undivided and covered with a tegument of living cells, forming a type of epidermis; one or more suckers; digestive tract mostly two branched; one ovary; life cycles may be simple and direct or complex with alternation of generations; mostly hermaphroditic but some dioecious; almost exclusively ectoparasitic or endoparasitic.

        Order Monogenea (Heterocotylea). *Polystoma.*

        Order Aspidobothria (Aspidogastraea). *Aspidogaster.*

        Order Digenea (Malacocotylea). *Opisthorchis.*

    Class Cestoda. Body with outer covering of living

cells (tegument) connected to deeper cells and permeated by pore canals; no digestive system at any stage; usually an anterior organ (scolex) specialized for attachment with bothria, bothridia, acetabula, or hooks; body usually of few to many proglottids, or segments; each proglottid with one or two complete hermaphroditic reproductive systems; embryo hooked; complicated life cycles, usually involving two or more hosts.

Subclass Cestodaria. Body undivided.

　Order Amphilinidea. *Amphilina.*

　Order Gyrocotylidea. *Gyrocotyle.*

Subclass Eucestoda. Body usually divided into proglottids.

　Order Tetraphyllidea. *Phyllobothrium.*

　Order Proteocephaloidea. *Proteocephalus.*

　Order Trypanorhyncha. *Haplobothrium.*

　Order Pseudophyllidea. *Dibothriocephalus.*

　Order Cyclophyllidea (Taenioidea). *Taenia.*

GENERAL MORPHOLOGIC FEATURES

Although flatworms are basically flattened dorsoventrally, as the name suggests, and in general have a vermiform appearance, considerable diversity of form exists within the phylum. Their organization is more complex than that of the coelenterates and can exploit a greater range of habitats. This diversity is not restricted to differences between the three classes but also applies to members of the same class. Some species are only moderately elongated, with segments (proglottids) that form tapering or ribbonlike organisms. Still others have a broad leaflike form. In addition to these typical shapes, many odd and bizarre forms are also found in a group that counts many thousands of species. Since many flatworms are generally transparent, their immediate color may be due to ingested food; but in some species the color is normally brown, gray, or blackish, with occasional brighter colors. The bilateral vermiform condition has stressed a head end in contrast to a less differentiated posterior region, but exceptions exist in which posterior features are more prominent than those of the head.

Most flatworms have organs of adhesion that may be glandular, muscular, or in the form of suckers and hooks. Most organs of attachment and adhesion are found in the head region, although they may also occur in the posterior

FIG. 6-2

Cross section through pharyngeal region of planarian. (From Hickman: Integrated principles of zoology, The C. V. Mosby Co.)

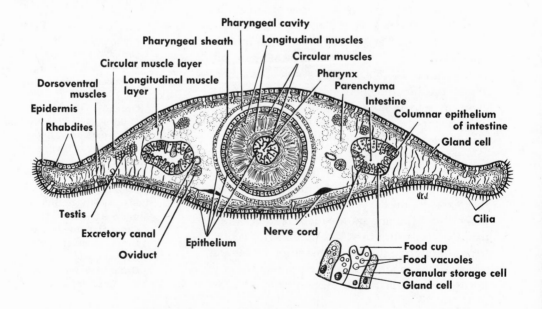

region or in other regions of the body. Parasitic members especially are usually well provided with such adhesive devices, but they may also be found among free-living individuals.

The body covering varies with the class. Turbellarians have a definite, one-layered epidermis, which is cellular or syncytial and usually bears cilia in tracts or over the entire body (Fig. 6-2). Rod-shaped bodies (rhabdites) are found in the cells and, when discharged in the water, may form gelatinous sheaths around the body for protection. Irritants may be responsible for their discharge. Epithelial cells may be large and flat or columnar with round nuclei.

Below the epidermis is a thin, noncellular basement membrane that may serve for the attachment of muscles. This basement membrane is scanty or absent in certain parts of the body (sensory organs). In some cases the epidermal cell nuclei sink into the parenchyma and retain connection with the epidermis by long strands of cytoplasm.

In the parasitic classes Trematoda and Cestoda, studies with the electron microscope have completely changed the classic concepts of their cuticle. It is now known that their body covering (tegument) consists of syncytial, living protoplasm (not an inert, lifeless secretion) that represents extensions of deeper-lying nucleated protoplasm or cells (Fig. 6-3). The outer surface of the tegument is folded into many small extensions (microtriches), and pore canals, which open to the exterior and penetrate through it into the parenchyma. These newer morphologic discoveries explain many functional aspects once considered obscure. The

FIG. 6-3

Section through typical tegument of cestode as revealed by electron microscope (diagrammatic).

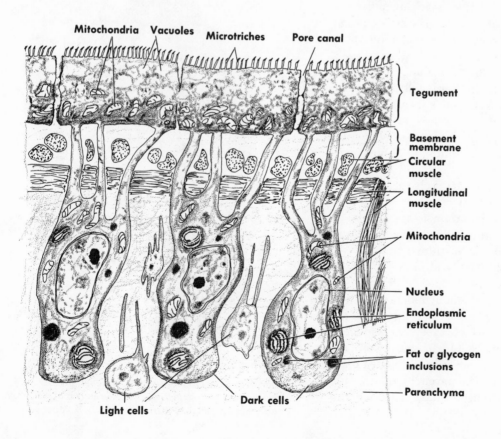

microtriches facilitate absorption of food by increasing the surface area, and by interlocking with the microvilli of the host's mucosa, they may partially act as holdfast organs (A. H. Rothman, 1963). Pore canals could also help absorb substances in solution. Finally, the presence of living protoplasm may explain the ability of the worm to resist digestion by the host because of the constant renewal of the outer tegument by its deeper layer of cells. It will be noted that the tegument lies on a basement membrane, below which are the circular and longitudinal muscle layers. Beneath the muscle layers are dark, elongated cells, the extensions of which supposedly form the tegument. Variations in the structure of the tegument are known to occur in the different regions of the body, both within the same species as well as between species (B. J. Bogitsh, 1968).

A cuticle is almost entirely lacking in turbellarians because of the cilia. Subepidermal unicellular glands of many kinds are common in free-living flatworms and secrete slime trails and adhesive devices.

A skeleton of any kind is lacking in the phylum, although relatively hardened parts exist, such as spines, hooks, and suckers, that are differentiations of the epidermis, or tegument. Such hardened parts are made up of scleroprotein (cuticularized), but chitin is almost nonexistent.

The parenchyma, or mesenchyme, that fills the body spaces of all flatworms varies in different platyhelminths. In origin it appears to be chiefly endomesoderm. Its general structure is that of a netlike syncytium of cells with long, irregular processes and small granules. The syncytium may be formed by the anastomosis of branched mesenchymal cells. Numerous fluid-filled interstices are scattered in the syncytium through which free ameboid cells wander. These ameboid cells are useful as formative cells in regeneration of injured parts and in reproduction. The parenchyma, in addition, serves in the transportation of food and in excretory functions.

The nervous system appears to be derived directly from coelenterates, at least in primitive flatworms; it consists essentially of a subepidermal network and an aggregation of nerve cells at the anterior end to form cerebral ganglia (brain) in free-living forms. Other modifications of the nerve net pattern involve the sinking of the nervous system into deeper regions (mesenchyme) of the body and the formation of longitudinal cords, of which the 2 ventral ones are the most prominent (Fig. 6-4). Transverse connections between the longitudinal cords form the characteristic ladder type of nervous system among flatworms (Figs. 6-4 and 6-13).

Sense organs are common in free-living flat-

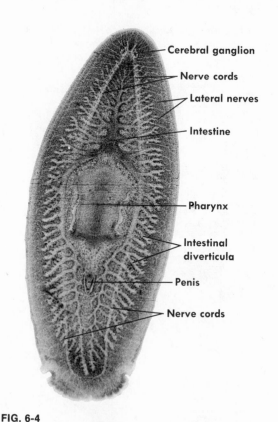

FIG. 6-4

Ladder-type nervous system of turbellarian *Bdelloura,* a marine triclad (stained preparation). (From Hickman: Integrated principles of zoology, The C. V. Mosby Co.)

worms and in the larval stages of the parasitic classes. In the adult trematodes and cestodes they are restricted to a few tangoreceptors and free sensory nerve endings. In turbellarians the sense organs take the form of eyes, otocysts, ciliated pits, tentacles, chemosensory receptors, and statocysts. The most common sensory organ is the tangoreceptor, or tactile

cell, that terminates in bristles or hairlike structures. Taste and rheotactic receptors have a wide body distribution and enable the animal to detect food juices and changes in the flow of water over the surface of the body. Ciliated pits with long cilia and bristles are the location of sensory cells. Certain areas on the head region are provided with ciliated grooves that bear special chemosensory receptors (auricular organs).

Eyes, or ocelli, are very common among the turbellarians (Fig. 6-5); usually there is a single pair, but there may be more. They usually consist of pigment cups sunk into the mesenchyme, in which photosensitive cells (Fig. 6-6) are located. The retina (photosensitive) consists of bipolar nerve cells with rounded, club-shaped distal endings projecting into the cup, their proximal endings joining the central nervous system. A lens is absent, but the epidermis over the eye is not pigmented and may serve as a type of cornea. Many modifications of this pattern exist. In this inverse type of eye the light has to first pass through the mass of nerve fibers before reaching the sensitive retina.

A digestive system is present in all flatworms

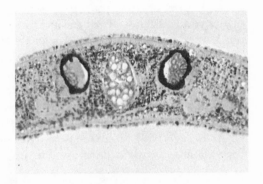

FIG. 6-5

Cross section through anterior end of planarian showing eyes as pigmented cups. Structure between eyes is anterior branch of intestine. (From Andrew: Textbook of comparative histology, Oxford University Press, Inc.)

FIG. 6-6

Ocelli, or eyes, of planarian (diagrammatic). Pigment cup allows light to enter open side parallel to long axis of light-sensitive cells. Planarian determines light direction from stimulation of light-sensitive cells. (From Hickman: Integrated principles of zoology, The C. V. Mosby Co.)

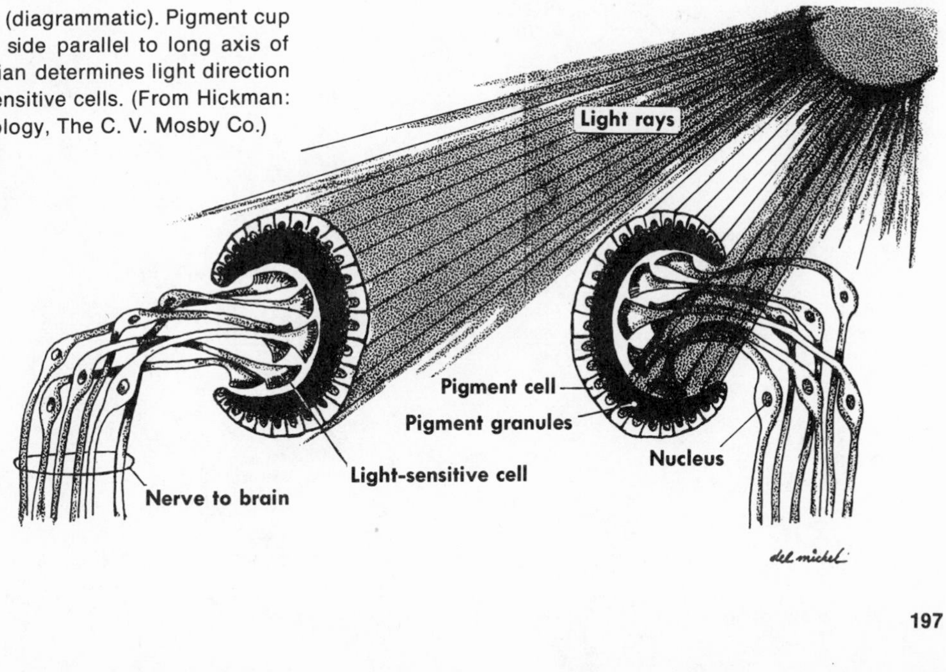

Light rays

Pigment cell

Pigment granules

Nucleus

Light-sensitive cell

Nerve to brain

del michel

except those of the order Acoela and the class Cestoda. The mouth, which serves for both ingestion and egestion, is found either anteriorly or ventrally. In simple forms the digestive sac is straight without diverticula, but in others it is branched.

The stomodeal pharynx is typically a muscular tube and may or may not be protrusible (Figs. 6-9 and 6-13). It occurs in a variety of types such as simple, bulbous, and plicate. The *simple* pharynx is a short tube of inturned ciliated epidermis with subepidermal musculature and is not delimited from the mesenchyme. It is found in acoels and in some of the suborders of the alloeocoels and rhabdocoels. The *bulbous* pharynx is sharply delimited from the mesenchyme and is much more muscular than the simple type. Its distal part is nonmuscular and forms a pharyngeal cavity into which the bulbous part fits. This type of pharynx occurs in two variations, the rosulate (spheric-shaped) type oriented at right angles to the body axis and the dolioform type (cask-shaped) oriented parallel to the body axis. The bulbous pharynx is found in some of the rhabdocoels and alloeocoels. The *plicate* pharynx, characteristic of triclads and some of the alloeocoels and rhabdocoels, has a fold, cylindric or ruffled, projecting into the pharyngeal cavity from the anterior end or roof of the cavity. It is protruded through the mouth in feeding, and the food is ingested through the orifice of the pharynx. There may be many mouth openings, each provided with a plicate pharynx (polypharyngy) in some freshwater and land planarians. The plicate pharynx is the most complicated of the three types, having muscular, glandular, and nervous tissue.

The digestive tract is lined with a single, nonciliated layer of cells that are phagocytic and digestive in function. With the exception of the pharynx, the digestive system appears to have little or no muscular layer. Digestion is known to be both extracellular, occurring in the gut or outside the animal, in which case enzymes are poured out on the food, and intracellular in the phagocytic cells lining the gut. G. G. Berg and O. A. Berg (1968) discovered that the enzyme trimetaphosphatase is decreased during starva-

FIG. 6-7

Adenoplea, a soil turbellarian (order Rhabdocoela). **A,** Feeding on a nematode worm *(Panagrellus);* turbellarian curls over prey, attaches pharynx, and draws in prey whole. **B,** Adult with fully developed egg; when laid, eggs are enclosed in a protective capsule similar to that of planarian eggs. (From Sayre, R. M. 1970. Amer. Biol. Teacher **32:**487-490.)

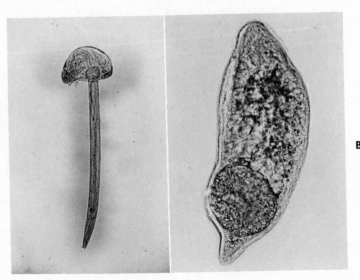

A   B

tion and greatly increased at feeding. They are the first to describe this particular enzyme in an invertebrate. The acoels have no intestine, and the phagocytic cells are found in the mesenchyme.

Turbellarians as a rule are carnivorous, living on rotifers, copepods, nematodes, small annelids, and others (Fig. 6-7, A). The larger species feed on earthworms (Fig. 6-17), snails, crustaceans, insect larvae, etc. Detecting the prey either by chemoreception or by noting movements of the prey as reflected in water disturbance, the turbellarian grips the prey with its anterior end and wraps itself about the prey, using mucous and adhesive secretions to help entangle and hold it. Most turbellarians swallow their prey whole by means of muscular waves of the pharynx (Fig. 6-7). Triclads can extend the pharynx from the mouth and, instead of swallowing the food whole, may insert the

pharynx into the body of the prey or into carrion, aided by secretions of the pharyngeal glands. These secretions may also soften the prey's tissue for ingestion but have no digestive action. The food is then sucked into the gut by peristaltic action. The suctorial action of the pharynx may tear the food into fragments, and a little digestion is continued by proteolytic enzymes from gland cells in the gut (extracellular digestion). Food particles are taken into the gastroderm cells by phagocytosis, and digestion continues in the food vacuoles by intracellular digestion (J. B. Jennings, 1962).

Throughout the phylum, with the exception of the order Acoela, the typical protonephridial excretory system is found. It usually consists of main canals, which run down either side of the body and which break up into smaller and smaller branches that terminate in flame bulbs. The main canals open to the exterior by nephridiopores, the positions of which vary (Fig. 6-8, A). The terminal part of a main canal may be enlarged to form a bladder just before the pore. A flame bulb, or protonephridium (Fig. 6-8, C), is made up of a cell with branched processes and a capillary in which cilia from the bulbous free end of the cell vibrate to produce the characteristic "flame." A single cell may supply cilia to several capillaries in some forms. Flame bulbs are best seen in the cercaria of trematodes. The tubes and flame bulbs run throughout the mesenchyme. The smaller branches of the tubular . excretory system are ciliated, whereas the main canals are usually not. In the process of excretion the cells of the capillary wall appear to pass fluids and substances into the lumen of the tube, and the action of the flame bulb is to maintain a hydrostatic pressure that keeps excreted substances moving down the tube. The function of the excretory system seems to be osmoregulation of body fluids and possibly some excretion as well.

Flatworms lack respiratory and circulatory systems. Exchange of respiratory gases takes place by diffusion through the body wall in turbellarians. Among parasitic members

FIG. 6-8

Excretory system of flatworm such as planarian. **A,** Two canals with their network of tubules in anterior part of body. **B,** Portion of one canal system enlarged to show network of tubules and protonephridia (flame cells). **C,** Two protonephridia enlarged.

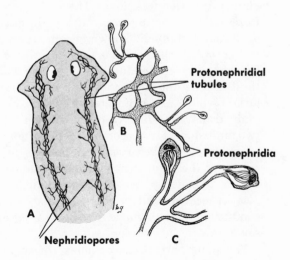

Protonephridial tubules

Protonephridia

A

B

C

Nephridiopores

anoxybiotic respiration occurs in which glycogen is used and carbon dioxide and fatty acids are given off. Some facultative parasites may be oxybiotic and make use of the small amount of oxygen in the gut.

Platyhelminthes reproduce both asexually and sexually. Asexual reproduction usually involves fragmentation or binary fission. By such methods some turbellarians produce, temporarily, long chains of individuals. Trematodes in their larval stages undergo polyembryony, by which individuals are asexually increased. Tapeworms bud off proglottids asexually from the neck region of the scolex, and some (e.g., *Echinococcus*) produce numerous larvae by internal or external budding.

Most flatworms are hermaphroditic, and their reproductive systems are the most complex in the animal kingdom. Sex cells are developed from certain free cells of mesenchyme and migrate to regions where gonads are formed. (In the primitive form Acoela no gonads or oviducts exist, and the germ cells are arranged in longitudinal rows of eggs and sperm.) In the reproductive system (Fig. 6-9), testes are usually numerous, but only one or two ovaries are present as a rule. One unique feature of the female system is that the yolk is added by vitelline glands to the eggs after the latter leave the ovary, instead of yolk material becoming a part of the egg structure within the ovary as in other animals. The yolk, or vitelline, glands may discharge into the oviducts either directly or by special yolk ducts.

To ensure cross-fertilization, a copulatory apparatus or complex is usually present (Figs. 6-4 and 6-9). This consists mainly of a genital antrum for receiving the male penis and sacs to store the sperm from the sex partner. The oviducts may or may not unite before entering the complex. The female complex, therefore, usually consists of a branched or rounded ovary from which runs a short oviduct that is joined by the common yolk duct and a short duct from the seminal receptacle. The muscular Laurer's canal also joins the oviduct between the ovary

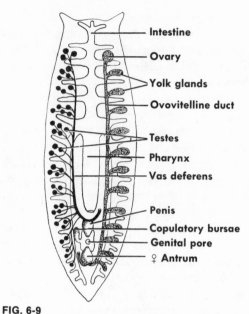

FIG. 6-9

Reproductive system of turbellarian. Male structures are shown on left, female on right; but in life, both systems are paired.

and the entrance of the common yolk duct. Laurer's canal (often absent) may open to the outside and may function in copulation in some unknown manner. The oviduct widens to form the ootype, which is surrounded by Mehlis' glands (function unknown). These glands were formerly thought to secrete the shell, but the shell material is now known to come from yolk cell granules. The ootype is the place where the finished eggs are produced. Along each side of the body are the numerous yolk glands (vitellaria), which are connected by ducts to one main transverse duct from either side. These two main ducts form a common duct (often with an enlarged yolk reservoir) before entering the ootype or the oviduct nearby. From the ootype the much-coiled uterus (a continuation of the oviduct or ovovitelline duct) extends to the common genital antrum, or the uterus may open separately to the exterior.

The male part of the complex consists of a protrusible penis or eversible cirrus and single or paired vesicles (spermidual) for storing sperm. The sperm ducts, which carry the sperm from the testes, may unite before entering the copulatory apparatus. The sperm tails in many flatworms have a $9 + 1$ arrangement of fibrils instead of $9 + 2$ (B. R. Hershenov and others, 1966). The spermiducal vesicles are dilated portions of the sperm ducts; when they are present, seminal vesicles are usually absent and vice versa. The lumen of the common sperm duct is called the ejaculatory duct and opens at the end of the penis. In addition, a prostatic apparatus may be present, consisting of gland cells that open directly or by way of the prostatic vesicle into the ejaculatory duct. The penis is pear shaped and muscular; at rest it opens into the genital atrium, but in copulation it is pushed through its own gonopore into the gonopore of its sex partner. The penis often bears a stylet or spines; sometimes in its place is an eversible cirrus formed from the terminal part of the ejaculatory duct. In trematodes the male copulatory organ is usually a cirrus.

Both male and female systems are highly variable in the flatworms, and modifications of the plan just described are numerous throughout the group. For instance, a common gonopore or separate gonopores for the male and female systems may exist. Gonopores may be located ventrally (the usual place), or they may be lateral or even dorsal. A common genital antrum or separate ones may also exist. In some turbellarians a genitointestinal canal is found between the reproductive system and the intestine, and the sex cells are discharged through the mouth. A separate vaginal or bursal pore for copulation only is found in some members.

Although cross-fertilization by mutual insemination is the rule among flatworms, self-fertilization sometimes occurs. During copulation the ventral surfaces of the sex partners are closely applied together so that their gonopores are opposite to each other. Then the penes or cirri are each pushed into the gonopore of the other for the mutual exchange of sperm. No preliminary courtship is involved (M. M. Jenkins and H. P. Brown, 1964). Hypodermic im-

pregnation, in which sperm are injected directly through the epidermis, may occur in some turbellarians and cestodes. Either eggs are laid in gelatinous masses (Acoela and Polycladida), or the egg mass is enclosed in a capsule. Capsules are often deposited in cocoons within which juveniles develop (most Turbellaria). Eggs of trematodes and cestodes usually emerge from the hosts and hatch into free-swimming larvae that enter the intermediate host. In some trematodes and many cestodes the eggs must be eaten before hatching occurs. Some turbellarians produce thin-shelled (summer) eggs with few yolk cells and thick-shelled, dormant (winter) eggs that can withstand the rigors of winter. The summer eggs are self-fertilized and may develop in the body of the parent.

Except in the Acoela and the Polycladida, which lack yolk glands, the eggs are ectolecithal, and the yolk is contained within yolk cells; however, in the Acoela and Polycladida the egg is endolecithal, with the yolk enclosed in the ova. The type of development is determined primarily by the differences between these two types of eggs. The most typical and least modified development is that of the endolecithal egg. In this egg the cleavage is of the spiral determinate type similar to that of annelids and mollusks.

In some flatworms, development is direct, but many (marine polyclads) produce characteristic larvae such as Müller's or Gotté's larvae. These larvae will be discussed under descriptions of the various groups. (See discussion on embryology of the metazoans in Chapter 3.

■ CLASS TURBELLARIA
■ ORDER ACOELA

Acoel marine worms are considered to be the most primitive members of the phylum. They range in length from 1 to 12 mm. and are somewhat oval to elongate in form. They have little color except what they acquire from symbiotic algae. The ciliated epidermis is now known to be

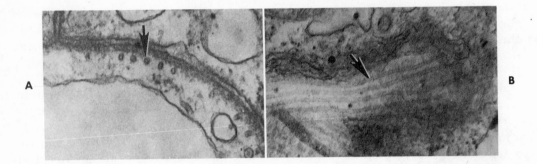

FIG. 6-10

Microtubules of a subdermal mucous gland cell in the acoel *Convoluta roscoffensis,* as seen with electron microscope. **A,** Transverse section of a mucous gland cell cut interior to surface of animal. Microtubules (transverse cut) are just inside gland cell membrane but exterior to membrane bounding a secretion granule. (×67,000.) **B,** A partial section of a mucous gland cell cut tangential to surface membrane showing microtubules cut longitudinally. These microtubules may be useful in the attachment of the animal to its substratum. (×58,000.) (From Oschman, J. L. 1967. Trans. Amer. Microscop. Soc. **86:**159-162.)

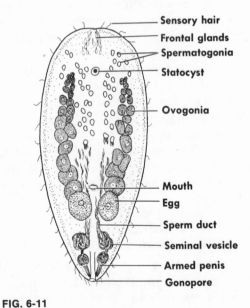

FIG. 6-11

Childia spinosa (order Acoela). This primitive form has epidermal musculature similar to that of coelenterates. (Modified from Graff, 1911.)

cellular and not syncytial, and the bases of the epidermal cells may bear contractile elements that form a muscle layer. Acoels have many kinds of subepidermal glands on the ventral and marginal surfaces of the posterior third of the organism. These glands may have an adhesive function, although the chemical nature of their secretions is unknown. The glands may protrude through the epidermis to the level of the tips of cilia that cover the epidermal surfaces. J. L. Oschman (1967) has shown that the striations on the protruded necks of the glands are microtubules (in *Convoluta*), which may give rigidity to the glands for attachment to the substrate on which the worm moves (Fig. 6-10). One or two statocysts and some frontal glands are usually found at the anterior tip (Fig. 6-11). Sense organs may also include sensory bristles and one or two pairs of eyes. The mouth is usually found near the middle of the ventral side and leads into a parenchymal mass that makes up the interior. In feeding, the acoel *Convoluta* forms a solid mass of its endoderm, which is protruded through the mouth, and the food is engulfed into food vacuoles (E. J. W. Barrington, 1967). According to the electron microscopic studies of J. L. Oschman and P. Gray (1965), the acoel *Convoluta roscoffensis* (as an adult) depends on its endosymbiotic algae (zooxanthellae) for its major nutrients. These green algae *(Carteria)* lie in the intercellular spaces interior to the subepidermal musculature. Extensions of the algal cells pass through the musculature and into the ciliated layer of the epidermis, thus facilitating transfer of stored carbohydrates (pyrenoids and starch grains) to the region of the greatest energy requirements.

The nervous system consists of poorly developed ganglia and several pairs of longitudinal nerves that lie subepidermally or deeper.

As mentioned before, the hermaphroditic reproductive system contains no gonads or oviducts, but sperm ducts are present. Scattered clusters of mesenchyme cells directly form sex cells that are arranged in longitudinal tracts of eggs and sperm. Both male and female copulatory mechanisms are present. Single or paired seminal or spermiducal vesicles open into the male canal of the penis (often 2 or 4 in number), which may be provided with prostatic glands. Sperm is discharged through the male gonopore near the posterior end, or the male canal may unite with the female canal. A female canal, or vagina, may join the common antrum, or a separate female gonopore may be present. In the absence of female structures, copulation may occur by hypodermic impregnation, in which sperm are injected directly through the epidermis of the sex partner. Fertilized eggs are discharged through the mouth, gonopore, or body rupture. Development is direct, and the young worms are similar to the adult in form.

The Acoela live under stones, in the bottom muck, or among algae. Some are pelagic and others are ectocommensal. *Convoluta* and *Amphiscolops* are two well-known genera.

The phylogenetic significance of Acoela has already been mentioned (p. 191).

■ ORDER RHABDOCOELA

The rhabdocoel turbellarians are broken up into three suborders—Catenulida, Macrostomida, and Neorhabdocoela—which in the newer classification are raised to the rank of orders. The length of these worms may range from 0.3 to 15 mm., and they include freshwater, marine, and terrestrial species. The body is elongated to an oval, plump shape. The ciliated epidermis is cellular or syncytial and is provided

with rhabdites. The worms usually lack the frontal glands, but their body surface has sensory bristles or hairs. A statocyst is usually lacking. They often bear one or more eyes of the epidermal pigment-spot type. The nervous system usually consists of a bilobed brain and 2 main longitudinal nerves, with few transverse connectives. The rhabdocoels have a simple, unbranched digestive tract with few or no diverticulations and a simple or bulbous pharynx. The excretory system is of the protonephridian type. The reproductive system is made up of compact gonads (a pair of testes and 1 or 2 ovaries), a pair of yolk glands, and the copulatory complex. Mutual insemination normally occurs, and the eggs (single or multiple) are enclosed in a capsule when laid (Fig. 6-7, *B*, and 6-16). Yolk glands are lacking in many species.

The order Catenulida is made up of colorless freshwater forms with a simple pharynx, a single median protonephridium, no yolk glands, and usually no sex organs. The nervous system contains a brain and four pairs of longitudinal nerves. Reproduction is mainly asexual by binary fission and the formation of zooid chains. In some members one or more ovaries may exist without ducts, and a single testis with a duct modified into a copulatory organ that opens through an anterior, dorsal pore may be present. In such cases the endolecithal eggs are discharged by rupture of the body wall. *Stenostomum* is a common freshwater genus.

Members of the order Macrostomida have paired protonephridia and a single pair of longitudinal nerves. Yolk glands are lacking, but the usual hermaphroditic reproductive system is often present. Chains of zooids are formed by asexual methods. The common freshwater genus *Microstomum* is often provided with nematocysts obtained by eating hydras.

Most rhabdocoels belong to the order Neorhabdocoela and have a bulbous pharynx, paired protonephridia, sexual reproduction, yolk glands, ectolecithal eggs, and ventral gonopores. The mouth may be located at the anterior end or near the middle of the body. Some members are provided with a protrusible proboscis.

The extensive group of neorhabdocoels is broken up into subgroups or suborders such as Dalyellioida, Typhloplanoida, Kalyptorhynchia, and Temnocephalida.

The Dalyellioida are characterized by a dolioform (barrel-shaped) pharynx, with the mouth near the anterior end. The principal genus is *Dalyellia*, common in fresh waters.

The Typhloplanoida usually have a rosulate pharynx (sometimes dolioform), with the mouth near the middle of the body. *Mesostoma* (Fig. 6-12) is a common freshwater genus throughout the world.

The Kalyptorhynchia have a rosulate pharynx, but at the anterior end they are also provided with a pouch possessing a protrusible glandulomuscular proboscis by which they are able to capture their prey. *Gyratrix* is a common genus in both fresh and marine water.

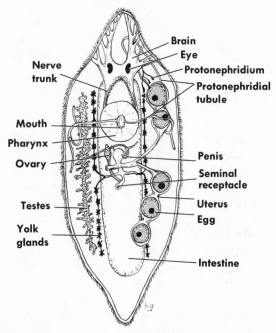

FIG. 6-12

Mesostoma (order Rhabdocoela), a common freshwater form whose summer eggs hatch within the body. Testes shown only on left; uterus only on right.

The Temnocephalida are ectocommensals on crayfish and other forms in tropic and subtropic waters. They differ from most rhabdocoels in possessing tentacles and adhesive organs; also, most species lack cilia. *Temnocephala* is a typical genus.

■ ORDER ALLOEOCOELA

As a group the alloeocoels are intermediate in their characteristics between the Acoela and the Tricladida. They range in length between 1 mm. and 40 mm. They are mostly of an elongated and cylindric shape and are found in freshwater, saltwater, and terrestrial habitats. The alloeocoels share with the rhabdocoels in being the most common turbellarians in soil habitats. Their chief characteristics are a plicate pharynx, a diverticulated enteron, three or four pairs of longitudinal nerves with transverse connections, a pair of protonephridia each with one to three branches, and a reproductive system of paired ovaries, many testes, and a penis. Eyes may or may not be present. The group ranges between the primitive characteristics of Acoela and the more specialized and complex structures of the triclads, which were supposedly derived from the alloeocoels.

A recent classification of the Turbellaria divides the Alloeocoela into a number of orders based on raising former suborders to ordinal rank. These new orders are Archoophora, Lecithoepitheliata, Holocoela, and Seriata. A brief description follows.

Order Archoophora possesses a frontal gland, buccal tube, plicate pharynx, and male copulatory apparatus but no female ducts. So far it is represented by a single species, *Proporoplana jenseni*.

In the order Lecithoepitheliata, yolk glands are absent, germovitellaria are present, a simple female duct may be present (sometimes absent), and usually also a typical male apparatus. A common freshwater species is *Prorhynchus stagnalis*.

Order Holocoela has yolk glands distinct from the ovaries and an unarmed penis. Their pharynx is plicate, and a frontal gland may be present. The most common genus is *Plagiostomum,* found in both fresh and marine waters.

The members of the order Seriata have a backward-directed plicate pharynx, numerous yolk glands with a pair of yolk ducts, and usually adhesive papillae. A seminal bursa is usually found. A common marine genus is *Monocelis,* found in littoral regions.

■ ORDER TRICLADIDA

Triclad worms range in length from 2 mm. to more than 50 cm. Their body is elongated, usually flattened, and provided with many mucous glands. The epidermis is ciliated on the ventral surface (rarely elsewhere) and bears many rhabdites. Adhesive glandulomuscular organs are common on the ventral side of the anterior end. Eyes are lacking in cave forms only. The mouth is located posterior to the ventral middle surface and opens into the pharyngeal cavity (Fig. 6-13). The worms possess a plicate and tubular pharynx attached anteriorly and a three-branched, diverticulated intestine, consisting of one anterior branch and two posterior ones. Some have more than one pharynx (poly-

pharyngy) and many mouth openings. The excretory system consists of a network of protonephridial tubules, with numerous nephridiopores on each side of the body. The female reproductive system (Fig. 6-9) consists of 2 small ovaries anteriorly located, a pair of ovovitelline ducts with numerous yolk glands, the female antrum, and usually 1 or 2 copulatory bursae. The male system (Fig. 6-9) has few to many testes (Fig. 6-14) laterally located, a sperm duct on each side connected to the testes by sperm ductules, a pair of spermiducal vesicles or enlarged parts of the sperm ducts, and a male antrum containing the copulatory organ (penis) into which the sperm ducts enter singly or after fusion (Fig. 6-15).

In copulation, which occurs by mutual insemination, the sperm first enter the copulatory bursa before traveling up the ovovitelline ducts to fertilize the eggs near the ovaries (Fig. 6-15).

FIG. 6-13

A, Digestive and nervous systems of planaria (diagrammatic); cutaway section shows ventral mouth. **B,** Pharynx extended through ventral mouth. (From Hickman: Integrated principles of zoology, The C. V. Mosby Co.)

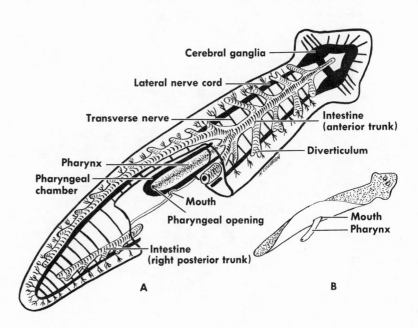

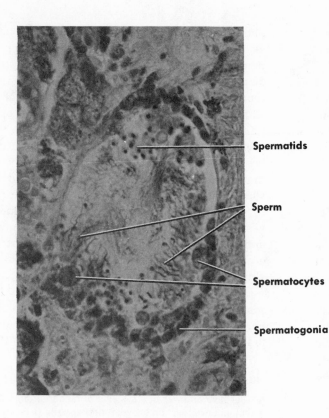

FIG. 6-14

Section through testis of *Dugesia* showing sperm cells in different stages of development.

Spermatids

Sperm

Spermatocytes

Spermatogonia

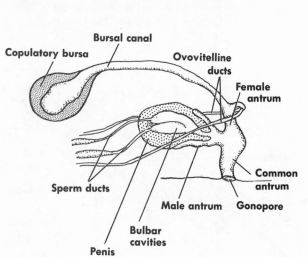

FIG. 6-15

Copulatory organs of planarian.

FIG. 6-16

Egg capsule (cocoon) of planarian. Capsule is formed in male antrum and contains 1 to several eggs and yolk cells. Eggs hatch into juvenile worms in 2 to 3 weeks. (Courtesy Carolina Biological Supply Co., Burlington, N. C.)

The fertilized eggs collect yolk from the yolk glands as they pass down the ovovitelline ducts (Fig. 6-15) and finally enter the male antrum, where a capsule, or cocoon, is formed around them. Each capsule (Fig. 6-16) has a stalk and an attachment disk which is extruded first through the gonopore followed by the cocoon. (M. M. Jenkins, 1966). The cocoon, containing one to several fertilized eggs, is attached by a stalk to some object after discharge and hatches in about 2 weeks into juvenile worms. Asexual reproduction by fission is common in some triclads, and in some, (the land planarian *Bipalium*), fragmentation and regeneration are the usual method.

Classified according to habitat, there are three well-recognized suborders of this order.

The Maricola are rather small marine forms that occur under stones along the coastlines of

FIG. 6-17

Bipalium kewense feeding on an earthworm. This land planarian (suborder Terricola, order Tricladida) may reach a length of a foot and reproduces asexually by fragmentation. In fragmentation a small portion of posterior end pinches off and assumes adult characteristics in a few weeks. In the laboratory it may release one or two fragments per month. (From Connella, J. V. and D. H. Stern. 1969. Trans. Amer. Microscop. Soc. **88**:309-311.)

most seas and are incapable of asexual reproduction. *Bdelloura,* a common ectocommensal on the horseshoe crab *Limulus (Xiphosura),* is a familiar genus.

The Paludicola, or freshwater forms, are usually larger than the Maricola and are found in streams, ponds, and lakes of the temperate zones. Lake Baikal in Siberia is unusually rich in species of this group. To this group belong the familiar planarians, of which there are many species in America, Europe, and Asia. *Dugesia tigrina* is well known to all zoologists. It reproduces both sexually and asexually, depending on weather and temperature conditions.

To the Terricola belong the land planarians that live on the forest floors under leaf mold, logs, etc. They cannot endure complete submersion in water, although they must have humid conditions. They are found mostly in tropic and subtropic regions, although a few have been introduced into the temperate zones. *Bipalium,* a large worm 10 to 12 inches long, has often been found in greenhouses in the United States. Some of its species reproduce entirely by fragmentation, but one or two species in California reproduce sexually. In their natural habitats of the tropics the Terricola reproduce both sexually and asexually. Fig. 6-17 shows *Bipalium* feeding on an earthworm (J. V. Connella and D. H. Stern, 1969).

■ ORDER POLYCLADIDA

Polyclad marine flatworms derive their name from their enteron, which has many diverticula radiating from the central main stem. In length they vary from 2 mm. to more than 15 cm. Their body is much flattened and often broadly leaflike. Their main habitat is on the bottom of littoral zones, although some members are pelagic and swim down to considerable depths. A few are commensals on shells of mollusks and hermit crabs. Their color varies from the transparency of pelagic species to the bright hues of those in warm waters. Many eyes and usually some tentacles are present. They lack statocysts and frontal glands; adhesive

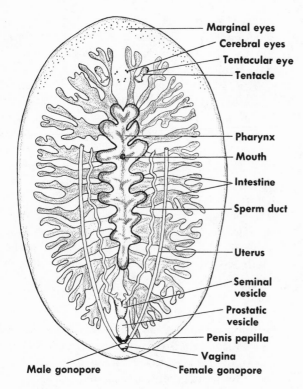

Marginal eyes
Cerebral eyes
Tentacular eye
Tentacle

Pharynx
Mouth
Intestine
Sperm duct

Uterus

Seminal vesicle
Prostatic vesicle
Penis papilla
Vagina
Male gonopore   Female gonopore

FIG. 6-18

Stylochus ellipticus, a common polyclad of both Atlantic and Pacific coasts. (Modified from Pearse, 1938.)

organs are rare. The ciliated cellular epidermis is provided with numerous glands and rhabdites. The nervous system is made up of a brain anterior to the pharynx and a number of radiating nerves, two of which are usually larger than the others. The mouth is located near the middle of the body and opens into the plicate pharynx. Yolk glands are absent, but testes and ovaries are numerous. Other parts of the reproductive system are similar to the flatworm plan (Fig. 6-18). Insemination may be hypodermic or regular, but self-fertilization does not occur. The endolecithal eggs are fertilized just before laying. Sperm usually collect first in Lang's vesicle, a bursa with a long stalk. Eggs are laid in gelatinous masses that become enclosed in capsules. In some polyclads the egg develops into Müller's larva, which

may indicate an evolutionary link between the turbellaria and the ctenophores on the one hand or with the annelids on the other. The larva is provided with a series of ciliated lobes that are lost when it assumes the creeping habits of the adult. As already described, the embryology of polyclads is of the spiral determinate type and resembles that of mollusks and annelids.

Three of the best known and most widely distributed genera are *Notoplana* and *Stylochus* (Fig. 6-18) found on both the Atlantic and Pacific coasts, and *Planocera*, a pelagic form.

■ CLASS TREMATODA
GENERAL MORPHOLOGY AND PHYSIOLOGY

The flatworms of class Trematoda are called flukes because of their flat shape. They are unsegmented, chiefly hermaphroditic animals that have a digestive system but are covered by a living tegument (as already explained), not by a nonliving cuticle. They are well adapted to their parasitic life, which involves all vertebrate classes as hosts. Along with the tapeworms, the flukes make up a majority of metazoan parasites (Table 5). Although parasitic, flukes resemble the Turbellaria more than they do the Cestoda. Their body assumes many shapes, but is usually oval to elongate. Some are broader than long and others are very slender. The anterior end is somewhat tapered and thinner than the posterior region. There is no head.

With the exception of life histories and the discovery of new species, no aspect of trematode investigation has received more attention than the nature of their integument (tegument). Investigations have stressed the electron microscope, which has caused altogether different interpretations of the trematode external structure from those made with the light microscope. Of the early investigators, W. Hein's (1904) careful work on the trematode integument, in which he described the fine protoplasmic strands to the integument as originating from

Table 5. Some common trematode parasites

Scientific name	Developmental stages	Hosts	Distribution
Polystoma nearcticum	Fresh water	Tadpole and frog	North America
Opisthorchis sinensis	Freshwater fish and snails	Man, cat, dog, and fish-eating mammals	Indo-China and Far East
Fasciola hepatica	Freshwater and aquatic vegetation	Sheep, cattle, pigs, and man	Most continents
Paragonimus westermani	Freshwater crayfish and snails	Various mammals, including man	Far East and North America
Schistosoma haematobium	Freshwater snails	Primates and some others	Asia and Africa
Schistosoma mansoni	Freshwater snails	Man and other mammals	West Indies, South America, and Africa
Schistosoma japonicum	Freshwater snails	Various mammals, including man	Far East
Aspidogaster conchicola	Life cycle in host	Clams, gastropods, and turtles	Cosmopolitan

cells below the muscle, has been frequently quoted. L. H. Hyman in her great work on the invertebrates (vol. 2, 1951) stated that the cuticle is secreted by special mesenchymal or parenchymal cells, or by ordinary mesenchyme. However in 1963 L. T. Threadgold demonstrated with the electron microscope that the integument in the classic *Fasciola hepatica* was cytoplasmic, with connections to underlying cells. R. F. Bils and W. E. Martin (1966) have studied with the electron microscope the integument of the various stages of a number of trematode species and have found the presence of mitochondria in cercariae, metacercariae, and adults. This indicates that the trematode integument is metabolically active. They also described microvilli or irregularities in sporocysts, rediae, and developing cercariae, which may be involved in increasing the surface for respiratory or other exchange. Briefly, the present status of the trematode cuticle as originally understood seems to be quite nebulous.

The three basic divisions of the integument are the surface membrane, cytoplasmic layer, and basement membrane. Most of the integument of adult parasitic flukes is cytoplasmic and metabolically active, and the evidence presented above indicates that most of the larval stages have about the same arrangement. Below the basement membrane are layers of circular and longitudinal muscles. The interior of the body contains the parenchyma, consisting of vacuolated granular, branched cells, embedded in which are the internal organs and bundles or separate strands of muscle. Adhesive gland cells are present in some trematodes.

The alimentary canal (which has many variations) partially resembles that of some rhabdocoels. The funnellike mouth cavity is usually at the anterior tip or slightly ventral to the tip. In most Digenea the mouth is encircled by the oval sucker. The mouth cavity empties into a highly muscular plicate pharynx and a short esophagus. From the esophagus the intestine usually gives rise to two branches or ceca, which extend posteriorly as blind tubes to the rear end of the body. These branches may have many lateral branches. The wall of the intestine is composed of cuboidal or columnar epithelium, usually underlaid by a very thin muscular layer. The food of trematodes consists of blood, tissue, tissue exudates, epithelial cells, or even food particles from the host. Digestion may be chiefly extracellular. Trematodes may absorb some food through the body surface.

When available, the fluke will use oxygen, depending on the oxygen tension and the location of the parasite within the host. Endoparasites are mostly anaerobic, and their glycolysis releases carbon dioxide and various organic acids. The ectoparasites are aerobic.

For their excretion trematodes have a protonephridial system similar to that of the turbellarians. The arrangement of the system varies with different groups. The system usually is made up of small tubules with terminal flame cells connected to paired protonephridial tubules that join medially to form a bladder and then continue separately to nephridiopores in the tail region. In addition to removing waste (fatty acids), the system may be involved in osmoregulation.

The nervous system is similar to that in the turbellarians. Sense organs are poorly developed.

NATURE OF PARASITISM

Although we have discussed many parasites before the trematodes, it is well at this point to consider the host-parasite relationship, for the trematodes and cestodes represent groups that not only are exclusively parasitic but also are among the most widespread parasites in the animal kingdom. A parasite is an organism that gets some essential food factor at the expense of another organism (the host) without actually destroying the host. Parasitism differs from predation, in which the prey is killed by the predator. However, among certain insects the host is always destroyed eventually by the parasitic forms so that there are varying degrees of this association of parasite and host.

Depending on the effect on the host there are a number of types of this intimate relationship between one organism and another. The broad term symbiosis may be used for all types of close association between different species of animals. In this broad sense, symbiosis includes parasitism, whereby a member lives in or on another at the expense of the host; com-

mensalism, whereby one member of the relationship derives all the benefit and the other member has neither an advantage or disadvantage; and mutualism, whereby each member derives some benefit. There may also be intermediate aspects of symbiosis.

Since parasitism is a universal feature of all organic communities, it must be considered an ecologic relationship of long evolutionary standing. Parasites in their evolutionary history have become modified and specialized for this kind of existence. They have lost organs no longer of value and have developed new ones that offer adaptive use for their mode of life. Parasites must not be considered degenerate; rather they have become specialized in both morphology and function to a degree that have no other counterparts in the animal kingdom. Their physiologic adaptations have advanced to a degree that some parasites can live successfully only in a single host species; in other cases the parasite has a wide range in organisms it can infect. Usually the older the association, the more strict is the relationship between host and parasite.

The complexity of the life cycle of some parasites is a striking feature of parasitism. This cycle may involve anywhere from one to three intermediate hosts, in each of which there is a specific larval stage before the adult takes up its residence in the final, or definitive, host. This complexity adds enormously to the hazards of completing the life cycle, a hazard that the parasite tries to overcome by excessive reproductive capacity or in other ways. A possible explanation for this complex life cycle is that the cycles have repeated the phylogenetic history of the parasite. Different hosts have been added as new groups have emerged in evolution. Organisms that are now considered intermediate hosts were at one time definitive hosts. This may be an advantage to the parasite because it prevents competition between larva and adult in the same organism.

Theories to account for parasitism are largely speculative, although many logical explanations can be advanced. Life in the body cavity of invertebrates may have been the earliest form of parasitism. No one can deny that pre-adaptation for such associations may have performed an important role. J. G. Baer (1952), speaking of nematodes, said he knew of no other group of organisms so perfectly adapted for becoming parasites, although most are non-parasitic. These worms have a generalized structure that enables them to adapt to varied environmental conditions. Not only do they have adaptive morphologic characteristics, but also their ceaseless activity as well as predatory and saprophagous habits enable them to seek food of all kinds. Some nematodes are known to feed on the decomposing bodies of their dead hosts, and thus become predators. The trematodes are also preadapted—for example, their creeping habits, flat body, powers of adhesion, etc. The role of mutation in bringing about morphologic and functional changes suitable for parasitic niches could also lead to the intimate dependence of parasitism in those that have had such mutations. Mutation could also lead to exacting nutritional requirements so that an animal, if given the opportunity, could adopt a wholly parasitic mode of existence to survive.

The general effects of parasitism have many variables. In some cases, parasites seem to have no effects on the host. Some hosts, however, are destroyed by parasites that are imperfectly adapted. Destruction of the host is not in the plan of most parasites because the death of the host would militate against the parasitic race. Some of our most destructive diseases are caused by parasites, but such diseases may be caused by some interference with the normal ecology of the parasite. Great aggregations of parasites in one animal may also result in harmful effects such as those of mechanical pressure, production of toxic substances, or interference with some vital function. There are also striking effects on the parasites themselves. A delicate physiologic balance must be maintained between the larval forms

that live in the intermediate hosts and the requirements of the adult that lives in the definitive host. Different requirements are involved in each host of the life cycle. Also, if the host by adaptive radiation becomes diversified, the parasite may have to diversify also to survive.

■ ORDER MONOGENEA (HETEROCOTYLEA)

The members of Monogenea are mostly ectoparasites on the gills, skin, and external orifices of fishes, a few amphibians, turtles, and a few other hosts. They are provided with distinctive anterior and posterior adhesive organs called, respectively, prohaptor and opisthaptor (posthaptor). The prohaptor consists of small lateral suckers and adhesive glands and sometimes an oral sucker surrounding the mouth; the opisthaptor usually has many hooks. The distinctive characteristics of the order are a single host in the life cycle, sexual reproduction, and direct development. Body shapes vary but are typically flattened and leaflike. The excretory pores are paired and lie near the anterior end. The pharynx is a muscular organ, and the enteron is bifid, often with diverticula. There are separate genital openings for the male and female reproductive systems. Cross-fertilization is the rule, although self-fertilization may also occur. They feed on both mucus and blood of the host. The egg capsules are provided with filaments for attachment to hosts. Monogeneids have a high host specificity and are very sensitive morphologically to the position of their microhabitats on the host.

The Monogenea are usually divided into two groups—Monopisthocotylea, in which the opisthaptor is a single sucker or disk with anchor and marginal hooks, and Polyopisthocotylea, with multiple suckers and cuticularized clamps. *Gyrodactylus,* a common genus on the gills of freshwater fishes, is an example of the Monopisthocotylea; *Polystoma,* which lives in the urinary bladder of frogs, is an example of the Polyopisthocotylea.

Polystoma (Fig. 6-19) has one of the best-

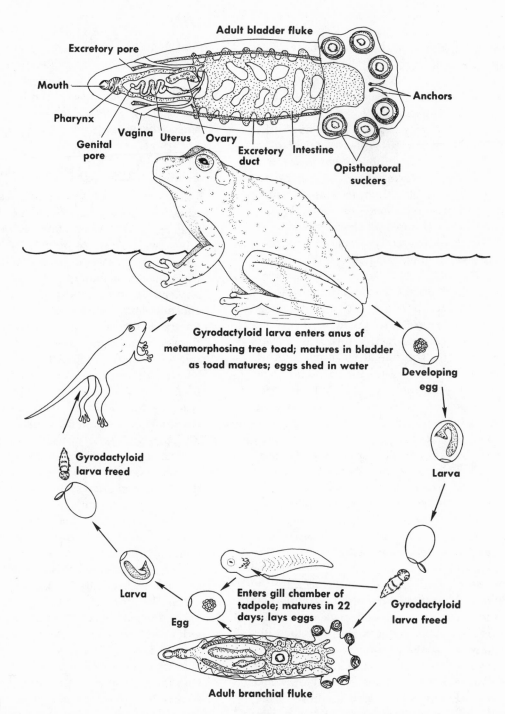

FIG. 6-19

Life history of *Polystoma nearcticum* (order Monogenea). This bladder fluke lives in urinary bladder of tree toads *(Hyla versicolor, H. cinerea);* sheds eggs into urine. Larvae develop in water and attach to external gills of young tadpoles where, in 3 weeks, they mature neotenically into branchial flukes. When larvae from eggs produced by either bladder fluke or neotenic branchial fluke attach to older tadpoles, neoteny does not occur; larvae enter bladder by way of cloaca, develop slowly, maturing as toad matures, in 2 or 3 years. (Modified from Olsen, 1962.)

known life histories, although it is not typical of all monogeneids. The adult fluke lives in the urinary bladder of frogs and breeds in the spring; its eggs in the urine are discharged into the water. The egg hatches into a larva with a hooked opisthaptor, attaches itself to the gills of a tadpole, and grows as the tadpole develops. When the frog metamorphoses, the young fluke may enter the host's esophagus and pass down to the bladder, where it completes its development and becomes sexually mature in about 2 to 3 years. The bladder fluke form may infect the frog by way of the anus instead of the esophagus. In some cases, however, the fluke may attach itself to a young tadpole, mature neotenically, and reproduce the first year. Such neotenic forms never get into the bladder and never fully mature.

■ ORDER ASPIDOGASTRAEA (ASPIDOBOTHRIA)

The Aspidogastraea were formerly included under the order Digenea. They are endoparasitic in mollusks, fishes, turtles, and skates and are mostly characterized by very large opisthaptors or suckers divided by septa into longitudinal rows of depressions (alveoli). Such a sucker may occupy most of the ventral surface. The alveoli are important in taxonomic procedures. An oral sucker may be absent or weakly developed. In general anatomy the Aspidogastraea resemble the Digenea, but the development of the former is direct, as in Monogenea.

Aspidogaster, common in the pericardial chamber of freshwater clams, is a well-known genus. Only one host is necessary in the life history.

■ ORDER DIGENEA (MALACOCOTYLEA)

Digenea, an extensive group of endoparasitic trematodes, is characterized by having more than one host in their life histories. The adults are sexual and usually have vertebrates of all classes except cyclostomes for their definitive hosts. The asexual larval stages are passed in intermediate hosts, usually mollusks, rarely annelids. The life history may be very complicated and involve two to four hosts for complete development. In most cases the adults are hermaphroditic, but a few are dioecious. In the definitive host the adults usually occupy the digestive tract and its derivatives or appendages, such as the gallbladder and its ducts, the liver, the lungs, the bladder, and kidneys. Some live in blood vessels and cavities (coelom and head). Aquatic animals harbor more digenetic trematodes than do terrestrial forms.

Certain unique morphologic features of the adults in this group exist. No opisthaptor is found at the posterior end as in other trematodes. The mouth is usually anterior and opens into an oral sucker that may be poorly developed or lacking. In some members, the mouth is ventral without a sucker. In place of the opisthaptor there is often a ventral sucker, the acetabulum, usually highly muscular with a raised rim. It operates mainly on the principle of a vacuum and not by glandular secretions. It may be found at various locations on the body, such as anterior to the middle of the ventral surface (Distomata) or at the posterior end or it may be absent. It is often of great diagnostic value in taxonomic determinations. The length of the adults is usually 20 to 25 mm., but some are smaller and others are larger than this. The Digenea as a whole show an enormous variety of modifications.

In their complex life histories, development is indirect. Development of the egg may begin before it leaves the host, or it may be delayed until later. The egg (Fig. 6-20, *B*) usually divides into propagation and somatic cells, forming a ball of cells. From the ball of cells the first larval stage, the ciliated miracidium, develops (Fig. 6-20, *C*). The miracidium enters an intermediate host and becomes a germinal sac, the sporocyst (Fig. 6-20, *D*), which contains, among other cells, germ balls or germinal cells. The germ balls may develop into daughter sporocysts, or they may produce a generation of the next larval stage, the rediae (Fig. 6-20, *E*),

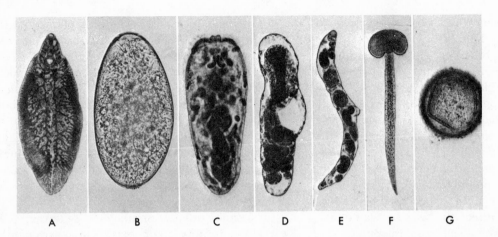

FIG. 6-20

Stages in life cycle of sheep liver fluke, *Fasciola hepatica* (class Trematoda, order Digenea). Not enlarged to same scale. **A,** Adult (up to 30 mm. long). **B,** Egg (130 to 150 microns long). **C,** Miracidium. **D,** Sporocyst. **E,** Redia. **F,** Cercaria (250 to 350 microns long). **G,** Metacercaria. (Courtesy Carolina Biological Supply Co., Burlington, N. C.)

which differ from sporocysts in having a pharynx and saclike gut. The rediae may produce a second generation of rediae, or the redial stage may be absent altogether. From the daughter sporocysts or from the rediae, the cercariae develop (Fig. 6-20, *F*). The number of cercariae that emerges from the intermediate host may be small or large. Reproduction in the sporocysts and rediae appears to be by mitosis of segregated germinal cells or balls without sexual phenomena, a process called polyembryony. The cercaria consists of an oval-shaped body and a simple or forked tail (lacking in some) and has many of the structures of the adult fluke. After leaving the intermediate host, it encysts on vegetation or on a second intermediate host. When the encysted metacercaria, or juvenile fluke (Fig. 6-20, *G*), is eaten by a vertebrate, it is transformed into the adult fluke. Some cercariae (blood flukes) are able to penetrate the skin directly. Many variations of the life cycle are found in this group in which (1) the number of species is constantly increased as a result of investi-

FIG. 6-21

Opisthorchis, the human liver fluke (order Digenea).

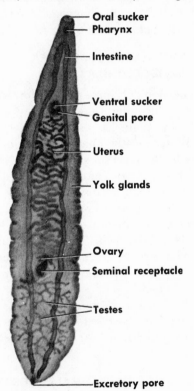

Oral sucker
Pharynx

Intestine

Ventral sucker
Genital pore

Uterus

Yolk glands

Ovary
Seminal receptacle

Testes

Excretory pore

gations and (2) the taxonomy is in a state of flux and confusion at the present time.

Of the many members of this extensive order, some of the more important parasitic genera follow: *Fasciola* (Fig. 6-20), found in

FIG. 6-22

Adult male and female *Schistosoma mansoni* (order Digenea) in copulation. Male has long sex canal that holds female (darker individual) during insemination and oviposition. Man is usually primary host; certain snails are intermediate hosts. (AFIP No. 56-3334.)

the bile ducts of sheep, goats, and other domestic animals, with pond snails and vegetation as the intermediate hosts; *Opisthorchis* (Fig. 6-21), in the bile ducts of mammals (including man), birds, and other vertebrates, with certain snails and fishes as the intermediate hosts; *Schistosoma* (Fig. 6-22), in the blood vessels of birds and mammals (including man) with pulmonate snails as the intermediate host; *Paragonimus,* in the lungs of mammals (Fig. 6-23), with operculate snails and crayfish or crabs as intermediate hosts; and *Fasciolopsis,* in the intestine of man, pigs, and dogs, with snails and water nuts *(Trapa natans)* as intermediate hosts. The first intermediate host is nearly always a mollusk, but W. E. Martin (1944, 1952) has found two examples of a

FIG. 6-23

Paragonimus fluke (order Digenea) in lung tissue. (AFIP-ACC No. 496822.)

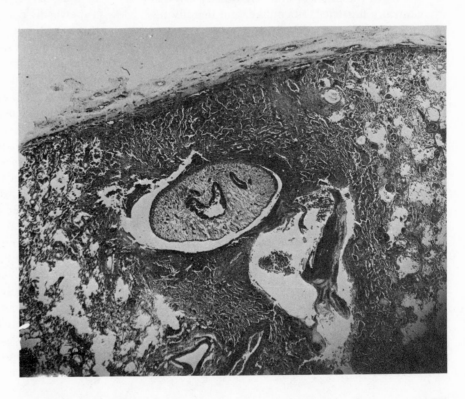

polychaete worm serving as the first intermediate host.

Classification of the trematodes has been based primarily on certain features, such as the arrangement of the reproductive system, the position and nature of the suckers, the type of miracidia and cercariae, the nature of the epithelial lining of the bladder, and the arrangement of the excretory system.

■ CLASS CESTODA (CESTOIDEA)
GENERAL MORPHOLOGY AND PHYSIOLOGY

Class Cestoda includes the tapeworms, the most highly specialized parasites in the animal kingdom. They are all endoparasites that may have come originally from a parasitic turbellarian of the former order Rhabdocoela. With few exceptions the adults are all parasitic in vertebrates, with their larval stages in the invertebrates. They have no mouth or digestive system, and their attachments are restricted chiefly to the anterior end. Their body is dorsoventrally flattened, ribbonlike, and usually has an elongated jointed body of segments, or proglottids, attached to a spherical or oval-shaped scolex (holdfast). Some tapeworms have undivided bodies that resemble flukes. Electron microscopy shows their bodies to be covered with a living tegument, formerly considered a nonliving cuticle.

The proglottids of the tapeworm originate in a neck or zone just behind the scolex and increase in size (in both width and length) as they are shoved farther from the neck by the continuous growth of new segments, forming a chain of proglottids called a strobila. At a variable distance from the scolex the proglottids acquire adult reproductive organs that usually develop progressively from the anterior to the posterior end of the chain, or strobila. There may be a succession of sex stages, the male organs appearing first, followed by the female organs, until mature proglottids contain fully developed systems of both sexes. In some tapeworms each proglottid may contain two complete reproductive systems bilaterally arranged. With few exceptions, all cestodes are hermaphroditic. The reduplication of the sexual organs may be an advantageous adaptation for the species in increasing the chance of impregnation and egg production. The posterior end of the strobila is composed of ripe, or gravid, proglottids that consist chiefly of uteri distended with eggs or developing embryos. Most of the other parts of the reproductive system are degenerate remnants at this stage. Since new proglottids are being formed continuously at the neck, the gravid proglottids are being shed unceasingly at the posterior end of the strobila so that the total length of a mature strobila remains more or less constant.

Much of our knowledge about tapeworm physiology is fragmentary and speculative because of the difficulty of keeping tapeworms alive outside their hosts for any length of time. However, many important breakthroughs have been made in duplicating life cycles in vitro in the past decade or so (J. F. Mueller, 1965). Tapeworms obtain their nutrition by absorbing through their exterior surface the split products of their host's digestion, or in certain cases by absorbing tissue fluid through the deeply-embedded scolex. They may be provided with proteolytic enzymes. Resistance to tapeworm infestation depends on age, sex, and immune reactions. The young are often more susceptible to infestation than are the old. True immunity refers to the formation of antibodies by the host in response to the invasion of tissues by the larval stages. Adult tapeworms in the intestine usually do not evoke immune reactions, but their presence may inhibit the presence of further specimens of the same species. This may definitely limit, in many cases, the number of tapeworms within a single host.

Tapeworms are not always injurious to their hosts. Ill effects are usually more apparent in the young, in whom the parasites may be responsible for serious digestive and nervous disturbances. Adult worms are known to secrete toxic substances. The migration of larvae within the body and secondary invasions are responsible for their most serious harm.

Subclass Cestodaria

The subclass Cestodaria is made up of a few endoparasites of primitive fishes. They differ mainly from other cestodes in being unsegmented, in not forming proglottids, and in having a larva (oncosphere or lycophore) with 10 hooks that is usually found in crustaceans. They also occur as parasites in the coelomic cavity as well as in the intestine. Instead of a scolex they may bear a protrusible proboscis with large frontal glands. In others the posterior end is provided with a holdfast organ, or rosette. They are hermaphroditic, with the sexual phase in the vertebrate host.

The best-known genera are *Amphilina,* a par-

asite in the coelom of the sturgeon *Acipenser,* and *Gyrocotyle* (Fig. 6-24) in the intestine of chimaeroid fishes.

Subclass Eucestoda

The Eucestoda are the true tapeworms whose bodies are chiefly divided into segments, or proglottids, and whose larvae bear 6 hooks. The adults are endoparasites in vertebrate intestines, and their life cycle involves one or two intermediate hosts that may be either invertebrate or vertebrate. The anterior end is usually modified into a scolex, or holdfast organ, followed by a short, undivided neck. Tapeworms range in length from 1 mm. to many feet, and the proglottids may number a few to many thousands. The chain of proglottids (the strobila) gradually increases in breadth and length posteriorly so that the broadest and longest proglottids are at the posterior end, and the smallest and least differentiated proglottids

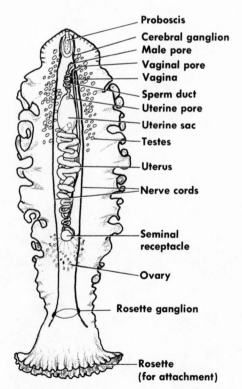

FIG. 6-24

Gyrocotyle (subclass Cestodaria). Adults are parasitic in chimaeroid fishes.

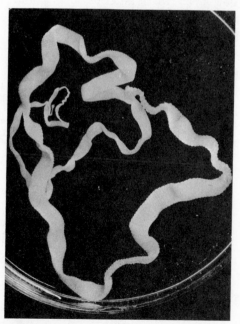

FIG. 6-25

Ruminant tapeworm *Moniezia expansa* (subclass Eucestoda, order Cyclophyllidea). Note small scolex and slender neck proglottids at upper left.

are just behind the neck, from which the proglottids proliferate asexually (Fig. 6-25). The scolex, which is modified in different species (Fig. 6-27), is a generally solid structure of muscle and mesenchyme and bears the organs of attachment (suckers, hooks, etc.), nephridial canals, and a type of brain. The body is covered with a living tegument, as already described. Their cuticle has more microvilli than does that of trematodes and is penetrated by tubules to facilitate absorption of food from the host. Through each proglottid and the length of the strobila runs a pair of longitudinal lateral nerves. The nerves are connected by commissures in each proglottid. A pair of longitudinal lateral excretory canals in dorsal and ventral arrangement on each side of the proglottid runs through the length of the body. Flame bulbs are connected to the 4 main excretory canals by a system of fine canals. With five or six exceptions all cestodes are hermaphroditic.

The reproductive systems are progressively developed from anterior to posterior ends of the strobila (Fig. 6-26), with the male system maturing before the female system (protandry). Each proglottid contains a full set of reproductive organs and, in some cases, two complete sets. The general plan of the reproductive system is similar to that found in other flatworms, with many diverse arrangements. The genital ducts (gonoducts) of both sexes usually open into a common antrum, which may be double if the reproductive systems are double. There is a single genital opening, or gonopore, for each common antrum.

Fertilization may take place within a single proglottid (self-fertilization), between proglottids in the same strobila, or between proglottids of different worms. Hypodermic insemination

FIG. 6-26

Proglottids from three sections of tapeworm *Taenia pisiformis* showing progressive sexual maturity. Left to right: immature proglottids from anterior region, mature glottids from middle portion, and gravid proglottid from posterior end (order Cyclophyllidea).

may occur in those that lack a vaginal opening. Copulation occurs by eversion of the cirrus into the vagina. Self-fertilization appears to be the most common method. The egg is fertilized in a region of the oviduct where the ovum, with yolk cells, becomes enclosed in a shell. Within the uterus the eggs develop into the oncosphere larval form with three pairs of hooks. The posterior proglottids become distended (gravid) with embryos (Fig. 6-26) and are shed from the strobila to the outside of the host. In some members only the shelled embryos are shed. After being eaten by an intermediate host, the oncosphere escapes from its membrane and undergoes further development. Two intermediate hosts are usually required, one of which is a crustacean and the other a fish or other vertebrate (Fig. 6-29). Many variations of this plan exist. The embryo attains entrance to the definitive vertebrate host when the latter eats the intermediate host. Usually the postoncosphere development involves a procercoid larval stage in the first intermediate host and a plerocercoid stage in the second intermediate host. The plerocercoid develops many of the structures of the adult, such as a scolex, thickened cuticle, and nervous and excretory systems. Strobilation, however, does not occur until it is eaten by the final vertebrate host. However, this pattern of development is modified in certain orders.

Tapeworms vary greatly in length of life. Some large, human tapeworms such as *Dibothriocephalus latus* (order Pseudophyllidea) and *Taenia saginata* (order Cyclophyllidea) may live 30 to 35 years, or sometimes for as long as the host lives. Many tapeworms live less than a year, and some live only a few days. Tapeworm infection may produce no visible symptoms, but diarrhea, abdominal pain, abnormal appetite, and even convulsions may occur. Diagnosis of infection is made by fecal examination for proglottids or eggs, x-ray films, skin tests, etc. Some progress has been made experimentally in cultivating larval and adult tapeworms outside of the body, but the process is very difficult.

■ ORDER TETRAPHYLLIDEA (PHYLLOBOTHRIOIDEA)

In these tapeworms the scolex is provided with 4 broad, leaflike muscular structures (bothridia) of many diverse forms symmetrically arranged around the anterior region of the elongated scolex. Hooks are also found on the scolex of some members. Tetraphyllids are small, intestinal parasites of elasmobranch fishes and are usually less than 3 or 4 inches long, with a few hundred proglottids. The sexually immature proglottids are often shed from the tapeworm and become sexually mature while motile in the intestine of the host. Each proglottid contains a single reproductive system of each sex that may open by a genital pore on each side. A complete life history is unknown, but larval forms have been found in various invertebrates and some marine fishes.

Phyllobothrium (Fig. 6-27, A) is one of the most familiar genera and has been extensively studied.

■ ORDER PROTEOCEPHALOIDEA

The tapeworms of Proteocephaloidea are intestinal parasites of freshwater fishes, amphibians, reptiles, and elasmobranchs. They are short (a few inches in length), slender, and are distinctly segmented. The scolex, or holdfast organ, bears 4 cup-shaped suckers (acetabula) of varying form, and many species have an apical organ or degenerate sucker at the tip of the holdfast. A rostellum with hooks is found in the genus *Gangesia*. The mesenchyme of the proglottid is divided into cortical and central regions, as in the order Tetraphyllidea, and there are other anatomic similarities between the two orders. The life cycle involves the ingestion of the oncosphere with its membranes by copepods (*Cyclops*), in the body cavity of which the larva develops into procercoid and plerocercoid stages, although the latter stage may occur in the definitive host. The definitive vertebrate host (freshwater fishes, amphibians, and reptiles) becomes infected by eating parasitized *Cyclops* or by eating small vertebrates

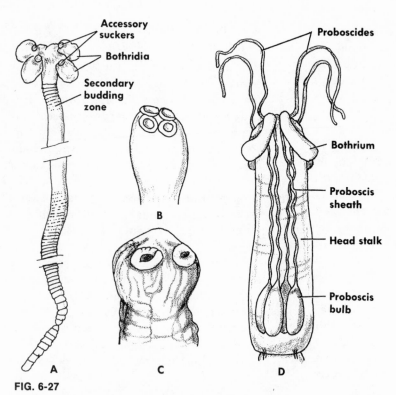

Accessory
suckers

Bothridia

Secondary
budding
zone

B

C

A

Proboscides

Bothrium

Proboscis
sheath

Head stalk

Proboscis
bulb

D

FIG. 6-27

A, *Phyllobothrium dornii* proliferates in usual manner at rear end of neck; then later starts forming proglottids in opposite direction just behind scolex. Small portions omitted from drawing (order Tetraphyllidea). **B,** *Proteocephalus coregoni* scolex; has no apical sucker though most species in this genus do; parasitic in coregonid fishes (order Proteocephaloidea). **C,** *Corallobothrium giganteum* scolex; its 4 suckers are facing forward; has been found in siluroid fishes in Mississippian drainage system (order Proteocephaloidea). **D,** *Gilquinia squali;* parasitic in spiny dogfish *Squalus.* This order (Trypanorhyncha) has eversible tentacles, or proboscides, as well as bothria. (Modified: **A,** Curtis, 1906; **B,** Wardle, 1932; **C,** Essex, 1927; **D,** Hyman, 1940.)

that have fed on parasitized copepods. *Proteocephalus* and *Corallobothrium* are among well-known species (Fig. 6-27, *B* and *C*).

■ ORDER TRYPANORHYNCHA (TETRARHYNCHOIDEA)

The order Trypanorhyncha is characterized by having 4 spiny, eversible proboscides of the scolex (Fig. 6-27, *D*). The elongated scolex is divisible into a bothrial region with 1 to 4 suckers (often called bothridia but which, because of their weak muscularity, should probably be considered bothria) and a head stalk with the proboscis apparatus consisting of 4 spiny proboscides. Each proboscide is contained in a sheath and ends with a muscular proboscis. The proboscides, or tentacles, are everted by muscular contractions of bulbs that force fluid against their inner ends and then are pulled back by retractor muscles. The tapeworms are usually under 4 inches in length and are chiefly parasites in the spiral valve of elasmobranch fishes and in the intestine of some ganoid fishes. The life history is incomplete for the marine species, but larval forms of the procercoid or cysticercoid types have been found in marine invertebrates and teleost fishes.

Haplobothrium, an aberrant genus found in the common freshwater fish *Amia,* lacks the suckers and has some other features not characteristic of the order. Its life history has been worked out and indicates that the larval stages

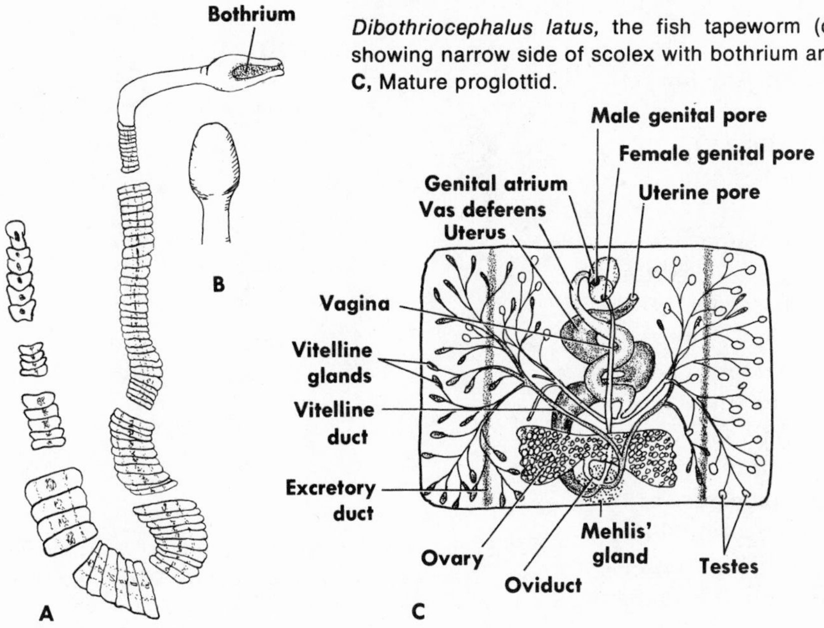

FIG. 6-28

Dibothriocephalus latus, the fish tapeworm (order Pseudophyllidea). **A,** Adult worm showing narrow side of scolex with bothrium and part of strobila. **B,** Flat side of scolex. **C,** Mature proglottid.

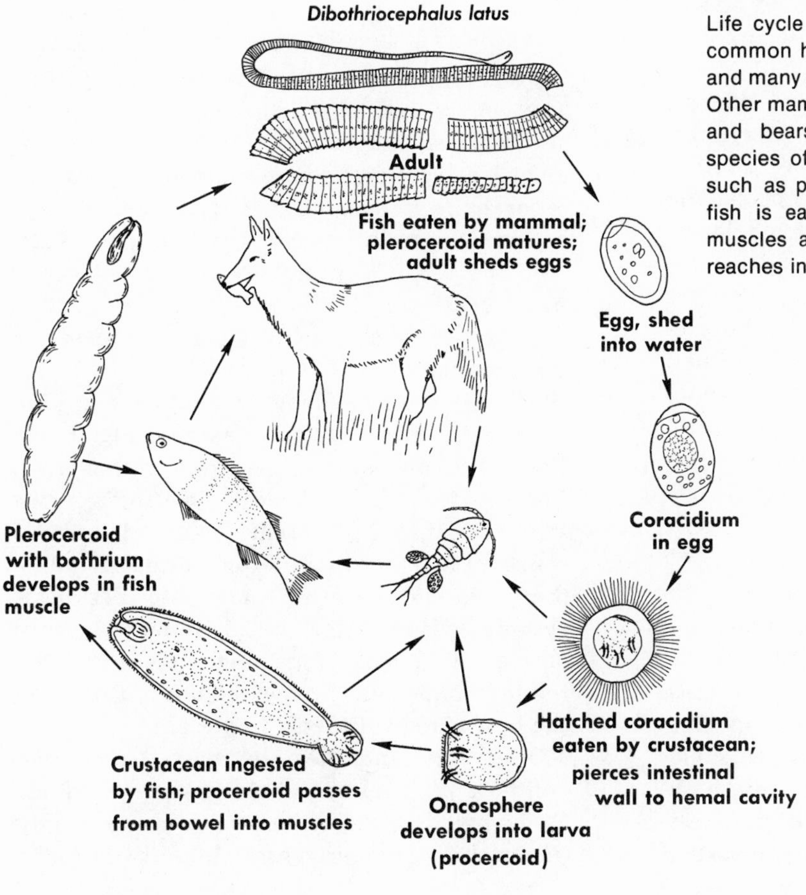

FIG. 6-29

Life cycle of broad tapeworm *Dibothriocephalus latus,* common human parasite in Russia and Siberia, Finland, and many other places, including our Great Lakes region. Other mammal hosts are usually dogs, cats, pigs (Russia), and bears (Canada). Intermediate hosts are certain species of *Cyclops* and *Diaptomus* and freshwater fish such as pike, perch, lawyer, trout, and ruff. If infected fish is eaten by another fish, plerocercoid travels to muscles again; it does not mature sexually unless it reaches intestine of mammal host.

involve *Cyclops* and the teleost fish *Ameiurus.*
(R. A. Wardle and J. A. McLeod consider *Haplo-bothrium* a member of Pseudophyllidea.)

■ ORDER PSEUDOPHYLLIDEA

Pseudophyllidea are widely distributed tapeworms and are parasitic in all classes of vertebrates. The chief diagnostic character is a scolex usually bearing 2 shallow grooves (bothria) arranged in a dorsal and ventral position to each other. In some species the scolex bears a circle of hooks. Some members are monozoic, with bodies not divided into proglottids. The genital openings, or gonopores, are often found near the ventral or dorsal midline instead of on the lateral margins. The uterine pore in some cases is separate from the male pore. One of the best known genera is *Dibothrio-cephalus latus* (*Diphyllobothrium*, Figs. 6-28 and 6-29), the broad tapeworm of man and of some fish-eating mammals. The life cycle of this tapeworm is well known (Fig. 6-29). The egg capsule hatches into a ciliated embryo (coracidium) that is eaten by the first intermediate host, a copepod (*Cyclops* or *Diaptomus*). In the body cavity of the copepod the coracidium develops into a procercoid larva. When the copepod host is ingested by the second intermediate host (usually a fish, but other vertebrate classes may be involved), the procercoid develops into a plerocercoid larva. When the definitive host eats the second intermediate host, the worm develops into the adult form, which may reach a length of 60 feet, with a strobila numbering up to 4,000 proglottids.

■ ORDER CYCLOPHYLLIDEA (TAENIOIDEA)

The tapeworms of order Cyclophyllidea are found in the intestine of warm-blooded vertebrates (birds and mammals), although they may also occur in reptiles and amphibians. The characteristic scolex bears 4 cup-shaped suckers (acetabula) and a muscular rostellum with a crown of hooks (Fig. 6-30). In some members the rostellum may be withdrawn into a rostel-

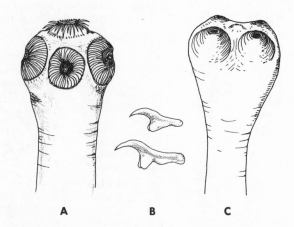

FIG. 6-30

A, Scolex (holdfast) of *Taenia solium* showing row of apical hooks in addition to suckers. **B,** Enlarged view of short and long hooks (side view) of *Taenia solium.* **C,** Scolex of *Taenia saginata,* which is provided only with suckers.

lar receptacle, or it may be absent altogether. The worms vary in length from 1 or 2 mm. to several feet, but most are only a few inches long. New proglottids are budded off just posterior to the scolex and, as they are pushed back, become sexually mature and larger (Figs. 6-25 and 6-26). With the exception of one genus (*Dioecocestus*), they are all hermaphroditic. One or two sets of sex organs are present in each proglottid, and both reproductive systems share a common gonopore and antrum that usually lie in the lateral margin (Fig. 6-31). Usually no uterine opening exists in this order. The ova may be fertilized by sperm from the same or from a different proglottid and are collected together in a uterus, ultimately filling the proglottid with ripe egg capsules. Ripe cestode eggs contain oncosphere or hexacanth larvae with 6 hooks. Ripe proglottids containing the infectious eggs are passed out of the gut of the host along with fecal material.

Only one intermediate host (sometimes omitted) is required in the life cycle of the Cyclophyllidea. After an intermediate host (invertebrate or vertebrate) ingests the eggs,

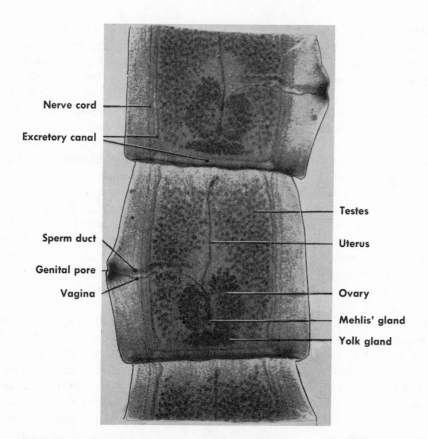

Nerve cord

Excretory canal

Testes

Sperm duct

Uterus

Genital pore

Vagina

Ovary

Mehlis' gland

Yolk gland

FIG. 6-31

Mature proglottid of dog tapeworm *Taenia pisiformis* (order Cyclophyllidea). (Courtesy General Biological Supply House, Inc., Chicago.)

FIG. 6-32

Piece of "measly pork" showing cysts of bladder worms (cysticerci) of *Taenia solium*. Bladder worm cysts of *Taenia saginata* have similar appearance.

the oncosphere is freed from its membranes and penetrates the intestinal wall into the coelom, hemocoel, liver, or other parts, where it develops into some type of metacestode such as a cysticercoid. A cysticercoid usually has the scolex retracted into a vesicle and is provided with a tail. Some metacestodes of this order have the scolex invaginated (introverted), lack the tail appendage, and form a bladder worm called a cysticercus (Fig. 6-32). This type of metacestode is most common in reptilian and mammalian intermediate hosts. A modified form of the single bladder worm is the hydatid cyst, which contains large numbers of attached and free-floating brood capsules, each of which buds many scolices internally or externally. Such hydatid cysts may reach the size of an orange and may be found in various parts of the body. The cyst greatly increases the asexual reproductive power of tapeworms.

Whatever the form of the metacestode, it usually remains quiescent until its intermediate host is eaten by the definitive host. Then the scolex of the cysticercoid or cysticercus emerges, attaches itself to the intestinal wall of the host, and begins to proliferate proglottids as a mature tapeworm. (In cysticercus types the scolex evaginates for attachment.)

The majority of the common tapeworms belong to this extensive order. The most common tapeworm of man, at least in the United States, is the beef tapeworm (*Taenia saginata*, Fig. 6-30, *C*) whose intermediate host is cattle. *Dipylidium* is a common cestode of dogs and cats and may be found occasionally in children. Its intermediate hosts are fleas and biting lice. The dwarf tapeworm (*Hymenolepis*) occurs in children, mice, and rats. Its life cycle is noted for having no intermediate host. The embryos (hexacanths) from the adult worm bore into the intestinal wall, transform to cysticercoids, and emerge into the intestine as adults. In some cases an invertebrate intermediate host (fleas, beetles) may be involved. The ruminant tapeworm (*Moniezia*, Fig. 6-25) occurs in sheep and other ruminants and has certain mites for its intermediate host. *Taenia solium* (Fig. 6-30, *A*), or the pork tapeworm, is found in man; the pig is the intermediate host. Hydatid tapeworms (*Echinococcus*) occur in dogs and related species, with man and other vertebrates as the intermediate hosts.

■ PHYLUM RHYNCHOCOELA (NEMERTINA)

The Rhynchocoela worms, along with the Platyhelminthes, belong to the acoelomate Bilateria. The Rhynchocoela are commonly known as ribbon worms or nemertine worms. They are elongated, thread-shaped or ribbon-shaped worms that are often cylindric in cross section. Most species are only a few inches long, but some may reach a length of several meters. They are nonsegmented and are without external appendages. Their epidermis is ciliated, and they are provided with an eversible proboscis in a sheath just above the enteron. The proboscis is well suited for capturing prey, for defense, and for burrowing. It is a highly sensitive organ as well as a muscular one. Many nemertines flip the proboscis in an act much like a snake flipping its tongue in and out.

With one or two exceptions, all nemertines are marine forms that live mostly in muddy or sandy bottoms beneath shells and stones. They are common in coral habitats and often secrete gelatinous tubes in which they live. Many are brightly colored. About 570 species are classified so far.

Their general organization bears many resemblances to the Platyhelminthes, in which they have often been included as a separate class. They resemble the flatworms in having the space between enteron, epithelium, and epidermis filled with parenchyma and muscle. They have also, like the flatworms, developed the creeping habit of locomotion and have a ciliated epidermis; but in general organization, nemertines are more highly developed than Platyhelminthes. They have a complete digestive system (mouth and anus), a circulatory system, and a mouth at the anterior end instead of midventrally. They also possess the eversible tubular proboscis altogether lacking in the flatworms. This proboscis may be largely responsible for the lack of parasitism in this group. Nemertines, however, have a much simpler reproductive system than Platyhelminthes, and the sexes are usually separate. In contrast to the parasitic habits of most flatworms, ribbon worms are all free living, with the exception of a few commensal species.

CLASSIFICATION

The 500 to 600 or so species are divided into two classes (or subclasses by some authorities), each of which is subdivided into two orders.

The class distinction is based mainly on the structure of the proboscis and the location of the longitudinal nerves with respect to the muscular layers.

Phylum Rhynchocoela. Body bilaterally symmetric, dorsoventrally flattened or cylindric, and acoelomate; nonsegmented, vermiform shape; body covered by ciliated epidermis; body soft, slimy, and very elastic; eversible proboscis in sheath dorsal to enteron; complete digestive system; excretory system with flame cells; blood vascular system; nervous system of brain and lateral nerves; brain with cerebral organs; body wall musculature of two or three layers; separate sexes usually; gonads of simple sacs; spiral determinate cleavage; pilidium larva in some.

    Class Anopla. Proboscis unarmed.

        Order Paleonemertini. *Tubulanus, Cephalothrix.*
        Order Heteronemertini. *Lineus, Cerebratulus.*

    Class Enopla. Proboscis usually armed with stylet.

        Order Hoplonemertini. *Prostoma, Amphiporus.*
        Order Bdellonemertini. *Malacobdella.*

GENERAL MORPHOLOGIC FEATURES

Nemertines in their general appearance resemble the Platyhelminthes. They have soft, slender bodies that are highly elastic. The anterior part, which may be lancet shaped or heart shaped, is usually broader than the rest of the body but is really not a head because it lacks the brain. In the forepart of the body the worm is somewhat cylindric and becomes flattened posteriorly, but many variations exist. The outer investment in all species consists of an epidermis of ciliated cells. The epidermis is provided with many unicellular glands, which are principal sources of the mucus with which nemertines are all liberally supplied. Sometimes a cluster of glands is located in the subepidermis and has a common opening to the exterior. Pigments that produce the various colors of orange, red, brown, etc. are principally located in the epidermis, but in some cases the pigments may lie deeper. In some species these pigments may produce stripes or transverse bars. A basement membrane of connective tissue (dermis or cutis) is usually present between the epidermis and the body musculature. In the heteronemerteans this cutis is divided into an outer layer of glands and an inner layer of connective tissue.

The muscular layers that form most of the body wall are not arranged the same way in all orders or even in lower taxa. In those worms that belong to the orders Paleonemertini, Hoplonemertini, and Bdellonemertini, the musculature consists chiefly of outer circular and inner longitudinal muscle layers; in the order Heteronemertini an outer longitudinal layer, a middle circular layer, and an inner longitudinal layer exist (Fig. 6-33). Some slight variations from this plan also occur.

The region between the enteron and body wall musculature is usually filled with parenchyma or mesenchyme. This mesenchyme is directly continuous with that in which the muscle layers are embedded. Dorsoventral muscle bands course through the mesenchyme and connect ventral and dorsal walls as well as surround the digestive tube. Mesenchyme is made up of an intracellular gelatinous substance in which nuclei are everywhere present. In some cases it is fibrous and consists of branched cells. Mesenchyme may be scanty in forms where the musculature largely replaces it.

Clusters of cephalic, or frontal, glands that open by a common pore above the proboscis pore are found in most Rhynchocoela. These glands may be homologous to the frontal glands of flatworms and may have the same function.

The characteristic proboscis apparatus is made up of a long, eversible, tubular proboscis; a short, anterior canal, or rhynchodeum, that opens anteriorly; and a long cavity, or rhynchocoel, in which the proboscis lies (Fig. 6-34). The tubular proboscis is blind at its inner end and is usually much longer than the body. It opens anteriorly into the rhynchodeum, which, in turn, usually opens by a pore to the outside just above the mouth. The muscular wall surrounding the rhynchocoel constitutes the proboscis sheath. The anterior, or open, end of the proboscis is fastened to the inner end of the rhyn-

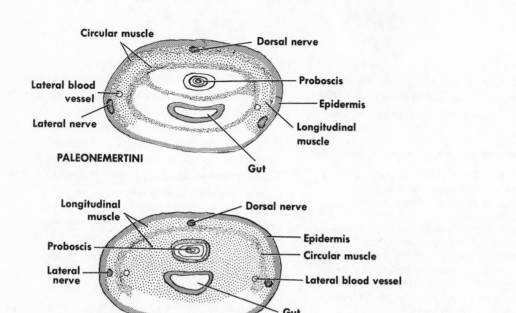

PALEONEMERTINI

Circular muscle — Dorsal nerve — Proboscis — Epidermis — Longitudinal muscle — Gut — Lateral blood vessel — Lateral nerve

HETERONEMERTINI

Longitudinal muscle — Dorsal nerve — Proboscis — Epidermis — Circular muscle — Lateral nerve — Lateral blood vessel — Gut

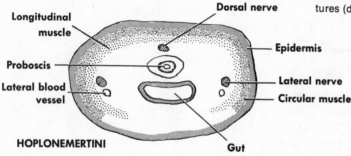

FIG. 6-33

Transverse sections of three orders of nemertines to show different arrangement of muscular layers and other structures (diagrammatic). Cilia not shown.

HOPLONEMERTINI

Longitudinal muscle — Dorsal nerve — Proboscis — Epidermis — Lateral nerve — Circular muscle — Lateral blood vessel — Gut

FIG. 6-34

A, Dorsal view of female *Amphiporus* (order Hoplonemertini) (from life). **B,** Circulatory and excretory system of *Amphiporus* (diagrammatic).

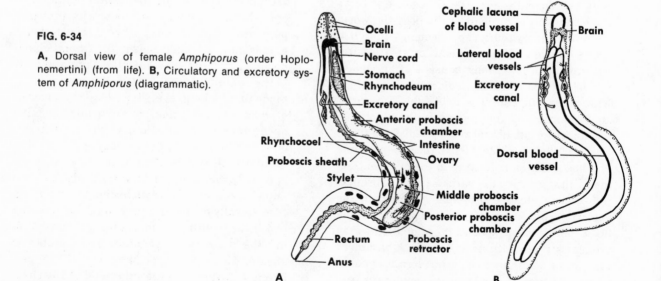

A

Ocelli — Brain — Nerve cord — Stomach — Rhynchodeum — Excretory canal — Anterior proboscis chamber — Intestine — Ovary — Middle proboscis chamber — Posterior proboscis chamber — Proboscis retractor — Rhynchocoel — Proboscis sheath — Stylet — Rectum — Anus

B

Cephalic lacuna of blood vessel — Brain — Lateral blood vessels — Excretory canal — Dorsal blood vessel

chodeum, and the posterior, or blind, end is attached by a retractor muscle to the posterior end of the proboscis sheath. The rhynchocoel, which is a completely closed cavity, is filled with a corpuscular fluid, and when the muscles of the proboscis sheath contract, the pressure on the fluid causes the proboscis to be everted and shot out of the proboscis pore (Fig. 6-35). The retractor muscle at the end of the proboscis then pulls back rapidly the extruded proboscis, aided somewhat by the musculature of the head. The proboscis is never completely everted because of the retractor muscle, and in the armed specimens or those with stylets there is a muscular diaphragm or bulb that closes off the anterior proboscis at the point of projection. This diaphragm is provided with a stylet and pouches of reserve stylets. When the proboscis is shot forward, or everted, it seizes the prey by coiling around it, by glandular secretions of the proboscis wall, or by the stabbing action of the stylet.

FIG. 6-35

Proboscis of nemertine extended in seizing prey.

In some species, nematocysts and rhabdites appear on the everted surface of the proboscis for overcoming the prey. Some differences exist in the structure of the proboscis in the different groups. In *Malacobdella* and a few others, for instance, there is a single opening for both the proboscis and the mouth (Fig. 6-36). The structure of the proboscis apparatus is similar to that of the body wall, with the layers in reverse (in the resting state) because it has originated as an ectodermal invagination. Since the rhynchocoel is formed by a split in the musculature, which is separated into proboscis muscles and proboscis sheath muscles, the rhynchocoel is considered a true schizocoel coelom.

The digestive system is complete in nemertines, with both mouth and anus (Fig. 6-37). The round or slitlike mouth is ventrally located near the tip of the head and in front of or behind the brain in most; in a few species, however, it has a more posterior location. The alimentary canal is a straight tube that is divided into foregut and intestine. The foregut consists of the buccal cavity, esophagus, and stomach, divisions not apparent in some. The lining epithelium of this portion is similar to the epidermis with many gland cells. The intestine (richly

FIG. 6-36

Longitudinal section through anterior region of hoplonemertine, a type in which esophagus enters rhynchodeum and separate mouth is lacking.

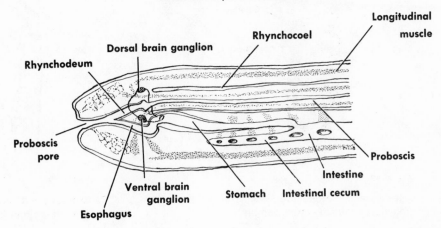

supplied with glands) usually lacks its own musculature and may be provided with diverticula and a cecum. Movement of the food is accomplished mainly by cilia. At the posterior end of the body is the rectum and anus. Digestion appears to be both extracellular and intracellular. Most nemertines are highly voracious, living on annelids, other worms, crustaceans, etc. (J. B. Jennings, 1963). Food seems to be digested, absorbed, and excreted very rapidly. Many nemertines can undergo starvation for months and still survive. In such cases the body may be reduced to one-twentieth of its original size.

The nemertines are the simplest animals that possess a circulatory system. The system is a closed type, and the blood is restricted to vessels and spaces (lacunae). The chief vessels are 3 longitudinal trunks—a median and two lateral ones—connected at both the anterior and posterior ends (Figs. 6-34, *B*, and 6-38). Various modifications of this plan exist. The lateral vessels often give off a network of lacunae to the wall of the foregut. The middorsal trunk, when present, runs anteriorly in the floor of the rhynchocoel and posteriorly beneath the proboscis sheath. It unites at the posterior end with the lateral vessels by means of anal lacunae. The blood fluid does not flow in any definite direction because its movements are influenced by the muscular body wall and to some extent by the muscular walls of the larger blood vessels. In most cases the blood fluid is colorless and contains definite nucleated corpuscles (oval or round) and amebocytes. Some nemertines have blood colored by pigments of orange, green, and red, which are carried in the corpuscles. In some species the red color is due to hemoglobin.

The excretory system is composed of a single pair of protonephridial tubules that are closely associated with the lateral blood vessels (Fig. 6-38). Usually the tubules are restricted to the anterior region, which may mark a tendency to

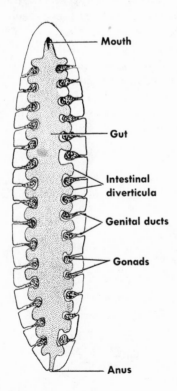

FIG. 6-37

Digestive and reproductive systems of nemertine (diagrammatic). (Modified from Lang.)

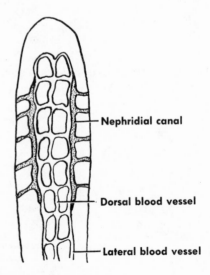

FIG. 6-38

Excretory and circulatory systems of anterior region of a nemertine worm (diagrammatic).

bring the products of excretion to a definite spot of the vascular system. On each side of the foregut region there is a nephridiopore from which a tubule extends anteriorly. On its inner surface each tubule gives rise to terminal flame bulbs that push in and indent the blood vessels, forming small blind ceca reminiscent of the glomeruli of higher forms. In some cases the blood vessel wall may disappear in that region so that the flame bulbs are directly bathed by blood. Many modifications of the system exist among different nemertines. In certain paleonemertines a ridge of protonephridial capillaries forms a nephridial gland that lacks flame bulbs; its capillaries open into the blood. In some species the protonephridia give rise to many capillaries and flame bulbs applied to the lateral blood vessels and foregut lacunae; in a few the excretory tubules extend into the alimentary canal. In the freshwater *Prostoma* and some others an extensive branching of the excretory system into capillaries and flame bulbs throughout the tissues occurs. A single protonephridium in these species may be broken up into smaller tubules, each with its nephridiopore to the exterior. The flame bulbs of nemertines may be a single cell or a cluster of cells. Large mesenchymal cells, known as athrocytes, may surround the flame bulbs and tubules and may be excretory in function.

With few exceptions, nemertines are dioecious. A notable exception is the hermaphroditic *Prostoma* that is also viviparous. In contrast with the complicated arrangement of Platyhelminthes, the nemertean reproductive system is quite simple. In the usual plan the gonads are a series of epithelium-lined sacs that alternate with the diverticula of the intestine (Fig. 6-37). When mature, each sac forms a genital duct to the exterior for the discharge of the gametes. These genital openings usually form a line dorsal to the lateral nerve. The whole sac, or gonad, with its epithelial wall and genital cells, arises from cells of the parenchyma. Sexual dimorphism is not evident, although the males of some species are provided with cirri

or copulatory projections of the male pores. Also, at sexual maturity each sex may have a different color because of the ripe sex cells seen through the transparent walls. Each ovary may produce only 1 egg (hoplonemertines) or as many as 50 (paleonemertines and heteronemertines). The eggs are usually discharged by body contractions. In spawning there may be external fertilization without contact of the sexes, or the male may discharge its sperm as it crawls over the female. Several worms may be enclosed in a mucous sheath where fertilization occurs. Self-fertilization often occurs in hermaphroditic species. The eggs are often laid in gelatinous strings produced by secretions of glands from the body walls.

The nervous system consists of a brain made up of lobes or ganglia, two longitudinal lateral cords, and many smaller nerves (Figs. 6-33 and 6-34). The ganglia are generally 4 in number, consisting of a dorsal and a ventral pair (Fig. 6-36). The dorsal ganglia are connected by a dorsal commissure above the rhynchodeum; the ventral pair are connected by a ventral commissure below the rhynchodeum. Through this ring of nervous tissue the proboscis apparatus passes. The ventral ganglia of the brain connect to the 2 main longitudinal nerve trunks, which are united by the anal commissure above or below the intestine. These longitudinal nerve trunks are made up of one continuous, unsegmented mass of fibrous and cellular nervous tissue and usually do not possess ganglionic swellings anywhere along their course. From the cerebral ganglia, nerves are given off anteriorly to the eyes and other sense organs, and from the lateral trunks small branches pass to the body wall; commissures connect the lateral nerves with each other. The nervous system of nemertines may be considered as local longitudinal accumulations of the less differentiated nerve plexus of primitive ancestors such as coelenterates. Various degrees of sinking of the nervous system from the epidermis to regions internal to the body wall musculature are found in the different orders, with the more primitive epidermal position found among the paleonemertines. In addition to the main lateral cords there also exist in some forms minor nerves such as a middorsal nerve, midventral nerve, paired esophageal nerves, and others. The proboscis is especially well supplied with nervous tissue of a primitive nerve plexus type.

A respiratory system is lacking, and exchange of gases occurs by direct diffusion.

Sense organs of a tactile function are widely scattered in the epidermis. These are slender cells with projecting bristles. Chemotactile organs in cephalic slits or grooves or in protrusible frontal organs are found. Eyes are of constant occurrence, although many heteronemertines and some others appear to be blind. Eyes are restricted to the anterior region in front of the brain and are located in the dermis, musculature, or parenchyma (Fig. 6-34). The eyes are mostly of the inverted type of pigment cell ocelli such as are found in flatworms. Simpler and more primitive types of eyes, some only of pigment specks without refracting apparatus, are present in some species. Eyes may number from 2 to many hundreds. Most nemertines also have a pair of cerebral organs forming 2 blind, invaginated canals from the epidermis. The inner ends of these canals are surrounded by glands and nerve cells near the dorsal brain ganglia. Their external openings are by way of cephalic grooves or on the body surface. Their function is probably chemotactile, as they are richly supplied with nerves from the brain.

Nemertines have spiral and determinate cleavage so that the potentialities of future development are fixed in the zygote before cleavage begins. In some cases the micromeres of the quartets are larger than the macromeres. Development involves the formation of a typical ciliated coeloblastula, and the endoderm is formed by either embolic or epibolic invagination of the macromeres with the fourth quartet of micromeres. In some nemertines, however, a stereoblastula is formed first and later acquires a cavity. In certain nemertines after early cleavage, development is direct without a larval

stage, and the embryo emerges from the egg membranes as a small, ciliated worm. In others, including the familiar *Cerebratulus*, the gastrula develops into a free-swimming, ciliated larva, the pilidium (Fig. 6-39). This helmet-shaped larva is formed by the downward growth of a pair of ciliated oral lobes at the sides of the mouth and bears an apical tuft of long cilia. The larval enteron with ectodermal foregut and endodermal midgut has a mouth but no anus (a flatworm characteristic). The blastocoel is filled with a gelatinous fluid and ameboid mesenchymal cells. A similar larva, but without the apical tuft and oral lobes, occurs in some other nemertines and is called a Desor larva. Both the pilidium and the Desor larva undergo metamorphosis by ectodermal invaginations called disks, which grow around the enteron, fuse together, and form the epidermis, nervous system, muscular body wall, and other structures of the future nemertine. The young nemertine becomes ciliated, frees itself from the surrounding pilidium investment, and develops into the adult worm.

Regeneration is usually quite common; nemertines have many enemies (annelids, crustaceans, etc.), and loss of body parts is frequent.

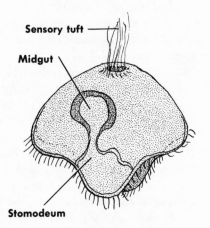

FIG. 6-39

Pilidium larva of Rhynchocoela (early stage). This larva has mouth but no anus, a flatworm characteristic.

Some worms may be broken into fragments, each of which may then give rise to a new worm (asexual reproduction). In others more restricted regeneration occurs. Posterior regions are usually easily regenerated, anterior or head regions rarely. The proboscis is often lost and is always regenerated.

■ CLASS ANOPLA
■ ORDER PALEONEMERTINI

The body wall of the Paleonemertini is either two-layered or three-layered. If two-layered, the musculature is composed of outer circular and inner longitudinal muscles; if three-layered, the inner and outer muscle layers are circular, with the longitudinal layer in between. The nervous system usually lies external to the outer circular layer, although it may lie deeper. A middorsal blood vessel and connections between the lateral vessels are usually lacking, and the intestine has a smooth contour. Eyes and cerebral organs are mostly absent. The dermis is

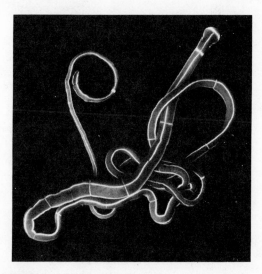

FIG. 6-40

Ribbon worm, *Tubulanus annulatus* (order Paleonemertini). May be several feet long; common along European shores. The anterior end has the enlarged, rounded cephalic lobe. (Courtesy Encyclopaedia Britannica Films, Inc.)

gelatinous. In general this order is the most primitively organized of all the nemertines.

Cephalothrix, a common nemertine along the New England coast, is a slender worm about 3 inches long with a long, pointed anterior end. The mouth is on the ventral side of the yellowish or flesh-colored body behind the brain. Other genera of this order are *Tubulanus* (Fig. 6-40), with a broad, rounded cephalic lobe at the anterior end; *Carinina*, with the cerebral organs in the form of canals; and *Procephalothrix*, similar to *Cephalothrix*.

■ ORDER HETERONEMERTINI

The Heteronemertini have a body wall composed of three muscular layers: an outer longitudinal layer, a middle circular layer, and an inner longitudinal layer. The nervous system lies between the first two muscular layers. A middorsal blood vessel, including transverse connections with the lateral blood vessels, is present. Cerebral organs usually occur, and the intestine is provided with diverticula but no cecum.

The most common genus along the Atlantic coast is the large form *Cerebratulus*, which may reach a length of 20 feet and a width of an inch. The body of this familiar flesh-colored nemertine is adapted for swimming by having a thick, narrow anterior end and an expanded, flattened posterior region with very thin margins. It is often found in burrows. Through the terminal proboscis pore the slender proboscis may be extended for more than 3 feet in the larger specimens. In the strange *Gorgono-rhynchus* the proboscis is divided into many proboscides (Fig. 6-41). *Lineus* is another well-known genus with a slender, rounded, highly contractile body. Six pairs of eyes are found in the head region of *L. socialis*, which grows to almost a foot in length with a width of $\frac{1}{4}$ inch. The members of this species are often found together in tangled masses. This species and some others can reproduce by fragmentation, and their regenerative powers are especially good. *Micrura* is similar to *Cerebratulus*, and

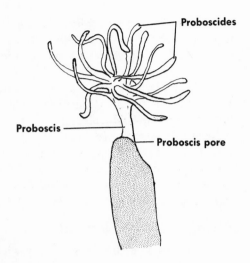

FIG. 6-41

Head region of *Gorgonorhynchus* (diagrammatic) (order Heteronemertini). Proboscis may be divided into 30 or more proboscides. Some zoologists think this genus arose suddenly by evolutionary saltation (Wheeler).

some of its species are considered transitional between the two genera. Some are bright red or orange in color.

◼ CLASS ENOPLA

◼ ORDER HOPLONEMERTINI

The members of this order are the only nemertines that possess a proboscis armed with a stylet apparatus (Fig. 6-34, *A*). Some have a single stylet and others have many stylets, a condition that causes the order to be divided into two suborders, Monostylifera and Polystylifera. The stylet or stylets are situated at the boundary between the eversible and noneversible portions of the proboscis. By this arrangement, when the proboscis reaches its maximum eversion, the stylet apparatus is terminally situated for puncturing the prey. Members of the order possess an intestine with lateral diverticula and a cecum. They have a middorsal blood vessel connected to the lateral vessels. Freshwater, terrestrial, marine, and commensal forms are found in this order.

Amphiporus (Fig. 6-34) is a common marine genus that is much studied. The members of

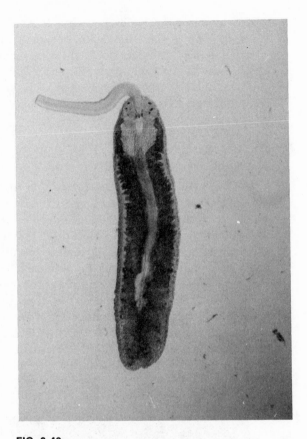

FIG. 6-42

Prostoma rubrum with proboscis extended. This freshwater form may be as long as 20 mm. and 2 mm. in diameter. (Phase contrast.) (Courtesy J. J. Poluhowich.)

this genus are usually short bodied; most are within the range of ½ to 5 inches in length. They have numerous eyes on the sides and top of the head, and shallow cerebral grooves are present. Their intestinal diverticula and long cecum are branched.

Prostoma (Stichostemma) (Fig. 6-42) is the only genus of nemertines that lives in fresh water (R. W. Pennak, 1953). Some species of this genus are strictly hermaphroditic with saclike ovotestis, although others are dioecious. *P. rubrum*, the most widely distributed species, is about 20 mm. long and is somewhat reddish in color, although other shades are found. It has a proboscis two or three times the length

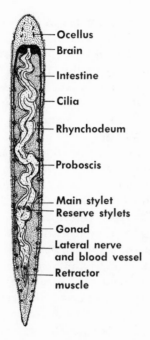

FIG. 6-43

Freshwater nemertine *Prostoma rubrum* (order Hoplonemertini). Anatomic relations (diagrammatic) as seen in
fixed whole mount. (From Poluhowich, J. J. 1968. Turtox
News **46**:2-6.)

of the body; in the quiescent stage it is coiled
within the rhynchocoel (Fig. 6-43). The single
stylet, when lost, is replaced by accessory stylets
from lateral pockets in the bulb region of the
proboscis (Fig. 6-44). This species appears to be
most common in the autumn, when it may be
found in masses of filamentous algae and other
aquatics. Under harsh weather conditions the
worm forms a cyst of hardened mucus where
it may live for several weeks, although the
cyst cannot withstand desiccation. Development of the eggs, which are laid in a mucous
sheath, is direct, and the young emerge as
juveniles. Several European species of similar
habits belong to this genus. *Prostoma* has
recently been studied by Poluhowich (1968) who
was able to collect them at all seasons in bottom
debris of mud and decaying aquatic plants of a
pond-marsh habitat in Connecticut.

FIG. 6-44

Large single stylet of *Prostoma,* with lateral reserve
pockets containing accessory stylets. (Phase contrast,
×500.) (From Poluhowich, J. J. 1968. Turtox News **46**:
2-6.)

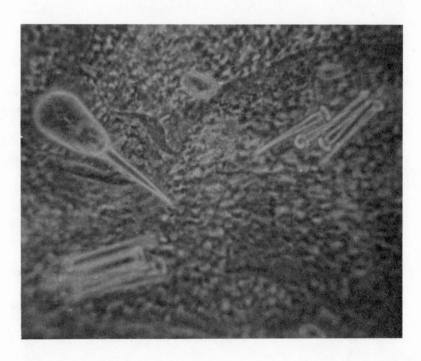

Other hoplonemertines include *Geonemertes,* a terrestrial genus widely distributed in tropic and subtropic regions; *Carcinonemertes,* an ectocommensal on the gills of crabs and provided with a common sperm duct into which all the testes discharge; and *Tetrastemma,* common along seashores and endocommensal in tunicates.

■ ORDER BDELLONEMERTINI

Bdellonemertini is an order of a single genus (*Malacobdella*) and consists of three species, all of which are commensals that feed on plankton in the mantle chamber or pulmonary sac of mollusks. The poorly developed proboscis does not bear a stylet, and the worms are devoid of cerebral organs, eyes, and other special sense structures. The proboscis pore and mouth form a common opening to the outside. At their posterior end, members bear an adhesive disk for attachment. Their vascular system contains a middorsal vessel and lateral vessels that are freely branched. The worm is usually 1 to 2 inches long, although it may be smaller.

MOVEMENT IN THE ACOELOMATES

To move about is one of the basic features of most animals, and it is done in various ways by different groups. Even among sessile animals, movement of some kind is a basic necessity. Some animals can move without muscles by the rhythmic beat of their cilia, but such a process is slow and suited only for very small forms. Larger organisms must have muscles. Muscle and body form are closely bound together. If contraction of muscle alters body form in any way, muscle fibers must relax to restore the body to its original form. Animal evolution has involved many improvements in the structure of muscular tissue and the mechanisms for exploiting the antagonistic action of muscles. Jointed skeletons with their system of levers make possible precise and restricted movements with the greatest economy of energy, but many invertebrates have no jointed limbs and must rely on other means. One of

these is the principle of the *hydrostatic skeleton.* Since animals have developed in water, which has formed so much of their general economy, it is understandable that they have made use of fluids in their mechanisms of movement. Water is incompressible and can transmit pressure changes equally in all directions. Such a system can function if a musculature is arranged so that it encloses a volume of fluid. Any contraction of the muscular system produces a pressure in the fluid that can be transmitted to other parts of the body. Also, water is easily deformed and body shape restored. The organization of the animal has a lot to do with the efficiency of the hydrostatic skeleton. It plays an important role in the form and movement of the radiate animals, and it will be seen later how effectively and precisely it is used in annelids and others. Such a system may involve general movements of the animal or simply certain parts that are employed for specific purposes.

In the acoelomates (Platyhelminthes and Rhynchocoela) the hydrostatic skeleton operates chiefly without the fluid component but makes use of the parenchymal arrangement found in this group. Here the mesenchyme of the parenchyma replaces the fluid as the chief deformable substance. Small members of the group may use their cilia for many movements, but muscles are needed for the larger members. The development of muscle is made possible by the emergence of the mesoderm as the definite originator of muscle tissue in place of the musculoepithelial cells found in coelenterates. In the acoelomates the skeletal system by means of which the muscles can act as antagonists to each other is made up of a thick basement membrane beneath the epidermis, consisting of fibers that form a type of lattice structure. These inextensible and nonelastic fibers run around the body as both left- and right-handed spirals to form alternate layers. This meshwork arrangement, similar to that of a trellis fence, allows the fibers to cross each other in such a way that the angles so formed can change as the length of the body changes. As the worm elongates, the angle between the fibers decreases; as the worm shortens, the angle between the fibers increases. The maximum volume would be attained when the lattice angle is about 54 degrees from the long axis of the body. Beneath the basement membrane, layers of longitudinal, circular, and diagonal muscles are located. To maintain a flattened form, dorsoventral muscles run between the dorsal and ventral surfaces. The two antagonistic systems of longitudinal and circular muscles can bring pressure to bear on the deformable parenchyma. By this relationship of the musculature to the parenchyma, acoelomates can undergo body deformities of twisting and flattening even though the parenchyma is less easily deformed than a fluid hydrostatic system such as the annelids have.

REFERENCES

Platyhelminthes

Ax, R. Relationships and phylogeny of the Turbellaria. In Dougherty, E. C. (editor). 1963. The lower Metazoa; comparative biology and phylogeny. Berkeley, University of California Press. Although the author believes in a monophyletic origin of the Bilateria, he removes the Turbellaria as direct ancestors of more recent groups except the other two classes of Platyhelminthes. According to his scheme, the early ancestor for most protostomes was provided with a coelom.

Baer, J. G. 1952. Ecology of animal parasites. Urbana, University of Illinois Press.

Barrington, E. J. W. 1967. Invertebrate structure and function. Boston, Houghton Mifflin Co.

Berg, G. G., and O. A. Berg. 1968. Presence of trimetaphosphatase in the Turbellarian *Dugesia tigrina* and the relation of enzyme distribution to food intake. Trans. Amer. Microscop. Soc. **87**:335-341.

Bils, R. F., and W. E. Martin. 1966. Fine structure and development of the trematode integument. Trans. Amer. Microscop. Soc. **85**:78-88.

Bogitsh, B. J. 1968. Cytochemical and ultrastructural observations on the tegument of the trematode *Megalodiscus temperatus*. Trans. Amer. Microscop. Soc. **87**:477-486.

Bullough, W. S. 1958. Practical invertebrate anatomy. London, The Macmillan Co. Practical descriptions of the anatomy of representatives of the Platyhelminthes are given in Chapter 5.

Cheng, T. C. 1964. The biology of animal parasites. Philadelphia, W. B. Saunders Co. A comprehensive treatise with special emphasis on the natural history and ecology of flatworms and other parasites.

Connella, J. V., and D. H. Stern. 1969. Land planarians:

sexuality and occurrence. Trans. Amer. Microscop. Soc. **88**:309-311.

Goodrich, E. S. 1945. The study of nephridia and genital ducts since 1895. Quart. J. Microscop. Sci. **86**:113-392. An important article that summarizes the problems of the gonocoel theory of coelom formation, genital ducts, and nephridia.

Graff, L. von. Acoela and Rhabdocoela. In Bronn, H. G. (editor). 1904-1908. Klassen und Ordnungen des Tierreichs, vol. 4, part 1. Leipzig, Akademische Verlagsgsellschaft. A definitive account of these two orders by a great specialist.

Grassé, P. P. (editor). 1961. Traité de zoologie, anatomie, systematique, biologie, Tome IV: Platyhelminthes, Mesozaires, Acanthocephales, Nemertiens. (Premier Fascicule). Paris, Masson et Cie. The most recent account of these four groups in one of the most authoritative treatises in the field of zoologic scholarship.

Hanson, E. D. 1958. On the origin of the Eumetazoa. Syst. Zool. **7**:16-47.

Hershenov, B. R., G. S. Tulloch, and A. D. Johnson. 1966. The fine structure of trematode sperm tails. Trans. Amer. Microscop. Soc. **85**:480-483.

Hyman, L. H. 1951. The invertebrates: Platyhelminthes and Rhynchocoela; the acoelomate Bilateria, vol. 2. New York, McGraw-Hill Book Co., Inc. The best general up-to-date account that is available at present. In volume 5 (1959) of this treatise a section is devoted to summarizing the more recent work on the flatworms.

Hyman, L. H., and E. R. Jones. Turbellaria. In Edmondson, W. T. (editor). 1959. Ward and Whipple's fresh-water biology, ed. 2. Methods of collecting and preparation of specimens as well as keys and figures of the freshwater forms are given.

Jenkins, M. M. 1966. Note on stalk formation in cocoons of *Dugesia dorotocephala* (Woodworth, 1897). Trans. Amer. Microscop. Soc. **85**:168.

Jenkins, M. M., and H. P. Brown. 1964. Copulation activity and behavior in the planarian *Dugesia dorotocephala* (Woodworth, 1897). Trans. Amer. Microscop. Soc. **83**:32-40.

Jennings, J. B. 1957. Studies on feeding, digestion, and food storage in free-living flatworms. Biol. Bull. **112**:571-581. This paper describes in detail how certain turbellarians ingest and digest their food. Digestion is both extracellular and intracellular, although the size of ingested particles is often the determining factor.

Jennings, J. B. 1962. Further studies on feeding and digestion in triclad turbellaria. Biol. Bull. **123**:571-579.

Karling, T. G. Some evolutionary trends in turbellarian morphology. In Dougherty, E. C. (editor). 1963. The lower Metazoa; comparative biology and phylogeny. Berkeley, University of California Press. The author shows how the evolution of the group is limited by the specific structure and organization of turbellarian tissue. Ecologic and functional principles indicate how evolution had to follow certain lines.

Kimmel, G. 1958. Das Terminolorgan der Protonephridien, Feinstrukter und Deutung der Funktion. Z. Naturforsch. [B] **136**:677-679. A description of the fine structure of flame cells.

Martin, W. E. 1952. Another annelid first intermediate host of a digenetic trematode. J. Parasitol. **38**:1-4. Two reports by Martin indicate the rare exceptions in which a group other than mollusks acts as the first intermediate host.

Mueller, J. F. 1965. Helminth life cycles. Amer. Zool. **5**:131-139.

Olsen, O. W. 1962. Animal parasites; their biology and life cycles. Minneapolis, Burgess Publishing Co. The author emphasizes illustrated schemes of parasitic life cycles. Concise descriptions are also given. An excellent treatise.

Oschman, J. L. 1967. Microtubules in the subepidermal glands of *Convoluta roscoffensis* (Acoela, Turbellaria). Trans. Amer. Microscop. Soc. **86**:159-162.

Oschman, J. L., and P. Gray. 1965. A study of the fine structure of *Convoluta roscoffensis* and its endosymbiotic algae. Trans. Amer. Microscop. Soc. **84**:368-375.

Rothman, A. H. 1963. Electron microscopic studies of tapeworms; the surface structures of *Hymenolepis diminuta* (Rudolphi, 1819) (Blanchard, 1891. Trans. Amer. Microscop. Soc. **82**:22-30.

Threadgold, L. T. 1962. An electron microscope study of the tegument and associated structures of *Dipylidium caninum*. Quart. J. Microscop. Sci. **103**:135-140. This study shows that the cuticle of cestodes is made up of living cells and must be considered an epidermis.

Threadgold, L. T. 1963. The tegument and associated structures of *Fasciola hepatica*. Quart. J. Microscop. Sci. **104**:505-512. The cuticle is considered to be a surface layer of protoplasm that is an extension of flasklike, nucleated cells of the interior.

Wardle, R. A., and J. A. McLeod. 1952. The zoology of tapeworms. Minneapolis, University of Minnesota Press. A pretentious and comprehensive treatise on these highly specialized parasites.

Rhynchocoela

Bohmig, L. Nemertini. In Kukenthal, W., and T. Krumbach (editors). 1929. Handbuch der Zoologie, vol. 2, part 1, section 3. Berlin, Walter de Gruyter & Co. An excellent technical account.

Bürger, O. 1890. Anatomie und Histologie der Nemertinen. Z. Wiss. Zool. **50**:1-279. This and other monographs are classic studies of the nemertines.

Coe, W. R. 1943. Biology of the nemerteans of the Atlantic coast of North America. Hartford, Trans. Conn. Acad. Arts Sci. **35**:129-328. One of many valuable papers on the Rhynchocoela by this author.

Grassé, P. P. (editor). 1961. Traité de zoologie, vol. IV, Platyhelminthes, Mesozaires, Acanthocephales, Nemertiens, fasc. 1. Paris, Masson et Cie.

Hyman, L. H. 1951. The invertebrates: Platyhelminthes and Rhynchocoela; the acoelomate Bilateria. New York, McGraw-Hill Book Co., Inc. The best general treatment in English of the Rhynchocoela.

Jennings, J. B. Some aspects of nutrition in the Turbellaria, Trematoda, and Rhynchocoela. In Dougherty, E. C. (editor). 1963. The lower Metazoa; comparative biology and phylogeny. Berkeley, University of California Press. A concise description of the methods by which nemertines capture and digest their prey. Their extracellular and intracellular methods of digestion are appropriate to either a predatory or parasitic mode of life, although the former method is mostly employed.

Pennak, R. W. 1953. Fresh-water invertebrates of the United States. New York, The Ronald Press Co. An excellent description of the structure and life history of the freshwater nemertine *Prostoma rubrum*.

Poluhowich, J. J. 1968. Notes on the freshwater nemertean *Prostoma rubrum*. Turtox News **46**:2-6.

Wells, M. 1968. Lower animals. New York, McGraw-Hill Book Co.

The pseudocoelomate animals

PHYLA ROTIFERA, GASTROTRICHA, KINORHYNCHA, GNATHOSTOMULIDA, NEMATODA, NEMATOMORPHA, ACANTHOCEPHALA, AND ENTOPROCTA

GENERAL CHARACTERISTICS

1. All the animals beyond the Rhynchocoela have a body cavity of some kind. This cavity may be divided into two types: (a) a peritoneal cavity lined with a modified mesoderm (peritoneum) or (b) a pseudocoel not lined with peritoneum—the inner surface of the body wall may be lined with mesoderm, but the endodermal gut is not covered with mesoderm.

2. The phylum Aschelminthes (cavity worms) formerly included a varied assemblage of common pseudocoelomate animals such as the rotifers, gastrotrichs, nematodes, and one or two other groups, each of which was called a class under the phylum. The aschelminth classes are now treated as separate phyla in their own right and are added to the other pseudocoelomate phyla. Altogether about eight phyla are now considered to be pseudocoelomates, of which the phylum Nematoda is by far the largest and most successful. Aschelminthes may now be considered as a superphylum, including such groups as were formerly considered classes within it.

3. In a pseudocoelomate animal no mesenteries hold the internal organs and no muscle layers surround the digestive tract. Thus, muscular peristalsis does not move food through the alimentary canal.

4. The pseudocoel in these phyla varies. In some it is filled with fluid; in others it may contain a gelatinous substance and mesenchyme cells. In all there is usually space enough, or the contained substance is so loosely arranged, that there is room for organ development and freedom of movement.

5. The advantages of a body cavity are that it affords space for the gut and other organs, for storage, and room for the gonads and sex cells. It also provides for a hydrostatic skeleton, similar to that found in some animals with a true coelom, e.g., the onychophorans and annelids.

6. The basic body plan of the pseudocoelomates is that of a roundworm, or a tube-within-a-tube arrangement. Parasitism may cause a modification of this tube-within-a-tube pattern.

7. In such a diversified grouping the pseudocoelomates all share in the unifying characteristic of a pseudocoel. Other characteristics are shared to some extent. Many have radial symmetry at the anterior end. The epidermis is commonly syncytial and secretes a cuticle, which may undergo specializations (spines, bristles, etc.). Only longitudinal muscles may occur in some, and constancy of cell numbers (or nuclei) is frequent.

8. Most members of the pseudocoelomate group are unsegmented or superficially segmented and are often cylindric in body form.

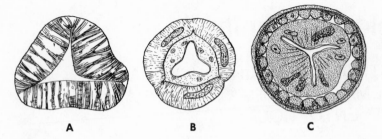

FIG. 7-1

Comparison of pharynx as seen in cross sections of, **A,** Gastrotricha; **B,** Kinorhyncha; and **C,** Nematoda. Similarity of pharynx in three groups may have phyletic significance.

Most are small, even microscopic, but some attain a considerable length. The group in general is made up of an odd assortment of animals that rarely share enough common characteristics to be placed together.

9. The body is covered with a scleroprotein cuticle that often bears spines, scales, and other skeletal structures. In many members the cuticle undergoes regular molting during the growth period of the life cycle. The epidermis may be cellular but is often syncytial. Underneath the epidermis the muscular layers are usually not well defined. Adhesive glands are common throughout the group.

10. Members possess a complete digestive system that is usually straight and without a muscular wall. The pharynx is very highly specialized in some (Fig. 7-1). An excretory system of protonephridia occurs in many. The circulatory and respiratory systems are absent. The group as a whole is often characterized by cell or nuclear constancy for a particular species, and cilia are either absent or greatly reduced. Most members are dioecious, and the reproductive systems are simple. Cleavage is mostly of the spiral determinate type, but in some cases it does not fit into either the spiral or radial types.

11. Both aquatic and terrestrial members are present in the group. Parasitism is fairly common.

PHYLOGENY AND ADAPTIVE RADIATION

1. Affinities are very confusing among such diverse groups. The most successful groups are the nematodes and rotifers, whereas the other phyla have modest numbers of species. Similarities in a few characteristics must be taken with great caution as evidence of relationship. Both kinorhynchs and priapulids, for instance, have an introvert at the anterior end but are quite different in other respects. Parasitism, so common with many, often brings about marked morphologic changes and obscures affinities. The flattened shape of the body and the presence of both circular and longitudinal muscles in the Acanthocephala may indicate relationship to the Platyhelminthes. Some may have come from an early offshoot of a line that led to other pseudocoelomates.

2. The Gnathostomulida are a new phylum whose evaluation is not yet completed. However, this group may be related to the Gastrotricha and Rotifera. Some consider this phylum to be related to the Turbellaria of the Platyhelminthes.

3. Each phylum of the pseudocoelomates may have its own unique basic pattern of adaptive radiation. The nematodes have been able to adapt themselves to almost every ecologic niche available to animal life because of the constant activity within the range of their restricted movement. The corona, body shape, and mastax trophi have guided the rotifers in their evolution. The kinorhynchs have kept within the restrictions of their zonite organization, and the entoprocts have been guided by

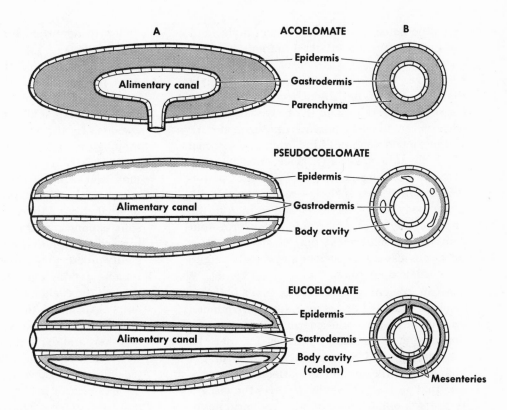

FIG. 7-2

Types of body cavity organization in Bilateria. In pseudocoelomate type, mesoderm may form lining around inner surface of body wall but not around enteron. (From Hickman: Integrated principles of zoology, The C. V. Mosby Co.)

their ciliary feeding and sessile habits. The small number of species of certain phyla indicates a very modest adaptive radiation or they may be the last remnants of a once much larger group. Many morphologic variations among the different phyla are difficult to explain.

4. Two members of the pseudocoelomates—nematodes and rotifers—may exhibit the condition known as cryptobiosis (formerly called anabiosis). This term refers to the ability of certain invertebrates to undergo a state of suspended animation under unfavorable environmental surroundings (extremes of heat and cold, ionizing radiation, etc.) at any stage of their life cycle. In this state the animal loses its water (partially or wholly), dries out, and contracts into a barrel shape (called a tun in rotifers) or into a tight spiral shape (nematodes). When returned to water, such animals rapidly assume ordinary life manifestations. Experiments reveal that there are several forms of

cryptobiosis among different species. Ordinarily a tiny oxidative metabolism is evident, but in an environment completely deficient in oxygen some cryptobiotic animals can switch to an anaerobic, or fermentative, type of metabolism and finally to a condition in which no metabolism can be detected. Such animals can be revived as long as they have structural integrity. Cryptobiosis greatly extends the life cycle of animals that normally have short life-spans, for an individual cryptobiotic state may last for many years.

THE BODY CAVITY

The bilateral animals include the acoelomates, pseudocoelomates, and eucoelomates.

It will be noted that this convenient division of the major groups of bilateral animals is based on the presence or absence of some form of body cavity (Fig. 7-2). In the acoelomates (phyla Platyhelminthes and Rhynchocoela) the space between the enteron and the body wall is filled with mesenchyme or parenchyma (mesoderm). In the pseudocoelomates the original blastocoel of the blastula makes up a space, or pseudocoel, between enteron and body wall without a peritoneal lining. The eucoelomates, or animals with a true body cavity, have a space formed in the endomesoderm or embryonic mesoderm that completely obliterates the blastocoel. Most phyla possess a true coelom and thus belong to the eucoelomate group.

A true coelom is therefore the space between the body wall and digestive system, surrounded on all sides by endomesoderm with a lining of peritoneum. Embryology teaches us that the mesoderm found on the inner surface of the body wall is the somatic or parietal mesoderm, and that part which is reflected over the viscera is the splanchnic mesoderm. The peritoneum is an epithelial specialization of the mesoderm that lines both the inside of the body wall (parietal peritoneum) and covers the exposed surfaces of protruding visceral organs (visceral peritoneum). The mesenteries that provide support for visceral organs are composed mostly of peritoneum.

The development of a coelomic cavity has been of the utmost importance in the evolutionary patterns of animals. The lack of such a cavity greatly restricts the differentiation of organs for special purposes. In a form such as hydra, for instance, transient gonads must lie on the outside surface of the body because there is no other place for them. When the space between body wall and enteron is filled with mesenchyme, the digestive system must increase absorptive surface by branching instead of by elongated coils. A coelom provides spaces for waste products and also for the development of sex cells. The coelom is correlated with genital ducts for eggs and sperm and with nephridia for the discharge of secretory products.

The formation of the coelom is correlated with the formation of endomesoderm. In this respect animals follow one or the other of two methods, the schizocoelous and the enterocoelous. In the schizocoelous method, which may be the primitive one, the coelom arises by a split in a solid mesodermal band that may or may not segment into somites, depending on whether or not the animal belongs to the segmented group. In whatever way formed, the split expands, pushing the mesoderm against the body wall and the wall of the enteron. The cavity produced between these two mesodermal regions is the coelom. In the enterocoelous method the coelom originates from mesodermal outpocketings of the primitive gut, or archenteron. These pockets are pinched off, fuse together, and expand until they touch the body wall and enteron, thus producing a space (coelom) enclosed within mesoderm. The schizocoelous method is characteristic of Protostomia, and the enterocoelous method characterizes the Deuterostomia.

■ PHYLUM ROTIFERA
CLASSIFICATION

Phylum Rotifera. Body somewhat cylindric with anterior ciliated disk (corona) and posterior forked tail or foot; body covered by a cuticle sometimes modified into hardened lorica; size microscopic to 3 mm. but most are not over 0.5 mm.; pharynx with movable jaws (mastax trophi); trunk often with sensory antennae, or palps; epidermis a syncytium; muscles mostly single and not forming layers; digestive tract mostly complete; excretory system of protonephridial tubes and flame bulbs; nervous system of ganglia and nerves; sense organs of membranelles, pits, and ocelli; reproduction dioecious with minute males (sometimes absent); parthenogenesis common; often oviparous; juvenile young; mostly freshwater; 1,500 species.

   Class Seisonacea. Body elongated with small head and long neck; corona reduced; body with joined cuticle; foot stalklike with adhesive disk; pharynx with jaws; sexes similar; paired ovaries without vitellaria; no parthenogenesis; epizoic or parasitic; size to 3 mm. *Seison* (Fig. 7-7, *G*).

Class Bdelloidea. Body elongated with retractile head and foot; 15 to 18 cuticular joints or false segments; corona usually of a cingulum and two trochal disks on pedicles; mastax (ramate) present; lorica absent; paired germovitellaria; males unknown; parthenogenesis of egg; oviparous or viviparous. *Philodina* (Fig. 7-7, *I*), *Rotaria*.

Class Monogononta. Body highly variable; lorica present or absent; foot of not more than two toes; mastax trophi of many types; males usually small or degenerate; one germovitellarium; eggs of three types (amictic, mictic, dormant); sessile or free swimming; parasitic (some). *Collotheca*, *Asplanchna* (Fig. 7-8), *Epiphanes* (Fig. 7-4).

GENERAL MORPHOLOGIC FEATURES

The rotifers, or wheel animalcules, are microscopic animals that range in length from 40 microns to 3 mm. Most are under 500 microns. They derive their name from a ciliated corona that gives the appearance of a revolving wheel. Most species are freshwater forms, although a few are marine. Most are solitary, and some are sessile. They are found in a great variety of habitats from purely aquatic environments to damp soils, sand, and mosses. They are cosmopolitan in their distribution.

External features

In their morphology the body is usually made up of the head, trunk, and foot. The head consists of the corona, mouth, mastax, and brain (Fig. 7-3, *A*). In the trunk are found most of the organs—stomach, intestine, cloaca, excretory system, and reproductive organs. The terminal foot is tapering and consists of two or more superficial segmentlike parts that are usually highly retractile. The foot is absent in some species, but when present, it may consist of 1 to 4 toes. There are many morphologic variations within the group. The body is usually elongated and cylindric, but some are short, stout, and even spheric. In cross section the body may be round or flattened dorsally or ventrally. The body is covered with a cuticle secreted by the syncytial epidermis and consists of scleroproteins. When the cuticle is much thickened,

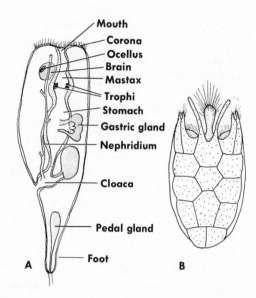

FIG. 7-3

A, Basic organization of rotifer, side view. **B**, *Keratella*, dorsal view, showing cuticle modified into sculptured, spiny lorica. (**A** modified from Lang.)

it is called a lorica and may bear characteristic patterns of plates and ornamentation (Fig. 7-3, *B*). In the bdelloid rotifers, superficial annuli may give the appearance of segmentation (Fig. 7-7, *G* and *I*), but true segmentation is lacking.

The corona, the most characteristic organ of rotifers, is of many types. The primitive form of the corona is primarily a ventral or anterior ciliated surface, or buccal field around the mouth, derived from a creeping, ventrally ciliated ancestor. From the buccal field a circumapical band, or ring, encircles the anterior margins of the head. The apical field is the unciliated region within the center of the ring and may bear outlets for the retrocerebral organ and sensory hairs. From such a ground plan the patterns characteristic of different groups of rotifers seem to have evolved. Structural modifications may involve either the buccal field or the circumapical ring or both. Parts of these regions may be lost altogether, cirri may replace short cilia, the circumapical ring may enlarge at the expense of the buccal field as in *Epiphanes (Hydatina)* (Fig. 7-4), and there may be an absence or reduction of cilia on the coronal surface (*Eosphora*). Ciliary tufts may be found on lateral processes, or auricles (*Notommata*). In *Collotheca* (Fig. 7-7, *B*) the corona forms a type of funnel, with its upper edges folded into lobes bearing stiff setae. In the familiar bdelloid rotifers (Fig. 7-7, *H* and *I*) the corona is made up of two trochal disks and the cingulum. This arrangement is a modification of two circles of cilia, anterior and posterior. The anterior circle (the trochus) is elevated on pedicels and divided into 2 trochal disks, each bearing on the edge a single row of membranelles. The posterior circle (cingulum) of small cilia encircles the bases of the pedicels and continues ventrally behind the mouth, which thus lies between the 2 rings in a ciliated groove. When such a corona is retracted, a projection (rostrum) covered with cilia and bristles is seen, which may be used in locomotion. This structure is retracted also when the corona is extended. In sessile types the corona may be scalloped into lobes that may be radially arranged (Figs. 7-5 and 7-7, *B*). Pigmented ocelli (eyes) may appear on various parts of the corona and are involved in reactions to light.

Among the characteristic structures of the trunk of rotifers are the sensory antennae, or palps. These are movable papillae, usually with tufts of sensory hairs at the tip. The dorsal antennae (single or paired) are found in the middorsal line near the anterior end of the trunk (Fig. 7-5); the paired lateral antennae (present only in the class Monogononta) vary in position on the trunk and may even be found dorsally or ventrally. A caudal antenna near the base of the toes is found in some rotifers. Long movable spines or skipping blades are present on the trunk of certain Monogononta.

The foot, or tail, is the terminal part of the

FIG. 7-4

Epiphanes (Hydatina) senta, ventral view of female (class Monogononta).

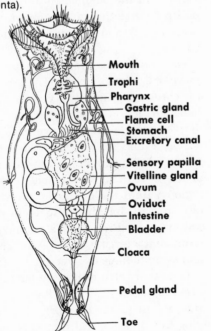

- Mouth
- Trophi
- Pharynx
- Gastric gland
- Flame cell
- Stomach
- Excretory canal
- Sensory papilla
- Vitelline gland
- Ovum
- Oviduct
- Intestine
- Bladder
- Cloaca
- Pedal gland
- Toe

FIG. 7-5

Floscularia, a sessile rotifer that lives in beautiful tube. Particles are pressed into pellets in its specialized pit and attached to tube in rows, or tube may be decorated with sand grains or fecal balls (class Monogononta).

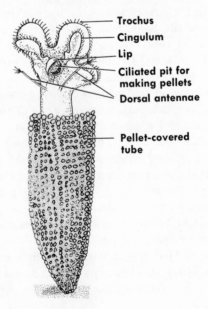

- Trochus
- Cingulum
- Lip
- Ciliated pit for making pellets
- Dorsal antennae
- Pellet-covered tube

rotifer. It may be a gradual extension of the body, or it may be sharply marked off from the trunk. Often its cuticle is ringed into joints that can telescope together when it shortens. The foot may be modified into a long stalk in sessile forms and may be absent altogether in others. It may bear an adhesive disk but is usually provided with 1 to 4 toes. Pedal glands (Figs. 7-3 and 7-4) open on some parts of the toes or foot in most rotifers. They secrete a sticky substance for attachment and construction of tubes or cases (Fig. 7-5). In swimming the foot acts as a rudder, and in creeping and looping methods it can anchor the animal temporarily to the substratum.

A change in body form, known as cyclomorphosis, occurs in many rotifers. This cyclic recurrence in relation to seasonal conditions may involve differences in length of spines, body, and head shape. In some parthenogenic females the spines become shorter in each succeeding generation so that individuals appearing early in the season may have longer spines than those appearing later in the season.

Internal features

The body wall of rotifers is made up of the cuticle, epidermis, and muscles. The epidermis is a syncytium of scattered nuclei usually constant in number for each species. *Epiphanes* (Fig. 7-4), for example, is supposed to have 960 cells or nuclei. The body cavity is not lined by a peritoneum, and mesenteries are lacking; it is thus a pseudocoel. It is filled with a fluid and a network of branched, ameboid cells. The musculature contains bands of circular muscles especially well developed in the head region; beneath the circular muscles are longitudinal muscles that act principally as retractors of the head and foot. In addition, visceral muscles extend from the body wall to the viscera for moving and holding the viscera in place. Single muscles, rather than layers, seem to be the rule. Group variations of muscular patterns are common. Both smooth and striated muscle fibers occur in rotifers; striated muscle is well adapted

for rapid movement of spines and appendages.

The nervous system consists of a bilobed cerebral ganglion, or brain, lying on the dorsal surface of the mastax, 2 main lateral nerves, and fine sensory and motor fibers from the brain to various sensory organs and to the muscles. A ganglion and plexus are found on or in the mastax wall and send nerves to the enteron. Sensory organs in the form of ocelli, antennae, bristles, ciliated pits, and membranelles are common. The controversial retrocerebral organ lying near the brain and opening on the corona is thought by some authorities to be a sense organ, but no conclusive evidence about its function has been determined.

A complete digestive system is present in most rotifers. The mouth, which is usually ventrally placed, may have a lower lip formed from the cingulum previously mentioned and usually opens into the pharynx by way of a ciliated tube. The pharynx is modified into a muscular mastax, or gizzard, lined with cuticle. A variable number of salivary glands open by ducts into the anterior end of the mastax.

The mastax is peculiar to rotifers and contains on its inner wall the masticatory apparatus, the trophi. The trophi are of many types, important in taxonomic diagnosis, and they are made up of hard, cuticularized pieces. The trophi consist of seven parts, three paired lateral and one median (Fig. 7-6). All these may be modified morphologically for specialized purposes in the different types. The unpaired piece is called the fulcrum, and the paired ones are rami, unci, and manubria. The fulcrum and two rami collectively make up the uncus; an uncus and its associated manubrium collectively form a malleus. Thus a typical trophi is made up of one uncus and two mallei. Eight basic types of mastax trophi are found and they are variously adapted to biting, cutting, holding, and crushing food particles (largely plankton, epiphyton, and particulate detritus). The eight basic types are malleate and ramate—chiefly for grinding, uncinate—for laceration of ingested plankton, virgate (Fig. 7-6, *C* and *D*)

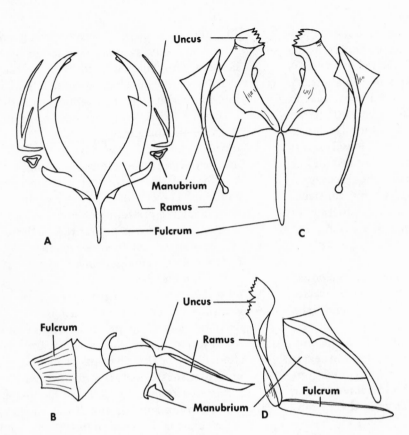

Uncus

Manubrium

Ramus

Fulcrum

A

C

Fulcrum

Uncus

Ramus

Manubrium

Fulcrum

B

D

FIG. 7-6

Two types of mastax trophi. **A** and **B,** Anterior and lateral views of forcepslike incudate trophi of *Asplanchna* used to grasp food organisms. **C** and **D,** Ventral and lateral views of virgate trophi of *Synchaeta*. This type is found in pelagic raptorial rotifers that grasp prey by unci and suck in its contents. (**A** and **B** after de Beauchamp, 1907; redrawn from Pennak: Freshwater invertebrates of the United States, The Ronald Press Co.; **C** and **D** redrawn from Hyman: The invertebrates: Acanthocephala, Aschelminthes and Entoprocta; the pseudocoelomate Bilateria, McGraw-Hill Book Co., Inc.)

and cardate—for sucking and oscillatory grinding, forcipate and incudate (Fig. 7-6, *A* and *B*)—protrusible forceps types for seizing prey, and fulcrate—a quite different type, found only in Seisonacea, and not fully understood regarding function.

In sessile and nonpredatory species the coronal cilia create currents of water that sweep the food particles to the region of the mouth.

If there are two trochal disks, such as are found in the Bdelloidea, the membranelles of each disk create a circular current opposite to that of the other disk. These two currents direct the food, by way of a groove between the trochus and cingulum, into the mouth. Most members that feed in this manner are omnivorous and will ingest most particles of the right size. The predatory species feed on protozoans, rotifers, and other small metazoan organisms, and detect their prey by chemical stimuli or by touch. They seize their prey by grasping with forcepslike trophi (Fig. 7-6, *A*) that can be protruded from the mouth. Some species can literally suck out the contents of their prey, either outside the rotifer or within the mastax. Another method of ingesting food is found in those with widespreading coronal funnels (Fig. 7-7, *B*) which act as traps for small organisms that accidentally wander into the funnels.

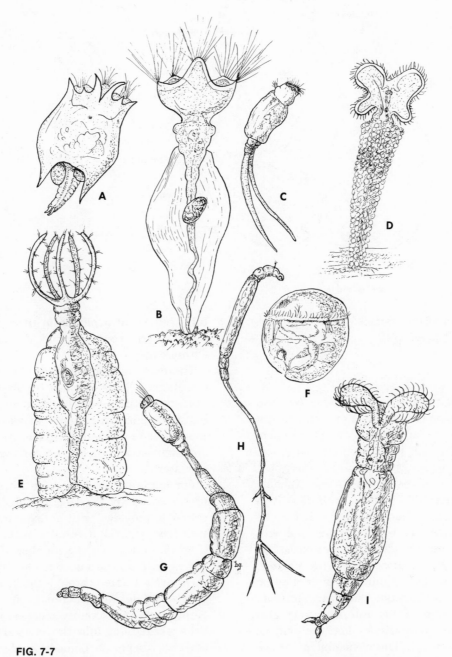

FIG. 7-7

Some rotifers showing many types of form. **A,** *Noteus.* **B,** *Collotheca.* **C,** *Furcularia (Cephalodeila).* **D,** *Floscularia.* **E,** *Stephanoceros.* **F,** *Trochosphaera.* (**A** to **F** class Monogononta.) **G,** *Seison* (class Seisonacea). **H,** *Rotaria.* **I,** *Philodina.* (**H** and **I** class Bdelloidea.)

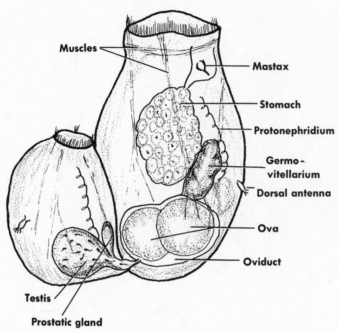

Muscles

Mastax

Stomach

Protonephridium

Germo-
vitellarium

Dorsal antenna

Ova

Oviduct

Testis

Prostatic gland

FIG. 7-8

Asplanchna in copulation (class Monogononta). (Modified from Wesenburg-Lund.)

After leaving the mastax, the food passes into the tubular esophagus, ciliated or cuticle lined, and thence to the large, thick-walled stomach, where most digestion and absorption take place. The stomach bears at its anterior margin a pair of bean-shaped gastric glands that probably furnish enzymes. The stomach wall usually is composed of large ciliated cells, but it may be syncytial in some members. It is provided with both circular and longitudinal muscles. The stomach may be separated from the narrow, ciliated intestine by a sphincter, or it may be indistinctly set off from that part of the enteron. The posterior end of the intestine may also receive the excretory tubules and the oviduct and is thus a cloaca. The cloaca opens to the outside dorsally at the base of the foot. In some rotifers, including *Asplanchna* (Fig. 7-8) and other sucking forms, an anus or cloacal opening is absent. Digestion is primarily extracel-

lular but is intracellular in those that possess large gastric ceca and lack an anus, such as *Chromogaster*.

There are no special organs of respiration and circulation. Respiration is by direct body surface, but rotifers vary in the amount of oxygen required. Some are found in oxygen-deficient mud in aquatic habitats, but most obtain a sufficient supply of oxygen carried by currents of water.

For maintaining a constant internal osmotic pressure and getting rid of metabolic waste, the rotifer is provided with an excretory system of two protonephridial tubules with flame bulbs (Figs. 7-3 and 7-4). The number of flame bulbs varies with the group, but usually 4 to 50 of them are scattered through the body. The 2 collective tubules are much convoluted and often empty into an excretory bladder, which opens by a short duct into the ventral side of the cloaca. A separate bladder is absent in those rotifers that have the posterior part of the cloaca modified into a bladder. The contents of the bladder are emptied through the cloacal opening four or five times per minute.

REPRODUCTION AND EMBRYOLOGY

Rotifers are dioecious, and the sexes are dimorphic. In some groups males have never been found and in others they are produced sporadically. In only a few groups are the males as large as the females, and most males have degenerate structures. Parthenogenesis is the only reproductive method in the class Bdelloidea; males are unknown in this group. In the class Monogononta, which has the most complex reproductive methods, males have been described for most species, although parthenogenesis is the usual method of reproduction. When males occur, they are usually found only during a few weeks of the year.

In most species the female has a single, bulky complex consisting of the ovary proper (germarium) with its developing oocytes, and the larger syncytial vitellarium that furnishes yolk to the egg (Figs. 7-4 and 7-8). When mature, the egg passes down the short oviduct into the cloaca, from which it is expelled to the outside. The bdelloids, however, have 2 ovaries, 2 vitellaria, and 2 oviducts. The egg is constricted into an oval shape as it passes down the oviduct, and it may be laid on a substratum or attached to the body of the female. In male rotifers a single, saclike testis exists, and a ciliated sperm duct usually opens into a gonopore because a cloaca is often absent. A copulatory organ and prostate glands are also present. Males commonly produce two kinds of sperm: a typical kind to fertilize the egg and a smaller, rod-shaped kind to aid the others in penetrating the female body wall. Hypodermic insemination, by which the male penis penetrates the female body wall and deposits sperm in the pseudocoel, is the common method of copulation (Fig. 7-8). The number of eggs produced by a female is restricted to the number of nuclei in her ovary at birth, usually 10 to 20.

In the Monogononta two types of females, which are usually morphologically indistinguishable, occur. One of these types is called amictic, and the females reproduce by parthenogenesis. Their amictic eggs undergo only one nonreductional division in the ovary, have the diploid number of chromosomes, and cannot be fertilized. They give rise only to females. The second type of female is called mictic and appears only at certain times of the year. Their eggs (mictic) are haploid because they are produced by the double meiotic division. If these mictic eggs are not fertilized, they develop by parthenogenesis into haploid males. If fertilized, they develop a thick-walled, resistant shell with more nutritive material and become winter, or dormant, eggs. Upon hatching, these dormant eggs always give rise to amictic females. A mictic female may give rise to both fertilized eggs and male eggs, but an amictic female produces only amictic eggs. Usually there are only two or three mictic generations per year, whereas there may be 30 to 40 amictic generations. In Seisonacea the eggs are all mictic and hatch into either sex. The Bdelloidea lack males entirely. Parthenogenetic females have a lifespan of 1 to 2 weeks as a general rule. Ecologic factors responsible for bisexual periods as well as abundance in number in the life cycle of rotifer populations are not well known, and many conflicting theories have been advanced.

In their embryology, rotifers have a modified spiral determinate type of cleavage.

The early embryology of *Monostyla cornuta* has been studied by F. A. Pray (1965) (Fig. 7-9). He found in the 4-cell stage that there are 3 small micromeres A, B, and C and one large D macromere. The D cell divides unequally and gives rise to the small d^1 cell, which with the micromeres A, B, and C forms the germ-ring cells. The gastrulation is epibolic. When the heavily granular A, B, and C cells divide, 3 of the resulting cells are heavily granular, with 3 clear cells arranged above them. The granular cells are the only cells to involute and eventually form the digestive tract (endoderm); the clear cells form the ectoderm. The D macromere and its descendants are shoved by the involuted micromeres to the end of the embryo opposite the blastopore. The D macromere and its de-

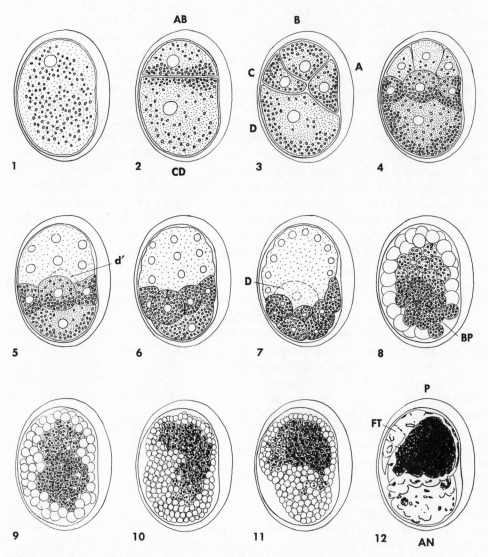

FIG. 7-9

Stages in the early development of the rotifer *Monostyla cornuta* Müller. **1-3,** Early cleavage, **AB** micromere dividing into **A** and **B** micromeres, **CD** macromere forming **C** micromere and **D** macromere. **4,** Eight-cell stage. **5,** Position of **d'** cell, which cannot be seen in **4. 6-7,** Epiboly and beginning of involution of micromeres. **8,** Involution of yolk-laden cells and position of blastopore **(BP).** Progeny of **D** cells have smaller amounts of yolk than do involuted micromeres. **9-11,** Internal migration of derivatives of **D** cell, which will form ovary and vitellarium. **12,** Orientation of anterior-posterior **(AN-P)** axis of embryo relative to position of blastopore, and location of foot **(FT).** (From Pray, F. A. 1965. Trans. Amer. Microscop. Soc. **84:**210-216.)

scendants give rise to the reproductive system (vitellarium and ovary).

Although a blastopore is formed, it is soon covered by the ungranulated ectodermal blastomeres moving over it. The mastax appears to be derived from the stomadeum, whereas the nervous, excretory, and muscular systems may be formed from cells that infiltrate from the overlying ectodermal layer.

■ PHYLUM GASTROTRICHA
CLASSIFICATION

Phylum Gastrotricha. Pseudocoelomate body somewhat vermiform and unsegmented, with differentiated cuticle and restricted tracts of cilia; microscopic (minute to 1.5 mm.); ventrally flattened; syncytial epidermis with adhesive tube glands; no corona; pharynx without trophi but with bulbous enlargements; muscles mostly longitudinal and not in definite layers; enteron a straight tube with circular muscle fibers; protonephridia (in some); nervous system of a brain and a pair of lateral nerves; parthenogenetic females with two types of eggs in some, hermaphroditic individuals in others; development direct with total cleavage; short life cycle (3 to 21 days); marine and fresh water; 500 species.

   Order (or class) Macrodasyoidea. *Macrodasys.*
   Order (or class) Chaetonotoidea. *Chaetonotus.*

GENERAL MORPHOLOGY

The 500 species of the microscopic Gastrotricha are about equally divided between those that live in fresh water and those that are marine. The order Macrodasyoidea is mostly marine; the order Chaetonotoidea is all freshwater with few exceptions. Both kinds are found in similar habitats, being especially common in littoral zones, where they crawl over the bottom or on masses of vegetation. They are usually less than 1 mm. in length (most are around 500 microns) and are about the same size as rotifers. The body is usually elongated, with a convex dorsal surface and a flattened ventral surface. The anterior end may be modified as a lobelike head set off by a constricted neck from the trunk, and the posterior end may be forked, pointed, or other shapes.

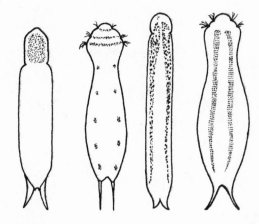

FIG. 7-10

Some patterns of ciliation found in gastrotrichs, ventral views.

Ciliation is mostly restricted to certain tracts such as the ventral region and the head lobe. The cilia may be arranged in longitudinal bands, in patches, in transverse rows, or in tufts (Fig. 7-10). Very long cilia or bristles are common and may function as sensory organs. The ventral cilia are used for gliding over a substratum. The animals can also swim by ciliary movement.

The body surface is covered with a thin cuticle that may be smooth but is usually provided with a pattern of scales, plates, and spines (Fig. 7-11). The scales or plates may overlap on the head lobe and ventral surface, or they may cover the entire body. Their color is variable but is usually light green to reddish brown; some are transparent and colorless. Some may have pigment spots (ocelli) and a few are provided with palps (Fig. 7-11, *B*). Adhesive tubes, similar to those of rotifers, are found in both orders and are used for attachment to objects (Fig. 7-11, *A*). Each is a cuticular tube supplied by a gland and movable by muscles. The tubes may be very numerous and are often arranged in rows along the lateral margin or on the ventral side of the head; the tail in some species may also bear them.

Beneath the cuticle the body wall is made up of syncytial epidermis and muscle bands of

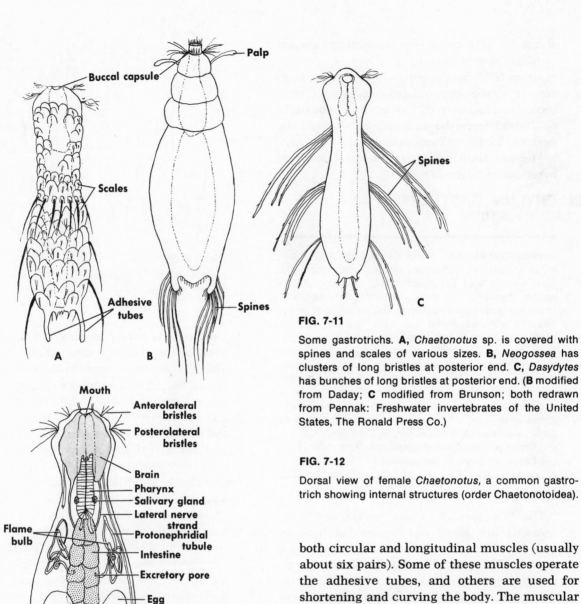

FIG. 7-11

Some gastrotrichs. **A,** *Chaetonotus* sp. is covered with spines and scales of various sizes. **B,** *Neogossea* has clusters of long bristles at posterior end. **C,** *Dasydytes* has bunches of long bristles at posterior end. (**B** modified from Daday; **C** modified from Brunson; both redrawn from Pennak: Freshwater invertebrates of the United States, The Ronald Press Co.)

FIG. 7-12

Dorsal view of female *Chaetonotus,* a common gastrotrich showing internal structures (order Chaetonotoidea).

both circular and longitudinal muscles (usually about six pairs). Some of these muscles operate the adhesive tubes, and others are used for shortening and curving the body. The muscular system is similar to the rotifer type.

Gastrotrichs are pseudocoelomate animals, but the pseudocoel is mostly in the form of spaces and is devoid of ameboid cells.

Excretion and osmoregulation are accomplished by 2 flame bulbs that are fastened to 2 protonephridial tubules (Fig. 7-12). This system, which is restricted to the freshwater species (Chaetonotoidea), has no bladder. The protonephridia lie coiled in the pseudocoel and open near the midventral line.

The digestive system consists of a terminal mouth, often with small hooks, a short buccal cavity (sometimes with projecting teeth), a nematode-like pharynx (Figs. 7-2, C, and 7-12) with bulbous enlargements and glands, a straight intestine or midgut, a rectum, and an anus with a sphincter muscle at the posterior end of the trunk. Most of the enteron, with the exception of the midgut, is lined with cuticle. Circular muscle is also found in the wall of parts of the alimentary canal.

Gastrotrichs live on small organisms such as bacteria, protozoans, algae, and organic detritus. They make use of their head cilia in the ingestion of food. They are most common in standing fresh water where there is decaying vegetation, and they may be found at all seasons of the year. It is thought that they can withstand low oxygen concentrations. Their chief enemies are aquatic insects, ameboid protozoa, annelid worms, and hydras.

The nervous system contains 2 lateral nerve strands connected to a large, dorsal, saddle-shaped brain covering the anterior part of the pharynx (Fig. 7-12). Fine nerve branches extend to the body wall and viscera. Sensory receptors are represented by tufts of cilia, ciliated pits, and scattered tactile bristles.

REPRODUCTION AND EMBRYOLOGY

Most marine species (Macrodasyoidea) are hermaphroditic, and a few are protandrous (in which male and female gametes are formed at different times). Only parthenogenetic females are known in freshwater species. The 1 or 2 ovaries consist of cell masses without a capsule and lie lateral to the intestine. They connect to a tiny oviduct that opens to the outside by a gonopore near or in common with the anus. The male system, when present, is made up of 2 testes, from each of which a sperm duct carries the sperm to a single or double male gonopore. A vestigial copulatory bursa may also be found. Most females produce 1 to 5 eggs, which are of two types according to whether or not they have immediate or de-layed cleavage. The tachyblastic eggs begin cleavage as soon as deposited and hatch usually within 12 to 20 hours; opsiblastic eggs are larger and are found in older cultures where they may survive harsh environmental conditions of temperature. M. Sacks (1964) found no opsiblastic eggs in his study of *Lepidodermella squammata*.

Gastrotrichs have direct development. Their total cleavage pattern in early stages is similar to that of nematodes (Fig. 7-25), producing equal blastomeres. Members of Macrodasyoidea have more cells than do those of Chaetonotoidea, and cell constancy is found in the group. Their life-span is apparently short; in cultures it is about 3 to 21 days, depending on ecologic conditions. Like rotifers they do not shed their cuticle (M. Sacks, 1955).

Chaetonotus (Fig. 7-11, A, and 7-12), which also occurs in marine waters, is one of the most familiar gastrotrichs as well as one of the most cosmopolitan. *Macrodasys* is a common marine genus.

■ PHYLUM KINORHYNCHA (ECHINODERA)
CLASSIFICATION

Phylum Kinorhyncha. Body vermiform with superficial, nonciliated segments (13 to 14 zonites); head with circles of spines or scalids, completely retractable; neck with plates (placids); cuticle covered with bristles; epidermis mostly syncytial forming longitudinal cords; one pair of ventral adhesive tubes on third or fourth zonite; musculature segmentally arranged, with a pair each of dorsolateral and ventrolateral longitudinal bands; muscles mostly cross-striated; muscular pharynx lined with syncytial epithelium; enteron straight with muscle layer; pseudocoel with fluid and amebocytes; one pair of protonephridia with flame bulbs; nervous system of a brain encircling the pharynx, a ventral ganglionated cord with a ganglion in each trunk zonite, and ganglia in the lateral and dorsal epidermal cords; reproduction dioecious, each sex with one pair of gonads; male with penial spicules; development largely unknown, but larval stages with molts occur.

Suborder Cyclorhagae. *Echinoderella* (Fig. 7-13, A), *Centroderes*.

Suborder Conchorhagae. *Somnoderes.*
Suborder Homalorhagae. *Pycnophyes* (Figs. 7-13, *B* and *C*, 7-14).

These three suborders are based chiefly on the way the head is protected when withdrawn. Among the Cyclorhagae the placids of the second zonite close over the opening. In the Conchorhagae the closing mechanism over the first 2 zonites consists of a pair of lateral plates

on the third zonite, whereas in the Homalorhagae (which also retracts the first 2 zonites) the opening is closed by a single dorsal plate and 3 ventral plates on the third zonite.

There were 53 described kinorhynch species in 1964, but other species are discovered from year to year (R. P. Higgins, 1964).

GENERAL MORPHOLOGY AND PHYSIOLOGY

The small marine kinorhynchs are superficially segmented and lack external ciliation. They are found mostly in the littoral zone of shallow waters, where they dwell in muddy bottoms.

The body is divided into three regions—the

FIG. 7-13

Phylum Kinorhyncha. **A,** *Echinoderella,* dorsolateral view, with head projected. **B,** *Pycnophyes,* ventral view, with head and neck extended showing sternal plates. **C,** Digestive system and protonephridia of *Pycnophyes.*

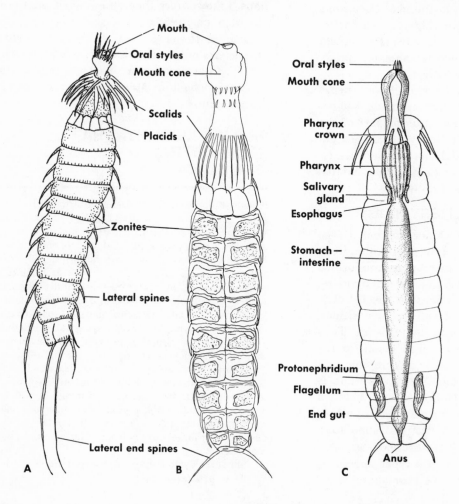

head, neck, and trunk. The head is rounded with a terminal mouth, bears five to seven circles of spines (scalids), and is completely retractile (Fig. 7-13). The neck is enclosed by a number of plates (placids) that may close over the head when the latter is withdrawn. (See description under classification.)

The trunk makes up the remainder of the body. Throughout the body the cuticle is divided into apparent segments called zonites. Head and neck each consist of 1 zonite, and the trunk usually has 11 zonites (12 in *Campyloderes*). Each zonite is usually flattened ventrally with a midventral groove and is covered with a dorsal, or tergal, plate and 2 ventral, or sternal, plates. Between the zonites the cuticle is thin, allowing flexibility. A pair of ventral adhesive tubes is present in the third or fourth zonite. Each of the remaining trunk zonites bears a recurved dorsal spine and a pair of lateral spines. The terminal zonite may also bear a pair of large, movable, lateral spines. Spines are hollow and filled with epidermis and, together with the zonites, are among the features characteristic of the group.

The body wall is made up of the cuticle and a syncytial epidermis, which forms middorsal and lateral longitudinal thickenings or cords that bulge into the pseudocoel. In each zonite each cord is slightly enlarged. There is no subepidermal sheath, but the musculature is segmentally arranged. Two pairs of longitudinal muscle bands run along the dorsolateral and ventrolateral regions of the body. The head is protruded by using these bands and ring muscles in the first 2 zonites to compress the body. Locomotion is produced by protruding the head and attaching it to the substratum by the scalids; then by means of the longitudinal muscles, the head is retracted and the trunk is pulled forward. Repetition of this process produces a characteristic creeping movement. Kinorhynchs cannot swim.

The nervous system is in close contact with the epidermis and consists of a brain encircling the anterior end of the pharynx, a ventral ganglionated nerve cord, and ganglion cells in the dorsal and lateral epidermal cords of each zonite.

The digestive system is a straight tube that is differentiated into certain specialized regions. The mouth is placed terminally on the protrusible mouth cone that encloses a buccal cavity. Posterior to the buccal cavity is the muscular pharynx (Fig. 7-13, C) lined with cuticle and syncytial epithelium. On its exterior the pharynx is covered by several bands of protractor muscles. Following the pharynx is a short, slender esophagus, which has lining continuous with that of the pharynx. To the esophagus are attached 2 dorsolateral and 2 ventrolateral salivary glands. The pharynx and esophagus are considered a stomodeum because their lining is cuticular. The stomach-intestine is covered by a network of muscles and lacks cuticular lining and gland cells. Pancreatic glands, however, are present at the junction of the esophagus and stomach. A sphincter separates the intestine from the short hindgut, and another, more posterior sphincter is also present. The posterior end of the hindgut is lined with cuticle and is a proctodeum. The terminal anus is found in the last zonite.

The pseudocoel is fluid-filled and lies between the body wall and enteron. Numerous amebocytes are present in the cavity. The excretory system consists of a single pair of protonephridia, one on each side of the intestine in the tenth zonite. The flame bulb in each protonephridium is multinucleated and usually contains a single flagellum. The protonephridia open by separate nephridiopores on the tergal plate of the eleventh zonite.

The kinorhynchs are dioecious, and the two sexes are similar externally. Each sex contains one pair of saclike gonads that open separately on the thirteenth zonite. The ovary contains ova and nutritive cells and is provided with a short oviduct. The genital pore, or gonopore, of the male is armed with 2 or 3 penial spicules that are useful in copulation.

Not much is known about the embryology

of the kinorhynchs. In some the eggs hatch into small larvae that lack external segmentation, spines, head, and other structures. By successive molts the larva acquires zonites and adult morphology.

American genera are best represented by *Echinoderella* (Fig. 7-13, *A*) and *Pycnophyes* (Figs. 7-13, *B* and *C,* and 7-14), which, so far, are known only from the northern New England coast.

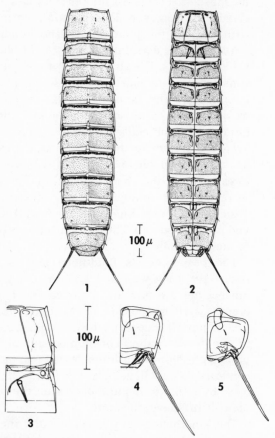

FIG. 7-14

Pycnophyes frequens (Blake, 1930), a homalorhagid kinorhynch. **1,** Dorsal view of neck and trunk segments (male). **2,** Ventral view of neck and trunk segments (male). **3,** Ventral view of left half of segments 2-4 (male). **4,** Ventral view of left half of segments 12-14 (male). **5,** Ventral view of left half of segments 12-13 (female). (From Higgins, R. P. 1965. Trans. Amer. Microscop. Soc. **84:**65-72.)

■ PHYLUM GNATHOSTOMULIDA

Gnathostomulida is the latest phylum to be discovered in the animal kingdom. It may not belong to the pseudocoelomates, but is placed here as one of the miscellaneous phyla. The first specimen appears to have been discovered in Kiel Bay of the Baltic Sea by A. Remane in 1928, but it was not until 1956 that the first publication of the new phylum was made by P. Ax (1963), another German investigator. Since that time the discovery of other species has been made in many seas, including those of North America, so that the phylum probably has a worldwide distribution. At the present time some 100 species in 15 genera have already been reported. In North America they have been found abundantly along the eastern coast, especially along the North Carolina coast. Their small size and their habit of adhering closely to the substrate, rather than scarcity of numbers, may account for the delay in their discovery. They are so minute and delicate that they could easily be overlooked.

Their common habitat is in the coastal sediments of the intertidal region, where they have a preference for extremely fine sandy sediments containing a small percentage of organic matter. They occupy the interstitial spaces between grains of very fine sand (125 to 250 microns), usually the black sediment that may be found several feet beneath the sand surface and often contains iron bacteria. This may indicate that the gnathostomulids may live under partially anaerobic conditions. A liter of such sediment may contain thousands of these worms. They occupy, it is thought, the minimal-sized interstitial environment occupied by metazoans. Associated with them are often found gastrotrichs, nematodes, ciliates, tardigrades, and other small forms. In isolated cases they have been found clinging to algae and other marine vegetation.

Evolutionary divergence, as far as an imperfect evaluation permits, seems to center mainly around the mouth region and size dimensions. Considerable variation is found among the different genera in respect to the lateral jaws that are present in all specimens. The jaws range from simple pincers or forceps to complicated lamellar jaws with comblike rows of numerous teeth. This differentiation may be associated with their food intake. Size variations among the different species may enable the phylum to make the fullest efficient use of their unique interstitial environment.

The phylogeny of the phylum is very obscure and awaits further evaluation of the entire group. R. J. Riedl (1969) thinks they may show some relationship to turbellarians, gastrotrichs, and rotifers. They have no pseudocoel spaces but are placed among the pseudocoel group because of other features.

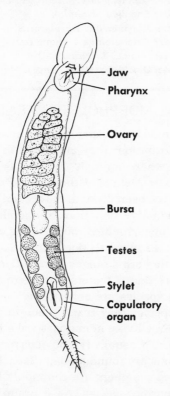

FIG. 7-15

Gnathostomula paradoxa, a hermaphroditic species, showing general shape and internal organization (diagrammatic). The recently described phylum Gnathostomulida awaits further evaluation. Its lack of a coelom and an anus would fit it for a group other than the pseudocoelomates. (Modified from Ax and Riedl.)

GENERAL FEATURES AND PHYSIOLOGY

Considerable variations in size occur among the different species. Some are as long as 3.5 mm.; others are as small as 0.7 mm. They are worm-shaped, mostly cylindric, and are semitransparent (Fig. 7-15). In many there are definite external divisions of head, trunk, and tail. The skin is a thin single-layered epithelium and bears one cilium on each of its polygonal epidermal cells, thus showing with the adults of sponges, coelenterates, brachiopods, and amphioxus a monociliated organization of the skin. A small amount of parenchyma may be found between the gut and the skin wall.

Some concentration of nervous tissue occurs in the anterior end, and a peripheral septum of nervous tissue lies just under the epithelium. Several stiff cilia (sensory in function) are found at the anterior end, as well as an anterior sensorium of 2 stiff cilia in some of the genera. A cushion of thick skin and nervous tissue on the dorsal side bears a cluster of stiff cilia.

The relatively large mouth is located ventrally in the collar region of the head and is surrounded by a specialized muscular apparatus. The lateral jaws, which give the name to the phylum, are diverse in structure, as already mentioned. Bacteria and fungi are scraped off the substrate and passed into the midgut by rapid snapping movements of the jaws, accompanied by jaw retraction. The simple midgut has no anus.

Sexual stages in some species include males, females, and hermaphrodites. In the same specimen there may be a dorsal ovarium followed by two posterior groups of testes follicles. The male system may include a tubular copulatory stylet of muscles or rods. The female system also includes a bursa for sperm storage. In copulation the male stylet of one worm injects sperm in packets surrounded by a mucous ball into another worm in a region between the skin and gut just back of the bursa. The mucous ball (prebursa) with the enclosed sperm then extends over the bursa and discharges the sperm, which migrate through the bursa mouthpiece and fertilize a mature egg. The fertilized egg is discharged through the dorsal body wall just behind the bursa and adheres to the substratum. The development of the egg follows, more or less, the spiral plan of cleavage, without the formation of a blastocoel or blastopore.

Gnathostomulids, as far as is known, move much as other lower worms do. They are able to glide, swim in loops and spirals, coil, nod the head from side to side, and adhere closely to their substratum. They are gregarious by nature and often are found together in large numbers. In a mixture of gnathostomulids and other small organisms (nematodes, turbellarians, gastrotrichs, etc.) in sand-water samples they are usually the last to emigrate to the surface of the sediment.

■ PHYLUM NEMATODA
CLASSIFICATION

Phylum Nematoda (Nemata). Body vermiform, cylindric, unsegmented, and pseudocoelomate; pointed at both ends; complex cuticle of many layers with various surface differentiations; cilia in some; epidermis cellular or syncytial with 4 longitudinal cords; subepidermal musculature of a single layer of longitudinal fibers arranged in 4 quadrants between the epidermal cords; a pair of amphids (chemoreceptors) at anterior end; enteron complete with triradiate pharynx and esophagus adapted for suction; excretory system of gland cells (renette) or of intracellular canals without flame bulbs in lateral cords; nervous system of circumenteric ring (brain), with anterior and posterior longitudinal nerves; mostly dioecious, with paired or single tubular gonads continuous with ducts; female genital opening (vulva) separate, male opening into cloaca; development direct, with determinate and modified spiral cleavage; molts of cuticle and parts of enteron; young juvenile.

Classification of nematodes varies greatly. The following scheme assumes that whether the nematodes are considered a separate phylum or a class in the phylum Aschelminthes, there is only one class of nematodes.

> **Subclass or superorder Phasmidia (Secernentea).** Body usually with a pair of phasmids near posterior end; caudal glands absent; amphids reduced or porelike on lips; excretory system of lateral canals in cords; accessory copulatory organs mostly absent; mostly parasitic.

> **Order Rhabditida.** *Rhabditis.*
> **Order Tylenchida.** *Tylenchus.*
> **Order Strongylida.** *Necator.*
> **Order Ascaridida.** *Ascaris.*
> **Order Oxyuroidea.** (Sometimes included with Ascaridida.) *Enterobius.*
> **Order Spirurida.** *Gnathostoma.*
> **Order Dracunculoida.** (Sometimes included with Camallanida.) *Dracunculus.*
> **Order Filarioidea.** (Sometimes included with Spirurida) *Filaria.*
> **Order Camallanida.** *Camallanus.*

> **Subclass or superorder Aphasmidia (Adenophorea).** Body without phasmids; caudal glands present; amphids spiral or disk types; excretory system not of lateral canals but glandular of one or more cells (renette); accessory copulatory structures present; mostly free living.

> **Order Chromadorida.** *Desmodora.*
> **Order Monhysterida.** *Monhystera.*
> **Order Enoplida.** *Enoplus.*
> **Order Dorylaimida.** *Dorylaimus.*
> **Order Trichuroidea.** (Sometimes included with Dorylaimida.) *Trichuris.*
> **Order Dioctophymatida.** *Dioctophyme.*

GENERAL MORPHOLOGIC FEATURES

Nematodes are an extensive group, worldwide in distribution; they are found in nearly every place where animals (free-living or parasitic) can exist. Details of their varied ecologic habitats have been given in the section on phylogeny and adaptive radiation. Nematodes are vermiform, unsegmented roundworms circular or cylindric in cross section (Fig. 7-16). The length of adult worms ranges from 100 to 200 microns to more than 7 meters, but the majority of free-living species are only about 1 mm. long. Marine species may reach a length several times this. Free-living nematodes are a homogeneous group, and usually few obvious morphologic distinctions are found among them. The more specialized members are commonly found among the marine species. As a group the ecologic and taxonomic status of roundworms is poorly worked out. Species identification is very difficult, and multitudes of species are undescribed.

According to some authorities, the marine nematodes are more primitive than freshwater

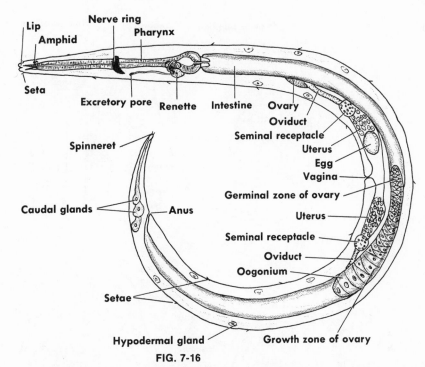

FIG. 7-16

Plectus parietinus, a chromadorid nematode (female), generalized. (After Hirschmann; modified from Sasser and Jenkins.)

FIG. 7-17

Marine nematode, *Metepoilonema* sp. (lateral view). Collected from beach sand, it may be characteristic of psammon habitats. Transverse cuticular markings (annulations) give the worm a ringed, segmented appearance and some rigidity. The few scattered bristles may be tangoreceptors. These cuticular sensory bristles are usually scanty in marine forms. (Courtesy R. P. Higgins.)

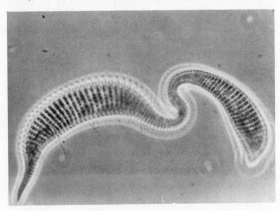

and terrestrial forms. They may have been at one time more or less sessile, fastened at the tail tip with cement from the caudal glands (which are mostly absent in terrestrial forms). Many marine forms are provided with rings of cuticle that give them rigidity (Fig. 7-17).

External features

The body of nematodes is covered with a cuticle consisting mainly of scleroproteins but no chitin and is not divided into regions. The cuticle may bear protuberances (Fig. 7-18, A), pits, ridges, and striations (Fig. 7-18, D) or it may be smooth. The body usually tapers at both ends.

Structures around the mouth at the anterior end have a definite radial or biradial arrangement (Fig. 7-1, C), which may indicate a sessile ancestor. The mouth is surrounded by liplike lobes (3 on each side in primitive forms; three fused pairs in terrestrial and parasitic species), each bearing an inner and outer labial papilla or bristle (Fig. 7-18). Thus there are two circles, each of 6 papillae or bristles. Outside the

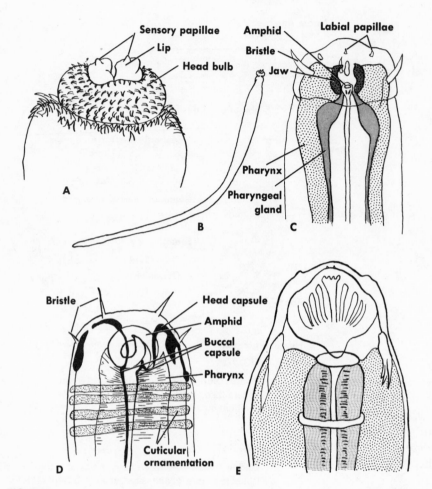

FIG. 7-18

Phylum Nematoda. **A,** *Gnathostoma,* anterior end with large protruding lips. **B,** *Gnathostoma,* whole specimen (order Spirurida). **C,** *Enoplus,* anterior end showing 2 of 3 large, movable jaws (order Enoplida). **D,** *Desmodora,* anterior end showing amphid characteristic of aphasmids (order Chromadorida). **E,** *Camallanus,* lateral view showing shell-shaped half of buccal capsule; has no lips (order Camallanida).

lips a third circle of 4 cephalic papillae may also be present. Many familiar nematodes have 1 dorsal and 2 ventrolateral lips. Many modifications of the lip structures exist in the various worms by loss, fusion, etc. The lips may also bear cuticular projections such as head shields and toothlike structures (J. E. Harris and H. D. Crofton, 1957). In some animal and plant parasitic nematodes the mouth cavity is provided with a hollow hypodermic needle for penetrating and extracting tissue contents. It will be noted that none of the lips or sense organs is found in strict medial position.

Free-living marine nematodes usually have caudal glands (Fig. 7-16) in their tail region, which secrete an adhesive cement of sticky threads for attachment.

Internal features

In nematodes the body wall consists of cuticle, epidermis or hypodermis, and a muscular layer (Fig. 7-19). The cuticle may be very complex in such nematodes as *Ascaris.* Nine layers

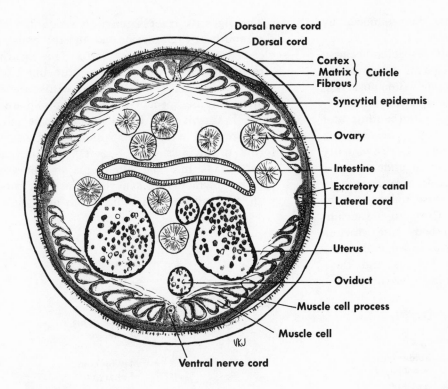

Dorsal nerve cord
Dorsal cord
Cortex
Matrix } **Cuticle**
Fibrous
Syncytial epidermis
Ovary
Intestine
Excretory canal
Lateral cord
Uterus
Oviduct
Muscle cell process
Muscle cell
Ventral nerve cord

FIG. 7-19

Cross section of *Ascaris* showing cuticle and internal structure. (From Hickman: Integrated principles of zoology, The C. V. Mosby Co.)

have been identified in this form by some investigators. Many substances have been found in it, such as lipoids, proteins, and carbohydrates, but keratin appears to be absent. The thick cuticle has three strata of collagenous fibers that are oriented spirally. The outermost protein layer is thought to undergo tanning. The cuticle usually undergoes four molts in its development. Beneath the cuticle the epidermis, which may be cellular or syncytial, forms two median and two lateral enlargements that project into the pseudocoel as longitudinal cords. Four subsidiary cords without nuclei may be present in some members. The cords run the body length but are usually better developed anteriorly. In rows within the cords are the

epidermal nuclei that are absent from the epidermis between the cords. Unicellular hypodermal glands (Fig. 7-16) may be present in the lateral cords of free-living nematodes.

The musculature of the body wall is made up entirely of longitudinal fibers arranged in bands, each of which is found in the space between 2 longitudinal cords. The muscle cells are uninucleated, nonstriated, spindle shaped, and interlocking. Each muscle cell has an outer fibrillar zone and an inner protoplasmic zone containing a nucleus, and each cell is innervated by a protoplasmic process from a medial nerve (Fig. 7-19). Nematodes have no circular muscles. Muscles with special functions include male copulatory muscles, female vulvar muscles, somatointestinal muscles, and rectal muscles.

The pseudocoel is filled with fluid and fixed cells or nuclei and is usually lined with a pseudocoelomic membrane that is formed from the growing together of protoplasmic processes of mesenchyme cells during embryology. Visceral organs are also supported in the pseudocoel by these mesenchymal processes.

Nematode movements are restricted because of the lack of circular muscles and the presence of a generally inflexible cuticle. When observed, they are frequently undergoing thrashing movements produced by the alternate bending and straightening of the body. Some can swim by undulations in a dorsoventral plane, movements that also may enable them to glide through thick vegetation. Much of their thrashing seems to be without definite purpose unless it is to explore the environment. Crawling is accomplished by throwing the body into sinuous curves or by gripping the substratum with cuticular specializations (stilt bristles and their secretions) and pulling the body forward.

The digestive tract is generally straight, with the mouth at the anterior end and the anus or cloaca at the other (Fig. 7-16). The mouth, already described, opens into the tubelike buccal capsule (Fig. 7-17, *D*) lined with cuticle, which may be specialized into ridges, plates, or teeth. Carnivorous nematodes are especially well provided with teeth, which are often referred to as jaws (Fig. 7-18, *C*). The buccal cavity leads into the esophagus or pharynx (foregut), which is a syncytial muscular tube with a triradiate lumen in which one angle,

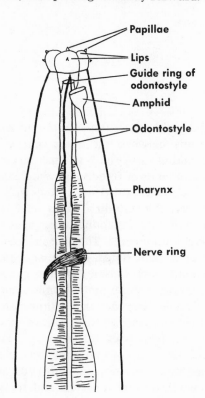

FIG. 7-20

Dorylaimus, an aphasmid nematode. Pharynx provided with enlarged tooth (odontostyle) that serves as stylet; amphid of cyathiform (cup-shaped) type.

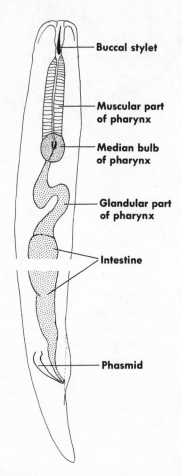

FIG. 7-21

Tylenchus, a phasmid nematode, anterior and posterior ends. Portion of middle part of worm omitted. Total length of worm, 1-2 mm. Pharynx is elongated and muscular (order Tylenchida).

or ray, is directed toward the midventral line. It is also lined with cuticle, and its walls usually contain 1 dorsal, and 2 subventral esophageal glands that may open into the buccal cavity or into the esophagus. The esophagus is provided with a small, complex esophagosympathetic nervous system that coordinates the sucking or pumping action of this highly muscular organ. To aid in the ingestion of food the esophagus may also have bulbs (Fig. 7-21) and valves. It may also be provided with a spear or stylet for puncturing prey (Figs. 7-20 and 7-21). An esophagointestinal valve usually separates the esophagus from the long, straight intestine (midgut) formed of a one-layered cellular epithelium. The intestine is not lined with cuticle, but it may bear ceca and its cells may be multinucleated. Sometimes its cells have stored food inclusions and waste materials. Its posterior end may also be provided with a thin muscle net. The rectum (cloaca in male), or hindgut, is a short, dorsoventrally flattened tube that is set off from the intestine by an intestinorectal valve. Unicellular rectal glands in parasitic

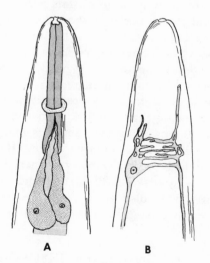

FIG. 7-22

Two types of excretory systems in nematodes. **A,** Glandular type with 2 large renette cells *(Rhabdias).* **B,** Tubular or H-type showing network of canals *(Ascaris).*

nematodes open into the rectum anteriorly. The rectum is lined with cuticle (being a proctodeum) and its wall contains a few large, flat epithelial cells covered externally by muscle tissue. Dilator muscles may run from the rectum to the body wall. The slitlike anus may be terminal or posteriorly ventral in position. The sperm duct in the male usually enters the cloaca ventrally.

Digestion of food is probably extracellular because many nematodes live on food that has been simplified by bacterial action or by enzymes of the host.

The plan of the excretory system in nematodes is not shared with the other pseudocoelomates. Flame bulbs have never been discovered in them. Two types of excretory systems, glandular and tubular, are recognized in the class. The glandular type is characteristic of the free-living marine nematodes and is made up of 1 or 2 large cells (renette gland) located at the ventral side of the junction between the esophagus (pharynx) and midgut (Figs. 7-22, *A,* and 7-24). The duct from this gland runs forward and opens by an excretory pore in the midventral region of the body. The more specialized tubular type of excretory system is often called the H type because of the arrangement of the canals of which it consists (Fig. 7-22, *B*). This H type consists of 3 canals; a lateral excretory canal in each lateral cord and a single transverse canal that connects the 2 lateral canals. A short, common excretory canal leads from the transverse canal to the excretory pore located in the midventral line of the body behind the mouth. The renette system is considered to be the primitive plan from which the H type has evolved by intracellular outgrowths.

The nervous system consists of a circumenteric nerve ring around the esophagus with paired lateral ganglia, a dorsal ganglion, and a ventral ganglion; other ganglia directly or indirectly associated with the rings; and various nerves that run anteriorly and posteriorly from the nerve ring (Fig. 7-23). Six sensory papillary nerves run anteriorly from the ring to the oral

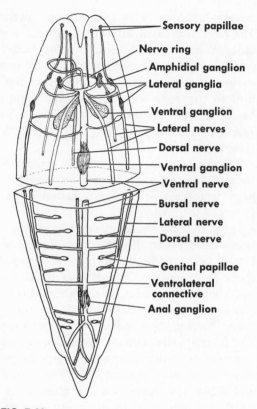

FIG. 7-23

Nervous system of *Ascaris* as found in head and tail regions.

Labels on figure:
- Sensory papillae
- Nerve ring
- Amphidial ganglion
- Lateral ganglia
- Ventral ganglion
- Lateral nerves
- Dorsal nerve
- Ventral ganglion
- Ventral nerve
- Bursal nerve
- Lateral nerve
- Dorsal nerve
- Genital papillae
- Ventrolateral connective
- Anal ganglion

papillae or bristles; a pair of dorsolateral nerves run to the amphidial glands; and 4 major longitudinal nerves (dorsal, 2 lateral, ventral) run posteriorly from the circumenteric ring in the corresponding longitudinal cord. The most important of these 4 nerves is the ventral nerve, which is a partly paired and partly ganglionated nerve chain that gives off branches to the enteron, copulatory organs, etc. and terminates in the anal ganglion or ganglia. The ventral nerve has both sensory and motor functions, whereas the dorsal nerve is motor with few or no ganglia along its course. A sympathetic nervous system is present in both the esophagus and rectum and connects with the nerve ring and anal ganglia respectively.

Free-living nematodes have many sense or-

gans; parasitic forms have lost many of theirs. Most sense organs are in the form of papillae and bristles. Cuticular structures such as the labial and cephalic bristles are each supplied with a nerve fiber from the nearest papillary nerve. Tactile receptors, mostly in the form of bristles, are also scattered on the body surface. Some bristles are supplied with gland cells. Males, especially, have cuticular sensory papillae around their copulatory mechanism.

Unique nematode structures are the amphids, which are best developed in aquatic nematodes but are greatly reduced in terrestrial and parasitic forms. An amphid (Figs. 7-16, 7-18, *D*, and 7-20) is a cuticular excavation of a pocket, spiral, or circular shape and provided with a slit-like aperture to the outside. One is present on each side of the body at the anterior end in or near the circlet of cephalic bristles. They are probably chemoreceptors; each has a gland and nerve ending.

A pair of unicellular glands, the phasmids (Fig. 7-24), open by a pore on each side of the tail in one group of nematodes (Phasmidia, or Secernentea). Unlike the amphids, they possess nearby neurons, whose axons run through the lateral caudal nerves to the large ventral nerve. These organs are supposed to be chemoreceptors. A few aquatic nematodes have ocelli on the sides of the anterior region. Free nerve endings are also found in nematodes.

In 1966 it was discovered (with the electron microscope) that dendritic nerve processes of certain sensory organs (amphids and papillae) are ciliary in structure, having the typical basal body and symmetry of peripheral and central fibrils (D. R. Roggen and associates, 1966). This discovery brings the nematodes (formerly thought to have no cilia) in line with other groups.

REPRODUCTION AND EMBRYOLOGY

Sexes are chiefly separate in nematodes, and the females are usually larger than the males, which may have curved ends, spicules, etc. The tubular gonads may be paired or single and

FIG. 7-24

Rhabditis (subclass Phasmidia [Secernentea], order Rhabditida). **A,** Male. **B,** Female. **C,** Posterior end showing phasmids. (**A** and **B** after Hirschmann; modified from Sasser and Jenkins.)

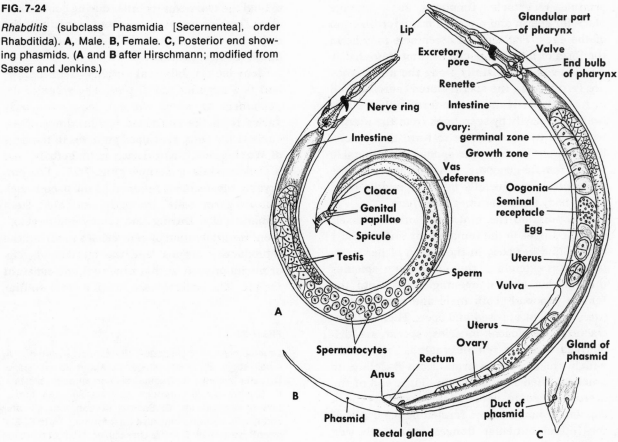

are often long and coiled. The germ cells arise either at the blind distal end from a terminal cell in the germinal zone or along the sides of the gonads (Figs. 7-16 and 7-24). In hermaphroditic nematodes both sperm cells and oocytes are formed from the same primordial germ cells at different times.

In the male the gonad, or testis, is usually single and extends anteriorly (Fig. 7-24). Its free end is often much coiled. The sperm duct is a continuation of the testis and toward its posterior end it widens to form the muscularized seminal vesicle. Posterior to the sperm duct is a heavier, muscularized ejaculatory duct that opens posteriorly into the ventral side of the rectum or cloaca and bears prostatic glands that furnish secretions in copulation. Most male

nematodes are provided with 1 or 2 cuticular nonhollow spicules of varying forms that originate from dorsal cloacal pockets. These spicules have special muscles and are pushed into the female vagina during copulation; they aid in the transfer of sperm by spreading the female gonopore. To direct the spicules in the right direction, an accessory cuticular piece, the gubernaculum, is sometimes found in the dorsal wall of the spicule pouch.

In the females there are usually 2 tubular ovaries (Figs. 7-16 and 7-24), but reduction from 2 to 1 ovary is not rare. An ovary may be straight or coiled. It continues at first as an oviduct with tall epithelium and then widens into the uterus with flat or cuboidal epithelium and muscular walls. A portion of the uterus, the seminal receptacle, functions as a storage place for sperm and a place for the fertilization of the eggs. The seminal receptacle may be in the form of a lateral pouch in some nematodes. As the eggs pass distally along the uterus they are fertilized by the sperm stored there. A shell is formed partly from the fertilization membrane and partly by secretions from the uterine wall. The two uteri unite to form the cuticle-lined single vagina that opens to the outside by the female gonopore, or vulva, found in the midventral line, usually near the middle third of the body but sometimes elsewhere.

In copulation the male may orient himself at right angles to the female body and partially coil around her body in the region of her gonopore. By extending the copulatory spicules through his cloacal opening (anus) into the female gonopore, both male and female apertures are kept apposed and open. Then, by contracting his cloacal muscles, sperm are discharged into the female vagina. The sperm, which are of various types from flagellate to ameboid, then migrate to the upper end of the uterus. In some nematodes there may be a copulatory bursa that is formed by the union of the lateral cuticular flanges from each side of the cloacal aperture. This structure is used to hold the two sexes together during copulation. Most females are oviparous, but sometimes the eggs are retained in the uterus until they hatch (ovoviviparous).

Cleavage is bilateral, highly determinate, and is a modified spiral type. The stage of development at which the egg leaves the body varies from the undivided fertilized egg of *Ascaris* to the fully developed embryos in the eggs of *Wuchereria*. Gastrulation is by epiboly, and a coeloblastula is formed (Fig. 7-25). The embryonic blastocoel is retained as the pseudocoel. Future germ cells are early separated from somatic cells. During late embryonic development multiplication of cells ceases in all but the reproductive organs, and the number of cells or nuclei present at that time remains constant for life. The embryo soon takes a form similar

FIG. 7-25

Development in nematodes *(Parascaris equorum)*. **A,** 4-cell stage. **B,** 18-cell stage. **C,** About 30-cell stage. **D,** Later embryo. **E,** Sagittal section through blastula. **F,** Sagittal section through gastrula showing formation of stomodeum. **G,** Sagittal section through later embryo, **D,** showing digestive tract forming. (**A** to **C, E, F** after Boveri, 1899; **D** and **G** after Müller, 1903; all modified from Hyman.)

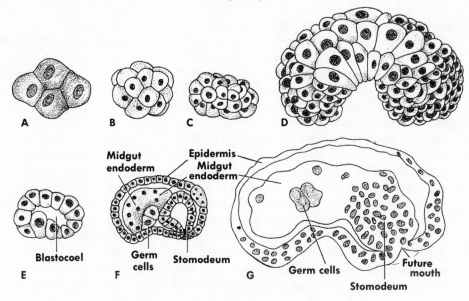

to that of the adult. Stages of growth are marked by four molts, of which one or two may occur within the egg. Regeneration is not found in nematodes.

No animal group has more patterns of sexual reproduction than have nematodes. Almost all types known in the animal kingdom are found among them. Two or three types may be combined in the same species. Perhaps one of the most curious types of reproduction is found in one species of *Rhabditis* (*R. monhystera*), in which, after normal copulation occurs between the two sexes, the sperm merely initiate the cleavage of those eggs that develop into females and do not fuse with the egg nucleus to form a zygote.

RÉSUMÉ OF ORDERS

Nematodes are divided into two great groups, the Phasmidia (Secernentia) and the Aphasmidia (Adenophorea). These groups are designated as classes, subclasses, or superorders by various authorities. Only some of the more important orders are described in the following.

■ ORDER RHABDITIDA

The members of Rhabditida are small to moderate size and are characterized by having an esophagus usually with 2 bulbs; the amphids are reduced to small pitlike pockets. There are no caudal glands. Most members are terrestrial and live in rich, decayed organic matter for which their saprophagous habits are best suited. Some are partial or complete parasites. Their life cycle may be direct, or it may include an infective larval stage in an insect or other invertebrate.

The earthworm nematode *Rhabditis maupasi* undergoes part of its development as a parasite in the nephridia or in the coelomic cavity of the earthworm and completes its development by feeding on the carcass of the dead earthworm. The familiar vinegar eel, *Turbatrix aceti,* is found in old vinegar. A similar form occurs in beer mats in certain parts of Germany. *Hetero-*

dera rostochiensis, or the golden potato eelworm, has done much damage to potatoes.

■ ORDER STRONGYLIDA

Among the common parasites of this economically important group are the hookworms *Ancylostoma,* found chiefly in the Old World and South America, *Necator,* common in the United States and the West Indies, and *Uncinaria* (Fig. 7-26), which is generally cosmopolitan. Most of these worms infect both man and domestic mammals, producing severe hookworm anemia. *Syngamus,* or gapeworm, is another common strongylid nematode that causes gapes in domestic poultry by blocking the trachea. Many other members of this order are responsible for severe damage to internal tissues, such as ulcerations, pulmonary

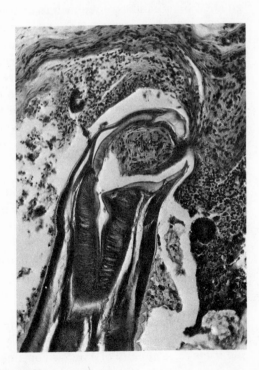

FIG. 7-26

Anterior end of hookworm *Uncinaria americana*. Note cutting plates pinching off mucosa and thick muscular pharynx that sucks up blood; parasitizes dogs, cats, foxes, and swine; about 8 mm. long. (AFIP No. 33810.)

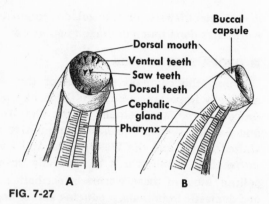

FIG. 7-27

Anterior end of hookworm showing buccal capsule with dorsal mouth and cutting plates. **A,** Dorsal view. **B,** View from left side. (Order Strongylida.)

irritations, and inflammation of various organs in both mammals and other vertebrates.

The male worms of Strongylida are distinguished by a caudal cuticular bursa supported by rays. The esophagus is without a spherical bulb and valvular apparatus, and the digestive tract lacks ceca. Members are endoparasites of the digestive tract. The life cycle is direct and usually includes two free-living juvenile stages that are passed in the soil. In size the length of the adult worm may be as short as 1 mm. or as long as 100 mm. in the various species. They are mostly oviparous, although some are ovoviviparous.

Infection of the host occurs by larval penetration of the skin, ingestion of larvae by mouth, or ingestion of an intermediate host. After entering the host, the larvae pass to their definitive habitats by way of the blood or directly through the tissues. Damage to the host is produced by their migratory habits and by blood sucking and tissue damage (Figs. 7-26 and 7-27). The host usually shows the effect of this damage by emaciation, loss of weight, anemia, and general lassitude. The amount of damage depends upon the host, number of worms, age of host, etc.

■ ORDER ASCARIDIDA

Order Ascaridida includes roundworms that may reach a considerable size and have a cos-

mopolitan distribution as endoparasites of vertebrate intestines. They are characterized by having 3 large lips, often with interlobia and other cephalic specializations, 8 submedian cephalic papillae and 2 lateral papillae, a muscular cylindric esophagus or pharynx but no buccal capsule, and an intestine, sometimes with an anterior cecum or ceca. Males are nearly always provided with 2 equal genital spicules. Females are oviparous, may have more than 2 uteri, and lay fertilized but uncleaved eggs. The worms may have a direct life history, or some may require an intermediate host as well as a definitive host to complete their development. They produce the disease ascariasis.

Common ascarids are the familiar *Ascaris lumbricoides,* the intestinal roundworm of man, which may reach a length of 40 cm.; *A. suum,* the pig ascarid, which is similar to *A. lumbricoides* but is considered a separate species; *Toxocara cati,* the cat ascarid, which may also infect man; and *T. canis,* common in dogs, especially puppies.

Infection with *A. lumbricoides* occurs when the eggs are ingested. The eggs hatch in the intestine; then the larvae burrow into the blood vessels, make their way to the lungs by way of the heart, are coughed up and swallowed, and finally become adults in the intestine. Damage occurs mostly in the lungs as a result of this peculiar migration.

■ ORDER OXYUROIDEA

The members of Oxyuroidea are parasites. They are usually small with a life cycle involving one host and are found in all classes of vertebrates and in some invertebrates. Their head papillae are usually in a circle and their lips, when present, are 3 or 6 in number. The esophagus is commonly divided into a corpus, isthmus, and posterior bulb. The reproductive system may be single or double, and the males may or may not have genital spicules. Females are sometimes viviparous but are usually oviparous. A precloacal sucker is found in some species. The life cycle and method of infection

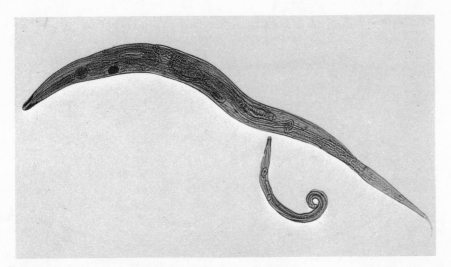

FIG. 7-28

Pinworms, *Enterobius vermicularis,* female above and smaller male below (order Oxyuroidea). (Courtesy Indiana University School of Medicine, Indianapolis.)

in some species are similar in many ways to that of *Ascaris.*

The familiar pinworm, or seatworm *Enterobius vermicularis* (Fig. 7-28), is very common in children and produces enterobiasis, characterized chiefly by restlessness and insomnia. The female worm (10 mm. long) deposits eggs around the anal region of the host. These may be carried to the mouth by the hands. Although it is easy to rid the child of the worms, reinfection is also easy because infective ova contaminate surroundings.

■ ORDER DRACUNCULOIDA

Dracunculoids are parasitic nematodes without definite lips and with an esophagus divided into a thin anterior muscular portion and a wider posterior glandular part. The vulva is near the middle of the body but is reduced in the mature worm. They are characterized also by their habitat in the deeper subcutaneous tissues and by the way the larvae, which are produced by viviparous females, pass from the body through lesions in the skin.

The classic example of this order is the guinea

worm (*Dracunculus medinensis*), or the fiery serpent of Biblical times. The female, which may be a meter or more in length, is found in the subcutaneous tissues of man in Africa and Asia. When mature, she produces via a toxic substance a small lesion in the region over the worm's head. When this part of the host's body comes in contact with water, the worm contracts and expels a large number of juvenile worms through the lesion into the water. These may be swallowed by a *Cyclops* water flea, the intermediate host, and there undergo development to the infective stage. Man is infected by drinking water containing *Cyclops*. It takes 8 to 12 months for the females to mature in the body cavities before migrating to the adult habitat. Rolling the worm out little by little on a stick is the ancient method for getting rid of it, but getting rid of *Cyclops* by chemical treatment is a more modern method.

■ ORDER FILARIOIDEA

Filarial worms have no lips or buccal cavity, and the esophagus is similar to that in Dracunculoida. Their cuticle is chiefly devoid of superficial ornamentation, and the male spicules are unequal and dissimilar. The cephalic papillae are reduced in size and number. The worms are parasites of the vascular system,

267

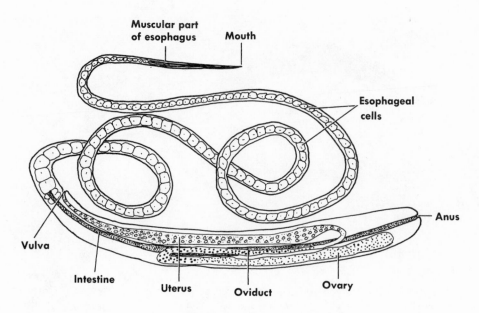

FIG. 7-29

Trichuris trichiura, the whipworm (order Trichuroidea). (Modified from Olsen, 1962.)

lymphatic system, coelomic spaces, and some other places in vertebrates. The life cycles of all are similar and require a bloodsucking insect for transmission.

Wuchereria bancrofti, which produces elephantiasis in man, is one of the best-known members of the order. The adults, which may be 4 to 5 inches long, live in the lymphatic system and, if present in large numbers, may block the flow of lymph, producing enormous swellings (elephantiasis) of the legs, scrotum, or other parts of the body. The females are ovoviviparous and release their microfilaria into the lymphatics, from which they make their way to the blood. At night the immature worms (150 to 400 microns long) reside mainly in the peripheral vessels, where they may be ingested by a bloodsucking mosquito. In the mosquito's stomach the larvae lose their sheath and migrate to the thoracic muscles, where they undergo further development and complete two molts. They then pass to the proboscis of the mosquito and into the definitive host (man) through the wound made when a mosquito bites a person.

The life history of the eye worm (*Loa loa*) is similar in many ways to that of *Wuchereria*

except that the eye worm is found sometimes in the eye; transmission is made through the deerfly (*Chrysops*), and the microfilariae appear in peripheral blood vessels during the day.

■ ORDER TRICHUROIDEA

The members of Trichuroidea are endoparasites of the digestive tract of birds, mammals, and other vertebrates. Their chief characteristic is the unique structure of the anterior part of the body, in which the pharynx or esophagus is located. This part of the body is usually very slender and in some members is very much longer than the thicker posterior part. It is known as the stichosome and is made up of a longitudinal row of large esophageal cells (stichocytes) that represent pharyngeal gland cells.

Two very common genera are parasites of man—*Trichuris* and *Trichinella.* The whipworm (*Trichuris trichiura,* Fig. 7-29) has a very long anterior portion (whip) and is a common parasite of man in warm climates. Its eggs

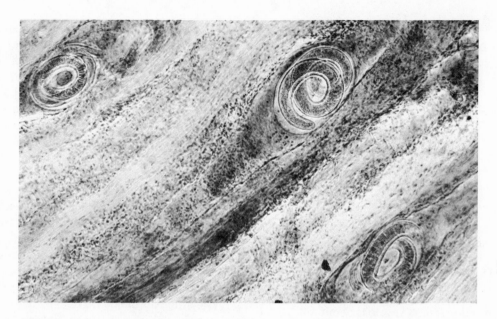

FIG. 7-30

Trichinella spiralis, larvae encysted in pig muscle. Larvae may live 10 to 20 years in these cysts; if eaten by another host, larvae are liberated in intestine of host, where they mature and release many more larvae into host's blood (order Trichuroidea).

develop in the soil but do not hatch unless they are ingested by the host, usually on contaminated vegetables. The trichina worm *(Trichinella spiralis,* Fig. 7-30) is one of the most common nematode parasites of Americans. Infection occurs when man eats poorly cooked pork, in which the worms are encysted. In either man or pig the female worms burrow into the intestinal wall and give birth to larvae that are carried by the blood and lymph to muscles, where they encyst. The symptoms of trichinosis infection occur mainly at the time of migration of the larvae. Pigs become infected by eating uncooked garbage containing contaminated meat scraps, or by eating infected rats.

■ **ORDER DIOCTOPHYMATIDA**

The male members of Dioctophymatida are characterized by the peculiar copulatory bursa without rays at their caudal end. They have no lips, but the mouth region is provided with 6

to 18 papillae. Males have only 1 genital spicule. Some have a sucker around the mouth. The various species are parasites of fish, aquatic birds, and mammals.

Perhaps the largest nematode, free living or parasitic, belongs to this order. This is the giant kidney worm, *Dioctophyme renale* (Fig. 7-31), which is bright scarlet in color and up to 40 inches long (females). It is a parasite of mammals, especially dogs and minks, where it may be found in the kidney, abdominal cavity, and other organs. It is thought that the worm does not become fully mature until it parasitizes the kidney. The complete life cycle, although much investigated, was first worked out by A. Woodhead. The eggs are discharged in the urine of the definitive host and become infective several months later when they are swallowed by branchiobdellid worms (epizoic oligochaetes on crayfish), the first intermediate host. After a period of development, the worm encysts in the annelid, which may be eaten by certain fish, the second intermediate host. After further development within the fish, the kidney worm encysts in the mesenteries and becomes infective to the definitive host (mammal) that

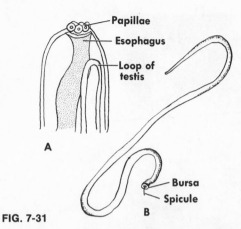

FIG. 7-31

Dioctophyme renale, the giant kidney worm, is the largest nematode (up to 40 inches long). Parasitic in kidney and elsewhere (order Dioctophymatida). **A,** Anterior end of male. **B,** Adult male.

devours the fish. The complete life cycle is about 2 years. The giant kidney worm eats away the interior tissue of the kidney, where it may remain for 2 to 3 years.

Another member of this order is *Hystrichis;* it lives in the proventriculus of aquatic birds, which become infected by eating fish.

■ PHYLUM NEMATOMORPHA
CLASSIFICATION

Phylum Nematomorpha (Gordiacea). Body long, filiform, and cylindric with rounded ends; cuticles may bear thickenings (areoles); lateral epidermal cords usually absent; anterior tip (calotte) bounded posteriorly by dark ring; epidermis of a single layer of cells; pseudocoel mostly filled with mesenchyme; excretory canal absent; enteron mostly degenerate; muscles of longitudinal fibers; nervous system of ventral cerebral mass and nerve cord in or near epidermal layer; dioecious; genital ducts open into intestine; gonads usually of paired cylindric bodies; parasitic as juveniles, free living as adults; holoblastic cleavage with coeloblastula.

    Order Gordioidea. *Gordius.*
    Order Nectonematoidea. *Nectonema.*

MORPHOLOGY AND PHYSIOLOGY

Nematomorpha, "horsehair" worms, are usually long, threadlike worms that are closely allied to the nematodes. The adults are free living, whereas the juvenile worms are all parasitic in arthropods such as insects and crustaceans. Nematomorphs differ from nematodes in a number of ways such as in having an absence of lateral cords and excretory canal, a more or less degenerate enteron, and the presence of a cloaca in both sexes. They may reach a maximum length of 1.5 meters with a diameter of 3 mm., but the majority are smaller than this, with diameters not exceeding 0.5 mm. They are cosmopolitan and are found in almost every kind of aquatic habitat. Adults do not feed and require only wet or moist surroundings and sufficient oxygen.

The two orders, Gordioidea and Nectonematoidea, total about 225 species. Of the two orders, all the genera with the exception of one belong to the order Gordioidea. The Nectonematoidea has only the single genus *Nectonema,* which are planktonic worms anywhere from 10 to 800 mm. long.

Nematomorphs (Fig. 7-32, *A*) are mostly dark brown or of some darkish shade except for the anterior tip (calotte), which is white with a dark pigmented ring and terminal mouth. The posterior end may be rounded or forked, having 2 or 3 lobes with a terminal or subterminal cloaca. The male is shorter than the female in the Gordioidea, but the reverse is true with the Nectonematoidea. The body is made up of three layers: (1) a thin cuticle of homogeneous, fibrous layers, often provided with prominences (areolae), papillae, and hairs; (2) an epidermis of a single layer of cells or a syncytium; and (3) an inner muscle layer of longitudinal fibers only (Fig. 7-32, *B*). In Gordioidea the pseudocoel is filled with mesenchyme and other cells, so only certain longitudinal spaces are found; in *Nectonema* the pseudocoel is more evident.

The digestive system in the adult is relatively degenerate, and the anterior part may consist of a row of cells. Ingestion of food is absent in the adults. No specific circulatory, respiratory, or excretory organs exist.

The nervous system consists of a circumenteric nerve ring of ganglia lying ventrally in the

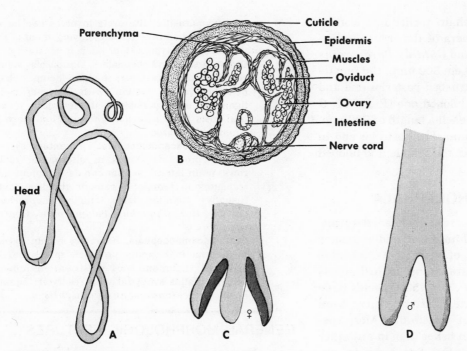

FIG. 7-32

Class Nematomorpha. **A,** Gordian worm. **B,** Cross section of female nematomorph. Posterior end of *Paragordius* female, **C,** and of male, **D.**

head calotte and a midventral nerve cord that originates in the epidermis. Sensory functions may be restricted to the bristles and areoles of the cuticle. A rudimentary eye in the form of a large sac is found in *Paragordius.*

Nematomorphs are all dioecious, and the paired gonads are cylindric strings that extend the length of the body. Each gonad of the male enters the cloaca by a sperm duct. There are no genital spicules. In the female the ovaries contain lateral diverticula in which the eggs mature. The eggs pass from the diverticula into the ovarian tubes or uteri and enter the cuticle-lined cloaca separately. A saclike seminal receptacle extends anteriorly from the cloaca.

In copulation the male coils around the female and places a drop (spermatophore) of sperm near the cloacal opening of the female. The sperm then actively enter the seminal receptacle. Sometimes a number of worms of both sexes may be found together in a tangled skein while copulating. Eggs are laid in stringy masses in the water, after which the adults die. When the eggs hatch, the larvae swim to an arthropod, which they enter by means of their proboscis armed with hooks and stylets. In some cases the larvae may secrete mucus, in which they encyst until the proper terrestrial host ingests them. In the hemocoel of the host the juvenile worms undergo gradual development without metamorphosis, a process that may last several months. Only one host is involved in the parasitic stage. This host may be a cockroach, cricket, grasshopper, or other arthropod. The hosts of *Nectonema* are various decapod crustacea. Within the host, nematomorphs take all their nutrition by absorption through the body surface because no food enters the enteron during their life cycle. It is possible that their epidermis secretes enzymes for digesting the host's tissues.

When the worms are mature, they molt, discard unnecessary larval characters, and leave the host, usually when the latter is near the water. In the past this often took place in water troughs, which gave rise to the popular superstition that horsehairs turned into worms.

The best known genera of the order Gordioidea are *Paragordius* and *Gordius*. *Paragordius* grows to a length of about 300 mm.; the female is distinguished by a trilobed posterior end and the male by a deeply bilobed one (Fig. 7-32, *C* and *D*). *Gordius* may reach a length of 900 mm. with a diameter of 1 mm. The posterior end in this genus is unlobed in the female and bilobed in the male.

■ PHYLUM ACANTHOCEPHALA

The Acanthocephala are spiny-headed worms. The adults of this phylum are all endoparasites in the digestive tract of vertebrates, whereas the juvenile stages are found in arthropods (the intermediate host). Over 500 species have been described to date, and they have been found in all classes of vertebrates. More species have been found in fishes than in any other class of vertebrates; only a few occur in amphibians and reptiles. Most of the small species are found in fishes; the largest are found in mammals. Host specificity is found in some but not in others. However, their host tolerance never extends to both cold-blooded and warm-blooded animals for the same species of acanthocephalan. The distribution of acanthocephs is worldwide, but not all species are cosmopolitan; some are restricted to rather small areas.

CLASSIFICATION

Phylum Acanthocephala. Body cylindric or laterally flattened in form and divided into proboscis, neck, and trunk; spiny proboscis retractile into sheath; epidermis syncytial, covered with cuticle; fluid-filled canal system (lacunae); muscles of both longitudinal and circular types; body cavity a pseudocoel; digestive tract absent; excretory system of branched, ciliated protonephridia; circulatory and respiratory systems absent; nervous system of a brain on the proboscis sheath and two lateral cords; sensory papillae on body surface; separate sexes; male organs of paired testes on ligament, ducts, cement glands, and penis apparatus; female organs of paired ovaries in a ligament sac, uterine bell, uterine tube, and uterus; endoparasitic; life cycle involves two hosts, a vertebrate and an invertebrate; development of the spiral determinate type; cell or nuclear constancy.

Order Archiacanthocephala. Proboscis spines concentrically arranged; median lacunar channels; excretory system with protonephridia; ligament sacs persistent in females; terrestrial hosts. *Macracanthorhynchus, Moniliformis, Gigantorhynchus.*

Order Palaeacanthocephala. Excretory system absent; proboscis spines in alternating radial rows; main lateral lacunar canals; ligament sacs transitory in female; 6 cement glands in male (usually); aquatic hosts (chiefly). *Leptorhynchoides, Gorgorhynchus, Polymorphus, Centrorhynchus.*

Order Eocanthocephala. Excretory system absent; median lacunar canals; proboscis spines radially arranged; ligament sacs persistent in female; cement glands syncytial with reservoir; aquatic hosts. *Neoechinorhynchus, Pallisentis.*

GENERAL MORPHOLOGIC FEATURES

Most species are about an inch long; some may be smaller, and others may be as long as 2 feet. Females are nearly always longer than the males of the same species. In the host the body is often flattened, but after removal it becomes turgid and cylindric (Fig. 7-33). Many different shapes may be found, such as curved or spirally coiled. The surface may be smooth or wrinkled, suggestive of pseudometamerism. The color is mostly translucent or white, but the body often assumes the color of the intestinal contents. The body is divided into the presoma (proboscis and neck) and trunk. Trunk spines are found on the body surface of some species.

The body wall is composed of cuticle, a hypodermis, and an inner muscular layer. The syncytial epidermis is provided with a branching and anastomosing lacunar system of channels without walls, which are arranged in definite patterns (Fig. 7-34). The system may be in the form of a network of lacunar channels, but usually it consists of a pair of longitudinal channels that are connected by a series of smaller vessels, the lacunae. The longitudinal channels may lie in dorsal and ventral median positions, or they may occupy lateral positions.

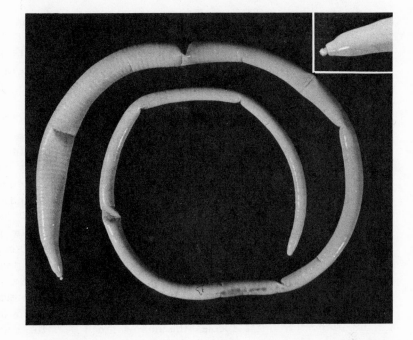

FIG. 7-33

Macracanthorhynchus hirudinaceus, the spiny-headed worm of pigs (preserved specimen). Inset is enlarged view of anterior end with proboscis. (From Hickman: Integrated principles of zoology, The C. V. Mosby Co.)

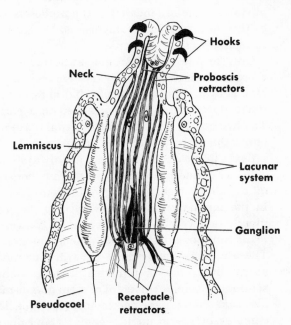

FIG. 7-34

Longitudinal section (diagrammatic) through anterior end of acanthoceph with proboscis partially everted. Recurved hooks or spines are numerous; their arrangement has taxonomic significance. (After Hamann; modified from Hyman.)

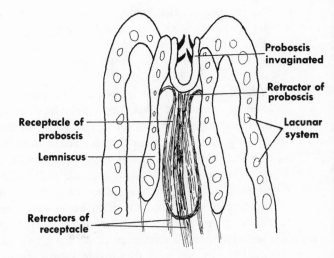

FIG. 7-35

Longitudinal section (diagrammatic) through anterior end of acanthoceph showing relations of invaginated proboscis and other structures. Note spines or hooks pointed forward in this position.

In some species only the dorsal channel is present. The lacunar system does not communicate with the outside or the inside except in the case of the lemnisci, which are supplied by lacunar vessels. The presoma contains lacunae but no longitudinal vessels. The lacunae contain a nutritive fluid and are a means of distributing food absorbed from the host because of the absence of digestive organs. Round or oval giant nuclei are also found in the hypodermis, which rests on a thin basement membrane or dermis. The muscle part of the body wall is made up of outer circular and inner longitudinal fibers syncytial in nature.

The characteristic proboscis is the main organ of attachment. It may be globular, elongated, or cylindric in shape and is always armed with sharp, recurved hooks, spines, or both. These hooks are arranged in patterns of taxonomic importance. In most species the hollow proboscis can be retracted or everted into a muscular, saclike structure, the proboscis receptacle, that is attached to the inner surface of the proboscis wall and hangs suspended in the pseudocoel. When the proboscis is introverted into the receptacle, the hooks are on the inside of the proboscis with their tips directed anteriorly (Fig. 7-35); when everted, the tips of the hooks point posteriorly and fasten into the host's intestinal wall like a thorn (Fig. 7-34). The neck and proboscis receptacle can also be retracted into the pseudocoel, but they cannot be everted. Special muscles (retractor or invertor) operate the proboscis and its receptacle. These muscles are attached to the inside tip of the proboscis and are inserted on the receptacle wall; also, passing through the wall, they are fastened to the inside of the body wall to become the receptacle retractors. The neck is retracted by special muscles that run from the posterior edge of the neck to the inside of the trunk. From the posterior end of the neck region are 2 lateral diverticula, or lemnisci, which hang down in the pseudocoel and may act as reservoirs for the lacunar fluid when the proboscis is retracted. They communicate with the la-

cunar system by special channels (Figs. 7-34 and 7-35).

The pseudocoel, or body cavity, extends also into the presoma and contains the internal organs (Fig. 7-36). The chief organs are the hollow connective tissue ligament sacs that enclose the reproductive organs and run the length of the body. The ligament sac may be single or double and may represent separate parts of the body cavity. Alongside the ligament sacs lies an endodermal ligament strand to which the gonads are attached. A digestive system is entirely lacking in this phylum.

The nervous system consists of a cerebral ganglion (Figs. 7-34 and 7-36) located within the proboscis receptacle; 2 lateral longitudinal nerve branches, the retinacula, pass posteriorly to innervate the trunk wall. The cerebral ganglion also gives off small nerves to such structures as the musculature of the proboscis and sensory papillae. Sense organs are scanty in this phylum. Some are located in the proboscis and in the male copulatory apparatus. Perhaps all the sense organs are tactile.

Only in one order (Archiacanthocephala) is there a definite excretory system consisting of a pair of small protonephridial organs each made up of flame bulbs attached to a canal. The two canals form a common canal that joins either the sperm duct or the uterus.

Sexes in the Acanthocephala are always separate. The male reproductive system consists of a pair of testes arranged tandem fashion in the ligament sac and fastened to the ligament strand (Fig. 7-36). A sperm duct from each testis runs posteriorly through the ligament sac. These 2 ducts finally unite to form a common sperm duct that is provided with a saclike seminal vesicle. A cluster of cement glands may also send ducts to the common sperm duct. The complex of sperm ducts, cement gland ducts, and protonephridia (in one order) are all enclosed in a genital sheath that is continuous with the ligament sac. The genital sheath terminates posteriorly on the bell-shaped copulatory bursa, which is eversible and holds the posterior

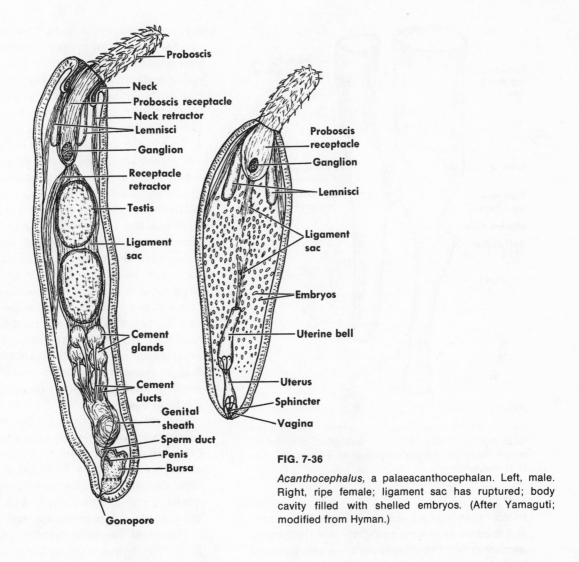

FIG. 7-36

Acanthocephalus, a palaeacanthocephalan. Left, male. Right, ripe female; ligament sac has ruptured; body cavity filled with shelled embryos. (After Yamaguti; modified from Hyman.)

end of the female in copulation. In some acanthocephalans a muscular elongated pouch (Saefftigen's pouch) contracts and injects fluid to assist in the eversion of the bursa. The sperm and secretions of the cement glands are discharged through a penis that projects into the bursa. The secretions of the cement glands hold the bursa firmly in place while the penis enters the female vagina and discharges its sperm in copulation.

The female reproductive system consists of an ovary (single or double) that is present in larval stages but later dissociates into egg balls. These egg balls float free in the ligament sac and, as the eggs are released from the egg balls, they are fertilized. The ligament sac may rupture and the pseudocoel may be filled with ovarian balls and eggs. Only one quota of eggs can therefore be produced in the life cycle of one female. From the ligament sac the eggs pass into the uterine bell, a funnel-shaped structure that opens into the ligament sac or the pseudocoel (Fig. 7-37). During embryonic development in the pseudocoel, the eggs acquire

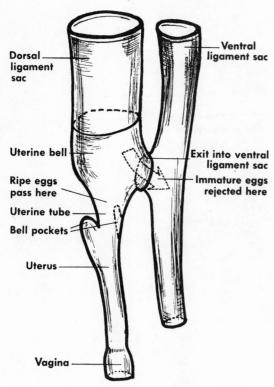

Dorsal
ligament
sac

Ventral
ligament sac

Uterine bell

Ripe eggs
pass here

Uterine tube

Bell pockets

Uterus

Exit into ventral
ligament sac

Immature eggs
rejected here

Vagina

FIG. 7-37

Genital selective apparatus of female acanthoceph
(diagrammatic). See text for functional description.
(From Hickman: Integrated principles of zoology, The
C. V. Mosby Co.)

membranes and a hard shell. From the uterine
bell the ripe or mature eggs pass into the narrow
uterine tube; the immature ones are detoured
back into the body cavity by openings in the
selector apparatus. The uterine tube enters the
muscular uterus, which opens to the outside by
a short vagina. The eggs are thus discharged
through the vaginal, or genital, orifice into the
intestine of the host.

The eggs, which are discharged in the feces of
the host, can withstand severe environmental
conditions for months. At this stage the egg
contains a mature embryo, the acanthor, that is
surrounded by 3 membranes. No further development occurs until the acanthors are ingested by a suitable intermediate host, usually

an arthropod. Within the intermediate host
(grubs, roaches, or crustaceans) the eggs hatch
and release the spindle-shaped acanthor with a
spiny body and retractor hooks. The acanthor
has an inner nuclear mass of undifferentiated
nuclei, the endoblast, that gives rise to the structures of following stages. The acanthor is found
at first in the intestine of the intermediate host
but eventually make its way into the hemocoel
where its further transformation occurs. Its
next stage is called the acanthella, in which
occur the beginnings of certain adult structures
such as the proboscis, reproductive organs,
lemnisci, etc. When the larva becomes infective,
it is called the cystacanth, which contains all
the adult structures in an immature form (a
juvenile worm). From the acanthor to the cystacanth may require 2 to 3 months. The juvenile
usually remains quiescent in the intermediate
host until ingested by the definitive vertebrate
host, although sometimes encysted juveniles
may be found in a vertebrate as a second intermediate host or transport host before reaching
the final host.

The cleavage pattern in the embryology of
acanthocephalans seems to be a modified spiral
determinate type. Cell or nuclear constancy
(eutely) seems to be fixed for each tissue. In
their development, changes in the form of the
nuclei are correlated with a favorable ratio
between the nuclear surface and cytoplasm.

The most familiar member of the phylum
is *Macracanthorhynchus hirudinaceus*, which
is cosmopolitan in pigs and some other mammals. A common parasite of fish is *Leptorhynchoides thecatus*, whose larval forms are passed
in the familiar amphipod *Hyalella knicker-
bockeri*. *Moniliformis dubius* is a common
parasite of rats, and cockroaches serve as intermediate hosts.

■ PHYLUM ENTOPROCTA

The entoprocts are sessile pseudocoelomates
and were formerly included in the moss animals
Bryozoa or Polyzoa, which also include the
coelomate Ectoprocta. Entoprocts are solitary,

colonial, stalked animals that are always sessile and that have the anus within the circle of tentacles, not outside as in Ectoprocta. The Entoprocta are all small animals and most live in marine waters, with only one or two fresh-water representatives. They are usually attached to a hard substratum or to other animals. The phylum comprises three families and about ninety species. Within the phylum itself the Entoprocta have not been classed in higher taxa than families. The three families are Loxosomatidae, with the calyx pseudocoel not separated from that of the stalk by a septum and with buds arising from the sides of the calyx; Pedicellinidae, with the pseudocoel separated by a septum from that of the stalk and with buds arising from stolons; and the Urnatellidae (freshwater), with small colonies formed by a few chitinized stalks from a basal plate and with deciduous calyces. Only one genus,

FIG. 7-38

Typical entoproct calyx (partly cut away in front) showing internal structures.

Urnatella (Fig. 7-39), with two species is found in the last family.

MORPHOLOGIC FEATURES AND PHYSIOLOGY

An individual entoproct shows four chief divisions: the crown of tentacles, the calyx, the stalk, and the substratum to which the stalk is attached. Most entoprocts are not more than 5 mm. in length. The internal organs are contained within the calyx, a bulbous-shaped structure that bears a circle of 8 to 36 ciliated tentacles around the upper margin (Fig. 7-38). Tentacles are outgrowths of the body wall and can move individually or together. The depression surrounded by the circle of tentacles is the vestibule containing the mouth at one end and the anus at the other. The crown of tentacles can be rolled inward over the mouth and anus but cannot be retracted into the calyx. Ciliated grooves in the sides of the vestibule run to the corners of the mouth. The calyx is slightly compressed laterally and may be at right angles to the stalk or oriented obliquely to it.

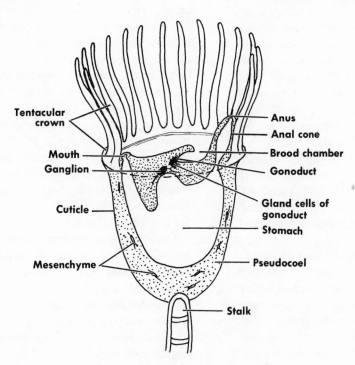

Tentacular crown
Mouth
Ganglion
Cuticle
Mesenchyme
Anus
Anal cone
Brood chamber
Gonoduct
Gland cells of gonoduct
Stomach
Pseudocoel
Stalk

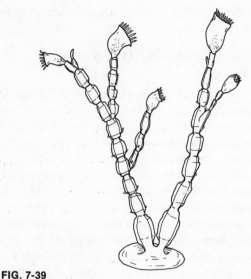

FIG. 7-39

Urnatella, the only freshwater entoproct. (After Cori; modified from Hyman.)

The stalk is an elongation of the calyx, but usually the two parts are partially separated by an incomplete septum. The stalk may have a beaded appearance because of the regular constrictions on it (Fig. 7-39), but many other variations are also found in the stalks of the different species. Smooth muscle fibers are found in the stalks of all species. Although they are sessile forms, entoprocts can move their stalks, curl their tentacles, bend their calyces, etc. In some the basal plate of the stalk sends out slender stolons that run over a substratum and, by branching, give rise to many stalks.

The body wall consists of a cuticle, epidermis, and smooth muscle fibers. The wall is of about the same structure in all parts of the animal. The epidermis is cellular, of cuboidal cells, and is thicker in the more heavily ciliated regions such as the tentacles.

The body wall musculature consists of mostly longitudinal fibers and is thinnest in the wall of the calyx. Muscle fibers are common along the inner wall of the tentacles, and their contraction makes it possible for the tentacles to be pulled toward the vestibule. A sphincter composed of a muscle band contracts the ten-tacular membrane over the tentacles when the latter curl into the vestibule. The jerky movements of the stalks of some species is brought about by the muscular swellings that can contract independently of each other.

The pseudocoel, or body cavity, is found in the tentacles, stalks, stolons, and calyx between the body wall and viscera. It is filled with a gelatinous mesenchymal tissue. It also contains both free ameboid and fixed cells. The calyx of most species has its pseudocoel separated from that of the stalk by a septum.

The U-shaped alimentary tract is composed of a mouth, a short buccal cavity, a tubular esophagus, a large saclike stomach, a narrow intestine, and a rectum that opens to the outside by the anus within the circle of tentacles. An anal cone is present in some species. Parts of the digestive system are lined with cilia. Glandular regions in the stomach may produce enzymes for extracellular digestion. Absorption may occur in both the stomach and intestine.

Entoprocts are filter feeders. Food particles are directed toward the mouth by the beating of the cilia on the tentacles. Water currents are carried down on the inner ciliated surfaces of the tentacles into the vestibule, where a certain amount of rejection or straining may occur by the closure of the esophageal sphincter so that rejected items are sent back in the outgoing current of water. The food is carried through the alimentary canal by ciliary movement because the walls of the digestive tract lack muscles. Most of the food consists of protozoans, diatoms, and other small organisms. When feeding, the tentacles of the calyx are well extended and turn about in different directions for the location of food.

The excretory system consists of a pair of flame bulbs in the calyx placed ventral to the stomach near the esophagus. Each bulb is drained by a nephridial tube that unites with its corresponding mate to form a common tubule before opening by a single median nephridiopore in the vestibule between the mouth and anus.

The nervous system is made up of a large, bilobed ganglion in the pseudocoel between the stomach and vestibule, and nerve fibers that radiate through the body to the tentacles, stalk, and parts of the calyx. Tactile sense organs of one or more bristles are found on the margin of the calyx and on the tentacles. The tentacular crown is especially sensitive to touch and retracts inward when disturbed. No special respiratory or circulatory organs are present.

Entoprocts reproduce both sexually and asexually. Sexual reproduction produces new colonies, and asexual budding adds new members to existing colonies. Sexually, some entoprocts are dioecious, some are hermaphroditic, and some are protandrous. An entire colony may be of one sex. The gonads consist of a pair of testes and a pair of ovaries. In the hermaphroditic forms the paired testes lie posterior to the paired ovaries. In dioecious species each gonad has a short gonoduct that unites with the other before opening by a common gonopore on the vestibule near the nephridiopore; in hermaphrodites the sperm duct joins the oviduct of that side before the formation of the common gonoduct.

Fertilization occurs in the ovaries, and the developing egg is released through the gonopore. By means of a secreted membrane the egg is attached in the female brood pouch, which lies in a depression of the vestibule between the gonopore and the anus. Here the egg is retained during the period of embryonic development. Cleavage is of the spiral determinate type, and a coeloblastula is formed by the 67-cell stage; a blastula is formed by the 90-cell stage.

A free-swimming, ciliated larva of a modified trochophore type is hatched and is incubated for a time in the brood pouch. After its release, the larva swims about for a short time before its permanent attachment to a substratum and its metamorphosis into the initial individual of a colony. The colony then is formed by asexual budding from the calyx or from the stalk.

Entoprocts have a wide distribution in marine waters from the poles to the tropics. Some have

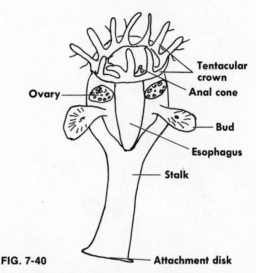

FIG. 7-40

Loxosoma with two buds. This common marine genus is considered to be most primitive of Entoprocta. (Modified from Nickerson.)

been found to a depth of 150 fathoms. Of the marine species, the solitary *Loxosoma* is fairly common, usually attached to worm tubes in shallow water (Fig. 7-40). It is not colonial in habit because the asexual buds separate from the parent after maturing and attach their pedal gland independently to a substratum, where they lead a solitary existence. The animals vary from 1/12 to 1/50 inch in height. Some authorities consider this genus to be one of the most primitive of the entoprocts. *Pedicellina* is another familiar entoproct of our eastern seashore. Its colonies are attached to the substratum by slender branching stolons, from which the stalked individuals are budded. The calyx is separated from the stalk by a septum. Its buds arise from the stolons. This genus is often found on the piles of old wharfs. *Barentsia* is another genus that is found along both eastern and western seacoasts; it is characterized by muscular enlargements that may give rise to branches along the stalk. *Gonypodaria* (Fig. 7-41) has a similar arrangement of muscular swellings along the stalk. The freshwater species *Urnatella gracilis* (Fig. 7-39) seems to have a limited distribution in the streams of the eastern half

FIG. 7-41

Gonypodaria, a pedicellinid entoproct, has slender stalks set in muscular sockets.

of the United States. It is found in small colonies, the 1 to 6 members of a colony all arising from one basal plate. The stalks, branched or unbranched, are beaded by cuticular constrictions. Individuals do not exceed 5 mm. in length.

REFERENCES

Ax, P. 1963. Relationship and phylogeny of the Turbellaria. In Dougherty, E. C. (editor): The lower Metazoa; comparative biology and phylogeny, Berkeley, University of California Press.

Beauchamp, P. de. 1907. Morphologie et variations de l'appareil rotateur. Arch. Zool. Exp. Gen. **6**:1-29. In this and other papers Beauchamp was the first to point out that the ground plan of the corona of rotifers consisted of a large buccal field of short cilia around the mouth and a circumapical band encircling the head and that the animal was adapted for a creeping existence.

Chitwood, B. G., and M. B. Chitwood. 1949. An introduction to nematology. Baltimore, Monumental Press. The best general treatment of the nematodes yet published.

Cori, C. Kamptozoa. In Bronn, H. G. (editor): 1936. Klassen und Ordnungen des Tierreichs, vol. 4, part 2. Leipzig, Akademische Verlagsgesellschaft. An authoritative and technical treatise of the entoprocts.

Crowe, J. H., and A. F. Cooper, Jr. 1971. Cryptobiosis. Sci. Amer. **225**:30-36 (Dec.).

Davenport, C. 1893. On *Urnatella gracilis.* Bull. Mus. Comp. Zool. Harvard Univ. **24**:1-44.

Edmondson, W. T. (editor). 1959. Ward and Whipple's freshwater biology, ed. 2. New York, John Wiley & Sons, Inc. Many groups of pseudocoelomates are treated in this excellent handbook. Keys and figures are emphasized.

Goodey, T. 1951. Soil and freshwater nematodes. New York, John Wiley & Sons, Inc.

Harris, J. E., and H. D. Crofton. 1957. Structure and function of the nematodes. J. Exp. Biol. **34**:116-155.

Higgins, R. P. 1964. Redescription of the Kinoryhnch *Echinoderes remanei* (Blake, 1930; Karling, 1954). Trans. Amer. Microscop. Soc. **83**:243-247.

Higgins, R. P. 1965. The homalorhagid Kinorhyncha of Northeastern U. S. coastal waters. Trans. Amer. Microscop. Soc. **84**:65-72.

Hyman, L. H. 1951. The invertebrates: Acanthocephala, Aschelminthes and Entoprocta; the pseudocoelomate Bilateria, vol. 3. New York, McGraw-Hill Book Co., Inc. An authoritative treatise of the first rank, and one of the best treatments of these groups yet published.

Lang, K. 1963. The relation between the Kinorhyncha and Priapulida and their connection with the Aschelminthes. In Dougherty, E. C. (editor): The lower Metazoa; comparative biology and phylogeny. Berkeley, University of California Press.

Lee, D. L. 1965. The physiology of nematodes. San Francisco, W. H. Freeman & Co., Publishers. An up-to-date work on a difficult and much neglected field of study.

Maggenti, A. R. Comparative morphology in nemic phylogeny. In Dougherty, E. C. (editor). 1963. The lower Metazoa; comparative biology and phylogeny. Berkeley, University of California Press. This chapter attempts to trace the affinities between the different nemic taxa.

Meyer, A. Acanthocephala. In Bronn, H. G. (editor). 1933. Klassen und Ordnungen des Tierreichs, vol. 4, part 2, section 2. Leipzig. Akademische Verlagsgesellschaft. The definitive German account of Acanthocephala.

Nickerson, W. 1901. On Loxosoma davenporti. J. Morphol. **17**:357-376.

Olsen, O. W. 1962. Animal parasites; their biology and life cycles. Minneapolis, Burgess Publishing Co. Good schemes of the acanthocephalan parasites found in the pig, rat, and fish.

Pennak, R. W. 1953. Fresh-water invertebrates of the United States. New York, The Ronald Press Co. Good sections with keys are devoted to gastrotrichs, rotifers, and nematodes.

Pray, F. A. 1965. Studies on the early development of the rotifer *Monostyla cornuta.* Trans. Amer. Microscop. Soc. **84**:210-216.

Remane, A. Rotatoria. In Bronn, H. G. (editor). 1933. Klassen und Ordnungen des Tierreichs, vol. 4, pp. 1-4. Leipzig, Akademische Verlagsgesellschaft. One of the best general accounts of rotifers.

Remane, A. 1963. The systematic position and phylogeny of the pseudocoelomates. In Dougherty, E. C. (editor). The lower Metazoa; comparative biology and phylogeny. Berkeley, University of California Press. The author regards the pseudocoelomates as playing minor roles in the production and evolution of higher forms. No phyletic relationships of the different members with each other or with other phyla have been established.

Riedl, R. J. 1969. Gnathostomulida from America. Science **163**:445-452.

Roggen, D. R., D. J. Rask, and N. O. Jones. 1966. Cilia in nematode sensory organs. Science **152**:515-516.

Rogick, M. D. 1948. Studies on marine Bryozoa. Part II. *Barentsia laxa*. Biol. Bull. **94**:128-142. A good account of the larva is included in this study. Miss Rogick considers the Entoprocta as a class under the Bryozoa, not a separate phylum (Hyman).

Sacks, M. 1955. Observations on the embryology of an aquatic gastrotrich, *Lepidodermella squammata*. J. Morphol. **96**: 473-495.

Sacks, M. 1964. Life history of an aquatic gastrotrich. Trans. Amer. Microscop. Soc. **83**:358-362.

Sasser, J. N., and W. R. Jenkins (editors). 1960. Nematology. Chapel Hill, University of North Carolina Press. Many specialists have contributed to this important treatise on nearly every aspect of nematodes.

Swellengrebel, N. H., and M. M. Sterman. 1961. Animal parasites in man. Princeton, D. Van Nostrand Co., Inc. An English edition of a standard Dutch work. Part 4 is devoted to nematodes.

Van Cleve, H. J. 1941. Relationships of the Acanthocephala. Amer. Natur. **75**:31-45. This and other papers on this phylum are by a lifelong student of the group. He strongly favors a platyhelminthic affinity of these parasitic worms.

Part IV □ The protostome eucoelomate animals

THE LOPHOPHORATE PHYLA

THE MINOR PROTOSTOME PHYLA

PHYLUM MOLLUSCA

PHYLUM ANNELIDA

PHYLUM ARTHROPODA

INTRODUCTION TO PHYLUM ARTHROPODA

THE EXTINCT TRILOBITES AND THE CHELICERATES

THE CRUSTACEANS

THE MYRIAPODS AND INSECTS

8

The lophophorate phyla

PHYLA PHORONIDA, ECTOPROCTA, AND BRACHIOPODA

GENERAL CHARACTERISTICS

1. The three phyla of eucoelomate, sessile animals—Phoronida, Ectoprocta, and Brachiopoda—have a crown of ciliated, hollow tentacles on a horseshoe-shaped or circular fold of the anterior body wall, the lophophore. The lophophore is an extension of the coelom, and its tentacles surround the mouth but not the anus. The tentacles of the lophophore are specialized for ciliary feeding organs and may also assist in respiration by providing a large surface for gaseous exchange. All three phyla (with one or two exceptions) have a recurved (U-shaped) alimentary canal with the anus placed near the mouth. A vascular system occurs in the Phoronida and the Brachiopoda.

2. No lophophorate has a distinct head. The body is divided into a lophophore region and a trunk. Each of these divisions encloses a coelom. The body may be considered regionalized into protosome, mesosome, and metasome with the consequent coelomic divisions of protocoel, mesocoel, and metacoel. The lack of a head may be associated with their sessile mode of life. The three phyla may have come from a common ancestral type whose body was composed of head, lophophore, and trunk; but due to the degeneration or reduction of the head associated with a sessile, sedentary existence, the protocoel has disappeared, leaving only the mesocoel and metacoel divisions of the coelomic cavity. A septum may separate the

mesocoel from the metacoel. The protosome may be absent because of the lack of a head.

3. Although all three phyla typically develop a free-swimming larva (sometimes called a modified trochophore larva), each phylum has a unique larval form.

4. Both ectoprocts and lamp shells (Brachiopoda) have extensive fossil records of thousands of species. Ectoprocts still have many existing species, but only about 300 lampshell species are now living. There are only a few species of phoronids now living, and they have left no fossil record.

PHYLOGENY AND ADAPTIVE RADIATION

1. All three phyla appear to be closely related, as all have a lophophore and the same body plan, but their relationships are not obvious on superficial examination. In all of them there is a nervous plexus in the base of the epidermis, which indicates a primitive characteristic. In addition, they have some centralization of a nervous system, as indicated by ganglionic enlargements around the digestive system.

2. Although classified as protostomes, they may represent a connecting link between the protostomes and the deuterostomes. Both lophophorates and deuterostomes have the same regionalization of protosome, mesosome, and metasome except that the protosome is chiefly suppressed in the lophophorates. Brachiopods have an enterocoelous formation of the

coelom, a deuterostome characteristic. Ectoproct cleavage may be highly modified or even radial in pattern, as is also cleavage in the brachiopods.

3. The evolutionary diversification of the lophophorates has been guided to a great extent by the varied patterns of the lophophore, which often takes the form of lobulations, arm formations, and spiral coils. Some brachiopods have undergone few changes in their long history because their environment has remained fairly stable. The skeleton in ectoprocts as well as in brachiopods has played a part in their adaptive radiation.

■ PHYLUM PHORONIDA

Phoronida is a small phylum of wormlike animals. Sixteen species have been named. Along with the other lophophorate coelomates, the phoronids possess a lophophore, which is a variously-shaped distal fold of the body surrounding the mouth, specialized as a food-collecting mechanism.

L. H. Hyman has emphasized the definition of a lophophore and excludes those animals that have clusters of tentacles in similar positions but lack other features of a true lophophore. Her definition of a lophophore involves a tentaculated extension of the mesosome, with a coelomic lumen surrounding the mouth but not the anus. The primitive shape of the lophophore is considered to be horseshoelike, but it is circular in some forms. Generally the lophophore is flexible and bears muscles and nerves. In some species it can be retracted into the body cavity; in others it can be pulled back for protection into the secreted tube in which the animal lives. When expanded, the lophophore forms a funnel of tentacles surrounding the mouth. Each ciliated tentacle bears 2 lateral

FIG. 8-1

A phoronid and some of its distinctive characteristics (diagrammatic). Tube may have in its wall pebbles, sand, shell fragments, and other environmental objects that cling to secreted part of tube.

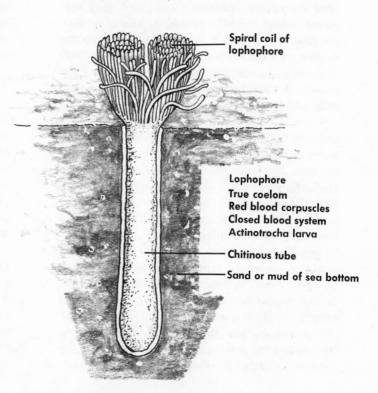

Spiral coil of lophophore

Lophophore
True coelom
Red blood corpuscles
Closed blood system
Actinotrocha larva

Chitinous tube

Sand or mud of sea bottom

tracts of cilia and 1 on the median inner surface. The retracted lophophore may carry the captured food to the mouth, or the trapped food may be carried by the ciliary tracts. Protrusion of the lophophore occurs by hydraulic pressure, produced by compression of the coelomic fluid.

The tube-dwelling phoronids are benthonic and live just below the surface in the upper littoral zone in intertidal or subtidal mud flats (Fig. 8-1). They are strictly marine and are rare in brackish water. They often form masses of intertwined tubes attached to rocks and other substrata in fairly shallow water (Fig. 8-2). Some have the habit of burrowing in calcareous shells and rocks in which the worms are also enclosed by their secreted tubes. Phoronids are restricted to the tropical and temperate zones where, in favorable environmental conditions, they may be abundant. They occur both along the Atlantic coast south of Chesapeake Bay and along parts of the Pacific coast. In Europe most species have been found from Norway to the Mediterranean region. Certain species live on the coasts of Japan, Australia, and elsewhere. So far, no records of phoronids have been reported from the polar regions.

The major activity of a phoronid consists in protruding the anterior end of the body, erecting and expanding the tentacular crown, and moving the cilia vigorously to create water currents. At the same time, the mouth is kept open to catch the food brought to it. When danger threatens, the body is withdrawn into the tube by contraction of the trunk longitudinal muscles. Protrusion and expansion are brought about by contraction of trunk circular muscles, which force the coelomic fluid forward to the lophophore. The giant fiber or fibers may be responsible for rapid retraction into the tube.

GENERAL MORPHOLOGIC FEATURES AND PHYSIOLOGY
External features

The body of the phoronid is cylindric and ranges in length from about ¼ inch to 8 inches, with an average of about 4 inches. The chief features of the body are the tentacular crown (lophophore) at the anterior end and the long slender trunk without appendages but with a slightly enlarged posterior end (the end bulb). Each worm is always found in a chitinous tube, which it secretes. L. H. Hyman (1958) found chitin in the tube but not in the body of the phoronid. The tube is somewhat longer than the worm and is open at the anterior end, from which the forepart of the body with its lophophore can be protruded. The parchmentlike tube with sandy encrustations may be erect and buried in sand or mud, or it may, with others, form masses of intertwining tubes attached to rocks and shells (Fig. 8-1 and 8-2).

The most conspicuous part of the worm is the anterior lophophore. This characteristic organ consists of 2 body wall ridges curved in the shape of a horseshoe whose limbs may each be rolled up into a spiral. The convex part of the horseshoe is located ventrally and the concave side is dorsal. The outer, or external, ridge of the horseshoe runs ventral to the mouth, and the inner, or internal, ridge is dorsal to it. Each ridge of the lophophore bears a variable number of flattened tentacles, which are hollow, ciliated extensions of the body wall. The lumen in the tentacles is a continuation of the coelomic cavity and is lined with peritoneum. The tentacles are rather stiff and are supported by a thick basement membrane that acts as a skeleton. The number of tentacles varies in the different species from 18 to more than 300.

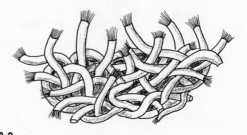

FIG. 8-2

Part of colony of *Phoronis.*

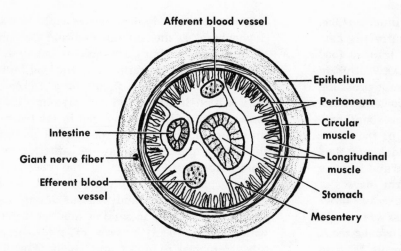

FIG. 8-3

Cross section through muscular part of trunk of phoronid (diagrammatic).

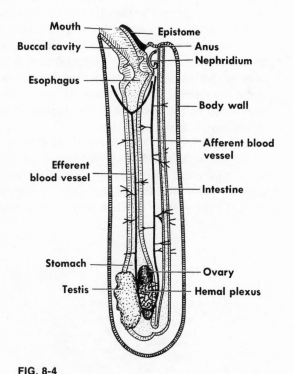

FIG. 8-4

Principal internal structures of a phoronid (diagrammatic). Lophophore has been omitted.

Internal features

The body wall consists of an outer epidermis, which may bear a cuticle in certain regions; a musculature of circular and longitudinal layers; and a peritoneal lining (Fig. 8-3). Cilia are restricted to the tentacles and anterior part of the trunk. The epidermis contains many gland cells, especially on the tentacular crown and anterior part of the trunk. The coelomic cavity is made up of the mesocoel, which forms the coelom of the tentacular crown, and the metacoel, which is found in the trunk. The mesocoel and metacoel are separated by a septum. Mesenteries supporting the intestine occur in the metacoel.

The digestive tract is U-shaped, with the mouth and anus opening close together at the anterior end of the animal (Fig. 8-4). The mouth between the 2 ridges of the lophophore is overhung dorsally by a crescent-shaped fold, the epistome, which has a cavity that communicates with the lophophoral coelom. The mouth, which can be closed by the epistome, opens into a type of buccal cavity and then into the long gullet, or esophagus. The latter passes into the globular stomach, which extends into the end bulb of the trunk region. Here the stomach constricts and joins the intestine, which turns sharply, ascends along the dorsal side of the trunk, and terminates in a short rectum emptying at the anal papilla on the concave side of the horseshoe lophophore. The digestive tract is lined with mostly ciliated epithelium, and its walls have muscular investments in certain regions. Gland cells are found in the esophagus and prestomach. Digestion is probably intracellular.

Phoronids are filter feeders, and the tentacular cilia create water currents that are directed toward the groove between the two ridges of the lophophore. This groove enters the angles of the mouth on each side. Food particles of plankton are entangled by the mucus of the tentacles and groove and are carried by the cilia to the mouth, while the water current passes dorsally over the anus to the exterior.

Two U-shaped metanephridia open on each side of the anal papillae. Each metanephridium consists of a coiled duct with a funnellike, ciliated opening at each end. They get rid of excess water and excreta and serve as gonoducts for the gametes.

The closed circulatory system is well developed. Basically it consists of 2 longitudinal blood vessels — an afferent vessel that runs dorsally, and an efferent vessel that is ventral. The vessels are connected with each other anteriorly by a pair of semicircular vessels (W. K. Brooks and R. Cowles, 1905). From one member of this pair a single blood vessel supplies each tentacle. Blood ebbs and flows in the tentacle by the same vessel and is carried away posteriorly by the efferent, or ventral, vessel lying in the lophophore ridge. The dorsal and ventral vessels are connected by a hemal plexus on the stomach wall. No heart is present, and the blood is driven by pulsation of the muscular walls of the vessels. The blood contains nucleated corpuscles with hemoglobin, which imparts a red color to the blood.

The nervous system consists of a general nerve layer of fibers in the base of the epidermis that thickens into a nerve ring at the outer base of the lophophore. This ring sends nerves to the tentacles, motor nerves to the longitudinal muscles of the body wall, and a lateral or giant fiber through the epidermis on the left side (Fig. 8-3). A smaller lateral nerve may occur in some forms on the right side. The lateral nerves give off branches to the trunk musculature and produce rapid contraction of those muscles. The sensory system is restricted to the neurosensory cells that occur on the tentacles and muscular part of the trunk. A lophophoral sense organ or a cluster of neurosensory cells is found on a flap in some phoronids.

REPRODUCTION AND EMBRYOLOGY

Most phoronids are hermaphroditic, but some are dioecious (e.g., *Phoronis architecta*). The sex cells originate from cells of the thin peritoneum of the blood capillary ceca. The gonads consist of indefinite masses of these cells closely applied to the efferent vessel and capillary ceca. In some hermaphrodites the ovary is on the dorsal side of the efferent vessel and the testis is on its ventral side. The gametes are shed into the body cavity and pass through the metanephridia (which act as gonoducts) to the outside. J. Rattenbury (1953) studied reproduction in *Phoronopsis viridis* and found sperm in metanephridia of both male and female animals and also sperm in the body cavity of the female from January to May. In *Phoronis hippocrepia* the fertilized eggs are retained on the tentacles until the larval stage. Fertilization is usually external. Some worms brood the eggs in a brood chamber formed by the inner tentacles and the concavity of the lophophore, which holds the eggs in place by mucous secretions. Asexual reproduction by transverse fission occurs in *Phoronis ovalis* in the nonmuscular part of the trunk.

Cleavage in phoronids is holoblastic, but there is no regular radial or spiral pattern. Indications of spiral cleavage exist in *Phoronopsis viridis*. Cleavage produces a coeloblastula with cilia at its apical pole. The gastrula formation is embolic with a very large blastopore that becomes small, ventral, and displaced inward by an overhanging preoral lobe of ectoderm. In this way a mouth is formed at this

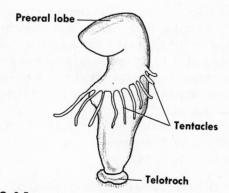

FIG. 8-5

Actinotroch larva of phoronid.

position with the blastopore still connected to the archenteron. An anus is established by the fusion of the elongated archenteron with the ectoderm. Protonephridia are later established with openings near the anus. When the larva (actinotroch) hatches, it is completely covered with cilia and assumes a planktonic existence (Fig. 8-5). The larva is somewhat oval in shape with a large preoral hood over the mouth, behind which is an oblique ridge of hollow tentacles that vary in number with the different species. At the posterior end of the larva a ciliated telotroch is formed for locomotion. After its planktonic existence of some weeks, the larva undergoes metamorphosis. The rapid metamorphosis, which occurs after the larva settles to the bottom, first involves eversion of a metasome pouch or future body wall of the trunk, to which a fold of intestine is attached, thus producing the characteristic adult digestive tract loop. The ciliated tentacles are also cast off, but new ones regenerate from their stumps. The preoral lobe and its nervous ganglion are shed along with the tentacles. After several internal changes and the completion of metamorphosis, the worm secretes a tube in which it will spend the rest of its existence.

■ PHYLUM ECTOPROCTA (BRYOZOA)

The ectoprocts are small marine or freshwater animals that are chiefly colonial (a few

are solitary), sessile, and usually live in some form of case of their own secretion. The colony (zoarium) is nearly always attached to a solid substratum such as rock, vegetation, debris, or the body of another animal. Most zoaria are immobile, but some colonies can move slightly over a substratum. The individuals are called zooids, each of which usually lives in its own case, often maintaining organic continuity with the other zooids of the colony. The case in which the zooid lives is called a zooecium; its walls may be in contact with or in common with those of others. The opening of the case to the exterior for the protrusion of the lophophore is termed the orifice, which may have a closing device.

The ectoprocts have a worldwide distribution in fresh water and the sea. Freshwater species are found in all altitudes from sea level to high mountain lakes except in the north and south polar regions. W. R. Tenney and W. S. Woolcott (1966) made an extensive investigation of the distribution of freshwater bryozoans in the headwaters of certain southern rivers. They found that the Bryozoa grew best and occurred more frequently in streams below impoundments. Their favorite habitat was underneath the downstream side of rocks, where they were attached and were always in the shade. The north temperate zone contains about the same species everywhere. Marine ectoprocts, which are far more common than freshwater ones, are most abundant in the littoral zone of the continental shelf but have been found in the deep abyssal regions of the ocean at depths of several thousand meters. They occur in the polar regions as well as in tropical waters. They are scarce in brackish water such as the Baltic Sea. Many of their species are cosmopolitan, and others are localized in their distribution. Their general adaptability is very great, and many factors account for their methods of dispersal such as ocean currents, attachment to floating seaweeds, the bodies of other animals, and (in fresh water) by statoblasts. Their fossil record (Fig. 8-6) is impressive, beginning

A

B

FIG. 8-6

Fossil bryozoans common from Ordovician to Devonian times. **A,** *Fenestrellina,* a cryptostome found in Indiana limestone, dating from middle Mississippian. (×2.) **B,** *Hallopora,* a treptostome, from Middle Devonian. (Class Gymnolaemata.)

with the Ordovician period. Those with calcified skeletons have left the best record, but those with a membranous zooecium are also found. Their fossils are most abundant in impure calcareous limestone rocks alternating with shale or shell marl. Their paleoecology indicates that they were most abundant in neritic littoral zones. Most bryozoan fossils are fragmentary because of the fragile nature of the colonies.

CLASSIFICATION

The Ectoprocta and Entoprocta are two groups that have long been classified together as a larger group, the Bryozoa, or moss animals. There has been considerable disagreement among authorities regarding the taxonomic status of these two groups. One classification still held by many students considers the phylum Bryozoa (or Polyzoa of British students) to consist of two subphyla, the Ectoprocta and the Entoprocta.

An alternative classification assigns phylum rank to both the Ectoprocta and Entoprocta. The term "Bryozoa" is now often restricted to the phylum Ectoprocta. Arguments for such a separation are the coelomate nature of ectoprocts against the pseudocoelomate entoprocts, the differences in embryologic development of the two groups, and the presence of a true lophophore in ectoprocts but not in entoprocts.

The present work is based on L. H. Hyman's classification. The phylum Ectoprocta has about 4,000 species, of which only about fifty are phylactolaemates.

Phylum Ectoprocta. A protostomial group of coelomate and unsegmented animals; mostly sessile and colonial; animals minute and each in a separate case (zooecium); digestive system U-shaped with anus outside the circumoral ring (lophophore) of tentacles; lophophore retractile into a sheath; coelom with peritoneum; no excretory, circulatory, or respiratory organs; usually hermaphroditic; asexual reproduction by budding; statoblasts in freshwater species; larva a trochophore type; marine and freshwater.

Class Gymnolaemata. Zooid with circular lophophore; no epistome (upper lip); no body wall musculature; no direct coelomic communications between zooids; three existing orders; mostly marine.

Order Ctenostomata. Zooids usually with uncalcified zooecia of chitinous cuticle; orifice with pleated collar for a closing apparatus; without brood chambers; zooids may or may not bud from each other; two suborders. *Victorella, Clavopora, Pherusella.*

Order Cheilostomata. Boxlike zooids with zooecia calcified wholly or in part; ventral orifice usually with operculum; brood chambers present (ovicells); vibracula; two suborders. *Bugula* (Fig. 8-7), *Microporella, Membranipora.*

Order Cyclostomata (Stenostomata). Zooids tubular with fully calcified zooecia; circular terminal orifice without operculum; brood chambers in form of gonozooids or modified zooids; membranous sac from tentacular sheath to posterior end of zooid; five suborders. *Tubulipora, Crisia.*

Class Phylactolaemata. Zooids usually with horseshoe-shaped lophophores; epistome present; body wall with musculature; zooecia gelatinous or chitinous (not calcified); zooids with continuous coeloms; no polymorphism; freshwater. *Cristatella, Pectinatella* (Fig. 8-8), *Lophopus, Plumatella* (Fig. 8-9).

GENERAL MORPHOLOGIC FEATURES AND PHYSIOLOGY

As described under the section on classification, two major groups or classes of ectoprocts exist, the gymnolaemates and the phylactolaemates. Many striking morphologic differences occur between the two classes, and what applies to one may not apply to the other, although a basic, uniform type of structure is present in both. Evolutionary divergence has greatly modified basic patterns because one class has evolved in fresh water and the other chiefly in the sea.

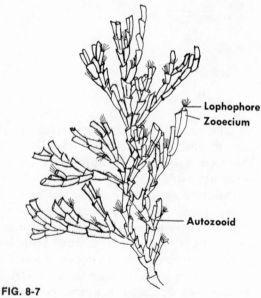

FIG. 8-7

Arrangement of zooids in colony of *Bugula* (order Cheilostomata of class Gymnolaemata).

FIG. 8-8

A, *Pectinatella* colony on stem showing rosette arrangement. Zooids form single-layered patches (rosettes) over gelatinous base which may reach size of football or larger. **B,** Part of rosette greatly enlarged showing zooids of colony (class Phylactolaemata).

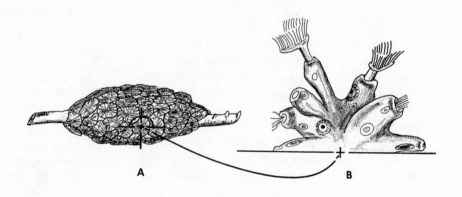

The bryozoan colony

Most ectoprocts live in colonies (zoaria) that vary greatly in shape, size, and organization. The taxonomy of many ectoprocts, especially marine species, depends on these features. Some colonies are small microscopic patches, and others are huge masses of zooids. The colony is nearly always attached to a great variety of substrata where it remains and grows, but a few colonies can creep about on their substrata. Some colonies are stoloniferous; the zooids are borne on creeping or erect stolons composed of modified zooids, which are divided by nodes or septa into internodes, giving a jointed appearance to the colony. Most colonies are formed by a succession of adjacent zooids fused to each other by their outer parts, or zooecia. Some colonies form a meshwork of

regular pattern. Many of the calcified ectoprocts form encrusting masses in the form of flat sheets (laminar) that may be unilaminar or multilaminar when formed by continued superposition of new layers. The most common type is the unilaminar, in which the dorsal surface of the zooids is attached to the substrate with their orifices facing away from the substrate. The freshwater *Pectinatella* consists of a slimy, gelatinous mass with the zooids in the surface. The colonies are in the form of a rosette (Fig. 8-8). The common marine *Bugula* forms branching colonies via a biserial attachment of zooids and has a plantlike appearance (Fig 8-7). When colonies are highly calcified, they often assume compact structures as corals do. The arrangement of the zooids in the colonies is also highly variable. A common organization is the attachment of the dorsal surfaces of the zooids to the substratum with the lateral surfaces attached to adjacent zooids and the ventral surface (with orifices) exposed.

The class Gymnolaemata usually illustrates polymorphism, in which several types of zooids

FIG. 8-9

Zooid of *Plumatella*, a phylactolaemate ectoproct. (From Hickman: Integrated principles of zoology, The C. V. Mosby Co.)

or zooid modifications exist. The typical zooid is the autozooid, or feeding individual. The others, which are modified zooids, are called heterozooids. Heterozooids are represented by stolons, rhizoids (rootlike attachments), avicularia, vibracula, or any cavity enclosed by walls other than those of the autozooids. Of these heterozooids, the avicularium and vibraculum are of special interest. The avicularia resemble bird heads and are characteristic of some of the cheilostomes. These modified zooids may occupy a place in the colony normally taken by an autozooid (vicarious), or they may be found anywhere (adventitious). They may be sessile or stalked. The operculum in this modified zooid forms a mandible that can close and open by muscles. The avicularia help keep the colony free from debris, as well as

entrap small organisms. In vibracula the opercula are modified into long bristles or setae (in cheilostomes only), which may keep the colony clean by sweeping back and forth.

Communication between adjacent zooids takes place by interzooidal pores often located in pore plates. The pores are not open holes but are filled with epidermal cells. These pores

FIG. 8-10

Four mechanisms of lophophore protrusion in ectoprocts, all essentially hydraulic, in which muscle contraction causes decrease in volume of zooecium, compressing coelomic fluid and thus forcing lophophore to extrude. **A,** Membranous frontal wall is pulled down by muscle contraction. **B,** Muscles pass through openings in a calcareous shelf, or cryptocyst, to frontal membrane. **C,** When entire zooecium is calcified, muscles attach to floor of a compensation sac, which dilates sac and compresses coelomic fluid. **D,** In the cyclostome with a tubular zooecium, dilator muscles enlarge vestibule, compressing that part of the coelom, and forcing coelomic fluid downward into the lower part, the outer coelomic sac, where it presses against inner coelomic sac to extrude the lophophore. Retractor muscles aid in lophophore retraction. (Adapted from several sources.)

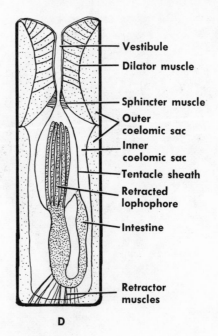

may be located in the transverse end walls or in the lateral walls. A slow diffusion of fluid between the zooids occurs through these pores.

Morphology of the zooid

The zooids, or individuals, of a particular colony have a uniform structure. Each zooid is very small (0.25 to 2 mm. in length) with an average length of less than 1 mm. The shape of the zooid may be boxlike, tubular, vaselike, or oval. It is made up of a body wall that encloses the lophophore and the polypide (visceral mass) (Fig. 8-10). The body wall varies in the two classes of ectoprocts. In the phylactolaemata and a few marine forms the body wall consists of an outer covering, the zooecium, which may be gelatinous, membranous, or chitinous, and an inner layer of epidermis with an underlying musculature of circular and longitudinal muscles. In the calcareous gymnolaemates the zooecium has a calcareous layer between the chitinous cuticle and the epidermis, thus forming a rigid exoskeleton. No muscles are present in the wall, and the only other part of the body wall is the peritoneal lining. The frontal membrane may remain uncalcified and may be membranous to permit expansion when the lophophore is retracted (Fig. 8-10, A). If the frontal membrane is calcified, a compensation sac opening by a pore to the outside permits

entrance of water, thus increasing coelomic fluid and aiding in the protrusion of the lophophore (Fig. 8-10, C). An oral aperture, the orifice, in the zooecium allows the lophophore to protrude. This orifice is often provided with a lid (operculum) that can be closed when the animal withdraws into the zooecium. The body wall is often turned in at the orifice as far as a constriction (diaphragm) provided with a sphincter muscle. The region between the orifice and diaphragm is called the vestibule (Fig. 8-10, D). At the bottom of the vestibule is a circular collar through which the lophophore can be projected or retracted. A tentacle sheath covers the necklike region of the protruded animal and extends from the base of the lophophore to the diaphragm. The sheath consists of a thin cuticle, an epidermis, a muscular layer, and a peritoneal lining in that order outside in. The sheath contains the anterior portion of the digestive tract and the anus. When the lophophore is retracted, the sheath encloses the tentacles bunched together.

Within the body cavity, or the coelom enclosed by the body wall, are the lophophore and the

FIG. 8-11

Comparison of lophophore arrangement. **A,** Class Gymnolaemata. **B,** Class Phylactolaemata. (Modified from Harmer.)

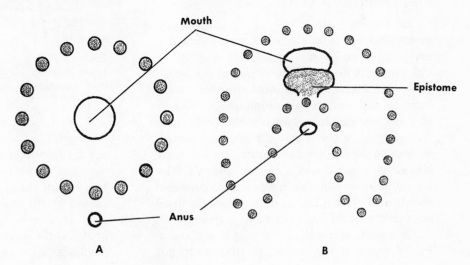

Mouth

Epistome

Anus

A

B

polypide. The lophophore may be circular (gymnolaemates) or horseshoe shaped (phylactolaemates) and surrounds the oral aperture (Fig. 8-11). It bears a series of hollow tentacles on 1 ridge (marine species) or 2 ridges (freshwater species). The number of tentacles varies from 16 to 106 in the freshwater forms and from 8 to 34 in marine forms. The tentacles are ciliated on the median and lateral surfaces. The cavities of the hollow tentacles open into a circular canal at the base of the lophophore. The circular canal may open into the coelomic or body cavity or may become separated from it.

The lophophore can be retracted into the tentacle sheath or introvert, or it may be extended through the diaphragm, vestibule, and orifice when the animal is feeding. Protrusion is mostly a matter of the evagination of the tentacle sheath, which, being attached to the lophophore base, pushes out the tentacular crown as the sheath everts (Fig. 8-10). The eversion of the tentacle sheath may be brought about by coelomic fluid pressure due to muscular contraction of several types. Retraction is brought about by the retractor muscles of the lophophore, the most powerful muscles in the zooid. The tentacles create water movements and bring food to the mouth.

Internal anatomy of the zooid

The digestive tract has the U shape commonly found in sessile animals, with the mouth and anus both on the dorsal side (Fig. 8-12). The mouth is located in the center of the lophophore and opens into a ciliated pharynx, which may be followed by an esophagus, although pharynx and esophagus are often not delimited. The esophagus opens into the stomach, which is divided into a tubular cardia, a posteriorly elongated cecum, and a narrow pylorus. The cardia and cecum are lined with glandular, nonciliated epithelium, and the pylorus is lined with ciliated epithelium. Circular and longitudinal muscles occur in the stomach walls and in the walls of the entire digestive tract, but

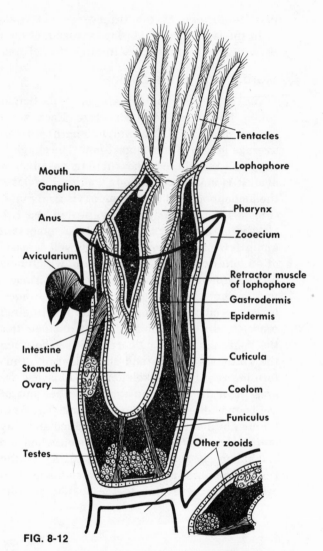

FIG. 8-12

Zooid of *Bugula* (diagrammatic), a gymnolaemate ectoproct, showing principal internal structures. (From Hickman: Integrated principles of zoology, The C. V. Mosby Co.)

not uniformly. The stomach empties into the intestine or rectum, which opens at the anus just outside the circle of tentacles. Attaching the cecum of the stomach to the posterior zooidal wall is a cord called the funiculus, which anchors the digestive tract in position. Food particles may be carried down the digestive tract singly or in mucous cords. Both intracellular and extracellular digestion may occur in the marine species, but only extracellular digestion occurs in the freshwater forms. Food in ectoprocts is often stored in the epithelial lining of the digestive tract.

The excretory, vascular, and respiratory organs are absent in ectoprocts. Gaseous exchange occurs through the body surface, and the coelomic fluid in the spaceous coelom and its divisions circulates the products of digestion. Waste materials are picked up by coelomocytes and other cells and are discharged to the outside.

The nervous center is a ganglion of small nerve cells and fibers and is located on the dorsal side of the pharynx. The ganglion forms part of a nerve ring around the pharynx. From the ganglion and nerve ring a network of nerves runs to the tentacles and the tentacle sheath, the digestive tract, and the muscles. A subepithelial nerve network may extend throughout the wall of the zooid except in that part cemented to a substratum. Nervous connections between zooids are uncertain. No special sense organs exist, but neurosensory cells with projecting bristles are known to occur.

REPRODUCTION AND EMBRYOLOGY

Most ectoprocts are hermaphroditic. Asexual reproduction also occurs, chiefly budding. Perhaps no group exhibits a greater variety of reproductive methods than do the bryozoans. Not only hermaphroditism, but dioecism, polyembryony, protandry, and the curious brown body regeneration are found in the group. The gonads show a wide range of locations, but the ovaries are usually found in the upper, or distal, end of the zooid and the testes are found in the lower, or basal, end. In freshwater species the funiculus may bear the testes, although they may be found on the body wall, where the ovaries also occur (Fig. 8-12). A gonad of either sex is simply a mass of developing gametes enclosed by peritoneum from which the sex cells differentiate. Gonoducts are entirely absent, and the gametes are shed into the spacious coelom. The male cells may be shed in balls of sperm. Protandry, or protogyny, in which sperm and eggs ripen at different times, may also occur. Many cheilostomes have separate sexes.

Fertilization is mostly internal, and most ectoprocts brood their eggs in some manner. The ova usually develop into embryos in special brood chambers, the ooecia or ovicells. In most cheilostomes, including *Bugula*, the body wall at the distal end of the autozooid forms an outgrowth or hood called the ovicell, which consists of two evaginations, one inside the other. The cavity of the inner evagination is continuous with the coelom of the zooid. In the space between the inner and outer evaginations the egg develops and is supplied with nutrition from the inner evagination. Many species brood their eggs in the coelomic cavity in special evaginations of the body wall, the embryo sacs. Generally, when the embryo in the sac increases in size, the polypide of that zooid degenerates into a dark reddish mass, the brown body, and a new polypide is regenerated. The fate of the brown bodies varies. They may accumulate in the coelomic cavity by successive degenerations, or they may be taken into the digestive tract of the regenerated polypide and voided through the anus. The same body wall and its zooecium (which do not degenerate in the process) can thus house several successive polypides as embryos are formed. Polyembryony, in which several embryos are produced from one zygote by the separation of the cells in early cleavage, is known to occur in some cyclostomes.

Self-fertilization is common in hermaphroditic species that produce sperm and eggs in the same individual, but in dioecious species sperm in some way not yet known must enter the coelom of another individual to fertilize the eggs.

Cleavage is of the holoblastic, radial, or biradial type and results in a coeloblastula. Gastrulation is by primary delamination at the vegetal pole. The larva is a form of trochophore provided with a ciliated girdle for swimming and an anterior tuft of cilia or flagella. An apical nervous organ and an adhesive sac or sucker are also present in the larva. After escaping from the brood chamber or embryo sac, the larva swims around, and if hatched from a nonbrooding species, it may feed during its larval existence. When it is ready for attachment, the larva everts the adhesive sac, which becomes fastened to a substratum. It then undergoes metamorphosis, resulting in the first zooid of the colony called an ancestrula. Then, by budding, the ancestrula gives rise to many other zooids. Repeated asexual reproduction of these zooids results in the formation of a new colony. The budding pattern of a colony varies with the species. The budding process usually involves the constriction of a part of the parent zooid, which includes an evagination of the parent body wall.

W. Lynch (1947) studied many of the factors that influenced the metamorphosis of *Bugula neritina* and found that sodium induces a rapid onset of metamorphosis, but potassium has the opposite effect. Excess of calcium tends to inhibit metamorphosis altogether and prolongs the swimming period. Magnesium in excess also inhibits metamorphosis but does not prolong the swimming.

The development of freshwater species differs in certain particulars from the foregoing account, which applies to the gymnolaemates. In the freshwater forms, development leads to a larva that is really a juvenile colony of primary and secondary polypides; these initiate a new colony after the attachment and degeneration of the larva.

The Phylactolaemata, or freshwater ectoprocts, also have a form of asexual reproduction other than that of budding. This method of propagation is the formation of statoblasts (Fig. 8-13). These resistant bodies are masses of epidermal and peritoneal cells that have come from basal epidermal cells forming the funiculus. The peritoneal cells store reserve food.

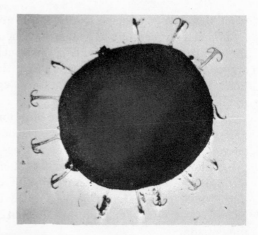

FIG. 8-13

Statoblast of *Pectinatella* (about 1 mm. in diameter). (From Hickman: Integrated principles of zoology, The C. V. Mosby Co.)

The germinal mass forms a hollow ball or disk protected by a chitinous shell, the outer part of which is modified into a pneumatic ring (annulus) of air that keeps the statoblast afloat. Statoblasts vary with different species and have taxonomic diagnostic value. Some of them bear hooks around their margins (spinoblasts). Statoblast formation is correlated with favorable conditions of temperature and medium. All must remain dormant for some time before they germinate; they are able to withstand harsh conditions. They generally accumulate in the zooids during the summer and are released when the colony disintegrates before winter begins. In the spring, if conditions are favorable, the statoblast germinates into a zooid and starts a new colony.

■ PHYLUM BRACHIOPODA

Phylum Brachiopoda consists of the lamp shells, which reached their greatest abundance during the Paleozoic era. The present number of 250 or so species is quite modest in view of the more than 25,000 species that existed during that remote geologic time. They were perhaps one of the most abundant forms to be found in the ancient seas. At present they are

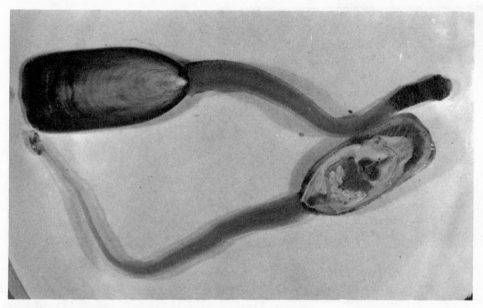

FIG. 8-14

Lingula, inarticulate brachiopods with long pedicles. One valve has been removed from lower specimen exposing soft parts.

FIG. 8-15

Examples of an inarticulate brachiopod, *Lingula,* and an articulate brachiopod, *Terebratulina.*

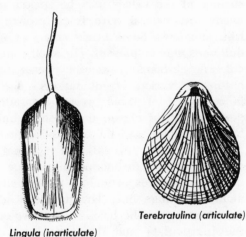

Lingula (inarticulate)

Terebratulina (articulate)

only locally abundant in such places as the Sea of Japan and Australian waters, although they are found in all seas and at all latitudes.

The brachiopods are exclusively benthonic, marine forms that live from the lowest tide levels to abyssal depths, although most are restricted to the continental shelf regions. None is found in the great abyssal trenches. They are almost entirely sessile forms, being attached by a pedicle to some object. When a pedicle is absent, they attach themselves by the ventral valve. They are commonly found on well-oxygenated and relatively firm sea bottoms. However, the lingulid genera, *Lingula* and *Glottidia,* live in vertical burrows in sandy shores or in mudflats and are the most active of all brachiopods. Some forms are restricted to very specific habitats, and others, such as *Lingula,* are cosmopolitan.

CLASSIFICATION

Phylum Brachiopoda. Coelomate Bilateria with Protostomia and Deuterostomia characteristics; unsegmented, with shells of dorsal and ventral valves somewhat dissimilar; valves may be hinged by articulation or unhinged; valves with mantle

FIG. 8-16

External view of articulate brachiopod, *Terebratella*. Left, specimen with dorsal valve up; right, ventral valve up.

originate from body wall; lophophore present; heart present; excretory system of metanephridia; mostly dioecious; free-swimming larva.

   Class Inarticulata. Valves held together by muscles only; lophophore without internal skeleton; anus present.

      Order Atremata. Shell of calcium phosphate; pedicle attached to ventral valve through notch of both valves. *Lingula* (Fig. 8-14), *Glottidia*.

      Order Neotremata. Pedicle attached through notch of ventral valve only; shell of calcium phosphate or calcium carbonate. *Crania*, *Discina*.

   Class Articulata. Valves hinged together by interlocking device; lophophore with internal skeleton; pedicle emitted through ventral valve; no anus; order status unsettled; many superfamilies recognized. *Orthis*, *Terebratula*, *Terebratulina* (Fig. 8-15), *Terebratella* (Fig. 8-16).

GENERAL MORPHOLOGIC FEATURES AND PHYSIOLOGY
External anatomy

The ventral valve is usually larger than the dorsal valve and is provided with a fleshy stalk, the pedicle or peduncle, by which it attaches itself to a substratum. The pedicle may extend through an opening (foramen) between both valves (*Lingula*, Fig. 8-14), or only through the ventral valve (*Laqueus*, Fig. 8-17). In *Crania* the ventral valve is cemented directly to the substratum. The valves may be oval, circular, or triangular in general contour and may be elongated with rounded anterior margins. Both valves are usually convex externally, although the dorsal one may be flat or concave.

Most existing shells fall within the range of 5 to 80 mm. along the long axis, but extinct ones were often much larger. The external surface of the valves may be smooth or variously ornamented with folds, growth lines, ribs, or spines. Some bright colors as well as dull ones may be present. The shell consists of calcium carbonate, calcium phosphate, and chitin in various combinations. The shell is composed of three layers: the outermost periostracum of chitin; the middle laminated layer of calcite; and the inner prismatic layer of fibers. Both internal and external shell characters are of taxonomic importance.

The two valves articulate with each other along the hinge line. The nature of the articulation forms the basis for the two classes, the Inarticulata and the Articulata. In the inarticulates (the more primitive forms) the two valves are held together at the hinge line by muscles; in the articulates the valves are

FIG. 8-17

Two fossil brachiopods. Bottom, *Kingena* of lower Cretaceous period. Top, *Laqueus* of Recent epoch.

bound together by an interlocking device of teeth-and-socket arrangement. The movement of the valves is controlled by two to five pairs of internal muscles between the valves. The inarticulates can open their valves much wider than can the articulates.

The posterior end of the ventral valve in most brachiopods is tapered to form a beak, which is perforated for the passage of the pedicle. The dorsal valve may send out projections that form an internal skeleton in the lophophore.

The body occupies only the posterior part of the space between the shells. From the body wall two double folds extend anteriorly as the dorsal and ventral mantle lobes, which secrete the valves and line their underlying surfaces. The lobes give off fine papillae that penetrate the shell surface.

The pedicle differs in the various groups. In the lingulids (e.g., *Lingula*) it is long and flexible, arises from the ventral mantle lobe, and includes a coelomic evagination continuous with the main body coelom (Fig. 8-14). The wall of the pedicle consists of a cuticle, an epidermis, a connective tissue layer, a muscle layer, and a peritoneal lining. When the animal is in its burrow, the pedicle is stuck in the bottom sand, with the anterior end of the valves directed toward the surface opening. Retraction of the pedicle can pull the animal back into its burrow. In the articulates the pedicle is rather short, is made up mostly of connective tissue with no coelom, and is very restricted in its movements. L. H. Hyman (1958) found chitin in the pedicle of lingulids but not in the valves of the articulates.

Internal anatomy

The body of the animal is attached to the inner surface of the two valves by its dorsal and ventral walls (Fig. 8-18). The other parts of its walls are largely free. Since the animal occupies only the posterior region of the chamber between the valves, the anterior body wall sends extensions dorsally and ventrally to cover those parts of the valves not occupied by the animal's body.

The body wall is covered with a one-layered epidermis. Beneath the epidermis is connective tissue and, in those parts of the body wall not adherent to the valves, a muscle layer is found beneath the connective tissue. Ciliated peritoneum lines the body wall. The mantle lobes are extensions of the anterior body wall and thus consist of a double wall with a coelom between. The coelomic cavity consists of the embryonic mesocoel and metacoel, the protocoel being absent, as already mentioned.

The lophophore is an anterior outgrowth of the body wall and lies in the space between the mantle lobes, filling the greater part of that cavity. The lophophore is supported by a calcareous loop from the dorsal valve. The lophophore is fringed with long, ciliated tentacles that

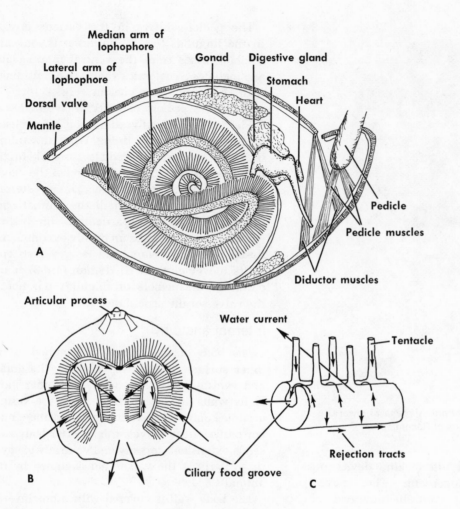

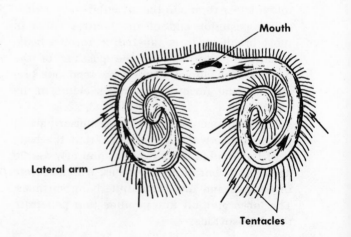

FIG. 8-18

A, An articulate brachiopod (longitudinal section). **B,** Branchial and feeding currents of brachiopod. Large arrows show water currents moving over lophophore and smaller arrows indicate ciliary movement in food grooves. **C,** Portion of lophophore arm and tentacles showing water current, ciliary feeding tracts, and rejection tracts. (**B** and **C** adapted from Russell-Hunter.)

FIG. 8-19

Pattern of lophophore in *Lingula,* showing direction of movement of food particles. Many other types of lophophores are also found in brachiopods.

occur in a row along its entire edge (Fig. 8-19). Extensions from the mesocoel extend into the tentacles, which are covered with a tall epidermis of ciliated cells.

The lophophorates are ciliary feeders. Water with food particles entering the gaping valves flows over the lophophore. Water enters the anterior gape by two currents, one on each side, produced by the lateral tracts of tentacular cilia that beat at right angles to the long axis of the tentacles. The two currents converge posteriorly to form a median outgoing current (Fig. 8-18, *B*). The inner surfaces of the tentacles have mucous glands and cilia that beat toward the food groove in the arm of the lophophore (Fig. 8-18, *C*). Food is caught in the mucus and is carried by the cilia to the brachial food groove, whose cilia transport the food to the mouth. Heavy or unwanted particles are carried down ciliated rejection tracts (Fig. 8-18, *C*), dropped on the ventral mantle lobe, and carried away by outgoing ciliary currents.

From the mouth the food is conveyed into a ciliated alimentary canal made up of an esophagus, a dilated stomach provided with a digestive gland that opens into it, and an intestine that, in the articulates (Fig. 8-18, *A*), ends blindly but in the inarticulates ends with an anus. Digestion may be intracellular within the digestive glands.

The circulatory system is an open one, consisting of a longitudinal vessel in the dorsal mesentery over the stomach. One part of this longitudinal vessel is enlarged into a contractile vesicle, or heart, from which extend an anterior channel to the sinuses of the intestine and lophophore and a posterior channel to the mantle, nephridia, and gonads. The colorless, coagulable blood may contain coelomocytes.

A respiratory system is absent in brachiopods, and gaseous exchange may occur principally in the mantle and lophophore.

All brachiopods have one or two pairs of metanephridia located on each side of the digestive tract. Each metanephridium has a fringed nephrostome that opens into the coelomic cavity, a tubule that extends anteriorly to the mantle cavity, and a nephridiopore that opens on each side of the mouth. Waste is picked up by coelomocytes from the coelomic cavity and carried into the nephrostomes. The ciliated tubules carry the waste to the mantle cavity for expulsion.

The nervous system consists of an esophageal nerve ring with a small ganglion on its dorsal side and a larger one ventrally, and a number of nerves from these centers to the lophophore, mantle lobes, and valve muscles. The nervous system is located mainly in the base of the epidermis. Nerve plexuses have been found in the lining of the digestive tract of inarticulates. Specialized sense organs have not been identified, except for a pair of statocysts near the anterior adductor muscle in a species of *Lingula*.

REPRODUCTION AND EMBRYOLOGY

Brachiopods are mostly dioecious, and their gonads are localized in regions of the coelomic peritoneum. Usually 4 gonads, which are quite large when filled with mature gametes, are present. In inarticulates the gonads are found in the free ends of the peritoneal bands among the viscera in the metacoel spaces. In the articulates the gonads occur in the coelomic canals of the mantle. The ripe gametes are discharged into the metacoel and are carried to the outside by the metanephridia, which act as gonoducts. A few brachiopods may brood their young in modified parts of the metanephridia or in a special pouch of the ventral valve. Fertilization is external in those that discharge their gametes to the outside or internal in those that have brood pouches.

Although classed as Protostomia, the development of brachiopods is similar in many ways to that of the Deuterostomia. Their cleavage is radial, mostly equal, and holoblastic, with some variations in the early blastomeres. The coeloblastula undergoes gastrulation chiefly by the embolic method. Mesoderm formation occurs by the enterocoelous method, another deuterostome characteristic. In inarticulates the free-swimming larva, which may last several weeks, bears a pair of mantle lobes and valves that enclose the body. The pedicle develops from the ventral mantle lobe. Locomotion of the larva is by the U-shaped ciliated lophophore. In *Lingula* no metamorphosis occurs; the larva develops directly into the adult. In articulates, after the formation of the coelomic sacs, the embryo shows three regions: an anterior lobe that will form most of the body, a median lobe (the mantle), and a posterior lobe (the pedicle). The larva swims around for a few hours by means of the ciliated anterior lobe. After it becomes attached by the pedicle, metamorphosis occurs, in which the mantle, formerly directed backward, now shifts its position forward around the body, where it secretes the valves. In most instances, the final stages of embryonic development and metamorphosis in brachiopods have not been clearly worked out.

REFERENCES

Benham, W. B. 1889. The anatomy of *Phoronis australis*. Quart. J. Microscop. Sci. **30**:125-158. One of the early detailed studies of a phoronid.

Bronn, H. G. (editor). 1939. Phoronidea. Klassen und Ordnungen des Tierreichs, vol. 4, part 4. Leipzig, Akademische Verlagsgesellschaft. An authoritative monograph.

Brooks, W. K., and R. Cowles. 1905. *Phoronis architecta*: its life history, anatomy, and breeding habits. Mem. Nat. Acad. Sci. **10**:28-45.

Chuang, S. 1959. Structure and function of the alimentary canal in *Lingula unguis*. Proc. Zool. Soc. (London) **132**:293-311. An account of the digestive system in the "oldest living fossil in the world."

Cori, C. 1941. Bryozoa. In Kukenthal, W., and T. Krumbach (editors). Handbuch der Zoologie, vol. 3, part 2, sect. 15.

Berlin, Walter de Gruyter & Co. The best general account of the phylum.

Cori, C. J. 1937. Phoronidea. In Kukenthal, W., and T. Krumbach (editors): Handbuch der Zoologie, vol. 3, part 2, sect. 5. Berlin, Walter de Gruyter & Co.

Edmondson, W. T. (editor). 1959. Ward and Whipple's freshwater biology, ed. 2. New York, John Wiley & Sons, Inc. The section on freshwater Bryozoa (which includes both entoprocts and ectoprocts) is by the late M. D. Rogick, foremost authority on the group.

Helmcke, J. G. 1939. Brachiopoda. In Kukenthal, W., and T. Krumbach (editors). Handbuch der Zoologie, vol. 3, part 2, sect. 5, Berlin, Walter de Gruyter & Co.

Hyman, L. H. 1958. Occurrence of chitin in lophophorates. Biol. Bull. **114**:106-112.

Hyman, L. H. 1959. The invertebrates; smaller coelomate groups, vol. 5. New York, McGraw-Hill Book Co., Inc. One of the best treatises yet published on phoronids.

Lynch, W. 1947. Behavior and metamorphosis of the larva of *Bugula neritina* (L.). Biol. Bull. **92**:115-150.

MacGinitie, G. E., and N. MacGinitie, 1968. Natural history of marine animals. ed. 2. New York, McGraw-Hill Book Co. This book describes the two species. *Phoronopsis californica* and *P. viridis*, which are common in spots along the California coast.

Moore, R. C., C. G. Lalicker, and A. G. Fischer. 1952. Invertebrate fossils. New York, McGraw-Hill Book Co., Inc. Section 5 of this excellent work is devoted to a good general description of the bryozoans and a detailed account of their fossil record dating back to the Middle Ordovician.

Pennak, R. W. 1953. Fresh-water invertebrates of the United States. New York, The Ronald Press Co. One of the best general accounts of the freshwater species, with keys.

Rattenbury, J. 1953. Reproduction in *Phoronopsis viridis* Biol. Bull. **104**:182-196.

Smith, R. I., F. A. Pitelka, D. P. Abbott, and F. M. Weesner. 1954. Intertidal invertebrates of the central California coast. Berkeley, University of California Press. Keys are given for the two genera (*Phoronopsis* and *Phoronis*) found on the California coast. Only one species is listed, although a number are known to occur along the Pacifiic coast.

Tenney, W. R., and W. S. Woolcott. 1966. The occurrence and ecology of freshwater Bryozoans in the headwaters of the Tennessee, Savannah, and Saluda River systems. Trans. Amer. Microscop. Soc. **85**:241-245.

The minor protostome phyla

PHYLA PRIAPULIDA, ECHIURIDA, SIPUNCULIDA, TARDIGRADA, PENTASTOMIDA, AND ONYCHOPHORA

GENERAL CHARACTERISTICS

1. All these animals are eucoelomates, or those that have a body cavity formed between layers of mesoderm and lined with an epithelium of mesoderm origin.

2. During their development, part of the mesoderm (somatic) is applied against the inner surface of the ectoderm, and another part (splanchnic) surrounds the gut. In this way the body wall is made up of ectodermal and mesodermal layers, whereas the gut wall has mesodermal and endodermal layers. The free space between the two mesodermal parts (the coelom, or body cavity) is lined by a mesodermal membrane, the peritoneum. Vertical portions of the peritoneum form the mesenteries that suspend the gut from the body wall. A pseudocoel may be lined by ectoderm, endoderm (sometimes partly by mesoderm), but never by peritoneum.

3. The phyla presented in this group are more or less distinct from each other and have puzzling affinities to each other and to other groups. Their grouping together is arbitrary, mainly for convenience.

4. These phyla have relatively minor economic and ecologic importance. All have undergone modest evolutionary diversification and are represented in most phyla by few species. They may have considerable interest in the evolutionary blueprint of life. Their few species may

be due to modest evolutionary development, or they may be remnants of groups that were larger and more widely distributed at one time.

5. The Priapulida have long been considered a pseudocoelomate and given a class rank under the Aschelminthes because of their close relation to the class Kinorhyncha and their supposed pseudocoel, but recent evaluations assign them true coelomate status.

6. The animals within this group have many odd morphologic and physiologic mechanisms for performing their functions, such as the introvert of the Sipunculida and the proboscis of the Echiurida for burrowing and feeding, the complex feeding apparatus of tardigrades, the caudal appendages of the priapulids, the state of suspended animation (cryptobiosis) also in tardigrades, and sex determination in certain echiuroids. Not the least of the oddities are the aggregations of cells called urns that may have excretory and other functions in the Sipunculida.

7. Although ranked among the protostomes, some have deuterostome characteristics in their embryologic development. A type of trochophore larva is found in some.

PHYLOGENY AND ADAPTIVE RADIATION

1. The Sipunculida and the Echiurida in their development show relations to the Annelida.

Sipunculids have annelid-like nervous systems but have no segmentation either as adults or as larvae. Echiuroids have a degree of segmentation in early stages but are unsegmented as adults. Both phyla may have arisen from protostome lines that led to the annelids, mollusks, and arthropods. The Pentastomida and Tardigrada (together with the Onychophora) are placed in a group called Pararthropoda, or Oncopoda, because they have unjointed limbs and claws (at some stage) and a molting cuticle. However, there is little evidence that they are related to arthropods. The onychophorans may serve as a connecting link between the annelids and arthropods, for the onychophorans have characteristics of both these major phyla.

2. The evolutionary development of the Sipunculida, the Priapulida, and the Echiurida has been guided chiefly by their proboscis devices, which in some cases may be a chemoreception center for exploring their environment, as well as for burrowing and food getting. Tardigrades have varied their mechanisms of

claws and feeding apparatus among their different species, whereas the pentastomidan adaptations have been mainly those that have to do with attachments to their hosts. Onychophorans have a primitive method of locomotion involving hydrostatic pressure for extending the leg, flexors for raising the leg, and muscles for producing thrust. Number of legs (14 to 43 pairs) probably represents their greatest adaptive feature.

■ PHYLUM PRIAPULIDA

The Priapulida consist of a small group of vermiform marine animals that are found in the colder waters of both northern and southern hemispheres. They occur from Massachusetts to Greenland on the Atlantic coast and from California to Point Barrow, Alaska, on the Pacific coast of North America, and from the coast of Uruguay in South America to certain islands of the Antarctic region in the southern hemisphere. They are also found in similar regions of Europe and Asia, especially in the Baltic Sea and the region around Norway. Some species have been found in deeper, colder waters of tropic and subtropic regions. They live buried in mud and sand on the sea floor and range from intertidal zones to depths of

FIG. 9-1

Cross section of portion of pharynx of *Priapulus* showing peritoneum. (Courtesy W. L. Shapeero, University of Washington, Seattle.)

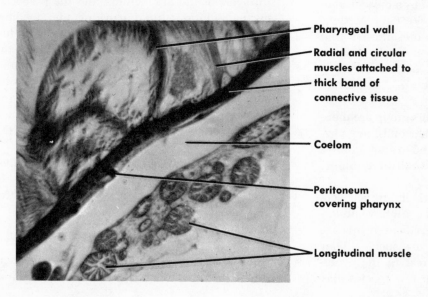

Pharyngeal wall

Radial and circular
muscles attached to
thick band of
connective tissue

Coelom

Peritoneum
covering pharynx

Longitudinal muscle

thousands of meters. Although they have the power to plow through the bottom muck in seeking their prey, they usually remain quietly buried in a vertical position, the mouth on a level with the bottom surface. They are predaceous in their eating habits.

Taxonomically the Priapulida have been linked in the past with various other groups such as the Echiurida and the Sipunculida. Since they were supposed to have a pseudocoel and for various other reasons, some zoologists have considered them a class of the Aschelminthes. However, it has been found (W. L. Shapeero, 1961) that they are true eucoelomate animals because their body cavity is lined with a thin peritoneum (Fig. 9-1). Because of this and other structural features, the group is now assigned the rank of phylum.

CLASSIFICATION

In his extensive and excellent monograph on the Priapulida, J. Van der Land (1970) has given a systematic résumé of the phylum somewhat different from the one given in the first edition of this text. His classification is followed here, but where the names of the various taxa have been changed, the ones formerly given are in parentheses. It will be noted that Van der Land has proposed a new family for the species (*Tubiluchus corallicola*) he has recently named.

Family Priapulidae
Priapulus caudatus
Priapulus tuberculatospinosus
Priapulopsis bicaudatus (Priapulus bicaudatus, P. atlantisi)
Priapulopsis australis (Priapulus australis)
Acanthopriapulus horridus (Priapulus horridus)
Halicryptus spinulosus
Family Tubiluchidae
Tubiluchus corallicola

In a particular habitat the animals may be fairly common, but usually they are solitary and uncommon. W. L. Shapeero collected 13 specimens of *Priapulus caudatus* in one locality along the coast of Washington. However,

more spectacular finds have been recorded. For example, a subspecies of *P. tuberculatospinosus* has been reported to occur at densities of up to 250 individuals per meter.

MORPHOLOGIC FEATURES

The Priapulida are mostly medium-sized animals, the largest being 7 to 8 inches long (Fig. 9-2), although some are much smaller. The warty body is cylindric, and the trunk is annulated into several unsegmented rings (30 to 40, *Priapulus;* 100 or more, *Halicryptus*). There are three divisions of the body: the proboscis, or introvert; the trunk; and the caudal appendage (lacking in *Halicryptus*). The proboscis, which is separated from the trunk by a constriction, is as large in diameter as the trunk and sometimes may be larger. It bears 25 longitudinal ridges (ribs), each made up of rows of papillae or spines that lead toward the mouth. The proboscis is eversible and is used both in capturing prey and in sampling the surroundings. At the terminal end of the proboscis is the mouth, which is surrounded with concentric rows of teeth (Figs. 9-4 and 9-5, *A*). This circumoral region is usually invaginated into the interior, but when this region is everted in capturing the prey, the teeth point forward.

Scattered over the trunk region are many tubercles and spines that give a distinctly warty appearance to the animal. At the posterior end of the trunk are 2 urogenital pores and the anus.

Most priapulids have a tail composed of 1 or 2 caudal appendages (Fig. 9-2), which are hollow stems containing many oval vesicles and communicating with the coelomic cavity. The function of these appendages, according to R. Fange and A. Mattisson (1961), is respiratory as formerly supposed because their thin walls facilitate the rapid uptake of oxygen and their rhythmic movements force coelomic fluid into the coelom. K. Lang (1948) considers them to act as chemoreceptors or accessory excretory organs. He proved that when the tail is removed the animal can get along without

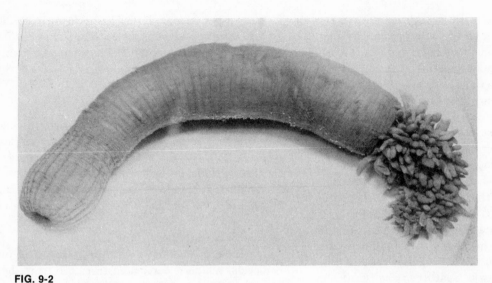

FIG. 9-2

Preserved specimen of *Priapulus caudatus* Lamarck.
(From Hickman: Integrated principles of zoology, The
C. V. Mosby Co.; photograph by W. L. Shapeero.)

it, although a new tail regenerates rapidly to
replace one cut off. The tail may play important
roles in conditions of low oxygen concentrations.
But in general, the morphologic differences
of the tail among the species may indicate
different functions. No respiratory or circulatory organs are found.

The thick, muscular body wall (Fig. 9-3) consists of a two-layered cuticle, an epidermis,
both circular and longitudinal muscle layers,
and the thin peritoneal lining. The epidermis
is made up of a single layer of definite cuboidal
or columnar cells and is not a syncytium (W. L.
Shapeero). The cuticle and epidermis are modified in certain regions to form such integu-
mentary specializations as the papillae of the
proboscis and trunk, the pharyngeal teeth, the
caudal appendage vesicles, and others. Molting
occurs at intervals throughout life. Before
the molting process the epidermis secretes a
new cuticle under the old one, which is then
shed. The cuticle is partly composed of chitin
during the intermolting periods but of protein
only in the newly molted individual (D. B.
Carlisle).

Next to the epidermis is the circular muscle layer arranged in separate rings; inside
the circular muscles are found the longitudinal
muscles (in part striated) either in separate
bundles or forming a continuous layer. In the
proboscis only one pair of longitudinal muscles
is found in the space between each rib so that
the circular muscle layer is attached to the
epidermis only between the ribs. These spaces
of the proboscis are anterior continuations of
the body cavity. The true coelom is lined internally with a thin, nucleated membrane (peritoneum, Figs. 9-1 and 9-4), which also forms the
outermost layer of the intestine and mesenteries
for the visceral organs and covers the retractor
muscles. The body cavity, the cavity of the
caudal appendages, and the spaces of the musculature are filled with a fluid containing cells
(coelomocytes). These coelomocytes have a
respiratory pigment (hemerythrin).

The digestive system consists of mouth,
pharynx, intestine, and rectum (Fig. 9-4). The
invaginable pharynx is composed of circular
and radial fibers and is lined with an epidermis
and cuticle. The cuticle bears teeth (spines)
continuous with those around the mouth (Fig.
9-5, *B*). A circle of short and long retractor
muscles is attached around the anterior end of
the pharynx and originates on the body wall

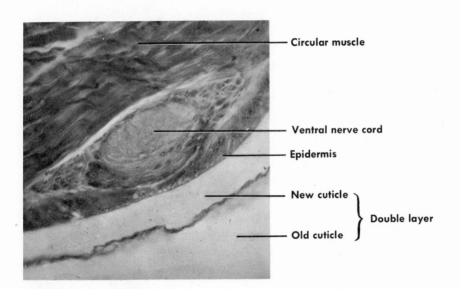

FIG. 9-3

Cross section through ventral nerve cord of *Priapulus*. Note double-layered cuticle of body wall and relation of nerve cord to epidermal layer. (Courtesy W. L. Shapeero, University of Washington, Seattle.)

(Fig. 9-4). They are used in the invagination of the proboscis and the inversion of the pharynx into the midgut. The intestine is mostly a straight tube in the center of the trunk, with much of its columnar epithelium thrown into folds. The lining epithelium is provided with a brush border (Fig. 9-6). Circular and longitudinal muscle layers are found in its walls. The short rectum is lined with cuticle and opens to the outside by the anus.

An interesting structure (in *Tubiluchus*) is the polythyridium at the entrance of the intestine. It consists of circlets of hollow cuticular plates (valvulae) bearing small tubercles and delicate hairs and may serve as a filtering apparatus by preventing large particles from entering the intestine.

The Priapulida live on polychaetes, small crustaceans, and even other priapulids, which they swallow whole.

The nervous system consists of a circumenteric nerve ring around the pharynx and a midventral cord (Fig. 9-3), which can be seen

FIG. 9-4

Major internal structures of *Priapulus* (diagrammatic).

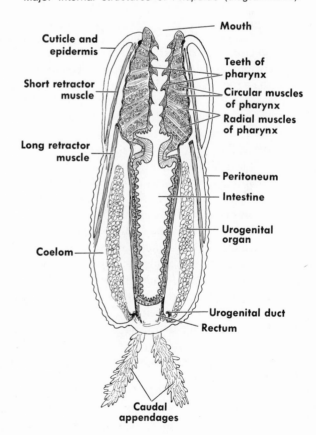

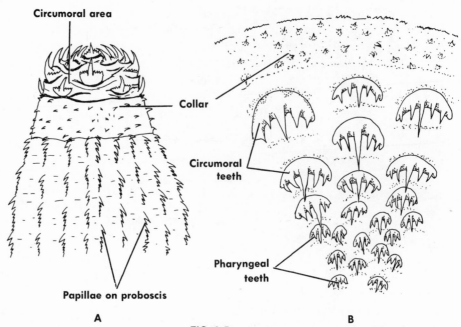

A

B

FIG. 9-5

A, Anterior end of proboscis of *Priapulus tuberculato-spinosus* showing recurved teeth around mouth region. **B,** Small portion from upper part of pharynx of *P. caudatus* showing pharyngeal teeth, large in circumoral area but becoming gradually smaller toward gut. (Modified from H. Theel, 1906.)

FIG. 9-6

Portion of lining epithelium of *Priapulus* gut showing brush border of epithelial cells. (Courtesy W. L. Shapeero, University of Washington, Seattle.)

as a longitudinal band in the undissected specimen. Several pairs of nerves from the nerve ring run to the proboscis ribs and to the pharynx. The nervous system is in very close association with the epidermis, of which it seems to be a thickening. Little is known about the sensory system, but the papillae and spines may be sense organs.

The paired urogenital organs are each made up of a gonad on one side and clusters of solenocytes on the other (Fig. 9-4). A central urogenital or protonephridial tubule serves to carry both excretory and genital products. This common duct is really an excretory duct that picks up the genital duct and then branches into a number of smaller ducts. These smaller ducts expand into nephridial sacs, into which the solenocytes open (A. Norrevang, W. L. Shapeero). The two urogenital canals open by separate pores at the posterior end of the trunk. Sex cells are shed into the sea, where fertilization occurs.

Little is known about the embryology of this phylum. Early cleavage stages are radial instead of the spiral pattern of most proto-

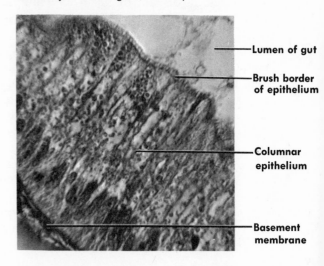

stomes. The egg hatches into a small post-gastrula, the larval trunk of which is later encased in a cuticularized lorica similar to that of the rotifers. The lorica is made up of plates that vary in the two genera. The spiny anterior end of the larva is invaginable into the interior. A foot with toes and caudal glands may also be present. Larval internal structures are similar to those of the adult. The larvae live in the same habitat along with the adults and feed on detritus. They molt in about 2 years and emerge as juveniles without the lorica and with the same structures as adults. As already mentioned, priapulids molt throughout both juvenile and adult life.

The most common species of the Priapulida is *Priapulus caudatus,* which has a wide but spotty distribution in northern waters; a closely related species, *P. tuberculatospinosus,* is found in antarctic regions. These species have only 1 caudal appendage, whereas *Priapulopsis bicaudatus* has 2. *Halicryptus spinulosus,* also found in northern regions, has no caudal appendage.

■ PHYLUM ECHIURIDA

The members of the marine phylum Echiurida are vermiform, protostome animals that live on ocean bottoms in habitats similar to those of the sipunculids. They are commonly found in U-shaped tubes on sandy or mucky areas, although their habitats may also include crevices in rocks, empty shells, and others. They are found from intertidal zones to very great depths in the ocean and are more common in tropic and subtropic water. Most members live in rather shallow water. Some are called spoon worms from the shape of their pro-stomium or proboscis. The smallest vary in length from 15 to 25 mm., and the largest from 400 to 500 mm.

CLASSIFICATION

Some authorities classify Echiurida into three orders: Echiuroidea, Xenopneusta, and Hetero-myota. Another classification places most members in three families: Echiuridae, Thalas-sematidae, and Bonnelliidae. The classification within this phylum is very uncertain at present. Some eighty species have been named.

MORPHOLOGIC FEATURES AND PHYSIOLOGY

The body of echiurids consists of a cylindric, saclike, or sausage-shaped trunk and a preoral lobe (prostomium or proboscis) that is flattened and of varying length, depending on the species. Although the proboscis cannot be retracted into the body, it is capable of great extension. Its edges are flared so that a ventral ciliated groove is formed along its surface, and its basal edges are so fused as to form a funnel around the mouth. In the common spoonworm

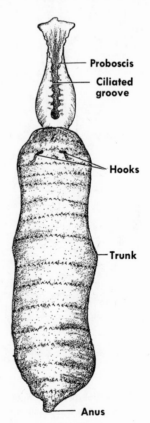

FIG. 9-7

Echiurus, an echiurid common on both Atlantic and Pacific coasts.

(*Echiurus*, Fig. 9-7), the proboscis, when extended, is not half as long as the trunk; but in *Bonellia* (Fig. 9-8) the ribbonlike, bifurcated proboscis may extend to the length of a meter, although the trunk is usually about 2 inches

long. The proboscis is solid and consists of connective tissue and muscles covered with a columnar epithelium, which secretes a cuticle (except in the ciliated grooves). It also bears the nerve ring, or brain, and the vascular ring.

The surface of the trunk may be relatively smooth, or it may bear rings of papillae that secrete slime to line the walls of the burrows. Just behind the mouth on the ventral surface are 2 large, curved, chitinous setae, which are used in the burrowing process (Fig. 9-7). Some species have 1 or 2 rows of small setae at the posterior end of the trunk. These may be used in cleaning out the burrow and in anchoring the body when burrowing, as already described. As in annelids, each seta is produced by a single basal cell and is moved by muscles. Replacements for worn-out setae are kept in reserve sacs.

The body wall consists of a simple columnar epithelium that secretes a thin cuticle and is modified into glandular cells, papillae, and pits; a subepidermal connective tissue containing pigment cells, which give the characteristic brown, green, or red color to the worm; and a muscular coat of circular, longitudinal, and oblique fiber layers. The longitudinal fibers are very pronounced, and the muscles may be arranged in sheets or in bundles.

The body cavity is spacious, unsegmented, and lined with peritoneum. In the proboscis are found some canals and lacunae that are separated from the true body cavity by a diaphragm. However, coelomic fluid can move between the proboscis cavities and the coelom. Embryologically the cavities of the proboscis are derived from the embryonic blastocoel, whereas the coelom is a schizocoel. The coelomic fluid, which circulates freely in the body cavities, has respiratory, nutritive, and excretory functions. It consists of spherical erythrocytes that, in some species, contain hemoglobin; ameboid leukocytes with pigment granules; and a fluid part that contains no plasma proteins and is isotonic with sea water.

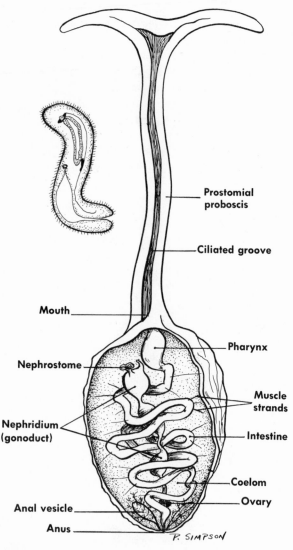

Prostomial proboscis

Ciliated groove

Mouth

Pharynx

Nephrostome

Muscle strands

Nephridium (gonoduct)

Intestine

Coelom

Anal vesicle

Ovary

Anus

P. SIMPSON

FIG. 9-8

Bonellia, internal structure. Left, male. Right, female. When fully expanded, the proboscis may be only a thin thread many times as long as when in the contracted condition. Males are very small (1 mm.) and are parasitic in or on the larger females (up to a meter long, when extended).

The digestive system (Figs. 9-8 and 9-11) is a thin-walled tube with many convolutions and irregular loops that are supported by fine muscular strands running from the body wall. It consists of an ectodermal foregut of mouth, pharynx, and esophagus, and an endodermal midgut and hindgut comprising the intestine. A sphincter separates the foregut and midgut. The mouth at the base of the proboscis opens into a muscular pharynx that is followed by a thick-walled esophagus. The posterior part of the esophagus is dilated into a bulbous crop, which molds the food bolus. A ciliated groove runs along one side of the long, much-coiled intestine. A collateral tube known as the siphon, of unknown function, arises from the ciliated groove near the anterior end of the intestine and reenters the groove farther down the intestine. The intestine is terminated by a short rectum that opens through the anus at the posterior end of the body. The intestine secretes an alkaline digestive juice that contains proteolytic and lipase enzymes.

At least two methods of feeding are found among the echiurids. Most are deposit feeders (deposit refers to organic detritus that accumulates on the substratum). The edges of the proboscis roll up to form a ciliated groove. Food particles are caught on mucus secreted by the proboscis and then are carried back to the mouth along the ciliated groove. In this process of feeding the proboscis is extended out and the ventral face is spread out over the substratum where the food particles adhere to the mucus on the proboscis (Fig. 9-9). A different method is employed by *Urechis*, a Pacific coast form that lives in burrows (Fig. 9-10). Just behind the short proboscis in the body of this form is a ring of mucous glands that secrete mucus as the worm backs up in its burrow. This process forms a funnel-shaped mucous collar around the anterior end of the worm. Then, by

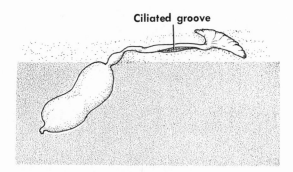

FIG. 9-9

An echiurid, *Tatjanellia,* feeding. Being a detritus feeder, the echiurid extends its proboscis over bottom surface, picks up detritus, and carries it along a ciliated food groove to the mouth. (After Zenkevitch; modified from Dawydoff.)

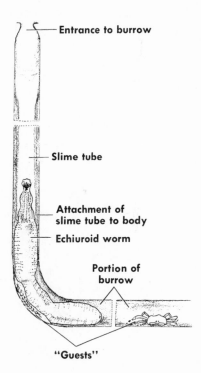

FIG. 9-10

Urechis, a burrowing echiurid common to California mudflats. Spins slime tube in U-shaped burrow. Food is collected in slime tube from water currents pumped through it; tube and food are then swallowed. Often called "innkeeper" because of characteristic "guests" safely harbored in its tube. (Modified from Fisher and MacGinitie.)

peristaltic movement of the body, water is pumped through the burrow, and as the water passes through the mucous tube, the food particles in the water are strained out. When filled with food, the funnel tube is detached from the body, and with the aid of the proboscis the worm swallows the mucous tube and the food within it. The time required to fill a mucous tube naturally varies with the amount of food material in water; usually the tube formation occurs in a steady succession. Echiuroids have about the same ecologic niche in the marine environment as earthworms do in terrestrial ones.

The excretory system is made up of metanephridia (Figs. 9-8 and 9-11), varying in number in the different species. *Bonellia* has only one. A metanephridium contains one or more openings (nephrostomes) into the body cavity and a nephridiopore to the outside. In some species the metanephridia are located anteriorly. Highly characteristic of the phylum is a pair of anal sacs, or vesicles, which open into the intestine close to the anus. They may be much branched; the free ends may be provided with ciliated tunnels, or nephrostomes, that open into the coelom. These anal vesicles are considered to be modified metanephridia and have both excretory and respiratory functions.

The closed vascular system (Fig. 9-11) consists of a ventral vessel that runs along the trunk above the nerve cord and below the digestive tract and bifurcates to form a dilated ring around the margin of the proboscis. A short dorsal vessel receives blood from the ventral vessel by way of peri-intestinal vessels or sinuses surrounding the intestine and propels the blood forward through a median vessel, which communicates anteriorly with the proboscis marginal loop of the ventral vessel. The circulating blood is colorless and contains only phagocytic amebocytes, but not the same kind of corpuscles as those found in the coelomic fluid.

The nervous system consists of a ventral nerve cord, which separates anteriorly into circumpharyngeal connectives. These latter form a loop at the margin of the proboscis

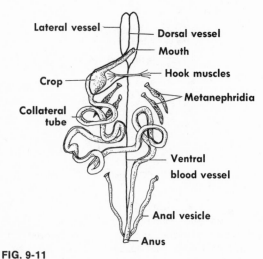

FIG. 9-11

Digestive, vascular, and excretory systems of *Echiurus* (diagrammatic).

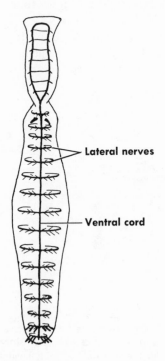

FIG. 9-12

Nervous system of *Echiurus* (diagrammatic).

and are united at the tip (Fig. 9-12). No separate ganglia can be recognized in the ventral nerve cord, and there are no cerebral ganglia. Lateral nerve fibers are given off along the ventral cord and unite dorsally to form rings in the skin of some species. Tactile or chemoreceptor sensory cells are found in the epidermis, especially in the proboscis, but special sense organs are lacking in the adult.

REPRODUCTION AND DEVELOPMENT

Echiurids are dioecious, and sexual dimorphism is conspicuous in some, such as *Bonellia*. The gonad of each sex is attached to the ventral vessel, and the gametes are shed into the coelomic cavity while still immature. When mature, the gametes pass from the body through the nephridia. Fertilization occurs in the seawater, except in *Bonellia* and one or two other species, in which it occurs in the nephridium of the female. In the common *Echiurus* two pairs of modified metanephridia, known as gonoducts or storage organs, open to the outside behind the ventral setae. These serve as storage chambers for the mature gametes.

Bonellia is noteworthy in two particulars —its extreme sexual dimorphism and its unique sex determination. The males are only about 1 mm. long and live as parasites upon or within the bodies of the females, which may reach a length of 1 meter. Mature males (Fig. 9-8, left) are usually found in the single metanephridium of the female; most are equipped with functional reproductive organs, a few degenerative or larval structures, and a ciliated surface. According to the work of F. Baltzer and some others, all larvae are potentially bisexual. Those that develop independently become females; those that become attached to the female proboscis undergo a transformation into a male. Various degrees of intersexuality are correlated with different periods of attachment to the female. It may be that the male absorbs from the female proboscis secretions that serve as masculinizing agents by causing genes for maleness to be expressed

while repressing those of femaleness, although the active principle of such a secretion has not been identified as yet.

In their early embryology spiral cleavage, so characteristic of annelids and others, is the echiurid type. In later stages of development certain molluscan features appear. A transitory segmentation of the mesoderm is shown in some species by the development of several pairs of rudimentary coelomic pouches. The larva is a free-swimming trochophore, and in the adult the anterior part of the animal is of a persistent larval type.

The Echiurida are best represented by the genera *Echiurus, Urechis, Thalessema*, and *Bonellia*. *Echiurus* is found on both the Atlantic and Pacific coasts as well as elsewhere and is characterized by a relatively short, spoon-shaped proboscis and a trunk bearing spiney rings. *Urechis* (Fig. 9-10) lives along the California coast in U-shaped burrows and has a short proboscis and an anterior band of mucous glands that secrete a mucous net for entrapping food particles swept into the net by peristaltic contractions of the worm. Some of them may be 15 to 20 inches long. In *Thalessema*, found along the Atlantic coast and tropic regions, the body is small with no anal setae, and the proboscis may be very long. Many echiurids may be found in the shells of sea urchins and sand dollars.

■ PHYLUM SIPUNCULIDA

Sipunculida, a phylum of unsegmented worms, belongs to the protostomatous coelomates. The worms vary in length from a few millimeters to 2 or more feet; most are less than a foot long. There are 275 species of them (N. M. V. Rothschild), and they have been found in all seas from intertidal zones to depths of several thousand meters. They live on the ocean floor (benthonic) where they make mucus-lined burrows or mucus-sand tubes in the sand or mud. Sometimes they are found in empty mollusk shells and in various kinds of crevices. Some have developed special de-

315

vices at their posterior end for anchoring their position. In some of their localities a spadeful of intertidal mud may yield several specimens.

These worms, together with the priapulids and the echiurids, have long been included under a hodgepodge phylum called Gephyrea. Since there are no close affinities between these three groups, many zoologists (but not all) have assigned each of them to phylum status. Sipunculids are often called peanut worms because of their shape.

CLASSIFICATION

Only genera are recognized in the phylum; they have not been assembled into higher taxa. The species are distributed among thirteen genera, the most familiar of which is *Sipunculus*, containing many species. *Golfingia* is the largest genus in number of species—more than seventy-five. In general the phylum is greatly in need of taxonomic revision.

MORPHOLOGIC FEATURES AND PHYSIOLOGY

The body of the sipunculid is cylindric in shape and is divided into a posterior trunk and an anterior eversible introvert, or proboscis, which bears the terminal mouth. The introvert may be shorter or longer than the trunk in the different species. The mouth may be entirely or partly surrounded by ciliated tentacles or folds, which may be in single or multiple circlets. In *Phascolosoma* (Fig. 9-13, *A*) the crescentic series of tentacles are dorsal to the mouth. Retractor muscles from the body wall of the trunk to the base of the tentacles are used to withdraw or invaginate the proboscis into the body. Evagination of the introvert is performed by the hydraulic pressure of the coelomic fluid when the circular muscles of the body wall contract. This pressure may be equal to almost 1 atmosphere. The introvert may be smooth, but often bears spines, hooks, and papillae or other rough projections. In a few cases the anus may be placed on the introvert. The proboscis base in one or two species is encircled by a collar

formed of calcified plates. *Aspidosiphon* (Fig. 9-13, *C*), found in crevices of coral, has a shield-shape or caplike plate to close the opening into its burrow.

The trunk, which may be dull whitish to grayish brown in color, varies from a short, saclike form to a long cylinder. The shape of the trunk may depend on the nature of the worm's habitat. Those found in gastropod shells may even have a twisted body to conform to the shape of the shell. The trunk of those that live in shells may also be provided with modified papillae in the form of chitinous holdfast organs at the ventral extremity. The trunk of most sipunculids bears papillae or scalelike thickenings but lacks spines and hooks. The trunk in *Sipunculus* is marked off

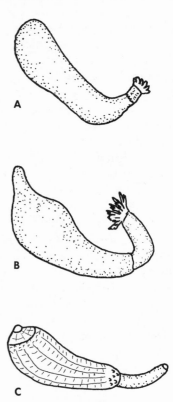

FIG. 9-13

Some types of sipunculids. **A**, *Phascolosoma*, most common genus along Pacific coast. **B**, *Dendrostomum*. **C**, *Aspidosiphon*.

into a pattern of small squares or rectangles by the circular and longitudinal muscle layers, which produce external grooves in the body wall (Fig. 9-14). Glands are also found in the body wall and open to the outside to furnish secretions for the tube linings. The anus is a middorsal opening on the anterior end of the trunk.

The body wall consists of a chitinous cuticle secreted by a simple epithelium, a dermis of connective tissue, a circular muscle layer, an oblique muscle layer (in some), a longitudinal muscle layer, and a peritoneum. The longitudinal muscle layer may be in bundles or it may be continuous. The retractor muscles of the introvert come from the longitudinal layer. Smooth muscle fibers are characteristic of sipunculids. Many glands are found in the epidermis; some are unicellular, but others have 2 or more cells.

The large coelom is a schizocoel filled with coelomic fluid and traversed by muscular fibers and mesenteric supports of the viscera. In some coeloms septal partitions may be present, but true segmentation is absent. Coelomic fluid contains some protein and is isotonic with sea water. The coelomic fluid contains many kinds of cells or corpuscles. The abundant red corpuscle is a biconvex nucleated disk and bears the respiratory pigment, the iron-containing hemerythrin, which is pink when oxidized and colorless when reduced. This kind of pigment is also found in a few polychaetes. Some species may have as many as 100,000 red corpuscles per cubic millimeter. The chief function of the respiratory pigment is to serve as a reservoir for oxygen, which hemerythrin releases in times of great need or deficiency. Other types of cells in the coelomic fluid are phagocytic amebocytes, chlorogogue cells, and peculiar cellular bodies known as urns, which may be fixed or free swimming (Fig. 9-15). The urns are budded off from the peritoneum, from the intestinal wall, or from the surface of compensation sacs. An urn is composed of a vesicle enclosed in flat peritoneal cells and is filled with a fluid and some cytoplasmic strands. It also has an apical ciliated cell and a circlet of cilia. The urns assist in the circulation of the coelomic fluid and help clear the coelomic fluid by collecting foreign substances and degenerating cells. Such accumulated debris may be carried to the nephridia for disposal.

A kind of vascular system is the lacunar, or tentacular, system that is found in connection with the tentacles and the anterior part of the esophagus. Around the beginning of the esoph-

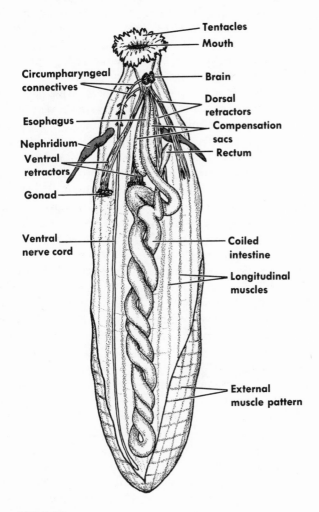

FIG. 9-14

Internal structure of *Sipunculus.*

Tentacles
Mouth
Circumpharyngeal connectives
Brain
Dorsal retractors
Esophagus
Compensation sacs
Nephridium
Rectum
Ventral retractors
Gonad
Ventral nerve cord
Coiled intestine
Longitudinal muscles
External muscle pattern

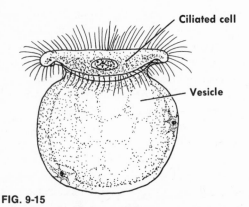

FIG. 9-15

Free urn of *Sipunculus*. These curious bodies originate from peritoneum and are useful for ridding body of waste debris. (Modified from Selensky.)

agus is a circular ring space, which sends canals forward into the hollow tentacles and sends 1, 2, or several blind tubes called compensation sacs (formerly called polian vesicles) backward alongside the walls of the esophagus. The fluid in this system contains corpuscles similar to those in the body cavity. The system is considered a part of the coelomic cavity, and the hydrostatic pressure initiated on the body cavity in the eversion of the introvert also presses on the blind sacs. This pressure, together with the contraction of the sacs, forces fluid into the tentacles and causes them to expand; when the tentacles contract, the compensation sacs receive the fluid.

No special organs of circulation are present because the functions of the blood are largely assumed by the coelomic fluid. Respiration occurs through the skin and the tentacles.

The digestive system of the Sipunculida is fitted for the sedentary life of the animal, being somewhat U-shaped (Fig. 9-14). Sipunculids live on organic matter collected from the muck and sand by the tentacles and taken to the mouth by ciliary action. From the wide but jawless mouth a ciliated esophagus lies in the introvert and leads to a long, coiled intestine, or midgut. The intestine passes to the end of the trunk and then forward to the anus, with the limbs of the loop generally twisted around

each other. The intestine is lined with ciliated epithelium and has a ciliated groove that assists in moving food along. Most digestion and absorption probably occur in the descending arm, and feces formation in the ascending arm of the intestine. The intestine opens into a short rectum, or hindgut, which runs to the anus. A blind diverticulum associated with rectal glands may be attached to the rectum; it may be secretory in nature.

The excretory system consists of a pair of metanephridia (only 1 in some) that open on the ventral surface of the body near the anus (Fig. 9-14). A nephridium may be straight or V-shaped and is partly saclike and partly a ciliated tube. These organs remove the waste brought to the nephrostome by the coelomic amebocytes and the urns previously mentioned.

The nervous system consists of paired cerebral ganglia located above the pharynx and connected by circumenteric connectives to a ventral nonsegmented nerve cord that runs to the posterior tip of the animal. The brain receives sensory nerves from the tentacles and various sense organs and sends motor nerves to the retractor muscles of the introvert. The ventral nerve cord gives off many lateral nerves to the muscles of the introvert wall and elsewhere. The structure of the nerve cord is complex, consisting of an outer peritoneal covering, a connective tissue layer with longitudinal muscles, and an inner sheath around the nerve elements. There is some evidence that the brain has an inhibitory effect on other parts of the motor nervous system, as is the case in annelids.

Numerous sense organs, especially for tactile and chemical stimuli, are present, particularly on the tentacles, introvert, and skin papillae. There are also frontal organs whose functions are not well understood. A cerebral organ found in some species is a ciliated pit that may be sensory in function; it extends back to the brain and has been found to contain neurosecretory granules. The nuchal organ, a ciliated area on the oral disk,

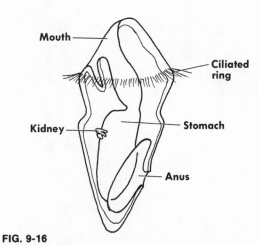

FIG. 9-16

Larva of *Sipunculus*. Cilia omitted except for metatroch (ciliated ring).

is probably a chemoreceptor. Pigment spots, or ocelli, are also found in various species.

REPRODUCTION AND DEVELOPMENT

Sexes are separate in this phylum, but externally they are similar. The gonads are attached on the coelomic wall at the origin of the retractor muscles. Immature sex cells are shed into the coelomic fluid, where they become mature before they pass to the outside through the nephridia, which thus serve as gonoducts. Females are stimulated to shed their gametes when they are touched by the sperm of the males. Fertilization occurs in the sea. In some sipunculids, females greatly outnumber the males.

In sipunculid development, cleavage is of the spiral type characteristic of annelids, mollusks, and others of the protostome pattern. Gastrulation is partly by epiboly and partly by invagination. The larval form (Fig. 9-16) is a type of free-swimming trochophore provided with a prototroch and a metatroch (bands of cilia). After swimming around in plankton for a number of days, the larva, depending on the species, metamorphoses into a juvenile worm by elongation of the trunk, loss of the ciliary bands, and other changes. In this metamorphosis the anus is shifted to an anterior position,

and the intestine becomes spirally twisted as it elongates. Throughout its development no visible evidence of segmentation occurs.

Although they have no asexual method of reproduction, these worms can regenerate parts such as tentacles, portions of the introvert, posterior ends, etc. Ameboid coelomocytes supposedly play an important part in the regeneration processes.

The most familiar sipunculid is, without doubt, *Sipunculus nudus*, which has a very wide distribution. The most characteristic feature of this genus is the rectangular pattern of the trunk surface produced by raised areas of the circular and longitudinal muscles (Fig. 9-14). *Phascolosoma*, with horseshoe-shaped arrangement of tentacles, is common along certain parts of the Atlantic coast and also in European waters. Certain species of this genus are also common along the Pacific coast, and some species may be 6 to 7 inches long. The genus *Dendrostomum* has many-branched tentacles. *Golfingia* is another genus on which much work has been done. *Aspidosiphon* (Fig. 9-13, *C*) has shields of plates in front of the anus and also at the end of the body.

■ PHYLUM TARDIGRADA

Water bears, or tardigrades (Fig. 9-17), are minute forms, usually less than 1 mm. in length, and are found in both fresh water and salt water, as well as wet moss, sand, damp soils, pond debris, liverworts, and lichens. Not more than 10% of the known species have been reported from the sea. Marine species are usually found in the sandy beaches of shallow water, where they may occur several inches under the surface. Tardigrades are found in all latitudes from the tropics to the Arctic. Only forty to fifty species of tardigrades have been found in North America, but about 180 named species occur in other parts of the world.

CLASSIFICATION

E. Marcus, a prominent authority on the Tardigrada, considers this group as a class under

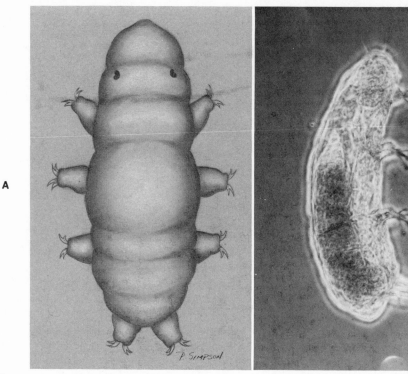

FIG. 9-17

A, Dorsal view of typical freshwater tardigrade. **B,** Lateral view of typical marine tardigrade, *Batillipes mirus.* Note the modified claws with disk endings. Marine species are relatively few compared with terrestrial and freshwater forms. Some are found in interstitial spaces of marine as well as freshwater sandy beaches. (**B** courtesy R. P. Higgins.)

the phylum Arthropoda. He divides the tardigrades into three orders and several families.

Order Eutardigrada. *Macrobiotus.*
Order Mesotardigrada. (Only one species from hot springs of Japan.) *Thermozodium.*
Order Heterotardigrada. *Echiniscus.*

ADAPTIVE FEATURES

Tardigrades have varied chiefly their claws and feeding apparatus of tubular pharynx and needle-like stylets in their adaptive radiation. The stylets, which can be protruded through the mouth, are useful in puncturing the cells of mosses, algae, and even other microscopic animals (predaceous tardigrades). Morphologic differences (spines, plates, papillae, etc.) among the species may indicate physiologic adaptations to various kinds of habitats. The terrestrial forms (which make up the majority of species among the tardigrades) do not as a general rule have well-defined ecologic limits; the same species may occur on lichens, mosses, and liverworts. Some are adapted as psammobionts and live in interstitial habitats.

RELATIONSHIPS

The affinities of tardigrades are among the most puzzling of all animal groups. Some authorities consider them a class under the Arthropoda; others consider them more closely related to the mites (Arachnida). There are some who believe that they have certain annelid characteristics. Their anatomy reflects similarities to other animal groups, but these likenesses may not indicate real relationships. Although they have a holoblastic cleavage pattern, their development does not conform to any of the other patterns found in other phyla. Most zoologists at present are inclined to regard them as a separate phylum. Because of certain characteristics, they may be highly modified protostomes.

MORPHOLOGIC FEATURES AND PHYSIOLOGY

The body is elongated, cylindric, or a long oval and is unsegmented. The head is merely the anterior part of the trunk. The body bears four pairs of short, stubby, unjointed legs armed with four single or two double claws that may be unequal in size and are used for clinging to the substrate (Fig. 9-17). Tardigrades cannot swim. The last pair of appendages lies at the posterior end of the body. The body is covered by a nonchitinous cuticle that may be smooth or ornamented with spines or plates. It is secreted by the hypodermis. Cilia are absent. The cuticle is shed (ecdysis) four or more times in its life history. The body consists of a prostomium and five indefinite segments. Cell constancy is characteristic of the group, and the number of epidermal cells is the same for all members of a genus.

The mouth (surrounded by cuticular rings) is at the anterior end of the prostomium and opens into a tubular and oval pharynx (Fig. 9-19, A), which is adapted for sucking by radial muscle fibers. The oval pharynx contains 6 sclerotized macroplacoids for support. The tubular pharynx is supplied with 2 needlelike stylets that can be protruded through the mouth or withdrawn by muscular action. These stylets are used for piercing the cellulose walls of the plants and body walls of animals that tardigrades live on, and the liquid contents are sucked in by the pharynx. A pair of glands, which secrete the stylets at each molt, open into the tubular pharynx. The pharynx has a triradial lumen. The rest of the digestive tube consists of a short esophagus, a large stomach, a short rectum, and a cloacal aperture or anus

that opens between the posterior pair of legs. Glands empty into both the pharynx and the intestinal tract. At the junction of the stomach and rectum two paired, ventral Malpighian tubes and a single dorsal gland empty into the digestive system and may be excretory in function (Fig. 9-19, B). Although they live primarily on plant juices, which they suck out with their pharyngeal pumping action, some tardigrades may suck out the body fluids of nematodes (Fig. 9-21, A), rotifers, and other small animals. Most of the body cavity is a hemocoel, with the true coelom restricted to the gonadal cavity. No circulatory or respiratory systems are present, as fluids freely circulate through the body spaces, and gaseous exchange can occur through the body surface.

The muscular system consists of a number of long muscle bands, each of which is composed of a single fibrillar cell that may be a syncytium. These muscles have most of their origins and insertions on the body wall, and they manipulate body and leg movements. Circular muscles are absent, but the hydrostatic pressure of the body fluid may act as a type of skeleton. The legs are manipulated by a set of muscles fastened to the body wall and inserted near the tip of the leg.

The lobed brain is large and covers most of the dorsal surface of the pharynx (Fig. 9-18). By way of circumpharyngeal connectives around the pharynx, it connects to the subpharyngeal ganglion, from which the double ventral nerve cord extends posteriorly as a chain of 4 ganglia. Sense organs usually consist of a pair of pigmented eyespots and various tactile organs such as bristles and spines.

Tardigrades are dioecious, but females are usually more common than males and may make up an entire population. Parthenogenesis is common. The ovary or testis (Fig. 9-19, B) lies dorsal to the intestine. A single oviduct in the female or 2 sperm ducts in the male open into the cloaca or through a separate gonopore near it. Both thin-shelled and thick-shelled eggs similar to the summer and winter eggs

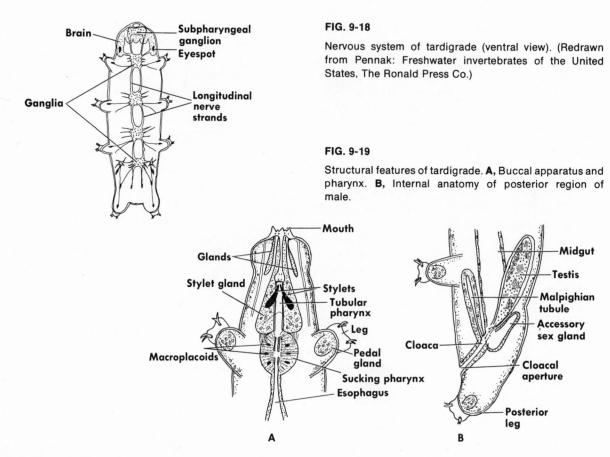

FIG. 9-18

Nervous system of tardigrade (ventral view). (Redrawn from Pennak: Freshwater invertebrates of the United States, The Ronald Press Co.)

FIG. 9-19

Structural features of tardigrade. **A,** Buccal apparatus and pharynx. **B,** Internal anatomy of posterior region of male.

of rotifers occur. Eggs are often found in the shed cuticle (Fig. 9-20). Up to 30 eggs may be deposited by the female at one time. Eggs may be up to 200 microns in diameter. Fertilization may occur in the ovary or in eggs deposited in the old cuticle. The young are juveniles.

Development is direct and may last for a month or less, depending on the species and temperature. Cleavage is holoblastic but is irregular because of the asynchronous division of the blastomeres. The endomesoderm has an enterocoelous origin, and five pairs of coelomic sacs are formed, of which the four anterior pairs form muscles and storage cells, whereas the posterior pair merge to form the gonad. When the young juveniles hatch, they are rarely more than 50 microns long. They grow by water intake and increase in cell size, rather than in cell number.

Like rotifers, tardigrades have great ability to withstand harsh environmental conditions (desiccation, freezing, etc.). Under such conditions the tardigrade loses water, shrivels, and contracts into a generally rounded condition with low metabolism (cryptobiosis) (Fig. 9-21, B and C). When placed in water, it becomes normal again even after being in the cryptobiotic (anabiotic) state for years. By such a method the life-span of these little animals may be extended several years.

SOME RECENT TARDIGRADE INVESTIGATIONS

More and more investigations on tardigrades are being done chiefly because of the enigma of their taxonomic relationships, the scanty knowledge about the American species, and the intrinsic charm of the little animals them-

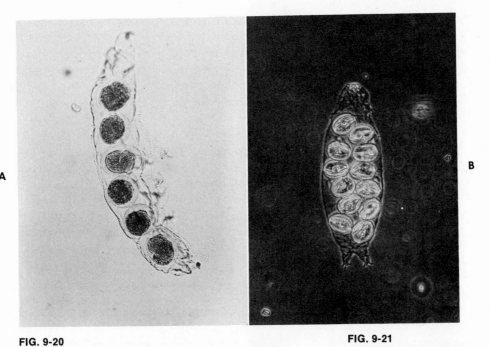

FIG. 9-20

Shed exoskeletons of freshwater tardigrade *Hypsibius* often contain eggs. **A,** Containing developing eggs. **B,** Containing eggs with developing embryos. (**A** from Sayre, S. M. 1969. Trans. Amer. Microscop. Soc. **88:** 266-274. **B** courtesy R. P. Higgins.)

FIG. 9-21

Tardigrades. **A,** *Hypsibius,* a freshwater form, grasping and feeding on a nematode worm. **B,** *Hypsibius* encysted. **C,** *Macrobiotus* in a cryptobiotic state. (**A** and **B** from Sayre, S. M. 1969. Trans. Amer. Microscop. Soc. **88:** 266-274. **C** courtesy R. P. Higgins.)

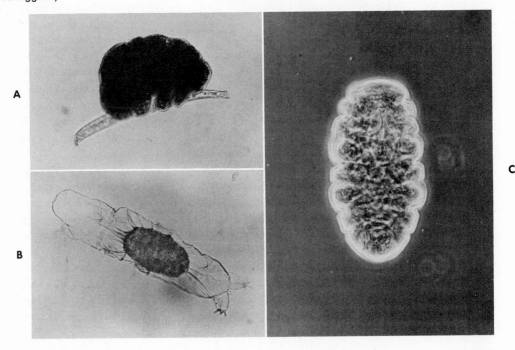

selves. Their remarkable ability to withstand desiccation, their widespread distribution, and their ability to lay eggs early in their life cycle are other interesting features.

Although E. C. Dougherty (1964) found only the species *Hypsibius arcticus* on the subantarctic Ross Island (of the Ross Dependency, Antarctica), he mentions that four or five other species had been found there by others. In his experiments he was able to revive desiccated specimens obtained from algal felt. He cultivated this species for several generations in a medium of blue-green algal felt infusion, but found that there was a deterioration that he ascribed to lack of suitable nutrients. J. H. Crowe and R. P. Higgins (1967) investigated the factors that affected the revival of the tardigrade *(Macrobiotus areolatus)* from a cryptobiotic state. They found the optimum recovery conditions from the cryptobiotic state to be a pH of 6 to 8, a temperature range of 5° to 27° C., and a dissolved oxygen range of 5 to 9 mg. per liter. The revival time was directly proportional to the duration of the cryptobiotic state. They believe that the metabolism of the cryptobiotic condition never reaches zero. They state that as a rule, marine and freshwater tardigrades do not undergo cryptobiosis.

Because of the present state of knowledge of the distribution of tardigrades, many investigations are concerned mainly with surveys of the species found in various regions. C. R. Puglia (1964) made an extensive survey of five microhabitats (lichens, moss, water, grass, and soil) in Illinois and found thirteen species, although only one species had been reported before. G. T. Riggin, Jr. (1964) collected thirteen species of semiaquatic tardigrades in moss and lichens of the Appalachian Mountains in North and South Carolina. M. M. McGinity and R. P. Higgins (1968) report that marine tardigrades are found mostly in psammolittoral regions, but some do occur in the upper limits of abyssal substrata. In 1968 twenty-eight species of marine tardigrades and fourteen different genera were known.

R. M. Sayre (1969) makes the interesting observation that certain predaceous tardigrades may be developed as biologic control agents for plant nematodes on which they normally feed. He experimented with the predaceous species *Hypsibius myrops* (Fig. 9-21, *A*) using the nematode *Panagrellus redivivus* as the prey. This tardigrade forms cryptobiotic cysts (Fig. 9-21, *B*) that could be stored and used when needed. However, a great deal more knowledge of factors governing both their entry into and revival from the cryptobiotic state would be necessary, as well as more information on specific prey-predator relationships between nematodes and tardigrades.

Common American genera of tardigrades are *Macrobiotus, Echiniscus, Milnesium,* and *Hypsibius.*

■ PHYLUM PENTASTOMIDA

The wormlike Pentastomida are bloodsucking internal parasites of carnivorous vertebrates. They live in the lungs or nasal passageways. Another name for the phylum is Linguatulida, and they are often placed in a class or subphylum under the Arthropoda. Their common name is tongue worm. Adults resemble small worms, but the larval forms are mitelike with stumpy legs, which may indicate their relationship to the arachnids.

They are found chiefly in tropic regions, where they may parasitize a large variety of vertebrates such as snakes, crocodiles, amphibians, and sometimes birds and mammals. Most adults are found in reptiles, especially snakes and lizards. Their life history usually includes an intermediate host for the larval stages (fish, rabbit, dog, etc.), and they complete their life history when the intermediate host is eaten by the definitive host. Among those that have mammals as the definitive host, the most common intermediate hosts are fishes, reptiles, and amphibians. Some have a host preference. Human infection is known in Africa and southern Europe, where man may act as an intermediate host for one stage in the life history of pentastomids.

A typical life history is illustrated by *Lingua-*

tula taenioides, which lives in the nasal passageways of carnivorous mammals. When the eggs of the adult worm leave the host in the feces or mucous secretions and hatch, the larvae climb on vegetation, where they are eaten by rabbits. In the rabbit the larvae encyst in the liver. After larval development involving several molts, the rabbit may be eaten, whereupon the parasite takes up residence in the nasal passages of the definitive host.

ADAPTATIONS AND PHYLOGENY

Most pentastomidan adaptations are about the same as those of many other metazoan parasites, such as hooks in both adults and larvae for attachment to the host. As in most internal parasites, the reproductive system is stressed above other systems, which may be absent. They are bloodsuckers, and the anterior part of the digestive tract is adapted as a pump.

The three phyla—Pentastomida, Onychophora, and Tardigrada—are often placed together as a group called Pararthropoda because they have unjointed limbs with claws (at some stage), a cuticle that molts, and some other common features. They may bridge the gap between annelids and arthropods, although this theory is disputed. The pentastomids show several arthropodan characters such as jointed appendages in the larva, skin with breathing pores (stigmata), molting cuticle, striated muscle, lobular gonads, and lack of ciliated epithelium. R. Heymons thinks the pentastomids have evolved from polychaetes because they show some features of the Myzostomidae, an aberrant group of polychaetes usually commensal or parasitic on echinoderms. *Myzostoma* has a rounded body with a fringe of limblike appendages and a protrusible proboscis for sucking food from ambulacral grooves. G. Osche suggests that the pentastomids are offshoots of the myriapods.

CLASSIFICATION

Phylum Pentastomida. Vermiform and unsegmented; body cylindric or flattened; chitinous cuticle undergoes molting; mouth with two pairs of retractile hooks; adults legless; no antennae; no respiratory or excretory system; some with complex life histories; dioecious; parasitic in respiratory passages of vertebrates; fifty to sixty species in phylum.

Order Cephalobaenida. With 6-legged larvae; adults live in mammals, birds, and reptiles. *Reighardia, Cephalobaena.*

Order Porocephalida. With 4-legged larvae; adults in snakes, larvae in mammals. *Porocephalus, Armillifer.*

GENERAL FEATURES
External anatomy

The adults of pentastomids vary from 20 to 130 mm. in length. They are elongated with cylindric or flattened bodies. There is no differentiation of head, thorax, or abdomen. The body has a thick, chitinous cuticle, which is molted during development; it is transversely ringed to give the appearance of superficial segmentation (Fig. 9-22). The anterior end bears on each side of the mouth region 4 retractile hooks, each of which is located on a short protuberance and provided with strong muscles. The type of hook and the number of body rings are important in the diagnosis of species. Body wall muscles are striated and arranged in layers very much like those of annelids. The genital pore and the anus are on the ventral side of the body. Tiny pores (stigmata) are found in the skin.

Internal anatomy and life history

The digestive tract is a straight, undifferentiated tube, with the mouth end adapted for sucking the host's blood, on which these worms live (Fig. 9-22, *A*). They have no respiratory, excretory, or circulatory organs. The simple nervous system is similar to that of annelids and arthropods, with three pairs of ganglia located along the ventral nerve cord.

The sexes are separate, and the females are much larger than the males (Fig. 9-22, *A*). The genital pore in each sex may be located ventrally near the anterior end, especially in the primitive species, or sometimes it is found posteriorly. Fertilization is internal, and the ripe

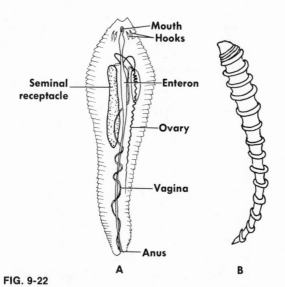

FIG. 9-22

A, *Pentastomum,* a common pentastomid in lungs and nasal cavities of snakes and other vertebrates. Female specimen is shown with certain internal structures. **B,** Female *Armillifer,* a pentastomid with pronounced body rings. In Africa, man is frequently parasitized by immature stages of this pentastomid, adults of which are usually found in lungs of snakes. Human infection may occur from eating snakes or from contaminated food or water. Adult of this genus grows to be 4 inches or more in length.

eggs of the female are discharged through the mouth of the host in the mucous secretions. In some life cycles the eggs may pass with the feces. Most pentastomids require an intermediate host in their life cycle, such as fishes or mammals. The larva has 4 or 6 legs; the adult has two pairs of hooks in the head region for its appendages. When the intermediate host is eaten by the final host, the parasitic larva is taken to the stomach and migrates by way of the esophagus and trachea to the lungs, or if they live in the nasal passages, by way of the esophagus and the pharynx. The larva matures in a few months and begins to release eggs. In the order Porocephalida the adults are found mostly in snakes that are infected by eating parasitized mammals. In the order Cephalobaenida the adults commonly occur in birds and lizards.

■ PHYLUM ONYCHOPHORA

In the past, onychophorans have been placed in the Arthropoda as a class or subphylum but are now often classified as a phylum in their own right. The group has been of special interest because it has characteristics of both the Annelida and the Arthropoda and therefore has been considered as the "missing link" between these two great phyla, which have an unmistakable relationship (Table 6). However, onychophorans are neither worms nor arthropods but have distinctive characteristics of their own. Although they comprise a small group of few species that are so much alike that the description of one applies to all, the unique organization and characters of both a primitive and specialized nature have given the little animal an unusual evolutionary significance.

The wormlike onychophorans are often cited as classic examples of discontinuous distribution. They are found in tropic and subtropic regions but only in restricted parts of those regions. Various species are found in Mexico, Central America, northern South America, South Africa, the West Indies, Malay Archipelago, India, Australia, and New Zealand. This discontinuity of occurrence indicates that they were once very widespread. It may also suggest that they are on the verge of extinction because they have very poor powers of distribution.

In regions where they are found, their habitats are in moist situations, although under dry conditions they have ways of protecting themselves and surviving. They live under logs, stones, tangled vegetation, bark, in crevices in rotting stumps, or near stream banks. They are nocturnal in their habits and are not seen out in the open during daylight.

They have a fossil record as far back as the Cambrian period. In the famous Burgess Shale deposit found in British Columbia (1911) one of the soft-bodied forms found in this unique collection was *Aysheaia,* a marine species that had a general body form and other characteristics very much like the modern onychophoran.

Table 6. Characteristics of *Peripatus* showing arthropod and annelid characteristics

Characteristics peculiar to *Peripatus*	Arthropod characteristics	Annelid characteristics
Tracheal apertures or spiracles diffused	Appendages modified as jaws	Paired segmentally arranged nephridia
Scanty metamerism	Hemocoelic body cavity	Cilia in reproductive organs
Single pair of jaws	Tubular heart with ostia	Muscular body wall
Texture of skin with transverse wrinkles and conical papillae	Presence of tracheae	Structure of eye
	Cuticle similar	Organs arranged as in annelids
Distribution and organization of reproductive organs	No coelom around gut	General nervous system
	Large size of brain	

FIG. 9-23

Peripatus in natural habitat. (Courtesy Ward's Natural Science Establishment, Inc., Rochester, N. Y.)

This fossil history plus studies on their comparative anatomy and embryology suggest that the Onychophora may represent a persistent specialized relic of an ancestral line that gave rise to many of the various groups of arthropods.

CLASSIFICATION

Phylum Onychophora. Body vermiform and somewhat cylindric; body of adult not segmented; head not distinct; skin with soft cuticle and transverse rows of fine papillae; anterior end (head) with a pair of antennae, a pair of oral papillae, and a pair of single eyes; ventral mouth with jaws; double row of short, thick legs segmentally arranged; spiracles numerous and tracheal system of unbranched tracheae; one pair of nephridia per each pair of legs; body cavity a hemocoel; nervous system of dorsal brain and a pair of ventral nerve cords; dioecious; oviparous, ovoviviparous, or viviparous; juvenile young; terrestrial.

1. Two families: Peripatidae (equatorial regions) and Peripatopsidae (subtropic or south temperate regions).
2. Seven genera: *Peripatus, Peripatoides, Eoperipatus, Opisthopatus, Paraperipatus, Peripatopsis, Oöperipatus.*
3. Seventy-eight species.

In popular usage the term *Peripatus* is often employed as the generic name for all species.

GENERAL MORPHOLOGIC FEATURES AND PHYSIOLOGY
External features

Onychophorans are vermiform with a generally cylindric body that is from 15 to 100 mm. or more in length (Fig. 9-23). The head is not distinct from the body, and the only external evidence of segmentation is the paired arrangement of the legs. The head, or anterior end, bears a pair of large, annulated preoral antennae and a pair of simple eyes. Below and slightly behind each antenna is a blunt oral papilla, between which is the buccal cavity. The mouth is located at the posterior end of the buccal cavity, which is ventral and surrounded by a circumoral fold. The buccal cavity contains a pair of horny jaws, each of which terminates in a 2-hooked structure. The jaws are modified appendages. The surface of the body bears transverse ringlike ridges or bands of small spine-bearing tubercles (Fig. 9-24).

The legs (14 to 43 pairs) are conical, lateroventral outgrowths of the body wall, and each leg terminates in a pair of sickle-shaped claws. Each sole between the claws is made up of a

FIG. 9-24

Ventral view of anterior end of *Peripatus*. (Courtesy Ward's Natural Science Establishment, Inc., Rochester, N. Y.)

varying number (3 to 6) of pads in contact with the substratum. Like the trunk, the surface of the legs bears rings of tubercles. The anus is located at the extreme posterior end, and a genital pore is present on the ventral surface between the last pair of legs.

The skin is somewhat dry and velvety in texture. The color of the different species may be gray, olive green, reddish above and light below, uniform in color, or of variegated pattern.

Locomotion in onychophorans is slow. A few segments, with their attached legs, are used at one time for movement. The legs are shortened by intrinsic muscles, swung forward, and then given a propulsive backward stroke by muscles. Successive waves of contraction and extension pass posteriorly, involving both the legs and body.

When the speed of locomotion is modified by changes of gait, faster movements involve an extension of the body (by hydrostatic pressure) and slower movements, a shortened body. As it walks, each leg undergoes a cycle similar to the parapodium of an annelid (although in *Peripatus* the legs are located more ventrally) — a protraction, a thrust, and a recovery with the leg elevated. In a faster speed the thrust stroke is faster, but a longer time is spent in raising and moving each leg forward for the next step (Fig. 9-25).

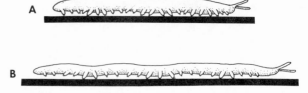

FIG. 9-25

How the onychophoran *Peripatus* moves. **A,** When worm is moving slowly, most of the legs are thrusting at any one time, by hydrostatic pressure. **B,** When moving rapidly thrust strokes are faster than flexor muscles in raising limbs and fewer legs are on the ground at one time. (Modified from M. Wells.)

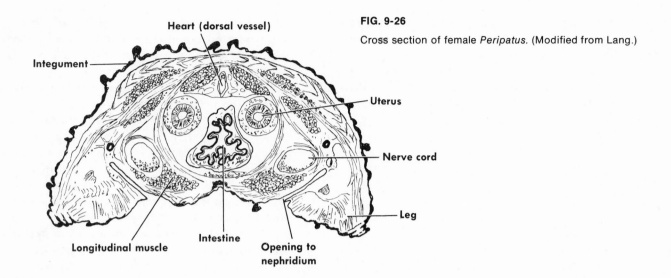

FIG. 9-26

Cross section of female *Peripatus*. (Modified from Lang.)

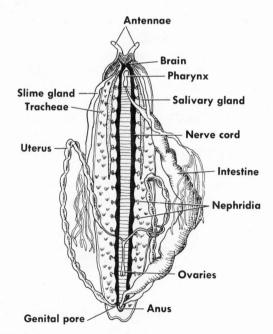

FIG. 9-27

Internal anatomy of female *Peripatus*.

Internal features

The body wall from outside to inside consists of a thin, chitinous cuticle that is not made up of movable plates; a single layer of epidermis; a thin basement membrane (dermis) of connective tissue; a musculature of circular, diagonal, and longitudinal layers; and a lining periteum. The body wall muscles are involved in changes in shape and length of the body, and not in locomotion (S. M. Manton, 1952).

Like the arthropods, the major body cavity is a hemocoel, and the coelom is restricted to cavities associated with the gonads and nephridia (Fig. 9-26). The hemocoel is divided by longitudinal partitions of tissue sheets into a dorsal pericardial sinus containing the heart; a central sinus containing the intestinal tract, slime glands, and reproductive organs; and 2 ventrolateral sinuses, each with a nerve cord, nephridia, and salivary glands.

The alimentary canal is usually a straight tube from mouth to anus (Fig. 9-27). A mouth, located at the back of the buccal cavity, is provided with a pair of jaws or mandibles, already referred to. The jaws, which are really cutting plates, move backward and forward instead of sideways and thus are an adaptation for narrow passageways. The mouth passes into a short,

329

muscular pharynx, from which an esophagus opens into an enlarged intestine, or stomach, according to some authors. The pharynx and esophagus make up the foregut, which is lined with chitin. The intestine passes back to a short, narrow rectum that opens at the anus on the ventral posterior end of the body. A pair of salivary glands, which open by a common duct into the buccal cavity, represent the only glands known with certainty to be in the alimentary canal. The large intestine is the site of digestion and absorption.

Onychophorans are mostly carnivorous and feed on insects, worms, snails, and other small forms, which they capture directly. When feeding, the animal presses its prebuccal lobes against its prey, tears it with its jaws, and sucks in juices and particles with its muscular pharynx. For defense they can discharge from their oral papillae two streams of milky fluid that hardens on contact with the air. Enemies and sometimes prey are immobilized when this fluid is directed upon them, sometimes from as much as 50 cm. away. The fluid is produced by a pair of slime glands that open at the ends of the papillae and extend posteriorly in the body (Fig. 9-27). The animals will also eat decomposing vegetation when hard pressed for food. They are capable of enduring a long fast.

They breathe by means of a tracheal system characteristic of many arthropods. However, the tracheal system in onychophorans is different from that in arthropods and probably has evolved independently. The external openings called spiracles are small pits, or epidermal pockets, which are numerous all over the surface of the body between the bands of tubercles. These spiracles are too small to be seen externally with the unaided eye. From the inside of each spiracle a bunch of fine, unbranched tracheae run parallel with each other to the tissues they supply. The tracheal tubes are supplied with tiny spiral thickenings, the same as those found in insects and other arthropods. In some species of onychophorans the spiracles are arranged in definite rows. In the

legs of some a small vesicle, which opens near the nephridiopore of the nephridium, is present and may serve as a type of gill under humid conditions.

The tracheae provide efficient respiration, but there is no means of closing the spiracles and thus preventing water loss. This has been an important factor in restricting onychophorans to moist habitats and noctornal activity and contrasts with most tracheate arthropods, which can close the spiracles. For example, water loss in an onychophoran may run eighty times as great as that in a cockroach.

The excretory system consists of a pair of nephridia for each segment or each pair of legs (Fig. 9-27). Each nephridium is made up of a tube that opens into a reduced coelomic vesicle at one end and a nephridiopore at the other end. Near the nephridiopore, or external aperture, the tubule enlarges into a bladderlike collecting vesicle, which is contractile. The nephridiopore is located on the inner base of each leg except the fourth and fifth pairs, where it opens on a papilla more distally. Cilia are present in the excretory tube near the coelomic vesicle, thus sharing with the reproductive female ducts the only place where cilia are present in onychophorans. Cilia are characteristically absent in the great arthropod group.

Not much is known about the excretory physiology of onychophorans, but there is some evidence that they are predominantly uricotelic (produce uric acid as a chief excretory product), having a water-salvaging device utilized by lizards, birds, most insects, and land snails.

The vascular system has about the same pattern as that of arthropods. A tubular heart in the form of a long, pulsating dorsal blood vessel is open at both ends and is provided with a pair of lateral ostia for each segment (Fig. 9-26). The heart lies in the pericardial sinus, and its contraction keeps the blood from the vast hemocoel moving forward and in circulation. Partitions between the 4 or 5 major sinuses have openings through which the blood passes as it circulates. The blood carries no respiratory pigments and is colorless. Some formed elements, such as amebocytes, occur in it.

The nervous system includes a well-developed bilobed brain located over the pharynx and a subesophageal ganglion, from which extend 2 widely separated ventral nerve cords connected with commissures (Figs. 9-26 and 9-27). From the brain, nerves are given off to the antennae, the eyes, and the jaws, and circumesophageal commissures run to the subesophageal ganglion. The ventral nerve cords in each segment form ganglionic swellings, from which paired nerves are given off to the legs and body wall. The antennae are abundantly supplied with tactile endings. Paired ocelli of the direct pigment-cup type are located near the base of the antennae, and each has a chitinous lens and simple retina. Sensory taste spines are found on the circumoral ridges of the mouth, and sensory receptors on the large tubercles respond to vibrations and tactile stimuli. Hygroscopic receptors on the antennae and body surface are important in orienting the animal to water vapor. Some onychophorans, such as *Peripatus,* are very sensitive to chemical stimuli.

REPRODUCTION AND EMBRYOLOGY

Onychophorans are all dioecious, the males being somewhat smaller than the females; the males may also have fewer legs. A saclike crural gland found near the nephridium of most legs in the male may play a part in reproduction. The gonads in each sex are paired and lie in the posterior part of the body above the intestine. The paired, elongated ovaries are partially fused, and each is connected to a nephridium that has been modified to form a genital tract, consisting of a seminal receptacle and a uterus. The 2 tracts, or uteri, merge together and open to the exterior by a common genital pore on the ventral posterior end of the body. The paired male testes remain separated, and each is connected to a modified nephridium that becomes an ejaculatory tube in its lower part. Both ducts join to form a common median opening in front of the anus. In the

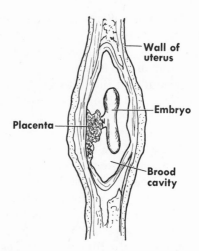

FIG. 9-28

Uterus of *Peripatus* with young and placenta. (Modified from Lang.)

common median tube the sperm are packaged into elongated spermatophores. Fertilization is internal, and sperm is deposited in the seminal receptacles of the female. In one or two species the female has no seminal receptacle, and the spermatophores are injected through the body wall into the hemocoel, where the sperm make their way to the ovaries.

Most species produce living young, except a few in Australia, which lay large yolk-laden eggs in sculptured shells. The oviducts of females that produce living young are usually enlarged into a number of uterine chambers in which the embryos develop. The young may simply develop in the uteri without maternal connection (ovoviviparous), or they may have a placental attachment (viviparous) (Fig. 9-28).

The basic pattern of embryonic development in onychophorans is similar to that of arthropods, especially to those of the lesser groups (Pauropoda, Chilopoda, and Symphyla). The coelomic cavity is well developed in the embryo. After the gastrula stage the blastopore elongates and gives rise to the mouth at its anterior end and the anus at its posterior end, while the median margins fuse together. The paired mesoblastic somites arise behind the blastopore.

331

The first pair become the preoral somites and form the mesoderm of the segment bearing the antennae. All the somites acquire a coelomic cavity and divide into ventral and dorsal parts. The ventral part migrates to the region of the appendage and becomes a part of the nephridium, but the dorsal portion approaches its partner from the opposite side at the mid-dorsal line to form the tubular heart between them. In the posterior region of the body the fusion of the dorsal coelomic cavities of the somites form the paired gonads. The remaining parts between the organs are filled with blood and become partitioned into sinuses as the hemocoel. In this way the coelom is restricted to the segmented excretory organs and the cavities of the gonads.

REFERENCES

Baltzer, F. 1931. Echiurida. In Kukenthal, W., and T. Krumbach (editors): Handbuch der Zoologie, vol. 2, part 2, sect. 9. Berlin, Walter de Gruyter & Co. An excellent authoritative account.

Baltzer, F. 1931. Sipunculida. In Kukenthal, W., and T. Krumbach (editors): Handbuch der Zoologie, vol. 2, part 2, sect. 9. Berlin, Walter de Gruyter & Co. One of the best technical discussions.

Carlisle, D. B. 1958. On the exuvia of *Priapulus caudatus*. Arkiv. Zool. series 2, **12**(5):79-81.

Chandler, A. C. 1952. Introduction to parasitology, ed. 8. New York, John Wiley & Sons, Inc. A short section with several illustrations is given to the Pentastomida in this well-known textbook.

Crowe, J. H., and R. P. Higgins. 1967. The revival of *Macrobiotus areolatus* Murray (Tardigrada) from the cryptobiotic state. Trans. Amer. Microscop. Soc. **86**:286-294.

Cuénot, L. 1932. Tardigrades. Faune de France **24**:1-96. An excellent French account of the tardigrades.

Cuénot, L. 1949. Les Onychophores. In Grassé, P. P. (editor). Traité de zoologie, vol. 6. Paris, Masson et Cie.

Dawydoff, C. 1959. In Grassé, P. P. (editor). Traité de zoologie, vol. 5, part 1. Paris, Masson et Cie. One of the best general treatments of the echiurids. Extensive bibliography.

Dougherty, E. C. 1964. Cultivation and nutrition of micrometazoa. II. An anarctic strain of tardigrade *Hypsibius arcticus* (Murray, 1907; Marcus 1928). Trans. Amer. Microscop. Soc. **83**:7-11.

Fange, R., and A. Mattisson. 1961. Function of the caudal appendage of *Priapulus caudatus*. Nature (London) **190**: 1216-1217. This paper presents evidence to show that the appendages have respiratory functions.

Fisher, W. K., and G. E. MacGintie. 1928. The natural history of an echiuroid worm. Ann. Magaz. Natur. Hist., series 10, **1**:204-213. A good presentation of echiurid habits and behavior.

Grassé, P. P. (editor). 1949. Traité de zoologie, vol. 4. Paris, Masson et Cie. An authoritative account of the onychophorans.

Gregory, W. K. 1951. Evolution emerging, vol. 1. New York, The Macmillan Co. This work compares the structure of *Peripatus* with that of insects.

Heymons, R. 1935. Pentastomida. In Bronn, H. G. (editor). Klassen und Ordnungen des Tierreichs, vol. 5, part 4. Leipzig, Akademische Verlagsgesellschaft. Perhaps the best account of the pentastomids is given in this monograph.

Hill, H. R., 1960. Pentastomida. In McGraw-Hill encyclopedia of science and technology, vol. 9. New York, McGraw-Hill Book Co., Inc. A concise account of the group.

Hyman, L. H. 1959. The invertebrates; smaller coelomate groups, vol. 5. New York, McGraw-Hill Book Co., Inc. The most up-to-date account. Excellent illustrations and bibliography.

Hyman, L. H. 1951. The invertebrates: Acanthocephala, Aschelminthes, and Entoprocta; the pseudocoelomate Bilateria, vol. 3. New York, McGraw-Hill Book Co., Inc. The author places the Priapulida as a class under the pseudocoelomate Aschelminthes, but more recent investigation shows the priapulids to be true coelomate animals. Dr. Hyman recognizes three species, but others have since been found.

Kukenthal, W., and T. Krumbach (editors). 1926. Handbuch der Zoologie; Onychophora, vol. 3, part 2. Berlin, Walter de Gruyter & Co. A comprehensive account of the group.

Lang, K. 1948. Contribution to the ecology of *Priapulus caudatus* Lamarck. Arkiv. Zool. **41A**(5):1-12.

MacGinitie, G. E. and N. MacGinitie. 1968. Natural history of marine animals. ed. 2. New York, McGraw-Hill Book Co. A fine section on the Echiurida, with descriptions of the physiology and behavior of *Echiurus* and *Urechis*.

Manton, S. M. 1970. Onychophora. In Gray, P. (editor). Encyclopedia of the biological sciences, ed. 2. New York, Reinhold Publishing Corp. This author considers their basic pattern as definitely that of an arthropod.

Manton, S. M. 1952-1961. The evolution of arthropodan locomotory mechanisms (several papers). J. Linn. Soc. (zool.), part 3.

Marcus, E. 1929. Tardigrada. In Bronn, H. G. (editor). Klassen und Ordnungen des Tierreichs, vol. 5, part 4. Leipzig, Akademische Verlagsgesellschaft. The most important monograph on the tardigrades.

Marcus, E. 1959. Tardigrada. In Edmondson, W. T. (editor). Ward and Whipple's fresh-water biology, ed. 2. New York, John Wiley & Sons, Inc.

McGinity, M. M., and R. P. Higgins. 1968. Ontogenetic variation of taxonomic characters of two marine tardigrades with the description of *Batillipes bullacaudatus* N. sp. Trans. Amer. Microscop. Soc. **87**:252-262.

Newby, W. W. 1940. The embryology of the echiuroid worm *Urechis caupo*. Mem. Amer. Phil. Soc. **16**:1-219. An important embryologic study with evidences for considering the Echiurida as a separate phylum.

Norrevang, A. 1963. Fine structure of the solenocyte tree in *Priapulus caudatus* Lamarck. Nature (London) **198:** 700-701.

Osche, G. 1963. Die systematische Stellung und Phylogenie der Pentastomida. Zeitschr. Morphol. Okol. Tiere **52:** 487-596.

Peebles, F., and D. L. Fox. 1933. The structure, functions, and general reactions of the marine sipunculid worm *Dendrostomum*. Bull. Tech. Ser. **3:**201-224. The burrowing habits and general reactions of the animal to experimental stimuli are described.

Pennak, R. W. 1953. Fresh-water invertebrates of the United States. New York, The Ronald Press Co. The best American account of the group is given on pp. 240 to 255. A good general description and keys to the common species.

Pickford, G. 1947. Sipunculida. Encyclopaedia Britannica, vol. 20. A concise, well-written account.

Puglia, C. R. 1964. Some tardigrades from Illinois. Trans. Amer. Microscop. Soc. **83:**300-310.

Ricketts, E. F., and J. Calvin. 1952. Between Pacific tides, ed. 3. Stanford, Stanford University Press. Summarizes the movements of *Urechis caupo* in its burrow and its use of the slime net in capturing its prey.

Riggin, G. T., Jr. 1964. Tardigrades from the southern Appalachian Mountains. Trans. Amer. Microscop. Soc. **83:** 277-282.

Sayre, R. M. 1969. A method of culturing a predaceous tardigrade on the nematode *Panagrellus redivivus*. Trans. Amer. Microscop. Soc. **88:**266-274.

Shapeero, W. L. 1961. Phylogeny of Priapulida. Science **133** (3456):879-880. The author presents evidence why the Priapulida should be placed among the true coelomates and why they should be considered a separate phylum.

Shapeero, W. L. 1962. The epidermis and cuticle of *Priapulus caudatus* Lamarck. Trans. Amer. Microscop. Soc. **81:** 352-355.

Smith, R. I., E. A. Pitelka, D. P. Abbott, and F. M. Weesner. 1957. Intertidal invertebrates of the central California coast. Berkeley, University of California Press. A short section with a key is devoted to the sipunculids of the Pacific coast.

Swedmark, B. 1964. The interstitial fauna of marine sand. Biol. Rev. **39:**1-42.

Swellengrebel, N. H., and M. M. Sterman. 1961. Animal parasites in man. Princeton, D. Van Nostrand Co., Inc. An account of *Linguatula* and *Armillifer* as pentastomid parasites of man.

Tétry, A. 1959. Classe des Sipunculiens. In Grassé, P. P. (editor). Traité de zoologie, vol. 5, part 1. Paris, Mason et Cie.

Tiegs, O. W., and S. M. Manton. 1958. The evolution of the Arthropoda. Biol. Rev. **33:**255-337.

Van der Land, J. 1970. Systematics, zoogeography, and ecology of the Priapulida. No. 112 Zoologische Verhandelingen Uitgegeven door het Rijksmuseum van Natuurlijke Historie te Leiden. Leiden, E. J. Brill. An excellent and up-to-date monograph.

Phylum Mollusca

A PROTOSTOME EUCOELOMATE PHYLUM

GENERAL CHARACTERISTICS

1. Next to the arthropods, Mollusca is the largest described invertebrate phylum, numbering perhaps 100,000 or more species. At least 35,000 species of fossils are known. There may be more species existing among the nematodes and protozoans, but as yet not as many species have been described in those groups.

2. The plan of structure involves a body divided into two functionally distinct regions – the upper region (visceropallium) composed of a visceral hump and an overhanging mantle specialized for mucous secretion and ciliary action and the lower region (head-foot complex) composed of the head and a foot whose activity is chiefly muscular.

3. Although mollusks have diverged into numerous evolutionary lines, the body plan of all can usually be traced back to a single ancestral pattern. The plan of structure and function is uniform throughout the phylum, but there is no standard shape. Homologies involving the mantle and gills and functional homologies stressing ciliary and mucous mechanisms in feeding and other processes are easily established.

4. As a group, mollusks have added or stressed a soft extension of the outer layer of body wall (the many-functioned *mantle*), the flexible file-like tongue (the *radula*), the highly protective *shell*, the ventral muscular *foot*, an effective respiratory system of *gills* or *lungs*, and efficient *eyes*.

5. Although most mollusks are marine, one group – the gastropods – has been successful also on land and in fresh water. Bivalves are basically filter-feeders and so are unknown on land. Cephalopods differ from other mollusks in their very high level of efficiency in locomotion and behavior. They stand high in invertebrate evolution.

6. Mollusks range from forms that are only a few millimeters in diameter to the largest of all invertebrates many meters long.

PHYLOGENY AND ADAPTIVE RADIATION

1. Little is known with certainty about the relationship of mollusks to other phyla. A possible connection between flatworms (Turbellaria) and mollusks is seen in the shell-less Aplacophora (solenogastrids), which have some of the features of chitons and also certain reproductive tubes similar to those of primitive flatworms. A relationship with annelids is evidenced by a similar trochophore larva found in both phyla. The original ancestral stock of mollusks may have given rise to two series of schizocoelomates; one of these was unsegmented (mollusks) and the other was arranged metamerically (annelids). The discovery in 1952 of *Neopilina* with its somewhat superficial segmentation has strengthened the belief that annelids and mollusks came from a common stem. However, segmentation is absent in molluscan embryos and in larvae of *Neopilina*.

Many zoologists, moreover, think the segmentation in *Neopilina* is not the same type as that in annelids but represents an aberrant example of convergent evolution.

2. Mollusks have a basic design of two interacting symmetries—a visceropallium composed of viscera and overhanging mantle, and a head-foot complex. The head-foot part has a definite bilateral symmetry; the visceropallium at first had radial symmetry, with a tendency later toward biradial symmetry. This basic design has been modified in various ways by different groups of mollusks. For instance, the head and foot may be fused together (cephalopods), or the head may be absent altogether (pelecypods).

The mantle has been one of the most versatile structures in molluscan evolution and is modified to form the mantle cavity, gills or lungs, siphons or apertures, locomotory structures, feeding processes, etc. Torsion of the body and coiling of the shell (gastropods) produce an asymmetric growth and changes in the location of many basic structures. The shell has many evolutionary adaptations, such as heavy shells for withstanding stress and pressure, boring and burrowing shells for penetration, swimming shells for locomotion, and reduced (or absent) shells for preventing encumbrance. In cephalopods the body is lengthened along the dorsoventral axis, which becomes the functional anteroposterior axis. Whereas the foot is the chief organ of locomotion in gastropods and pelecypods, locomotion in cephalopods involves jet propulsion of water from the mantle cavity through a funnel.

GENERAL SURVEY OF MOLLUSKS

The mollusks include such forms as chitons, clams, mussels, oysters, snails, slugs, octopuses, squids, nautiluses and some others. The group has enormous diversity, both in shape and size.

Their ecology is also very diversified, for they have been able to live in nearly all kinds of habitats—from the depths of the oceans to some of the highest altitudes recorded for animal life.

The shell-bearing forms are restricted somewhat by the calcium content of their environments, but the shell-less mollusks can live even in regions that are more or less devoid of lime. As a consequence, mollusks are found in the sea, brackish water, fresh water, and on land. They share with the arthropods the capacity to adapt themselves to terrestrial life better than most groups of invertebrates. They include some of the slowest and most sluggish of animals and some of the swiftest invertebrates. In structural and functional adaptations, mollusks rank with the arthropods because some of the most unique invertebrate behavior patterns are found among them. Nutritionally they range from primitive filter feeders to highly carnivorous predators.

Most mollusks live in the sea. Three of their classes—Amphineura, Scaphopoda, and Cephalopoda—are not represented at all on land or in fresh water. Both gastropods and bivalves have colonized fresh water; however, they are fewer in number than their representatives in the sea. The chief physiologic requirements for conquest of the land involve air breathing, temperature regulation, and water control. Bivalves have successfully invaded fresh water but their ciliary feeding method has prevented their colonization of land. Only the gastropods have successfully invaded terrestrial habitats. Although both prosobranch and pulmonate gastropods are found in fresh water, the larger number belong to the pulmonates.

Mollusks are coelomate, unsegmented (except class Monoplacophora), and bilaterally symmetric (modified in some) animals. The mollusk typically has a head, a visceral mass, a mantle, a shell, and a foot. The head is located at the anterior end; the compact visceral mass is enclosed in a soft integument; the mantle is a modified dorsal part of the integument and hangs down as free folds or flaps around the body to form a mantle cavity; the calcareous shell is secreted by the mantle; and the ventral muscular foot, modified from the integument, is the organ of locomotion. All these structures

FIG. 10-1

Some of many shell varieties among mollusks. (From Encyclopaedia Britannica film, "Mollusks.")

FIG. 10-2

Neopilina, a monoplacophoran, ventral view. Note mouth, anus, and five pairs of gills surrounding foot. (Modified from Lemche.)

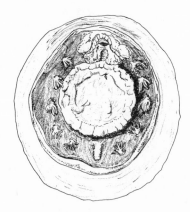

may be modified in various ways in the different groups of mollusks. A head, shell, or mantle may be lacking altogether in some. The shell may be bivalved, univalved, spiral, or cone shaped (Fig. 10-1). Most mollusks have a rasping organ, the radula. This is a kind of toothed tongue that bears denticles or teeth.

The exact status of the monoplacophorans is very much in doubt (Fig. 10-2). Their metamerism is slightly superficial, involving only some of the systems and lacking external segmentation of the body wall altogether. At present they cannot be considered a basic mollusk type because of their irregular metamerism and isolated case. If metamerism were a primitive molluscan feature, to account for its total absence in other mollusks would be difficult. All this may be interpreted to mean that metamerism in the monoplacophorans has been

secondarily acquired. More embryologic evidence may shed some light on the problem.

Stronger evidence for the annelid affinity is shown perhaps by the resemblance between the embryology of mollusks and that of polychaetes. A similar spiral cleavage is found in both groups, with the exception of the cephalopods.

The relations between the various molluscan classes are also obscure. Although logical explanations can be made for the modifications of most groups from the hypothetical ancestral type, these throw little light on possible interrelationships. Each class seems to have distinct evolutionary tendencies. The amphineurans have departed least from the primitive plan, although their fossil record is not so old as that of more specialized groups. Convergence of characters may well add to the confusion about relationships. The gastropods and the cephalopods share many characters, although there is a wide gulf in basic plan between the two classes. At present there are few logical reasons for grouping the classes according to relationships. The discovery (1959) of a bivalved gastropod, *Tamanovalva limax*, off the coast of Japan, and similar ones elsewhere, may suggest aspects of relationship of great interest. This type has the characteristic valves of a bivalve and the sluglike head and foot of a gastropod. The evolutionary appraisal of this and related species awaits further study.

Within the phylum two classes, the gastropods and the pelecypods (bivalves), each demonstrate some rather striking phylogenetic relationships and evolutionary differentiations among the members of the class. The most primitive basal gastropods are the Prosobranchia with torsion and a helical shell. Prosobranchs exhibit two grades—those with bipectinate (aspidobranch) gills (2 rows of gill leaflets on the branchial axis) and those with pectinibranch gills (1 row of gill filaments). From the pectinibranchs the evolutionary lines lead to two subclasses—the Opisthobranchia and the Pulmonata. The evolutionary trends in the Opisthobranchia have led to a reduction or loss of the shell, detorsion, loss of streptoneury, and reduction of visceral mass. The pulmonates retain many of the prosobranch characteristics but have lost streptoneury, and some are shell-less. Most evident evolutionary changes in gastropods have been associated with torsion and modifications of the

respiratory organs and foot. The fossil record indicates that the prosobranchs date from the Paleozoic era, whereas the Opisthobranchia and the Pulmonata appeared in the Carboniferous period.

The bivalves have been a slowly evolving group of organisms. Many date back as far as the early Paleozoic, whereas others have arisen in the Recent epoch. The Protobranchia are considered the most primitive group of existing bivalves, although they have some specialized characters. The most primitive of the living bivalves is the protobranch *Nucula*. *Nucula* has a pair of equal adductors and two pairs each of protractor and retractor muscles. Its gill structure is not unlike that of the gastropod *Haliotis* (Fig. 10-23).

Except in the Septibranchia, there are 2 gills in all bivalves. The primitive form of the gill consists of a vascular axis carrying a series of filaments on opposite sides. This is the type found in the Protobranchia. From this type of gill has emerged the various modifications found in the different groups (Fig. 10-40). These will be described later.

Ancestral form

Students of mollusks have reconstructed a hypothetical primitive mollusk (Fig. 10-3) which is supposed to have existed in Precambrian seas. Such a generalized ancestor was bilaterally symmetric with a dorsally arched body. Its head bore a pair of tentacles with ocelli at their bases, and its mouth was provided with a protrusible radula similar in structure and arrangement to that in modern snails. A dorsal mantle (pallium) extended over the body and secreted the arched, horny shell. Ventrally, a flat, muscular creeping foot was attached by muscles that could draw the foot into the protective shell.

Between the mantle and foot at the posterior end was a mantle cavity containing paired gills, openings to the nephridia (kidney), and the anus. Within the axis of each gill, blood vessels provided for the exchange of respiratory

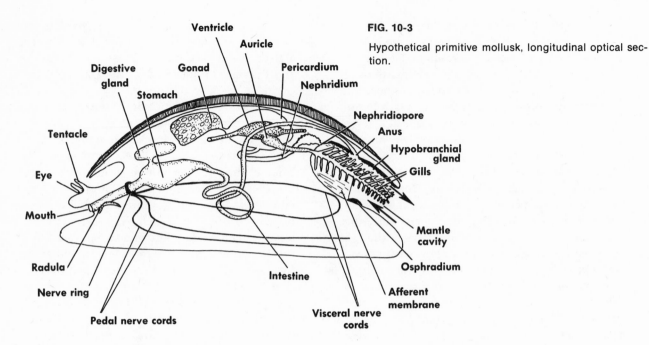

FIG. 10-3

Hypothetical primitive mollusk, longitudinal optical section.

gases. Water currents were drawn into the cavity by beating cilia on the gill filaments. A hypobranchial gland between each gill and the rectum helped carry waste particles into the dorsal branchial chamber for rejection. A small sensory patch, the osphradium, at the base of each gill tested the chemical nature of the incoming current.

The digestive system was lined with cilia and mucous glands. The posterior part of the stomach was differentiated into a style sac and a ciliated groove extended along the stomach floor. The food of this early ancestor was probably algae and other material scraped from rocks. Digestion may have been primarily intracellular. The early coelomic cavity was mostly the pericardium, surrounding the three-chambered heart. Blood circulation was similar to that of present-day mollusks. There were a pair of gonads and a pair of metanephridia, which emptied their products into the mantle cavity. The nervous system was made up of a central nerve ring around the esophagus, with two pairs of longitudinal nerve cords, one running to the

mantle and viscera and the other to the muscles of the foot. Nerve cells were probably diffused and not gathered into ganglia. Sense organs were represented by the tentacles, ocelli, statocysts in the foot, and osphradia.

From such a hypothetical ancestral form the various groups of mollusks could have evolved by adaptive radiation into the diversity demonstrated by both the fossil record (Fig. 10-4) and the existing classes (Fig. 10-5).

FIG. 10-4

Some fossil mollusks. **A** to **C**, Gastropods. **A**, *Hormotoma,* middle Ordovician. **B**, *Trepospira* and, **C**, *Worthenia,* both Pennsylvanian. **D**, Pelecypod *Grammysia,* middle Devonian. **E**, Cephalopod *Scaphites,* an Upper Cretaceous ammonoid. All shown at approximately natural size.

FIG. 10-5

Molluscan types. **A**, Class Amphineura. **B**, Class Monoplacophora. **C**, Class Scaphopoda. **D**, Class Pelecypoda. **E**, Class Gastropoda. **F**, Class Cephalopoda. Intestinal tract, gills, heart, shell, and cerebral ganglia are indicated in diagrams.

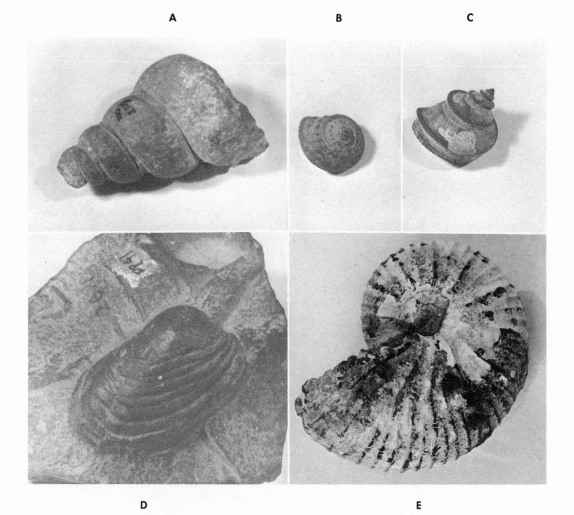

FIG. 10-4

For legend see opposite page.

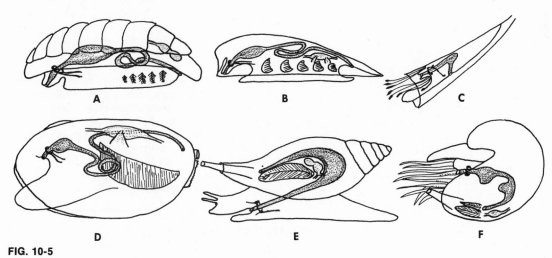

FIG. 10-5

For legend see opposite page.

CLASSIFICATION.

Phylum Mollusca. Body bilaterally symmetric with some modifications; chiefly unsegmented; coelomic cavity reduced; body soft and covered by a mantle or extension of the outermost layer of the body wall, which often secretes a calcareous shell of one to several parts; usually provided with a head and ventral muscular foot; mantle cavity, produced by mantle flaps, encloses gills (ctenidia) or lunglike structures; digestive system complete, U-shaped or coiled, and provided with a radula (usually), digestive glands, and an anus opening into the mantle cavity; excretory system of one or two pairs of nephridia (kidney) connecting pericardium and mantle cavity; circulatory system of a heart consisting of a ventricle and paired auricles, a pericardium, an arterial system, and an open venous system expanded into a hemocoel; nervous system of a circumesophageal ring modified usually into three pairs of ganglia (cerebral, pedal, and visceral) with longitudinal and cross connectives; sense organs of statocysts, touch, vision, smell, and variously modified; sexes separate (usually) or monoecious (a few); 1 or 2 gonads with ducts; fertilization internal or external; determinate and unequal total cleavage; development with larval stages (veliger, trochophore, or glochidium) or direct; terrestrial, freshwater, or marine.

Class Monoplacophora. Single symmetric shell; replication of certain organs (apparent metamerism). *Neopilina.*

Class Amphineura. Body elongated, with shell of 8 plates or none; mouth and anus terminal.

Subclass Polyplacophora. Large, flat foot and convex dorsal surface of 8 transverse calcareous plates. *Chiton.*

Subclass Aplacophora (Solenogastres). Wormlike body with mantle completely investing the body; no shell. *Chaetoderma.*

Class Gastropoda. Body asymmetric with spiral shell (usually); visceropallium typically with torsion of 180 degrees.

Subclass Prosobranchia (Streptoneura). Body with visceral torsion, shell and operculum; visceral loop in form of figure 8; gills anterior to heart.

Order Archaeogastropoda (Aspidobranchia). *Patella, Haliotis.*

Order Mesogastropoda (Pectinibranchia). *Littorina, Crepidula.*

Order Neogastropoda. *Urosalpinx, Busycon.*

Subclass Opisthobranchia (Euthyneura). Body with 1 auricle and 1 kidney; gill posterior to heart; shell reduced or absent; uncrossing of visceral loop.

Order Cephalaspidea. *Bulla, Philine.*

Order Anaspidea. *Aplysia.*

Order Thecosomata. *Cavolina, Limacina.*

Order Gymnosomata. *Pneumoderma.*

Order Notaspidea. *Pleurobranchus.*

Order Acochlidiacea. *Acochlidium.*

Order Sacoglossa. *Elysia.*

Order Nudibranchia. *Doris, Aeolis.*

Subclass Pulmonata. Gills absent; mantle cavity fitted for lung; nervous system concentrated; shell a simple spiral or absent; no operculum.

Order Basommatophora. *Planorbis, Physa.*

Order Stylommatophora. *Helix, Polygyra, Limax.*

Class Pelecypoda (Bivalvia). Body with head reduced and no radula; enlarged gills; shell of 2 calcified valves; compressed foot.

Subclass Protobranchia. Gill filaments not folded; foot with flat ventral surface; feeding proboscides common. *Nucula, Yoldia.*

Subclass Lamellibranchia. Gill filaments folded and reflected forming two-sided lamellae; adjacent filaments united by lamellar junctions.

Order Taxodonta. *Arca.*

Order Anisomyaria. *Mytilus, Ostrea, Pecten.*

Order Heterodonta. *Cardium, Mercenaria, Donax.*

Order Schizodonta. *Unio, Lampsilis, Anodonta.*

Order Adapedonta. *Ensis, Mya.*

Order Anomalodesmata. *Pandora.*

Subclass Septibranchia. Gills transformed into muscle pumping septum between inhalent chamber and suprabranchial cavity. *Cuspidaria, Poromya.*

Class Scaphopoda. Body in tubular shell open at both ends; head with many captacula; no gills; foot cylindric and pointed. *Dentalium, Cadulus.*

Class Cephalopoda. Body with bilateral symmetry and a foot divided into arms or tentacles provided with suckers; body greatly lengthened along dorsoventral axis; pallial funnel for jet propulsion.

Subclass Nautiloidea. A many-chambered shell, coiled or straight; head with tentaculate appendages, but lacking suckers; two pairs of gills and two pairs of nephridia; "pinhole" eye; mostly fossil. *Nautilus.*

Subclass Ammonoidea. Extinct group; external shells with complex septa and sutures. *Baculites.*

Subclass Coleoidea. Shells internal or absent; mantle a sac covering the viscera; head with 8 sucker-bearing arms and (sometimes in addition) a pair of retractile tentaculate arms.

Order Decapoda. *Sepia, Loligo, Architeuthis.*

Order Octopoda. *Octopus, Argonauta.*

Order Vampyromorpha. *Vampyroteuthis.*

SOME ECOLOGIC RESEARCH

The most primitive mollusks are aquatic, and the fossil record shows that they originated in the sea. Most mollusks are still marine, but about a third of all species live on land and in fresh water as a result of their invasion of those regions. This terrestrial invasion began in the Devonian period and has continued ever since. Although evolutionary diversification in those invaders has taken place more rapidly among strictly land forms, there has also been a slow evolution of freshwater types, producing considerable numbers but less variety of form and behavior than among the land dwellers. It is thought that this invasion occurred through littoral zones and estuaries.

Of the two mollusk classes that were able to overcome the handicaps of a radically new type of habitat (land and fresh water), the Pelecypoda must have entered by the estuarine route alone. The estuary is not an easy route by which to gain access to fresh water. Some authorities believe that only a restricted number of animals have succeeded in passing through the estuaries, which support few types but often considerable populations of animals.

Estuaries are characterized by brackish water of varying salinity and tide cycle fluctuations. There are also many changes in estuarine substrates because of the constant deposition of sediment, silt, mud, etc. by both sea tides and continental streams. Oxygen tensions and acidity ranges are also wider in brackish regions. One of the chief problems of those forms that live in estuaries is osmoregulation. Experimental work (L. C. Beadle and J. B. Cragg, 1940) shows that marine animals, in making the transition to fresh water, at first retain blood salts of sea water concentrations and gradually shift to lower blood salt concentrations as their tissues build up a tolerance and hypotonic urine regulation. Bivalves are probably less subjected to the fluctuations of an estuary because of their semiburrowing habits in the estuarine substratum. However, it may be stated that no marine species could pass into a land or freshwater habitat without an "internship" in littoral or estuarine existence (C. W. Cooke).

Although the mollusks are one of the most adaptive of all animal groups, as evidenced by their wide distribution in altitudinal ranges and varied habitats, they do have marked limitations in acclimation to rapid environmental changes. Recently C. M. Yonge has given an account of the changes in molluscan life brought about by the great dyke across the Zuider Zee in Holland. One of the effects of the construction of this dyke was to transform a shallow region of brackish water, considerably lower in salinity than sea water, into fresh water. Brackish water has its own unique physiologic properties, to which animals that live there must adjust themselves. Within 2 years after the erection of the dyke, all the marine mollusks (periwinkles, mussels, and cockles) were dead in the transformed region. In their place a number of freshwater mollusks from rivers invaded the region formerly occupied by the marine forms. Two bivalves belonging to the genera *Dreissena* and *Anodonta* were extremely mobile in entering the new region. Freshwater snails have been slower, although some parthenogenetic species have a foothold. Sediment carried down to the freshwater lakes formed by the reclaimed land was actually an advantage to molluscan life, especially to those whose habits are adapted to substrata of a semisolid medium.*

On the other hand, certain dykes in Holland were destroyed in World War II, and sea water flowed into regions formerly occupied by fresh water. Before the dykes were rebuilt, a number of marine bivalves such as *Mytilus*, *Cardium*, and *Mya* established themselves in the new saline region. Mollusks without pelagic larvae or those that could not swim, such as certain gastropods, had little opportunity to establish themselves. Some mollusks (e.g., *Urosalpinx*) have, however, a wide range of tolerance to different concentrations of salinity (H. Federighi).

Although accounts of ecologic experimentation on mollusks are numerous, they are scattered; and few attempts have been made to integrate them into general summaries.

---

*C. M. Yonge: Zuider Zee to Ijsselmeer, Times Sci. Rev., Winter, 1965, p. 9.

Temperature ranges of various mollusks have been determined by many investigators. R. G. Evans (1948) found that British littoral species could withstand temperature ranges well beyond those they would encounter under natural conditions. Lethal temperature effects show many variations in the different mollusks. There seems to be a correlation between the normal living temperatures of bivalves and the temperatures that are lethal to them on heating, although some experiments are not in agreement with this. A freshwater mollusk often can withstand the same temperature range as that of general metazoan metabolism. Ordinarily, temperatures beyond 45° C. are lethal to many mollusks. The time factor of exposure is also important in thermal determinations. Some species will react much more quickly than others.

Size and growth ranges also depend on many factors. C. R. Boettger (1950) found that molluscan populations had smaller-sized individuals in lower salinities, and J. S. Gutsell (1931) showed that *Pecten* grew much faster in swifter currents than in slower ones. P. A. Dehnel (1955) demonstrated that rates of gastropod growth were correlated with the latitudes in which the populations were normally found. Embryos and larvae of certain prosobranchs from northern latitudes (Alaska) grew two to nine times as great as those from southern populations (California) at a comparable temperature. It is interesting to note that, although biologic processes are determined chiefly by temperature, yet arctic or subarctic poikilotherms grow and metabolize more rapidly than do those in warmer regions. W. R. Coe (1942) showed that dwarf forms of *Crepidula* will grow into much larger ones when placed in suitable environments. No species of animal has an arctic-to-tropic distribution, but some have a wider latitudinal range than others. Arctic mollusks are chiefly infauna but show compensatory changes in growth and metabolic rates at different latitudes.

Land snails perhaps show the greatest range of physiologic factors that determine their existence and survival. The chief factors that restrict their lives are temperature, humidity, and physiochemical aspects of their substrata. However, little is known about the limits to which an animal can be acclimated to temperature. Those with shells require calcium, but many without shells (e.g., slugs) may be found where the soil is deficient in this mineral. Poikilothermous slugs and snails often maintain an internal temperature below that of their surroundings by water loss from their surfaces. E. Segal (1961) finds that the slug is able to adapt to different levels of environmental temperatures (2° to 30° C.) by increasing or decreasing oxygen consumption. This investigator, working on the pulmonate gastropod *Limax flavus*, maintained animals at 2°, 5°, 10°, 20°, and 30° C.; after they were given time to acclimate at each temperature, their oxygen consumptions were measured at the same temperatures. He found that the animals that were acclimated to temperatures of 2° and 5° C. consumed more oxygen than those kept at 20° and 30° C., but that those at 10° C. had the highest rate of oxygen consumption. He therefore concluded that 10° C. represented the optimal acclimation temperature for this species. He also established the limits of acclimation for this pulmonate at 2° and 30° C. He has also summarized many of the problems of acclimation in other mollusks subjected to specific tests.

The wide range of physiologic variations that have been reported may be due not only to phenotypic differences but to genetic differences as well (C. L. Prosser).

GENERAL MORPHOLOGIC FEATURES AND PHYSIOLOGY

Although each class of mollusks has its own distinctive features of morphology, the basic molluscan plan makes it possible to present certain generalized descriptions of their characteristic organs. More details of group modifications will be presented in the discussion of the different classes.

External anatomy

The body wall plays an important role in all animals, but especially so in mollusks. Modifications of the body wall structure are chiefly responsible for the varied assortment of pattern structures in this extensive phylum. Its varied modifications involve such structures as the mantle, the tentacles, and the foot. Each of these may have its special adaptations, such as the sensory or prehensile nature of the tentacles, the locomotory aspects of the foot, and the varied secretory mechanisms of the mantle for shell formation, swimming devices, food collecting, etc. Each of these adaptations depends on morphologic, or structural, modifications of the various tissues that make up the body wall.

The shell is one of the most characteristic structures of mollusks, although it may be reduced or absent altogether in some. Its formation has been the subject of many investigations because the process of its origin and development involves many problems of crystal formation and deposition, calcification, and synthesis of components. In its structural organization there are many variations among the different taxa. Although the shell is secreted by the mantle in all mollusks, there are variations in the way the components are put together.

The formation of the shell involves three major processes: the synthesis of the organic matrix of conchiolin (an albuminoid material) and calcium carbonate formation; the secretory activity of the mantle in forming the shell components; and the arrangement of the crystalline layers and the deposition of crystals. According to K. M. Wilbur, the organic matrix is first secreted in soluble form before being deposited as a sheet layer on the inner shell surface. Calcium carbonate is laid down in layers, alternating with the organic matrix. First laid down as small crystals, these grow and unite to form a crystalline layer that eventually assumes the character of columns—the so-called middle prismatic layer (Fig. 10-6). The crystalline columns are laid down at right angles to the inner laminated layer. The calcium

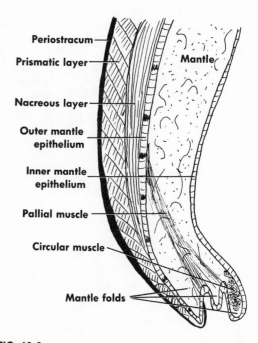

FIG. 10-6

Vertical section of shell and mantle of bivalve (semidiagrammatic).

carbonate may occur as calcite or aragonite, which are chemically identical but have different systems of crystallization. The forward edge of the mantle, which secretes both organic and inorganic components, forms the growing edge of the shell. The thickness of the shell is formed by the epithelial cells on the outer side of the mantle (Fig. 10-6). The innermost, or nacreous, layer (mostly in the form of aragonite of calcium carbonate) may be laid down as plates forming prisms that reflect light. The nacreous layer is secreted by scattered cells all over the outer lobe of the mantle. If a foreign substance gets between the mantle and shell, concentric layers of narcreous shell may be formed around it, producing a pearl. The outer layer, the periostracum, is composed of horny conchiolin and protects the shell from the corroding action of acid. The periostracum is secreted between the outer and middle lobes of the mantle. The shell reaches its greatest development in the gas-

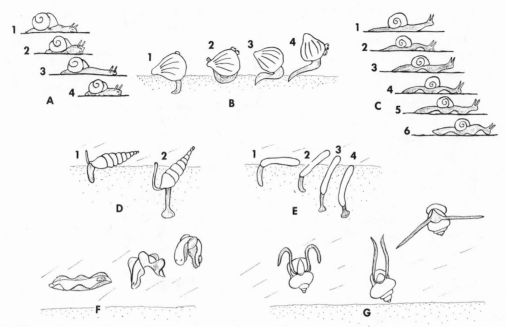

FIG. 10-7

Locomotion in mollusks. **A,** Creeping *(Helix aspersa).* **B,** Leaping *(Laevicardium).* **C,** "Galloping" *(Helix dupetith-ouarsi).* **D,** Burrowing *(Terebra).* **E,** Burrowing *(Ensis).* **F,** Swimming by means of muscular waves of notal margin and flexion of body in sagittal plane *(Hexabranchus).* **G,** Swimming by means of parapodia, using upward recovery stroke and downward effective stroke *(Limacina,* a pelagic pteropod). (Modified from Morton.)

tropods and the bivalves, both groups showing a similarity of plan.

The molluscan foot is adapted for attachment to a substratum, for locomotion, or for both these functions. The foot has been exploited in various ways and is found in diverse forms and shapes among mollusks. Its primitive form has been generally an elongated, broad, solelike structure with successive waves of contraction for creeping locomotion. Among its modifications are the attachment disk of the limpets, the compressed hatchet foot of the bivalves, and the many divisions of the anterior foot of the cephalopods. Besides these types there are many other adaptive forms of the foot. In most mollusks the foot is the only organ of locomo-

tion. *It corresponds to the ventral body wall of most other animals.* The foot in most mollusks has great powers of expansion and contraction, using the hemoskeleton of blood turgidity, together with an extensive system of muscles to change the shape of the foot for adaptive purposes. For attachment and locomotion of many of the mollusks, the action of the foot is aided by the abundant mucus, which can act either as an adhesive or as a mucous sheet over which the foot can glide by ciliary action.

Methods of locomotion are as diverse as the form of the foot (Fig. 10-7). In some snails, instead of a creeping movement brought about by waves of contraction and relaxation, there may be a type of bipedal locomotion involving an alternate extension and contraction of first one half of the foot and then the other half. Many are able to make leaping movements by vigorous muscle contractions. Swimming is performed in many by means of lobes or parapodia of the foot, by the whole body, by mantle appendages such as fins, or by the highly efficient jet propulsion of cephalopods, with lateral fins acting as keels or gliding surfaces.

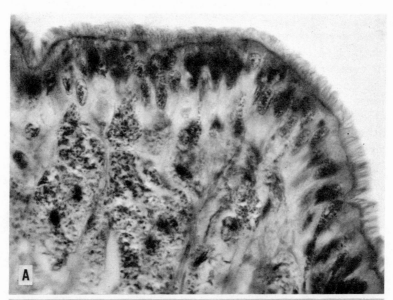

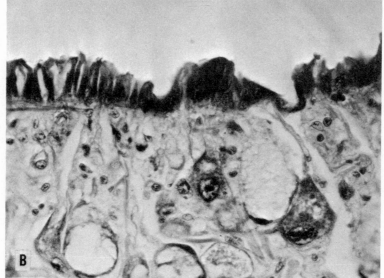

FIG. 10-8

Integument of mollusks. **A,** *Anodonta* (class Pelecypoda). Note abundant, long cilia of epidermis and mucus-secreting cells. (×810.) **B,** *Limax* (class Gastropoda). Note large mucous glands beneath epidermis. (×480.) (From Andrew: Textbook of comparative histology, Oxford University Press, Inc.)

The epidermis of mollusks is chiefly cellular and ciliated. Much of it, especially outside the mantle, contains mucigenous cells or mucus-secreting glands (Fig. 10-8). These cells and glands produce slime for locomotion and food-collecting strings. Some mollusks, like the cephalopods, have many chromatophores, whereby they can produce the fastest color changes in the animal kingdom. Each chromatophore consists of a thin elastic membrane that surrounds the fluid-filled space containing the pigment. Radially disposed muscle fibers are attached to the elastic membrane, and by contracting and relaxing, the sac of pigment is dilated or contracted, thus dispersing the pigment or concentrating it. There are many variations in the epidermis among the different molluscan groups.

The body wall below the epidermis consists of connective tissue and muscle. Mollusks tend to differ from other animals in lacking the regular arrangement of these tissues into definite layers, such as one finds in the vermiform animals that have definite circular and longitudinal layers. This lack of regular arrangement may be associated with the varied functions the body wall is called on to perform in the variety of forms having mollusk organization. Connective tissue, for instance, may be very compact and firm (chordoid tissue) in those with reduced shells or absent shells; it may be spongy and loose in those well protected by a shell. Chordoid tissue, described in many mollusks and some other invertebrates, is made up of large vesicular cells with a border of cytoplasm, and when turgid with a semifluid content, these cells become very firm and supporting. Loose, spongy tissue is of functional significance in the hydraulic systems of mollusks when fluid or blood causes tissue to expand or extend.

Muscle tissue in mollusks has a structural arrangement of antagonistic sheets of muscle and is irregularly distributed in the mollusk body. Only a few muscles are discrete units such as the adductors of shell valves and the retractors of byssal threads, foot, or radula.

In some mollusks, muscles are very important in locomotion; others have little or no movement. The nature of muscle contraction depends on the nature of their innervation. A given muscle must contract at widely different speeds. If a muscle is exposed to a series of stimuli, it may contract with each stimulus, but if the stimuli occur rapidly, the contractions fuse. Mollusks have the power to maintain high tensions for prolonged periods without excessive levels of energy expenditure. Some mollusks can swim by opening and closing the shell valves rapidly and do not need the prolonged period of keeping the valves closed. The speed of muscle contraction and relaxation and the frequency of stimulation necessary for fused contractions are important for fitting muscles to specific functions.

Most of the musculature in mollusks is composed of small nonstriated and uninucleated cells. Some striated fibers are also found in the adductors of lamellibranchs and a few others. Most slowly contracting smooth muscles are tough and opaque. Fast muscles are usually soft and translucent. In some lamellibranchs the adductors are divided by a connective tissue partition into smooth muscle fibers and striated fibers, but those two types of fibers may be intermingled in others. The striated portion contracts and relaxes rapidly; the smooth part contracts slowly and may maintain contraction for a long time. Fast muscles may also be spirally or obliquely banded instead of transversely. There is also the possibility that smooth muscle with peripheral, twisted fibers may give the deception of being striated fibers. The fast muscles of some mollusks present an appearance identical with the skeletal muscles of arthropods, being divided into sarcomeres by Z-bands and each sarcomere being provided with I- and A-bands.

The striated muscles of some mollusks may have a contraction time of 0.046 second and a half-relaxation time of 0.04 second, compared with the unstriated muscle (in the retractors of the tentacles of snails) of 2.5 seconds for contraction and 25 seconds for half-relaxation time.

The "catch mechanism" proposed by J. von Uexküll in 1912 based on the *Pecten* adductor suggests that a molecular ratchet mechanism within the muscle fibers sets itself automatically and remains so without the expenditure of energy. Later investigators indicate that these slow muscles could be kept in a state of contraction by a very low frequency of motor excitation with the expenditure of a small amount of energy.

Internal anatomy

The alimentary canal varies in the different mollusks, depending on nutritional habits. In general it is composed of an anterior buccal

FIG. 10-9

Sagittal section of anterior end of *Patella* showing relationship of radula and other buccal organs. (Modified from Davis and Fleure.)

mass and esophagus (of ectodermal origin) in which are the radula and jaws (Fig. 10-9), the midgut (endodermal) of stomach and liver, and the hindgut or intestine. Mollusks have different feeding habits. Those living on coarser food use the radula as a rasp. Carnivorous mollusks such as the cephalopods have powerful jaws for seizing and tearing their prey and a radula to aid in the process. There are many modifications of the feeding habits, especially in bivalves, which have no mandibles, radula, or pharyngeal musculature but have efficient ciliary-feeding devices.

In many bivalves and some gastropods that depend on filtering organic materials from suspensions or deposits, there is a special rod-like crystalline style that protrudes into the stomach (Fig. 10-10). This rod may lie free in the first part of the intestine or in a special pouch or sac. The rod is composed of a gelatinous substance containing a digestive enzyme. The crystalline style is kept rotating by means of

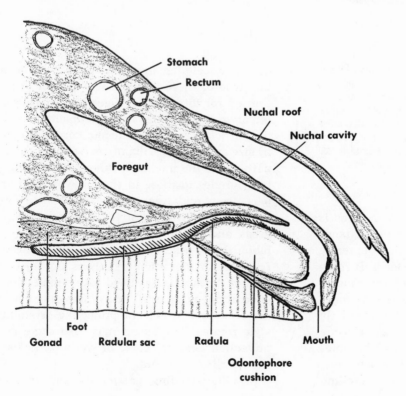

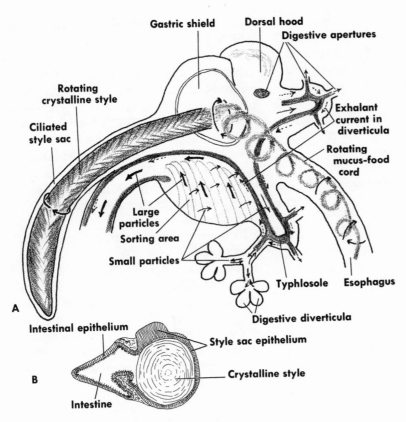

FIG. 10-10

A, Stomach and crystalline style of lamellibranch ciliary feeder. From volumes of water taken in, particles are engaged in cord of mucus, which is kept rotating by crystalline style. Ridged and ciliated sorting area directs large particles to intestine and small particles to digestive diverticula. **B,** Transverse section through straight portion of intestine of *Cardium* showing crystalline style. (**A** modified from Morton; **B** modified from Johnstone.)

cilia in the style sac or pouch. Its free end is attached to a string of mucus, which extends into the esophagus and in which entering particles of food become embedded. Turning of the style causes a twisting of the mucus-food cord into spiral coils and keeps it under tension, thus regulating the gradual and orderly entry of food into the stomach. At the same time, the surface of the style head becomes softened and dissolves in the stomach fluid, setting free an amylase that is active in prelimi-

nary extracellular digestion. Since most of the fine food of these filter feeders is later phagocytosed for intracellular digestion, ciliary sorting areas are provided in the stomach in which the particles freed from the food cord are graded by size and weight. Cilia in grooves direct larger particles to the intestine. Other cilia beating at right angles to those in the grooves keep fine particles circulating till they find their way to the cells of the digestive diverticula. Sorting areas also occur on the feeding organs (e.g., labial palps) where sand grains are separated from the smaller microorganisms and rejected. Part of the stomach is lined with cuticle, thickened at one point into a heavy flange or gastric shield against which the projecting crystalline style rotates. The shield may also serve as a protection against coarse or sharp particles in the whirling food cord.

In ciliary-feeding gastropods and in most

bivalves the finest particles released from the food chain are diverted into the apertures of the digestive diverticula in the anterior part of the stomach (Fig. 10-10, *A*). These apertures (usually 2 in number) open into ciliated ducts that branch into the digestive gland. The digestive gland is an organ of absorption and intracellular digestion. Some preliminary extracellular digestion may occur in the stomach, but digestion is chiefly intracellular. The digestive gland consists of many blindly ending tubules, each made up of one or more cell types. Many of the cells may bear several long cilia. Food particles in mucoid suspensions are carried by the incurrent ducts into the lumen of the tubules, where the food particles are engulfed by the cells of the tubules. Food particles are drawn in by the lowered pressure within the lumen of the tubules induced by the progressive absorption of food or by the squeezing action of the muscles of the stomach (J. L. Campbell, 1965). The digestive cells may also absorb products of extracellular digestion. In the digestive gland such enzymes as protease, lipase, and amylase are provided. Waste or rejected materials from the tubules are carried by excurrent ciliary tracts back to the intestinal groove and added to the feces of the intestine.

The dorsal heart is enclosed in the pericardium. Usually 2 auricles receive blood from the gills, and the single muscular ventricle pumps the blood through the body. The blood vessels are made up of arteries, veins, and sinuses. The integument of cephalopods may also contain capillaries. In some a single anterior aorta leaves the heart and breaks up into smaller arteries that pass to the mantle and viscera. The arteries form a network of vessels that pass into veins and sinuses. In its passage back to the heart the blood goes through the kidneys and gills for purification and gaseous exchange. The formed elements of the colorless blood are represented by a variety of amebocytes and corpuscles. Many mollusks have copper-bearing hemocyanin as a respiratory pigment, although a few have hemoglobin.

The types of respiration in mollusks are cutaneous respiration, branchial respiration, and pulmonary respiration, but there may be modifications of each of these.

Cutaneous respiration occurs in those Mollusca that have no differentiated respiratory organs (Scaphopoda, some Gastropoda, and small Aplacophora). Some cutaneous respiration may occur also through the skin of those with respiratory mechanisms (pulmonate gastropods). However, cutaneous respiration must be very restricted when the body surface is covered with a shell or when the body is thick and bulky. The necessity for some respiratory mechanism must have arisen early in the evolution of the molluscan structural requirements, which are about the same for all animals. An effective surface where exchange of gases must be made; a blood circulatory system, with a respiratory pigment for carrying off the oxygen as it enters the surface and for carrying carbon dioxide away from the surface; and mechanisms for ventilating the surface to ensure rapid exchange are basic to all animals.

Most aquatic mollusks have gills (ctenidia) for breathing (the branchial respiratory type). Molluscan gills have rows of lateral cilia on each side of filament rows alternately arranged along both sides of the gill axis. Blood passes through the gill filaments (by incurrent and excurrent blood vessels) while the cilia maintain a water current running in an opposite direction to that taken by the blood within the filaments, a countercurrent principle for efficient respiratory exchange. As water with its oxygen enters among the gill filaments, oxygen passes through the respiratory exchange surface of the gill into the blood; in return the carbon dioxide waste of the blood passes into the water. The amount of these gases that diffuses across a membrane is proportional to the surface area of the respiratory membrane and to the concentration gradient. In this respiratory exchange no active-transport mechanisms are involved so far as is known, for it is a matter of simple diffusion. Since oxygen diffuses slowly, organisms larger than 1 mm. in diameter cannot get enough oxygen supply by diffusion and must have the assistance of some respiratory mechanism or organ such as a gill or lung. Flat animals have an advantage over spheroid animals in this respect.

The gills in mollusks have arisen from modifications of the mantle. In the bivalves these gills lie in the branchial cavity covered by the mantle. The lung cavity of pulmonates is a simple pouch or sac that has an extensive plexus of blood vessels arranged netlike over its walls. The lungs in pulmonates are an adaptation for terrestrial existence. The muscular dome-shaped floor, by raising and lowering due to alternate contraction and relaxation, causes air to rush in and out through the pneumostome opening, which is provided with a valve. Compression of the mantle cavity increases the partial pressure of oxygen and facilitates oxygen absorption. Some pulmonates derive their oxygen from the water in the mantle cavity and do not come to the surface to breathe.

Efficient respiration requires regulatory mechanisms for adjusting the ventilation of respiratory surfaces to the changes of respiratory demands. In cephalopods the respiratory movements are made possible by the rhythmic contractions of the mantle, which brings water into contact with the gills. In those animals the gills are attached to the upper (anterior) wall of the mantle cavity. Water is kept flowing by the coordinated activities of the mantle, the funnel, and the inlet valves.

Small cephalopods tend to breathe more rapidly than larger ones. In the cuttlefish *Sepia*, when the animal is resting under normal conditions, the rate of inspiration is about 55 times per minute. It may be much lower (12 to 14 times per minute) in the octopus *Eledone* (K. M. Wilbur and C. M. Yonge, 1966). Respiratory movements in the cephalopods are controlled by nerves from the visceropallial lobe of the brain. The paired nerves (one on either side) pass by way of the stellate ganglion to the mantle musculature. In the mantle of the squid a rapid maximal contraction is mediated by impulses in the giant fiber system and a graduated response by the small, slow-conducting fibers. Chemical control with carbon dioxide in excess can produce a great increase in respiratory movements in the octopus.

Ciliary activity that produces water currents in bivalves appears to be autonomous, and cilia will beat without nervous connections. However, there are many factors that influence ciliary activity, such as temperature (frequency of beat of cilia increased by rise) in the range of 0° to 33° C. At 40° C. the cilia will stop. Vernberg and others (1963) found that the effect of salinity on ciliary activity of excised gill tissue varied with different species. The gill cilia of *Aequipecten irradians* (scallop) ceased to beat with reduced salinity; those of *Crassostrea virginica* (the American oyster) did not. In fresh water, pulmonate gastropods must come to the surface when the oxygen supply reaches a certain level. In water of 6 cc. of oxygen per liter they remain submerged three times as long as they do in water of 2 cc. per liter. Generally, pulmonates—both terrestrial and aquatic forms—are sensitive to the levels of oxygen and carbon dioxide. The pneumostome opening with its valve may control the snail's reaction to those gases.

The coelom is very much restricted in most mollusks, being confined mainly to the pericardium and gonadal cavity. Most of the body spaces represent the hemocoel, which is extensive in the Mollusca. In cephalopods the gonadal coelomic cavity is large and communicates directly with the pericardium. In most other mollusks the 2 cavities are separate. Closely connected to the coelomic cavities are the kidneys, which open internally into the pericardium in nearly all molluscans and may by considered a part of the coelom. In many the kidneys also serve as coelomoducts through which sperm and eggs are discharged; in the advanced cephalopods the genital ducts are separate from the kidneys. Each kidney is a tubelike structure that undergoes various differentiations such as bladder and glandular regions in the higher mollusks.

Few mollusks have more than one pair of nephridia. In *Nautilus* and the Monoplacophora there are more than one pair. Members of the Archaeogastropoda have one pair of nephridia, but the right nephridium is absent in other Mollusca. When only one nephridium is present, excretory functions may be assumed by other parts of the body, or biochemical changes may enable the mollusk to handle nitrogenous compounds with 1 kidney. Nitrogen is often eliminated as amino acids and uric acid in land forms. It is well known that aquatic species can get rid of ammonia through the surface of the gills and the body. Mollusks, moreover, can tolerate more of the toxic ammonia than can most other animals.

A nephridium is saclike, and its walls are much folded to increase the surface area. In some gastropods (freshwater forms and land pulmonates) there is a short ectodermal ureter, which is placed along the right wall of the mantle and opens to the outside near the anus and pneumostome. The distal end of the nephridium may connect with the pericardial cavity, or there may be a renopericardial canal that serves to connect the nephridium with the pericardial cavity near the proximal part of the nephridium. The nephrostome, which conveys the coelomic fluid and its nitrogenous waste into the nephridium, plays a very minor part in mollusks because of the reduced coelom.

The mollusk urine may be formed by ultrafiltration from the heart into the pericardium, and the ultrafiltrate is drained into the nephridium through the renopericardial tube. In terrestrial pulmonates, however, ultrafiltration occurs directly from the renal veins to the kidney sac. In many cephalopods, ultrafiltration takes place from the branchial heart appendage into the branchial heart pericardium and then into the sac (W. T. W. Potts, 1967). The ultrafiltration undergoes some change by the secretion of certain compounds and reabsorption of ions into the kidney sac and ureter (Potts, 1967).

Most marine Mollusca have a blood that is isosmotic with sea water, but in brackish water with a lower concentration of salt, mollusks simply reduce their blood concentration to that of the brackish water. Freshwater forms have a low concentration of blood salts and excrete a hyposmotic urine by reabsorbing some of the salt.

The nervous system of mollusks is made up of certain ganglionic masses, nerve cords, and sense organs. Although there are several variations, in general the nervous system is simpler than that in other forms as advanced as this phylum. The primitive plan already described is followed to some extent by many mollusks, but there is a tendency to concentrate the ganglia in the head region, usually in the form of a circumesophageal ring. The longitudinal pallial and pedal cords on each side are connected by commissures but are modified here and there. The system is usually differentiated into paired ganglia, which are connected by nerve cords that do not bear ganglia. In cephalopods the nerves to the muscles of the mantle contain giant fibers for rapid escape. The chief ganglia are the cerebral, pallial, and pedal. From these ganglia the head region, mantle, viscera, and foot receive nerves. Sense organs are represented by a pair of eyes (variable in plan) on the head region, accessory eyes, statocysts, osphradia, and possibly others.

Most mollusks are dioecious, although the members of the subclasses Pulmonata, Opisthobranchia, and a few others are hermaphroditic. Pulmonates, however, like earthworms, usually practice cross-fertilization. Copulation with a penis occurs in cephalopods and certain gastropods. Fertilization may be external or internal when there is no penis, as in pelecypods. The eggs may be laid outside the body in some, but in others they are incubated in the parent's body (P. S. Galtsoff, 1961). Certain types of viviparity occur only in some aquatic and pulmonate snails, but placental viviparity probably is never found. Parthenogenesis is almost nonexistent. Eggs vary in the amount of food

FIG. 10-11

Molluscan larvae. **A,** Trochophore of *Yoldia* (class Pelecypoda). **B,** Trochophore of *Dentalium* (class Scaphopoda). **C,** Young trochophore of limpet *Patella* (class Gastropoda). **D,** Veliger of *Patella*. (Redrawn from MacBride: Textbook of embryology, The Macmillan Co.)

they contain. The larvae of many aquatic mollusks pass through trochophore and veliger stages that are free swimming; but in all pulmonates and some others the trochophore and veliger stages are passed in the egg, and the young emerge as juvenile snails.

The embryologic development of mollusks

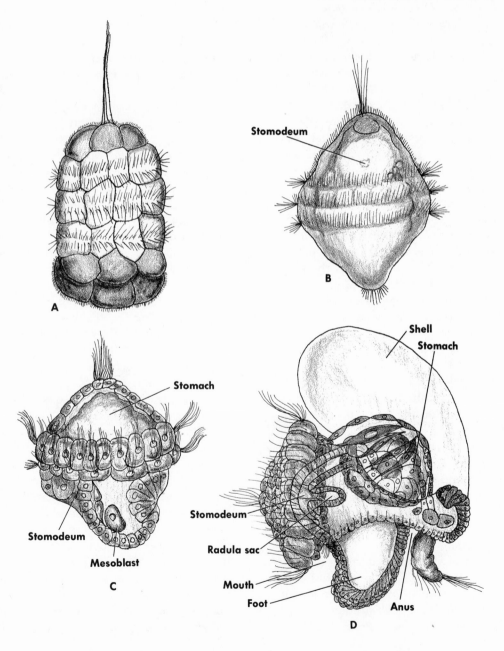

resembles that of polychaete worms. Cleavage is spiral and total, although its pattern may be influenced by the amount of yolk present in the egg. According to those who have made a careful study of the embryology of certain mollusks, the first two cleavages produce about the same size of blastomeres, which then divide to form 4 small micromeres and 4 large macromeres. By further divisions, four quartets of cells are formed. Each quartet has about the same fate as those described in a previous section. Gastrulation occurs in several ways: some by invagination, some by epiboly, and others by ingression (Fig. 10-34). Much of the mesoderm arises from endoderm. The coelom may arise as a single cavity, with the renal and gonadal cavities originating from its walls. A circlet of cilia forms around the equatorial region of the young embryo. One hemisphere remains small as growth takes place; the other enlarges. A mouth and anus appear in the larger hemisphere, and such a larval form is called the trochophore or trochosphere (Fig. 10-11, *A* to *C*). Later a second stage, the veliger (Fig. 10-11, *D*), which bears a velum of lobes for locomotion, appears.

■ CLASS AMPHINEURA

Amphineurans are elongated, bilaterally symmetric animals that represent the most primitive of existing mollusks. They vary in size from less than an inch to over 12 inches in length. They are divided into two subclasses, Polyplacophora and Aplacophora. Most of them live in rather shallow marine water and are sedentary in their habits. Some of them have a worldwide distribution, and their fossil record goes back to the Ordovician period. They have undergone little evolutionary change in their long history.

Subclass Polyplacophora

The most characteristic of the amphineurans belong to the subclass Polyplacophora, the most familiar members of which are the chitons. There are about 600 species existing at the present time. The chitons have a flattened, ovoid body with reduced head and no cephalic eyes or tentacles. The flat foot is broad, occupies the whole ventral surface of the body, and is fitted for close application to a substratum. The thick mantle or girdle may be scaly or fleshy and encircles the whole body. It is commonly provided with bristles and spines. The mantle bears embedded within it 8 transverse and

FIG. 10-12

Chaetopleura, a chiton (subclass Polyplacophora). (From Hickman: Integrated principles of zoology, The C. V. Mosby Co.)

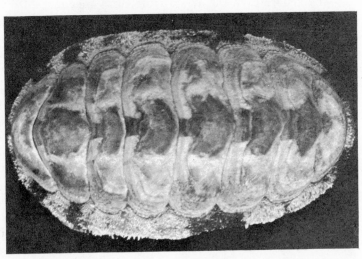

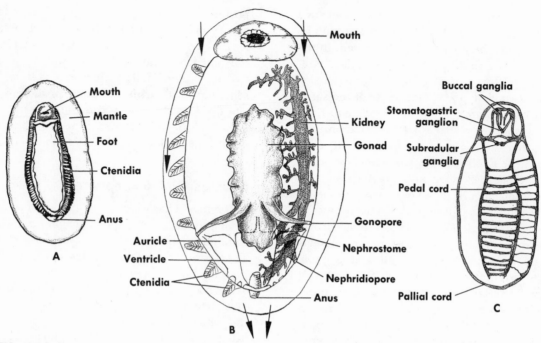

FIG. 10-13

Subclass Polyplacophora. **A,** Ventral view of *Acantho-pleura*. **B,** Internal structure (diagrammatic) of chiton (gills and kidneys shown on one side only). **C,** Nervous system of *Acanthochiton*. (**A** and **C** modified from Pelseneer.)

arched calcareous plates so arranged that each plate overlaps the one behind (Fig. 10-12). The separate valves are articulated by means of anterior and posterior marginal projections and held in place by the reflexed mantle folds. This articulation enables the animal to roll up into a ball. In some chitons the entire shell is covered by the mantle. The reduced head and mouth are in front of the foot and beneath the mantle (Fig. 10-13, *A* and *B*). The anus lies on the ventral surface behind the foot. The mouth is usually provided with a radula. Each shell plate has a sculptured tegmentum, or upper layer, and an articulamentum, or lower layer. The upper layer is partly calcareous and partly organic matrix; the lower is entirely calcareous. The shell bears little branching canals through which nerves pass to aesthetes, or shell eyes.

Running completely around the animal is the shallow mantle groove between the mantle edge and the foot. This groove contains on each side a varying number (two to forty pairs) of gills which may be clustered close together or may extend along most of the groove on each side (Fig. 10-13, *A* and *B*). The large number of gills may be a multiplication of the single ancestral pair because a large mantle chamber is no longer available. The large marine limpets, such as *Lottia* (a West coast form), have passed through an evolution similar to that of chitons (parallel evolution); however, limpets lose their original gills and from mantle folds form secondary gills that project into the pallial groove along the side of the body. In chitons each gill is made up of a central axis and many short filaments, which are held together by interlocking cilia. The inhalant chamber lies outside the row of gills and the exhalant chamber inside the gills. When the margin of the mantle is pressed close against the substratum, the mantle groove is a closed chamber. Inhalant apertures are formed by the mantle margins near the anterior end. Water entering at these openings passes through the gills into the dorsal exhalant chamber on each side. The two exhalant chambers may converge and discharge through single or paired exhalant apertures.

The chitons feed chiefly on algae and other plant life found on rocks. Food may be scraped off the rocks by the projection of the radula and odontophore to the substratum; when the radula is retracted, the food particles are carried back into the mouth, where salivary glands empty a secretion for collecting the food and lubricating the radula. No digestion occurs in the mouth. The alimentary canal is differentiated into an esophagus, stomach, and intestine. The intestine may be coiled in some chitons and may be divided into anterior and posterior parts separated by an intestinal valve. The digestive gland discharges proteolytic enzymes into the stomach through paired ducts. Digestion is mostly extracellular and occurs in both the stomach and intestine. Phagocytic cells may play a minor role in some intracellular digestion. Ciliary tracts are present throughout the alimentary canal and aid in the passage of blood and fecal material.

The blood is discharged from the posterior heart through a single anterior aorta into a pair of sinuses. The blood passes to the gills for gaseous exchange and then is collected by a pair of auricles, which empty into the single ventricle. The 2 kidneys, one on each side, open into the pericardium, and the nephridiopores empty into the mantle groove near the posterior gills (Fig. 10-13, B). As the kidneys pass through the hemocoel, they form many diverticula and windings.

Amphineurans largely retain the primitive ancestral plan of nervous system with 4 main trunks and connecting commissures (Fig. 10-13, C). Nerve cells are distributed largely in the nerve cords, and ganglia are mainly absent. The esophageal nerve ring gives off nerves to the buccal region, radula, pedal region, mantle, and viscera. Besides the shell eyes already mentioned, sense organs are restricted chiefly to tactile and photoreceptor cells. The shell eyes, or aesthetes, vary. They may consist of bundles of sensory cells, or they may be made of cornea, lens and retina. Chemoreceptors are also located in the epithelium around the mouth.

The sexes are separate in chitons. There is a single median gonad located in front of the pericardium, with which it has no communication. A pair of gonoducts opens separately in the mantle cavity near the openings of the kidneys and carries sperm or eggs to the outside. Fertilization may occur in the female mantle cavity, where the sperm are carried by inhalant branchial currents. Eggs, either singly or in strings, are discharged into the sea. Usually the free-swimming pelagic larva, or trochophore, develops directly into the adult without an intervening veliger stage.

Subclass Aplacophora

The members of subclass Aplacophora are commonly known as Solenogastres. They are wormlike forms without shell plates, but calcareous spicules may be embedded in the thick integument. The vestigial foot is represented by a median ridge in a small ventral groove (Fig. 10-14, A) in one of the two orders (Neomeniomorpha); in the other order (Chaetodermomorpha) this pedal groove is lacking.

In the Neomeniomorpha there is a large mucous gland that discharges its secretion into the anterior part of the pedal groove. The only remnant of the mantle cavity is represented by a deep posterior depression, the cloaca, into which the anus and the nephridiopores from the kidneys open. In the cloaca a circlet of epithelial folds, through which the blood circulates, serves as gills. There are no true blood vessels, but there is a ventral sinus between the foot and the gut; a posterior part of a tubular dorsal sinus serves as a contractile heart. A radula is sometimes present in the simple alimentary canal. The kidneys are represented by a pair of simple coelomoducts, each of which opens into the pericardium at the

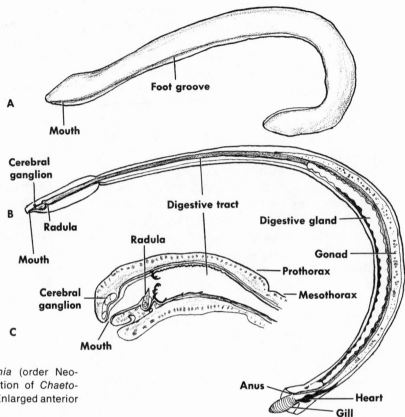

FIG. 10-14

Subclass Aplacophora. **A,** *Proneomenia* (order Neomeniomorpha). **B,** Median sagittal section of *Chaetoderma* (order Chaetodermomorpha). **C,** Enlarged anterior end of same. (Modified from Pelseneer.)

inner end and the cloaca at the outer end (nephridiopores). In this order a pair of hermaphroditic gonads opens separately into the pericardium, and the gametes are discharged through the coelomoducts, which serve as gonoducts. Glands at the terminal parts of the gonoducts secrete material for an eggshell. Seminal vesicles, or cecal outgrowths of the gonoducts, are also present. The eggs develop into free-swimming trochophore larvae. The nervous system is similar to that in chitons, with the same arrangement of the two sets of longitudinal nerve tracts—a dorsal pallial pair and a ventral pedal pair.

In the Chaetodermomorpha, besides the differences already mentioned, the margins of the mantle cavity have fused to form a bell-shaped, contractile cloaca, which bears a pair of bi-

pectinate gills (Fig. 10-14, *B*). They also have separate sexes, each with a single dorsal gonad that opens into the pericardium. Gametes, however, are carried to the outside by the coelomoducts or renal tubules.

There are fewer than 100 species of these aberrant aplacophoran mollusks. All dwell in deep water, and some may burrow in the sand. They live on small invertebrates and microorganisms. Some live on corals and hydroids. Since some chitons have lost their shell, this may indicate how the aplacophorans could have originated from them. Several authorities consider the Solenogastres to be so aberrant that they are placed with such primitive worms as the flatworms.

The common genera are *Neomenia, Chaetoderma,* and *Proneomenia;* since they live in deep water, they are most commonly found only in deep-sea dredgings.

■ CLASS GASTROPODA

Class Gastropoda is by far the largest of all molluscan classes in number of species and includes perhaps the most familiar members in the phylum. Often the members of the class, because of their single shell, are referred to as univalves in contrast to the bivalves. Their single shell, which is usually coiled, is one of their most distinguishing characteristics. The shell is very much modified, however, especially in the adult state, and assumes many forms. In some it is entirely absent. The gastropods also have a unique, asymmetric body organization brought about by a torsion process that is found nowhere else among the mollusks. A well-defined head, which may be scanty or absent altogether in many other mollusks, is found in all gastropods.

This group has had an amazing adaptive radiation, which has enabled them to move into most kinds of ecologic niches and into nearly every area of the earth. They are found in marine and fresh water and on land. They exist from the greatest depths of the sea to mountainous altitudes of 15,000 to 20,000 feet. They are adapted to all types of aquatic bottoms, both littoral and bathyal. Some are pelagic and swim with ease. Some are adapted for living in desert regions, although most land forms prefer fairly moist habitats. They can occupy such specialized habitats as trees, exposed limestones, and coral formations; or they can live a parasitic

existence. Their food habits are equally wide ranging; they feed on vegetation of all kinds, on other animals as predators, and on dead animals as scavengers.

Their fossil record is impressive, extending throughout the Paleozoic era and becoming increasingly abundant during the Mesozoic and Cenozoic eras (Fig. 10-3, A to C). Many families have become extinct, and there have been fluctuations in the relative abundance of the three subclasses — Prosobranchia, Opisthobranchia, and Pulmonata — during their long geologic history. It is generally agreed that the terrestrial forms evolved in a marine habitat spread to fresh water and then to land. As fossils the first nonmarine snails are known from the Pennsylvanian period. Gastropods are apparently enjoying the heyday of their evolutionary divergence at present.

Gastropods number between 35,000 and 40,000 species. In size they range from tiny species not more than 1 mm. long to the giant whelks that may be more than 2 feet long. Some fossil forms were much longer.

Within this extensive class are included the snails, slugs, whelks, periwinkles, nudibranchs, cowries, cones, and sea butterflies.

FIG. 10-15

Four basic stages in process of body torsion of gastropods. Alimentary canal is stippled in all figures. **A,** Embryo before torsion. **B** and **C,** Ventral flexure with anus and mantle cavity shifted to forward position. **D,** Embryo with lateral torsion in which organs on right have shifted to left, and vice versa.

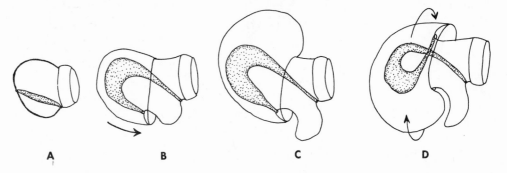

A   B   C   D

SIGNIFICANCE OF TORSION

No aspect of gastropod organization and morphology has played a more important role in their evolutionary tendencies than the phenomenon of torsion and a related modification, the coiling of the shell. The body of a gastropod, like that of most mollusks, is composed of four parts—the visceral sac, the mantle, the head, and the foot. The basic plan of the molluscan body is bilaterally symmetric, with the mouth at one end and the anus at the other. In the gastropods, however, this bilateral symmetry is modified by a torsion, or twisting, of 180 degrees (Fig. 10-15) so that the anus is brought forward toward the mouth and the nervous system forms a figure 8. The body of most gastropods is generally wormlike, with the head and foot at one end and most of the body spirally coiled. The coiled visceral sac is covered by a mantle, which projects as a free fold at the anterior end and hangs down like a skirt around the head and foot. Thus the mantle cavity is in a position opposite to that in other mollusks. In the cavity between the mantle and head is the lung or gill. The site of the torsion is the neck behind the head-foot, through which the esophagus, rectum, aorta, visceral loop, and muscles of the shell pass. The actual twisting thus involves this neck tissue and the structures within it.

The development of the asymmetric condition begins in the veliger stage and may be very rapid (2 to 3 minutes in *Acmaea*), although slow in some snails. In the archaeogastropod (aspidobranchian) *Haliotis*, torsion occurs in two unequally timed phases of 90 degrees each. The initial cause of the torsion is the asymmetric growth of the right and left muscles attaching the shell to the head-foot, which forces the twisting process. During the first stage of torsion, there is a ventral flexure in the sagittal plane so that the anus and the openings of the mantle cavity are brought forward ventrally (Fig. 10-15, *A* and *B*). A second torsion

FIG. 10-16

Effects of torsion upon position of gills, digestive tract, and nervous system (diagrammatic). Note that nervous system in **C** has twisted into figure 8 (streptoneury).

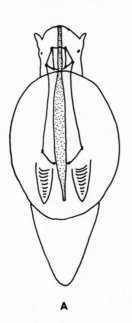
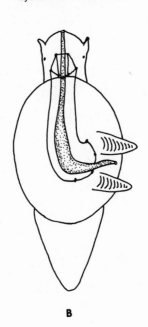
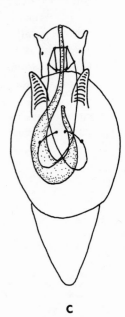

A   B   C

movement at right angles to the first involves the rotation of the visceropallium 180 degrees upon the head-foot so that the mantle and other structures that were lying ventrally are shifted up the right side to a dorsal position where they face forward (Fig. 10-15, C and D). At the same time, structures that were dorsally located pass down the left side to lie ventrally. In this way the left structures (gill, renal organ, etc.) are shifted to the right and the right structures to the left (Fig. 10-16). The mantle cavity that originally faced backward now faces anteriorly. Such a torsion allows the sensitive head end of the animal to be drawn first into the protection of the mantle cavity and shell, with the tougher foot forming a barrier to the outside.

The torsion process as described applies to right-handed (dextral) snails, in which the

organs on the left side eventually disappear, and to left-handed (sinistral) snails, in which the organs on the left side persist, whereas those on the right side atrophy.

The spiral coiling of the shell and viscero-pallium is an entirely separate process from that of torsion and does not seem to result from torsion. In fact, it is believed that coiling evolved before torsion. Some gastropods that have torsion exhibit no spiral coiling. There may be no causal relation between the two processes. Coiling is achieved by a more rapid growth of one side of the visceral mass than the other. When viewed from the apex of the shell, the direction of the coils is clockwise in dextral snails; counterclockwise in sinistral snails. In the spiral cleavage of the snail's egg during development the direction of the cleavage planes in dextral coiling is the reverse of that in the sinistral type. Early gastropods had a planospiral shell that was bilaterally symmetric instead of conically spiral (Fig. 10-17). A plano-spiral shell has certain disadvantages, such as a more unwieldly shape; some bilateral re-

FIG. 10-17

Types of coiling (diagrammatic). **A,** Planospiral coiling with coils all in one plane. **B,** Planospirally coiled shell in (left) normal, upright position and (right) lying on side. Shell has bilateral symmetry but is topheavy and has limited viceral space. **C,** Helical (conispiral) coiling with spirals in different planes. **D,** Shell with helical coiling is asymmetric, compact, and easily balanced.

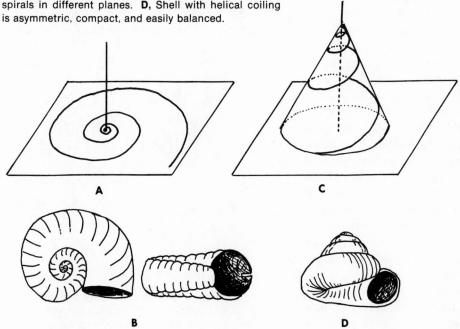

A

C

B

D

striction of the mantle cavity; and a sharp reduction in the diameter of the inner coils, with consequent reduction in available visceral space. Growth of the planospiral shell in an anteroposterior direction is restricted by the forward position of the anus; nor can it grow dorsally without becoming top-heavy. Helical (conispiral) coiling provides better balance for the visceral sac as it grows. By twisting into a spiral, the shell and its contents acquire a more compact form and a diminished diameter, as well as better balance.

The effects of torsion are not equally evident in all gastropods. In the opisthobranchs and some others, although normal torsion takes place in the veliger larva, some detorsion occurs in later development, usually a juvenile stage. This frequently involves the carrying back of the anus and mantle cavity to a posterior position. In extreme cases, detorsion may involve a disappearance or reduction of the mantle and visceral mass, with the body becoming an elongated form (*Pterotrachea*). In the order Nudibranchia there is no mantle cavity or external shell in the adult, although they appear in the veliger. There are various degrees of detorsion, always accompanied by some degeneration of the shell. Many different morphologic patterns are produced, according to the degree of detorsion (e.g., 90 to 180 degrees). In the Opisthobranchia there is a marked evolutionary sequence of patterns, depending on the amount of torsion. When detorsion is only partial, the mantle aperture may be shifted to the right side. In *Aplysia*, in which there is more detorsion, the reduced shell is enclosed by the mantle and moved backward along with the anus and the gill (Fig. 10-26, *A*). In the Nudibranchia, in which detorsion is complete, the shell, mantle cavity, gill, and visceral hump all disappear; the body assumes an elongated form with secondary bilateral symmetry (Fig. 10-26, *D*).

In pulmonates the detorsion is suppressed. They retain the posttorsion anus and mantle cavity but have no gills. The mantle cavity is modified into a lung. The secondary symmetry of the nervous system is due to a shortening of connectives and concentration of ganglia into a circumesophageal ganglionic complex.

There are several theories regarding the relative advantage or disadvantage of torsion to the gastropod. Some authorities believe that the main advantage must have been to the larva, and others believe that torsion would not have survived had it not been directly advantageous to the adult. One theory stresses the advantage of torsion to the larva in its adaptation to pelagic life. When the mantle cavity is twisted anteriorly, the larva is able to retract its head and its ciliary mechanism, the velum, into its mantle cavity and thus drop away from potential predators. Other pelagic larvae, however, such as those of the lamellibranchs, have no torsion; and in *Haliotis,* torsion is not completed until after the animal has left its pelagic life and settled to the bottom to live. According to the other theory, the main advantage of torsion must be to the adult. Certainly torsion promotes stability in the adult by placing the bulky mass of the animal nearer the substratum. Other advantages may be obtained from facing the inhalant currents, which flush out the mantle cavity, and from the strategic forward placing of

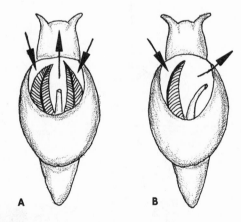

Fig. 10-18

Direction of currents in mantle cavity. **A,** Archaeogastropod. **B,** Mesogastropod; one-way current carries feces away from gill. (Modified from Yonge.)

the sensory elements of the mantle cavity. Whatever the advantages are, torsion has undoubtedly also resulted in some problems to the adult. The diversity of gastropod morphologic patterns indicates some of the ways these difficulties have been met. For instance, torsion has caused a tendency for the incurrent water to sweep the feces onto the gills, thus introducing a sanitation problem (Fig. 10-18, A). In primitive gastropods this was solved by a slit in the mantle and shell near the location of the anus. *Haliotis* and others have a series of holes for this purpose (Fig. 10-23). In Mesogastropoda (pectinibranchs) the problem is met by an arrangement of a single gill with a row of filaments that produce a lateral through current which carries the feces away from the gill (Fig. 10-18, B). Some have lost the gills altogether and use an adaptation of the mantle for respiration, such as the lung in pulmonates and the pallial gills in *Patella*.

GENERAL MORPHOLOGY
External anatomy

The major external features of a gastropod are the head, foot, mantle, and shell (which is absent in some). The head is one of the best developed among the mollusks. It usually bears one or two pairs of tentacles and is connected to the viscera by a mobile neck. When eyes are present, they are located on the tips of the tentacles or at their base. Provided with a snout and armed with a radula in the buccal cavity, the head is the active agent in food getting. When disturbed, the animal can retract its head and foot into the mantle cavity and shell.

The foot in most snails is usually elongated, flat, highly muscular, and fitted for creeping. However, it may be modified for digging or for swimming (Figs. 10-19 and 10-20). Extension of the foot, as mentioned in a former section, is produced by blood pressure turgor in hemocoelic spaces with the cooperation of muscles; retraction of the foot is by powerful muscles. Muscles are of both the circular and longitudinal type and extend from columellar shell

attachments into the foot. The foot is not only a locomotor organ (Fig. 10-7) but also acts as an adhesive organ by suckerlike action. Ciliary movement also takes place over a slime sheet. In general the creeping movement occurs by locomotor waves, which involve longitudinal muscle contraction followed by relaxation from behind forward. Various modifications of the locomotor wave are also found. Some are able to produce forward movement by having the waves start at the front of the foot and pass posteriorly. Muscles involved in the locomotion of the sole also serve in the postural maintenance of the heavy shell and visceral mass. Leaping in the Strombidae may take place by extending the foot, attaching its opercular blade to a substratum, and then producing a quick, powerful contraction of the column muscles.

On the posterior surface of the foot in many snails there is a horny plate, the operculum, which forms a closing lid over the shell aperture when the head and foot are retracted into the shell. Certain gastropods also have a ridge or fold called the epipodium along the lower edge on each side of the foot. It bears appendages and sense organs and may be homologous to the funnel of cephalopods. The foot along with the head is connected to the visceral mass by the neck because there is no boundary between foot and head.

The mantle shows some variations that are correlated with the shape of the shell and the nature of respiration. The visceral coiled mass is enclosed in the mantle, which projects as a free fold at the anterior end down over the head and foot like a skirt. The cavity between the mantle and head serves to carry out respiration by gills or lungs. Those snails with gills usually have a wide opening to the outside, although in some a part of the mantle forms a siphon through which water is drawn by ciliary action for breathing. In air-breathing gastropods, when the cavity forms a lung, the edge of the mantle forms a small opening (pneumostome) for the admission of air by fitting tightly against the neck or foot. The mantle edge, by

FIG. 10-19

A, *Carinaria,* a free-swimming, pelagic pteropod. Shell is reduced, and foot is modified into strong ventral fin. It swims ventral side up. **B,** *Cavolina,* the sea butterfly. Transparent shell is uncoiled and vase shaped. Ciliated filter tracts on fins direct small organisms to mouth where radula assists in swallowing. (**A** modified from Cambridge Natural History; **B** modified from Yonge.)

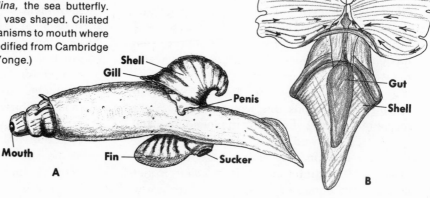

FIG. 10-20

Sea butterfly, a swimming pteropod. Long fins, or parapodia, are lateral projections of foot. (Photograph by Roman Vishniac, courtesy Encyclopaedia Britannica Films, Inc.)

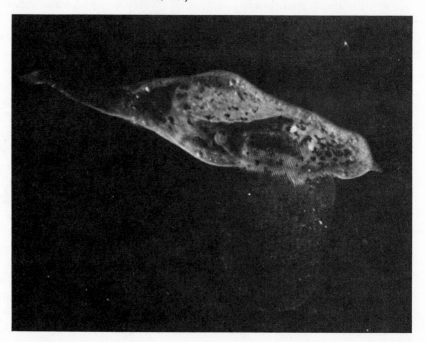

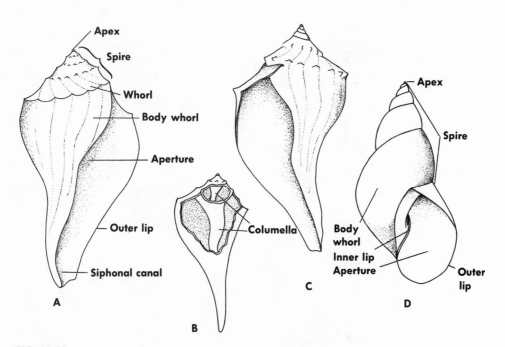

FIG. 10-21

Gastropod shell. **A** and **B**, *Busycon carica*, a dextral (right-handed) marine snail. **C**, *Busycon contrarium (perversum)*, a sinistral (left-handed) snail. **D**, *Lymnaea*, a freshwater snail.

not fusing completely, may form a slit in some gastropods.

The mantle secretes the shell, which is of one piece but highly modified into many shapes. The typical shell is spirally coiled into more or less a cone shape. Such a shell is made of tubular whorls, with the smallest whorl at the apex and successively larger ones to the aperture of the shell (Fig. 10-21). The whorls are coiled around a central axis called the columella. If, when the shell is held with the spire pointed upward and the aperture facing the observer, the aperture is to the right of the columella, or shell axis, the shell is said to be right-handed (dextral, or clockwise); if the aperture is to the left, it is said to be left-handed (sinistral, or counterclockwise) (Fig. 10-21, *A*, and *C*). Only a relatively few shells are sinistral, although some species may have either a sinistral or dextral type.

The larval shell of gastropods, called a protoconch, is usually a plate or is caplike. The protoconch is represented in the adults by the small apex whorl and is usually smooth and not sculptured. In its development the protoconch becomes coiled by the deposition of mineral salts round the mantle edge at the aperture end. The gastropod shell has the three layers characteristic of most mollusk shells—the thin outer periostracum of horny, organic structure (conchiolin); the middle prismatic layer of vertical crystals of calcium carbonate; and the inner nacreous layer of calcium carbonate laid down in lamellar sheets.

Shells exist in a great variety of shape, color, and pattern (Fig. 10-22). Although primitively spiral, the shell in some has become uncoiled and cap shaped (*Patella*). Detorsion, or the untwisting of the original embryonic twisting, is always associated with reduced or complete loss of the shell (Anaspidea, Gymnostomata). The edges of the mantle may grow over the shell and cover most or all of its surface. This causes the shell to be internal and degenerate (slugs). Shell reduction is especially pronounced

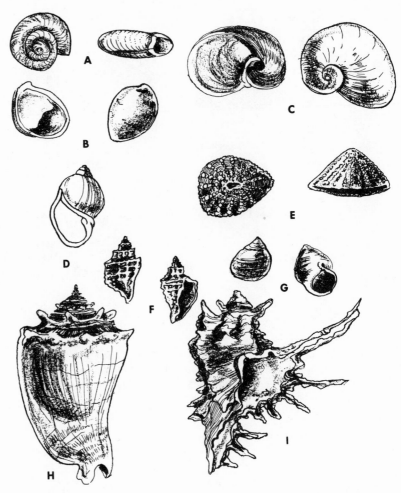

FIG. 10-22

Some gastropod shells. **A,** *Gyraulus (Planorbis)*. **B,** *Crepidula fornicata*. **C,** *Sinum perspectivum*. **D,** *Physa gyrina*. **E,** *Diodora*. **F,** *Urosalpinx*. **G,** *Littorina*. **H,** *Strombus*. **I,** *Murex*.

in opisthobranchs. In the Nudibranchia and some others the shell may disappear completely. In *Vermicularia* the whorls are so arranged that the shell looks like a corkscrew. Sculpturing on the whorls also has many patterns; in some these patterns are quite rough and horny, in others very smooth. Many tree snails in the tropics are of a bright, uniform color, whereas others are striped and varicolored. Mollusk shells illustrate the principle of allometry, or disproportionate rate of growth of various parts of a structure in relation to the whole structure. Since the shell increases by marginal increments from the mantle, growth rates at different points on the margin vary during the shell's

development. Many ecologic factors can alter this allometric relationship of the shell's growth. In oysters (*Crassostrea*) the length-width ratio of the shell is greater when the animal grows on a soft bottom than when it grows on a hard substratum of the same area. When *Mytilus* lives below low-water levels, it has a narrower and thinner shell than when it is exposed between tides.

Most gastropods have aquatic respiration

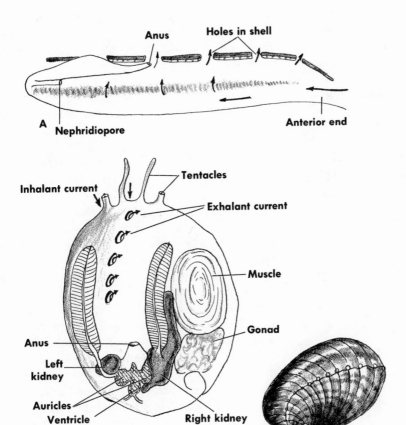

FIG. 10-23

Haliotis, the abalone (order Archaeogastropoda). **A,**
Water current through mantle cavity (diagrammatic).
B, Mantle cavity (diagrammatic). **C,** Shell. (**B** modified
from Kerkut.)

carried out by gills. Some aquatic mollusks have
to rely on cutaneous respiration because they
have no gills. Some prosobranchs of a primitive
nature have 2 gills of the bipectinate type, with
2 rows of filaments and cleft shell (Figs. 10-18,
A, and 10-23). In more specialized forms only
the left gill persists, with 1 row of filaments
along the gill (Figs. 10-18, *B,* and 10-27, *A*).
The single pectinbranchiate gill pattern of
Mesogastropoda, with the inhalant current en-

tering the mantle cavity at one side of the head
and the exhalant current leaving on the other
side where the anus is located, is often found
in prosobranchs. This arrangement helps solve
the problem of waste fouling caused by torsion.
Many shore-dwelling prosobranchs have a mantle that can carry on aerial respiration. All
kinds of modifications of respiration can be
found between those snails with gills and those
that are strictly aerial breathing (pulmonates).
Aerial respiration may exist along with gill
breathing. Some have eliminated regular gills
altogether and have respiratory outgrowths;
or true gills and outgrowths may be found together. *Littorina* has part of its gill modified

into a vascular network over the mantle roof similar to that of pulmonates. Such snails can live out of water for long periods of time. The keyhole limpet (*Diodora*) (Fig. 10-22, *E*) has a cone-shaped shell with a slot at the summit of the shell that curves forward. Although the limpet begins life in a spiral shell with an elongated slit along the shell margin, the slit during growth gradually closes except at the apex. At the same time, the apical spire becomes reduced or lost altogether. A siphon in the form of a modified mantle may project through the slot. Although the pulmonates have a mantle specialized for a lung, the pulmonate condition has also arisen separately in such groups as the Helicinidae and Cyclophoridae. Certain pulmonates (e.g., *Physa* and *Lymnaea*) have secondarily returned to fresh water; they must come to the surface to expel a bubble of air from the lung and protrude the edge of the mantle around the pneumostome into a long, muscular siphon to take in a fresh supply of air.

Internal anatomy

The alimentary canal varies somewhat with the food habits of gastropods, which are extremely varied. Modifications of the gut have thus been correlated with their nutrition. The alimentary canal usually consists of a mouth at the snoutlike end of the head; a buccal or pharyngeal cavity that is nearly always provided with a radula and often with cuticular mandibles; a long, slender esophagus, much folded within and sometimes provided with a muscular gizzard; a thin-walled stomach, which may have a cuticular lining, and a liver that opens into the stomach by 2 ducts; an intestine that may be long (herbivorous forms) or short (carnivorous forms) and bears a typhlosole ridge; and an anus that is usually found on the right side of the body near the head.

Because of torsion, the stomach is turned around so that the esophagus enters at the posterior end and the intestine leaves at the anterior end. Digestion is mostly extracellular. Digestive enzymes are produced by the digestive glands and esophageal pouches. Salivary glands, which open on each side of the radula, mostly secrete mucus for lubricating the radula and for easing food passage.

The radula, one of the most characteristic organs of gastropods, is a bandlike organ consisting of a horny basement membrane on which are rows of recurved teeth, usually arranged into two symmetric sets (Fig. 10-24). The number of teeth varies from a few to many thousands. The arrangement of the teeth is of great taxonomic value. The radula is located in a ventral diverticulum of the buccal cavity called the radula sac (Fig. 10-9). As old parts are worn away, new surfaces with teeth replace the old. A subradular membrane and odontophore cartilage support the radula. By special muscles and movement of an attached epithelial

FIG. 10-24

Some patterns of radula teeth. **A,** *Busycon carica.* **B,** *Murex regius.* **C,** *Cypraea tigris.* **D,** *Elysia viridis* (side view). **E,** *Scaphander lignarius.* These represent only a few of rows of teeth on radula and only a few of types. (**B** to **E** modified from Cambridge Natural History.)

A

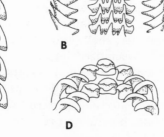

B

D

C

E

layer, the radula moves forward continuously on the surface of the buccal cavity or pharynx (N. W. Runham) and may be used for a variety of purposes, such as raking, scraping, tearing, drilling holes, and macerating, depending on the species. The radula is secreted continuously by permanent odontoblasts located posteriorly in the radula sac or gland. Runham (1963) has shown that in the pond snail *Lymnaea stagnalis* the radula moves forward at a rate of about 2.9 rows of teeth a day. At the anterior end of the radula the attached epithelium becomes detached from the radula, secretes a buccal cuticle, and disintegrates along with that end of the radula and its subradular membrane.

The stomach has no digestive function of its own but serves to hold the food that is being digested by enzymes from other sources. The products of digestion pass into the ducts of the digestive gland for absorption. The digestive gland is a bilobed structure, with the left lobe usually larger. It may be composed of ciliated tubes and glandular alveoli. In the alveoli are various cells that are secretory and resorptive in function. The secretory cells produce enzymes for digestion, and the resorptive cells also have proteolytic enzymes for intracellular digestion. In carnivorous gastropods (many prosobranchs and opisthobranchs) the radula has fewer but larger teeth. Jaws may be absent, but an extensible proboscis formed from the buccal cavity lies within a special proboscis sheath. With the aid of the radula, food is taken into the proboscis, which is retracted into the proboscis sheath. Digestion in carnivorous snails takes a different course from that in herbivorous forms. Proteolytic enzymes are discharged into the stomach, where digestion takes place, rather than in the cells of the liver.

Prosobranchs such as *Urosalpinx* (Fig. 10-22, *F*) drill holes with their radula into the shells of bivalves such as oysters. The boring mechanism involves a chemomechanical theory of penetration in which the abrasive action of the radula removes shell softened by secretions (of unknown nature) from a specific boring organ (although it may be absent in some), which may be located near the tip of the proboscis or in the sole of the foot. The boring mechanism is found in six families of prosobranchs and pulmonates. The snail sucks out the contents of the prey with its proboscis after the radula has torn the tissues apart. Another prosobranch, *Conus,* mainly a tropic form, has the radula teeth modified into long, hollow structures for stabbing its prey. A duct from a poison gland opens into the buccal cavity near the opening of the radula sac, and poison may flow into the wound made by the stabbing action of the tooth, which is attached by a slender tissue thread to the basement membrane and thrusts into the prey like a harpoon. The poison acts on the neuromuscular junction, thus paralyzing the prey. Some species of this genus are dangerous to man.

The carnivorous Nudibranchia, which feed mainly on hydroids and sea anemones, have digestive glands in the numerous papillae or cerata that cover their backs (Figs. 10-25 and 10-26, *D*). The digestive tubule of each ceras joins that of others to form a branching, diffuse digestive gland, which opens into the stomach, where digestive enzymes are discharged for extracellular digestion. The digested products are then absorbed by the stomach wall and digestive glands. Undischarged nematocysts from their coelenterate prey are carried by the digestive gland ducts to the tips (cnidosacs) of the cerata (Fig. 10-26, *E*), where they are discharged by the nudibranchs for their defense.

Ciliary feeders such as the sessile form *Crepidula* are provided with a crystalline style, a solid gelatinous protein rod that is located in a style sac at the intestinal end of the stomach and projects into the other end of the stomach. (It is recalled that, because of torsion, the stomach is turned around so that its anterior end is the intestinal one and its posterior end the esophageal.) The crystalline style (Fig. 10-10), which has evolved several times in the mollusks, is a special modification of the style

FIG. 10-25

Aeolis, a nudibranch; ventral view of preserved specimen.

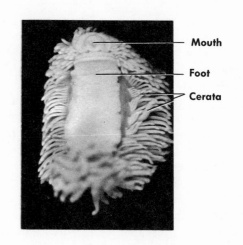

Mouth

Foot

Cerata

FIG. 10-26

Subclass Opisthobranchia. **A** to **C,** *Aplysia,* sea hare, a tectibranch. **A,** Living animal in expanded condition (dorsal side). Small thin shell lies between parapodia and is covered by mantle. **B,** Chain of 3; middle one acts as male for female below and as female for male above. **C,** Internal structure. **D,** *Aeolis,* sea slug, a nudibranch snail; cerata contain branches of digestive gland. **E,** Section of ceras of *Aeolis.* (**A** to **C** modified from Eales; **E** modified from Lankester.)

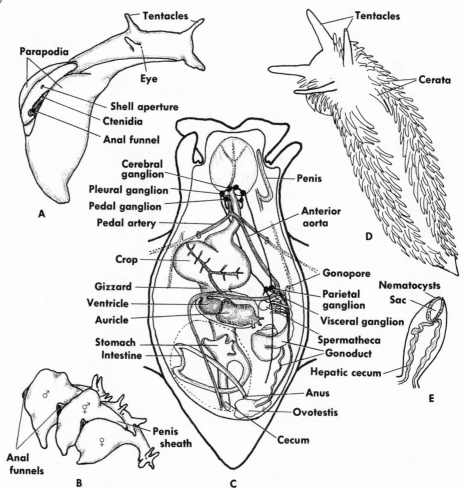

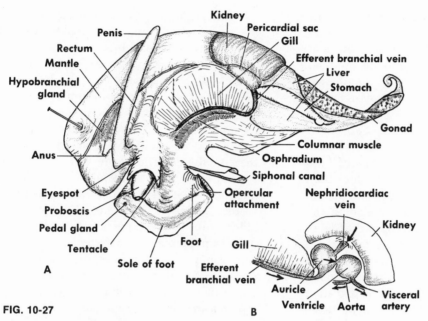

FIG. 10-27

A, Surface anatomy of male *Busycon* removed from shell
(order Neogastropoda). **B,** Dissection of heart and chief
vessels of *Busycon.* (Drawn from life.)

FIG. 10-28

Side view of pulmonate land snail with shell removed
(semidiagrammatic). (Modified from Hegner.)

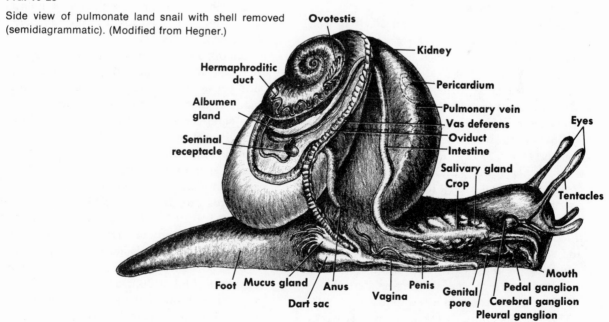

sac. Amylolytic enzymes are adsorbed on the protein rod. The rod is rotated by the ciliary action of the style sac, releasing the enzymes of the rod into the stomach and mixing the food contents. In such ciliary feeders the lengthened gill filaments aid in trapping the plankton as the water current passes through the mantle cavity. The particles of food are carried by gill cilia into a longitudinal groove alongside the gill filaments, where they are formed into mucous strings. The strings of food are then pulled into the mouth by the radula and passed to the stomach. A special ciliary field in the floor of the stomach carries the food to the region of the gastric plate, where the enzymes released by the crystalline rod digest the food. Digested products and small particles are carried into the ducts of the digestive gland where the digestion is completed intracellularly. The larger particles are passed directly by a ciliary groove to the intestine. Many other modifications of the style sac are found among the gastropods, depending on feeding habits.

A few gastropods are parasitic on mollusks and echinoderms. Many of these ectoparasites or endoparasites belong to the Cephalaspidea of the opisthobranchs.

The circulatory system in gastropods is well developed for a mollusk. The heart is usually in front of the visceral mass because of torsion (Figs. 10-27 and 10-28), but may be located posteriorly in some of the Opisthobranchia that have completed detorsion (Fig. 10-26). The primitive members have 2 auricles (Fig. 10-23, B), but only the left auricle is found in other gastropods (Fig. 10-27). From the single ventricle an aorta arises and branches to form the posterior visceral artery to the viscera and an anterior cephalic artery to the head and foot. The venous blood is carried directly to the gills through sinuses, or it may first go to the kidneys and then into the branchial circulation before returning to the heart. The blood is usually colorless but may contain hemoglobin (*Planorbis*) or hemocyanin in a few genera.

The coelom in gastropods is restricted to the pericardium, kidneys, and to some extent, the reproductive gland. The large body cavity is the hemocoel. The kidneys are mostly paired in the Archaeogastropoda (Aspidobranchia), but in the other gastropods only the left is functional. The kidney, or nephridium, connects with the pericardium and is much folded to increase its surface. The connection with the pericardial cavity may be made by a renopericardial canal that is ciliated. The waste from the kidney is carried away by a thin-walled ureter. In mesogastropods the single left kidney has only an excretory function; whereas the former right kidney tube is now a gonoduct for genital products. In pulmonates the ureter opens to the outside behind the pneumostome and above the anus; in opisthobranchs and prosobranchs it opens inside the mantle cavity where waste is swept away by the branchial current. Aquatic gastropods excrete ammonia or its compounds as their chief waste, whereas land forms, to conserve water, excrete the insoluble uric acid. Terrestrial forms usually live in fairly humid surroundings, or they are nocturnal. Under very dry conditions they may aestivate in the soil and remain inactive. To protect themselves they may secrete a special calcareous membrane over the shell aperture during the period of inactivity.

Gastropods in general have a well-developed nervous system consisting of nerves, ganglia, and sense organs. Its basic organization is the same as that of other mollusks, although torsion has complicated certain aspects of it. The molluscan nervous system is mainly a diffuse one in its primitive plan (Fig. 10-13, C), and in many higher gastropods this plan has been little altered. Ganglionic nerve centers and commissures are the same as those in other mollusks (Figs. 10-26, C, 10-28, and 10-29, C). A pair of cerebral ganglia lies over the posterior esophagus and sends nerves to the tentacles, eyes, and buccal ganglia. The buccal ganglia innervate the radula muscles and adjacent structures. From each cerebral ganglion a pedal nerve cord passes on each side of the esophagus

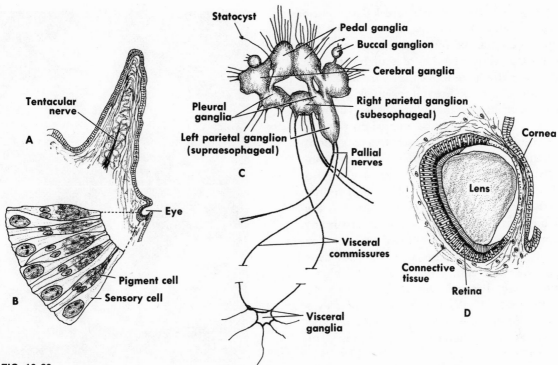

FIG. 10-29

Nervous and sensory organs. **A,** Cephalic tentacle of *Patella,* longitudinal section showing eye (order Archaeogastropoda). **B,** Enlarged section of eye epithelium of *Patella.* **C,** Nerve centers of *Buccinum* (order Neogastropoda). Commissure connecting cerebral ganglia has been severed and ganglia laid back. View is from inside nerve collar. **D,** Transverse section of eye of *Buccinum.* (A and B modified from Davis and Fleure; C and D modified from Dakin.)

to a pair of pedal ganglia in the middle of the foot. The pedal ganglia innervate the muscles of the foot. Also from the cerebral ganglia, a pair of visceral nerves passes posteriorly to the paired visceral ganglia in the visceral mass. The visceral ganglia innervate the visceral mass. Along the visceral nerves are located two other pairs of ganglia, the pleural ganglia near the cerebral ganglia and the parietal ganglia more posteriorly toward the visceral ganglia. The pleural ganglia send nerves to the mantle and the muscles of the columella. Commissures from these ganglia also pass to the pedal gan-

glia. The parietal ganglia innervate the mantle, gills, and osphradia.

In the brain cells (some of which are a millimeter in diameter) of the large Mesogastropod (*Tritonia*) it has been possible to pinpoint with precision (with fine electrodes) the brain cells involved in escape behavior (A. O. D. Willows, 1971).

Torsion has produced a twisting of the basic molluscan type of nervous system into a figure 8 (streptoneury) (Fig. 10-29, *C*). Because of their position, the right parietal ganglion becomes the left or supraparietal ganglion, and the left ganglion the right or infraparietal ganglion after torsion (Figs. 10-16 and 10-29, *C*). The locations of the pleural and cerebral ganglia are not changed by the torsion process. In most gastropods the ganglia tend to be concentrated, and the commissures and connectives between ganglia shortened. A secondary bilateral symmetry of the nervous system is an evolutionary trend in most gastropods. In the whelks such as *Busycon* and *Buccinum* (Fig. 10-29, *C*), most ganglia, except the visceral, are located close together around the esophagus just below the cerebral ganglia. In pulmonates there is an even greater concentration of ganglia, for the visceral ganglia have moved forward also.

The sense organs are represented by eyes, statocysts, osphradia, tactile organs, and chemoreceptors. The eyes may be located on the tips of the tentacles or at their bases. Eyes may be of two or three types. In the simplest or more primitive types (e.g., limpets) the eye is merely a cuplike invagination with a pigmented retinal layer that is exposed to the water through the open aperture (Fig. 10-29, *A* and *B*). In many gastropods the aperture is closed by a cornea and lens formed from the fusion of the edges of the aperture (Fig. 10-29, *D*). The photoreceptors are directed toward the source of light (direct eyes). Perhaps most of their eyes function merely to detect light. Tentacles also bear tactile and chemoreceptor organs. The tentacles (rhinophores) of nudibranchs may have olfactory as well as other functions. Usually the statocysts (Fig. 10-29, *C*) are found in the foot near the pedal ganglia, but in nudibranchs and some others they may be located near the cerebral ganglia. The Archaeogastropoda have an osphradium located anterior to and above the attachment of each of its 2 gills; a single osphradium is found in most other gastropods that have only 1 gill (Fig. 10-27). This sense organ may test the chemical quality of the branchial water or the sediment that the water contains. In terrestrial gastropods the osphradia have been replaced by rhinophores on the tentacles.

A. McClary (1968) has studied the change in form, size, and number of statoliths in the statocyst of the prosobranch snail *Pomacea paludosa* under controlled laboratory conditions. He finds the two types of statoliths shown in Fig. 10-30 and that both size and number increase with age up to 120 days. Statoliths are calcareous and are formed by the crystallization of the vesicle fluid secreted by the wall of the statocyst.

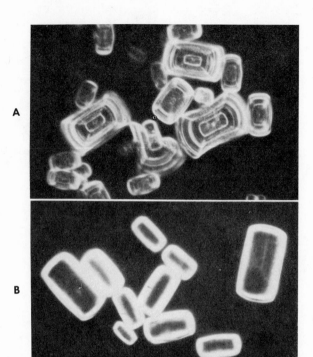

FIG. 10-30

Statoliths from a statocyst of the gastropod *Pomacea paludosa* (about 25 to 100 microns long). **A,** Lamellate statoliths from a snail 70 days old. **B,** Nonlamellate statoliths. Lamellate type is more common. In most snails a pair of statocysts, or organs of equilibrium, are found in foot near pedal ganglia. (From McClary, A. 1968. Trans. Amer. Microscop. Soc. **87:**322-328.)

Gastropods may be dioecious or monoecious (Fig. 10-32, *A* and *B*). In all gastropods the gonad (ovary, testis, or ovotestis) is unpaired and located near the apex of the visceral hump. In the Archaeogastropoda the kidney acts as a gonoduct for the discharge of the gametes into the mantle cavity. Fertilization in such a primitive condition takes place in sea water. In most other gastropods there is a separate gonoduct, which is partly made up of that portion of the right kidney remaining after the kidney has degenerated. The gonoduct is lengthened by the addition of a pallial duct formed from the mantle. Differentiation of this pallial portion of the gonoduct provides for sperm storage and egg membrane formation. A gonoduct separate and distinct from the kidney, with its provision for forming membranes, has made possible internal fertilization in contrast to the external fertilization of the primitive Archaeogastropoda. A penis is found in most prosobranchs other than the Archaeogastropoda (Fig. 10-27). The penis is an elongated, tubular structure formed by a fold or extension of the body wall. In most gastropods with a penis the vas deferens is a closed tube and opens at the penial tip (Fig. 10-32, A). A prostate gland for the production of seminal fluid is formed by part of the vas deferens. That part of the genital duct leading directly from the testis is usually very much coiled for the storage of sperm. The pallial section of the female oviduct is modified to form an albumen gland and a jelly or capsule gland. Some species lay their eggs enclosed in jelly from the jelly gland; others have the eggs within a capsule. A seminal receptacle for receiving sperm from the penis is found in that part of the oviduct proximal to the ovary (Fig. 10-28). There the eggs are fertilized before they receive their accessory coats or membranes in the secretory parts of the oviduct. A more complicated method of sperm transfer to the seminal receptacles is found in some (e.g., *Urosalpinx*, *Busycon*), in which the sperm are discharged first into a bursa at the distal end of of the oviduct and then make their way posteriorly along a ventral groove to the seminal receptacle.

Most of the hermaphroditic species of gastropods (pulmonates and opisthobranchs) are simultaneously male- and female-gamete producing (W. R. Coe, 1936 and 1942), although some are protandrous; i.e., they first form functional sperm and later produce the mature eggs (*Crepidula*, *Aplysia*, Fig. 10-26, B). Coe has shown in *Crepidula* the effect of association of individuals of the opposite sex. This snail forms a group of individuals attached in the form of a chain with the youngest and smallest male on top of the chain. At the bottom of the chain are the mature females, and hermaphro-

dites occupy the intermediate position. He found no evidence of a masculinizing hormone from the female but suggested that the male may have received via its sense organs stimuli that, through the mediation of hormones, may influence male characteristics. The hermaphroditic system of the pulmonates may be taken as typical of such a system (Fig. 10-28). The single gonad, called an ovotestis, is located at the apex of the visceral mass (Fig. 10-31). Both eggs and sperm are produced in the same follicle of the ovotestis. A common hermaphroditic duct, formed from the gonadal and renal portions of the genital duct, carries both eggs and sperm to the albumen gland. At the terminal part of the hermaphroditic duct is a seminal receptacle or fertilization pouch where sperm is stored and fertilization may occur. In the region of the albumen gland the hermaphroditic gland begins to separate into 2 ducts, the oviduct and the sperm duct (vas deferens). This separation of the 2 ducts at first is only partly complete morphologically, but eventually the vas deferens breaks apart from the oviduct and passes to the tip of the penis; the oviduct continues straight to the vagina, which communicates with the common genital atrium on the right side of the snail's body (Fig. 10-31). The anterior part of the female tract contains many accessory reproductive structures. First of all, the female tract in this region is divided into the oviduct and a thick-walled vagina. The vagina communicates with an elongated spermatheca, which is supposed to receive sperm directly from the penis. Connected to the vaginal region also is a mucous gland for lubricating purposes and a dart sac (Fig. 10-28), which secretes calcareous darts (Fig. 10-32, D) for sex-stimulating purposes.

Cross-fertilization is probably the rule in most pulmonates. When two snails prepare for mutual insemination, they approach each other with their genital atria everted and by muscular action of the dart sac each shoots a dart into the other. New darts are formed by the epithelium of the dart sac. The two snails then copulate

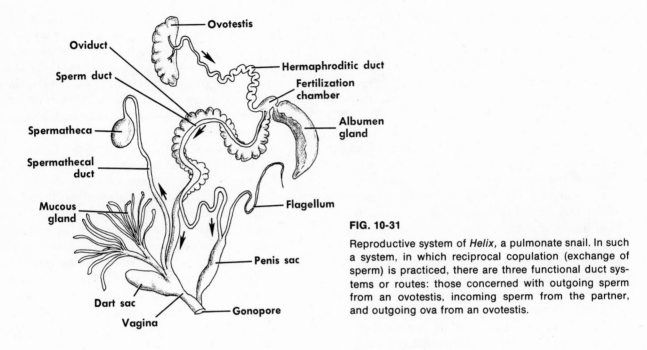

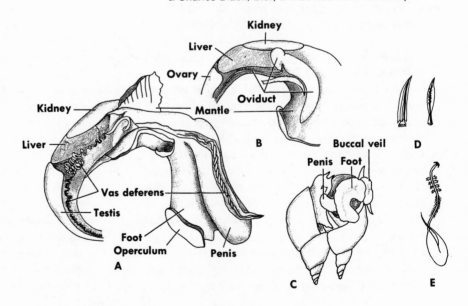

FIG. 10-31

Reproductive system of *Helix,* a pulmonate snail. In such a system, in which reciprocal copulation (exchange of sperm) is practiced, there are three functional duct systems or routes: those concerned with outgoing sperm from an ovotestis, incoming sperm from the partner, and outgoing ova from an ovotestis.

FIG. 10-32

Reproductive system. **A,** *Buccinum,* male. **B,** *Buccinum,* female. **C,** Copulation in *Lymnaea stagnalis,* left one acting as male. **D,** Darts of (left) *Helix hortensis* and (right) *Helix aspersa.* (About $\frac{5}{16}$ inch long.) **E,** Spermatophore of *Nanina.* (**A** and **B** modified from Dakin; **C** and **E** redrawn from Lankester: Treatise on zoology, Adam & Charles Black, Ltd.; **D** modified from Ashford.)

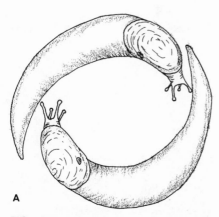

A   B

FIG. 10-33

Courtship and copulation of the naked pulmonate *Arion*.
A, Courtship, in which they trail each other in circles,
each licking caudal gland of the other. **B,** Copulation.
Each snail makes contact with the other with their genital
atria and exchange spermatophores. (After Ferrussac, redrawn from Hyman.)

(Fig. 10-33), each inserting its penis into the
vagina of the other. The sperm are bound together as spermatophores (Fig. 10-31), and
each snail discharges its spermatophores into
the reproductive tract of the other. The spermatophores pass up the spermathecal duct to bulb-
shaped spermatheca, where the sperm are set
free. Later the sperm pass to the oviduct and
up to the fertilization pouch, where fertilization
occurs. After fertilization the eggs take on
large quantities of albumen before being discharged from the body.

Some pulmonates may undergo a courtship
behavior before copulating (Fig. 10-33).

DEVELOPMENT

The eggs of primitive gastropods are fertilized
externally and may be laid singly or in masses
held together by gelatinous material. Higher
gastropods usually have some form of membrane around their eggs. Fertilization is usually
internal, and eggs are laid soon after. Some lay
their eggs in a capsule containing many eggs
embedded in albumen (certain fresh-water pulmonates). The eggs may be in a simple gelati-

nous mass (*Lymnaea*) (B. M. McCraw, 1961),
or they may each have a shell enclosing the
egg and albumen to prevent evaporation (terrestrial pulmonates). Most land snails lay their
eggs in holes within the ground or under logs.
Many prosobranchs attach their eggs to snail
shells, their own or those of adjacent ones.
Aquatic snails often attach their egg masses
to vegetation.

Cleavage in gastropods is spiral and total. The
plan of this cleavage has already been described
(see pp. 99 to 101). Quartets are formed in
the usual way. Radial symmetry is characteristic of the 32-cell stage, and bilateral symmetry
of the 64-cell stage (*Patella*, Fig. 10-34, *A*). Gastrulation is usually by epiboly (Fig. 10-34, *C*),
and the macromeres give rise to endoderm and
mesoderm, whereas the ectoderm comes from
the micromeres. Only in the primitive Aspidobranchia is the trochophore larval stage free
swimming (Fig. 10-11, *C*). In others this stage
is passed in the egg membrane. Most marine
gastropods have a free-swimming veliger larva
derived from the trochophore (Fig. 10-11, *D*).
The veliger is provided with a special swimming
structure of 2 ciliated lobes or folds, the velum.
The velum has arisen as an outgrowth of the
girdle of long cilia (the prototroch) characteristic of the trochophore. Structural details that
started in the trochophore stage undergo a
further development in the veliger, and some
structures now appear for the first time. The

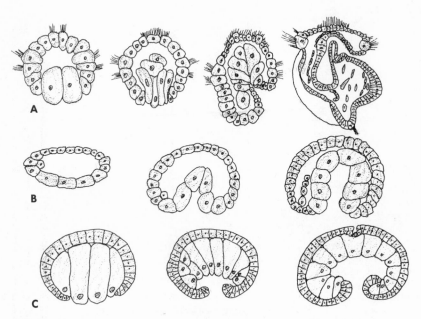

FIG. 10-34

Cleavage and gastrulation in, **A,** *Patella* (ingression); **B,** *Littorina* (invagination); **C,** *Crepidula* (epiboly).

foot, tentacles, and eyes now emerge; and the shell (cap shaped in the trochophore) assumes a spiral shape. Torsion also occurs in the late veliger stage, in which the visceral mass twists about 180 degrees in relation to the foot and head (see pp. 358 to 361). Torsion may occur in a few minutes or it may involve many days of development. The growth of the foot is one of the characteristic aspects of the veliger stage; the foot actually assumes a creeping form, and the larva relies less on the receding velum for locomotion. In the subclass Opisthobranchia, in which detorsion is a marked tendency, the torsion occurs to a degree in the veliger before detorsion sets in. In the shell-less nudibranchs a shell appears in the veliger stage but is lost in its metamorphosis. All pulmonates and some prosobranchs (marine and freshwater) have no free-swimming larvae; but the trochophore and veliger stages occur, at least partly, within the egg. In these a small juvenile snail is hatched from the egg. Only a few gas-

tropods are ovoviviparous, such as the freshwater prosobranchs (*Viviparus* and *Campeloma*) and certain species of *Littorina*.

■ CLASS PELECYPODA (BIVALVIA)

The pelecypods represent one of the three largest classes of mollusks. They have a worldwide distribution and are mostly marine, although some have invaded brackish and fresh water. There are no terrestrial species. They number more than 10,000 living species and many thousands of fossil forms. They are characterized by a bilaterally symmetric body that is also laterally compressed. The body is completely enclosed in a mantle, which is subdivided into lateral lobes that secrete a shell of 2 calcareous valves joined together dorsally by an elastic cuticular ligament and hinge. Adductor muscles can close the shell ventrally and work in opposition to the elastic ligament that tends to open the valves. The head is reduced to a mouth without a radula or buccal mass. The mouth is surrounded by labial palps. The mantle cavity is unusually capacious and bears 2 large gills (ctenidia), 1 on each side of the body. The ventral foot is wedge shaped and adapted for locomotion through mud or sand. Most pelecypods are ciliary feeders, and feeding is accomplished mainly by the ciliated gills and labial palps, which form a mechanism of collecting mucous strands of food brought in by the incurrent water.

Although they have undergone some modifications of structure and habits, most pelecypods are somewhat sedentary and are found mainly in soft-bottom habitats. Their morphology has emphasized such a mode of living because their food of plankton and organic detritus is brought to them directly by water current rather than being sought.

The fossil record shows that they first appeared in the early Cambrian period. During the Ordovician period they became differentiated and widely distributed. It is thought by some authorities that pelecypods arose from limpetlike monoplacophorans when calcifications occurred at two places in the shell and became hinged by the outer layer of periostracum. Their gill structure suggests that they may have come from a primitive gastropod such as *Haliotis*. By the Carboniferous period, bivalves became very numerous and displaced the brachiopods. Unlike the gastropods, pelecypods probably reached their peak in the ancient past and may be said now to be undergoing a slow decline. Fossil bivalves are important in the study of paleoecology because the sensitivity of pelecypods to temperature and salinity gives a clue to ecologic conditions under which they flourished.

The evolution of the pelecypods is not so striking as that of the gastropods. The group has undergone adaptive radiation and convergence. Some show a tendency toward fusion of the mantle margins to produce long siphons for burrowing forms. Many seem to have acquired these trends independently. Each subclass usually has certain morphologic characteristics that are emphasized for a special function. Thus, the Protobranchia have become superficial burrowers, and the Lamellibranchia utilize byssal attachment to a substratum.

GENERAL MORPHOLOGY
External anatomy

The body of a pelecypod is composed of a mantle (with a shell), foot, and visceral mass. There is no head in a strict sense. The main axis of the bilaterally symmetric body is the visceral mass. The mouth and labial palps located at the anterior end of the visceral hump are usually not set off from the rest of the body. The visceral mass is found mainly in the dorsal part of the body and is covered by the mantle that hangs down on each side like a skirt and

A

B

FIG. 10-35

A, *Lima,* the file shell. **B,** Posterior end of freshwater clam showing exhalant (dorsal) and inhalant (ventral) apertures formed by posterior edges of mantle lobes. (From Encyclopaedia Britannica films: **A,** "Pond Life"; **B,** "Adaptive Radiation—the Mollusks.")

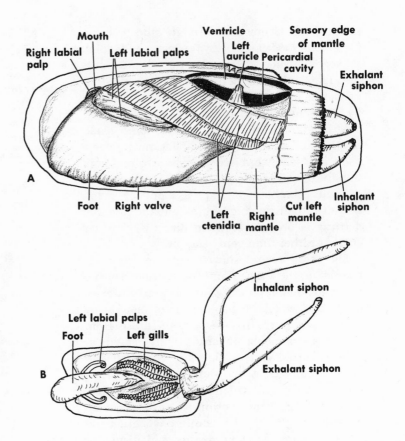

FIG. 10-36

A, *Tagelus gibbus* (order Heterodonta); left valve, mantle, and pericardial wall removed. **B,** *Tagelus,* ventral view with valves open, as seen in small aquarium where oxygenation was inadequate. Gaping valves afford good view of gills and siphons.

forms a sheet of tissue between the valves of the shell. Between the visceral mass and the free side of the mantle is the mantle cavity, which is unusually large in most bivalves. The ventral margins of the 2 loose folds of the mantle may be free and unattached. In most bivalves the 2 mantle lobes are united with each other at one or more points below the ventral surface of the visceral mass. Some have only one posterior junction between the inhalant and exhalant orifices. In the freshwater clams *Anodonta* and *Unio* the mantle edges are not completely fused to form the inhalant aperture, although a functional orifice with frilled edges is present (Fig. 10-35, *B*). In many bivalves the orifices are prolonged as siphons so that the animal can retain communication with the water while burrowing beneath the surface of the mud (*Tagelus,* Figs. 10-36 and 10-37;

Pholas, Fig. 10-39, *C*). Shell impressions of sharply recurved, inward pallial lines often determine whether or not siphons have been present in the living animal (Fig. 10-37, *B*).

Many modifications of mantle fusion are found among pelecypods. In *Teredo* the siphons secrete a calcareous tube (C. E. Lane, 1961). The foot projects down as a highly muscular extension of the visceral mass and can be projected beyond the ventral margin of the valves. A common type of mantle fusion produces 3 apertures: a posterior exhalant aperture, a posterior inhalant aperture, and a large anterior pedal orifice through which the foot protrudes (*Ensis, Cardium,* and *Solemya*). A fourth orifice posterior to the pedal orifice is formed in some razor clams by a third fusion of the mantle edges.

The mantle and its derivatives are important in the basic organization plan of bivalves. Besides having its margins modified into apertures and siphons, the mantle secretes the shell, and its gills are modified chiefly for secondary feeding and developmental functions. The margin of the mantle bears 3 folds: an outer secretory fold, a middle sensory fold (sometimes with tentacles and eyes), and an inner muscular fold (Fig. 10-7). The outer fold contains an inner surface that secretes the noncalcareous periostracum and an outer surface that secretes

FIG. 10-37

A, *Tagelus gibbus,* the stubby razor clam, from right side. **B,** Inside of left valve of *Tagelus* showing muscle scars. **C** and **D,** Frontal sections of bivalve (diagrammatic) showing function of adductors and hinge ligament. **C,** Adductor is contracted, stretching hinge ligament and closing valves. **D,** Adductor is relaxed, allowing hinge ligament to pull valves apart.

the outer calcareous layer of the shell. The inner calcareous (narcreous) layer is also secreted by the outer surface of the mantle. The mantle is connected to the shell by pallial muscles (absent in gastropods) a short distance from the edge of the shell. This line of mantle attachment to the shell is represented on the inner surface of the shell as the pallial scar line (Fig. 10-37, *B*). The mantle margins have undergone a fusion pattern ventrally, posteriorly and dorsally in the evolution of the bivalves, although in the primitive condition they are free. The inner fold fuses first, followed by the middle fold, and then by the inner surface of the outer folds.

The shell of pelecypods is typically made up of 2 convex valves of various shapes and contours (Fig. 10-38). It first appears in the embryo as a single shell. Later in its development, 2 separate calcareous plates are secreted by the left and right regions of the shell gland. The 2 valves of the shell are joined dorsally by an elastic hinge ligament, which acts as a spring by exerting tension on the valves, forcing them apart when the adductors relax (Fig. 10-37). The ligament (a persistent rudiment of the

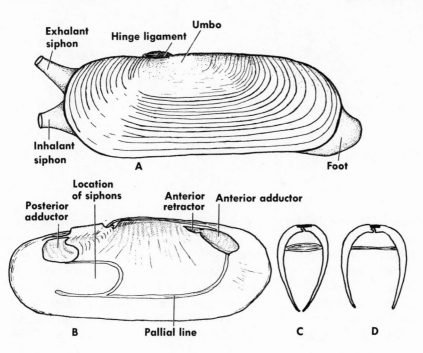

FIG. 10-38

Some pelecypod shells. **A,** *Dosinia discus.* **B,** *Chione cancellata.* **C,** *Dinocardium robustum.* **D,** *Tridacna gigas.* **E,** *Lampsilis ventricosa.* **F,** *Pecten irradians.* **B,** *Atrina rigida.* **H,** *Yoldia.* **I,** *Mya.*

larval shell) is made up of conchiolin, or the horny substances of the periostracum. The valves are also usually articulated by a series of interlocking teeth located on the hinge line just beneath the hinge ligament. The shell is composed of the same three layers described in the discussion of gastropods—periostracum, prismatic, and nacreous (Fig. 10-6). The dorsal margin of each valve bears a prominence point near the hinge ligament called the umbo, or beak, which is the oldest part of the shell and around which are the concentric lines of shell growth. Each umbo may point slightly anteriorly so that it is usually possible to determine right and left valves. On the external surface of each valve the different areas may be called the anterior slope, the posterior slope, and the central disk. The shell valves are pulled together by 2, usually transverse, adductor muscles (anterior and posterior) attached to the inner surfaces of the valves. Scars on the valves indicate where these adductor muscles are attached (Fig. 10-37, B). The adductor muscles act in opposition to the hinge ligament, which pulls the top edges of the 2 valves together, causing the lower edges of the valves to gape apart unless they are closed by the strong adductors (Fig. 10-37, D). Although in primitive bivalves both adductors are of about the same size, the anterior adductor in some becomes reduced (Fig. 10-45); and in oysters and scallops (Fig. 10-44) it is completely absent. To compensate for its absence, the posterior adductor occupies a more central position. Adductor muscles have developed from the inner muscular fold of the mantle. In the scallop and some other bivalves the single adductor muscle consists of two kinds of fibers—quick striated fibers for quick closing of the valves and the catch smooth muscle fibers for holding the valves together.

Pelecypod shells are found in a great variety of shapes, sizes, colors, and surface sculpturings. In some genera the 2 valves are of unequal size and of different shape (Ostrea, Pecten). One of the valves may be permanently attached to a substratum. The shell is long and somewhat tubular in the razor clams (Ensis [Fig. 10-7, E], Tagelus [Fig. 10-37]). In the shipworm the shell is rudimentary and adapted for drilling. In some the mantle may be reflected over the entire surface of the shell (Scioberetia). Ornamentation of the shell is often brought about through the interruption of growth lines by ribs radiating from the umbo. Shell size varies from certain tiny, freshwater forms smaller than a fingernail to the giant clam Tridacna (Fig. 10-38, D), which may weigh several hundred pounds.

Whenever a foreign object becomes lodged between the mantle and the shell, concentric layers of nacreous shell are laid down around it. Pearls are formed in this manner. The most valuable pearls are produced by the pearl oysters (Pteriidae), with which artificial pearl culture is conducted by introducing a tiny piece of shell between the mantle and shell. Pearl buttons are made from shells of the freshwater family Unionidae.

The foot is highly muscular, and in a few protobranchs (Yoldia, Nucula) it may be used for creeping like a snail. However, most bivalves use it as a burrowing organ. The highly specialized Solemya uses its foot as both a burrowing and a swimming organ. Even in those bivalves that creep the foot can be folded to form a bladelike edge. In most cases the foot is bilaterally compressed, with the lower edge keeled to form a bladelike structure (hatchet footed). E. R. Trueman (1966) has described the mechanism of burrowing in Anodonta, Cardium, Mercenaria, and other bivalves. The foot is first extended by the contraction of a pair of pedal protractor muscles, which run from each side of the foot to the adjacent valve. Then the extended foot probes around in the substrate. As the foot is protruded, the valves begin to close (by adductor muscles) and hydrostatic pressure is increased both in the foot and in the whole body within the shell. Water from the mantle cavity and blood are ejected into the pedal hemocoel. To facilitate easier penetration, the water tends to liquify the sand around the foot. The dilation of the foot occurs and gives a firm pedal anchorage. The shell is pulled downward and toward the swollen end of the foot by the contraction of a pair of anterior and a pair of posterior retractor muscles. After pedal contraction, relaxation follows and the valves gape. Some variations of this method are involved, especially in those bivalves that burrow deeply.

Some clams such as Ensis and Laevicardium can leap and swim as well as burrow. The leaping movement is achieved by first extending the foot, looping it back beneath the shell, and suddenly straightening it (Fig. 10-7, B). Swimming is effected by alternately thrusting out and withdrawing the foot, with a simultaneous closing of the shell valves to expel water from the centrally closed valve through an anterior pedal gap. This modified version of jet propulsion forces the clam backward by jerks. The scallop *Pecten*, which swims by jerky movements, also uses jet propulsion. By rapidly shutting and opening its valves and forcing water out dorsally from each side of the hinge, the animal moves forward. File shells (Fig. 10-35, *A*) can swim slowly by clapping their valves, which are held vertically. Sometimes the water current passes out ventrally and the hinge joint becomes anterior. The adductor muscle in *Pecten* is single (Fig. 10-44), but it is divided into two unequal parts. The larger of these parts produces the rapid movement of swimming; the smaller keeps the valves closed by continual contraction. *Pecten* has a very reduced foot and cannot burrow.

Instead of being fastened down by the attachment of 1 valve, as in oysters, many marine bivalves are attached to the substratum by strong byssal threads (Fig. 10-39). These threads are secreted by a gland located in the middle line of the surface of the foot in a byssus pit (Fig. 10-39, *B*), which may correspond to the ventral pedal mucous gland of gastropods. The threads of *Mytilus* are composed of conchiolin and are pushed to the exterior singly when secreted. When the tip of the foot is in contact with the substratum, the fluidlike thread flows down a groove to the tip, hardens on contact with water, and attaches the clam to the substratum. The mussel withdraws its foot, changes its position, and attaches another thread in the same manner. In this way many threads are attached.

All pelecypods, except the Septibranchia, have a pair of gills. In some cases the single gill on each side has been folded to form a double gill. The gills have in some ways kept pace with the formation of the large, lateral mantle cavity and have extended anteriorly toward the mouth with a great increase in number and length of filaments. The mantle cavity on each side of the body is divided into a small dorsal exhalant (suprabranchial) chamber and a large ventral inhalant chamber.

In the primitive Protobranchia (*Nucula*) each gill is composed of a central axis bearing on its opposite sides a series of filaments that are loosely connected by tufts of cilia (Fig. 10-40, *A*). Ciliary connections also lock the filaments to the sides of the foot and to the mantle. This basic plan has undergone many modifications among the different pelecypods. In one common type the rows of filaments become parallel to each other and hang downward in the mantle cavity, with the extremities of the filaments turned up at the ends to form the descending (attached to the axis) and ascending limbs for each filament. The gill is thus made up of four series (lamellae) of filaments, or each gill is a W-shaped structure with each half of 2 thin platelike lamellae (Fig. 10-40, *C*). At the ventral margin, where each filament is upturned, is a food groove. Adjacent limbs are

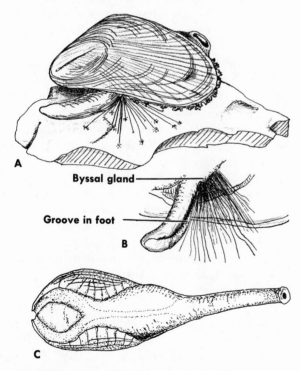

Byssal gland

Groove in foot

FIG. 10-39

A, *Mytilus* attached by byssal threads (order Anisomyaria). **B,** Foot and byssal gland of *Mytilus*. **C,** *Pholas (Barnea)* (order Adapedonta), a rock borer.

connected by interlamellar junctions, and adjacent filaments are attached by tufts of cilia or disks. Each gill of 2 lamellae thus becomes a pair of demibranchs, with each demibranch composed of 1 descending and 1 ascending lamella.

The degree to which the gills are fused together to form continuous plates varies with the different subclasses, reaching its greatest development in the Eulamellibranchia. In this group ciliary junctions between the filaments have been replaced by tissue fusion, and the lamellae are sheets of tissues (Fig. 10-41). The

interlamellar junctions extend dorsoventrally the entire length of the lamellae and partition the interlamellar space into vertical water tubes (Fig. 10-41). Blood is carried through the afferent and efferent vessels in the interlamellar junctions for aeration. By fusing with the upper surface of the mantle on the outside and the foot on the inside, the tips of the ascending limbs have separated the mantle chamber horizontally into an upper suprabranchial (exhalant) chamber and a lower (inhalant) cavity. A current of water entering by the ventral aperture, or siphon, circulates among the original filaments and passes through ostia in the lamellae to the water tubes. Passing dorsally through the water tubes, the water then flows into the suprabranchial chamber and out the exhalant opening (aperture or siphon) (Fig. 10-42). Some pelecypods such as *Ensis* have increased the lamellar surface area by secondary folds so that the gill surface has an undulated aspect.

Most oxygenation occurs in the gills, but some also takes place in the inner mantle surface.

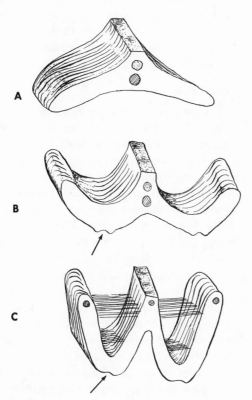

FIG. 10-40

Evolution of bivalve ctenidium by folding. **A,** Primitive protobranch gill. **B,** Hypothetical intermediate condition in which food grooves are formed (indicated by arrow). **C,** Gill in which filaments are folded at food groove to form ascending and descending limbs, connected by interlamellar junctions (anisomyarian condition) or interfilamentary tissue junctions (lamellibranch condition).

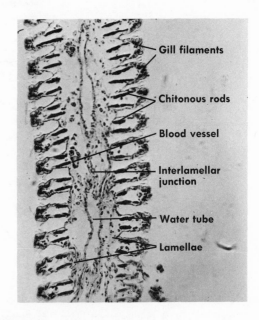

Gill filaments

Chitonous rods

Blood vessel

Interlamellar junction

Water tube

Lamellae

FIG. 10-41

Cross section through gill of mussel.

The function of oxygenation is assumed entirely by the mantle surface in the subclass Septibranchia, in which the gills are absent (Fig. 10-42, *C*).

The ecologic habits of bivalves have brought great changes to the basic primitive plan of water circulation through the mantle chamber. In the primitive protobranch *Nucula,* water passes through an anterior inhalant opening, then over the gills, and leaves the body by a posterior exhalant opening (Fig. 10-42, *A*). Lateral cilia sweep the water currents upward between the gill filaments into the dorsal exhalant chamber. Cleansing (frontal) cilia on the gill and mantle carry the heavier sediment to the ventral mantle edge, where it is expelled from the inhalant cavity with other rejected particles by the sudden contraction of the adductor muscles.

Although *Nucula* is a burrowing form and a deposit feeder, its plan of water circulation is not sufficient for the filter-feeding bivalves. Burrowing in soft silt and mud involves the problem of bringing in much inorganic sediment with the inhalant current. Most pelecypods have the inhalant as well as the exhalant openings at the posterior end so that the water current makes a U-turn as it passes through the gills into the suprabranchial chamber and out the exhalant opening (Fig. 10-42, *B* and *C*). The mantle edges of *Nucula* are free and do not form distinct apertures for the inhalant and exhalant currents.

Internal anatomy

Most bivalves are filter feeders, a process in which the gills play an important role. The primitive *Nucula* and related forms use what is called deposit feeding, by which they extend the 2 palp proboscides (Fig. 10-42, *A*) into the mud to pick up, by their ciliated grooves, organic deposits. From here the material is transferred to the base of the proboscis. It is then carried between the palpal lamellae, the inner surfaces of which are ridged with ciliary tracts that form a sorting and rejecting device. Ingested particles are conveyed in a groove to the mouth, whereas rejected particles are swept by cilia to the tips of the palpal lamellae and transferred to the mantle cavity by way of the mantle surface. Helping in the circulation of water currents through the gill chamber is the "pumping" action of the gills and dorsal membrane. The gills in *Nucula* are formed of modified filaments united by extensive ciliary junctions, which make them good pumping organs. Similar pumping devices are found in the Septibranchia, in which the gills are converted into a muscular septum, perforated with ciliated pores, that can be raised and lowered to draw in the tiny ani-

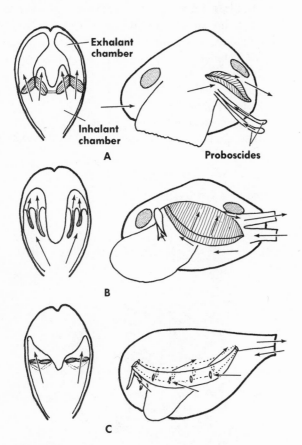

FIG. 10-42

Water currents in pelecypod types. **A,** Protobranchia. **B,** Lamellibranchia. **C,** Septibranchia; dotted lines indicate pumping movement of muscular septa.

mals on which the organism feeds (Fig. 10-42, C).

In the filter feeders the inhalant current, on entering the large mantle chamber, is checked somewhat to allow the large particles to sink to the bottom of the chamber, whence they are carried by cilia toward the posterior part of the mantle chamber. The inhalant current with the smaller food particles is drawn over the gills by the action of cilia covering the gills and mantle. The food is trapped by mucus on the gills and carried ventrally to the food groove along the ventral edge of the gill and thence to the palps around the mouth. The cilia involved in these processes differ in structure and function. Lateral cilia on the gills (Figs. 10-43 and 10-44, D and E) draw the main current of water into the mantle cavity. The frontal cilia carry the trapped food down to the food groove by producing a constant stream over the surface of the gill toward its ventral margin. Laterofrontal cilia along the angles between the lateral and frontal cilia (Fig. 10-44, D) prevent large particles from clogging the gills and deflect the small food particles toward the frontal cilia. Trapped food is entangled in the mucus in stringlike masses and

is directed by cilia partly into the food groove and toward the palps around the mouth (Fig. 10-45). In the labial palps the food may pass directly to the mouth, whereas rejected material may be carried posteriorly by ciliary paths toward the outgoing circulation.

In *Nucula* and a few related forms an enlarged buccal cavity with 2 lateral glandular pouches is present, but in most other pelecypods the mouth is a simple transverse aperture more or less compressed between a dorso-anterior and a ventroposterior lip. The lips are scalloped in *Pecten*, and the labial palps are actually modifications of the lips.

Except in carnivorous forms such as the Septibranchia, the alimentary canal of bivalves is adapted for dealing with a constant supply of small food particles, mostly of plant origin. In the mechanism of handling food there are some differences between the primitive proto-

FIG. 10-43

Simple gill filaments of *Donax* variabilis (Coquina) (order Heterodonta). Epithelium has extremely long cilia to maintain currents of water. (×810.) (From Andrew: Textbook of comparative histology, Oxford University Press, Inc.)

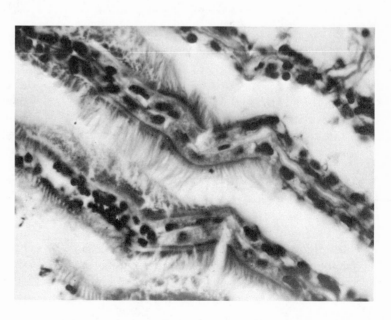

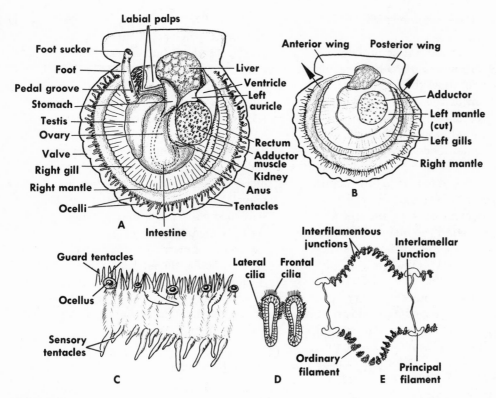

FIG. 10-44

Pecten irradians (order Anisomyaria). **A,** Structure as seen with left valve, mantle, and gills removed. **B,** Left valve and part of left mantle removed. **C,** Portion of velar fold showing tentacles and steely blue eyes. **D,** Two ordinary filaments of *Pecten* enlarged from **E. E,** Section of gill transverse to filaments at level of top ascending filaments. (×130.) (**A** to **C** drawn from life; **D** and **E** modified from Dakin.)

branchs and the higher bivalves. The method in protobranchs will be considered first.

In protobranchs, after food enters the mouth, it is conducted by a short esophagus into a thin-walled and laterally compressed stomach. The stomach walls contain no muscle, except in the septibranchs, and may be partly buried in the visceropedal mass. The stomach is divided into a dorsal portion, into which the esophagus and digestive glands open, and a ventral portion made up of the style sac. Surrounding the stomach is the liver or paired digestive glands made up of one kind of phagocytic cells. The

dorsal part of the stomach is mainly a ciliated sorting organ, consisting of a series of longitudinal grooves formed by ridges (Fig. 10-46). It contains a semicircular, chitinous plate called the gastric shield that is provided with a tooth-like elevation. When food enters the dorsal part of the stomach, it is carried along a ciliated tract to the cecum. In its transfer to the cecum the food passes the 3 ducts of the digestive glands, and some of the finer particles enter. After the rest of the food passes into and out of the cecum, it is entangled in a mass of feces and mucus called a protostyle, which is rotated counterclockwise by long cilia lining the walls of the style sac. As the protostyle with food at its anterior end rotates, it is pushed dorsally into the stomach by muscular contraction and by the action of the cilia. The food mass at the tip of the protostyle is now wound around the elevation of the gastric shield by the rotation of the protostyle. This process causes large parti-

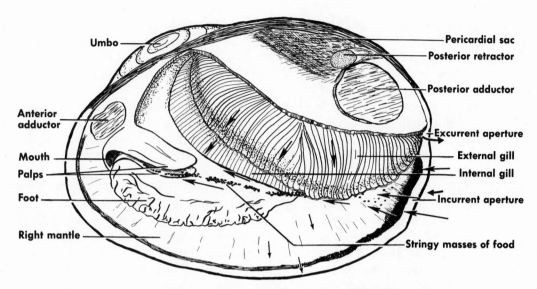

FIG. 10-45

Feeding mechanism of lamellibranch as seen in fresh-water mussel. (From Hickman: Integrated principles of zoology, The C. V. Mosby Co.)

cles of food to be freed and ground into smaller pieces by the pressure against the gastric shield. The smaller, freed particles are sent by the ciliated sorting grooves into the ducts of the digestive glands, whereas the larger particles are passed directly into the intestine by a groove from the anterior wall of the style sac. No extracellular digestion occurs by this method, but the food taken into the digestive gland is engulfed and digested intracellularly by cells in the tubular walls. Some digestion may also occur in phagocytic blood cells, which enter the gut. Wastes are returned by ciliary tracts to the stomach and thence to the style sac and intestine.

The long intestine forms a loop or two before passing dorsally between the anterior adductor muscle and the stomach. The posterior part of the intestine, or rectum, passes through the heart and pericardial chamber and over the posterior adductor muscle before opening by the anus into the suprabranchial chamber. An internal ridge (often called a typhlosole) extends the length of the intestine and is continuous with one of the ridges found beside the style sac groove. The intestine has no digestive or absorptive function but is concerned only in forming the feces.

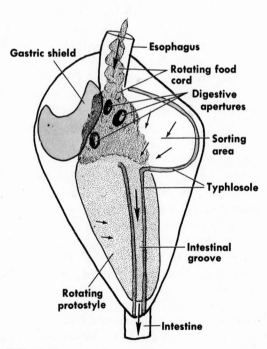

FIG. 10-46

Protobranch stomach, with protostyle (diagrammatic). Compare with lamellibranch stomach, Fig. 10-10. (Modified from Owen.)

387

The more advanced pelecypods such as the Eulamellibranchia differ from the protobranchs by having a longer style sac, which is always provided with a crystalline style (Fig. 10-10). They also lack proboscides, characteristic of the deposit feeders. The advanced bivalves are strictly filter feeders, living mainly on phytoplankton. The plankton is trapped by the mucus on the surface of the gills and passed by the frontal cilia to the food grooves along the ventral edge of the gills, as described earlier. There are several patterns of food currents over the gills among the different genera. Some may have both dorsal and ventral food grooves (*Pecten* and Unionidae). The palps have about the same function as those in the protobranchs, sorting out the accepted food and passing it to the mouth; the rejected particles are carried by cilia to the ventral edge of the mantle, then posteriorly to be expelled just below the inhalant apertures.

The crystalline style consists of a protein matrix and bears at its anterior end the carbohydrate-splitting enzyme amylase (and in some a lipase, or fat-splitting, enzyme as well). Cilia in the style sac rotate the crystalline style several turns per minute and drive its free end against the gastric shield, where the enzymes are rubbed off and mixed with the stomach contents. The style also forms a stirring mechanism for circulating the particles into the sorting areas. As the tip of the style in many pelecypods is eroded by rubbing against the gastric shield, the style at its base is continually being reformed. In some the style is replaced two or three times per day. The length of the crystalline style varies, usually between 1 and 2 inches, although it may be longer in large species.

The location of the style shows several variations. In some it is found in the intestine or in a sac derived from the intestine (Fig. 10-10), whereas in others it is derived as a pocket from the stomach. In oysters and some others the crystalline style acts as a capstan, and as it rotates, it draws in the mucous strings of food that are shed in the sorting areas of the stomach.

In the sorting areas of the stomach of higher bivalves, cilia pass the finer particles of food into the numerous digestive gland ducts, which convey the food to the tubules of the gland. The digestion here is intracellular, and enzymes for all the classes of food may be found here. Some extracellular digestion occurs in the stomach when the crystalline style releases its enzymes, as already described.

The carnivorous septibranchs, with their muscular pumping septum, draw their prey into the mantle chamber and to the mouth (Fig. 10-42, C). Their stomach is muscular, lined with chitin, and acts as a gizzard to macerate the prey. Their reduced style may furnish mucus for coating sharp and injurious particles of food.

The circulatory system consists of a heart, arteries, and sinuses. The heart consists of a ventricle, wrapped around the rectum, and 2 auricles, each of which opens into the ventricle by a canal (Figs. 10-36 and 10-44). Surrounding the heart and part of the digestive tract is the pericardial sac, to which the broad bases of the auricles are attached. Usually only the anterior aorta leaves the heart (Fig. 10-47, B), but in some (*Anodonta* and others) a posterior aorta is found (Fig. 10-47, A). After it leaves the heart, the anterior aorta enlarges into an aortic bulb, from which many arteries emerge. These include the pallial arteries to the mantle and the visceral arteries, the most important of which are the gastrointestinal to the digestive tract, the hepatic to the digestive glands, and the terminal arteries to the foot and anterior regions. In the different tissues the arteries break up into a network of sinuses, which join together to form larger sinuses mostly on the inner side of the mantle and superficial parts of the body. The course of the blood varies somewhat with different species. In some, such as *Mytilus*, the blood from the tissue sinuses flows through the kidneys; part of the blood from the kidney then passes to the auricles through a longitudinal vein; the other part of the blood circulates through the gills before returning to the auricles by the longitudinal vessel (Fig. 10-47, *B*). In *Anodonta* most of the blood passes through the gills for oxygenation before returning to the heart, and only a small part of it goes directly from the kidneys to the heart (Fig. 10-47, *A*). Much oxygenation occurs in the mantle.

The blood is usually colorless, but hemoglobin is found in some (*Solen*) and hemocyanin in others (*Venus*).

In those bivalves with a well-developed foot the blood plays an important part in the extension and manipulation of the foot. Blood is pumped into the foot sinuses and prevented from returning by a sphincter valve while the foot is being extended. When the foot is retracted, the blood retreats into the mantle sinuses.

Excretion in bivalves occurs through a pair of U-shaped nephridial tubules near the pericardium. The glandular part of each tubule

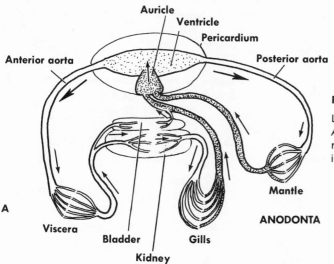

A ANODONTA

FIG. 10-47

Lamellibranch circulation. **A,** *Anodonta.* **B,** *Mytilus.* In *Anodonta,* most of blood from viscera goes to gills before returning to heart. In *Mytilus,* branchial circulation is less important and mantle circulation plays greater role.

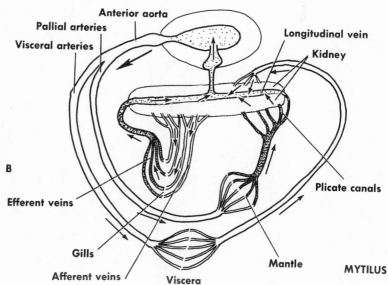

B MYTILUS

opens into the anterior pericardium, whereas the bladder portion opens by a nephridiopore into the suprabranchial chamber. Modifications of the excretory system involve the absence of glandular and bladder differentiations in primitive forms and many foldings and ramifications in higher pelecypods. In *Pecten* and other Filibranchia the reproductive ducts also open into the kidneys instead of having separate openings.

The nervous system has in general followed the simple basic plan of mollusks, such as three pairs of ganglia and two pairs of longitudinal nerve cords (Fig. 10-48). In the Protobranchia there are four distinct pairs of ganglia; in others the cerebral and pleural ganglia are fused. Commissures connect the members of paired ganglia with each other. The cerebropleural ganglia are located on each side of the

esophagus and send paired nerve cords to the visceral ganglia below the posterior adductor muscle. A second pair of nerve cords from the cerebropleural ganglia runs to the pedal ganglia in the foot.

Sense organs are represented in the class as a whole by ocelli, tactile organs, statocysts, and osphradia (Figs. 10-44, 10-48, and 10-49).

FIG. 10-48

Nervous and sensory system of pelecypods. **A,** General view of nervous system of *Pecten.* **B,** Cerebropleural and pedal ganglia of *Pecten.* **C,** Visceral ganglion of *Pecten;* this is the largest ganglion in *Pecten* because posterior part of body is most developed and mantle (pallial) nerves arise from it. **D,** Section through tip of siphonal tentacle of *Cardium,* passing through an ocellus. **E,** Section through right statocyst of *Cardium.* **F,** Section through osphradium of *Pecten.* (**A** to **C** modified from Dakin; **D** and **E** modified from Johnstone.)

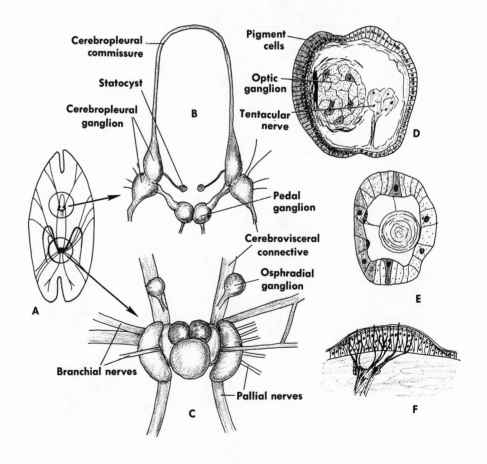

Most sense organs are located on the edge of the mantle. The edge in some species bears pallial tentacles with tactile and chemoreceptor sense organs (*Pecten*) (Fig. 10-44, *C*). The ocelli are best represented in such forms as *Pecten*, in which each consists of cornea, lens, retina, and pigmented layer (Fig. 10-49). They are steel blue in color and are found along the edge of the mantle. Their retinal cells are directed away from the source of light. In most pelecypods the ocelli are simply pigment spots. Statocysts are usually located in or near the pedal ganglia. Osphradia, for testing incoming water or other functions, are located below the posterior adductor muscles in the exhalant canal.

Separate sexes are the rule in most pelecypods, although some species, e.g., *Ostrea* and *Pecten* as well as certain freshwater species (family Sphaeriidae), are hermaphroditic. The paired gonads are practically fused together in their location around the intestinal loops. In the Protobranchia and some others the gonads open into the kidneys, which discharge the gametes to the exterior. In the Lamellibranchia the gonoducts are separate from but close to the kidneys. The renal ducts and gonoducts may open on a common papilla (*Mercen-*

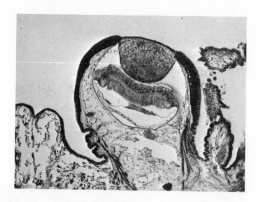

FIG. 10-49

One of numerous eyes from margin of mantle of *Pecten*. Large lens, abundant pigment, and complex retina (here partly detached) are present. (×83.) (From Andrew: Textbook of comparative histology, Oxford University Press, Inc.)

aria). In some hemaphroditic species the ovary and testis are separate from each other and discharge their gametes by separate openings.

Fertilization occurs externally in the surrounding water in most marine pelecypods. After the gametes are shed into the suprabranchial chamber, they are carried through the exhalant aperture, or siphon, to the outside. In some marine species and freshwater forms the eggs are fertilized in the suprabranchial chamber by sperm that have entered by inhalant water currents. The fertilized eggs are then carried into the water tubes of the gills, where they are incubated. Incubation is a feature of most freshwater bivalves. Self-fertilization in the gonoducts may occur in some hermaphroditic species, such as Sphaeriidae. Fertilization of the eggs may also occur in the water tubes (e.g., family Unionidae). In some species of *Ostrea* the incubation takes place in the mantle cavity outside of the gills.

A. L. Edgar (1965) describes in the pelecypod *Anodontoides ferussacianus* sperm packed in such a way that the heads were buried in a spherical mass with their tails extended radially. These tails beat in synchrony, rotating the entire mass as it moves through the water. He considered these sperm balls to be motile spermatophores that later broke up into freed sperm for fertilization (Fig. 10-50).

Pelecypods have spiral cleavage, which may vary in certain details among the different groups. After the first two cleavage divisions and the formation of 4 macromeres, successive divisions of the macromeres produce quartets of micromeres. After the formation of certain body organs, a shell gland produces a dorsal horny plate, which grows in extent. Later, 2 calcareous plates are formed by the folding of the single plate and are joined together by the hinge ligament, thus giving rise to the 2 valves of the shell. Around the prototroch a girdle of long cilia arises with a tuft of cilia at the apex. This first free-swimming stage is called the trochophore (Fig. 10-11, *A*) and is followed by the veliger stage, which is usually of short dura-

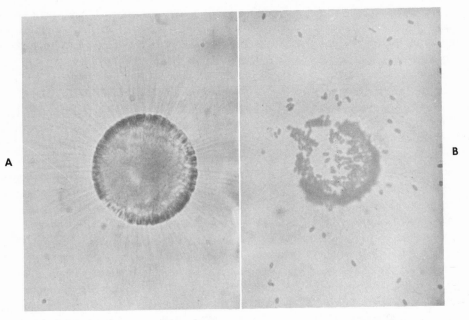

FIG. 10-50

A, Mass of living sperm in pelecypod *Anodontoides ferussacianus.* Sperm heads are packed close together in surface of sphere. Sperm tails extend radially from mass. Synchronous beat of tails enables the mass to rotate and move. **B,** Disintegration of the mass of sperm. When the sperm are detached, the sphere in which they are embedded disappears. (Courtesy Edgar, A. L. 1965. Trans. Amer. Microscop. Soc. **84:**228-230.)

FIG. 10-51

Bivalved glochidia larvae.

tion. During the veliger stage many body structures are developed, such as the nervous system musculature, gills, and foot. The band of cilia (the velum) is lost; the larva musculature disappears; and the apex moves downward above the mouth, which assumes the characteristics of the adult mouth. In the Protobranchs the velum is unusually large with rows of ciliated cells, the whole forming a type of test. Inside this test the shell gland and shell develop. In the freshwater *Sphaerium* and *Unio* there is no free-swimming stage; development is direct, although both a reduced trochophore and veliger stage can be detected in their embryonic development. When development is completed, the young clam is discharged from the gills.

Development in the freshwater Unionidae is unique in having a veliger stage that is represented by a parasitic glochidium (Fig. 10-51). The glochidium (less than 0.5 mm. long) is a bivalve shell usually provided with hooks or teeth and a rudimentary foot bearing a long adhesive byssus thread. Its reduced digestive system is nonfunctional. When discharged from the gills through the suprabranchial chamber and exhalant aperture, the glochidium sinks to the bottom. Here it may by chance come in contact with a fish, to which it attaches itself by its hooks either to the fins or to other body parts. Those glochidia without hooks usually attach themselves to the gills to which they are carried by respiratory currents of water. The phagocytic cells of the larval mantle are able to obtain nutrition from the host tissues.

During the month or less of its parasitic existence the glochidium, in a cyst, undergoes considerable transformation, losing most of its larval structures and developing adult structures. When the immature clam breaks out of its cyst, it falls to the bottom mud, where it takes up its existence for the rest of its life cycle. Infected fish can be detected by the small cysts of the glochidia, called "blackheads."

The number of glochidia produced by a freshwater clam may exceed 2 or 3 million, most of which die without an opportunity to parasitize a fish.

■ CLASS SCAPHOPODA

The class Scaphopoda, or elephant's-tusk shells, is represented by a small group of about 300 to 400 species. Many fossil species dating back to the Devonian period are known. Most of their evolutionary development has occurred, however, within the Cretaceous period.

The shell is an elongated, cylindric or tube-shaped structure, which may be slightly curved and open at both ends. It may resemble a miniature elephant's tusk or the canine tooth of a carnivore, with a gradual reduction in diameter posteriorly. The shell may also assume other shapes in a few species. It may be sculptured with longitudinal or annular ribs, or it may be smooth like a tooth. The size of the shell varies from 4 mm. to about 5 inches in length, although the average length is usually less than 2 inches. Fossil specimens reached a much greater size.

The class is divided into two families—Dentaliidae with pointed feet and Siphonodentaliidae with elongated feet terminating in a crenulated disk. They are exclusively marine and are adapted for burrowing. They are usually

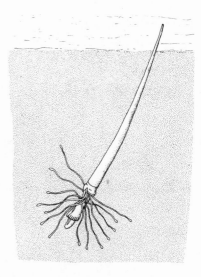

FIG. 10-52

Dentalium (class Scaphopoda) burrows into soft bottom to feed, leaving posterior end projecting into water. Its capitate oral tentacles (captacula) are used in feeding, and cilia in mantle cavity maintain water currents. (Modified from Light and others.)

found buried in the sand, with the posterior end projecting from the surface (Fig. 10-52). The body fits snugly into its shell tube and is attached along the concave or dorsal side by special columellar muscles. The mantle cavity is continuous throughout the shell on the ventral side. The mantle, originally of 2 lobes, is fused into a tubelike structure to conform with the shell and forms an anterior aperture, into which the head can be protruded or withdrawn (Fig. 10-53). A narrow posterior aperture serves for the inhalant water current (by ciliary action) and the exhalant current (by muscular contraction).

The head is cylindric and made up of two lobes, each with several slender tentacles or captacula with adhesive knobs or prehensile disks at their ends. These captacula, along with the mantle cilia, aid the animal in getting its prey, mostly microscopic plankton. The head has no special sense organs except the captacula, which apparently have sensory functions.

Gills are lacking in this class; respiration takes place in ciliated folds lining a portion of the mantle cavity.

The mouth, which is located in a projection of the pharynx or proboscis, is found at the base of the foot. It contains a well-developed radula and mandibles. A narrow esophagus leads into the stomach, which receives the ducts of the bilobed digestive glands and a pyloric cecum. The V-shaped intestine leaves at the anterior end of the stomach and opens through the anus into the mantle cavity.

On each side of the anus is an opening, or nephridiopore, from each of the paired kidneys. Renopericardial openings are not present, since there is no heart.

The blood system is made up of irregular sinuses, and the blood is kept circulating by the rhythmic protrusion and withdrawal of the foot.

The nervous system is typically molluscan (cerebral, pleural, and pedal ganglia with connectives). The only sense organs besides the captacula are the subradula organ of taste and the statocysts in the foot.

The sexes are separate, and the gametes from the unpaired gonads are discharged through the right nephridium into the mantle cavity and to the outside. The eggs, which are laid singly, are fertilized externally.

Segmentation is irregular and unequal. Gastrulation is by invagination. A floating trochopore stage (Fig. 10-11, *B*) is followed by a bilaterally symmetric veliger larva, which is transformed after a few days into a creeping organism. Elongation of the body and shell takes place gradually.

Scaphopods are mostly sedentary animals and are chiefly found in fairly deep water, although a few species may live in shallow shore water. They burrow their way through the mud or sand with the foot, which is provided with retractor muscles that are fastened to the distal wall of the shell tube. Locomotion in the Siphinodentaliidae may take place by extending and anchoring the foot by means of the foot disk.

Dentalium (Fig. 10-52) is by far the most common genus and consists of several species in the coastal waters of North America, mostly in southern waters. Shells of this genus are easily found along the beaches where they have been washed ashore.

■ CLASS CEPHALOPODA

The cephalopods represent the most specialized of the molluscan group and in many ways are the most advanced of all the invertebrate groups. Although they possess the basic molluscan pattern of structure, they surpass all other mollusks in organization, cerebral de-

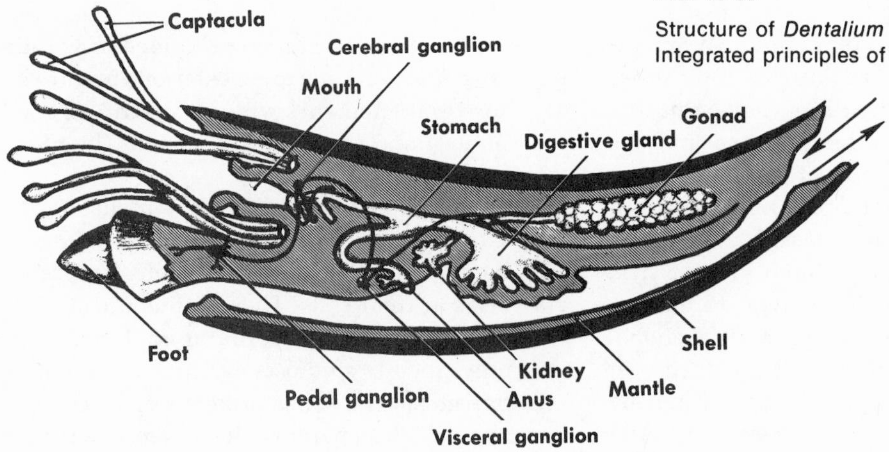

FIG. 10-53

Structure of *Dentalium* (diagrammatic). (From Hickman: Integrated principles of zoology, The C. V. Mosby Co.)

velopment, more precise sensory information, learning processes, and locomotion. They include octopuses, cuttlefish, squids (Fig. 10-54), and nautiluses. Some have attained the largest size of all invertebrates. The giant squid *Architeuthis princeps* reaches a length of more than 50 feet (including the arms). On the other hand, the smallest cephalopod is probably the pigmy squid *Idiosepius,* which does not exceed a total length of 15 to 20 mm.

Their fossil record indicates that the peak of their evolutionary development was reached

FIG. 10-54

School of young squids in swimming formation. (Courtesy C. P. Hickman, Jr.)

during the Paleozoic and Mesozoic eras, at which time they numbered thousands of species. At present there are only about 600 species of cephalopods found. In their early history they contained such important groups as the Ammonoidea (Figs. 10-3, *E,* and 10-57), now wholly extinct. Only a single nautiloid, *Nautilus* (Fig. 10-55), now exists, although most of the fossil cephalopods belong to this subclass. In contrast to most mollusks, the cephalopods are chiefly active swimmers instead of bottom dwellers. They seem to be more closely related to the Gastropoda than to other classes of mollusks.

The cephalopods are bilaterally symmetric animals, but they have undergone profound rearrangement of their body axes in comparison with other mollusks. The anteroposterior axis has become greatly shortened and the dorsoventral axis lengthened. This causes the latter axis to become the dominant functional anteroposterior axis. As a result of this shift of axes, the former ventral surface is now anterior, the anterior surface is dorsal, the dorsal surface is posterior, and the posterior surface is ventral. The posterior mantle cavity has become ventral. The mouth is now located in the middle of the foot, the edges of which are modified to form

FIG. 10-55

Chambered nautilus, optical section (diagrammatic). (From Hickman: Integrated principles of zoology, The C. V. Mosby Co.)

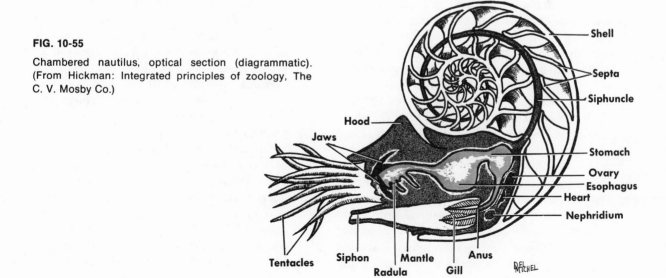

the tentacles or arms. Part of the foot (probably the region of the epipodium of gastropods) is represented by the funnel.

Most existing cephalopods have a very reduced shell (squids), one that is vestigial, or none at all (octopuses). The relict *Nautilus* has a completely developed shell that is also well represented by fossil cephalopods.

The evolutionary trends of the class have emphasized a vigorous, aggressive mode of life, fast locomotion, and predaceous habits. Molluscan body modifications have thus meant the elimination of the heavy calcareous shell and the formation, in many instances, of fins as an aid to active locomotion. By the forward movement of the foot the viscera forms a hump that is now posterior and covered by a dome-shaped mantle.

GENERAL MORPHOLOGY AND PHYSIOLOGY
External anatomy

Present-day cephalopods are divided into two chief groups—the Nautiloidea (Tetrabranchia) with an external shell and two pairs of gills and the Coleoidea (Dibranchia) with a shell that is internal or absent and a single pair of gills.

The Nautiloidea are represented today by a single genus, *Nautilus*. Their external shell is coiled in a planospiral and is partitioned into gas-filled chambers by successive, posterior, concave septa, except the last and largest chamber, which is occupied by the living animal (Fig. 10-55). Each of the chambers represents a stage in the animal's growth. In its growth and as the shell is enlarged, the animal moves forward while the mantle secretes a new septum, which is indicated on the shell wall by a suture. Each new successive chamber is larger than the preceding one. All the chambers, except the last, contain gas (mostly air with more nitrogen and less oxygen), which is secreted by a siphuncle or tubule extension of the visceral hump that extends through perforations in the middle of the septa. These gas-filled chambers make the heavy shell relatively buoyant so that the animal can swim more freely. The shell consists of an organic matrix, with an outer layer containing calcium carbonate crystals and an inner nacreous layer. The outer surface may bear alternating bands of orange and white, although in some the surface is a uniform pearly white.

The Nautiloidea are divided into the nautiloids and the ammonoids (extinct). The nautiloids probably reached their greatest development in the Paleozoic era. Many of them such as *Orthoceras* had straight cone-shaped or slightly curved shells. Some were 10 to 15 feet long. No general evolutionary pattern can be formulated about their coiling. Some forms tended to coil, culminating in the closely coiled *Nautilus*. Others tended to become uncoiled. All degrees of coiling are found among fossil forms (Fig. 10-56). The second group, the ammonoids (Fig. 10-57), reached their peak in the Mesozoic era and became extinct at the end of the Cretaceous period. They differed from the nautiloids mainly in the shape and sculpture of their whorls, the wrinkled septa, and the highly complex suture lines. Their morphology was highly varied. In some of these the shell may have been coiled at first and straightened later. Uncoiling was especially pronounced as they neared extinction. Their size ranges were equally varied, extending from tiny shells of $\frac{1}{2}$ inch diameter to some with a diameter of 6 to 7 feet. The nature of the sutures forms an important index to the taxonomic studies of the group. Complex sutures may have been involved in the strengthening of the shell, and the abandonment of the coiling pattern may have been related to their return to a benthonic existence. Of course, nothing is known about the soft parts of these fossils.

The Coleoidea, or Dibranchia, to which most of the existing cephalopods belong, have lost the rigid shell, which has become internal or absent. The mantle, which encloses the viscera and secretes the shell in the nautiloids, has assumed an active role in the locomotion of modern cephalopods. With the loss of the shell and presence of a highly muscular mantle, the

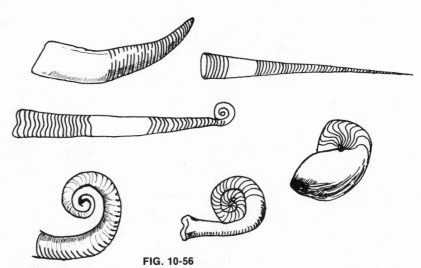

FIG. 10-56

Fossil cephalopod shells showing some varied shapes. Earliest fossil shells were slightly curved, and straight and coiled shells developed from these. (Modified from Moore, Lalicker, and Fischer.)

FIG. 10-57

Suture patterns in ammonoid shells. **A,** *Baculites.* **B,** *Ceratites.*

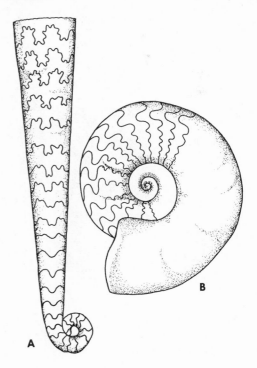

jet propulsion mechanism reaches its highest perfection.

The Coleoidea may have evolved from the nautiloids; and in their evolution the shell became enclosed in the mantle. As they evolved in the direction of an active swimming existence, the shell was partly or wholly discarded for more efficient locomotion. The fossil group Belemnites may represent a link between nautiloids and dibranchs. These forms are considered dibranchs, as they possessed the ink sac and the sucker type of tentacle. The belemnoids also possessed an anterior dorsal plate (pro-ostracum), which projected from the anterior end of the shell for protection, and an external calcified sheath, the guard, at the apex of the shell. Modifications of this plan gave rise to several lines of descent leading to modern forms. Reduction of the guard and coiling of the shell led to the present-day *Spirula* (Fig. 10-66, *A* and *B*), in which the shell is still partly external. Another evolutionary line developed into the squids, in which all the shell has disappeared except for the horny pen. A third line of descent involves the cuttlefish *Sepia*, in which the uppermost side of the chambered shell with numerous septa persists (Fig. 10-66, *C* and *D*). The cuttlefish has many thin chambers that are spaced and separated by numerous pillars and contain gas for buoyancy. In the octopuses, another evolutionary line, the shell has completely disappeared or is a mere vestige.

All cephalopods swim by some form of jet propulsion. They do this by drawing water into the mantle cavity by active contraction of radial muscles. Water enters the mantle cavity in the neck region between the mantle and head. When the mantle cavity is filled by the inhalant current, the longitudinal muscles relax, the circular muscles contract, draw the mantle margin tightly around the neck, and expel the water through the funnel. The snugness of the collar around the neck is furthered by mantle cartilage knobs that fit into sockets near the head (Fig. 10-58). As the water is

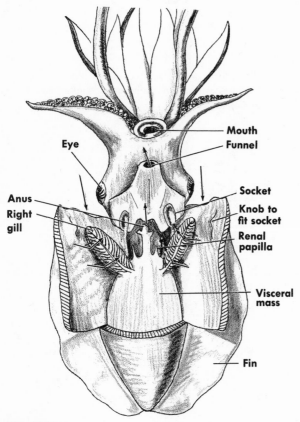

FIG. 10-58

Sepia, a cuttlefish. Arrows indicate direction of water current in mantle cavity.

expelled in one direction from the ventral funnel, the animal is propelled in the opposite direction. The funnel can be turned either forward or backward so that the animal can go in either direction. The body is stabilized during movement by lateral fins when such are present. Some animals perform slow swimming movements with their fins. Differentiated movements by jet propulsion are controlled by the force with which water is expelled from the mantle cavity.

In *Nautilus* the mantle is thin and plays no part in movement. Water is expelled from the funnel in these forms by drawing the body into the shell and contracting the funnel. The *Nautilus* shell is kept upright by the gas-filled chambers within it. Movement is slower because of the hindrance of the large shell.

Octopuses are more adapted for crawling and walking than for active swimming. They have no fins and their jet propulsion movements are jerky. They use their arms, which are provided with suction disks, to anchor and pull themselves along. Their movements in pursuing prey or escaping from enemies are skillfully and swiftly executed.

As the water circulates through the mantle cavity, it provides power for movement and oxygen for the gills. *Nautilus* has 4 gills, and the dibranchs only 2. The gills are featherlike, consisting of a central axis with numerous side filaments. Cilia are absent from the gills because nautiluses are not filter feeders, and the removal of sediment in such pelagic forms is no problem. Since the funnel is ventral, the exhalant current is ventral to the inhalant current—the reverse of that found in most other mollusks. Respiration may occur through the body surface in some cephalopods with vestigial gills.

Internal anatomy

The Coleoidea have an internal cartilaginous skeleton in the form of cartilage plates that are found around the brain, in the neck region, at the base of the fin, in the gills, and often in the arms. In *Nautilus* the cartilaginous skeleton is not so prominent, although it does offer some cartilage support.

The digestive system (Fig. 10-59) is mainly a U-shaped tube that runs from the mouth to the anus, which opens into the mantle cavity near the siphon. The tract consists of a pair of chitinous jaws, a typical mollusk radula (Fig. 10-59, *B* and *C*), two pairs of salivary glands, a long esophagus, a stomach, a spiral cecum, an intestine, and a rectum. The jaws are located in the buccal mass in the center of the ring of arms. Modifications of the tract are found among the various groups of cephalopods. A crop or expanded portion of the esophagus

occurs in the Octopoda. Some deep-sea forms lack a radula. Two pairs of salivary glands open into the buccal mass, except in *Nautilus*. The posterior pair of salivary glands opens by ducts just behind the lower jaw and often contains a poisonous secretion, which is injected into the prey when seized. The muscular, nonglandular stomach has a spiral cecum attached near its anterior end. This unique organ is provided with a ciliary mechanism that removes solid particles from the cecal contents so that only fluid products are absorbed. The solid particles are carried back into the stomach and into the intestine by a cilated groove. The digestive gland is made up of two parts, a solid bilobed

FIG. 10-59

Digestive system of *Eledone,* an octopod cuttlefish. **A,** Alimentary canal, ventral view; crop is turned to right, buccal mass to left, and liver forward. **B,** Buccal mass showing radula. **C,** Two rows of teeth from radula. (Redrawn from Isgrove: Eledone, L.M.B.C. Memoirs XVIII, Williams & Norgate, Ltd.)

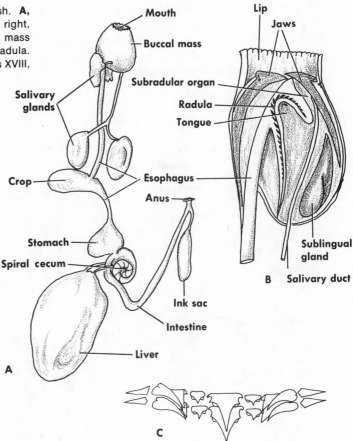

liver and a smaller spongy pancreas. Both parts of the digestive gland open by a common duct, which may be single or paired, into the cecum. In *Sepia* the liver portion of the digestive gland is paired. Enzymes from both the liver and pancreas are poured into the stomach when food is present there. Digestion, which is wholly extracellular, starts in the stomach and is completed in the cecum. Some absorption occurs in the intestine as well as in the cecum. The intestine discharges its waste into the mantle cavity through the anus.

A diverticulum of the intestine is located near the anus and forms the characteristic ink sac (absent in *Nautilus*) (Fig. 10-59). Its duct opens into the rectum behind the anus. The ink sac secretes a dark fluid of melanin pigment. Recent work on octopuses indicates that these animals, when threatened by a predator (e.g., moray eel), give off a smoke screen of sepia that assumes roughly the shape of a large animal. Behind this screen the octopus takes off at right angles to its original direction. A product of the ink secre-

tion is also thought to deaden the sense of smell of potential enemies.

Cephalopods are entirely carnivorous, feeding on crustaceans, fishes, and even their own kind. The prey is seized by the suckers or clawlike hooks of the arms. The food is then bitten into pieces by the jaws and passed down the digestive tract to the stomach.

Squids have 10 arms. Two from the dorsal side are much longer than the other 8 and are called tentacles. The arms bear on their flattened inner surface many cup-shaped adhesive suckers, which are located on short stalks. Horny rims are often found around the suction cups for reinforcements. Muscle fibers in the floor of the cups can produce a suction effect on the seized object. The 2 longer tentacles have suckers only at the flattened ends and are mainly responsible for the active seizing of prey. Octopuses have only 8 arms, all of which are equal in length. Their suckers are sessile and lack horny rims and hooks.

Nautilus has 4 kidneys, whereas the Di-

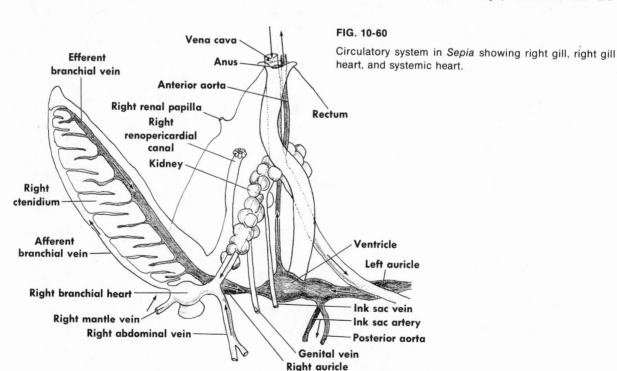

FIG. 10-60

Circulatory system in *Sepia* showing right gill, right gill heart, and systemic heart.

Vena cava
Efferent branchial vein
Anus
Anterior aorta
Right renal papilla
Right renopericardial canal
Kidney
Rectum
Right ctenidium
Right Afferent branchial vein
Ventricle
Left auricle
Right branchial heart
Right mantle vein
Ink sac vein
Right abdominal vein
Ink sac artery
Posterior aorta
Genital vein
Right auricle

branchia have only 2. The kidneys are compact, spongy, saclike structures, each of which surrounds a large vein (Fig. 10-60). Within the outer wall of each kidney is the long renopericardial canal, which opens posteriorly into the visceropericardial cavity and anteriorly into the kidney cavity close to the renal papillae. Near the anus the nephridiopore or renal papilla of each nephridium opens into the mantle cavity. Excretory tissue around the vein selects waste from the blood and carries it to the kidney cavity. Nitrogenous waste is excreted by the kidneys mainly in the form of uric acid. The 2 kidneys (Decapoda) are connected anteriorly and usually communicate with each other. In *Nautilus* the 4 nephridia have no communication with each other or with the pericardium.

The circulatory system is mostly of the efficient closed type, except in *Nautilus*, where a great part of it is of the primitive lacunar type. The system consists of closed vessels and 3 hearts, 2 branchial and 1 systemic (Fig. 10-60). The systemic heart consists of a median ventricle and 2 lateral auricles. An anterior and a posterior aorta carry blood pumped by the ventricle to the various parts of the body. Blood is returned from the head by way of the vena cava, which divides into 2 branches (branchial veins) near the kidney. Each branchial vein runs through a kidney to a branchial heart located at the base of a gill. The muscular branchial hearts facilitate unoxygenated blood flow through the capillaries of the gill. Oxygenated blood from the gills is returned to the auricles, which then pass the blood to the ventricle. The right branchial vein also receives blood from veins coming from the ink sac and gonads, and each of the branchial hearts receives blood from the mantle and abdominal veins.

The blood contains the respiratory pigment hemocyanin characteristic of the mollusks. Venous blood is blue in color, but becomes colorless when oxygenated.

The brain of cephalopods is the most complex among the invertebrates. It is made up of cerebral, branchial, pedal, and visceral ganglia, which are fused to form a ring around the esophagus (Fig. 10-61). Localized regions have been detected in the various ganglia so that certain areas or centers are linked with definite functions. Only a few ganglia lie outside the brain. Around the brain is a type of skull in the form of cartilage-like tissues. The large cerebral ganglia make up the part of the brain lying above the esophagus. From them a pair of buccal nerves passes by way of the paired superior buccal ganglia and commissures around the esophagus to the paired inferior buccal ganglia just behind the buccal mass. From each side of the cerebral ganglia an optic nerve extends to the well-developed eye. The funnel is supplied with nerves from the infundibular center, and the tentacles with nerves from the branchial center of the pedal ganglia. The

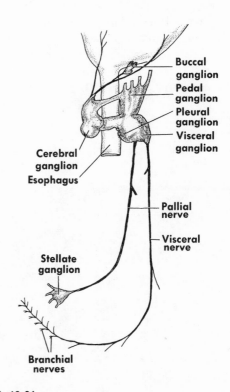

FIG. 10-61

Central nervous system of *Octopus*, from right side. (Modified from Lankester.)

401

visceral ganglia have many centers that give off nerves to the gills, sympathetic nerves to the stomach, and a pair of large nerves to the mantle. Branchial and gastric ganglia have centers associated with localized functions in the organs they innervate.

In the mantle nerves are giant motor neurons that are involved in the rapid locomotion of such forms as the squids. These fibers are absent in the giant squid *Architeuthis* and are best developed in *Ommastrephes*. The giant fiber system of cephalopods involves a pair of giant cells in the visceral ganglia of the brain, processess from these cells that make synapse in the visceral part of the brain, and giant neurons that pass to the stellate or mantle ganglia, where they synapse with other giant fibers. From the stellate ganglion, located anterior to each gill, many nerves containing giant axons, each made up of the fusion of processes from many stellate ganglionic cells, pass to the mantle. A single giant fiber may reach a diameter of 800 microns. Nerve impulses in the giant fiber are very rapid and ensure a quick contraction of the mantle in jet propulsion.

FIG. 10-62

Section through eyes of *Loligo*.

The best developed sense organs in cephalopods are the eyes, which are able to form images. They are similar in structure to the vertebrate type. The eye is provided with muscles for a limited amount of movement, a cornea, a lens suspended by a ciliary process, an iris diaphragm, a vitreous humor, and a retina with photoreceptors directed toward the source of light, not away from it as in vertebrates (Fig. 10-62). Focusing is performed by moving the lens forward and backward, not by changing the shape of the lens. Photoreceptors are connected to retinal cells, which send processes to an optic ganglion at the outer ends of the optic nerve. The eye of *Nautilus* is of the pinhole camera type, with the optic cavity opened to the exterior (Fig. 10-63).

Other cephalopod sense organs are represented by statocysts on each side of the brain, as well as tactile and chemoreceptor cells on the tentacles.

The sexes are separate in cephalopods, and sexual dimorphism, when it occurs, is chiefly a matter of size. Most males are distinguished by having one or more arms modified as copulatory organs; some males have longer prolongations of fins than females. The ovaries or testes are specialized parts of the coelomic wall and are located at the posterior end of the body.

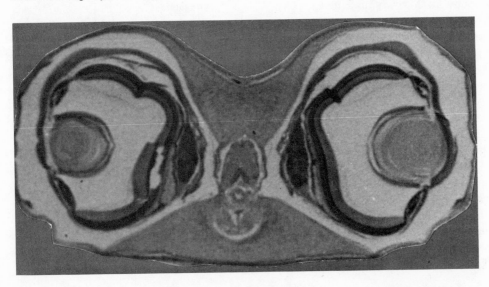

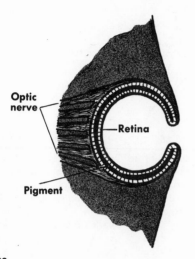

FIG. 10-63

Section through eye of *Nautilus.* (From Hickman: Integrated principles of zoology, The C. V. Mosby Co.)

The female system, which is simpler than that of the male, consists of a single, saccular ovary and a single (paired in *Octopus*) looped oviduct that opens into the mantle cavity. At the terminal end of the oviduct is an oviducal gland that furnishes part of the egg capsule. The eggs are fairly large (up to 15 mm. in diameter) and provided with a great deal

FIG. 10-64

A, Male reproductive organs of *Loligo.* **B,** *Loligo* in copulation. Male grasps spermatophores as they emerge through his funnel (as shown) and thrusts them between ventral arms of female, where sperm reservoirs become attached to depression in buccal membrane. In another copulatory position, male inserts spermatophore into female's mantle chamber near end of oviduct. **C,** Female reaching for egg string as it emerges from funnel. **D,** Several egg strings attached to rock by female. **E** to **G,** Early cleavage and larval stages of *Loligo.* (**A** after Duvernoy, modified from Lankester; **B** to **D** modified from Drew.)

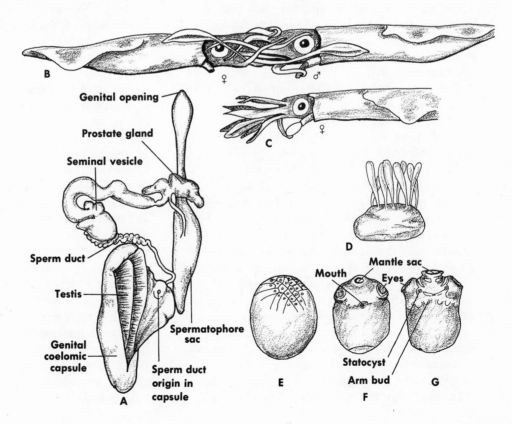

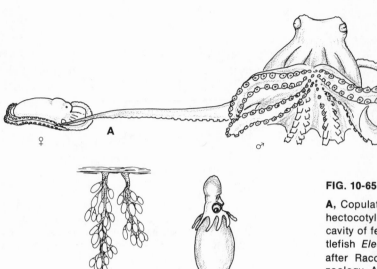

FIG. 10-65

A, Copulation in *Octopus*. Larger male uses third right hectocotylized arm to deposit spermatophores in mantle cavity of female in end of oviduct. **B,** Egg strings of cuttlefish *Eledone*. **C,** Unhatched embryo of *Eledone*. (**A** after Racovitza, redrawn from Lankester: Treatise on zoology, Adam & Charles Black, Ltd.; **B** and **C** modified from Isgrove.)

of yolk. In the ovary the eggs are nourished by follicle cells that are derived from the peritoneum. After passing through the oviduct, the eggs are covered by membranes from the oviducal gland and from the nidamental gland, which opens independently of the oviduct into the mantle cavity.

The male system (Fig. 10-64) consists of a saccular testis and a much-coiled sperm duct modified to form a seminal vesicle, a prostate gland joined to the seminal vesicle, and a storage reservoir (Needham's sac). Needham's sac opens into the left side of the mantle cavity. Sperm are produced in the wall of the testis, pass into the genital coelom, and enter the sperm duct through a small ciliated opening. In the seminal vesicle the sperm are encased or packed into the complicated spermatophores. From the seminal vesicle the spermatophores pass into the large Needham's sac, where they are stored until copulation occurs. A spermatophore is an elongated structure, consisting of a sperm tube, a cement sac, a coiled ejaculatory apparatus, a cap, and an outer sheath with a terminal filament. It may attain a length of several inches, although it is usually smaller.

In copulation the male makes use of a modi-fied arm for the transference of spermatophores to the female (Figs. 10-64 and 10-65). This arm, formerly thought to be a parasite, is called a hectocotylus ("the arm of a hundred suckers"). The arm modified for this purpose is the right or left fourth arm in squids (Fig. 10-64, *B*) and the right third arm in octopuses (Fig. 10-65, *A*). It varies in length and structure in different species. In some octopuses it may exceed 16 inches. When not in use the arm is coiled in a saclike sheath. Structural modifications usually involve some kind of groove with small suckers or a terminal depression or cavity. In the actual process of sperm transference, the male withdraws a bundle of spermatophores from Needham's sac with his hectocotylus (Fig. 10-64, *B*), which he then inserts into the mantle cavity of the female or into a fold (seminal receptacle) near the mouth (*Loligo*). In most species the spermatophores are deposited near the openings of the Oviduct. The hectocotylus arm is completely detached and left in the mantle cavity of the female. Some cephalopods (*Vampyroteuthis*) have no hectocotylus. When the spermatophores are picked up by the hectocotylus, the cap is removed and the ejaculatory apparatus is loosened.

After deposition the spermatophores adhere to the seminal receptacles or the mantle walls. They can be discharged by mechanical pressure on the terminal filament or by imbibition of water. This results in an inversion of the ejaculatory apparatus, a process that pulls the sperm tube forward and releases the sperm. The sperm are activated by the sea water and are ready for fertilization, which may occur in the mantle or in the egg mass held by the arms.

The female grasps the strings of fertilized eggs as they emerge from her funnel and attaches them to a substratum in clusters of elongated capsules, each containing 50 to 100 eggs (Fig. 10-64, C). The gelatinous egg capsules absorb water and swell to several times their original size, forming what are called "deadman's fingers" (Figs. 10-64, D, and 10-65, B). At the time of spawning the cephalopods are gregarious and often deposit their eggs in a common area. The primitive Nautilus has 4 tentacles modified as copulatory organs, and its eggs are laid singly on a substratum. Vampyroteuthis simply releases its eggs into the sea. The female of the paper nautilus (Argonauta) secretes a thin shell by means of 2 membranes on the dorsal pair of arms; in it, she lives and incubates her eggs. The smaller male without its own shell often lives in the shell with her. In some cases after the eggs have been deposited they are brooded over by the female (Octopus).

DEVELOPMENT

The eggs of cephalopods contain a great deal of yolk and are telolecithal. Cleavage is therefore incomplete and restricted to one end of the egg (animal pole) (Fig. 10-64, E). This causes the embryo to be placed at one end with the ectoderm spread over a large mass of yolk, forming a yolk sac. When the mouth is first formed, it is not surrounded by arms; these develop posteriorly and later migrate forward during development and encircle the mouth (Figs. 10-64, F and G, and 10-65, C). The funnel is formed from double outgrowths, which re-

main separate in Nautilus but fuse in other cephalopods to form a single funnel. A trochophore or veliger stage is lacking; but eggs hatch into larvae, which at first may differ widely from adults in form (J. M. Arnold, 1965). For a time the larvae may form a part of plankton populations. Embryologic development varies among cephalopods, but in every case it differs from that of other mollusks. In only a few cephalopods has it been completely explained.

ECOLOGY

Cephalopods are exclusively marine forms and occupy most ecologic habitats, such as littoral, surface, pelagic, and abyssal. They are absent from brackish and fresh water and do not tolerate water of reduced salinity, although some are found in high-salinity waters. The active squids are found mostly in the open sea, whereas the octopuses live in crevices along rocky shores. They have excellent protective devices against their potential enemies, mostly sperm whales and certain fishes. In addition to the black secretions of their ink gland and their swift locomotion, they have remarkable powers to change colors.

The pigment cells are of two kinds, chromatophores and iridocytes, which are found in the dermis. The method of changing color in the chromatophore has been described earlier. Being under nervous control, chromatophore camouflage can take place rapidly in cephalopods. Pigment colors are represented by red, yellow, orange, red-brown, blue, violet, and black. Many cephalopods have a great variety of differently colored chromatophores. The ability of a chromatophore to change color is very rapid because of its quickness to contract and expand. Patterns of color often match the animal's surroundings in a highly effective way.

The cuttlefish (Sepia) is of special interest ecologically because of the buoyancy of its calcareous shell. The chambered shell (Fig. 10-66, C and D) is reminiscent of the great group of cephalopods that is now represented only by Nautilus. It has been demonstrated that the

FIG. 10-66

A, *Spirula.* **B,** Internal shell of *Spirula.* **C,** *Sepia.* **D,** Internal shell (cuttlebone) of *Sepia.* (**A** and **B** modified from Cambridge Natural History; **C** and **D** modified from Lankester.)

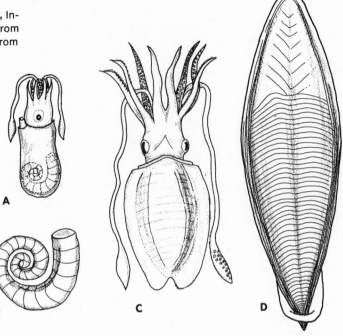

cuttlefish can vary its specific gravity by varying the proportions of liquid and gas in the cuttlebone. By pumping liquid out of the cuttlebone, the cuttlefish increases its volume of gas space and becomes less dense. It becomes more dense by pumping liquid in again. A special membrane at the posterior ends of the cuttlebone chambers seems to be responsible for pumping the liquid in and out. The gas, mostly nitrogen, always retains its pressure of about 4/5 atmosphere at all depths. The cuttlefish can change depth by regulating the osmotic pressure difference between cuttlebone liquid and the blood; and it can reestablish as needed an equilibrium between the hydrostatic pressure of the sea and the osmotic difference that keeps the sea water from entering the cuttlebone. *Spirula,* which lives at much greater depths, uses its shell device in the same way (Fig. 10-66, *A* and *B*).

LEARNING EXPERIMENTS

Training experiments in cephalopods, especially in octopuses, have revealed that these animals have a high level of intelligence, perhaps the greatest among all invertebrates. Many competent investigators, such as J. Z. Young (1961) and M. J. Wells (1966) have given extensive reports on learning in these animals. Sophisticated experiments have involved the tracing of those parts of the nervous system concerned in learning by removing parts of the brain. Such studies have given knowledge about the organization of the brain in this advanced invertebrate as compared with similar studies on vertebrates. It has been possible to ascertain what an octopus can and cannot learn and how this animal reacts to the various objects it can see and touch.

Some tactile discriminations are made by the octopus without difficulty, but there are others they are unable to make. Some discriminations they can be trained to make. For instance, Wells has shown that octopuses can be taught to discriminate among many types of cylinders, but to do so they make their discriminations on the basis of texture, or roughness of the surfaces of the cylinders concerned. They are unable to detect different patterns of grooves cut into the surface of cylinders, because all cylinders to which they were exposed have the quality of roughness. They can be taught to distinguish cylinders of different sizes by the basis of distortion detected by the individual sucker, and not on the degree it must bend its arms around the objects.

Wells also found that octopuses can detect taste with their suckers. In them this sense is very acute so that they are able to detect certain solutions as sugar or acid dilutions far below the range of the most sensitive tongue of man. This chemotactile sense has no doubt developed from their close ecologic relationship with their natural habitat and the source of their food. Weight discriminations apparently cannot be learned, for much investigation has shown that they cannot learn to distinguish between light and heavy objects. Some of their shortcomings in this respect may be ascribed to the way in which the nervous system of octopuses is constructed. Some investigators believe that the ability of octopuses to distinguish between horizontal and vertical rectangles and their inability to distinguish oblique rectangles may be due to the vertical and horizontal orientation of anatomic structures of their visual system, such as retinal cells, dendritic fields of the optic nerve, etc. Conformation of this is strikingly shown in the case of plane-polarized light; octopuses can distinguish sources of plane-polarized light vibrating in vertical and horizontal planes but are unable to do so if the planes are oblique.

In general, investigations on the behavior of octopuses have shown that soft-bodied, flexible invertebrates such as the octopuses are quick to learn if the experiments are suitable for their particular type of organization. However, much investigation with more sophisticated experiments must be done before the secrets of invertebrate learning are revealed.

REFERENCES

Abbott, R. T. 1954. American sea shells. Princeton, D. Van Nostrand Co., Inc. More than a thousand marine shells are described in this semipopular treatise.

Arnold, J. M. 1965. Normal embryonic stages of the squid, *Loligo pealii* (Lesueur). Biol. Bull. **128**:24-32.

Beadle, L. C., and J. B. Cragg. 1940. Studies on adaptation to salinity in *Gammarus* spp., part 1. Regulation of blood and tissues and the problem of adaptation to fresh water. J. Exp. Biol. **17**:153-163.

Bidder, A. M. 1950. Digestive mechanism of European squids. Quart. J. Microscop. Sci. **91**:1-43.

Campbell, J. L. 1965. The structure and function of the alimentary canal of the black abalone *Haliotis cracherodii* (Leach). Trans. Amer. Microscop. Soc. **84**:376-395.

Carriker, M. R. 1959. Comparative functional morphology of the drilling mechanism in *Urosalopinx* and *Eupleura*. Proceedings of the Fifteenth International Congress of Zoology, London. This author describes how these forms bore their way into the shells of mollusks.

Coe, W. R. 1942. Influence of natural and experimental conditions in determining shape of shell and rate of growth in gastropods of the genus *Crepidula*. J. Morphol. **71**:35-47.

Dakin, W. J. 1912. *Buccinum* (the whelk). L.M.B.C. Memoirs. London, Williams & Norgate, Ltd. A technical treatise with numerous detailed illustrations.

Davis, J. R. A., and H. J. Fleure. 1903. *Patella* (the common limpet). L.M.B.C. Memoirs. London, Williams & Norgate, Ltd. A general description of the adult structures of the limpet and its development; well illustrated.

Dehnel, P. A. 1955. Rates of growth of gastropods as a function of latitude. Physiol. Zool. **28**:115-144. This paper shows how northern populations grow faster than southern ones at the same temperature. The author was one of the first to point out how poikilotherms vary physiologically with latitude at the interspecific and intraspecific level.

Drew, G. A. 1911. Sexual activities of the squid. J. Morphol. **22**:327-359.

Eales, N. B. 1921. *Aplysia*. L.M.B.C. Memoirs. Liverpool, University Press of Liverpool.

Edgar, A. L. 1965. Observations on the sperm of the pelecypod *Anodontoides ferussacianus* (Lea). Trans. Amer. Microscop. Soc. **84**:228-230.

Edmondson, W. T. (editor). 1959. Ward and Whipple's freshwater biology, ed. 2. New York, John Wiley & Sons, Inc. Keys are given of the freshwater mollusks of the United States.

Evans, R. G. 1948. The lethal temperatures of some common British littoral molluscs. J. Anim. Ecol. **17**:165-173.

Fretter, V. 1937. The structure and function of the alimentary canal of some species of Polyplacophora, part 1. Trans. Roy. Soc. Edinburgh **59**:119-164. A detailed description of structure and physiology of the digestive tube in chitons.

Galtsoff, P. S. 1961. Physiology of reproduction in molluscs. Amer. Zool. **1**:273-289.

Graham, A. 1949. The molluscan stomach, part 3. Trans. Roy. Soc. Edinburgh **61**:737-778. An especially fine treatise on the comparative morphology and histology of the complicated molluscan stomach.

Gutsell, J. S. 1931. Natural history of the bay scallop. Bull. U. S. Bur. Fish. **46**:569-632.

Haas, F. 1929-1935. Bivalvia (Pelecypoda). In Bronn, H. G. (editor). Klassen und Ordnungen des Tierreichs, vol. 3, part 1. Leipzig, Akademische Verlagsgsellschaft. One of the most comprehensive treatises ever published on this group of mollusks.

Isgrove, A. 1909. *Eledone.* L.M.B.C. Memoirs. London, Williams & Norgate, Ltd.

Johnstone, J. 1899. *Cardium.* L.M.B.C. Memoirs. Liverpool, T. Dobb & Co.

Keen, A. M. 1963. Marine molluscan genera of western North America. Stanford, Stanford University Press. An up-to-date systematic account of the mollusks in the western part of our continent.

Lane, C. E. 1961. The teredo. Sci. Amer. **204**:132-142 (February).

Lemche, H. 1957. A new living deep-sea mollusc of the Cambro-Devonian class Monoplacophora. Nature (London) **179**:413-416. A description of the newly found *Neopilina*, a mollusk with internal segmentation.

MacBride, E. W. 1914. Textbook of embryology; Invertebrata, vol. 1. London, The Macmillan Co.

McCraw, B. M. 1961. Life history and growth of the snail, *Lymnaea humilis* (Say). Trans. Amer. Microscop. Soc. **80**:16-27.

MacGinitie, G. E., and N. MacGinitie. 1949. Natural history of marine animals. New York, McGraw-Hill Book Co., Inc. A large section of this popular book is devoted to the mollusks of the Pacific coast.

Moore, R. C., C. G. Lalicker, and A. G. Fischer. 1952. Invertebrate fossils. New York, McGraw-Hill Book Co., Inc. About 200 pages are devoted to the mollusks in this excellent treatise on paleontology; well illustrated.

Morton, J. E. 1960. Function of the gut in ciliary feeders. Biol. Rev. **35**:92-139. An up-to-date treatise on this method of feeding in the different groups of animals. The author points out the convergent resemblances that have arisen because of similar problems facing animals that use this method. An excellent and comprehensive account is devoted to ciliary feeding in Mollusca, where this method reaches its greater complexity.

Pelseneer, P. Mollusca. In Lankester, E. R. (editor). 1906. Treatise on zoology, vol. 5. London, Adam & Charles Black, Ltd. In many respects this treatise will never become out of date.

Pennak, R. W. 1953. Fresh-water invertebrates of the United States. New York, The Ronald Press Co. Freshwater mollusks are discussed and keyed.

Pilsbry, H. A. 1930. Land mollusks of North America (north of Mexico): Monograph 3, vol. 1, parts 1 and 2. Philadelphia, Academy of Natural Sciences. The most important work on the land snails of North America.

Potts, W. T. W. 1967. Excretion in molluscs. Biol. Rev. **42**:1-42.

Raven, C. P. 1958. Morphogenesis; the analysis of molluscan development. New York, Pergamon Press, Inc. This monograph is of great value in understanding the embryology of the mollusks. Extensive bibliography.

Robsin, G. C. 1932. A monograph of the recent Cephalopoda, 2 vols. London, The British Museum. An outstanding account of this group.

Runham, N. W. 1963. A study of the replacement mechanism of the pulmonate radula. Quart. J. Microscop. Sci. **104**:271-277.

Segal, E. 1961. Acclimation in mollusks. Amer. Zool. **1**:235-244.

Thiele, G. Solenogastres, Mollusca. In Kukenthal, W. and T. Krumbach (editors). 1924. Handbuch der Zoologie, vol. 5. Berlin, Walter de Gruyter & Co. A complete account of this group.

Trueman, E. R. 1966. Bivalve mollusk: fluid dynamics of burrowing. Science **152**:523-525.

Wells, M. J. 1962. Brain and behavior in cephalopods. Stanford, Stanford University Press. A fascinating study of this highly developed group of invertebrates.

Wilbur, K. M., and C. M. Yonge (editors). 1966. The physiology of Mollusca, vols. 1 and 2. New York, Academic Press, Inc. Represents the most important recent treatise on mollusks.

Willows, A. O. D. 1971. Giant brain cells in mollusks. Sci. Amer. **224**:68-75 (Feb.).

Younge, C. M. 1932. The crystalline style of the Mollusca. Sci. Progr. **26**:643-653. In this and other papers one of the greatest students of mollusks analyzes the functional aspects of molluscan structures.

Young, J. Z. 1961. Learning and discrimination in the octopus. Biol. Rev. **36**:32-96. An analysis of the learning ability of a form that is near the peak of learning capacity among the invertebrates.

Phylum Annelida

SEGMENTED PROTOSTOME ANIMALS

GENERAL CHARACTERISTICS

1. The annelids are the segmented worms, the most highly developed of the wormlike invertebrates, and, in general, the largest. Here metamerism, or segmentation, has evolved in its most elaborate form, is reflected internally as well as externally, and involves basically the mesoderm. The structure of the body wall and of the body cavity consists of a longitudinal series of segments—the oldest segments being anterior and the youngest posterior. The tendency has been toward a serial succession of segments, each containing unit subdivisions of the various organ systems. The functional significance of metamerism in coelomate animals is mainly concerned with development and locomotion.

2. A true coelomic cavity subdivided into segmental compartments provides space for the development of the body systems found in an advanced type of animal life.

3. Specialization of the head region, with differentiated organs such as tentacles, palps, and eyespots, has advanced further in annelids than in other invertebrates so far considered.

4. The centralization of the nervous system is highly developed, with cerebral ganglia (brain), closely fused ventral nerve cords, and segmental ganglia.

5. Digestion is extracellular, with a complete mouth-to-anus alimentary canal held in place by mesenteries and septal partitions. Suspension feeding in some has been carried to a fine art and involves several mechanisms.

6. The closed circulatory system of dorsal and ventral longitudinal vessels, segmentally arranged connecting vessels, and a system of capillaries ensures good and rapid blood supply for active organisms.

7. Respiration may occur as simple diffusion through the body surface or may involve the parapodia, specialized gills, or the radioles of the crown of fanworms.

8. Excretory organs (nephridia) are typically arranged with a pair to each segment for removal of nitrogenous waste and for osmotic and ionic regulation.

9. For the removal of gametes from the gonads to the exterior, coelomoducts grow outward through the body wall from the coelomic cavity.

PHYLOGENY AND ADAPTIVE RADIATION

1. Annelid ancestors with creeping and burrowing habits may have produced the origin and early evolution of segmentation, for early development is greatly subject to evolutionary modifications. Animals lower than annelids have a marked tendency to repeat organ systems. The hydrostatic skeleton of segmented forms, by subdividing the coelomic cavity into compartments, could pinpoint pressure changes. The pressure in the fluid of one segment is isolated from the fluid of other segments, resulting in more conservation of energy and faster and more powerful forces for propulsion.

Pseudometamerism, or even true metamerism, according to some interpretations, exists

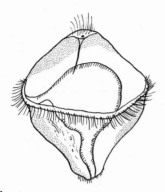

FIG. 11-1

Trochophore larva (diagrammatic). Trochophore of some form is found in several phyla and may indicate relationship.

in certain flatworms, although such metamerism is mainly ectodermal. Although sipunculids are not segmented, they have many characteristics of annelids, such as body wall structure, type of nervous system, and pattern of embryology. In general, the similarities in the early development of flatworms, annelids, and mollusks indicate a close relationship. Also the trochophore larvae (Fig. 11-1) found in polychaetes is also common in mollusks, bryozoans, nemertines, phoronids, and others. But at present it is impossible to assign the origin of annelids to any particular stock of flatworms or to ribbonworms (Rhynchocoela). The well-developed segmental organization of annelids represents a functional climax of putting together many features found here and there in other groups. This is also evident in their functional alimentary canal, which is similar in many respects to that of vertebrates.

Within the phylum Annelida it is thought that the Clitellata (oligochaetes and leeches) have arisen from a polychaete stem, even though polychaetes are more specialized in many ways. It is impossible from present knowledge to indicate the polychaete groups from which oligochaetes have come, although some oligochaete families such as the Aelosomatidae possess certain primitive characters of prostomium (ciliated pits, nervous system, the double func-

tion of nephridia for excretion and reproduction, etc.) that fit them to some extent as transition types between polychaetes and oligochaetes. Many authorities think that the terrestrial oligochaetes are degenerate or have lost many polychaete characters in their adjustment to a different environment. Each functional pattern seems to have developed independently for meeting varied ecologic conditions. The leeches have a meshwork of fluid-filled tissue as their coelom, but it serves the same hydrostatic function as it does in other annelids. They have specialized attachment suckers for a bloodsucking existence but lack setae and septa. They differ from oligochaetes mostly in their adaptations for their specialized existence.

2. The evolutionary diversity among the annelids has centered primarily around the metameric arrangement of the fluid-filled compartments and parapodia. In the more highly evolved annelids the exact number of somites is less easily determined in the adult than in the larva. Fusion, or modification of somites, may obscure the regularity of repetition of the organ systems, or even of the serial succession of the metameres. In primitive or unspecialized worms the adult often shows metamerism of identical units. The coelomic cavity as a hydrostatic skeleton has evolved subdivisions (septa) providing greater locomotory efficiency and greater capacity for precise changes of shape.

Significance of metamerism

In three great phyla of animals—Annelida, Arthropoda, and Chordata—the bodies are serially divided into units (metameres or segments) that are generally similar. Two of these phyla, Annelida and Arthropoda, are definitely related, as revealed by morphologic and embryologic evidence. The annelidan ancestry of the arthropods seems to be well established. However, metamerism in chordates must be considered as having an independent origin, an example of convergent evolution. Thus metamerism has arisen at least twice, in annelids and in chordates, and probably has a different adaptation in each case. In annelids it is an adaptation for burrowing; in chordates, for swimming. These three phyla have been very successful by whatever criteria one may use to measure success in the animal world. Perhaps three fourths of all named animal species belong to these metameric groups. It is, however, a difficult matter to assess the degree to which they owe their success to metameric organization. One may speculate on this or that advantage accruing from metamerism, but the evolutionary pattern of animals is too complex, from our present knowledge, to assign their differential survival and success to any one contributory factor.

In the three groups a considerable degree of variation exists in the intensity or degree of metamerism. Annelids possess the generalized type of metamerism, and it is in them that one can best see the basic plan of a metamere. In the annelid the metameres are marked off from each other by external grooves and by internal partitions (septa). The inner longitudinal muscles form a continuous sheath throughout the worm, but the fibers of a metamere run only between intersegmental junctions. The outer circular muscles are arranged in segmental groups. Many internal organs such as nephridia, certain blood vessels, ganglionic enlargements, and lateral nerves are serially arranged according to segments. On the other hand, the alimentary canal and parts of the reproductive system show little or no metameric organization. Each metamere thus functions individually to some extent.

The chief advantage of metamerism, especially of the primitive plan, is that it allows controlled movement. Such a metameric arrangement largely restricts the coelomic fluid to separate septal compartments. By muscular action the coelomic fluid of a compartment can be compressed transversely to lengthen a metamere, or it may be compressed longitudinally to thicken and shorten the metamere. Compression of the fluid on one side of a metamere will cause the other side of the seg-

ment to lengthen and thus produce a curve. The septa are therefore of paramount importance in that they allow each segment to move separately from the others, and directional changes are easily effected.

It may be stated that the concept of segmentation is not so simple as the foregoing description of conditions, such as those in an earthworm, may indicate. As will be seen later, this generalized pattern is often greatly altered and modified. If the repeated segments are generally alike, as they are in the earthworm, the metamerism is called homonomous; when they are unlike, as in most arthropods, it is termed heteronomous. Fusion and specialization have made many alterations in the primitive basic plan. An additional fact is that many groups other than the three mentioned have some tendencies toward a segmental arrangement. Some authorities think that in the cestodes the proglottids may represent true segmentation, although in these flatworms new units are added just behind the anterior scolex, whereas in true metamerism new segments are formed just in front of the terminal segment, the pygidium. However, there may be many types of metamerism for different adaptive reasons —swimming, reproduction, and repetition of organs in elongated animals for effective control of all types of movement, especially in soft-bodied animals that are coelomate and unsegmented.

The origin of metamerism has been speculative among zoologists ever since the process has been studied. The fission theory explains its origin as the result of an incompletely divided chain of zooids, a common method of asexual reproduction. Another theory is that of pseudometamerism found in some forms in which organs were repeated and scattered through the body before segmentation occurred. The muscular theory of segmentation stresses the idea that metamerism first arose as a locomotory device in the somite arrangement of muscles for undulating movements such as are found in certain larvae and in the freshwater origin of the vertebrates. It is possible that segmentation could have arisen first from the repeated arrangement of certain body units, with the later development of interseptal partitions separating these units from each other.

CLASSIFICATION

The phylum Annelida consists of bilateral animals that belong to the Protostomia, which have spiral cleavage and a schizocoelic coelom. They are commonly divided into three main groups—Polychaeta, Oligochaeta, and Hirudinea. Of these, the polychaetes are mainly marine; the oligochaetes are chiefly terrestrial and freshwater; and the leeches (Hirudinea) are wholly or partly parasitic, chiefly in freshwater or on land.

Phylum Annelida. Body vermiform and metamerically segmented externally and internally; cuticle thin and secreted by an epidermis or hypodermis; chitinous setae paired, few or many in number; fleshy parapodia in some; body wall of circular and longitudinal muscles; coelom may be divided by septa; blood system closed, with respiratory pigments; digestive system complete and not metamerically arranged; nephridia paired; respiration by skin or gills; nervous system of brain and midventral nerve cord with ganglia and nerves; sensory system of tactile, photoreceptor, and equilibrial organs; hermaphroditic or dioecious sexes; larva, when present, a trochophore; terrestrial, freshwater, and marine.

Class Polychaeta. Body has numerous metameres with lateral parapodia that bear many setae; head distinct with eyes and tentacles; clitellum absent; sexes usually separate; gonads transitory; asexual budding in some; trochophore larva usually; mostly marine.

Subclass Errantia. Body of many segments, usually similar except in head and anal region; parapodia alike and with acicula; pharynx usually protrusible; head appendages usually present; free living, tube dwelling, pelagic, mostly marine. *Neanthes (Nereis)*, *Aphrodite, Manayunkia, Glycera* (Fig. 11-9).

Subclass Sedentaria. Body with unlike segments and parapodia, and with regional differentiation; prostomium small or indistinct; head appendages modified or absent; pharynx without jaws and mostly nonprotrusible; parapodia reduced and without acicula; gills anterior or absent; tube dwell-ing or in burrows. *Arenicola* (Fig. 11-10), *Chaetopterus* (Fig. 11-13), *Amphitrite* (Fig. 11-11).

Subclass Archiannelida. This ill-defined subclass is a small group of reduced polychaetes of unrelated ancestry. They are usually characterized by having a ciliated epidermis, no parapodia, reduced or missing septa, and indistinct segmentation. They may be primitive or highly aberrant and may have evolved from different polychaete lines. A form of trochophore larva links them to the polychaetes. *Polygordius.*

Class Clitellata. Body with clitellum; segmentation conspicuous; segments with or without annuli; segments definite or indefinite in number; parapodia absent; hermaphroditic; eggs usually in cocoons; mostly freshwater and terrestrial.

Order Oligochaeta.* Body with conspicuous segmentation; setae few per metamere; head absent; coelom spacious and usually divided by intersegmental septa; development direct, no larva; chiefly terrestrial and freshwater.

Suborder Plesiopora. Male pores or gonopores in segment following the testes; seminal receptacles in region of, or in front of, genital segments; mostly aquatic. *Aeolosoma, Nais, Tubifex* (Fig. 11-27), *Enchytraeus.*

Suborder Prosopora. Male gonopores in same segment with last pair of testes (if more than one pair); gonads and seminal receptacles variable in position and in number; aquatic. *Branchiobdella.*

Suborder Opisthopora. Male gonopores at least one segment or more behind last pair of testes. *Lumbricus* (Fig. 11-37), *Eisenia, Allolobophora.*

Order Hirudinea.* Body has definite number of segments (33) with many annuli; body with anterior and posterior suckers usually; setae usually absent; coelom closely packed with connective tissue and muscle; terrestrial freshwater, and marine.

Suborder Acanthobdellida. Body with posterior sucker only; no proboscis or jaws; segments 30 in number; setae present; body cavity spacious with segmental septa; primitive and intermediate between Oligochaeta and Hirudinea. *Acanthobdella.*

Suborder Rhynchobdellida. Body with or without anterior sucker; protrusible proboscis without jaws; blood colorless; marine and freshwater. *Placobdella* (Fig. 11-39), *Glossiphonia.*

*There is much discussion at present as to the classification of Oligochaeta and Hirudinea. Some authorities consider them as classes coordinate in rank with Polychaeta; others place them as orders under the class Clitellata.

Suborder Gnathobdellida. Body with both anterior and posterior suckers; proboscis lacking; three pairs of jaws; blood red; land and freshwater. *Hirudo* (Fig. 11-39), *Macrobdella*, *Haemopis*.
Suborder Pharyngobdellida. Without proboscis and jaws; teeth lacking; land and aquatic. *Erpobdella*, *Dina*.

CLASS POLYCHAETA

The polychaetes are the largest and oldest group of the Annelida. More than 10,000 species and 1,600 genera of these worms exist, and most of them are marine. They display an enormous size range, varying from less than a milli-

meter to about 3 meters in length, although the majority are around 50 to 100 mm. long. Many are highly colored (red, green, etc.), and some are very beautiful. An ecologic classification divides them into two subclasses, the errant, or free-moving, Errantia and the sedentary tube dwellers, or Sedentaria. The Errantia include active burrowers, pelagic forms, crawlers, and tube dwellers (Fig. 11-2) that may leave their tubes for food or for breeding or may even carry their tubes about with them. The Sedentaria live in tubes and rarely expose more than the head from the end of the tube (Figs. 11-3 and 11-4).

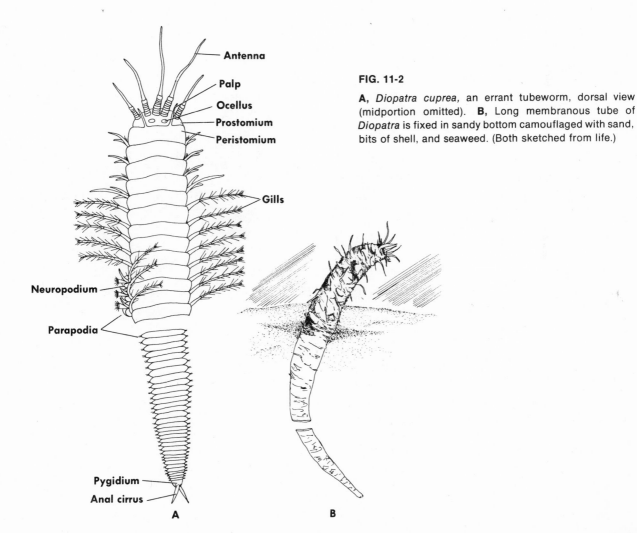

FIG. 11-2

A, *Diopatra cuprea,* an errant tubeworm, dorsal view (midportion omitted). **B,** Long membranous tube of *Diopatra* is fixed in sandy bottom camouflaged with sand, bits of shell, and seaweed. (Both sketched from life.)

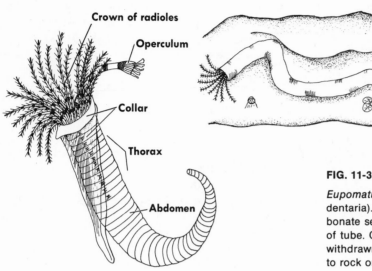

Crown of radioles

Operculum

Collar

Thorax

Abdomen

FIG. 11-3

Eupomatus dianthus, a serpulid fanworm (subclass Sedentaria). Glands beneath collar produce calcium carbonate secretions, which collar molds into place at end of tube. Operculum serves as plug in tube when head is withdrawn. Its calcareous tube (right) is usually cemented to rock or shell. (Sketched from life.)

FIG. 11-4

Some fanworms in their tubes. (From Encyclopaedia Britannica film, "Worms—the Annelida.")

TUBE STRUCTURE AND BUILDING

The tubes constructed by the polychaetes vary a great deal. Mucus secreted by special gland shields may simply harden to form a protective covering around most of the body, or in other cases the mucous secretions may be used to stick mud and sand particles together to construct a more or less cylindric tube. The tube itself may be smooth and regular, or it may be jagged and rough in appearance. Sometimes it is branched and very extensive. It may be fixed in position (*Sabella*), or it may be carried around with the worm (*Pectinaria*).

Materials used by the worms in making their tubes also show considerable variation. Some worms are very selective in their choice of materials, and others simply make use of whatever material is at hand. *Sabella,* for instance, chooses only sand particles of a very fine size, whereas *Terebella* often makes its irregular tube of sand and fragments of detritus. The serpulids secrete carbonate of lime from glands located under their collars and construct a tube of this mineral. Some species of *Eunice* form a parchmentlike tube that is translucent. Various substances of both organic and inorganic nature from tubes have been chemically analyzed.

In constructing their tubes, certain worms make use of a collar that is formed by the folding back of the peristomium (*Sabella, Serpula*). In *Sabella* the collar is employed in molding the additions of particles that are first stored in ventral sacs below the mouth and mixed with mucus. These sacs then deliver strings of mucus and sand to the collar folds located below the sacs. The collar is provided with lappets that mold the mucus-sand strings to the end of the tube as the worm rotates in the tube. In this way the tube is slowly built up. *Serpula,* which forms its tube entirely from calcium carbonate secretions, secretes the material into the space between the body wall and collar, where it hardens in the space as a ring so that the tube is elongated by the fusion of rings on top of each other. As the worm rotates, a mucous coating is laid down on the inner surface of the tube.

In nearly all tube-building worms part of the tube projects above the surface as a chimney. *Chaetopterus* with its U-shaped parchment tube has 2 chimneys, one at each end, that project above the sand surface (R. D. Barnes, 1965). Many worms have tubes located vertically in the bottom of the sea or in intertidal zones, but some have tubes placed at different angles to the surface. Usually only the head emerges from the tube opening to procure food. A number of tube-dwelling forms entrap their food in various ways by the use of mucus.

MORPHOLOGIC FEATURES

Polychaetes as a class show great diversity. They may be classified morphologically as scale-bearing, jaw-bearing, or crown-bearing forms.

Typically the polychaete worm has an anterior head or prostomium. It may be a simple lobe above or in front of the mouth, it may be pushed far back on the body, or it may be retractile into one of the first segments. The head may bear eyes, antennae, and a pair of palps (Fig. 11-5). The first segment, the peristomium, surrounds the mouth, forming the lower lip, and may bear setae and other complex structures such as the tentacular crown of the sabellid group. In some cases the peristomium may fuse with more posterior segments to form a food-

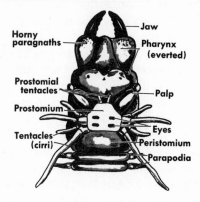

FIG. 11-5

Dorsal view of head and anterior segments of *Neanthes,* with pharynx and jaws protruding. (From Hickman: Integrated principles of zoology, The C. V. Mosby Co.)

FIG. 11-6

A to **D,** Some types of polychaete setae. **A,** *Hermoine,* a scale worm. **B,** *Amphitrite.* **C,** *Neanthes.* **D,** Oar-shaped swimming seta of *Heteronereis.* **E,** Diagram of polychaete parapodium.

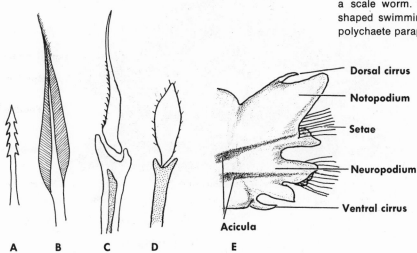

gathering structure. It also often bears a pair of palps.

The trunk consists of a definite or indefinite number of segments that may be quite similar or may be modified into specialized regions such as thorax, abdomen, and tail. The Sedentaria are usually the most modified worms. Most body segments have a pair of lateral, fleshy expansions of the body called parapodia (Fig. 11-6, *E*). Each parapodium may be single (uniramous) or double (biramous), consisting of a dorsal notopodium and a ventral neuropodium. Each may also have lobes or processes such as tentacles, cirri, or folds. Each parapodium may also have embedded rodlike acicula and emergent setae, which are secreted by special cells within their bases. Setae are hardened by a type of sclerotization similar to that found in arthropods. When lost or worn away, the setae are replaced by these cells. The type of setae and their arrangement have taxonomic value. Setae (Fig. 11-6) function in swimming, in gripping surfaces, in reproduction, in respiration, as organs of defense or offense, and as sensory organs. In many pelagic polychaetes the setae may be entirely absent,

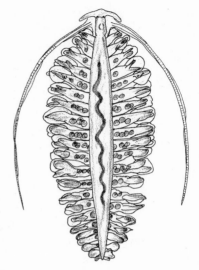

FIG. 11-7

Tomopteris sp., a planktonic polychaete, has few or no setae; parapodia are adapted for swimming. (Modified from Davis.)

FIG. 11-8

Transverse section through *Neanthes* (diagrammatic). (From Hickman: Integrated principles of zoology, The C. V. Mosby Co.)

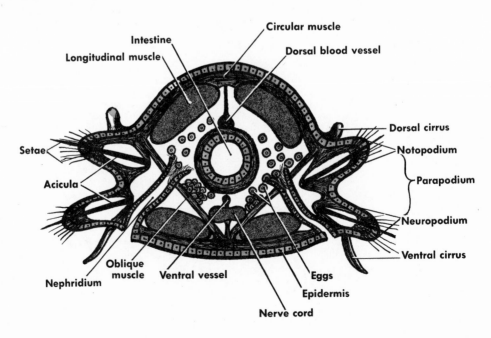

and the parapodia function as paddles in swimming (Fig. 11-7). The dorsal setae in some scaleworms form a dorsal mat or feltwork over most of the body.

The body wall of polychaetes consists of a dermomuscular tube that is separated from the gut by the coelom and the peritoneum (Fig. 11-8). The outer layer of the tube is formed of a single layer of columnar epithelium, which is covered by a thin cuticle. Some of the epithelial cells are modified into glands that secrete mucus. Sometimes restricted areas of the epidermis are ciliated. Below the epithelium is a basement membrane of connective tissue, followed by the smooth muscle layer of circular and longitudinal muscles, the latter of which is usually in the form of two dorsolateral and two ventrolateral bundles. Oblique muscles often extend from the midventral to the midlateral line on each side, usually at the level of the parapodia.

The coelom of polychaetes is spacious and is usually divided into compartments in accordance with the metamerism of the body. The successive compartments are separated by the intersegmental septa, each of a double layer of peritoneum, which extend from the gut to the body wall. The gut thus runs through the septa and is suspended dorsally and ventrally by longitudinal mesenteries. In this way each coelomic compartment is divided into left and right halves. However, the successive cavities are not completely closed off from each other because the septa may be deficient or may be entirely lacking. The coelomic fluid contains excretory granules as well as many kinds of cells and gametes. Communication between the coelomic cavity and the outside takes place through the nephridia, the coelomoducts, and the dorsal or head pores (in some).

The alimentary canal is usually a straight tube running from the anterior mouth to the anus in the terminal segment (pygidium). In a few worms the gut is coiled and may also bear branched ceca for increasing the surface. Regional specializations of the alimentary tube vary, but a buccal cavity or pharynx is usually present, as well as a short esophagus (sometimes with glandular ceca), a stomach, and a rectum. A type of gizzard is present in a few worms. The cuticle-lined pharynx or buccal cavity is armed with characteristic jaws in many errant polychaetes that are raptorial feeders. The pharynx is eversible, and one or more pairs of jaws are usually located on its lateral walls (Figs. 11-5 and 11-9). By everting the pharynx, the food is seized by the jaws and drawn into the body when the pharynx is retracted. Eversion of the pharynx is effected principally by the coelomic fluid pressure induced by the contraction of the body wall muscles; the pharynx is withdrawn by special retractor muscles fastened to its posterior end and to the body wall. The pharynx usually occupies the first 6 body segments, but it may extend pos-

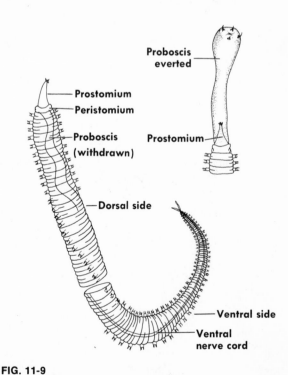

FIG. 11-9

Glycera dibranchiata, the "beak thrower." Pointed head is adapted for burrowing. At right, anterior end with proboscis everted (family Glyceridae, subclass Errantia). (From life.)

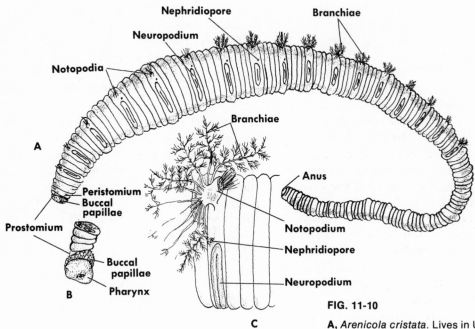

FIG. 11-10

A, *Arenicola cristata.* Lives in U-shaped burrow in sand or mudflat, usually identified by little mound of castings at one end of burrow. **B,** Everted pharynx. It everts and withdraws, carrying sand and organic substances into esophagus. **C,** One segment of *Arenicola* showing annuli, notopodium, and gills. (Subclass Sedentaria.) (Modified from Ashworth.)

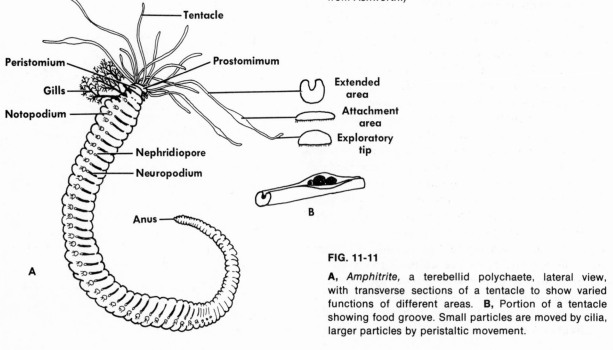

FIG. 11-11

A, *Amphitrite,* a terebellid polychaete, lateral view, with transverse sections of a tentacle to show varied functions of different areas. **B,** Portion of a tentacle showing food groove. Small particles are moved by cilia, larger particles by peristaltic movement.

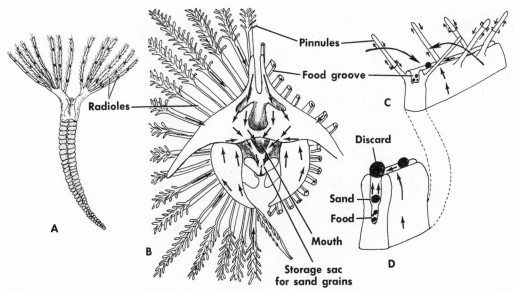

FIG. 11-12

Ciliary feeding in fan worms. **A,** *Sabella,* a tubicolous fanworm, with its crown of feeding radioles. **B,** Anterior view of crown showing bases of grooved and pinnate radioles. Cilia direct small food particles toward mouth and sand grains toward storage sacs, where they are saved for tube building. **C** and **D,** Enlarged portions of a radiole. **C,** Distal portion showing food groove and ciliary tracts of pinnules. **D,** A more proximal portion of radiole where particles are sorted for size. (**B** to **D** adapted from Nicol and others.)

teriorly a much greater distance in those worms with a long proboscis (*Glycera,* Fig. 11-9). Some active forms burrow through the sand and mud by swallowing quantities of soil with organic food by means of an eversible pharynx or proboscis without jaws (*Arenicola,* Fig. 11-10). Others, having a proboscis that is free at the anterior end, merely extend the proboscis when feeding and suck in food. A circle of teeth instead of jaws is usually found at the tip of the proboscis in such arrangements.

Sedentary, or tube-dwelling, forms are ciliary feeders and collect microscopic food by means of ciliated mucous grooves on numerous extensible and contractile tentacular filaments (*Amphitrite,* Fig. 11-11). Fanworms such as *Sabella* (Fig. 11-12) and *Serpula* have an enormous number of head appendages or bipinnate tentacles (radioles) that are rolled up when the worm is withdrawn into the tube. Ciliated grooves carry particles down the tentacles to their bases, where the small particles are sorted out and passed to the mouth. Large particles are rejected, but medium ones are used in tube construction.

The unique *Chaetopterus* (Fig. 11-13) has a different method of feeding (G. E. MacGinitie, 1939; R. D. Barnes, 1965). These worms live in U-shaped parchment tubes with 2 apertures. Their tentacles are rudimentary and play little part in food collecting. Instead, three pairs of modified notopodia near the middle of the body form fans and by rhythmic oscillation produce currents of water that pass through the tube from end to end. Food particles carried by these currents are entangled in a baglike sheet of mucus formed by the specialized winglike (aliform) notopodia of the segment just anterior to the fans. These notopodia are richly supplied with mucous glands for the formation of the mucous bags, which are continually being produced as successive boluses of food and are carried along a ciliated mid-dorsal groove to the mouth.

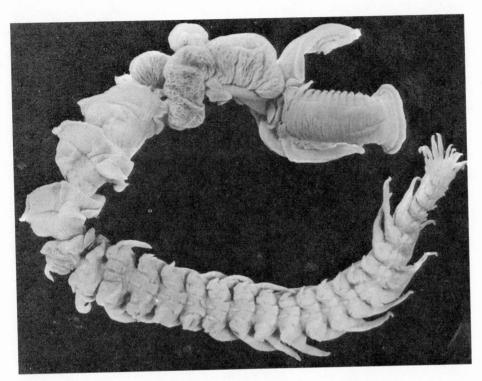

FIG. 11-13

Chaetopterus lives in mudflats in a leathery, open-ended, U-shaped tube. This genus has a complicated food-gathering device involving mucous bag secreted by aliform twelfth notopodia to trap detritus and plankton from water that is kept flowing by means of beating fans. Food and mucus are rolled into ball by ciliated cup below aliform notopodia, then carried up ciliated groove to mouth. (Subclass Sedentaria.) (From Hickman: Integrated principles of zoology, The C. V. Mosby Co.)

Esophageal glands found in many polychaetes, including *Neanthes*, probably produce proteolytic enzymes for digestion.

Food is moved through the digestive system by muscular contraction in the gut wall, by cilia lining the lumen, or by a ventral ciliated groove. More than one of these methods may be involved in the same worm. Since many polychaetes take in a large amount of indigestible substances along with organic food, extensive castings are often extruded through the anus. This is especially the case with *Arenicola*, which leaves its castings on the intertidal beaches. Sedentary

FIG. 11-14

Anterior end of nervous system of *Neanthes*. (After Turnbull; modified from Brown.)

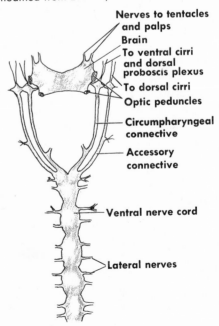

Nerves to tentacles and palps

Brain

To ventral cirri and dorsal proboscis plexus

To dorsal cirri

Optic peduncles

Circumpharyngeal connective

Accessory connective

Ventral nerve cord

Lateral nerves

forms living in tubes with 2 apertures have their egested waste removed by water currents, but those that live in deeply buried tubes may have to thrust out the anal end during defecation. Some worms such as the fanworms have a ciliated groove (partly ventral and partly dorsal) that extends from the anus to the head region for the expulsion of fecal material released from the anus.

The nervous system (Fig. 11-14) consists of a brain or dorsal ganglion mass located primarily in the prostomium; a pair of circumesophageal connectives; and paired ventral, ganglionated, or smooth cords that may be separated in primitive forms and fused into a single cord in others. The brain sends nerves to supply eyes, palpi, antennae, and nuchal organs. Lateral branches from the ventral cord supply parapodia and other segmental structures. In some forms the ventral cord is embedded in the epidermis; in others it lies in the coelomic cavity. Many variations of the nervous plan exist among the polychaetes. In general, the nervous system tends to move inward for better protection. In most polychaetes it lies in the muscle layer. Ganglionic swellings may be single or double per segment; lateral nerves vary from two to five pairs per ganglion; pedal ganglia separate from the ventral cord may innervate the parapodia; and giant fibers in the cord may vary from the characteristic number of 3. The giant fiber system is especially characteristic of the Sedentaria for rapid conduction, as in the withdrawal of tentacles for protection.

Although the errant polychaetes are primitive in many respects, they have the most evolved brains among the annelids. The cerebral ganglia are developed functionally into centers, each of which is specialized for a definite function. It may be stated that their brain sets a pattern for the annelid-mollusca-arthropod line. This pronounced development in errant forms may be associated with their activity in search of food, especially for active prey. However, the general annelid brain seems to be in transition toward a higher-developed one. Their associa-

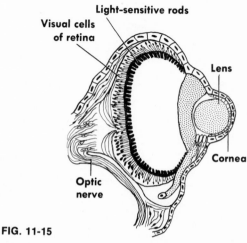

FIG. 11-15

Section through eye of polychaete *Alciope* (diagrammatic).

tional neurons relay sensory stimuli to motor centers, producing responses more or less stereotyped. Cerebral ganglia are not mere relay stations, however, but mediate to some extent definite responses for bodily welfare as the occasion demands.

In some pelagic polychaetes the sense organs are well developed; the eye is complex enough to form an image (Fig. 11-15). Simpler eyes are of the retinal-cup type (Fig. 11-16, *A*), in which rodlike photoreceptors, pigment, and refractile bodies are used for light detection in many errant polychaetes. They are usually located on the prostomium and may consist of several pairs. Eyespots may be present in forms such as fanworms. Statocysts (Fig. 11-16, *B* and *C*) occur in *Arenicola* and some others. Other sense organs include the head nuchal organs, tactile cells scattered over the body, and others. The nuchal organs are glandular sensory pits located on the first segment of certain polychaetes, and are comprised of chemoreceptors.

M. Wells (1968) distinguishes between tonic and phase receptors in sense organs. He describes tonic receptors as being those in which the energy of the stimuli induces in a nerve cell in close contact with the receptor a generator potential that may discharge steadily and re-

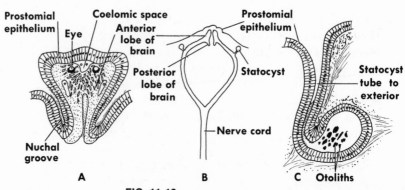

FIG. 11-16

A, Horizontal section (diagrammatic) of prostomium of *Arenicola* showing relationship of brain, eyes, and coelomic spaces. (×35.) **B,** Anterior end of nervous system of *Arenicola*; nerves to statocysts arise from esophageal connectives. (×3.) **C,** Section of statocyst and tube to exterior *(Arenicola)*; irregular otoliths are chiefly quartz grains. (×100.) (Modified from Ashworth.)

quires continual monitoring, as in the proprioceptors of muscle tension. On the other hand, phasic receptors also generate a generator potential, but the impulses it produces tend to disappear if the conditions remain steady. Phasic receptors are especially adapted to the changes that occur in an animal's environment and depend on the stored information the animal has of its past history (memory) regarding the meaningful significance of a particular stimulus.

Respiration in polychaetes may occur through the body surface (scale worms and some others) or by epithelial outgrowths (gills or branchiae). Gills are lobes or simple filaments, sometimes much divided for increase of surface, and are supplied with vascular loops. Gills may be found in various localities, but they are chiefly associated with the parapodia. The neuropodium or, more commonly, the dorsal cirrus may be modified into a kind of gill. Forms such as *Arenicola* (Fig. 11-10) have gills on several segments in the middle of the body, whereas *Amphitrite* (Fig. 11-11) has gills only on the first 3 segments. Forms with less regional differentiation are more likely to have gills on most of the segments. In the crown of fanworms the numerous radioles serve for gas exchange. Water currents over the gills are chiefly produced by bands of cilia on the gills or on the body surface. Water currents may also be produced by undulatory movements of the worms and by the devices employed for keeping the water circulating through the tubes of tubicolous worms.

Polychaetes in general have a closed vascular system (Fig. 11-17). The circulatory plan consists of median longitudinal dorsal and ventral vessels, with transverse connections segmentally arranged. The blood flows forward in the dorsal vessel, which lies over the digestive tract. At its anterior end the dorsal vessel is connected to the ventral vessel by means of one or more vessels. The ventral blood vessel carries the blood backward beneath the digestive tract. The propulsive power of blood flow is produced by the contractile walls of the dorsal vessel and to some extent by that of other vessels as well. The ventral vessel, however, seems to have little or no peristaltic contraction. A part of the dorsal vessel may be modified as a bulbous, muscular heart in some worms, especially in the Sedentaria. Blood passes from the ventral vessel to the dorsal vessel by indirect routes. In each segment the ventral vessel gives rise to vessels that supply the parapodia, the body wall, the nephridia, and the digestive system. The dorsal vessel receives vessels from the parapodia and the alimentary canal. The branches of these ventral and dorsal systems are interconnected by a meshwork of smaller vessels. Many variations of this plan are found, and in some Er-

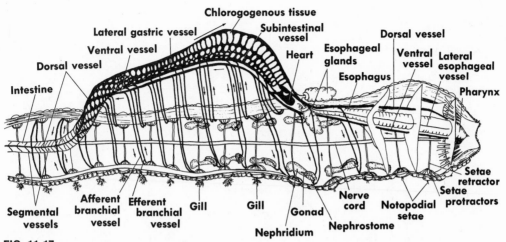

FIG. 11-17

Internal structure of *Arenicola,* dorsal view, with portion of digestive tract lifted out to show arrangement of circulatory system. (Modified from Ashworth.)

rantia the vascular system is an open one instead of closed.

The blood, in its course from the ventral to the dorsal vessel, undergoes many changes in its composition. In the body wall the blood loses carbon dioxide and picks up oxygen; in the nephridia it loses waste and picks up nutrients from the intestine. The blood also varies greatly in color. In some forms it is colorless (scale worms, syllids, phyllodocids), but it may also be greenish, reddish, or other hues. Most polychaetes have some kind of respiratory pigments dissolved in the plasma of their blood. These pigments aid in the transport of molecular oxygen. One respiratory pigment is chlorocruorin, which shows up as red in high concentrations or bright green in more dilute solutions. This pigment is common in the blood of the serpulid and sabellid fanworms. Red hemoglobin is one of the most common pigments, and some polychaetes such as *Serpula* have both chlorocruorin and hemoglobin. Hemoglobin molecules may be large or small. In *Arenicola* the molecular weight of the molecule is 3,000,000. Some polychaetes contain amebocytes in their blood, and in *Glycera,* which has no blood vessels, erythrocytes with hemoglobin

are present in the coelomic fluid. See the discussion on respiratory pigments at the end of this chapter.

A pair of nephridia is typically found in every segment that bears parapodia (Fig. 11-17). A nephridium is usually an invaginated ectodermal tube, although some may be mesodermal in origin. Two kinds occur—protonephridia and metanephridia (Fig. 11-18). In some polychaetes only one pair of nephridia may be present in the entire animal; in others a pair is found in each segment. The typical metanephridium extends into 2 segments. The nephridial tubule and its external opening, the nephridiopore, lie in one segment, and the other end of the tubule is a ciliated funnel (nephrostome) opening into the coelom of the next anterior segment (Fig. 11-18, *F*). The primitive protonephridial tubule does not open directly into the coelom but has at the preseptal end a slender solenocyte tubule provided with a long flagellum (Fig. 11-18, *A* and *B*). Fluid passing from the coelom through the walls of the solenocyte tubule is driven by the flagellum into the chief lumen of the nephridium. There may be several of these solenocytes to a single nephridium. Nephridial tubules are ciliated for driving the fluid and for discharging it through the nephridiopore. Metanephridia are generally coiled and are chiefly found in Sedentaria but also occur in Errantia. All types of nephridia

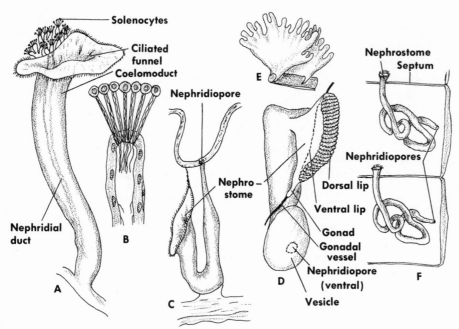

FIG. 11-18

Nephridia. **A,** Protonephridia and coelomoduct of *Phyllodoce*. **B,** Group of solenocytes and branch of protonephridium of same. **C,** Metanephridium of *Nerine*. **D,** Metanephridium of *Arenicola;* dorsal lip bears many highly vascular, lobed, ciliated processes; vesicle is shown distended with sperm. **E,** Two lobed processes from dorsal lip of nephridium of *Arenicola*. **F,** Metanephridia of *Neanthes*. (**A** to **C** modified from Goodrich; **D** and **E** modified from Ashworth.)

collect and transport waste from the coelom to the exterior through the nephridiopores. In some polychaetes the nephridia are surrounded by a network of blood vessels that furnish some of the waste to the tubules. Chloragogen tissue on the wall of the intestine (Fig. 11-17) and coelomocytes may also play a part in excretion.

Polychaetes have both asexual and sexual methods of reproduction. Asexual reproduction occurs by budding and transverse fission; sometimes it may happen by accidental fragmentation. In a few worms a chain of sexual individuals is formed by an asexual stolon (blastogamy or budding). This is most common among the fanworms, serpulids, and syllids. In sexual reproduction most polychaetes are dioecious,

but occasional cases of hermaphroditism, parthenogenesis, and protandry also exist. Gonads or masses of sex cells are usually widely scattered in the coelomic cavity. They usually appear first as swellings of the peritoneum on the septa of certain segments. In some forms the gonads may have a generally segmental arrangement (*Aphrodite*); in the familiar *Neanthes* the ventral septal peritoneum is the region of gonad formation. There is no differentiation of special sex ducts or receptacles such as uteri, seminal vesicles, and gonoducts in *Neanthes*. Unripe sex cells are shed into the coelom, where maturation occurs in the coelomic fluid. Special nurse cells may nourish the ova as the latter develop. Gametes are released from the body cavity through special gonoducts, through modified nephridia, or through a breakdown of the body wall. When the body wall ruptures, the adults die. Perhaps the majority of polychaetes shed their sex cells to the exterior through the metanephridia.

It is thought by some authorities that in the primitive plan of polychaetes each segment bore a pair of protonephridia, with solenocytes

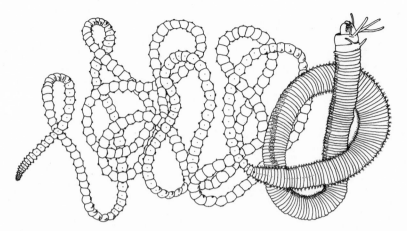

FIG. 11-19

Eunice viridis, the palolo worm of Samoa and Fiji. Epitokal region is long chain of posterior segments, each with eyespot on ventral side. (Modified from Woodworth.)

for excretion and a pair of gonoducts for carrying the gametes. The union of the ciliated opening of a gonoduct with the protonephridial tubule produced the metanephridium (the nephromixium of E. S. Goodrich). The degeneration of the solenocytes in this combination would result in the typical metanephridium with both genital and excretory functions. Other investigations, however, have shown that in some polychaetes the metanephridia may arise as new developments from the mesoderm and are not homologous to the protonephridia as the Goodrich theory implies. Moreover, such metanephridia may have both excretory and genital functions the same as does the nephromixium or may have only an excretory function.

Fertilization occurs in the sea outside of the worms. During the breeding season, worms of both sexes swim together near the surface of the water, and the proximity of the worms ensures contact of sperm and eggs.

The external form of the body of a sexual worm may involve radical changes in color, body texture, setae morphology, and parapodia.

This phenomenon is called epitoky and is characteristic of the Nereidae, the Syllidae, and the Eunicidae. The reproductive individual, which is usually pelagic, is called an epitoke; the nonsexual and nonpelagic form is an atoke. The most striking modifications in epitokal forms are found chiefly in the gamete-forming segments, which are much distended with gametes. Their parapodial lobes are greatly enlarged, with long swimming setae. The palolo worm (*Eunice viridis,* Fig. 11-19) is a classic example of epitoky. Once a year during the full moon in October and November the posterior segments, full of eggs or sperm, are cast off by the adult worms, which live in burrows of coral reefs. These segments rise and swarm at the surface, making the water milky with the discharged gametes. Similar behavior is found in the West Indian palolo *E. schemacephala* (L. B. Clark and W. N. Hess, 1940). In these worms, swarming occurs in July during the first or last quarter of the moon. The physiologic reasons for swarming apparently involve environmental factors that are poorly understood.

Another striking case of epitoky is found in *Autolytus* (Syllidae), which has marked dimorphic phases of male and female sexes. In the male epitoke of this form the body is divided into three distinct regions, the preepitokal, the epitokal, and the postepitokal. The female epitoke is more slender, with the prostomial antennae simple instead of forked, as in the male.

Polychaetes usually have marked regenerative powers. Most lost organs such as parapodia, tentacular crowns, and tails are replaced. In a few species an entire worm may regenerate from a single segment. Autotomy, or the voluntary discarding of damaged parts, is also a frequent occurrence.

DEVELOPMENT

In development the polychaete egg has typical spiral cleavage, with mosaic development (Fig. 11-20). Although the egg is telolecithal, the amount of yolk is variable and determines the type of gastrulation. In some polychaetes the egg has little yolk, and the macromeres and micromeres are about the same size. In such cases gastrulation takes place by invagination. Most polychaetes, however, have micromeres much smaller than macromeres in their spiral cleavage. Gastrulation in these forms occurs by epiboly. In some polychaetes, gastrulation may involve a combination of both invagination and epiboly. A solid blastula or stereoblastula, instead of a coeloblastula, is chiefly characteristic of those forms with unequal cleavage. In *Neanthes*, for instance, the first

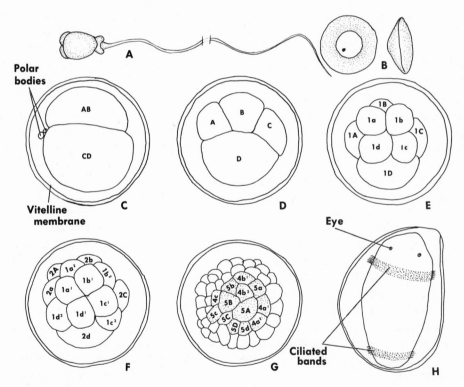

FIG. 11-20

Early development of *Arenicola*. **A,** Spermatozoa. (×3,000.) **B,** Two views of flat ovum. (×100.) **C,** Two-cell stage about $4\frac{1}{2}$ hours after fertilization. **D,** Four-cell stage, from anterior pole. **E,** Eight-cell stage, with first quartet of micromeres. **F,** Sixteen-cell stage about 2 hours after first cleavage; second quartet has formed. **G,** Blastula at end of 24 hours, seen from posterior pole; fourth quartet has formed; **4d,** mesoblast cell has sunk into segmentation cavity; **4a, 4b,** and **4c,** along with **4A, 4B, 4C,** and **4D,** are endoderm cells at posterior pole. **H,** Larva 72 hours after fertilization; has eyes and 2 ciliated bands and is ready to hatch. (**C** to **H** ×210.) (Redrawn from Ashworth.)

two divisions are equal, producing 4 similar blastomeres or macromeres. Each of these macromeres in its further cleavage gives rise to one of the quadrants of the embryos. The third, fourth, and fifth divisions are unequal and at right angles to the first two cleavage stages and produce from the macromeres three quartets of small micromeres. These three quartets of cells give rise to the ectoderm and its derivatives. The sixth division involves a fourth quartet and results in the separation from the macromeres of cells that produce the mesoderm and aid in the formation of the endoderm.

Characteristic of many polychaetes, but not of *Neanthes,* is the formation during development of a pelagic, ciliated larva called a trochophore (Fig. 11-21). In some forms this stage is found in the egg, and the larva has no opportunity for a free-swimming existence because the egg hatches into a posttrochophore form. In the typical free-swimming trochophore the chief band of cilia is the prototroch, which is preoral in position. Later another band of cilia, the telotroch, rings the larva near the anus, followed by a third girdle, the metatroch, located just below the mouth. Both the mouth

and the anus may arise from the elongated blastopore, or the anus may arise as a new opening. The alimentary tract, in addition to the mouth and anus, consists of an ectodermal esophagus (stomodeum), an endodermal stomach, and an ectodermal rectum. The body cavity is the old blastocoel and is the space between the gut and ectoderm. The cavity contains larval muscle bands, processes of mesenchymal cells, and 2 larval protonephridia. A group of ganglionic cells near the apical plate, together with several nerves that encircle the trochophore, may make up a nervous system in the fully-developed larva. An apical organ of a tuft of cilia and other sensory structures may also be present.

The trochophore varies in certain details among the different polychaetes that have such a larva. The most typical ones are perhaps found in *Eupomatus* (Fig. 11-21), *Polygordius* (Fig. 11-22), and some others, in which the larva has a free-swimming, planktonic existence. In their metamorphosis the trochophore larvae pass through stages in which larval structures are replaced by those characteristic of the adult. A polytroch stage with a few segments (Fig.

FIG. 11-21

Gastrula and two trochophore stages of *Eupomatus uncinatus.* (Modified from Hatschek.)

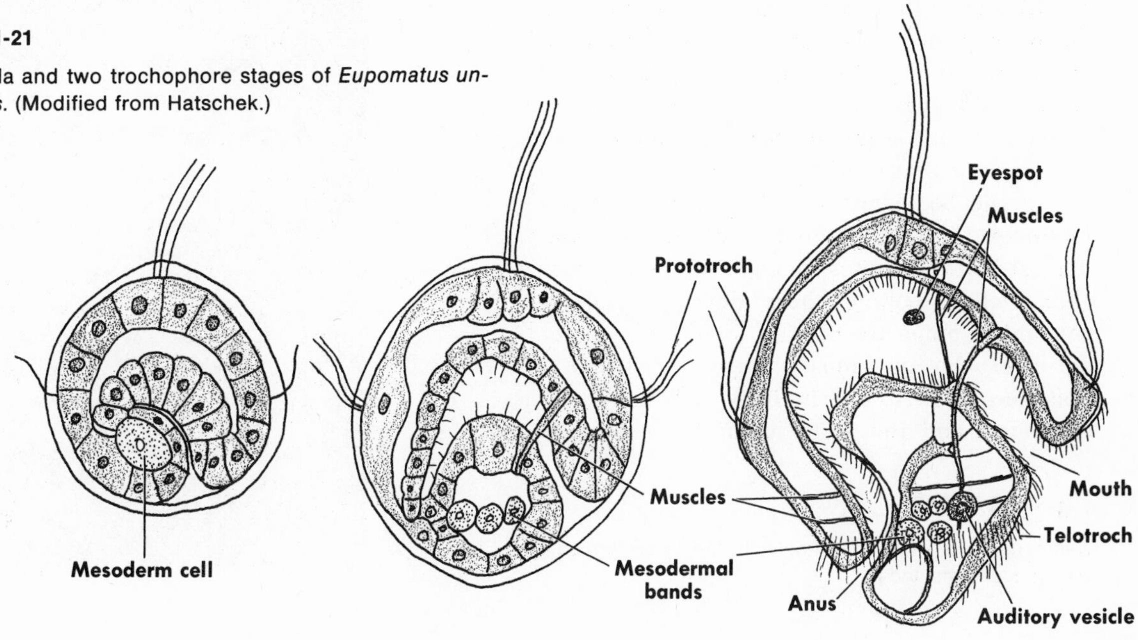

Prototroch

Mesoderm cell

Muscles

Mesodermal bands

Eyespot

Muscles

Mouth

Telotroch

Anus

Auditory vesicle

FIG. 11-22

Three larval stages of *Polygordius*. **A,** Trochophore. **B,** Polytroch with early segmentation. **C,** Free-swimming, segmented larva. (Modified from Hatschek.)

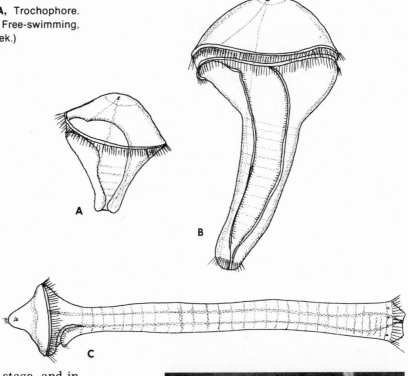

11-22, *B*) follows the trochophore stage, and in turn, the polytroch is succeeded by various free-swimming larval forms (Figs. 11-22, *C*, and 11-23). Some of the most characteristic changes during metamorphosis are the transformation of the apical organ into the adult prostomium with brain, eyes, and tentacles; the development of trunk segments between mouth and anal regions that results in a gradual lengthening of the body; and the arrangement of the metameric segmentation into a pair of mesodermal hollow somites for each segment. From each hollow somite there develops half a coelomic compartment, the spanchnic mesoderm of the gut musculature, the somatic mesoderm that forms the body wall under the ectoderm, and the adult nephridia and blood

FIG. 11-23

Nectochaete larva of polychaete. (From Encyclopaedia Britannica Films, Inc.)

vessels. In addition, mesenteries and septa are also laid down by the somites. This development results in the elimination of the old blastocoel, and its place is taken now by the peritoneum-lined cavities of the segments.

■ CLASS CLITELLATA
■ ORDER OLIGOCHAETA

The members of the order Oligochaeta were probably derived from some branch of the polychaetes, but they differ from the latter in several important particulars. They have both external and internal segmentation, but with the exception of one family (Branchiobdellidae) they have no parapodia, and their setae are comparatively few in number. External regionalization of the body is mostly nonexistent, and the pharynx is not eversible. They are hermaphroditic instead of dioecious, and mutual insemination is the rule. Gonads are usually one or two pairs in number, and special genital ducts (coelomoducts) exist for discharge of the gametes. A clitellum is found at sexual maturity, and the eggs are laid in a cocoon. Eggs develop directly into juvenile worms without larval stages. Ecologically, these worms are divided into two main types: burrowing terrestrial forms and aquatic dwellers (mostly freshwater). Some are marine, and several species are intertidal. They range in size from small aquatic species less than a millimeter in length to the large terrestrial earthworms, which may reach a length of 7 to 9 feet with a diameter of 1 inch. About 3,100 species have been named at the present time.

MORPHOLOGIC FEATURES

The body of oligochaetes is generally cylindric, although it is sometimes slightly flattened, with the anterior mouth usually overhung by a rather small prostomium that does not bear tentacles. The anus is terminal, although in a few forms it is displaced dorsally. The external segmentation is well marked by furrows, but the segments are not distinctive of species and are not fixed in number. In some aquatic species the segments may number only 6 or 7; in giant earthworms they may number 600 or more.

The setae are not as numerous as they are in the polychaetes; they display a variety of lengths from very short to very long, and they may be straight or curved, simple or branched. A characteristic arrangement of setae is a pair of ventrolateral bundles and a pair of ventral bundles on each segment (Figs. 11-24 and 11-25). The number of setae in a bundle varies from one to many. In the genus *Achaeta* setae are entirely absent. The setae are produced in setal sacs in the body wall (Figs. 11-25 and 11-26); in some species a separate setal sac occurs for each seta. A cell in the base of each sac secretes the seta and replaces lost ones, either in a new sac or in the old one. Setae are manipulated by protractor and retractor muscles and can be set at different angles for effective use.

Projecting gills are found in the body of some species. Besides the mouth and anal openings, reproductive pores on some segments, openings of the nephridia, and in many terrestrial forms, dorsal pores that open externally from the coelom occur. At sexual maturity the oligochaete has a clitellum. This glandular structure may consist of only a few cells restricted to 1 or 2 segments in aquatic forms, or it may consist of many cells that cover many segments, as in earthworms.

The body wall consists of a simple columnar epithelium interspersed with glandular and sensory cells, a delicate cuticle secreted by the epithelium, a muscular coat of outer circular and inner longitudinal layers, and a lining peritoneum (Fig. 11-26). A connective tissue dermis or basement membrane is found beneath the epidermis in some forms. The cuticle is ciliated in *Aeolosoma* (Fig. 11-27). The longitudinal muscle layer is very thin in aquatic forms and is a muscular syncytium in others.

Locomotion of terrestrial forms involves alternate contraction of the circular and longitudinal muscle layers, with the resulting extension and contraction of the body (Fig. 11-28).

This peristaltic progressive motion is aided by the gripping and anchoring action of the setae on the substratum. Undulatory movements, so characteristic of polychaetes, are absent in terrestrial oligochaetes.

The coelom is well developed and is segmentally partitioned by septa. It communicates with the exterior by dorsal pores regulated by sphincter muscles. In some aquatic forms a single pore, usually on the prostomium, is found. Coelomic fluid, which fills the septal compartments, may be discharged through the dorsal pores of the intersegmental furrows or through special openings in the floor of the pharynx. The fluid, which may be highly malodorous, plays an important role in the differential turgor of the septal compartments and in regional movement of the body for burrowing, as mentioned regarding the polychaetes. In composition the coelomic fluid may contain protein, coelomocytes, pigment, bacteria, and sarcolytes (fragments of muscle). Some of the formed elements—lymphocytes, leukocytes, and monocytes—resemble types found in vertebrates and are phagocytic. Eleocytes with oil droplets are also present. In some coelomic fluids phosphorescent discharges from special granules occur.

The digestive tract consists of a buccal cavity, pharynx, esophagus, and intestine. The pharynx is highly muscular and is used to suck in the

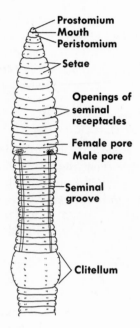

FIG. 11-24

Ventral view of anterior end of earthworm *Lumbricus*.

FIG. 11-25

Sections through setal sacs. **A,** *Tubifex*. Parietovaginal muscles regulate movement of whole follicle; intrafollicular muscles direct movement of individual setae. (×510.) **B,** *Lumbricus*. (**A** modified from Dixon; **B** modified from Stephenson.)

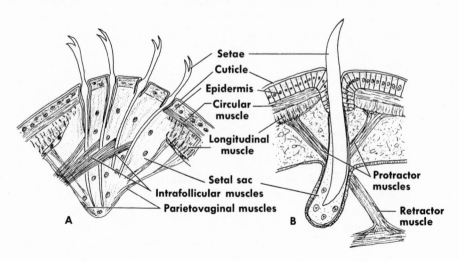

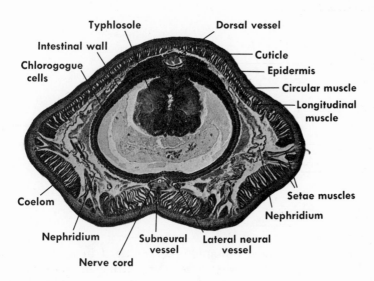

FIG. 11-26

Cross section through intestinal region of *Lumbricus*.

FIG. 11-27

A, *Tubifex,* with portions of body omitted. *T. rivulorum* is 30 to 50 mm. long; lives in tube in mud of pond or lake. **B,** *Aeolosoma* dividing by transverse fission; single zooids are usually less than 5 mm. long, but in some species up to 10 mm. (**B** modified from Lankester.)

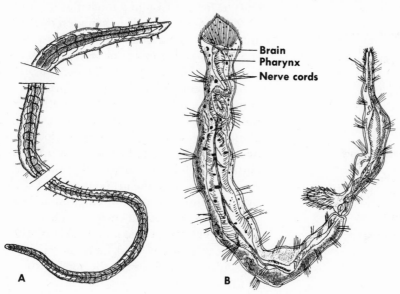

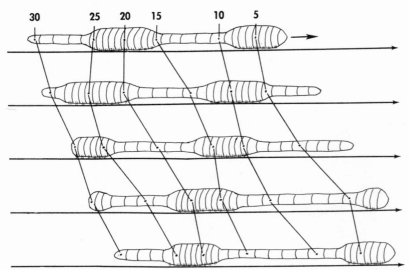

FIG. 11-28

Locomotion in an earthworm (diagrammatic) showing the progression of alternate contractions passing backward as the animal goes forward. Contractions of the longitudinal muscles produces short and thick segments, with extension of setae. Contractions of circular muscles cause segments to become thin and long, and setae are withdrawn. Every fifth segment is marked and linked by numbered lines. (Adapted from Gray and Lissman.)

FIG. 11-29

Dissection of *Lumbricus.*

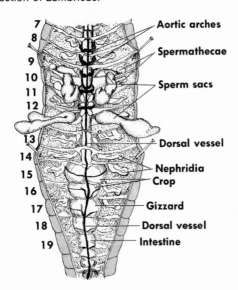

food by pumping contractions; in aquatic worms the pharynx may be protrusible and eversible, and food adhering by mucous contact is pulled in when the pharynx retracts. To assist in the action of the pharynx, muscle strands are present between the pharynx and body wall. Jaws are absent except in the family Branchiobdellidae. Pharyngeal glands secrete a saliva containing mucus and a proteolytic enzyme. A narrow, tubular esophagus follows the pharynx and may be modified (usually at its posterior end) to form a crop and muscular gizzard (Fig. 11-29). A dilated pouch or stomach may be present in some oligochaetes. In terrestrial forms such as earthworms one or more pairs of diverticula or evaginations of the esophageal wall known as calciferous glands are present. These glands may be separate but communicate with the lumen of the esophagus by a duct. The glands are well supplied with blood vessels and secrete calcium carbonate into the esophagus in the form of calcite crystals. The glands thus play a part in the pH regulation of the blood by controlling the level of calcium and carbonate ions in the blood. These glands are mostly absent in primitive and aquatic forms. The thin-walled crop stores food, and the cuticle-lined muscular gizzard grinds the food with the aid of soil particles.

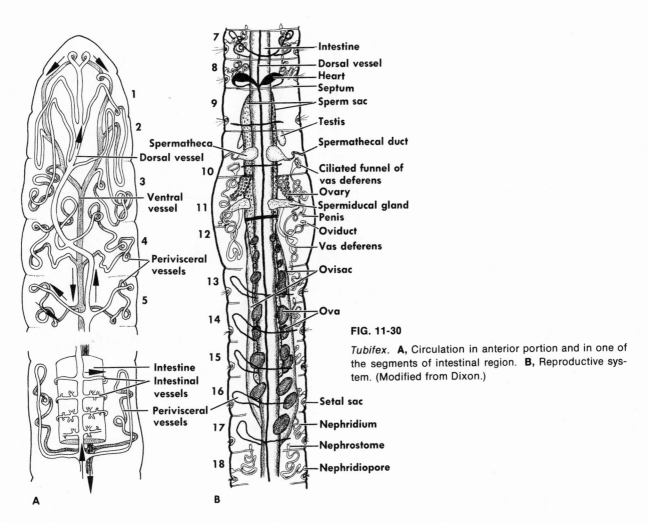

FIG. 11-30

Tubifex. **A,** Circulation in anterior portion and in one of the segments of intestinal region. **B,** Reproductive system. (Modified from Dixon.)

The intestine is a straight tube lined with cilia, secretory cells, and absorptive epithelial cells. To increase the absorptive surface, a longitudinal dorsal fold (the typhlosole) (Fig. 11-26) projects downward into the lumen of the esophagus. Digestive secretions contain enzymes for breaking down proteins, fats, and carbohydrates. The intestine has thin muscular layers and is invested by a modified peritoneum of yellowish cells, the chloragogen tissue, which serves for fat and glycogen synthesis and storage. It may also play a part in excretion, such as the formation of ammonia and urea. The intestine in earthworms may have an osmoregulatory function of assisting the nephridia

in eliminating excess water when the worm becomes water soaked.

Excretion in oligochaetes occurs by segmentally arranged, paired metanephridia (Fig. 11-30). The primitive plan seems to have been a pair of nephridia per segment, but this arrangement is slightly modified in some worms. A typical metanephridium (Fig. 11-31) consists of the funnellike preseptal nephrostome that opens into the coelom, a duct through the septum, and a coiled tube that discharges ventrolaterally to the exterior by an opening (the nephridiopore). The coiled tube has various differentiations such as a ciliated region, variations in the tubular bore, and a terminal

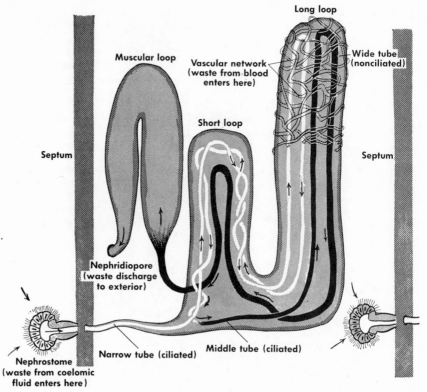

FIG. 11-31

Earthworm nephridium (diagrammatic).
(Modified from Moment.)

dilated vesicle. Nephridia are variously modified in the different species. Instead of the
typical paired nephridia per segment (holo-
nephridia), the nephridia may be multiple
(meronephridia). In some giant earthworms
many nephrostomes exist for each nephridium.
The cells composing the lips of the nephrostomes are of diagnostic value in taxonomic
determinations. The nephridia may open to the
exterior (exonephric), into the pharynx (pharyngeal), or into the intestine (enteronephric).
More than one type of nephridia may be found
in the same worm.

In addition to the waste fluid that is drawn
from the coelom into the nephrostome by
ciliary action, each nephridium is also richly
supplied with blood from branches of the chief
ventral artery; this blood is involved in active
reabsorption of fluid from the nephridial tubule
as well as excretion of waste into the tubule.
As the coelomic fluid passes down the tubule,

much of the water is reabsorbed back into the
blood, thus concentrating the waste. Some salts
are also reabsorbed back into the blood. Among
the chief wastes excreted are ammonia (mostly
in aquatic forms) and urea (chiefly in terrestrial
species). The nephridium in oligochaetes is
thus seen to have about the same functions as
has the nephron unit of the vertebrate kidney—
filtration, secretion, and reabsorption.

Gaseous exchange is chiefly a function of the
body surface in oligochaetes. The skin is usually
well vascularized, and when moist, the exchange of gases can readily occur through it.
In some aquatic species (Naididae) anal respiration by the circulation of water in and out of a
rectal chamber takes place. *Dero* and some
other aquatic genera have a circle of anal
gills (Fig. 11-32). A few oligochaetes have
tuftlike gills on certain segments, usually in the
posterior part of the body. Many oligochaetes,
including earthworms, can do without oxygen

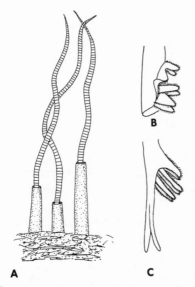

for some time and can build up an oxygen debt by the accumulation of lactic acid.

In oligochaetes the circulatory system is a closed system that includes a pulsating region for driving the blood, arteries to carry the blood away from this region, fine capillaries for the exchange of foods and wastes between blood and tissues, and veins for returning the blood to the propulsive structures, or heart. The primitive plan of the circulatory systems seems to be a contractile dorsal vessel arising posteriorly from a peri-intestinal blood sinus, a ventral vessel for blood distribution, and a few pairs of segmental, commissural connectives at the anterior end between the dorsal and

FIG. 11-32

A, *Tubifex*, a small red oligochaete, is tube builder in bottom mud of ponds or lakes. (×2½.) Exposed posterior end waves vigorously, possibly to increase oxygen supply to its thin-walled surface. Anal gills (lateral view) of *Dero limosa*, B, and *Aulophorus*, C. (Modified from Pennak.)

FIG. 11-33

Scheme of circulatory system of eathworm *Lumbricus*. A, Circulation in any segment posterior to esophagus. B, Circulation in segments 7 to 11 showing aortic arches around esophagus. Septa not shown. (From Hickman: Integrated principles of zoology, The C. V. Mosby Co.)

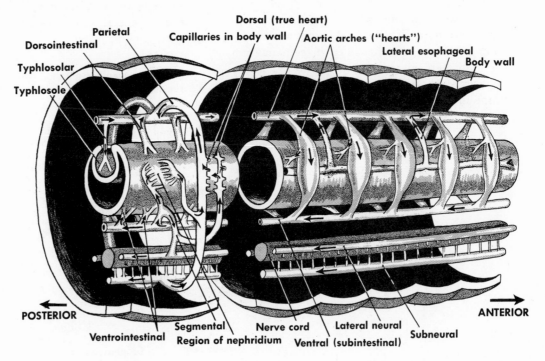

ventral vessels. Such a plan is found in *Aeolosoma* at the present time.

In more advanced oligochaetes the dorsal vessel has become separated from the intestinal sinus and extends to the posterior end (Figs. 11-30 and 11-33). In addition, a subneural vessel and 2 lateroneural vessels are present. Dorsal and ventral blood vessels are also connected by lateral vessels in every segment. The functional heart, or pumping organ, is really the contractile dorsal blood vessel, although the five or more pairs of aortic arches or "hearts" (in segments 7 to 11 in the earthworm) have been assigned this function in the past because of their vigorous contractions. They may assist the flow of blood and also regulate the pressure. The blood is propelled forward and downward by rhythmic contractions that begin near the posterior end of the dorsal blood vessel. The main distributing artery is the ventral vessel, which is found below the gut and above the nerve cord. This vessel receives blood from the paired aortic arches and carries the blood toward the tail. From the ventral vessel a pair of segmental arteries is given off in each segment and delivers blood to the body wall musculature. Small branches in each segment from the main ventral artery also send blood to the nephridia, to the lateral neural artery, and to the underside of the intestine. In the regions supplied by these branches from the ventral vessel the arteries break up into capillaries, where food and waste exchanges are made. The principal vein for the return of blood is the subneural blood vessel. It receives blood from the nerve cord and delivers it posteriorly toward the tail and upward through segmentally paired parietal vessels. The parietals also receive blood from the body wall before discharging into the dorsal vessel. Also, in each segment the dorsal blood vessel gives off a branch to the typhlosole and receives a pair of veins from the intestine.

The blood is colorless in some oligochaetes but usually carries hemoglobin or erythrocruorin in solution in the plasma, producing a red color. The formed elements, or amebocytes, are colorless but are phagocytic.

The typical central nervous system in oligochaetes consists of a bilobed cerebral ganglion (brain), circumpharyngeal connectives, and a ventral nerve cord. In some simple or primitive oligochaetes (Aeolosomatidae), 2 ventral nerve cords as well as the brain are united to the epidermis. In more advanced oligochaetes the 2 ventral cords are fused into 1, and the

FIG. 11-34

Nervous system of anterior end of earthworm. (From Hickman: Integrated principles of zoology, The C. V. Mosby Co.)

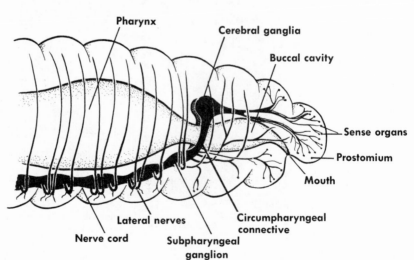

brain lies just above the pharynx in the third segment (Fig. 11-34), instead of at the base of the prostomium as in *Aeolosoma* (Fig. 11-27, *B*). In the nerve cord segmentally arranged ganglia give off lateral nerves to the body wall and intestine. Some aquatic oligochaetes have four pairs of lateral nerves per segment, but terrestrial forms usually have only three pairs per segment. These lateral or segmental nerves contain both sensory and motor fibers. The brain supplies a pair of prostomial nerves to the prostomium and first segment. The subpharyngeal ganglion, at the union of the circumpharyngeal connectives with the ventral nerve cord, also gives off nerves to the first as well as the second trunk segment. The subpharyngeal ganglion exerts a control over the motor activities and reflexes of the worm. The worm is unable to adapt to environmental conditions when the brain is removed; when the subpharyngeal ganglion is destroyed, all movements cease.

The nerve cord contains nerve fibers of all three types of neurons: motor neurons, with their cell bodies in the segmental ganglion; association neurons for carrying impulses from one cell to another; and sensory neurons, with their cell bodies in the body wall. In locomotion the ordinary rhythmic peristaltic movements are produced by a succession of tactile reflexes involving reflex arcs of sensory and motor neurons. Whenever a segment is stretched, receptors are stimulated, and sensory nerves carry impulses to the motor neurons of the cord. The motor neurons produce contractions of the longitudinal muscles, which cause the adjacent segment to stretch. The stretch receptors in this adjacent (second) segment are stimulated, causing this segment to contract; this stimulates the third segment, etc. Thus a wave of muscular contraction passes over the worm.

For rapid movements to avoid danger, nearly all oligochaetes have an escape mechanism in the giant fibers. Of the 5 giant fibers, the 3 large ones on the dorsal side of the ventral nerve cord are primarily concerned with the rapid movement of escape. In these large giant fibers the rate of nerve conduction is five times faster than that in the regular neurons. The median giant fiber conducts impulses toward the tail; the 2 lateral ones conduct impulses toward the head. Whenever a worm is overstimulated at one end, the rapid impulses in the giant fibers produce a simultaneous contraction of all the body segments that jerks the worm into its burrow and out of danger. Giant fibers differ from other neurons in being multicellular and in having the 2 lateral fibers connected with each other at regular intervals so that they react as a unit (J. A. C. Nicol, 1948).

Nerve plexuses found in the epidermis and muscle layers are connected with the central nervous system through the segmental nerves. Some neurons apparently are independent and initiate local responses of muscles. An enteric nervous system attached to the circumpharyngeal connectives innervates the muscle fibers of the digestive tract in earthworms. Recently neurosecretory cells have been found in annelida (F. I. Kamemoto and co-workers, 1966).

The sensory system consists of sense organs of several types located primarily in the epidermis (Fig. 11-35). Most oligochaetes show a wide range of reactions to stimuli, but they generally lack eyes, although a few aquatic forms have eyespots. Free nerve endings among the epithelial cells may respond to tactile,

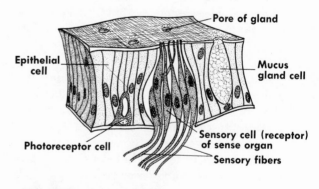

FIG. 11-35

Portion of epidermis of earthworm (diagrammatic) showing various sensory specializations. (From Hickman: Integrated principles of zoology, The C. V. Mosby Co.)

heat, and pain stimuli. The photosensitive cells in the anterior end of the worm are oval with a lenslike body. The work of W. N. Hess and others indicates that the prostomium bears many sense organs. Chemoreceptors are common, especially in the anterior end (C. L. Prosser). Earthworms are very sensitive to vibrations but do not respond to sound waves.

All oligochaetes are hermaphroditic, with distinct gonads that are derived from the septal peritoneum. The gonads are usually restricted to a relatively few segments. Members of the primitive family Aeolosomatidae resemble the polychaetes in lacking distinct gonads and in having masses of reproductive cells scattered through many segments. The testes in oligochaetes are located anterior to the ovaries. In some aquatic species one pair of testes and one pair of ovaries are located in successive segments (Fig. 11-30), but the typical plan in terrestrial species is two pairs of testes and two pairs of ovaries, with each pair in a different segment (Fig. 11-36). Most earthworms retain only the posterior pair of ovaries. The location of the reproductive segments is not the same in all genera. Gametes are released into the coelom or special coelomic pouches where maturation is completed, and they reach the outside through the differentiated gonoducts. The coelomic pouches (seminal vesicles and ovisacs) are produced by septal outpocketings in the reproductive segments. Many variations exist in the number and position of the pouches. In the common earthworm *Lumbricus* (Fig. 11-36) one pair of the testes is located in the tenth and the other is in the eleventh segment. The testes are very small and usually lobed. Each is enclosed in a sperm sac or seminal vesicle. Of the three pairs of seminal vesicles, one pair projects forward from the septum anterior to the first pair of testes; the other two pairs project from the septa posterior to the two pairs of testes. Each segment containing a pair of testes is provided with a pair of sperm ducts, or vasa deferentia, one on each side. The internal opening of each duct is a ciliated funnel located on

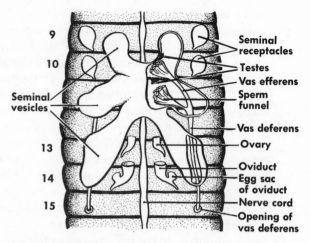

FIG. 11-36

Dorsal view of reproductive organs of earthworm, with seminal vesicles partly removed on right side to reveal testes. (From Hickman: Integrated principles of zoology, The C. V. Mosby Co.)

the posterior septum of each testicular segment. From the funnel the vas deferens extends backward through the septa of several segments. Before opening on the ventral side of the body, the two pairs of vasa deferentia on each side join so that there is only one pair of openings on the fifteenth segment. This outer opening may be enlarged in some oligochaetes (aquatic) to form a muscular antrum with an eversible penis (Fig. 11-30, *B*). The vasa deferentia are straight in the earthworm but may be coiled in other worms. A prostate gland may be attached to the posterior end of the vasa deferentia in some aquatic species, or it may be a separate structure.

A single pair of pear-shaped ovaries in *Lumbricus* is located on the anterior septum of the thirteenth segment. Ovisacs are small or entirely absent. The narrow end of the ovary projects into the coelomic cavity, into which the eggs are discharged in ovulation.

Each of the paired oviducts is provided with a funnel that is located on the posterior septum of the ovarian segment. The oviducts penetrate the septum to which the funnels are attached,

FIG. 11-37

Earthworms in copulation. (Photograph by Guy Carter.)

and each opens by an oviducal pore on the ventral surface of the segment following the ovarian segment. The female system also includes the seminal receptacles (spermathecae) that receive sperm from another worm. They are small, round structures that usually open on the ventral surface of the segments in which they are found. Two pairs of these receptacles are found in *Lumbricus,* but there may be more in some oligochaetes. Each pair is located in a separate segment, and in *Lumbricus* the openings are in the intersegmental furrows between segments 9 and 10, and 10 and 11.

Reciprocal copulation is the rule in oligochaetes, with heads in opposite directions and the anterior ventral surfaces of the mating pair in contact (Fig. 11-37). In some species (but not in *Lumbricus*) the male pores of one mate are directly opposite the seminal receptacles of the other, and the sperm from the male pores are transmitted directly into the seminal receptacles of the partner. In others including *Lum-*

bricus, the male pores and those of the seminal receptacles are not in apposition, and the sperm must be carried externally along the ventral seminal grooves from the male pores to the clitellar region where the openings of the seminal receptacles of the partner are found. Each worm is also enclosed in a slime tube. Sperm are propelled through the ventral grooves by muscular contractions. The mating pairs are held together in two zones by a common mucous coat secreted by the clitellum and by special ventral setae, which penetrate the body of the other. After the mutual exchange of sperm, the two worms separate. Copulation may last for more than 2 hours. After separation, each worm produces a series of cocoons for the eggs. For each cocoon a slime tube is first secreted around the anterior segments and the clitellum. Within this slime tube, the clitellum secretes around itself another tougher band that will form the cocoon. Eggs and albumen enter the cocoon before it slides off the clitellum. As the slime tube with the cocoon slides forward over the anterior end of the worm, sperm from the seminal receptacles enter the cocoon and fertilize the eggs. When the worm has backed completely out of the slime tube and the cocoon and freed itself, the mucous tube disintegrates, and the ends of the cocoon constrict to form a yellowish, lemon-shaped case. Cocoons of *Lumbricus* are about 7 mm. long, but those of the giant earthworm of Australia *(Megascolides)* are over 3 inches long. From each cocoon will hatch between 1 and 20 juvenile worms, depending on the species. Cocoons are formed every few days, and mating may occur frequently during the spring and fall.

Copulation in some oligochaetes is impossible, and eggs are self-fertilized. Parthenogenesis may also occur in a few. Asexual reproduction is common in the aquatic families Aeolosomatidae and Naididae. The members of these families reproduce asexually by transverse fission (Fig. 11-27, *B*). Sexual reproduction is unknown in some of these species. Chains of zooids are often found in *Aeolosoma* (Fig. 11-27, *B*) and *Chaetogaster,* in which a new fission occurs before the old division is completed.

Considerable regeneration occurs in oligochaetes. In earthworms, if the worm is bisected anterior to segment 18, a new anterior end of 1 to 5 segments will form. If the tail is cut off posterior to segment 18, a tail may grow forward from the cut, forming a worm with 2 tails. An anterior piece of 35 or more segments forms a new tail, but the number of segments regenerated will depend on the level of the cut. A cut at segment 50 will regenerate 10 fewer segments than one made at the level of segment 40. Grafting is also possible with earthworms.

DEVELOPMENT

The entire development of the egg into the juvenile worm takes place in the cocoon. There are no larval stages, for development is direct. The eggs of lower oligochaetes, especially aquatic species, are provided with a great deal of yolk. Eggs of terrestrial forms usually have less yolk, but sufficient albumen for growth is supplied to the cocoons by the clitellum before the cocoon is discharged from the body. The segmentation is a modified form of spiral determinate cleavage. There are, however, many divergent opinions about the origin of various structural features. For instance, some views hold that the nephridia are derived from ectoderm; others consider these organs to have originated from mesoderm. A few opinions maintain that nephridia are derived from both germ layers.

The holoblastic cleavage in earthworms is unequal (Fig. 11-38), producing a rather small blastocoel enclosed by small ectodermal cells above and larger endodermal cells below. At first spheric, the blastula elongates, with the endodermal cells becoming distinctly columnar (Fig. 11-38, *D*). During gastrulation the endoderm is invaginated as the future enteron, with the blastopore at the anterior end becoming the future mouth (Fig. 11-38, *E* and *F*). The somites and muscles are formed from 2 cords of mesoblastic cells that arise between the ectoderm and the enteron (Fig. 11-38, *G*). The anus is a new formation. The juvenile worms of *Lumbricus* hatch from cocoons in a few weeks, although the length of time depends on environmental conditions.

■ ORDER HIRUDINEA

The Hirudinea are commonly known as the leeches, of which more than 300 species have been classified at the present time. Although derived from the oligochaetes and included with them in the same class (Clitellata), leeches have evolved beyond the level of organization of all other annelids. With the oligochaetes they share certain common features of embryologic development: both groups have a clitellum for secreting a cocoon; both are hermaphroditic with well-defined gonads; and both have separate

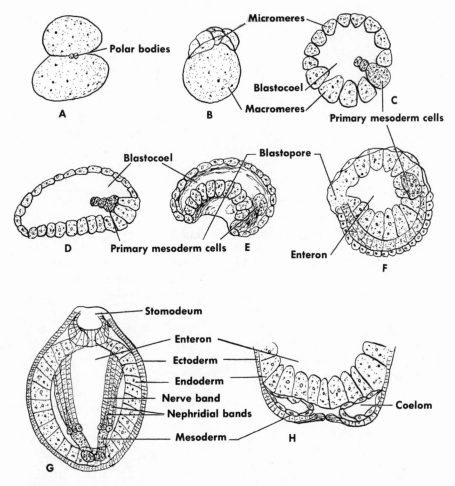

FIG. 11-38

Early development of *Lumbricus*. **A,** Two-cell stage. **B,** Eight-cell stage. **C,** Blastula.
D, Blastula elongating; macromeres forming flat ventral plate. **E,** Early gastrula, formed
by invaginating macromeres. **F,** Later gastrula, after formation of mesoblastic cords,
obliteration of blastocoel, and closure of blastopore. **G,** Ventral view of embryo after
formation of stomodeum and germ bands. **H,** Ventral part of embryo in transverse section
showing origin of coelom. (Modified from Wilson.)

nephridia and coelomoducts. However, leeches differ from oligochaetes in the lack of setae (with one or two exceptions), in having a reduced coelom mostly filled with mesenchyme, in having ventral suckers at the ends of their bodies, and in having a fixed number of segments with secondary external annulation. Many of their unique features are associated with their bloodsucking habits, such as the suckers for clinging to their host and the modifications of the enteron for storage of blood. Most are found in inland waters or in damp places on land. A few are marine. Although most of them are bloodsuckers on other animals, some of them are not parasitic. A few have even become predators and feed on earthworms and insects, which they swallow whole. Leeches show a considerable diversity in size. Some are less than

½ inch long, and the larger ones may exceed a foot in length when fully extended. All have great powers of extension and contraction.

MORPHOLOGIC FEATURES

The body of the leech is vermiform, subcylindric, and flattened ventrally (Fig. 11-39). The shape of the body, however, varies greatly with the amount of blood in its digestive tract and the state of its contraction. Leeches are usually not highly colored. Many have a greenish background of color with a diversified pattern of spots and stripes. Their ventral surface is often black interspersed with grayish markings. Because they lack parapodia and setae and have secondary rings (annuli), the 33 (or 34 according to some authorities) segments are not always clearly defined. Each segment usually bears a variable number of annuli. Leeches have no caudal growth zone similar to that of oligochaetes, and growth in length occurs by subdivision and growth of the annuli, especially in the middle of the body where the number of

annuli is usually constant for a species. A clitellum is usually present in segments 9, 10, and 11. Of the characteristic suckers, the anterior is usually smaller than the posterior and generally surrounds the mouth.

The body is commonly divided into five regions: (1) a head region of 4 segments bearing the ventral oral sucker, mouth, eyes, and jaws (in some), (2) a preclitellum of 4 segments (5 to 8), (3) a clitellum of three segments (9 to 11), (4) a middle region of 15 segments (12 to 26), and (5) the terminal region of 7 segments modified to form the posterior sucker. The number of annuli varies usually from 1 to 5 per segment, with the larger number in the middle of the body, but there are many variations in the different species. The pygidium, or terminal segment, is a part of the posterior sucker, and the anus opens dorsally in front of the posterior sucker. Segmental analysis is based mainly on the distributional pattern of the paired ganglia of the ventral nerve cord.

The body wall consists of a skin made up of a single layer of epidermal cells and unicellular glands; a dermis of connective tissue, pigment cells, and blood vessels; and a musculature of outer circular and inner longitudinal layers, between which are crossed oblique muscles (Fig. 11-40). Muscle strands pass from one side of the body to the other. Muscle fibers are

FIG. 11-39

Common leeches. **A,** *Placobdella,* commonly found on turtles; living specimen about 1 inch long. **B,** *Hirudo medicinalis,* a form once used in bloodletting; preserved specimen, about 3 inches long. (**A,** suborder Rhynchobdellida; **B,** suborder Gnathobdellida.)

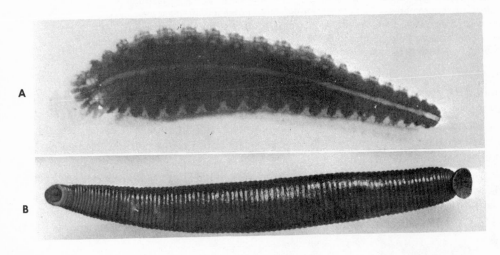

striated in the outer margins, but their centers consist of unmodified protoplasm.

With the exception of the Acanthobdellida, the body cavity is much reduced by the invasion of connective tissue (botryoidal tissue). The characteristic septa and coelomic compartments of the oligochaetes have entirely disappeared. Instead, the coelom in leeches consists of spaces or sinuses connected by capillary branches, forming a type of lymphatic system. The coelomic fluid of the spaces contains coelomocytes; in some forms the fluid is kept in circulation by special pulsatile vesicles. In some suborders the coelomic spaces are a part of the vascular system and contain red blood. The botryoidal tissue also contains yellow cells, similar to chloragogue cells, which are excretory in function. In some of the Rhynchobdellida longitudinal canals are generally connected with numerous transverse canals (Fig. 11-41). A network of sinuses between the epidermis and muscular layers in the wall connects with the longitudinal canals.

The blood-vascular system in leeches is of two main types: (1) a system of definite blood vessels and (2) a system of coelomic sinuses that serve to circulate the blood. The first type is characteristic of the suborder Rhynchobdellida (Fig. 11-41), in which dorsal and ventral longitudinal vessels (above and below the gut) are connected at their extremities with paired loops. These vessels lie in coelomic sinuses. One median and four pairs of lateral blind pouches extend from the posterior part of the dorsal vessel for increased absorptive areas. The contractile nature of the blood vessels, especially the anterior part of the dorsal vessel, serves as a heart in the propulsion of the blood. The other type of vascular system is characteristic of most leeches. It consists of sinuses that have become modified to form circulatory channels. The ventral sinus vessel is the principal channel, although a dorsal vessel is sometimes present. Lateral coelomic sinus vessels, one on each side, are also present and connect by transverse vessels to the ventral vessel. Paired contractile vesicles, or ampullae, are found on these transverse vessels in the posterior region of the body. These vesicles may have little to do with pumping the blood, which is carried on principally by contraction of the lateral longitudinal sinus vessels. A capillary network, formed chiefly from the botryoidal tissue and communicating with the sinus vessels, occurs in some leeches. Respiratory pigments are absent in the blood of some leeches (rhynchobdellids), but plasma pigments are found in others and produce a red color. The blood of leeches is really the same as the coelomic fluid.

Respiration in leeches occurs primarily through the general body surface, but a few have leaflike gills (*Branchellion*); others have

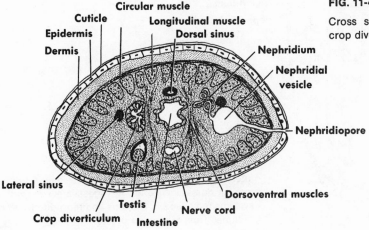

FIG. 11-40

Cross section of *Hirudo* (diagrammatic). Nephridium, crop diverticulum, and testis shown in one side only.

Circular muscle
Cuticle
Longitudinal muscle
Epidermis
Dorsal sinus
Dermis
Nephridium
Nephridial vesicle
Nephridiopore
Lateral sinus
Dorsoventral muscles
Testis
Crop diverticulum
Nerve cord
Intestine

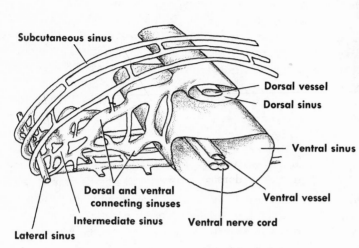

Subcutaneous sinus
Dorsal vessel
Dorsal sinus
Ventral sinus
Dorsal and ventral connecting sinuses
Ventral vessel
Intermediate sinus
Ventral nerve cord
Lateral sinus

FIG. 11-41

Portion of coelomic sinus system of *Glossiphonia,* a rhynchobdellid. Dorsal sinus contains contractile dorsal blood vessel; ventral sinus contains nerve cord, ventral vessel, female organs, and funnels of nephridia. Blood is carried forward in dorsal vessel and backward in ventral vessel. (After Oka; redrawn from Mann: Leeches [*Hirudinea*], their structure, physiology, ecology, and embryology, Pergamon Press, Inc.)

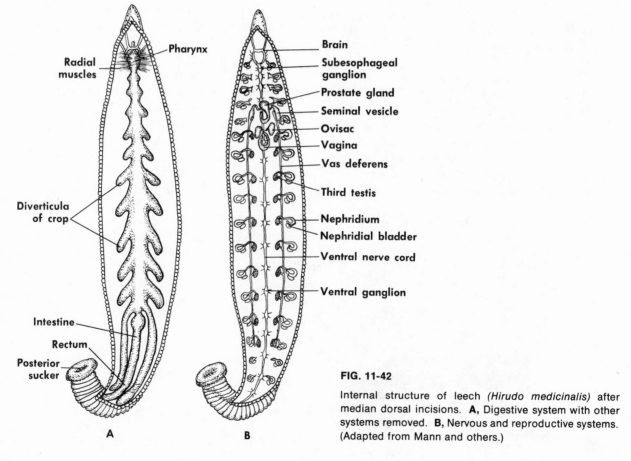

Radial muscles
Pharynx

Brain
Subesophageal ganglion
Prostate gland
Seminal vesicle
Ovisac
Vagina
Vas deferens
Third testis
Nephridium
Nephridial bladder
Ventral nerve cord
Ventral ganglion

Diverticula of crop

Intestine
Rectum
Posterior sucker

A   B

FIG. 11-42

Internal structure of leech *(Hirudo medicinalis)* after median dorsal incisions. **A,** Digestive system with other systems removed. **B,** Nervous and reproductive systems. (Adapted from Mann and others.)

vesicles that may serve for the exchange of gases. Investigation with leeches on the rate of oxygen uptake at different oxygen concentrations has yielded variable results. Some leeches maintain a constant rate of oxygen uptake independent of oxygen concentration, but in others uptake is proportional to the concentration. Many leeches are known to exist in fresh water with low amounts of oxygen.

The digestive system consists of a mouth, buccal chamber and sinus, pharynx, esophagus, stomach, intestine, rectum, and anus (Fig. 11-42, *A*). The mouth opens within the anterior sucker. In the rhynchobdellid leeches an eversible proboscis (muscular and lined with cuticle) is forced out of the mouth into the tissue of the host when the anterior sucker is attached. In those that lack a proboscis (most leeches), toothed jaws occur inside the mouth (Fig. 11-43). The muscular action of these 3 jaws produces a Y-shaped incision on the skin of the host when a leech attacks. The secretions of the salivary glands (in the walls of the pharynx) containing an anticoagulant (hirudin) are poured into the wound and prevent the clotting of blood. V. F. Lindemann, however, believes that the free flow of blood is due to the enlargement of blood vessels by histamine from the leech. The anticoagulant in such cases may prevent blood clotting in the leech's stomach. In those leeches that are not bloodsuckers, the jaws are reduced to muscular ridges for seizing the prey, which is then swallowed whole. Some bloodsucking leeches can store up to ten times their weight in blood, which may furnish them nutrition for many months before they need to feed again. The buccal chamber empties into the muscular pharynx, which exerts a powerful suction in drawing blood from the host. A short esophagus connects the pharynx to the stomach, which in bloodsucking leeches becomes a crop with paired lateral ceca or diverticula for the storage of the solid parts of the blood. In predaceous leeches the stomach is usually a simple chamber. A sphincter valve separates the stomach from the intestine, which may also bear

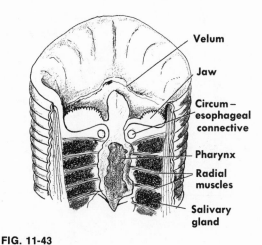

FIG. 11-43

Ventral view of dissected anterior region of leech showing toothed jaws. (Modified from Mann.)

some paired ceca. Most digestion and absorption probably occurs in the intestine, although the stomach may be involved. The digestive process is not well understood because it has been difficult to identify proteolytic enzymes (most of the leech's food is protein) in the gut wall. Bacterial decomposition of the food has been described in some leeches. The globin part of the hemoglobin seems to be the chief source of nutriment in the leech. From the intestine a short rectum leads to the dorsal anus in front of the posterior sucker.

The excretory system of leeches consists of paired metanephridia that are more specialized than those found in oligochaetes. Their typical arrangement is one pair to each segment in the middle region of the body. The number varies with different species, but 17 pairs occur in *Hirudo* (Figs. 11-42 and 11-44). A nephridium typically consists of a ciliated nephrostome that projects into a coelomic channel at one end and into a nonciliated capsule at the other, and a tubular structure composed of cells through which runs a canal that may drain a network of finer canals. The canal may or may not open into the capsule. Externally the nephridium opens by a vesicle or bladder and a nephridiopore on the ventral or lateral surface. The unique

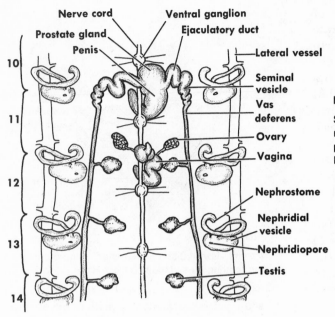

FIG. 11-44

Segments 10 to 14 of *Hirudo* (diagrammatic) showing reproductive organs and nephridia. This form has ten pairs of testes in segments 12 to 21. (Modified from Brown.)

capsule is supposed to receive and store excretory phagocytes and to manufacture coelomic amebocytes. This plan of the nephridium has many variations among the different species. From the coelomic fluid or blood the nephrostome receives excretory particles that are taken up by the capsular phagocytes. These phagocytes may carry the waste particles to the intestine or to the epidermis. The fluid urine is obtained by filtration through the walls of the tubule and eventually is discharged through the nephridiopore. Most of the urine is made up of ammonia, although small amounts of such constituents as purines, urea, and creatinine also occur. The excretory system is also concerned in maintaining the water balance of the body. Since the osmotic pressure of the body fluids is higher in most cases than that of the water in which they live, an excess of water must enter the leech's body at all times and must be eliminated by the nephridia. Experiments by A. Krogh indicate that leeches have special mechanisms in the epidermis for taking up sodium and chloride ions to replace those lost in metabolism and thus are able to maintain a constancy in the osmotic pressure of body fluids.

The nervous system of leeches follows the general annelid plan of a paired ventral nerve cord made up of paired ganglia (Figs. 11-42, B and 11-44). Of the 34 paired ganglia, 6 are in the head, 21 are along the trunk region, and 7 are fused in the posterior sucker. Each ganglion is composed of six groups (capsules or follicles) of nerve cells, which innervate that particular segment. The brain in segment 6 consists of a paired suprapharyngeal ganglion connected to a paired subpharyngeal ganglion by circumpharyngeal connectives (Fig. 11-42). The number of ganglia fused in the subpharyngeal ganglion can be determined by the number of capsules present in it. The capsules of the brain may be generally scattered and do not form a compact mass. From these various ganglia, peripheral nerves run to the prostomium and other regions of the head. From each trunk ganglion two pairs of mixed nerves innervate the wall and structures of the corresponding segment, the anterior pair going to the dorsal part of the segment and the posterior pair going to the ventral part. Sense organs are similar to those in oligochaetes and are grouped into three categories: (1) free nerve endings in the epidermis for temperature and touch, (2) epidermal sense organs (sensillae) of tall spindle-shaped cells with terminal hairs, probably tactile or chemoreceptor in function, and (3) photoreceptor cells (eyes). The eyes contain a pigment cup and are found on the dorsal surface of the anterior end.

Like oligochaetes, leeches are hermaphroditic, with a single pair of ovaries and many pairs of testes (Figs. 11-42, *B*, and 11-44). They are also provided with ducts for carrying the gametes. The four to ten pairs of spherical testes (usually ten pairs in *Hirudo*) are arranged on each side of the enteron, one pair to a segment, in the body region where they are found. They are enclosed in coelomic sacs, and those on each side join a vas deferens running anteriorly to a single midventral opening (the male gonopore), which is located about one third of the distance in back of the anterior sucker. The vas deferens is greatly coiled and forms a seminal vesicle, followed by a muscular ejaculatory duct that opens into the single antrum. The antrum is surrounded by glands and may form an eversible penis by which sperm are transferred directly to the female gonopore of another leech. When a penis is lacking, the sperm are formed into spermatophores, which are attached to the body of another worm. Here the sperm migrate (hypodermic impregnation) through the skin and tissue to fertilize the eggs in the ovisacs. The paired ovaries are located between the anterior pair of testes and the antrum (Fig. 11-44). Each ovary is enclosed in an ovisac, which is derived from the coelom. From each ovisac a short oviduct extends forward and joins its mate to form a common vagina that opens by a single female gonopore just behind the male gonopore. When the eggs are laid, a glandular clitellum enlarges and secretes a cocoon over the genital pores. Albumen and fertilized eggs are placed in the cocoon, which is then worked over the head, much as in the earthworm. The cocoons are deposited in various places—on stones, in damp places, on aquatic vegetation, and even on the ventral surfaces of their own bodies, where, on hatching, the leeches are carried about by the parent for some time. Cocoons may have few or many eggs and may be deposited soon after the eggs are fertilized, or there may be considerable delay between fertilization and cocoon formation. Self-fertilization probably does not occur.

DEVELOPMENT

Leeches have spiral determinate cleavage, but the pattern of cleavage depends to a great extent on the amount of yolk. In the development of leeches the eggs of some may have so much yolk that absorption of albumen from the cocoon is unnecessary; in others the eggs with little yolk absorb much of the fluid albumen. Young leeches are juveniles and are provided with both anterior and posterior suckers when hatched from the egg.

RESPIRATORY PIGMENTS

In most animals that require the transport and storage of oxygen, special colored proteins known as blood pigments are involved. Body fluids such as blood have a very low capacity to dissolve oxygen, and only a tiny fraction of oxygen can be carried this way (less than 1%). Only a few animals with low metabolism can transport oxygen this way. Some invertebrates may be sluggish enough to get by with such transportation, and even the vertebrate icefishes of Antarctica have vascular fluids that can combine this way with oxygen without the necessity of a respiratory pigment because of the very cold waters in which they live. In contrast, man's arterial blood can carry about 20 volume % of oxygen because some 98% of oxygen is carried in combination with the respiratory pigment hemoglobin. Many metazoans of any size therefore make use of a respiratory pigment of some kind for their storage and transportation of oxygen. The many kinds of respiratory pigments in the animal kingdom are thought to have been independently evolved in unrelated groups, and little phylogenetic significance can be attributed to them. Respiratory pigments made possible smaller respiratory structures, greater activity of animals, reduction of the circulatory system, and pickup of oxygen at lower oxygen tensions.

Perhaps no group of animals has a greater variety of respiratory pigments than the annelids. Most respiratory pigments are found in these forms, and more than one kind of pigment may be found in the same species. Wherever found, the pigment may carry oxygen continuously, or it may function only at low oxygen pressures. It may function in other ways, such as storage for oxygen, buffers in the transport of carbon dioxide, and in the maintenance of the colloid osmotic pressure of the blood. However, it is difficult to determine the function of these pigments in certain species.

All respiratory pigments contain a metal of some kind, usually iron or copper, combined with some protein. Some of the respiratory pigments have the porphyrin nucleus – the metalloporphyrins. The porphyrins are organic compounds in which four pyrrol nuclei are connected in ring structures by —CH groups. When an atom of ferrous iron is attached in the center to the pyrrol nitrogens of the protoporphyrin, the component is called a heme. Thus hemoglobin consists of an iron porphyrin compound, heme, associated with the protein globin. Porphyrins are found widely distributed in nature (H. M. Fox and G. Vevers, 1960). Compounds of metalloporphyrins are the prosthetic groups of proteins that perform many catalytic functions in the cell (W. B. Gratzer and A. C. Allison, 1960). In chlorophyll, for instance, the metal is magnesium. The heme part of the molecule is the same in all hemoglobins, but the globin component varies in different species. It is also possible for a number of hemoglobin units to unite and form polymers of different sizes. One unit of hemoglobin has a molecular weight of about 17,000, but in most vertebrates each molecule of hemoglobin has four units, or a molecular weight of around 68,000. In the earthworm the hemoglobin molecule is made up of about 180 units, producing a molecular weight of more than 3,000,000. In all these hemoglobins, the heme component is the same in each unit.

Among the invertebrates the amount of respiratory pigment determines the oxygen capacity. There are many individual variations within a species. The oxygen combining capacities of the different bloods are indicated in Table 7. Whatever variation of oxygen capacity is found depends on the property of the total molecule. In every case 1 atom of ferrous iron, or the heme unit, carries 1 molecule of oxygen to form oxyhemoglobin. The reaction readily occurs in the reverse, and the hemoglobin is thus reduced. Enzymes are not involved in these reactions, but the combination of the heme with the oxygen depends on such factors as the avail-

Table 7. Oxygen-combining capacities of different bloods in some invertebrates (at 0° C. and 1 atmosphere pressure)*†

Respiratory pigment	Color	Invertebrate and site of pigment	O_2 vol.% (in ml. O_2/100 ml. of blood)
Hemoglobin	Red	Annelids (plasma)	1−10
		Mollusks (plasma)	2−6
Chlorocruorin	Green to pink	Annelids (plasma)	5−9
Hemocyanin	Blue	Mollusks (plasma)	1−5
		Crustaceans (plasma)	1−4
Hemerythrin	Red	Sipunculids (coelomic corpuscles)	1−2

*From several sources.
†Compared with the oxygen volume capacities of invertebrates, those of vertebrates range from about 4 to 30. Other factors that influence the combining capacities are the pH and activity conditions.

ability of oxygen, pH and ionic content of the solution, carbon dioxide content, and structure of the total hemoglobin molecules. For instance, at high carbon-dioxide pressures the oxygen pressure at which hemoglobin is saturated with oxygen is higher than it is at low carbon-dioxide pressures.

Distribution and nature of respiratory pigments

Hemoglobin. Certain aspects of hemoglobin have already been presented, for it is the most widely distributed and the most efficient of all respiratory pigments. It is found in nearly every invertebrate phylum (C. Manwell, 1960); it may exist extracellularly dissolved in body fluids or intracellularly in corpuscles or tissue cells. It is especially characteristic of the Annelida and the entomostracan Crustacea, although it is found sporadically in a few other arthropods (e.g., *Chironomus*). It is found dissolved in the plasma of the snail *Planorbis*, and hemoglobin may occur in the blood corpuscles of a few lamellibranchs. The phoronid *Phoronis* has the pigment in its blood corpuscles. Hemoglobin may actually occur in the Protozoa (*Paramecium*). All heme compounds show characteristic absorption bands, as revealed by the spectroscope. Oxyhemoglobin has alpha and beta bands in the yellow regions and green regions and a large gamma band in the violet region. The position of the bands vary in the different hemoglobins. Invertebrate hemoglobin also differs from that of vertebrates in having more arginine and cystine amino acids and less histidine and lysine. In molecular weights they vary from 17,000 to many millions. This difference in kind of hemoglobin may be due to the variability of the environment in which the invertebrates live. Many invertebrates exist in regions where the concentration of oxygen is very low and even where the oxygen can be supplied without the aid of any pigment at all (M. L. Johnson, 1941). The polychaete *Arenicola* does not need hemoglobin for storage between low and high tides, as its hemoglobin functions only in oxygen transportation and its consump-

tion of oxygen is the same at both low and high tides. At both low and high tides the partial pressures of oxygen were high enough to provide sufficient oxygen.

By using carbon monoxide (which has a stronger affinity for hemoglobin than does oxygen), it is possible to ascertain at which percentage levels hemoglobin is able to transport oxygen. In leeches, for instance, oxygen consumption was reduced by carbon monoxide to 20% and 10% saturation but not at 3%. Species that had no hemoglobin were not affected by carbon monoxide.

Chlorocruorin. This pigment is closely related to hemoglobin but contains iron in a different porphyrin. Its prosthetic group is the same as that in cytochrome *a*. It does not occur in cells but is found in the plasma. It is distributed in four families of polychaetes, especially in the Sabellidae and the Serpulidae. The genus *Serpula* has both hemoglobin and chlorocruorin, and the amounts vary with the age of the organism—more in the younger worms. This pigment may have originated from hemoglobin by mutation (H. M. Fox and G. Vevers, 1960). The pigment is green in dilute solutions and pink reddish in concentrated solutions.

Hemerythrin. This respiratory pigment is the rarest of all the respiratory pigments, and its distribution is sporadic. It is characteristic of the sipunculid worms, the polychaete *Magelona*, the priapulids *Halicryptus* and *Priapulus*, and the brachiopod *Lingula*. The pigment always occurs in corpuscles, especially of the coelomic fluid, and is reddish violet in color.

Hemocyanin. Hemocyanin is next to hemoglobin in its distribution and importance. It also lacks the heme group and has copper instead of iron as the metal that combines with oxygen. The oxygen combines with 2 atoms of copper instead of the 1 iron atom of the hemoglobin unit and produces a blue color; when reduced, the pigment is colorless. A unit with 2 copper atoms has a molecular weight of 50,000 to 74,000 and forms multiple molecules of many units. Some of their molecular weights are more than 6,000,000 (*Helix*). Its distribution includes certain mollusks (amphineurans, cephalopods, and some gastropods), some malacostracans and crustaceans, *Limulus,* and other arachnids (Fox and Vevers, 1960). The pigment functions in both transport and storage. Its efficiency as an oxygen carrier is much lower than that of the vertebrate hemoglobin.

A principle of fundamental importance in all respiratory pigments is their oxygen affinity. This affinity is indicated by the partial pressure of oxygen at which the pigments are partially saturated with oxygen. A pigment with low affinity will give up its oxygen while there is still much oxygen to which it is exposed; one with a high affinity will retain its oxygen unless the partial pressure is very low. A low affinity usually indicates a transport function of the pigment; a high affinity, a storage function. Invertebrates as a rule can maintain their metabolic activities at very low internal oxygen tension, which would be fatal to most vertebrate species.

REFERENCES

Ashworth, J. H. 1904. Annelids. Liverpool Marine Biology Committee Memoirs XI. London, Williams & Norgate, Ltd.

Barnes, R. D. 1965. Tube-building and feeding in chaetopterid polychaetes. Biol. Bull. **129:**217-233.

Beddard, F. E. 1901. Earthworms and their allies. Cambridge, Cambridge University Press. This book is out of date but still useful.

Bell, A. W. 1947. The earthworm circulatory system. Turtox News **25:**89-94. Perhaps the best and most accurate description of the blood system in this worm.

Brown, F. A. 1950. Selected invertebrate types. New York, John Wiley & Sons, Inc.

Buchsbaum, R., and L. J. Milne, 1960. The lower animals. Garden City, Doubleday & Co., Inc. Superb illustrations of many marine worms.

Bullock, T. H. 1945. Functional organization of the giant fiber system of *Lumbricus.* J. Neurophysiol. **8:**55-71.

Clark, L. B., and W. N. Hess. 1940. Swarming of the Atlantic palolo worm, *Eunice schemacephala.* Tortugas Lab. Papers **33**(2):21-70. A study of the swarming behavior that occurs in July during the first or last quarter of the lunar cycle.

Dales, R. P. 1963. Annelida. New York, Hillary House Publishers, Ltd. An up-to-date treatise on this group.

Davis, C. C. 1955. The marine and freshwater plankton. East Lansing, Michigan State University Press.

Dixon, G. C. 1915. *Tubifex.* Liverpool Marine Biology Committee Memoirs XXIII. London, Williams & Norgate, Ltd.

Edmondson, W. T. (editor). 1959. Ward and Whipple's freshwater biology, ed. 2. New York, John Wiley & Sons, Inc. Sections devoted to Oligochaeta (C. J. Goodnight), Polychaeta (O. Hartman), and Hirudinea (J. P. Moore) are given in this important work. Keys are emphasized.

Fox, H. M., and G. Vevers. 1960. The nature of animal pigments. London, Sidgwick & Jackson, Ltd., Publishers.

Goodrich, E. S. 1945. The study of nephridia and genital ducts since 1895. Quart. J. Microscop. Sci. **86:**113-392.

Grassé, P. P. (editor). 1959. Traité de zoologie, vol. 5, part 1. Paris, Masson et Cie. An authoritative and comprehensive treatise of the annelids.

Gratzer, W. B., and A. C. Allison. 1960. Multiple haemoglobins. Biol. Rev. **35:**459-506.

Hanson, J. 1949. The histology of the blood system in Oligochaeta and Polychaeta. Biol. Rev. **24:**127-173.

Hess, W. N. 1925. Nervous system of the earthworm, *Lumbricus terrestris.* J. Morphol. Physiol. **40:**235-259. One of the best accounts of the nervous system in annelids.

Johnson, M. L. 1941. Hemoglobin function in *Lumbricus.* J. Exp. Biol. **18:**266-277.

Kamemoto, F. I., K. N. Kato, and L. E. Tucker. 1966. Neurosecretion and salt and water balance in the Annelida and Crustacea. Amer. Zool. **6:**213-219.

Krivanek, J. O. 1956. Habit formation in the earthworm, *Lumbricus terrestris.* Physiol. Zool. **29:**241-250. A study in animal behavior.

Laverack, M. S. 1963. The physiology of earthworms. New York, Pergamon Press, Inc. An excellent monograph on the physiologic aspects of the earthworm. A fine addition to the many treatises on this common animal.

MacGinitie, G. E. 1939. The method of feeding of *Chaetopterus.* Biol. Bull. **77:**115-118. This work describes how *Chaetopterus* in its tube gathers the food into mucous bags and forms pellets that are carried to the mouth.

Mann, K. H. 1962. Leeches (Hirudinea), their structure, physiology, ecology and embryology. New York, Pergamon Press, Inc. A useful, up-to-date monograph on leeches.

Manwell, C. 1960. Comparative physiology: blood pigments. Ann. Rev. Physiol. **22:**191-244.

Moment, G. B. 1953. On the way a common earthworm, *Eisenia foetida,* grows in length. J. Morphol. **93:**489-503.

Nicol, J. A. C. 1948. The giant axons of annelids. Quart. Rev. Biol. **23:**291-323. The morphology and functional significance of these fibers.

Pennak, R. W. 1953. Fresh-water invertebrates of the United States. New York, The Ronald Press Co. A good general account and useful keys to the freshwater annelids of all three classes.

Röhlich, P., S. Virágh, and B. Aros. 1970. Fine structure of photoreceptor cells in the earthworm, *Lumbricus terrestris.* Zeitschrift Zellforsch. **104:**345-361.

Stephenson, J. 1930. The Oligochaeta. New York, Oxford University Press, Inc. One of the best known treatises on this group and still very useful, although out of date in some particulars.

Wells, M. 1968. Lower animals. New York, McGraw-Hill Book Co.

CHAPTER 12

Introduction to Phylum Arthropoda

GENERAL CHARACTERISTICS

1. Phylum Arthropoda is the most extensive in the animal kingdom, both in number of species and in ecologic distribution. They make up about 80% of all named animal species.

2. Although closely related to the annelids, arthropods differ from annelids in many ways. They have lost much of the internal segmentation of annelids. Arthropods depend less on hydrostatic properties of locomotion, so characteristic of annelids. They have exploited the latent potentialities found in the annelid plan. Whereas annelids have soft bodies, most arthropods have developed a rigid exoskeleton that can resist predation and deformation and makes possible a system of levers for locomotion. In general, their exoskeleton is divided into two zones—the outer nonchitinous epicuticle, mostly nonwettable and protective, and the inner procuticle (endocuticle) of protein and chitin, the latter varying in the different arthropods. The uniformity of serial metameric units of annelids is replaced by grouping segments (tagmatization) into specialized structures for performing certain functions, thus producing a marked physiologic division of labor among them. (Some tagmatization occurs in a few annelids.) Cephalization is far more advanced among arthropods than it is among annelids. Much of the diversity in arthropod organization may have been independently evolved under the impact of environmental conditions. Although schizocoelomates, they have departed from spiral cleavage, but their eggs have remained mosaic and have become chiefly centrolecithal with superficial cleavage.

3. The basic body divisions of an adult arthropod are head, thorax, and abdomen. There are many arrangements of this organization. Head and thorax may be fused into a cephalothorax (often called a prosoma in contrast to the opisthosoma of the abdomen). A continuous exoskeletal shield (carapace) may cover the head and thorax. The head is composed of 6 embryonic metameres that fuse together, but the segmental arrangement is usually revealed. Head appendages vary, but are chiefly for sensory and ingestive purposes.

4. The three principal divisions of the arthropods are the Trilobita (extinct), the Chelicerata, and the Mandibulata. The head and limb characteristics may be considered the chief differences in these three divisions. The trilobites were the most generalized of arthropods in having the postoral appendages alike (biramous) and a distinct dorsal segmentation of the larva. In Chelicerata, mandibles are replaced by gnathobases of the anterior appendages (chelicerae), and they have four pairs of ambulatory limbs and no antennae. Mandibulates usually have true mandibular mouthparts, variously modified, as well as one or two pairs of antennae.

5. Arthropods owe their great success in the animal kingdom to many factors, such as their exoskeleton and jointed appendages, regional differentiation, division of labor, complexity of behavior patterns, ability to adapt to new ecologic niches, extensive histologic development of striated muscle, food diversities, and other factors. In general, arthropods have exploited the potentialities that are latent in the basic structural plan of annelids by the so-called arthropodization process.

PHYLOGENY AND ADAPTIVE RADIATION

1. Morphologic and embryologic evidence supports the view that arthropods are generally believed to have originated from annelids or annelidlike ancestors. The Precambrian fossil worm *Spriggina,* found in South Australia, had similarities to early trilobites, such as parapodia that could well be the forerunners of arthropod appendages. The nature of the arthropod ancestors must be largely speculative from present knowledge, but it may have been a form with an unsegmented head (acron) and a segmented trunk, with paired appendages on each segment except the telson. Whether the primitive appendage was biramous or uniramous is an unsettled matter. The three lines of evolution involved each one of the divisions already mentioned. According to some authorities, the Protoarthropoda, or ancestors of the arthropods, gave rise to three branches—the Prototrilobita, the Protocrustacea, and the Protomyriapoda. From the trilobites came the chelicerates, as the Prototrilobita are considered the oldest of the three branches. The Mandibulata differ from the chelicerates mainly by having antennae as the most anterior appendage and mandibles as biting and chewing jaws. The formation of ancestral archetypes is difficult because of the diversities and modifications found in such an extensive group.

2. Arthropods are the most highly specialized of all invertebrates. Certain features may have guided them in their great evolutionary development. In their segmentation and early embryo-

logic development, arthropods share a basic plan with that of annelids. Certain adaptive features common to the whole group have been modified and exploited in their diversification. Although segmented, arthropods have had the tendencies to suppress metamerism and to fuse segments together into specialized tagmata. The hydrostatic skeleton so useful to the annelids in their locomotion is little needed in a group with such a hard exoskeleton. Their evolutionary ranges of diversity have been guided chiefly by the exoskeleton, the jointed limbs, and the extensive specializations of their appendages for jaws and other feeding devices and for numerous other mechanisms, such as grasping organs, sense organs, accessory sex organs, and weapons of offense and defense. Locomotion and feeding methods have played an enormous role in the development of their morphologic patterns. The plasticity of their body systems in adapting them to different ecologic niches must be considered a primary factor in their diversity. In general, arthropods have great potentialities for adaptation to terrestrial life. Insects especially have made use of waterproofing their surfaces by a wax layer of the cuticle and their respiration by the unique tracheal system of spiracular closing mechanisms for adjusting to dry surroundings. For the conservation of water, the malpighian tubules are developed as outgrowths of the alimentary canal (and not from coelomoducts or nephridia as the coxal glands are) for the excretion of uric acid in a semifluid or solid form.

GENERAL SURVEY OF ARTHROPODA

The great phylum Arthropoda includes about 80 percent of all known species in the animal kingdom. About 1,000,000 species have already been described and some authorities think the number in time will exceed 5,000,000. Arthropods occupy about every conceivable ecologic niche from low ocean depths to very high altitudes, and from the tropics far into both north and south polar regions. Different species are adapted for life in the media of air,

Table 8. Time relations of invertebrate muscle*†

Muscle	Contraction time (seconds)	Time for 1/2 relaxation (seconds)
Arthropoda		
Schistocerca, wing muscle	0.025	0.250-0.340
Dectitus, leg muscle	0.200	0.200
Astacus, abdomen flexor	0.040	—
Limulus, abdominal muscle	0.197	0.435
Mollusca		
Mytilus, byssus retractor	1.000	2.0
Pecten, striated	0.046	0.04
Pecten, nonstriated	2.28	5.14
Helix, tentacle retractor	2.5	25.00
Squid, mantle	0.068	0.106
Echinodermata		
Thyone, lantern retractor	3.9	5.7
Thyone, longitudinal retractor	3.1	—
Stichopus, longitudinal retractor	1.0	—
Coelenterata		
Medusae	0.4-0.8	—
Sea anemone, sphincter	5.0	—
Sea anemone, slow circular muscle	120.0	—

*From several sources.
†This table does not show the effect of temperature, to which muscles are very sensitive.

land, fresh water, brackish water, marine water, and the bodies of other animals. Some species are known from only a few individuals, but most species are represented by thousands of individuals wherever found. In size they range from microscopic mites to the giant Japanese spider crab, which in its overall dimension may exceed 10 to 12 feet. The phylum includes insects, spiders, mites, crustaceans, ticks, centipedes, millipedes, and some smaller groups.

Both morphologic and embryologic evidences indicate that the arthropods have evolved from annelidan ancestors. Arthropods and annelids are segmented. Both phyla have a similar pattern of the nervous system—a dorsal anterior brain with a ventral nerve trunk and ganglionic swellings in each segment. Some arthropods have modified nephridia as excretory organs similar to the coelomoducts of polychaetes. In its primitive pattern each arthropod segment bears a pair of appendages, and polychaetes have the same arrangement of their parapodia. In their early development paired, segmental, coelomic compartments appear in the arthropods but become reduced as a hemocoel forms.

However, arthropods have distinctive characteristics of their own, such as a continuous cuticle variously modified, chitinized linings of the foregut and hindgut, an almost complete lack of cilia, a marked tagmosis of their segments, complex behavior patterns, and in many other ways. Not the least of their beneficial characteristics has been striated muscle, which has appeared only sporadically in lower groups. Their evolutionary divergence no doubt has depended on many factors, but one of the most significant is their mechanism of locomotion. A successful invasion of land and air environments, to a great extent, stresses meth-

Table 9. Appendages and some other characteristics of arthropods*

Segment	Arachnida	Crustacea (Malacostraca)	Myriapoda	Insecta
1	No append.	Embryonic	Embryonic	Embryonic
2	Chelicerae	Antenules	Antennae	Antennae
3	Pedipalps	Antennae	Embryonic	Embryonic
4	1st pair of legs	Mandibles	Mandibles	Mandibles
5	2nd pair of legs	1st maxillae	1st maxillae	1st maxillae
6	3rd pair of legs	2nd maxillae	2nd maxillae	2nd maxillae (labium)
7	4th pair of legs	1st maxillipeds	1st maxillipeds	1st pair of legs
8	Embryonic	2nd thor. append.	1st pair of legs	2nd pair of legs
9	Genital operc. ♀♂	3rd thor. append.	2nd pair of legs	3rd pair of legs
10	Pectines	4th thor. append.	3rd pair of legs	No append. and 11 seg.
11	1st lung books	5th thor. append.	4th pair of legs	11th abd. seg. (cerci)
12	2nd lung books	6th thor. append. ♀	5th pair of legs	
13	3rd lung books	7th thor. append.	6th pair of legs	
14	4th lung books	8th thor. append. ♂	7th pair of legs	
15	No append.	1st abd. append.	8th pair of legs	
16	1st seg. meta.	2nd abd. append.	9th pair of legs	
17	2nd seg. meta.	3rd abd. append.	10th pair of legs	
18	3rd seg. meta.	4th abd. append.	11th pair of legs	
19	4th seg. meta.	5th abd. append.	12th pair of legs	
20	5th seg. meta.	6th abd. append.	13th pair of legs	
			Additional legs	
			↓	
Gut	Complete with divisions; pumping pharynx; mesenteron with diverticula	Digestive glands; mostly straight intestine; gastric mill	No digestive glands	No digestive glands; usually gastric ceca
Excretion	Malpighian tubules into hind gut	Antennary glands	Malpighian tubules into hind gut	Malpighian tubules into hind gut
Respiration	Lung books plus trachea	Gills from biramous legs	Tracheae mostly unbranched	Branched tracheae

*Data from several sources.
Abd., abdominal; append., appendage; meta., metasoma; operc., operculum; seg., segment; thor., thoracic; ♂, position of male opening; ♀, position of female opening.

ods for getting around. Arthropods succeed in these methods to a remarkable degree; their dynamic nature is shown in their rapid locomotion when avoiding enemies and pursuing prey (Table 8). Their most extensive subdivision, the insects, have evolved flight, thus sharing with some vertebrates this most efficient method for wide distribution.

External anatomy

The typical arthropod body plan is a series of segments that have muscular movement, and each segment is provided with a pair of jointed appendages. This basic plan is greatly modified in their adaptive radiation because, unlike the annelids, the arthropods have their segments and appendages specialized for many varied functions (see Table 9 and also the section on tagmosis, p. 503). This trait has contributed to their evolutionary divergence into various ecologic niches.

One of the unique features of arthropods is their integument, which has been an important factor in the transition from water to land. The armorlike exoskeleton of arthropods has been the subject of much investigation (A. G. Richards and V. B. Wigglesworth). The general plan of the integument consists of a single layer of cells (hypodermis), an internal basement membrane, and an external cuticle secreted by the hypodermis. This external cuticle is made up of two basic layers: an outer thin epicuticle without chitin and an inner, thicker endocuticle, or procuticle, with chitin as its chief constituent. The epicuticle usually varies in thickness from 0.1 to 1 micron and is a waxlike substance in many arthropods. It is this wax that confers most of the impermeability of the cuticle to water. Two layers are usually found in the epicuticle: the outer lipoid (waxy) layer and an inner layer of proteins and lipoids. Other substances, especially in insects, have been identified in the epicuticle, which is very complex. The much thicker endocuticle (often in two layers) consists primarily of a chitin-protein complex that varies greatly in the differ-

ent arthropods. The characteristic chitin is a polymer (polysaccharide) of a high molecular weight and similar to cellulose. It is not found in nature in a pure condition but is always associated with proteins. Its metabolic source may be glycogen. It is often sclerotized into hard plates by the addition of other substances such as calcium salts (e.g., calcium carbonate and calcium phosphate). Flexible portions of the cuticle are called membranes and are found wherever movable articulations occur. Cuticle is also found in the ectodermal derivatives of the foregut and hindgut, tracheae, and parts of the reproductive ducts.

Arthropods often display a range of color patterns in the different groups, especially in insects and crustaceans. Both true pigments (biochromes) and structural colors (schemochromes) are represented. The biochromes are mostly carotenoids, although the darker melanin shades also occur, as well as some of the rarer pigments. The brilliant yellow, orange, and red colors are due to carotenoids, whereas shades of blue, green, and some other colors are primarily produced by the effects of light interference. Combinations of the two kinds of colors yield many shades. Pigment may be deposited directly in the cuticle or in chromatophores, which are multicellular and may or may not bear more than one kind of pigment.

The thick and generally inflexible cuticle or exoskeleton imposed on arthropods problems of locomotion and growth. Jointed limbs, however, made possible a wide variety of adaptations for specialized living. With the exception of the highly specialized flying insects, arthropods have evolved a segmented limb especially adapted for rapid movement on a substratum. For flexibility the appendages are provided with articular membranes, articular sockets, and systems of separate muscles. The propulsive force for exerting thrusts against the body is thus produced by extrinsic limb muscles. The division of the exoskeleton into separate sclerites, which are also connected by flexible articular membranes, also allows a movement of the body as well as of the jointed legs. Typically, each body segment is made up of a dorsal tergum, a ventral sternum, and 2 lateral pleura. Many modifications of this plan by fusion, elimination, or otherwise are found in the varied assortment of arthropods. Sclerotization of the integument also makes possible a greatly lengthened and stiffened limb for locomotor efficiency.

In tetrapod locomotion the stiffness of the limb segments is due to the bone elements, but in arthropods, stiffness is brought about by hollow cylinders enclosing the muscles. Arthropods have limb segments whose configuration and manipulation are controlled by antagonistic muscles that act reciprocally. In this way the muscles are converted into forces acting between the foot and the substratum. However, in some arachnids (spiders, scorpions, and others) the use of muscles in opposition to one another is done away with and another mechanical principle is used. Instead of having the joint axis passing through the center of the limb, thus allowing the attachment of extensor and flexor muscles above and below the axis, in spiders and some others the joint axis passes across the top of the limb, and only the flexor muscles are involved; extension of the limb is effected by fluid pressures within the leg. In the resting condition this fluid pressure is about 5 cm. of mercury but increases to about 40 cm. of mercury during vigorous activity. This pressure appears to rise in the body of the spider and is transmitted by this hydraulic system to the legs. Much of the evidence for this hydraulic system was obtained by D. A. Parry (1960) from the action of the jumping spider (*Sitticus*), which uses only its hind legs in jumping. The Onychophora uses a similar method of walking. It is suggested that the soft-bodied protoarthropod may have used this principle, and the hard cuticle was a later evolutionary development. This may indicate that at least two arthropod groups were evolved—a hard-cuticle, jointed animal with flexor and extensor muscles, and a flexi-ble-cuticle, jointed animal with the hydraulic system for limb extension.

S. M. Manton (1953) has made an extensive study of arthropodan locomotion and has described the mechanism of arthropodan movement. He has found that any arthropod exerts forces proportional to the volume of contractile fibers in its muscles and that the speed of movement depends on the length of stride, which increases as the length of the leg increases. Those requiring high speeds have quick thrust strokes, and most of the legs are off the ground for most of the time. Movement of legs on opposite sides of the body may coincide, or they may alternate with one another so that one limb of a given pair may be moving through its effective stroke while its mate is undergoing recovery —a process that induces body undulation.

During the growth of the arthropod the hard exoskeleton allows little room for expansion and must be shed periodically (molting, or ecdysis) to permit additional growth. A new cuticle then must be secreted to replace the one discarded. This process is under hormonal control and involves also the shedding of the ectodermal derivatives such as foregut and hindgut, tracheae, and some other parts. Just before molting most of the endocuticle is digested away by enzymes in the molting fluid secreted by the hypodermis. In this way the cuticle is freed from the hypodermis. Some constituents (limy salts) of the cuticle are also salvaged and stored temporarily in the liver or stomach gastroliths. While this is taking place, a new cuticle is secreted beneath the old by the hypodermis. The actual molting or shedding of the old cuticle now occurs, the cuticle being split typically along the middorsal line and other strategic places. This rupturing is aided by swelling induced by an absorption of water or air. The animal pulls out of the old cuticle and absorbs more water to increase its body size. In this enlarged state the new cuticle is hardened by the resorbed chitin and by the reserve of limy salts temporarily stored, as well as by the addition of new salts from the water.

During the growth period between molts the animal is called an instar, the number of instar periods varying with different species. Molting, or ecdysis, usually occurs frequently in the larval stages when growth is rapid, less often as the animal becomes older. In the postmolt period before the new cuticle has hardened the animal has little protection from enemies and usually seeks a hiding place. The many problems connected with ecdysis have no doubt restricted the size of many land arthropods.

Sclerotization refers to one of the hardening methods (the other being the deposition of calcium carbonate or calcite in some arthropods) that involves a group of proteins whose biochemistry has served for many investigations, but as yet is not fully resolved. When cuticle is freshly secreted, it is flabby and pale in color. In a short time it hardens and darkens. The rigidity of the cuticle is caused not by chitin, as formerly supposed, but by the sclerotin proteins. Sclerotization begins in the epicuticle and progresses inward. At the same time, tanning agents are operating. There is some experimental evidence that these tanning agents are derived from the amino acids tyrosine and tryptophane and are aromatic carboxylic acids. Some clue to sclerotization was discovered in the study of egg cases (ootheca) of cockroaches. Two colleterial glands (unequal in size) of the adult cockroach form secretions that are mixed and produce the walls of the egg case. The smaller gland produces a glucosidase that mixes with the presclerotin of the larger gland in a special sac where the egg-case is made. The glucosidase liberates phenol, which is oxidized by the oxidase. As a result, the oxidation product converts the presclerotin into sclerotin. However, the tanning agents of the cuticle and egg case are different, perhaps as a consequence of their being metabolized at different stages in their life history. One tanning agent (quinone) is thought to be the tanning agent in the puparium, for instance.

Sclerotization may be weakly developed in those with a flexible exoskeleton and is mainly restricted to the epicuticle. Here the hardening method may prevent the cuticle from being deformed too much in the arthropod's mode of life. The general method of hardening may vary among different groups of animals, for it is widely distributed in some form among non-arthropods.

The sclerotization of the cuticle and the formation of sclerites have resulted in the formation of distinct striated muscle groups. The muscles are supplied with few nerve fibers, of which usually at least one is inhibitory in its action. Each muscle fiber is also supplied with many motor endplates. Most of the body and limb muscles are attached directly on the inner surface of the integumental sclerites. Apodemes or endoskeletal ingrowths of the exoskeleton also serve for muscle attachments. Appendages are moved with a lever system the same as in vertebrates, except that in arthropods the muscles are attached to the inner surface of the exoskeleton and work across joints from the inside. Since the skeleton occupies the outside of the body and is heavy and bulky, this is another factor that has limited the size of arthropods.

Internal anatomy

The arthropods are all provided with the organ systems characteristic of higher forms. The coelomic cavity, however, is much reduced, being restricted to the gonocoel, where the gonads lie, and the space around the excretory coelomoducts. Most of the spaces form the hemocoel (blood sinuses) or swollen parts of the vascular system, which have generally obliterated the coelom proper.

The blood system is an open one, and the internal organs are contained in perivisceral cavities filled with blood. From these sinuses, blood is collected in a pericardial sinus (a part of the hemocoel) that surrounds the heart. The heart is an elongated dorsal structure and corresponds to the dorsal blood vessel of annelids. This heart is provided with lateral slits, or ostia, typically one pair to a segment. By pulsation the blood is forced forward into smaller vessels (arteries), which empty into the tissue sinuses. In Pauropoda and some other arthropods no heart is present, and the blood is circulated by the movements of the body or body organs. There are many modifications of the circulatory system within the phylum. The arthropodan blood bears some resemblance to that of annelids. The blood pigments, hemoglobin and hemocyanin, are represented in the plasma of many arthropods, hemoglobin being restricted mostly to the lower crustaceans. Several types of formed elements (leukocytes, amebocytes, and thigmocytes) are also found. The blood of some arthropods clots rapidly as a defense mechanism against wounds.

The digestive system may be straight or coiled, with the mouth in the head and the anus on the terminal segment. Embryologically it consists of three divisions: an anterior stomodeum, a posterior proctodeum (both of ectodermal origin and lined with chitin), and the mesenteron (midgut) of endodermal origin. The stomodeum, or foregut, is variously modified in many arthropods into a pharynx, esophagus, and storing or grinding organs. Lateral diverticula (blind pouches) may be found on the mesenteron, especially in the Arachnida and Pycnogonida, where they often extend into the legs. The proctodeum, or hindgut, may be divided into anterior intestine and posterior rectum. At the junction of the midgut and hindgut, the malpighian tubules (excretory) are present in some forms, including insects.

In some of the simpler arthropods (e.g., Pycnogonida) respiration takes place through the body wall; in most arthropods, however, the exchange of gases is effected by gills or tracheae. Some special breathing mechanisms such as book lungs are found in Arachnida. Gills are formed as branching outgrowths of the body wall or its appendages, whereas trachea have arisen as tubular invaginations of the ectoderm and cuticle. Tracheae may be unbranched in primitive forms, but usually they are branched and open to the outside by spiracles. Tracheae are lined with cuticle that is thickened in places, forming spiral staves to keep the tubes open. A tracheal system represents one of the most efficient breathing systems in the animal kingdom because it delivers oxygen directly to the tissues. Blood plays a very restricted role in the transportation of oxygen in insects.

Many kinds of excretory organs exist in arthropods. Some simple forms may have no organs at all, and excretory products are discarded at molting. Nephridia appear to be absent, but segmented excretory organs such as antennal, coxal, and maxillary glands, which are remnants of coelomic sacs, communicate with the exterior by coelomoducts. The coelomoducts in some forms open at the base of the legs. In arachnids, insects, centipedes, and millipedes, malpighian tubules of endodermal or ectodermal origin carry on the work of excretion. The chief excretory products include ammonia, amines, urates, and guanine. Fluids and solutes first enter the distal part of the malpighian tubule by active secretion, but as water and potassium bicarbonate are reabsorbed, the precipitated urates and other constituents are voided into the hindgut and discharged through the anus.

The nervous system reaches a high degree of complexity in higher arthropods. The basic plan, however, is similar to that of annelids. Typically, it consists of 2 ventral nerve cords (far apart in some primitive forms), a dorsal brain, and a pair of ganglia for each somite. Many of the ganglia may be fused together in accordance with the merging of the head segments. In the mandibulate arthropods the first 3 ganglia (protocerebrum, deutocerebrum, and tritocerebrum) are fused to form the brain. In insects especially, the ganglia of the thorax and abdomen undergo merging. From the ganglia, nerves run to the body wall and to the visceral organs. Sense organs include internal proprioceptors and many surface receptors such as olfactory, tactile, visual, and chemical.

Most arthropods are dioecious, but a few are hermaphroditic. Parthenogenesis is common in some groups. The gonads are nearly always in direct contact with the coelomoducts, and the genital openings are not constant in position. They may open near the posterior end of the body, near the middle of the body, and often toward the head end. Fertilization is internal in terrestrial forms but is often external in aquatic species.

EMBRYOLOGY

Eggs of arthropods usually contain a great deal of yolk and are of the type known as centrolecithal or some modification thereof. The nucleus is centrally located, and after fertilization it undergoes mitotic division without formation of cell membranes or cleavage of the yolk materials. In time the products of nuclear division migrate to the periphery, where cell membranes are formed around the nuclei in the layer of protoplasm on the surface of the yolk. This is the stage of the stereoblastula. Gastrulation may occur by invagination toward the yolk mass, by other methods involving migration of cells, or by delamination. From a germinal band along one side of the embryo, cells form a cell mass of mesoderm and endoderm by overgrowth or proliferation. A plate outside this mass forms the ectoderm. Numerous variations of this pattern exist. Immature stages of the arthropods vary from juvenile-like forms to larval forms wholly unlike the adults, such as the larval stages of insects. Many of the arachnids are anamorphic, that is, there is a general increase in complexity from larval form to adult. Horseshoe crabs resemble trilobites but a number of somites and appendages are added later to make the larval form an adult. The same occurs in mites and ticks, which hatch with three pairs of legs and add a fourth one at molting. Among the Crustacea the nauplius, or first, larva has only three pairs of appendages (the two pairs of antennae, and the third the mandibles). New segments and appendages are added at each molt.

Spiral cleavage, so characteristic of the protostomes, occurs in very few arthropods to any extent.

CLASSIFICATION

Phylum Arthropoda. Bilateral symmetry with body segmented and jointed externally; head somites fused with thorax, and abdomen distinct or fused; appendages usually one pair per somite, with hinge joints and variously differentiated; exoskeleton of a chitinous cuticle secreted by the epidermis and molted periodically; muscles striated and arranged in units; coelom much reduced and largely replaced by a hemocoel; mouthparts adapted for chewing or sucking; digestive system complete with terminal anus; excretory system of coelomoducts or tubules; circulatory system open, dorsal heart with vessels and tissue spaces; respiration direct or by gills, tracheae, or book lungs; nervous system of brain, paired ventral cords, ganglia, and connectives; sensory system of tactile, chemical, and visual receptors (simple or compound); parthenogenesis and dioecious sexes; eggs with much yolk (centrolecithal); cleavage superficial with gradual or abrupt metamorphosis; usually one or many larval stages; terrestrial, freshwater, and marine.

Subphylum Trilobita. Wholly extinct; body divided by 2 longitudinal furrows into 3 lobes; head distinct, with abdomen of varying number of somites and fused caudal plate; somites with jointed biramous appendages fringed by setae and present on most segments; one pair of uniramous antennae; development with larval stages; Cambrian to Permian periods; marine. *Triarthrus, Isotelus.*

Subphylum Chelicerata. Body usually divided into a cephalothorax (prosoma) and an abdomen (opisthosoma); cephalothorax unsegmented with paired preoral chelicerae bearing chelae or claws and paired postoral pedipalpi variously modified; antennae absent; four pairs of legs (usually); abdomen has varying number of segments, usually without appendages; respiration by gills, book lungs, or tracheae; excretion by coxal glands or Malpighian tubules; sexes separate, with single genital opening located anteriorly on the abdomen; chiefly terrestrial and predaceous.

Class Merostomata. Aquatic chelicerates; five or six pairs of abdominal appendages modified as gills;

continuous dorsal carapace; 12 to 18 somites and a spikelike telson at the end of the body.

Subclass Xiphosura. *Limulus.*

Subclass Eurypterida. All extinct.

Class Pycnogonida (Pantopoda). Body reduced to a series of cylindric segments (4 to 6), with vestigial opisthosoma; head or cephalon fused to first trunk segment; appendages of one pair of chelicerae, a pair of palps, a pair of ovigers, and usually four to six pairs of long walking legs; first four pairs of appendages usually on first (cephalic) segment; marine. Sea spiders—*Nymphon, Pycnogonum.*

Class Arachnida. Body divided into a prosoma and abdomen; prosoma (cephalothorax) usually covered dorsally with carapace and consisting of 6 fused segments; prosoma appendages of one pair of chelicerae, one pair of pedipalps, and four pairs of legs; opisthosoma (abdomen) of 13 segments generally fused and divided into a preabdomen and postabdomen; prosoma and opisthosoma fused in some; mostly terrestrial.

Order Scorpionida. *Centruroides.*

Order Amblypygi. *Acanthophrynus, Charinus.*

Order Uropygi. *Mastigoproctus, Thelyphonus.*

Order Palpigradi. *Prokoenenia, Koenenia.*

Order Araneae. *Miranda, Latrodectus, Lycosa, Argiope.*

Order Solpugida. *Eremobates, Galeodes.*

Order Pseudoscorpionida. *Chelifer.*

Order Ricinulei (Podogona). *Cryptocellus, Ricinoides.*

Order Phalangida (Opiliones). *Phalangium, Leiobunum.*

Order Acarina. *Dermacentor, Trombicula.*

Subphylum Mandibulata. Body of two or three divisions (cephalothorax-abdomen, head and trunk, or head-thorax-abdomen); antennae one or two pairs, mandible or jaws paired; maxillae one or two pairs; walking legs three or more pairs; respiration by gills or tracheae; excretion by glands or malpighian tubules; sexes usually separate; development usually with larval stages; terrestrial, freshwater, or marine.

Class Crustacea. Body segmented, with chitinous and limy exoskeleton and usually with carapace and telson; head usually with five pairs of appendages (antennules, antennae, mandibles, first and second maxillae); eyes compound; appendages typically biramous; segments of varying regional specializations and number; excretion by antennal glands; respiration by gills or body surface; usually dioecious; larval stages in development; cosmopolitan and mostly aquatic; free living or parasitic.

Subclass Cephalocarida. *Hutchinsoniella.*

Subclass Branchiopoda

Order Anostraca. *Branchinecta, Eubranchipus, Artemia.*

Order Notostraca. *Lepidurus, Triops.*

Order Conchostraca. *Cyzicus.*

Order Cladocera. *Daphnia.*

Subclass Ostracoda

Order Myodocopa. *Cypridina, Gigantocypris.*

Order Cladocopa. *Polycope.*

Order Podocopa. *Bairdia, Bythocypris.*

Order Platycopa. *Cytherella.*

Subclass Mystacocarida. *Derocheilocaris.*

Subclass Copepoda

Order Calanoida. *Diaptomus, Calanus.*

Order Cyclopoida. *Cyclops.*

Order Harpacticoida. *Harpacticus, Epactophanes.*

Order Monstrilloida. *Monstrilla.*

Order Notodelphyoida. *Notodelphys.*

Order Caligoida. *Caligus.*

Order Lernaeopodoida. *Lernaea.*

Subclass Branchiura

Order Arguloida. *Argulus.*

Subclass Cirripedia

Order Thoracica. *Lepas, Balanus.*

Order Acrothoracica. *Trypetesa.*

Order Ascothoracica. *Baccalaureus.*

Order Apoda. *Proteolepas.*

Order Rhizocephala. *Sacculina.*

Subclass Malacostraca

Series Leptostraca

Order Nebaliacea. *Nebalia.*

Series Eumalacostraca

Superorder Syncarida

Order Anaspidacea. *Anaspides.*

Order Bathynellacea. *Bathynella.*

Superorder Peracarida

Order Thermosbaenacea. *Thermosbaena.*

Order Spelaeogriphacea. *Spelaeogriphus.*

Order Mysidacea. *Mysis.*

Order Cumacea. *Diastylis.*

Order Tanaidacea. *Tanais.*

Order Isopoda. *Asellus, Porcellio, Oniscus.*

Order Amphipoda. *Hyalella, Gammarus.*

Superorder Eucarida

Order Euphausiacea. *Euphausia.*

Order Decapoda. *Cambarus, Palaemonetes, Homarus, Callinectes.*

Superorder Hoplocarida

Order Stomatopoda. *Squilla, Gonodactylus.*

Class Chilopoda. Body long, flattened dorsoventrally, and of many similar segments; head distinct with single pair of antennae; one pair of jaws and two pairs of maxillae; most segments with a pair of appendages; first pair of legs behind the head modified into poison claws; segments vary, up to 190; ocelli present or absent; dioecious; genital opening on mid-ventral next to last segment; terrestrial; cosmopolitan. Centipedes.

Order Scutigeromorpha. *Scutigera.*

Order Lithobiomorpha. *Lithobius.*

Order Geophilomorpha. *Geophilus.*

Order Scolopendromorpha. *Scolopendra.*

Class Diplopoda. Body long and cylindric or hemi-cylindric, with distinct head and trunk; segments double with single dorsal plate and two pairs of appendages per couple, except certain single anterior and posterior segments; one pair of antennae; mandibles and maxillae (gnathochilarium); tracheae two pairs per double segment and not anastomosing; genital opening on midventral line of third segment; dioecious; cosmopolitan. Millipedes.

Subclass Pselaphognatha. *Polyxenus.*

Subclass Chilognatha

Order Limacomorpha. *Glomeridesmus.*

Order Oniscomorpha. *Glomeris.*

Order Colobognatha. *Platydesmus.*

Order Nematomorpha. *Striaria.*

Order Polydesmoidea. *Polydesmus.*

Order Juliformia. *Julus, Spirobolus.*

Class Pauropoda. Body small, soft, elongated, and generally cylindric; trunk of 12 segments with 6 tergal plates; 9 to 10 pairs of legs; antennae two-branched with flagella; mouthparts of one pair of mandibles and one pair of maxillae; no respiratory organs or eyes; dioecious; anamorphic; cosmopolitan. *Pauropus.*

Class Symphyla. Body small and elongated; 15 to 22 similar segments, 12 of which usually bear legs; mouthparts of one pair of mandibles and two pairs of maxillae; moniliform antennae of many joints; cerci stout; no pigment; eyes absent; dioecious and anamorphic; cosmopolitan. *Scutigerella.*

Class Insecta (Hexapoda). Body divided into three distinct divisions: head, thorax, abdomen; head of 6 segments and one pair of antennae; mouthparts modified for chewing, sucking, or lapping; thorax of 3 segments with three pairs of legs and two pairs of wings (reduced or absent); wings dorsolateral outgrowths with characteristic venation; abdomen typically of 11 segments without ambulatory appendages; respiration chiefly by a tracheal system of tubes and spiracles; body cavity a hemocoel; excretion by malpighian tubules of the gut; nervous system of brain, ventral nerve cord, and segmental ganglia variously fused; eyes simple and compound; sexes separate with genital apertures near anus; development direct, gradual or complete metamorphosis; mostly terrestrial or freshwater. (A résumé of insect orders and other taxonomic arrangements is given under the section on class Insecta, pp. 574 to 626.)

BACKGROUND OF THE ARTHROPODS

A résumé of the background of the arthropods could have been some such line of phylogeny as suggested by the following. The acoelomate flatworms gave rise to bilateral symmetry, a marked tendency toward cephalization, and a pattern of determinate spiral cleavage, which has undergone many modifications. The formation of a complete digestive system with both mouth and anus (as in the rhynchocoels) freed this system for peristaltic movements and avoided many of the pitfalls of a gastrovascular system. This made possible a pseudocoel with perivisceral fluid. Because this particular line of organisms developed an elaborate cuticular covering, ciliation was lost, as cilia and cuticle do not mix to any extent. Of the several ancestral stocks that may have branched out of the flatworms, one of these could have been the coelomate group. The Precambrian rock fossils include worms and jellyfish, but no arthropods or echinoderms. Hard skeletons were unknown at this time. It was not until early Cambrian times that several animal groups began to have hard coverings, as indicated by the mollusks and echinoderms. The phylum Chordata was missing at this time because maybe their early members did not have hard parts. Coelomates seemed to have diverged into those with sedentary habits (such as the lophophorates with their mucus-ciliary feeding) and those with active motility (the arthropods). There is little evidence that mollusks have been segmented (*Neopilina* to the contrary), but the arthropods got their segmentation originally from their annelid ancestors and greatly advanced its potentialities (i.e., tagmosis) by their amazing adaptive radiation. Some consider the vast assemblage of arthropods as only remotely related to the annelids through a common annelid-like ancestor and that they do not form a natural group. Many of their characteristics of metamerism, hardened exoskeleton, and jointed appendages may be due to convergence. For instance, head segmentation varies from one class of arthropods to the next, and some have no distinct heads at all (M. Wells, 1968). Crustaceans and arachnids have digestive glands, but these structures are lacking in myriapods and insects. However, most of our ideas about the origin of arthropods are purely speculative and to find evidences for them in the animal kingdom is very difficult or impossible.

Phylum Arthropoda

THE EXTINCT TRILOBITES AND THE CHELICERATES

GENERAL CHARACTERISTICS

1. The extinct trilobites and the chelicerates are characterized by having chelicerae (jawlike structures) as the most anterior appendages and a complete absence of antennae. Their body is divided into two tagmata—an anterior cephalothorax or prosoma (head, mouthparts, and ambulatory legs) and a posterior opisthosoma (abdomen)—except in the order Acarina, in which these parts are fused. The cephalothorax is unsegmented except in the solpugids, or sunspiders. The second pair of appendages are the pedipalps, which are variously modified in structure and function.

2. The class Arachnida is by far the most common group in the subphylum. Ten orders are recognized in this class, of which the order Araneae includes the extensive group of spiders.

3. As a general group, Chelicerata have an organization so different from other arthropods that they may have followed an independent line of development. Throughout this group there is a tendency for few locomotor appendages, usually four pairs, except in sea spiders, which may have five or more pairs. Compound eyes are lacking in the class Arachnida.

4. The biting jaws are formed from the gnathobases of certain appendages and usually consist of 2 joints. The distal segment of the gnathobase may be fanglike and provided with poison glands. The jaws move transversely in some and vertically in others. Jaws that move transversely are the more effective for cutting and crushing. Most Chelicerata are fluid feeders, and the gnathobases are chiefly used for extracting juices from the prey.

5. In comparison with insects, arachnids, especially the spiders, furnish a striking example of how a group has overcome handicaps largely unknown in the insect group. Without antennae, compound eyes, or wings of flight, spiders, by ingenious methods and wily behavior patterns making use of clever devices, have managed to meet successfully severe competition. All spiders produce silk, which they use in various ways. Some species form webs, which may be constructed as horizontal webs or as vertical webs. Webs may be fashioned at the entrance to burrows, and some are sheetlike or funnel shaped. Some are merely irregular strings that become unsightly in houses, for dust collects on them. But silk may be used for other purposes, such as for the cocoons of eggs, for packaging sperm, linings for burrows, hinges for trapdoors, draglines, and gossamer thread for ballooning in air currents. Whenever a new situation arises, the spider usually finds some use for spinning silk threads.

6. The chelicerate line of arthropods gave rise to three main branches. One of these was the class Merostomata, which declined during the Ordovician, leaving only a few species of horseshoe crabs. A second line was the class Pycnogonida, or sea spiders, made up of about 600 species and found in all seas, especially

in polar seas. Little is known about its fossil record. The third branch is the highly successful class Arachnida, which is chiefly terrestrial and consists of spiders, scorpions, ticks, mites and some others not so well known.

7. The wholly extinct subphylum Trilobita inhabited shallow seas but not fresh waters. They became extinct before the Mesozoic, and the mineral matter of their exoskeleton may be preserved in rock with little change, may be replaced by other minerals, or may be dissolved, leaving only a mold. The study of their fossils serves as a good introduction to the Arthropoda.

Subphylum Trilobita (Trilobitomorpha)

Trilobites are considered the most primitive of all arthropods. They had a wide distribution in the Cambrian and later periods but are now wholly extinct, having died out by the end of the Paleozoic era. Their fossils are common in

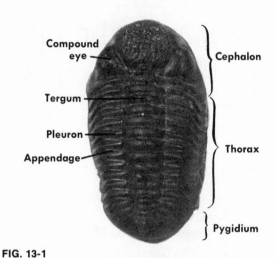

FIG. 13-1

Fossil cast of *Phacops* (middle Devonian period). These arthropods have been found in Silurian to Devonian rock in both North America and Europe.

FIG. 13-2

Cast of *Triarthrus,* an Ordovician trilobite. **A,** Ventral view. **B,** Dorsal view.

A   B

limestone and sandstone deposits, and they apparently inhabited the shallow Paleozoic seas. There were several kinds of trilobites, ranging in size from a fraction of an inch to 27 inches, although most were between 1 and 3 inches. About 4,000 fossil species are known.

The body was oval, flattened, and three lobed because of 2 lengthwise dorsal furrows (Figs. 13-1 and 13-2, *B*). The dorsal surface was covered by a chitinous and mineral exoskeleton that also extended over the side. This exoskeleton was molted at intervals to allow the trilobite to increase in size. For a long time only this dorsal surface was discovered, and little was known about the ventral surface and appendages. Specimens were eventually found with the ventral surface exposed, and the structure of the limbs has been ascertained (Fig. 13-2, *A*).

The body was divided into a head region, or cephalon, composed of 4 or 5 fused segments, with a head shield somewhat semicircular in shape; a thorax, or trunk region, of a varying number of free segments, each of which was made up of a convex tergal piece and a pair of lateral plates (pleura); and a terminal region

or tail (pygidium) of fused segments, also variable in number (Fig. 13-1).

The mouth was located on the midventral region of the cephalon (Fig. 13-3, *A*), which also usually bore a pair of compound eyes on the pleural region or cheeks. A pair of many-jointed antennae was placed on the sides of the forelip (hypostoma). The other appendages were all similar in structure. Four pairs of these were placed in the head region and one pair each on the segments of both the trunk and pygidium. Each appendage was two branched (biramous) and was made up of an inner seven-jointed branch or telopodite (walking leg) and an outer epipodite bearing a fringe of bristles (Fig. 13-3, *B*). Both branches of an appendage arose from a single coxal segment that bore a lobe (gnathobase). The gnathobases are thought by some authorities to have been used to seize and grind the food. In the region of the pygidium the appendages were smaller. The appendages lacked specializations, although the epipodites may have functioned as gills. The anus was situated at the posterior end of the pygidium. Many structural differences are found in trilobites.

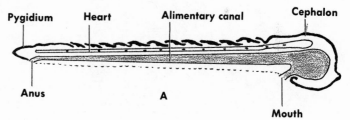

FIG. 13-3

A, Longitudinal midsection (diagrammatic) through axis of trilobite. **B,** Transverse section (diagrammatic) through trilobite showing attachment and segmentation of legs.

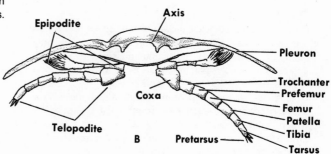

465

Trilobites appear to have been bottom dwellers, although some were adapted for swimming. Some trilobites may have been planktonic. They fed on small organisms they could seize and master. They may also have consumed mud and silt and used the organic part of it for their food. When disturbed, they could bury themselves in the mud or they could roll themselves into a ball. They have never been found in Mesozoic or more recent geologic deposits.

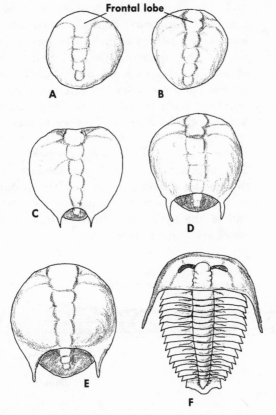

FIG. 13-4

Developmental stages of *Olenus,* an upper Cambrian trilobite. **A** and **B,** Protaspis stages, representing only head, or cephalon, of adult. Four divisions of glabella correspond to four pairs of head legs. **B,** Frontal lobe divides to form frontal lobe and free cheeks of adult. **C** to **E,** Small region behind cephalon generates new somatic segments, each new somite being added between last one and pygidium. **F,** Mature *Olenus.* (**A** to **E** after Störmer; redrawn from Snodgrass: Arthropod anatomy, Comstock Publishing Associates.)

Trilobites have had a varied evolution, with their greatest development in the late Cambrian and early Ordovician periods. Only one family survived to the end of the Paleozoic era. Theories of their relationships fluctuate between the crustaceans on the one hand and the Arachnida on the other. The discovery of *Hutchinsoniella* (order Cephalocarida) in 1954, with its trilobite characteristics of primitive, unspecialized appendages, may indicate a link between the crustaceans and the trilobites. As additional evidence, *Hutchinsoniella* is the only crustacean with a seven-jointed endopodite—a trilobite characteristic.

The early development of some trilobites is fairly well known. The early larval form has only indications of a head and pygidium (Fig. 13-4, *A* and *B*). Between these two regions new segments are developed from the front end of the pygidium in succession (Fig. 13-4, *C* to *F*). Several molts occur during this development. The evidence indicates that the pygidium is older than the thorax or trunk region.

Subphylum Chelicerata
CLASS MEROSTOMATA
Subclass Xiphosura

Fossils of horseshoe crabs have been found as early as the Devonian period, although some relatives of the group date as far back as the Cambrian period.

These ancient relics have the body divided into three divisions—the horseshoe-shaped prosoma (fused head and thorax), the segmented opisthosoma, and the spikelike tail (telson). The members of this group were characteristically found in shallow, brackish water, and their fossil record extends back to the Cambrian period. Four or five species and three genera are found living today. Their distribution is restricted to the Atlantic coast and to certain Indo-Pacific regions. The three genera are *Limulus* (Figs. 13-5 to 13-9), *Carcinoscorpius,* and *Tachypleus.*

Limulus polyphemus, the horseshoe crab, is the most familiar member of the group. Its

FIG. 13-5

Limulus, commonly called horseshoe crab, or king crab.

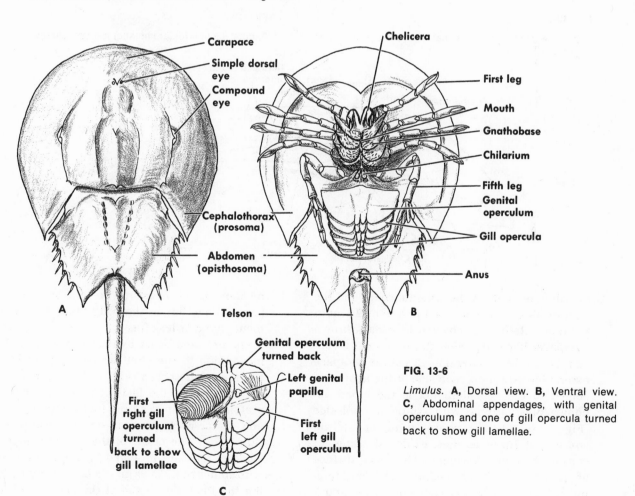

Carapace

Simple dorsal eye

Compound eye

Cephalothorax (prosoma)

Abdomen (opisthosoma)

Telson

Chelicera

First leg

Mouth

Gnathobase

Chilarium

Fifth leg

Genital operculum

Gill opercula

Anus

A

B

Genital operculum turned back

Left genital papilla

First right gill operculum turned back to show gill lamellae

First left gill operculum

C

FIG. 13-6

Limulus. **A,** Dorsal view. **B,** Ventral view. **C,** Abdominal appendages, with genital operculum and one of gill opercula turned back to show gill lamellae.

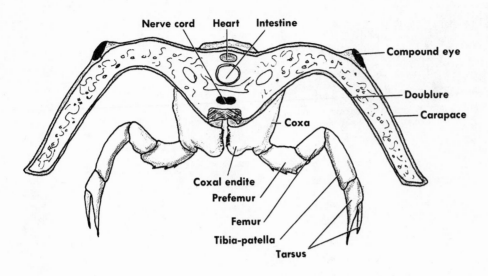

FIG. 13-7

Transverse section (diagrammatic) through prosoma of *Limulus*.

FIG. 13-8

Sagittal section (diagrammatic) through *Limulus*.

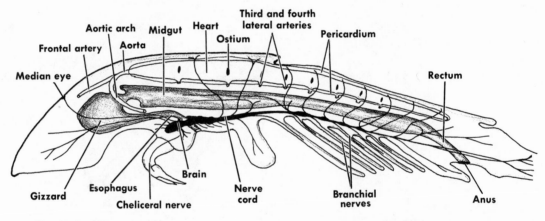

distribution is along the Atlantic coast and the Gulf of Mexico. It reaches a length of 2 feet or more and its body is covered by a dark brown carapace, horseshoe shaped and dorsally convex. The posterior lateral angles of the carapace extend backward on each side of the abdomen. Joined to the cephalothorax is the broad abdomen (opisthosoma), with 6 spines on each side (Fig. 13-6, *A*). The sharp, triangular telson consists of fused tergites, or dorsal plates, of certain abdominal segments. The dorsal surface of the carapace has a median and 2 longitudinal ridges. Two lateral compound eyes and 2 median simple eyes are present on the carapace. Altogether in the body there are 15 segments more or less fused.

The prosoma bears five pairs of walking legs in addition to the chelicerae on the underside of the cephalothorax (Fig. 13-6, *B*). Each walking leg (except the last) consists of a coxa, a prefemur, a femur, a tibia fused with a patella, and 2 tarsal segments, which form chelae (Fig. 13-7). The coxae of the first four pairs of walking legs are provided on the median side with spines that act as gnathobases for macerating the food. The coxae of the last pair of legs

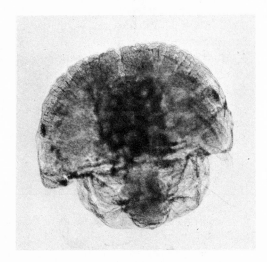

FIG. 13-9

Trilobite larva of *Limulus,* so called because of superficial resemblance to trilobites.

have flat processes (flabella) for cleaning the gills and for clearing away debris during burrowing.

The abdomen, or opisthosoma, bears seven pairs of appendages. The first pair, called chilaria and often considered to be the last pair of cephalothoracic appendages, is found between the coxae of the fifth pair of walking legs (Fig. 13-6, *B*). The second pair is fused medially to form a genital operculum, a membranous flap, on the underside of which are the 2 sexual apertures (Fig. 13-6, *C*). The last five pairs of appendages are modified as book gills. They also are flaplike and membranous, and on the underside each bears more than a hundred leaflike folds, or lamellae (Fig. 13-6, *C*). Movement of the gills circulates the water over the respiratory surfaces and also provides propulsive power for the animal, the gills serving as paddles.

The endoskeleton is made up of a connective tissue plate (endosternite) in the cephalothorax between the esophagus and the midgut, a number of endochondrites on the floor of the opisthosoma, and ectodermal invaginations (apodemes). Attached to these processes and to the carapace are paired and single muscles that move the cephalothoracic appendages. In addition, other muscles move the book gills and the telson.

The digestive system consists of a ventral mouth just behind the labrum or upper lip, an esophagus, a muscular gizzard, a midgut separated from the gizzard by a valve, and a short rectum that opens with an anus at the end of the abdomen (Fig. 13-8). A pair of large digestive glands (hepatic ceca) in the cephalothorax open by two pairs of ducts into the enlarged anterior part of the midgut. Horseshoe crabs live on mollusks and other organisms. Food is seized by the chelae of the appendages, macerated by the gnathobases, and passed to the mouth. Large indigestible particles are regurgitated through the esophagus and mouth. Ingested food is ground up in the gizzard, and the fine food particles are passed into the midgut, where they undergo enzymatic action. The food is then passed into the digestive glands, where further digestion and absorption occur. Digestion may be partly intracellular as well as extracellular. Indigestible particles are returned to the midgut and egested through the short rectum and anus.

Excretion occurs through two pairs of coxal glands or modified nephridia that are found in the cephalothorax near the gizzard. The glands open to the outside at the base of the coxae of the last pair of walking legs. Coxal glands can be distinguished by their red color.

Horseshoe crabs have a heart (Fig. 13-8), arteries, and veins. The dorsal tubular heart is situated in both the cephalothorax and abdomen. It has eight pairs of dorsoventral slits, or ostia, which serve as valves. The posterior end of the heart ends blindly. Blood is sent from the heart through a short aorta, which branches into 3 large arteries that carry blood to the carapace and to the sides of the gut, where they open into the ventral arterial sinus. From the ventral sinus, blood is supplied to the book gills, where it is oxygenated. Four pairs of lateral arteries also lead from the heart and supply blood to organs in the cephalothorax and abdomen. A median dorsal abdominal artery runs along the entire length of the telson. A number of paired veins carry the blood from the gills to the pericardium that surrounds the heart. From the pericardium the blood passes into the heart through ostia. Although provided with arteries and veins, the blood system is an open one.

Respiration occurs in the book gills through hundreds of thin, leaflike lamellae (Fig. 13-6, C). Here gaseous exchange occurs between the blood in the lamellae and the surrounding water. The blood contains the pigment hemocyanin, which imparts a blue color to it.

The entire nervous system of horseshoe crabs is enclosed within the ventral sinus (Fig. 13-8). The brain forms a compact ring around the esophagus. Only the protocerebrum forms the brain. Part of the nerve ring is composed of fused ganglionic masses that give off nerves to the thoracic appendages. From the nerve ring a ventral nerve cord, with paired ganglia enclosed in the longitudinal sinus, runs posteriorly. Each ganglion gives off an anterior motor and a posterior sensory nerve. Sense organs are represented by a pair of median ocelli and a pair of compound eyes. Rudimentary ocelli and compound eyes are also found. The compound eyes are uniquely constructed in that the ommatidia are made up of clusters of retinal cells with no intervening pigment. Taste buds may also be found in the gnathobases.

The sexes are separate, and the plan of the reproductive system is the same in both male and female. The reproductive glands form a type of network of tissues—a pair of vasa deferentia in the male and a pair of oviducts in the female. These ducts open to the outside at the base of the genital operculum. No copulatory organs are present, but the males can be distinguished by the enlarged basitarsus of the first pair of legs. Mating occurs by a type of amplexus, with the male holding onto the female carapace by the aid of hooks from his first pair of feet. As the eggs are discharged by the female into a scooped out depression in the sand, the male discharges his sperm over them. The eggs, which are about 3 mm. in diameter, are laid in batches of 200 to 300. In some places (e.g., Delaware Bay) spawning coincides with the new or full moon.

Eggs are centrolecithal and cleavage is total. Hatching usually occurs within 2 weeks, although it may be delayed. The newly hatched larva resembles a trilobite (Fig. 13-9) and is mostly incomplete in the development of its systems. Molting occurs many times during the first years of its life. Three years are required for the crabs to reach sexual maturity, although full size is not reached until much later. Like the adults, the young at all stages swim on their backs.

The body cavity of *Limulus* is a hemocoel and is derived from the fusion of several coelomic pouches. The cavity extends into all appendages.

Subclass Eurypterida

The primitive arthropods of subclass Eurypterida are wholly extinct, but they flourished from the Cambrian period to the end of the Paleozoic era. They are commonly known as giant water scorpions. Their elongated and lanceolate body, encased in a chitinous exoskeleton, was divided into a prosoma (cephalothorax) and opisthosoma (abdomen) (Figs. 13-10 and 13-11). The cephalothorax had a pair of marginal compound eyes and a pair of centrally located ocelli. They share many similar characteristics with the horseshoe crabs. Usually the appendages of the cephalothorax consisted of one pair of chelicerae, four pairs of walking legs, and one pair of paddlelike swimming legs. There were many variations among the different species. The bases of the appendages around the centrally located mouth were provided with masticatory organs.

Unlike that of the horseshoe crabs, the abdomen of eurypterids consisted of 12 separate segments, 7 of these forming the mesosoma (preabdomen), which bore appendages, and 5

FIG. 13-10

Eurypterus. Eurypterids flourished in Europe and North America from Ordovician to Permian periods. (From Hickman: Integrated principles of zoology, The C. V. Mosby Co.)

forming the metasoma (postabdomen) without appendages. The terminal telson probably had a poisonous sting. There were six pairs of abdominal appendages. The first pair formed the operculum and the other five pairs made up the gills.

Many eurypterids were less than 1 foot long, but some reached a length of 9 feet. Although much of their evolution occurred in marine and brackish water, they also invaded fresh water. Mostly they were bottom crawlers, but some could probably swim. They seemingly had a close affinity to the trilobites, from which both the xiphosurids and eurypterids apparently descended. Eurypterids may have left no direct descendants, but one theory holds that they were the ancestors of the great class Arachnida (Fig. 13-11).

Among the common fossil genera are *Pterygotus*, *Carcinosoma*, and *Eurypterus* (Fig. 13-10).

FIG. 13-11

Body divisions of chelicerates. **A,** Eurypterid. **B,** Xiphosurid. **C,** Spider. Shaded area denotes prosoma; white area shows opisthosoma.

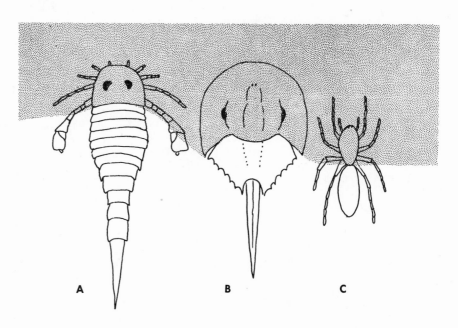

A   B   C

FIG. 13-12

Anoplodactylus lentus, a pycnogonid (preserved specimen).

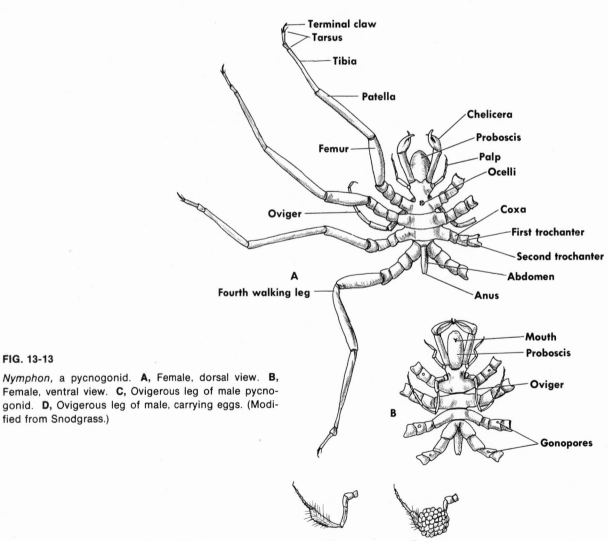

FIG. 13-13

Nymphon, a pycnogonid. **A,** Female, dorsal view. **B,** Female, ventral view. **C,** Ovigerous leg of male pycnogonid. **D,** Ovigerous leg of male, carrying eggs. (Modified from Snodgrass.)

■ CLASS PYCNOGONIDA (PANTOPODA)

Pycnogonids are commonly known as sea spiders because of their long legs and spiderlike form (Fig. 13-12). About 600 species are known (J. W. Hedgpeth, 1948, 1954). They are mostly small forms not over 10 mm. long, although some of the deep-sea forms may reach a length of 2 feet. Their body is greatly reduced, consisting of a cylindric trunk and segments that support the appendages. They are also provided with a proboscis on the head (cephalon), gonopores on the second joints of the legs, and a much reduced abdomen. The posterior part, or neck, of the cephalon has on its dorsal surface 4 eyes on a tubercle. The trunk usually consists of 4 to 6 segments, of which the first is fused to the cephalon (Fig. 13-13, A). In some pycnogonids there are seven pairs of legs, of which the first four—chelifores (chelicerae), palps, ovigers, and first pair of walking legs—are on the cephalic segment. Each of the other trunk segments bears a single pair of legs. Some species are known to have 12 legs. Some genera may lack chelifores, palps, and other structures. The palps are located just behind the chelicerae and bear sensory hairs. The ovigerous legs are used by the male to carry the eggs and are reduced or absent in the female.

Each of the walking legs has 8 segments. The legs are longer in those forms that have reduced bodies. The proboscis bears a three-cornered mouth at its end (Fig. 13-13, B).

Pycnogonids are carnivorous animals that suck out the juices of their prey, such as corals, bryozoans, and sponges. The chelicerae are used for seizing and carrying their prey to the mouth. The digestive and reproductive systems extend into the legs. Digestion is intracellular in the walls of the intestine, which opens into a short rectum, with the anus at the end of the abdomen. The nervous system is made up of a dorsal brain, circumesophageal ring, paired ventral ganglia, and ventral nerve cords. The circulatory system is open and consists of a dorsal tubular heart with a hemocoel divided into dorsal and ventral parts. Blood flows from the hemocoel into the legs and is returned to the heart by the dorsal hemocoel. Excretory and respiratory organs are absent. The appendages are provided with tiny muscles for movement, and strong exterior muscles run to the wall of the pharynx for producing a suction action by the pharyngeal lumen.

In reproduction the male fertilizes the eggs as they pass from the female and collects the eggs into masses on his ovigerous legs (Fig. 13-13, C and D). Glands on the femurs of these appendages form a secretion for attaching the egg masses (R. E. Snodgrass, 1952).

Sea spiders are more common in the cold waters of the Arctic and Antarctic seas than they are elsewhere, although they are found in all seas. They may be found from littoral or intertidal zones down to depths of several thousand meters. Some are ectoparasites on hydroids and other forms. Their exact relationships are very obscure, although their structures indicate that they are chelicerates.

Common genera of Pycnogonida include *Nymphon* (Fig. 13-13), *Pycnogonum*, and *Anoplodactylus* (Fig. 13-12). A Devonian fossil species, *Palaeopantopus*, is known.

■ CLASS ARACHNIDA
STRUCTURE AND PHYSIOLOGY

The arachnids are the most important of all the chelicerates. They are the spiders, scorpions, ticks and mites, harvestmen, and some lesser known types. Most are terrestrial forms, although a few mites are aquatic. Their general distinguishing characters (B. J. Kaston, 1953; T. H. Savory, 1964) are four pairs of thoracic legs, tracheae or book lungs for breathing (but no gills), simple eyes (ocelli), the first pair of appendages (chelicerae), and the second pair (pedipalps), which are variously modified. They have a chitinous exoskeleton and a body that is divided into two portions—an unsegmented cephalothorax (prosoma) and an abdomen. The prosoma is covered dorsally by a carapace, and on its ventral side are sternal plates and coxae of the appendages. In its primitive condition the abdomen is divided into a pre-abdomen and a postabdomen, but this distinction is lost in most forms. The abdomen is composed of 12 segments, which may be modified in various ways. A telson is found in some. In many arachnids, segmentation is more or less lost by fusion of segments. The segmentation of the cephalothorax is hard to determine but probably consists of 9 segments, thus making a total of 21 segments the typical number in Arachnida. The sternites, which cover the ventral surface of the abdomen, may also be fused so that they may be fewer in number than the dorsal tergites. There are many variations among the different groups. In some arachnids the tergites may be fused to form a single dorsal plate with the carapace. The cephalothorax may be broadly joined to the abdomen, or the two divisions may be joined by a narrow pedicel.

Of the six pairs of cephalothoracic appendages, the chelicerae are either two or three jointed. They may bear pincers (chelate) for holding and crushing prey, but they may also bear other structures, such as poison fangs, intromittent organs for transferring sperm, spinnerets, etc. The second pair (pedipalps) is built on the same plan as the legs but usually has only 6 joints. They may also vary and have different functions, serving as intromittent organs, for crushing food, as pincers, as sensory organs, and others. The four pairs of walking legs are typically seven jointed; and each is composed of a coxa, trochanter, femur, patella, tibia, metatarsus, and tarsus. In some arachnids the coxae are provided with gnathobases.

The endoskeleton consists of an endosternite (mesodermal in origin) and apodemes (ectodermal invaginations), variously modified. The complex muscular system is made up of many muscles for moving the numerous body parts, such as the cephalothoracic appendages, digestive specializations, abdomen, and spinnerets.

Although most arachnids are carnivorous, they possess no mandibles. While holding their prey with their chelicerae, they squirt enzymes from the midgut over the crushed parts, predigesting and softening them for swallowing. The juices are then sucked into the stomach. Their digestive system is made up of foregut, midgut, and hindgut (Figs. 13-16 and 13-23). The foregut and hindgut (ectodermal) are lined with chitin, which is shed at each molt. Digestion is completed in the midgut. The three digestive parts are modified in the different groups. In some forms the foregut consists of a pharynx, which may serve as a pump and a sieve; an esophagus; and a gizzard, which may also serve as a pump. The midgut, or mesenteron, has cells for absorption and for producing enzymes; it is provided with lateral diverticula, which become filled with partially digested food. Some food is stored in the interstitial tissue. The hindgut has a short rectum, which opens by an anus at the end of the abdomen.

Excretion in arachnids chiefly occurs through a pair (or sometimes two pairs) of coxal glands in the cephalothorax and also through malpighian tubules in the abdomen. Coxal glands are modified nephridia or saclike structures, which open at the coxae of the appendages. The position of the outlet varies. The waste is absorbed into the gland from the blood. Many arachnids have one or two pairs of long malpighian tubules that arise at the junction of the hindgut and the midgut. These tubules pick up guanine crystals (one of the end products of nitrogenous metabolism) and discharge them into the alimentary canal, or the crystals may be stored in special cells. Some arachnids may have one or the other or both types of excretory organs. Some excretory substances may be picked up by special cells (nephrocytes).

Most arachnids have a dorsal tubular heart within a pericardium (mostly in the abdomen) and arteries to various parts of the body (Figs. 13-16 and 13-23). The heart may be segmented to coincide with the segmentation of the abdomen. Ostia, or funnel-shaped openings with valves, allow blood to enter the heart from the pericardium. True capillaries and veins are absent. From the heart the anterior aorta follows the gut in the cephalothorax, bifurcates, and sends a branch on each side of the gut under the endosternite. Here they open into a sinus that feeds arteries to the appendages. In the abdomen the heart gives off lateral arteries and a posterior artery to the abdominal region. Blood collected in the ventral sinus bathes the book lungs and is conducted back to the heart by tissue sinuses. Hemocyanin, a copper-bearing compound, is the respiratory pigment in the blood plasma of some arachnids.

Arachnids have book lungs or tracheae (sometimes both) for breathing (Figs. 13-15 and 13-24). The number of book lungs varies from one to four pairs. The book lungs may be invaginations of the ventral abdominal wall and are actually groups of leaf tracheae, as compared with ordinary tubular tracheae. A book lung is made up of leaflike lamellar folds of one side of the wall through which air circulates and oxygenates the blood in the lamellae. One part of the book lung is in the form of an air chamber, which contracts and dilates by muscular action, thus sucking in and expelling air. The book lung opens to the outside by a spiracle. The tracheae are homologous to, and may be derived from, book lungs. Their spiracles are usually paired and may open directly into the main tracheal tubes, or they may open first into chambers. The main tracheal trunks divide and subdivide into smaller and smaller tubes, which eventually deliver the air to fluid-filled and branching tracheoles that pass the oxygen directly to the tissue cells. The tracheal tubes are lined with cuticle, usually in the form of spirals (Fig. 13-24, C).

The central nervous system is typically made up of a dorsal brain and a ventral chain of paired ganglia, usually one pair to each segment. The forebrain, or protocerebrum, furnishes nerves to the chelicerae. In many arachnids the abdominal ganglia of the ventral nerve cord migrate anteriorly and fuse with the subesophageal ganglia, forming a ganglionic mass or ringlike structure. Often some of the posterior ganglia are lost completely. In some arachnids with primitive nervous systems each fused ganglion of the ventral chain sends two pairs of nerves to each of its original segments, but such a pattern has been greatly modified in the various groups.

Sense organs in arachnids are represented by simple eyes, the corneal lens of which is lost or shed with the rest of the cuticle at each molt. The shape of the lens may be spherical, oval, or even triangular. The photoreceptors of the retina may be oriented toward the source of the light (direct eyes) or toward the postretinal membrane or reflector (indirect eyes) (Fig. 13-25). Photoreceptors may or may not be separated by pigment. The number of eyes varies greatly in the different orders. Many arachnids are totally blind. The sense of touch is well developed and is mediated by sensory hairs all over the body, especially on the appendages. One type of sensory hair (trichobothrium) can detect air currents and thus presumably may function in audition. Slit organs, or special pits in the cuticle, may have a function in olfaction, as mechanoreceptors, or vibration receptors. Some arachnids stridulate, thus producing sounds.

Arachnids nearly always have separate sexes, although there are rare exceptions in the order Acari. The genital opening in both sexes is located on the second abdominal sternite. The gonads are found in the abdomen and may be single or paired. Copulatory organs exist in a great variety of forms. The males of some arachnids have a tubular penis, and others make use of pedipalps or other appendages as copulatory organs (Fig. 13-20). Long, retractile ovipositors also occur. With the exception of the scorpions and some Acari, all arachnids are oviparous.

■ ORDER SCORPIONIDA

The distinguishing characteristics (J. L. Cloudsley-Thompson, 1958; W. S. Bristowe,

FIG. 13-14

Centruroides, a scorpion. The various species of the genus *Centruroides* occur in Mexico, Arizona and New Mexico, and are responsible for the death of many children in Mexico. The venom of the scorpion is a neurotoxin and may cause respiratory and cardiac failure.

anywhere that provides hiding places for them. In tropical countries they enter houses and often become a nuisance. As they are sensitive to moisture, they stay underground in dry seasons but emerge during a wet period and live under rocks and other hiding places. They are solitary in their habits and are never found in associations.

The body is made up of a prosoma, or cephalothorax, of a single piece and an abdomen, or opisthosoma. The abdomen consists of two divisions — an anterior preabdomen (mesosoma) of 7 segments and a postabdomen (metasoma) of 5 segments. The preabdomen is as wide as the cephalothorax, but the postabdomen is slender, with a curved sting at its end. The prosoma is relatively short and four sided. It is covered by an unsegmented carapace on which are located a median pair and two to five lateral pairs of eyes. The cephalothorax bears a pair of small chelicerae in front of the mouth, a pair of large pedipalps, and four pairs of walking legs. Each of the walking legs has 7 segments and two pairs of claws. The segments of a leg are a coxa, trochanter, prefemur, femur, tibia, basitarsus, and tarsus complex with the claws. The first two pairs of walking legs have their coxae provided with plates and glands that secrete digestive enzymes.

On the ventral surface of the second abdominal segment behind the last pair of legs are the paired pectines, each of which consists of 3 rows of chitinous plates and a series of toothlike projections (Fig. 13-15). The teeth of the pectines bear sense cells that may be used by the male to touch the ground and select a suitable mating site, or they may be used to explore the surface over which he crawls.

A single, slender sternite covers the ventral surface of the cephalothorax between the coxae of the walking legs. Each of the 7 preabdominal segments is covered by a tergal plate, and the ventral surface is covered by a sternal plate. The genital aperture, covered by the genital plate (Fig. 13-15), is located on the

1958) of order Scorpionida are the large chelate pedipalps, which are always kept in a forward position; the abdomen, which is divided into a broad portion and a slender, segmented portion; and the terminal caudal sting (Fig. 13-14). Although they vary in size from less than 1 inch up to 7 inches, they all have these basic characteristics.

Scorpions are tropic or semitropic forms, which in general prefer arid or desert regions, although they may be found in some humid areas. They are found on all continents but are absent from such islands as New Zealand. There are about 650 species, forty of which are found over a large portion of the United States. Arizona and California appear to have the largest number of species.

Scorpions in their natural habitat live under stones, in crevices, under bark of trees, under rubbish, in open fields and woods, and almost

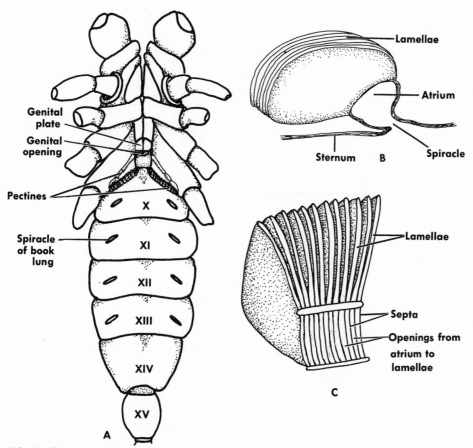

FIG. 13-15

A, Ventral view of scorpion *Pandinus* (appendages and most of postabdomen not shown). **B,** Vertical section of book lung of scorpion. **C,** Ventral view of book lung showing openings into lamellae from atrium. (Modified from Snodgrass.)

ventral side of the first abdominal segment. On the second segment a small basal piece articulates laterally with each of the pectines. On the undersurface of the preabdomen and along the lateral regions of the third to sixth sternites are four pairs of slits (spiracles), which serve as openings to the book lungs.

The sternite, pleurites, and tergites of each segment of the postabdomen are fused together in a ring. The anus is placed on the ventral side of the last segment. The base of the sting is enlarged and contains a pair of poison glands that open near the tip of the sharp, curved spine. The scorpion thrusts the tail forward over its back to inject the venom from the sting into its victim. The most dangerous scorpions in North America are members of the genus *Centruroides* (Fig. 13-14) found in Arizona, New Mexico, and Mexico, where they have been responsible for some deaths, especially among children. Other scorpions are poisonous enough to kill prey but are not harmful to man. Their poison is mostly neurotoxic.

Scorpions live mainly on small arthropods, especially insects. They seize their prey with the pincers of their pedipalps, sting it, and macerate it with their chelicerae. They then discharge digestive fluids with enzymes over the soft parts and suck the predigested food into the mouth by

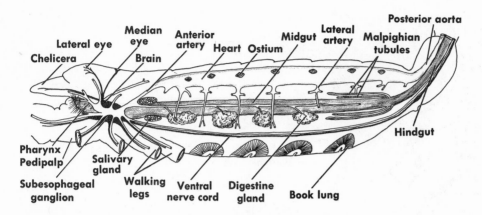

FIG. 13-16

Internal structure of scorpion. Sagittal section (diagrammatic) of cephalothorax and preabdomen. (Modified from Leuckart.)

the pharynx, which acts both as a pump and a strainer. After passing through the esophagus, the food enters the midgut and the large digestive glands (Fig. 13-16), where it is subjected to further digestion and absorbed by special cells. The indigestible part is discharged through the anus. Scorpions can exist many months without food or water.

The circulatory system (Fig. 13-16) is made up of a 7-segmented heart in the preabdomen, several pairs of ventrolateral arteries to regions of the preabdomen, and various venous channels. The book lungs (four pairs) are placed in the ventral blood sinus, where the gaseous exchange between oxygen and carbon dioxide takes place. The oxygenated blood then passes to the pericardium of the heart.

Excretion occurs through the paired coxal glands, which open on the coxae of the third pair of legs, and through the malpighian tubules, which discharge into the gut.

The nervous system of scorpions (Fig. 13-16) consists of a brain, a ventral nerve cord with 7 ganglia, and nerves that innervate various regions of the body. In addition to the eyes and the pectines already mentioned, scorpions have other sense organs such as the special

tactile hairs (trichobothria) on the femur, tibia, and pedipalps.

Although there are only slight differences between the male and female, the male is usually more slender and has a longer tail region. His copulatory organs (hooklike structures) are found on the opercular plates on the ventral side. The reproductive organs of each sex open by a ventral aperture on the second abdominal segment. The gonads are similar in both sexes and are located between the midgut diverticula in the preabdomen. From each group of ovarian tubules a short oviduct leads to a single genital atrium. Each oviduct may have a dilated seminal receptacle. The testes are also made up of tubules, and the paired vasa deferentia also empty into a common genital atrium. There is a complicated courtship pattern in which the male seizes the female pedipalps with his own pedipalps and undergoes a type of dance made famous first by J. H. Fabre, and later by Walt Disney in one of his nature films. The male then deposits a spermatophore on the ground, which is picked up by the female vulva. In some scorpions the two sexes may actually copulate by bringing their ventral surfaces together. As in spiders, the female has been known to devour the male after copulation.

The fertilized eggs develop inside the lumen of the ovarian tubules, and the young are born alive (ovoviviparous). The eggs are telolecithal with meroblastic cleavage. In some scorpions the young are nourished internally by maternal secretions from a tube attached to the diverticulum (viviparous). The eggs of viviparous forms have little yolk and show equal, or holoblastic, cleavage. As soon as they are born, the tiny young climb on the back of the mother, often completely covering her since they may number several dozen. While on her back they get their nourishment from embryonic yolk. The period of gestation may be from 6 to 15 months. Growth occurs by molting, the first molt occurring while the scorpion is still on its mother. Adulthood is attained in about a year.

■ ORDER AMBLYPYGI

The members of order Amblypygi are commonly known as the tailless whip scorpions. They are flattened, crablike animals that are found in tropic and the warmer temperate regions. They occur in the United States in some of the southern states and in the Southwest as well as in California. Altogether there are some fifty to sixty species. They are reddish brown in color and have a body divided into a prosoma, covered with an undivided carapace, and an abdomen of 12 segments. Their pedipalps are heavy and adapted for grasping and crushing prey (mostly insects). The first pair of legs is long and sensory in function. Their whiplike appearance is the basis for the common name for these animals. The other three pairs of legs (each of 7 segments) are used for locomotion. The chelicerae have a movable claw. No tail is present. The carapace bears a pair of median eyes and three small eyes on each side. Other sense organs are represented by lyriform organs or chemoreceptors for taste and smell found on the chelicerae. No poison glands or repellent organs are present. They have two pairs of book lungs, which lie in the second and third abdominal segments. Excretory organs are coxal glands and malpighian tubules. So far as is known, they have no elaborate courtship pattern. The 50 to 60 eggs are carried by the mother attached to the underside of her abdomen, where they hatch and where the young undergo a first molt.

The Amblypygi prefer semidesert and somewhat humid regions. They live under stones, logs, and leaves during the day. Some are found in caves. They vary in length from 4 to 45 mm. *Acanthophrynus* and *Charinus* are common genera.

■ ORDER UROPYGI

The uropygids are known as the tailed whip scorpions and have about seventy species (R. E. Snodgrass, 1952). They are found mostly in tropic and semitropic regions of Asia and the Americas. In the United States and Mexico a common species is *Mastigoproctus giganteus*. Some species do not exceed 2 to 3 mm. in length, and some, such as *Mastigoproctus* (Fig. 13-17, *A*), may reach 65 mm.

The prosoma of their flattened bodies may be single or divided. The chelicerae are small, two jointed, and provided with a hooklike fang. The six-jointed pedipalps are thickened, with strong spines, and adapted for grasping prey, whereas the first pair of legs is very long, secondarily segmented, and modified as feelers. Their long whiplike shape gives the common name to the group. The other three pairs of legs are unmodified.

The abdomen is made up of 12 segments, of which the first and the last 3 are not as wide as the others. At the terminal end of the abdomen is the many-jointed flagellum. In the smaller species the flagellum is short and composed of few segments. They have two pairs of book lungs on the ventral side of the second and third abdominal segments. At the base of the tail is a pair of large anal glands, which expel a liquid with the strong odor of acetic or formic acid when the animal is disturbed. This fluid acts as a repellent against potential enemies and serves also in catching insect prey. They do not have poisonous glands. Excretion is by coxal glands and malpighian tubules. Some have one pair of median and three pairs of lateral eyes; in others, median

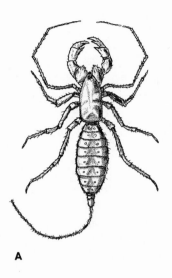

A

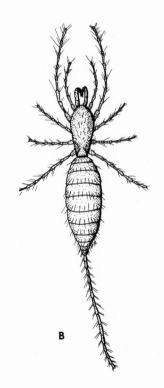

B

FIG. 13-17

A, American whip scorpion, *Mastigoproctus giganteus,* a uropygid. **B,** *Koenenia,* a palpigrade. (**B** after Hansen and Sörensen; redrawn from Snodgrass: Arthropod anatomy, Comstock Publishing Associates.)

eyes are absent and lateral eyes may or may not be present. The heart has five to nine pairs of ostia in the different species. A ganglionic mass is usually found in both the thoracic and abdominal regions.

Uropygids display an elaborate courtship pattern like that of scorpions. They are oviparous, and the female lays about 35 eggs in a subterranean brood chamber. The young remain with the mother until after the second molt.

They feed mostly on insects and are nocturnal in their habits. They prefer dark places and burrow in the ground, especially in damp humus soil. *Schizomus,* and *Thelyphonus,* in addition to *Mastigoproctus,* are well-known genera.

■ ORDER PALPIGRADI

Palpigrades are small, whitish, eyeless forms that never exceed 3 mm. in length (Fig. 13-17, *B*). They all belong to one family of about twenty species. They have been found in tropic and hot temperate regions in all the continents. A few species are present in Texas and California.

The body is made up of three divisions—prosoma, abdomen (opisthosoma), and flagellum. The carapace is divided into 3 plates (propeltidium, mesopeltidium, and metapeltidium). The chelicerae are three jointed and prominent. The pedipalps are used as walking legs, but the first pair of legs serves as sensory appendages and are vibrated to test the substratum, since these forms are eyeless.

The abdomen contains 11 segments, with transparent tergites and sternites. The last 3 segments are small, annular, and progressively reduced in size. The fifteen-jointed flagellum is provided with lateral setae. There are no book lungs. Some of the sternites bear eversible ventral sacs, which may serve in respiration. Excretion occurs by a pair of coxal glands. The nervous system contains an abdominal ganglionic mass in the second abdominal segment.

Palpigrades live mostly under stones, but in extreme drought they burrow deeply within the soil. Practically nothing is known about their feeding habits or their reproduction. Egg laying seems to occur at all seasons. *Koenenia* (Fig. 13-17, *B*) and *Prokoenenia* are two of the known genera.

■ ORDER ARANEAE

The order Araneae, which is made up of spiders, is one of the most extensive groups of arachnids (J. L. Cloudsley-Thompson, 1958; T. H. Savory, 1964). Spiders are found in most regions of the world and in almost every kind of habitat and represent one of the oldest of all groups. Their fossils, or evidences of their existence, date back to Devonian and Carboniferous times. Wherever found, their populations may reach enormous sizes in small areas. Because of their interesting habits and natural history, spiders have attracted the interest of biologists since man began to study animals. The number of species known to science is estimated between 28,000 and 50,000. In size they may be as small as 0.4 mm. or as large as 4 inches in length. They differ from other arachnids by having one or more of the following characteristics: chelicerae, usually with poison fangs; book lungs, tracheae, or both; spinnerets in all; pedipalps in the male as copulatory organs; and unsegmented abdomen in most.

General structure. The body of a spider is divided into a cephalothorax and an abdomen, which are connected by a slender pedicel. The head and thorax may be slightly separated by the cervical groove. The hard integument that

forms the dorsal wall of the cephalothorax is called the carapace. The body is covered with chitinous plates, which often bear accessory structures such as spines and tubercles.

The eyes, located near the front end of the head region, are all of the simple type, or ocelli. They are usually arranged in 2 rows of 4 each, although there may be fewer or none at all (Fig. 13-18).

Spiders have six pairs of cephalothoracic appendages. The first pair is called the chelicerae, or jaws, each of which consists of a basal piece and a slender distal fang. The duct of the poison gland opens at the tip of the fang. A few spiders have lost their poison glands. The second pair of appendages is the 6-segmented pedipalps. The pedipalps differ from the legs in having no metatarsus. The male pedipalps are modified in various ways as copulatory organs (Fig. 13-19). In male spiders the tarsus of the pedipalps contains a coiled tube, which is often found in a bulblike enlargement of the tarsus. In the female the pedipalp is used only as an organ of touch and manipulation. Each leg consists of 7 segments, and each of the four pairs of legs undergoes some variations in the different families. These differences of relative lengths, related sense organs, kinds and number of claws, etc. are of

FIG. 13-18

Front view of serveral spider carapaces showing chelicerae and eye arrangement. **A**, *Lycosa carolinensis.* **B**, *Araneus.* **C**, *Theridion.* **D**, *Argiope trifasciata.*

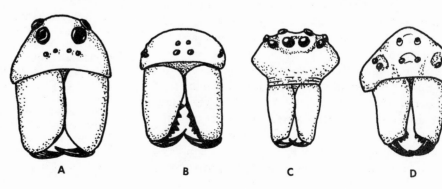

A   B   C   D

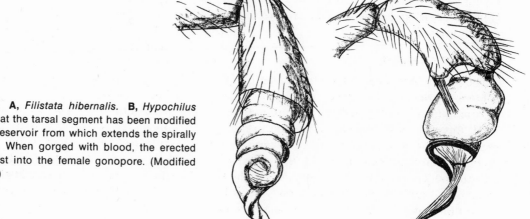

FIG. 13-19

Male pedipalps. **A,** *Filistata hibernalis.* **B,** *Hypochilus thorellii.* Note that the tarsal segment has been modified into a bulblike reservoir from which extends the spirally coiled embolus. When gorged with blood, the erected embolus is thrust into the female gonopore. (Modified from Comstock.)

adaptive significance, as well as of taxonomic importance.

The large saclike abdomen is usually without a trace of external segmentation, except in a few forms that have up to 12 segments. Many variations are found in the shape as well as size of the abdomen. The narrow pedicel that connects the abdomen to the thorax is usually concealed when viewed from above. On the ventral side of the abdomen the anterior, often densely chitinized, portion is called the epigastrum, and the crescent-shaped furrow that separates the epigastrum from the caudal part of the abdomen is known as the epigastric furrow. Openings of the reproductive organs are situated in the middle of this furrow, and a spiracle at each end of the furrow opens to a book lung.

The spinnerets are modified appendages of the fourth and fifth abdominal embryonic segments (Fig. 13-23 and 13-24, *A* and *B*). They are found on the ventral surface anterior to the anus. There are usually 6 spinnerets, but the number varies from 4 to 8; 8 is the primitive number. Each spinneret is a conical projection,

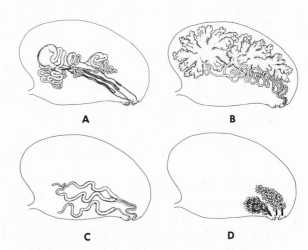

FIG. 13-20

Location of the various silk glands in the female abdomen of *Araneus sericatus.* **A,** Anterior, median, and posterior ampullate glands; they provide silk for dragline, frame threads, and some parts of orb web. **B,** Anterior and posterior aggregate glands produce viscid part of sticky spiral. **C,** Cylindric glands may produce silk for egg cocoon. **D,** Aciniform glands on median and posterior spinnerets and ventrally located glands on anterior spinnerets. Aciniform glands are source of swathing silk and fine silken threads of accessory frame and some others. (Courtesy Mullen, G. R. 1969. Trans. Amer. Microscop. Soc. **88:**232-240.)

FIG. 13-21

Web of orb-weaving spider. Starting with "bridge thread," framework of boundary threads is made of dry silk, usually an irregular four- or five-sided figure. Radii of dry silk are spun next, then functional spiral of sticky viscid silk. Spider working inward from rim to hub is careful to walk only on nonsticky radii. (From Hickman: Integrated principles of zoology, The C. V. Mosby Co.)

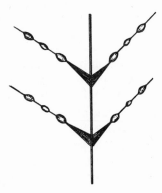

FIG. 13-22

Enlarged view of small portion of a nonviscid, inelastic radius and 2 viscid spiral threads of spider web.

enclosing many duct openings at the ends of the spinning tubes. Physically different kinds of silk are spun out of different types of glands within the abdomen. The silk of spiders is a scleroprotein of a fine, elastic, and strong character. Four amino acids (glycine, alanine, serine, and tyrosine) make up most of its molecule. Spiders spin several kinds of silk, each with special qualities. G. R. Mullen (1969) describes eight distinct types of glands in *Araneus sericatus* (orb weaver). These glands (and the chief kind of silk produced by them) are subdivided into four main groups, as shown in Fig. 13-20. No one spider spins all the kinds recognized, but most species spin at least four or five types. The silk is produced as a fluid, but hardens under the pressure of being drawn out as well as exposure to the air. The complex web of the orb-weaving spiders (Fig. 13-21) is made up first of a rectangular or irregular framework of silk, within which radial threads are spaced like spokes in a wheel from the center of the net. These threads are nonviscid. The spiral viscid threads are fastened in concentric patterns to these radiating threads (Fig. 13-22), although some of the spiral threads may be nonviscid. The spider carefully does its traveling on the nonviscid threads. In certain spiders a sieve-plate, the cribellum, lies anterior to the spinnerets; from this plate a banded silk is extruded. See E. L. Palmer (1961) for various types of spider webs.

B. J. Kaston (1964) has given an account of the evolutionary origin and method of construction of the spider's web. The web may have evolved from a mass of threads that covered the eggs or from a tube structure of silk in which the spider concealed its eggs. This may have led to the trap-door nest or the aerial net for catching prey. Some authorities consider that the criss-crossing of the dragline may have been the origin of the web as the spider moved about. Selective advantages may have led in time to the highly developed types used by the orb weavers. When constructing its net, the spider fastens the objects together between

483

which the web is to be formed. After testing the thread to determine when it has stuck on the object where it has been carried by wind currents, the spider draws the thread taut and carries across it a heavier thread, which becomes the bridge line. This new line may adhere to the original line, or the original line may be eaten as the heavier line is laid down. The spider next spins the foundation lines, which, together with the bridge line, form the general outline and plane of the web-to-be. The radial lines are now put in, running from the periphery to the hub like the spokes of a wheel. Just outside of the hub there is a spiral thread that connects the radii and holds them in place. In this way an attachment zone is laid down. The spider now spins a nonviscid spiral thread, starting at the attachment zone and proceeding with spiral turns to the periphery. The function of this scaffold spiral is to hold the radii in place while the web is being finished. The viscid thread is now placed in position beginning at the periphery and proceeding to the center attachment zone. As the spiral viscid threads are being laid down the scaffold spiral is cut away in advance. The hub threads may now be cut out, or the hub may be covered on both sides to form the stabilimentum. There are some modifications of this plan of structure among the spiders that form webs.

Spiders use silk in many ways. Although all spiders have spinnerets, not all of them make webs for ensnaring prey. Most of them use silk for other purposes, such as draglines for marking their course, sperm webs by the male before copulation, gossamer threads for ballooning, lining for burrows, hinges for trapdoors, cocoons, or egg sacs, attachment disks, swathing bands for binding prey, and many others. All spiders form draglines as they run about.

The alimentary canal (Fig. 13-23) exhibits the typical arthropod plan of foregut, midgut, and hindgut. The foregut is a stomodeum consisting of the pharynx, esophagus, and sucking stomach. The mouth cavity is between the bases of the pedipalps, the coxae of which form

FIG. 13-23

Internal structure (diagrammatic) of female 2-lunged spider. (Modified from Comstock.)

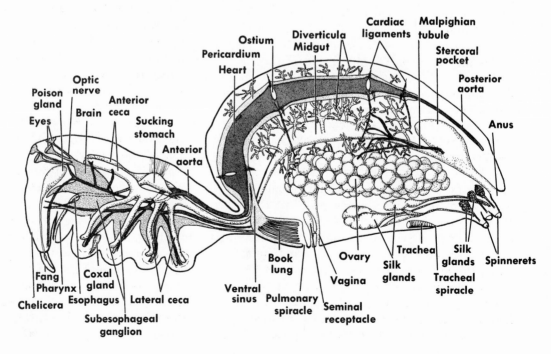

the sides of the cavity. A liplike structure, the so-called rostrum, lies at the top of the mouth region; and a lower lip, the labium, is at the bottom. The wall of the sucking stomach is strengthened by chitin as a firm support for the attachment of the powerful muscles that extend to the inner surface of the carapace and to the endosternite. Sphincter muscles are also found close to the stomach. The midgut (mesenteron), with its diverticula, or ceca, constitutes a large

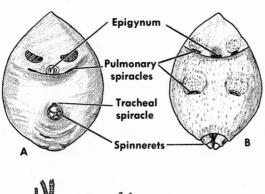

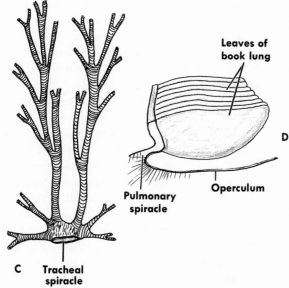

FIG. 13-24

Respiration in spiders. **A,** Abdomen of *Argiope,* with one pair of book lungs and one pair of tubular tracheae. **B,** Abdomen of tarantula, with two pairs of book lungs. **C,** Pair of tubular tracheae, with single spiracle. **D,** Vertical section of book lung. (From several sources.)

part of the abdomen and thorax. A bladderlike diverticulum, the stercoral pocket, is found on the dorsal side of the hindgut and serves as a reservoir for fecal material.

Spiders are carnivorous. Their food consists mostly of insects and other arthropods. Some of the larger members are also known to feed on birds, small fishes, and other vertebrates. The mouth lacks mandibles. When the prey is caught (by web or otherwise), it is bitten and in some spiders is macerated by the chelicerae. Hairs of the labium and endites of the pedipalps act as strainers for coarse particles. After some predigestion of food in the mouth, the pharynx and sucking stomach draw the liquefied food into the gut. Although voracious, many spiders can endure long fasts.

A much-branched malpighian tubule opening into the midgut on each side and a pair of reduced coxal glands serve for excretion (Fig. 13-23). Some primitive spiders have two pairs of coxal glands. The wall of the gut may also be involved in excretion.

Respiration takes place by book lungs and tracheal tubules (Figs. 13-23 and 13-24). Some spiders have only one of these types, but most have both. Usually there is only one pair of book lungs, but there are two pairs in tarantulas. The openings of the book lungs are slitlike clefts on the ventral side of the abdomen. In some spiders the anterior book lungs also may be transformed into tracheae. There is usually only a single tracheal spiracle, which is placed, in most cases, just in front of the spinnerets. There are two pairs of tracheal tubules, which are highly variable in their distribution and branching in the different species.

The circulatory system of spiders (Fig. 13-23) is an open one, for the blood courses partly in the hemocoel. The heart is in the dorsal part of the abdomen above the gut and has two to five pairs of ostia. The primitive condition appears to be five pairs, but tracheate spiders have two pairs exclusively. The heart is a simple tube that is not divided into chambers. Anterior and posterior aortae, as well as lateral abdom-

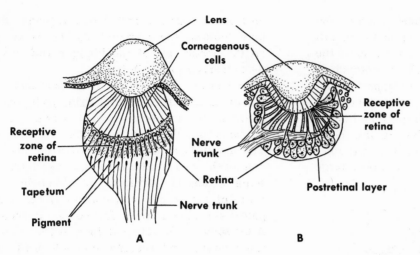

FIG. 13-25

A, Structure of indirect eye *(Lycosa agricola).* **B,** Structure of direct eye *(Tegenaria derhami).* (Modified from Widmann, 1908.)

inal arteries, extend from the heart. The heart is surrounded by a thin-walled sac, the pericardium. Cardiac ligaments hold the heart in position. Blood is carried from the book lungs to the pericardium by "pulmonary veins," or ventral sinuses, which are really hollow ligaments.

In most spiders the central nervous system is concentrated in the cephalothorax (Fig. 13-23). The ganglia in general are consolidated into a mass around the esophagus. The brain is the dorsal part of this mass, and the subesophageal ganglion is the part below. A visceral nervous system of scattered ganglia is connected with a median single nerve to the pharynx or paired nerves to the brain. Of the 8 eyes most spiders possess, the anterior median eyes are of the direct type, whereas the others have an indirect retina (Fig. 13-25). Indirect eyes have a tapetum for reflecting the light rays. The direct eyes may be provided with muscles. The eyes of hunting and jumping spiders are better developed for the formation of sharp images than those of most spiders.

Spiders differ sexually from most other animals in having a male copulatory apparatus separate from the reproductive system. The reproductive system of each sex is located in the abdomen (Fig. 13-23). The female has 2 elongated ovaries, a pair of oviducts, a uterus, and a vagina (Fig. 13-26). In addition, there are usually 2 pouches (seminal receptacles) that are associated with the vagina. In most spiders the seminal receptacles open by little tubes into the vagina and also to the outside by external openings so that the female has 3 external openings—the opening of the vagina and the 2 openings of the seminal receptacles. The ovaries are tubelike and bear ovarian follicles like bunches of grapes, especially during the breeding season when the ovaries are greatly distended. Ripe eggs rupture into the lumina of the ovaries and are carried by the oviducts to the common uterus and vagina. The vagina and 2 seminal receptacles open externally on a chitinized plate, the epigynum, which is found in front of the epigastric furrow on the midventral line of the body. Sperm from the male is stored in the seminal receptacles until the time of egg laying, at which time the eggs are fertilized.

In the male the reproductive organs consist of a pair of testes, a pair of sperm ducts (vasa deferentia), and a common ejaculatory duct, or seminal vesicle, which opens to the exterior at the same place on the body as the genital open-

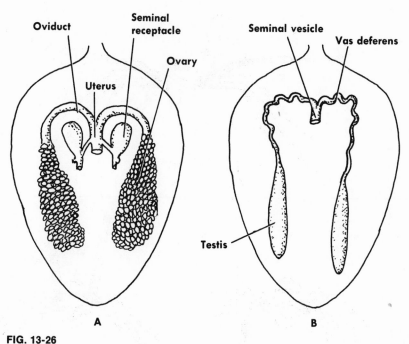

FIG. 13-26

Reproductive organs of spider. **A,** Female. **B,** Male.

ings of the female (Fig. 13-26). The male copulatory organ, as mentioned, is not connected with the reproductive system; the seminal fluid is transferred to the female by means of the specialized male pedipalps (Fig. 13-20). The tarsus of each pedipalp is modified in various ways in different spiders. One of the simplest types of copulatory organ is a bulb with a curved spout or embolus. Inside the bulb is a reservoir for the sperm, with a tube that opens at the end of the embolus. To fill its reservoir, the male first ejects a drop of its semen into a silk web, and then transfers it into the opening of the embolus. During the mating process the copulatory apparatus is distended by blood engorgement, and the male inserts the embolus into one of the seminal receptacles. That the male copulatory organ is fitted only for the epigynum of its species— the "lock-and-key" theory—is held by some authorities. Since the smaller male is often eaten by the female, the process of mating may be quite dangerous to the male so that elaborate

courtship patterns have evolved. Courtship patterns between the male and female vary with different spiders (Fig. 13-27).

Some species lay relatively few eggs, and others lay large numbers. Some females deposit all their eggs at one time, and others lay their eggs in several batches, each enclosed in an egg sac. The egg sac, or cocoon, is more or less peculiar to each species and may in some even serve to identify the species. Some females guard the cocoons in their retreats or burrows; others carry the sacs along with them; and others have nothing to do with the cocoon after it is formed. The young spiders hatch in the sac, where they stay until after the first molt. Various types of maternal care are found among spiders, from those that take no care to those that stay with the young for a long time and some who even carry their young on their backs.

There is no metamorphosis during growth, and the young differ from the parents mainly in size. Growth occurs by a series of molts, highly variable in number among the various species. At the final molt the sex organs are

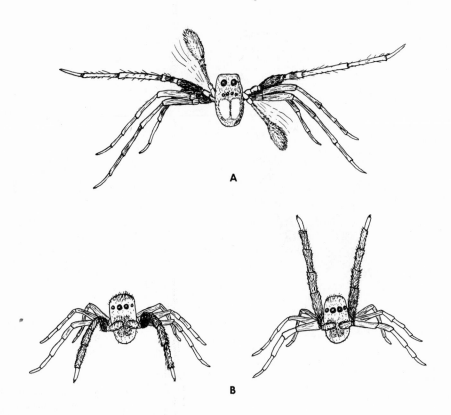

FIG. 13-27

Courtship display in some spiders. **A,** *Lycosa* stretches out his front legs, pauses, and raises first one pedipalp and then the other in a semaphore manner to attract female. **B,** *Euophrys* swings front legs up and down to attract attention of female. Such courtship displays may be necessary in precise fertilization involving use of spermatophores for transfer of sperm by male. Not all arthropods have courtship rituals; some mate directly. (Modified from W. S. Bristowe.)

mature. In a few of the larger, long-lived species the females may even molt after sexual maturity. Regeneration of lost parts is possible only if there are still molts to undergo. Some spiders live for only a single season, but others (some of those kept in captivity) live for many years.

With a few exceptions, the poison glands of all spiders are constructed on the same plan (Fig. 13-23). The gland is located either in the basal segment of the chelicera or else in the cephalothorax proper, with its duct opening at or near the tip of the fang. The flow of poison is voluntarily controlled by the spider. All spiders may use their poison in defense or when trying to escape. Only a few spiders are very dangerous to man. The black widow (*Latrodectus mactans*) is perhaps our most dangerous one. The poison seems to be neurotoxic in its action, bringing on severe pain, high pressure of the cerebrospinal fluid, muscular spasms, and finally respiratory paralysis. The venom of *Lycosa raptoria*, a Brazilian wolf spider, and of *Loxosceles reclusa* in our own southern states is hemolytic rather than neurotoxic, producing a necrosis of tissue around the bite. The large tarantulas are mainly inoffensive.

Students should read B. J. Kaston's (1965) paper on some of the little-known (and some unique) aspects of spider behavior.

Some characteristic families of spiders. Authorities are far from agreeing on the way spiders should be classified. Each revision of the order Araneae usually shows a marked tendency to increase the number of families. Most of the sixty-odd families are found in two suborders – Mygalomorphae and Araneomorphae.

Suborder Mygalomorphae

Suborder Mygalomorphae includes about eight families, to which the trap-door spiders, tarantulas, and purse-web spiders belong. The members of this suborder have paraxial chelicerae, i.e., with the fang movable in the anterior-posterior plane of the body (Fig. 13-28, *B*), two pairs of book lungs (Fig. 13-24, *B*), and usually 4 spinnerets.

FAMILY CTENIZIDAE – TRAP-DOOR SPIDERS

These spiders make burrows in the ground, line the burrow with silk, and close the entrance with a hinge lid (Fig. 13-29). Lid and entrance are artfully concealed by vegetation. Their chelicerae are adapted for digging. They are found in the southern and western United States. *Bothriocyrtum californicum* (Fig. 13-30) is a common species in California.

FAMILY THERAPHOSIDAE (AVICULARIIDAE) – TARANTULAS

These hairy spiders of large size are common in the western and southwestern states. Although venomous, their bites are relatively harmless to man. The tarantulas in America (Fig. 13-31) should not be mistaken for what is called a tarantula, actually a wolf spider, in southern Europe. Tarantulas have been known to live more than 20 years. Some species live in shallow holes, which they dig with their fangs and pedipalps. They are often associated in groups of a dozen or more. About thirty species are found in the United States.

Suborder Araneomorphae

Most of the common spiders in the United States belong to suborder Araneomorphae. About forty-nine families are included in this group. These spiders have diaxial chelicerae,

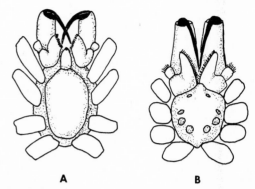

FIG. 13-28

Types of chelicerae. **A,** Diaxial, *Dysdera crocata.* **B,** Paraxial, *Atypus milberti.* (Modified from Kaston.)

FIG. 13-29

Entrance to burrow of trap-door spider. (From Hickman: Integrated principles of zoology, The C. V. Mosby Co.)

FIG. 13-30

Bothriocyrtum californicum, the common trap-door spider of California. (Redrawn from McCook: American spiders and their spinning work, Academy of Natural Science.)

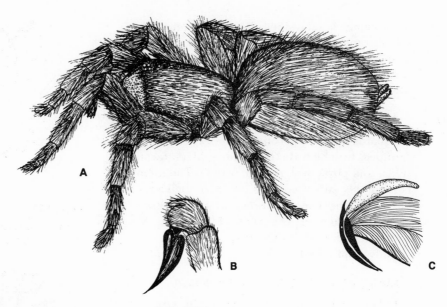

FIG. 13-31

A, *Dugesiella hentzi,* the tarantula common in our western and southwestern states. **B,** Male pedipalp of tarantula. **C,** Fang of tarantula with poison gland and muscles. (**A** redrawn from Kaston and Kaston: How to know the spiders, William C. Brown Co., Publishers; **B** and **C** modified from Baerg.)

i.e., articulating at right angles to the long axis of the body (Fig. 13-28, *A*); one pair of book lungs; usually one pair of tracheal spiracles; and 6 spinnerts (some with cribellum).

FAMILY ARGIOPIDAE (ARANEIDAE)—ORB WEAVERS

Most members of this family form circular or orb webs, often of beautiful geometric design (Fig. 13-21). The plane of the web may be vertical, horizontal, or at any angle. The frame and radii are constructed of nonviscid threads, whereas the main spiral threads are viscid (Fig. 13-22). Many of the species have bright patterns, such as the common garden spider, *Argiope aurantia,* which has yellow, reddish, and black colors (Fig. 13-32).

FAMILY AGELENIDAE—FUNNEL-WEB, OR GRASS, SPIDERS

These spiders form a web that has a tube or funnel with a platform on which the insect is seized. When the prey is caught, the spider retreats with it into the funnel. The common genus *Agelenopsis* (Fig. 13-33) can be recognized by the long hind spinnerets that extend posteriorly. In addition to the horizontal sheet or platform, the spider spins a type of labyrinth

FIG. 13-32

Argiope aurantia, the black and yellow garden spider. Legs not shown on left side.

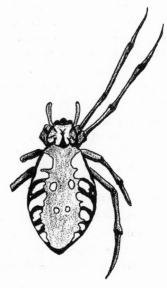

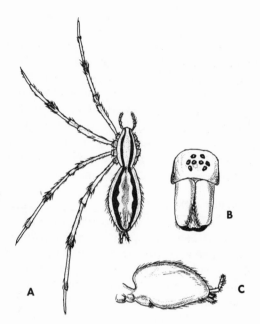

FIG. 13-33

Agelenopsis, the grass spider. **A,** Dorsal view. **B,** Front view of carapace. **C,** Abdomen, side view, showing spinnerets.

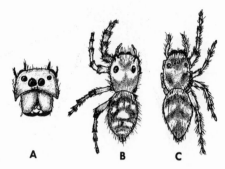

FIG. 13-34

Jumping spiders (family Salticidae). **A,** Front view of carapace of *Phidippus*. **B,** Dorsal view of *Phidippus*. **C,** Dorsal view of *Salticus*.

or stopping maze above the funnel for hindering the flight of insects and causing them to fall on the platform. The spider apparently catches the insect by its own rapid and deft movements, since the web is not composed of viscid threads.

FAMILY SALTICIDAE (ATTIDAE) – JUMPING SPIDERS

These spiders (Fig. 13-34) build no snare, but by their keen eyesight they stalk their prey and suddenly spring upon it. Two of their 8 eyes are very large and situated on the front of the head. They also spin retreats for molting, hiding, and other purposes. Their front legs are often unusually heavy and are used in pinning down the prey. There are many species in the family, especially in the southern climates. In the tropics many of them are brightly colored.

FAMILY PISAURIDAE – FISHING SPIDERS

These spiders do not spin webs for ensnaring but catch their prey by hunting. The genus *Dolomedes* (Fig. 13-35) lives near the water

and is known to catch small minnows as well as aquatic insects. The members of this family can usually be seen around ponds or else running over aquatic vegetation.

FAMILY LYCOSIDAE – WOLF SPIDERS

These common spiders are often found running through the grass of pastures or among the leaves of the forest floor. Wolf spiders (Fig. 13-36) are usually stoutly built, with eyes that are unequal in size (Fig. 13-18, *A*). They can be spotted at night with a flashlight because their light-colored eye tapetum reflects some of the light to the observer. The female carries her globular egg sac attached to the spinnerets. Some species (burrowing wolf spiders) build burrows. Three common genera are *Lycosa, Pirata,* and *Pardosa.*

FAMILY THERIDIIDAE

These spiders build irregular snares and suspend themselves upside down while awaiting their prey. The family includes one of the most

491

FIG. 13-35

Dolomedes tenebrosus (family Pisauridae).

poisonous members of the spider group—the black widow (*Latrodectus*, Fig. 13-37). The black widow has a globose, shiny black abdomen, with an hourglass-shaped reddish mark on the underside. Variations of abdominal markings are common. Although far more prevalent in the southern sections of the United States, this species has been found in every state and in most Canadian provinces as well.

FAMILY LOXOSCELIDAE

Some species of this family are very venomous. One of these species is the brown recluse (*Loxosceles reclusus*), which may exceed the black widow in this respect. *L. reclusus* is somewhat smaller than the black widow and is light fawn to darkish brown in color. It bears

FIG. 13-36

Huntsman spider (family Lycosidae). They are also sometimes called wolf spiders.

FIG. 13-37

Black widow, *Latrodectus* (family Theridiidae). Note "hourglass" on her abdoman. (From Hickman: Integrated principles of zoology, The C. V. Mosby Co.)

a fiddle-shaped dorsal stripe of darker brown in the forepart of the cephalothorax. It occurs in the eastern part of the United States, and its favorite lurking places are crevices and other protected regions.

■ ORDER SOLPUGIDA (SOLIFUGAE)

The members of order Solpugida are often called sun spiders and occur mostly in dry tropic and warm temperate regions. Most solpugids are from 10 to 50 mm. in size, although some may be larger. About sixty of the 200 known species are found in the southwestern states.

The most striking feature of these arachnids is the enormous chelicerae, which do not possess poison glands (Figs. 13-38 and 13-39, C). The chelicerae are two jointed, with the pincers articulating vertically. The pedipalps are simple, leglike, nonchelate, and terminate in an adhesive organ (Fig. 13-38). These appendages seize the prey and pass it to the chelicerae, which hold and crush the prey.

The carapace of the prosoma is made up of 3 plates, which represent the terga of the first 6 somites; the anterior of the 3 plates is the largest. The carapace bears anteriorly one pair of median eyes and obsolete lateral eyes. The first pair of legs is tactile, and the other three pairs are ambulatory. The abdomen is made up of 10 or 11 segments, which are plainly visible. The animal has no book lungs; but there are 7 ventral, slitlike spiracles, which open into a well-developed tracheal system. The paired spiracles are located between the second and third coxae and on the third and fourth abdominal segments. A single median spiracle may occur on the fifth segment. Excretion takes place by a pair of coxal glands in the prosoma and by a pair of malpighian tubules. A heart with eight pairs of ostia is placed partly in the cephalothorax and partly in the abdomen. The nervous system consists of both a thoracic and an abdominal ganglionic mass. On the last pair of legs are the characteristic racquet organs. These are T-shaped sense organs of which there are five on the ventral side of each leg (Fig. 13-39, *A* and *B*). The solpugids are among the most agile of all arachnids and are mostly nocturnal in their habits.

After a courtship ritual, in which the female is thrown into a trancelike condition, the male transfers his sperm into her genital orifice. The female lays her eggs in burrows, where she remains until after they are hatched.

Eremobates (Fig. 13-38) in the southwestern states and *Galeodes* (Fig. 13-39, *C*), an Old World form, are common genera.

■ ORDER PSEUDOSCORPIONIDA

The Pseudoscorpionida derive their name from their resemblance to scorpions; however, they have no postabdomen or sting (Fig. 13-40). Most of them are between 4 and 10 mm. in length. They live in forest debris, leaf mold, beneath bark, under stones, in barns, among books, and in nests of birds, mammals, and insects. One genus occurs on marine algae. Their food is chiefly small arthropods such as mites and small insects. Fossil pseudoscorpions similar to present-day forms have been found in Baltic amber. The form of their appendages may indicate a relationship to true scorpions, but the presence of a pregenital segment (which scorpions lack) indicates an ancestor common to both groups. About 800 to 1,000 species have been described.

The prosoma is covered by a single dorsal shield or carapace and contains 6 fused segments. One or two pairs of eyes are present, although some species are blind. The chelicerae are two jointed and chelated, and each finger of the pincers bears a serrula of plates. Near the end of the movable finger, ducts of silk glands lead to a spinneret, or galea, located in

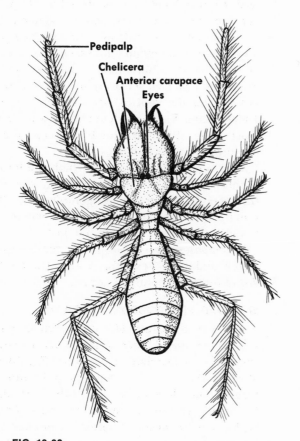

FIG. 13-38

Eremobates cinerea, a solpugid. (After Putnam; redrawn from Comstock: The spider book, Doubleday & Co. Inc.)

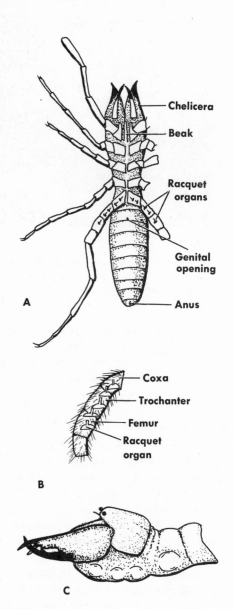

FIG. 13-39

A, Ventral view of *Eremobates formicaria* (order Solpugida). **B,** Portion of fourth leg of *Eremobates* showing racquet organs. **C,** Cephalothorax of *Galeodes,* side view, showing massive chelicerae. (**A** modified from Petrunkevitch.)

the prosoma. The pedipalps are the largest and most conspicuous of all the appendages. They are chelated, with a poison gland opening by a duct on one or both fingers. All the coxae meet in the median line. The four pairs of walking legs are all similar; each leg has 2 claws, between which is an adhesive pad. The legs have an additional joint (trochantia) between the trochanter and femur.

The abdomen is composed of 11 or 12 visible segments and is rounded at the terminal end. There are two pairs of tracheal spiracles on the ventral side of the third and fourth abdominal segments. Book lungs are lacking. Excretion takes place by a pair of coxal glands that open on the coxae of the third pair of walking legs. The heart has two or three pairs of ostia. The nervous system is concentrated in the cephalothorax.

The opening of the reproductive organs is between the second and third abdominal sterna. The gonads are single in both sexes. In the males, near the genital aperture, are genital sacs, which in some forms can be extended (the

FIG. 13-40

Pseudoscorpion. (From Hickman: Integrated principles of zoology, The C. V. Mosby Co.)

"ram's horn organs") and may be used to excite the female. In mating, the male deposits a spermatophore on the ground after an elaborate courtship display. The female is induced by the male to pick up the sperm with her genital orifice. The eggs and larval forms are carried on the underside of the female abdomen. Maturity is reached in about a year, after several molts.

The silk that is spun by the chelicerae is used to make a type of cocoon in which the pseudoscorpion molts and hibernates. The serrulae of the chelicerae are used in forming the cocoon.

Pseudoscorpions eat their prey as most arthropods do—by sucking the juices and predigested tissues into the mouth and esophagus after the chelicerae have torn open the exoskeleton and macerated the tissues. Among the typical genera are *Chthonius* (in caves) and *Chelifer* (in books and households).

■ ORDER RICINULEI (PODOGONA)

Order Ricinulei is a small arachnid group of some 20 species. It has been found in tropic Africa and in the New World from Texas to the Amazon River. Their fossil records indicate that they were more common and more widespread formerly than now. Their size range is around 4 to 10 mm. in length, and they resemble ticks in general appearance (Fig. 13-41).

The prosoma is covered by a shieldlike carapace in two divisions of different sizes, the smaller one in front. The anterior part articulates with the posterior one and can be raised and folded over the mouth and chelicerae like a hood. Both the chelicerae and pedipalps are chelate. The basal segments of the pedipalps are united in the middle line below the mouth. There are no eyes. There is no visible segmentation in the cephalothorax. The walking legs of the female are similar, except that the first pair is shorter and the second pair is longer than the others. The number of joints is variable, and their tarsi have 2 claws. In the male the third pair of legs is modified into

FIG. 13-41

Cryptocellus, male (order Ricinulei). (After Hansen and Sörensen; modified from Cambridge Natural History.)

FIG. 13-42

Member of order Phalangida, commonly called harvestmen, or daddy longlegs. **A** and **B,** Dorsal views. **C,** Ventral view.

peculiar copulatory organs that may be used in the transfer of spermatophores to the female. The cephalothorax at the posterior end also bears the tracheal tubes, which open by spiracles above the third coxae. There are no book lungs.

The abdomen is connected by a movable, membranous joint to the prosoma, although the narrow pedicel, which connects the two divisions, is not apparent; only a shallow groove appears to separate the two, as seen dorsally. The segments of the abdomen are fused.

The reproductive organs open between the first and second abdominal segments, but little is known about the copulation and development of ricinuleids except that they are oviparous and have a 6-legged larva at one stage. Their natural habitat is not known to any extent. Some have been collected from leaf molds and such places.

■ ORDER PHALANGIDA (OPILIONES)

The members of order Phalangida include the daddy longlegs, or harvestmen (Fig. 13-42). They range in length from about 1 to 15 mm. This order has a worldwide distribution but is most abundant in moist tropic regions. There are more than 3,000 species. The cephalothorax and abdomen are joined together without a furrow between them. The carapace is a single

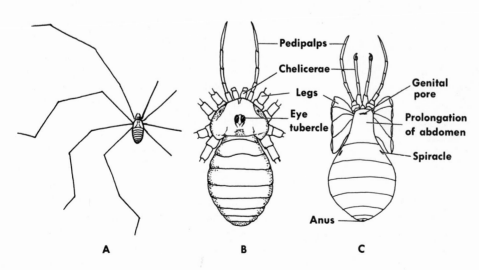

A   B   C

shield with a median tubercle, on each side of which is an eye. The stout chelicerae have 3 joints and are chelated, and in some forms they are comparatively long. The pedipalps are leglike but shorter and are also provided in some species with pincers. The legs are long and slender, which is their most distinctive feature. When lost, legs are not regenerated; some may be active with fewer than 8. In walking, the body appears to be suspended from the long arched legs. The legs are made up of 7 segments, but the tarsus in some forms is subdivided. The coxae of the pedipalps and first and second pairs of legs each bear an anterior process (endite). The abdomen appears to consist of about 10 segments, only some of which are visible. A single pair of tracheal tubes opens on the second abdominal sternite. Secondary spiracles are found on the legs of certain species. Book lungs are lacking.

Near the anterior margin of the cephalothorax close to the attachment of the first pair of legs are the openings to a pair of scent glands. They produce a material with a strong, pungent odor. Excretion is by means of a pair of coxal glands, which open on each side between the third and fourth coxae. The heart is provided with two pairs of ostia, and the nervous system is highly concentrated in the cephalothorax. Sense organs are scattered over the body and appendages; the 2 eyes are set in little turrets on top of the body.

The reproductive organs open in the region between the cephalothorax and the abdomen. The genital openings are on a plate (operculum), which in some cases covers the genital orifices. In the female there is an ovipositor, a tubular organ that is retracted into a sheath or sac when not in use. The ovipositor can be extended a long distance when laying eggs. The male is provided with a large penis, or large chitinous duct, which can be injected between the female chelicerae into the genital orifice at copulation. Unlike most arachnids, they have no form of courtship. Eggs are laid in the ground or in crevices of wood by the ovi-

positor. Eggs laid in the fall do not hatch until the following spring. After the first molt the young are white, but they gradually acquire the color of the adult.

The phalangids have a varied diet of soft-bodied insects, both living and dead, and will also eat vegetable matter. Plant lice seem to be one of their favorite foods. Food is seized by the pedipalps and passed to the chelicerae, which macerate it. In addition to fluid, the phalangids eat some coarse particles.

In temperate climates most harvestmen die in the autumn, although in warmer regions they undergo hibernation. They often get into human habitations and are common in barns and outbuildings. They are most active at twilight and usually remain in hiding during the day. They cannot endure deprivation of water and food for long periods of time as many other arachnids are able to do. Familiar genera include *Phalangium* and *Leiobunum*. T. H. Savory (1962) has described many features of the behavior of phalangids.

■ ORDER ACARINA

The diverse order Acarina includes ticks and mites; mites are the largest in number. There are so many species of them, of which only a fraction (about 20,000 species) has been described, that it is impossible to estimate the number even approximately. They exist in many morphologic patterns. In some ways they are the most important economic group of arachnids because of their parasitic habits, their destruction of crops useful to man, and their role as vectors of disease germs. They have a worldwide distribution. Several hundred species of Acarina have been described in the United States, but their taxonomy is a difficult one. Acarina vary in size from less than 1 to 30 mm. in length, with ticks as the largest members of the group.

Since the body divisions do not correspond to those of the other arachnids, different names have been assigned to them. The body is commonly divided into an anterior gnathosoma

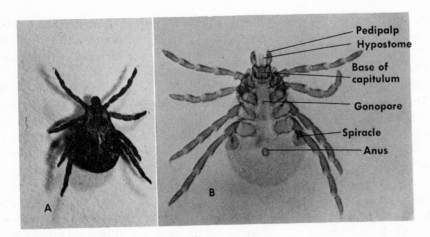

FIG. 13-43

Wood tick *Dermacentor* (order Acarina). **A,** Dorsal view. **B,** Ventral view.

and a posterior idiosoma. The gnathosoma is made up of the first 2 segments, which include the mouth region and its parts, and the idiosoma includes the rest of the body. The so-called abdomen of mites includes the typical arachnid abdomen, plus the last 2 thoracic segments. These two divisions are often imperceptibly fused together, producing a more or less saclike globular body. Many mites have transverse, fine lines that give a false impression of segmention.

The head region with the mouthparts is called the capitulum. A dorsal projection of the body wall forms a rostrum (tectum), or beak. The basal segments of the pedipalps may be fused together to form a lip or labium (hypostome), with a labrum above (Fig. 13-43). With other structures, these processes may form an oral tube, through which the chelicerae are projected. The chelicerae usually consist of 2 segments and may be chelate. However, these organs are variable. Many are needlelike for piercing. The pedipalps consist of 5 segments. They vary according to function. They may be simply tactile organs, or they may be large and chelate for predation. The four pairs of walking legs are each made up of 6 segments, with the

tarsus bearing a pair of claws. The legs may be modified in various ways, depending on the kind of mites and their habits.

Most mites breathe by tracheal tubes, but some lack tracheae and breathe through the general body surface. The spiracles (one to four pairs) are located near the base of the appendages in the anterior half of the body. The excretory organs consist of coxal glands, malpighian tubules, or both. Parts of the gut also serve for excretion. The circulation is much simplified, usually consisting of a network of sinuses without a definite heart. Most organs found on the cephalothorax of other arachnids are found on the idiosoma of mites. Many mites are blind, but when eyes occur, they are usually present in the idiosoma. Pits and slit sense organs are also present.

The male reproductive system consists of a pair of testes, a pair of vasa deferentia, and a chitinous penis. The penis can project through the genital orifice, which is variable in position. The female system usually has only 1 ovary that is connected to the genital orifice by an oviduct. The genital orifice is usually ventral but may be dorsal in some. Seminal receptacles are also present for the reception of sperm. The male usually deposits his sperm in a spermatophore, which is picked up by the female, or the sperm may be transferred by the male chelicerae or by the third pair of legs. In other species the male injects his penis into the female genital orifice.

Most species are oviparous but some are ovoviviparous. The female of some groups has an ovipositor. A 6-legged larva hatches from the egg; the fourth pair of legs is absent. After a molt all appendages are present. There may be a number of nymphal molts in passing through the nymph stage. After a resting period the nymph is gradually transformed into an adult. There are many variations in their metamorphosis.

Because of their small size and habitat preferences, mites and ticks are rarely considered important parts of faunal populations. However, they are abundant in the soil, humus, decaying vegetation, and other places. One group, often red in color, lives in water (mostly fresh water), but some groups are marine. Such forms may be parasitic on the gills of mollusks, or they may live on small aquatic organisms. Ticks are the largest members of the order and are parasitic on the higher vertebrates. Some ticks are responsible for transmitting serious diseases, such as Texas tick fever and Rocky Mountain spotted fever. The itch mites burrow into the skin of man and other animals, causing mange or scabies. Predaceous mites usually have typical arachnid digestive systems and ingest the juices of their prey, such as other small arthropods. Many live on plant juices. One group with only 4 legs, the gall mites, produces galls on leaves of plants. The 6-legged larvae of the family Trombidiidae, commonly called red legs or chiggers, cause severe dermatitis in man by burrowing under the skin.

REFERENCES

Arthur, D. R. 1962. Ticks and disease. New York, Harper & Row, Publishers. The role of ticks as vectors of disease.

Baerg, W. J. 1958. The tarantula. Lawrence, University of Kansas Press. A monograph on the habits and natural history of the tarantula.

Baker, E. W., and G. W. Wharton. 1952. An introduction to acarology. New York, The Macmillan Co. Stresses taxonomy of the group with keys to the families.

Brady, A. R. 1964. The lynx spiders of North America, north of Mexico (Araneae:Oxyopidae). Bull. Mus. Comp. Zool.

131(13):432-518. A systematic study of this active family of hunting spiders that fasten their egg sacs to shrubbery or other objects.

Bristowe, W. S. 1958. The world of spiders. New York, The Macmillan Co. A comprehensive treatise on this extensive group.

Bullock, T. H., and G. A. Horridge. 1965. Structure and function in the nervous system of invertebrates. San Francisco, W. H. Freeman & Co., Publishers. Certain aspects of the nervous system of arachnids are given in this comprehensive summary.

Butt, F. H. 1960. Head development in arthropods. Biol. Rev. 35:43-91.

Carthy, J. D. 1965. The behavior of arthropods. San Francisco, W. H. Freeman & Co., Publishers.

Cloudsley-Thompson, J. L. 1958. Spiders, scorpions, centipedes, and mites. New York, Pergamon Press, Inc. The natural history and ecology of the group.

Curtis, H. 1965. Spirals, spiders, and spinnerets. Amer. Sci. 53:52-58.

Edmondson, W. T. (editor). 1959. Ward and Whipple's freshwater biology, ed. 2. New York, John Wiley & Sons, Inc. A section is devoted to the aquatic mites (Acari) by I. M. Newell. Keys are given of this extensive group.

Gertsch, W. J. 1949. American spiders. Princeton, D. Van Nostrand Co., Inc. One of the best general works on spiders; excellent illustrations.

Grassé, P. P. (editor). 1949. Traité de zoologie, vol. 6, pp. 219-262. Paris, Masson et Cie. Good accounts of the Merostomata and the Pycnogonida by L. Fage. A definitive work with extensive bibliographies.

Hedgpeth, J. W. 1948. The Pycnogonida of the Western United States, North Atlantic and the Carribean. Proc. U. S. Nat. Mus. 97:157-342. An excellent treatment of the sea spiders by the foremost American student of the group.

Hedgpeth, J. W. 1954. On the phylogeny of the Pycnogonida. Acta Zool. (Stockholm) 35:193-213. Sea spiders have some characteristics that suggest affinities to the Chelicerata, and some superficial resemblances to crustaceans.

Hoff, C. C. 1949. The pseudoscorpions of Illinois. Bull. Ill. Natur. Hist. Surv. 24:407-498. This paper and the one by the same author on the pseudoscorpions north of Mexico afford an excellent taxonomic approach to the group.

Hughes, A. M. 1961. The mites of stored food. Techn. Bull. No. 9. London, British Information Service. A treatise on the economic importance of mites.

Kaston, B. J. 1964. The evolution of spider webs. Amer. Zool. 4:191-207.

Kaston, B. J. 1965. Some little known aspects of spider behavior. Amer. Midl. Nat. 73:336-356.

Levi. H. W. 1967. Adaptations of respiratory systems of spiders. Evolution 21:571-583.

Manton, S. M. 1953. Locomotion and evolution in arthropods. Symp. Soc. Exp. Biol. 7:339-376.

McCook, H. C. 1889-1893. American spiders and their spinning work. Philadelphia, Academy of Natural Science. The three volumes of this classic work may be outdated in some respects, but the beautiful plates and informal descriptions will never cease to interest all students of spiders.

Mullen, G. R. 1969. Morphology and histology of the silk

glands in *Araneus sericatus* CL. Trans. Amer. Microscop. Soc. **88**:232-240.

Palmer, E. L. 1961. Spiders and webs. Natur. Hist. **70**:33-43 (Oct.).

Parry, D. A. 1960. Spider hydraulics. Endeavour **29**:156-162.

Pennak, R. W. 1953. Fresh-water invertebrates of the United States. New York, The Ronald Press Co. A section is devoted to the water mites (Hydracarina). A good introduction is included as well as keys, figures, and bibliography.

Petrunkevitch, A. 1939. Catalogue of American spiders. Transactions of Connecticut Academy of Arts and Sciences, part 1, vol. 33, pp. 133-338. A valuable work for all students of spiders.

Savory, T. H. 1960. Spider webs. Sci. Amer. **202**:114-124 (April).

Savory, T. H. 1962. Daddy longlegs. Sci. Amer. **207**:119-128 (Oct.).

Savory, T. H. 1964. Arachnida. New York, Academic Press, Inc.

Snodgrass, R. E. 1952. A textbook of arthropod anatomy. Ithaca, N. Y., Cornell University Press. A detailed account of the morphology of this group. A work of established reputation and usefulness; well illustrated.

Wells, M. 1968. Lower animals. New York, McGraw-Hill Book Co.

Wilson, R. S. 1962. The control of dragline spinning in the garden spider. Quart. J. Microscop. Sci. **104**:557-571.

Phylum Arthropoda

THE CRUSTACEANS

GENERAL CHARACTERISTICS

1. It has been customary to classify together the Crustacea, Myriapoda, and Insecta under the subphylum Mandibulata, which does not form a natural group. Resemblances among the mandibulates are now thought to be due to convergence. Myriapods and insects seem to have a common plan of structure in the head region and in other ways, although they have some differences, whereas the crustacean head types are not comparable with those of the other two groups. However, the primitive crustacean body plan and its evolutionary adaptive radiation give a better insight into the mandibulate functional morphology than that of the other two groups.

2. Since the crustaceans are mostly aquatic and the myriapods and insects are mostly terrestrial, it may be well to call the crustaceans "aquatic mandibulates" and the other two "terrestrial mandibulates."

3. Crustaceans show an enormous diversity of structure. In general they have two pairs of oral antennae and at least three pairs of postoral appendages acting as jaws. The sharpest distinction of crustaceans is the second antennae because all other antennate arthropods have a single pair of antennae corresponding to the crustacean antennules. Crustaceans also have biramous appendages, consisting of the basal part (protopodite), an inner arm (endopodite), and an outer arm (exopodite). On the other

hand, terrestrial mandibulates have uniramous appendages with no indication of an endopodite and exopodite. Crustacean preoral appendages may also become locomotory or prehensile or sometimes absent. Parasitic forms undergo many modifications of crustacean structure. The basic arrangement of crustaceans has been modified in countless ways as their members have become adapted to different modes of life. Typically the head has 5 obvious segments (6 in the embryo), all of which bear appendages. Usually the thorax has 8 segments with limbs, and the abdomen has 6 segments, all of which can bear limbs.

4. In many crustaceans of diverse groups the larva that first hatches from the egg is called a nauplius. It has three pairs of jointed legs, which serve principally for locomotion and for feeding. During several molts the crustacean retains the nauplius form. Externally the nauplius shows 3 segments, but embryologically there may be 4 segments. The first pair of appendages (uniramous) corresponds in the adult to the first antennae. The other two pairs are biramous appendages and correspond in the adult to the second antennae and the mandibles. These biramous appendages have in their basal parts gnathobases with stout bristles. After later molts, the larva becomes a metanauplius with four pairs of appendages. In all primitive Crustacea and Arthropoda the limbs have a biramous pattern. More segments are added at

molting in an anteroposterior sequence, with the youngest segment just in front of the anus. A great deal of diversity is shown by different groups of crustaceans, and the process of growth and development is best seen in relatively primitive crustaceans, as there is increased tagmatization in the more advanced crustaceans.

5. The other mandibulates, insects and myriapods, are treated in a separate chapter because they form a natural group. The tagmosis of the insects usually differs markedly from that of other mandibulates. The first 6 somites are more or less fused to form a well-defined head. The next posterior region, the thorax, is composed of 3 somites somewhat fused into a compact structure and containing the muscles for operating the legs and wings. The remaining somites form the abdomen, which typically has 11 segments, but the terminal segment may be abbreviated or vestigial in the adult condition. Although the crustacean shows a threefold tagmatization, the segments involved do not correspond to those of the insects.

The Mandibulata

The mandibulate arthropods are characterized by jaws, maxillae, and antennae as head appendages. The body is made up of two or three divisions. In its simplest form it consists of a head and trunk, but it may be divided into a head, thorax, and abdomen. The head and thorax are often fused into a cephalothorax division (Table 10). Within this extensive group there are many variations in accordance with specializations for different purposes. This subphylum represents one of the two major lines of evolution of arthropods, the other being the Chelicerata. Evolutionary trends involve the composite head of fused segments and shifting posi-

Table 10. Comparison of the mandibulate arthropods

Organization of body divisions	Crustacea	Chilopoda	Diplopoda	Insecta
Tagmata	Head, thorax, and abdomen; carapace often over head and thorax	Head and trunk	Head and trunk, with first few segments each bearing a single pair of legs	Head, thorax, and abdomen
Head appendages	1st antennae, 2nd antennae, mandibles, 1st maxillae, and 2nd maxillae (often)	Antennae (of 2nd segment),* mandibles, 1st maxillae; and 2nd maxillae	Antennae (of 2nd segment), mandibles, 1st maxillae, and 2nd maxillae	Antennae (of 2nd segment), mandibles, 1st maxillae, and 2nd maxillae
Thorax	Segments vary in number	Not differentiated from trunk	Anterior trunk segments with a single pair of legs each	3 segments
Thoracic appendages	Several types with different functions	Maxilliped, followed by one pair of legs to each segment	First few segments with a single pair of legs each	One pair of walking legs for each segment
Abdominal appendages	Reduced or functionally differentiated	Appendages the same with one pair per segment	Two pairs of legs to each diplosomite; gonopods differentiated	Absent, except those for reproduction

*Mandibulates are considered by some to have an embryonic first segment without appendages. The first pair of antennae, or antennules (of 2nd segment), are homologous to the antennae of those mandibulate classes that have a single pair of antennae; hence the designation "of 2nd segment."

tions and the modifications or specializations of appendages by fusion and differentiation or otherwise. In general the mouth has passed backward, and the segments have pushed forward.

The antennae, which are absent in chelicerates, may be represented by two pairs in the mandibulates. The first pair of these is called the antennules in the crustaceans that have two pairs and simply antennae in those that have only one pair (insects, chilopods, diplopods). The antennule is homologous to the prostomial palps of polychaete worms. The second pair of antennae, when present, represents derivatives of trunk appendages borne by the first of the 4 trunk segments that have fused with the head region. This appendage may be vestigial or entirely lacking. The mandibles are the second postoral appendages and are modified appendages of the second fused trunk segment. Their evolution from primitive trunk appendages is evidenced by the small palpi they bear, since the major part of the mandible has evolved into its masticatory specialization from the original basal segments of the appendage. Some variations are found in the fate of the appendages of the third and fourth segments. In crustaceans they are the maxillae—two pairs of accessory feeding structures. In insects they become the maxillae and labium. In diplopods and chilopods they form the maxillae. The trunk appendages are used mainly for walking or swimming but vary in the different groups. In a simple form the head of the mandibulate may consist of only the eyes, antennae, and a preoral lobe (labrum), with the segments of the mandibles and maxillae united to the thorax. Usually these 2 latter structures are included in the head.

The mandibulates include the crustaceans, centipedes, millipedes, pauropods, symphylans, and insects (hexapods).

TAGMOSIS AND CARIDOID FACIES

The grouping of arthropod segments with their appendages for specializations of function is nowhere better developed than in the crustaceans and in the insects. This regional differentiation is a key to an understanding of the arthropod body and its organization. This demarcation of the arthropod body into specialized regions (tagmata) is called tagmosis. Thus for instance the head with its appendages performs sensory and alimentary functions, and the segments and appendages of the rest of the body are concerned mainly with locomotion, respiration, and reproduction. The outstanding example of tagmosis, in which a more or less uniform plan is emphasized throughout the group, is the insects. Here the first 6 segments form the head region, the next 3 segments the thorax, and the remaining segments the abdomen (although there are some modifications of this tagmosis). Crustaceans, however, display a greater range of modification of this regional distribution and have a greater overlap of functions between the several tagmata. Throughout their adaptive radiation each major group retains and is guided by its characteristic patterns of tagmosis in its evolution. In no crustacean are all the segments distinct. The head of arthropods, for instance, has evolved along several lines and each pattern of head is characteristic of a major group, but the possibility of convergent evolution may cloud important clues to an understanding of relationships.

Tagmosis has made possible the establishment of a plan of organization in the Malacostraca that is called caridoid facies. In this organization plan 5 or 6 segments form the head with their appendages; 8 segments the thorax, with appendages for walking and bearing accessory parts for swimming and respiration; and 6 segments the abdomen, with 5 bearing swimming pleopods and the sixth modified with the postsegmental telson to form the tail fin. There may be some modification of this plan, especially in terrestrial forms. Caridoid facies may have evolved as an adaptation to benthonic existence and walking movements but later independent evolution produced other modifications of the original pattern. However, crustaceans have done little to exploit the land and are mainly aquatic forms, for which they are well fitted. Perhaps physiologic reasons may explain their difficulties in conquering the land (water balance, excretion, etc.), or they may have become too specialized for aquatic existence to develop suitable adaptations for other kinds of habitats.

■ CLASS CRUSTACEA

The important arthropod class Crustacea comprises a group of such diversity of structure, function, and development that it is difficult to define them in characters that apply to all members of the group; the characters that apply to one subdivision of crustaceans may be highly modified or nonexistent in another subdivision. In general a crustacean is an arthropod of aquatic habits, with two pairs of antennae, three pairs of masticating and feeding appendages, and typical biramous appendages. However, this definition must be applied only with reservation.

Some members of this extensive group are familiar to everyone. They include crayfish, shrimps, crabs, lobsters, prawns, sow bugs, sandhoppers, wood lice, barnacles, water fleas, copepods, and others. Although typically aquatic, there are some members that are terrestrial or semiterrestrial. Crustaceans generally have avoided strictly terrestrial habitations.

A broad classification divides crustaceans into the smaller forms called entomostracans and the larger members called malacostracans. Entomostraca is not considered a valid taxonomic division because the members differ from each other in diverse ways. Malacostraca is more clearly defined because all members of this group have abdominal appendages, an 8-segmented thorax, a gastric mill, and an abdomen of 6 (sometimes 7 or 8) segments. About eight subclasses of crustaceans are recognized at present, although there is less agreement among carcinologists about lesser taxa such as orders and families. There are more than 30,000 known species in the class.

External morphology

All parts of the external anatomy of crustaceans are covered with a chitinous exoskeleton, which may be thick with limy salts or delicate and transparent. The exoskeleton is divided into segments (somites, or metameres) that may be more or less fused together, or in the more primitive forms they may be more or less distinct. No crustacean, however, has all of the somites distinct. Body segmentation may be absent altogether (Cladocera, Ostracoda). The three parts of the body—head, thorax, and abdomen—may be distinct, but often one or more thoracic segments are fused with the head to form a cephalothorax. The cephalothorax is often wholly or partly covered by a carapace that arises as a fold of the head and fuses with certain tergites behind it. However, in some forms the carapace may be an integumental fold that remains free of the thoracic segments, or it may coalesce with the thoracic terga in the middle but remain free at the sides, forming gill chambers. In other crustaceans the carapace is in the form of a bivalve, and in barnacles it is modified into a mantle bearing calcareous plates. Branchiopods have no carapace at all.

Typical segments are usually not as regular as circular rings but are somewhat compressed. Each is separated from the one in front and the one behind by a region of softer integument that forms a joint. The dorsal arched part of the segment is called the tergum; the lower narrow ventral part, the sternum; and each lateral part, the pleuron. The pleuron may extend over the attachment of the appendages to form a free plate, the epimeron.

The part of the body that bears the anus is usually called the telson. In nonmalacostracan forms the telson may be referred to as a true segment, although it never bears typical limbs. Various interpretations have been made about its nature. Some regard it as a median appendage or the fusion of a pair of appendages. In larval forms of primitive crustaceans the number of segments increases by addition of new segments between the last formed segment and the terminal unsegmented region that forms the telson.

The nauplius larva (Fig. 14-1), characteristic of crustaceans, has in its head three pairs of appendages—the antennules, antennae, and mandibles—which represent the appendages of 3 fused segments. To this primary head (protocephalon) are added 2 additional segments bearing the first and second maxillae to form the cephalon of later stages. The typical adult crustacean head thus has five pairs of appendages. In addition to these five pairs of appendages are the upper lip (labrum) from the body wall and the lower lip (labium) from the foregut. The compound eyes are not true segmental appendages but are simply specialized structures at the anterior ends of the nerve cord. Embryologically they do not develop in the regular segmental pattern.

The paired appendages of crustaceans are typically biramous, consisting of two branches, endopodite and exopodite (Fig. 14-2). These are attached to a basal protopodite composed

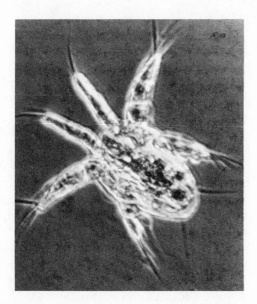

FIG. 14-1

Nauplius larva. (From Encyclopaedia Britannica film, "Life in a Drop of Water.")

of 2 segments, the coxopodite (coxa) articulated to the body and the basipodite (basis). The two branches exhibit many modifications in the different groups. Either the endopodite or the exopodite may be reduced or entirely absent, or they may coalesce into a foliaceous limb. The protopodite may bear on its outer margin a process called an exite or on its inner margin an endite. Exites in some crustaceans function as gills, whereas endites of appendages near the mouth function as gnathobases. The subdivisions of foliaceous limbs are usually considered as endites and exites. The appendages are all more or less similar in their primitive pattern, but the evolutionary trend has been toward reduction in number and specialization for different functions. In the more advanced crustacean the endopodite is segmented. The segments, named outward, are the ischiopodite, meropodite, carpopodite, propodite, and dactylopodite (Fig. 14-2). The endopodites in the different groups assume a variety of functions—perception, crawling, respiration, defense, offense, sperm transmission, and food handling. When present, the exopodite may be paddlelike for swimming or for respiration.

In a more specific way the functions of the various paired appendages may be summarized in the following manner. The antennules are mainly sensory in most crustaceans but may be used as clasping organs by some. In the higher crustaceans the antennae are chiefly sensory but may be used as swimming organs in some or as attachment organs in others. The mandibles are often heavy and short for grinding and biting but may also be used for swimming and other functions in the more primitive forms. The first and second maxillae are used mainly as accessory feeding appendages. They are often flattened and leaflike with gnathobasic lobes. The thoracic series of appendages (eight pairs in number) in malacostracans are modified into maxillipeds for accessory food handling and into walking legs. Usually only the endopodite is present in these, whereas the exopodite (if present) may be used as a swimming organ.

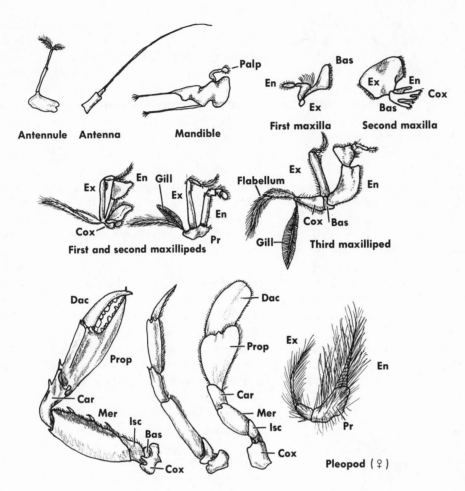

Antennule   Antenna   Mandible   First maxilla   Second maxilla

First and second maxillipeds   Third maxilliped

First, second, fifth pereiopods   Pleopod (♀)

FIG. 14-2

Crustacean appendages *(Callinectes sapidus)*. Key to labels is as follows:

Bas	Basipod	**Ex**	Exopodite
Car	Carpopod	**Isc**	Ischiopod
Cox	Coxopod	**Mer**	Meropod
Dac	Dactylopod	**Prop**	Propod
En	Endopodite	**Pr**	Protopodite

The abdominal limbs (six pairs) are generally biramous and are used mainly for swimming, although they are often sexually differentiated for the attachment of eggs and for copulatory organs.

The integument of the crustaceans consists of an inner, single-layered, cellular hypodermis or epidermis and an outer noncellular cuticle (Fig. 14-3). The cuticle shows many variations. In its typical form it is divided into two or more horizontal layers. The two chief cuticular layers are an outer, thin epicuticle and a thicker endocuticle (procuticle). The epicuticle is usually a waxy layer to shed water. The endocuticle has a chitin-protein complex, calcium salts, and scleroproteins. Pigmentation may occur in the inner part of the epicuticle and the outer part of the endocuticle. The innermost layer of the endocuticle does not contain minerals. The epidermis contains blood cells and cells for molt-

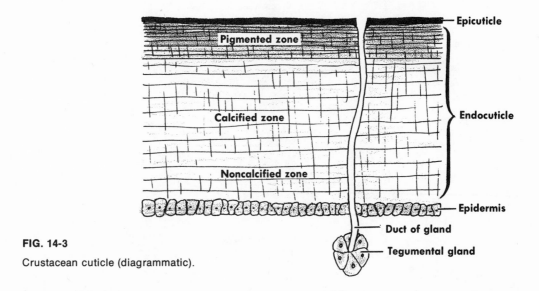

FIG. 14-3

Crustacean cuticle (diagrammatic).

ing and is largely glandular. Certain large glands (tegumental) just beneath the epidermis may be involved in the secretion of the epicuticle. They are provided with ducts that open on the surface of the epicuticle.

Internal morphology

Digestive system. The food canal is a relatively straight tube, except for the dorsal curve of the part that runs down anteriorly to the mouth. In some cases it is coiled upon itself. The mouth lies between the mandibles on the underside of the head. The fleshy upper lip of the mouth is called the labrum; the lower lip is called the metastoma, or labium. The metastoma is sometimes bilobed, each lobe being called a paragnatha. The anterior part of the foregut is the esophagus (stomodeum), whereas the posterior foregut is modified into some form of stomach. In malacostracans the stomach is modified into the gastric mill, which is provided with chitinous ridges or teeth and setae for churning and straining the food (Fig. 14-4). The teeth are manipulated by muscles to triturate the food as it passes into the stomach proper. The midgut, or mesenteron, is provided with pouchlike diverticula that serve as glands for secretion and for absorption of

food. In the decapods the glands are massive and are called the hepatopancreas, or liver. It is made up of ducts and blind secretory passages and is the chief source of digestive enzymes. The midgut may be very short and is the only division of the gut not lined with chitin. The hindgut (proctodeum) terminates at the anal opening. In filter-feeding crustaceans, fine setae on the appendages around the mouth collect food particles. Crustaceans vary in their food habits. Besides the filter feeders, many are scavengers, predators, or herbivores. Raptorial habits are characteristic mostly of the larger forms whose appendages are modified for seizing prey. Several crustaceans have multiple feeding methods. Parasitic forms are represented by both ectoparasites and endoparasites. In the highly modified parasites the alimentary canal is vestigial or absent.

Circulatory system. Circulation in crustaceans is similar to that in most arthropods. It is an open system consisting of heart, arteries, and sinuses (Fig. 14-5). The heart is often tubular and may extend the greater length of the body, but it may also be a compact organ. It lies in a pericardial blood sinus from which the blood passes by valvular ostia into the heart. In the more advanced crustaceans the blood is

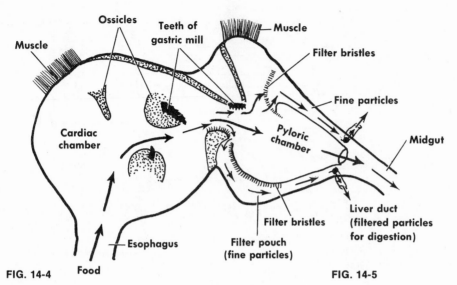

FIG. 14-4

Malacostracan stomach (diagrammatic) showing gastric mill and direction of food movements. (Modified from Yonge, 1924.)

FIG. 14-5

Circulation in crustaceans. **A,** Generalized plan in crustaceans. **B,** Circulation in amphipod. **C,** Circulation in decapod.

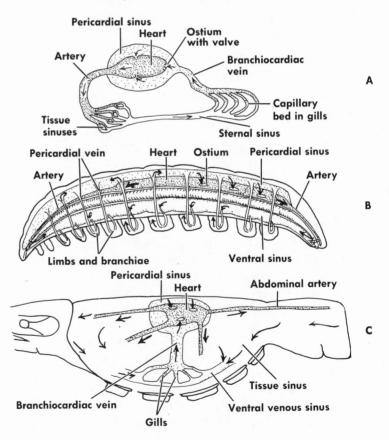

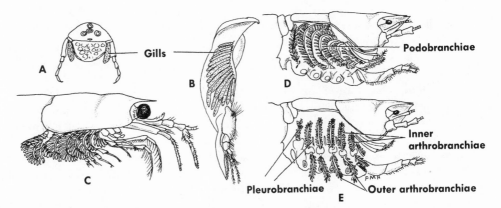

Gills

Podobranchiae

A

B

D

C

E

Inner arthrobranchiae

Pleurobranchiae

Outer arthrobranchiae

FIG. 14-6

Malacostracan gills. **A,** *Corophium,* an amphipod (transverse section), has pouchlike gills on protopodites of thoracic legs. **B,** *Diastylis,* a cumacean, has epipodites forming branchial tubes on first thoracic appendages. **C,** *Euphausia* has gills on protopodites of seven pairs of thoracic legs. **D,** *Astacus,* a decapod (branchiostegite removed), has podobranchiae arising from basal joints of thoracic limbs. **E,** *Astacus* (podobranchiae removed) also has two rows of arthrobranchiae attached to membranes between joints and the body and pleurobranchiae attached to the body itself.

pumped by certain arteries through the connecting sinuses of the tissues. Arteries are elastic but have little muscle; in small crustaceans arteries are greatly restricted. A heart may be absent in lower crustaceans, and the blood is then driven by body movements.

Many crustaceans have blue blood because of the respiratory pigment hemocyanin, but a few have the respiratory pigment erythrocruorin or hemoglobin. Hemoglobin is the pigment most common in lower crustaceans living in poorly oxygenated water because hemoglobin has a great affinity for oxygen. Crustaceans are usually represented by hemocyanins of smaller molecular weight. The blood has a variety of leukocytes and amebocytes, which may represent variations of a single basic type. Blood corpuscles may range from a few hundred to several thousand per cubic centimeter of plasma, depending on species, age, and physiologic states. Some of their corpuscles have phagocytic power. Some have no respiratory pigments, and those that do have only moderate oxygen-carrying capacity. Blood coagulation in crustaceans appears to take place by several methods. Certain amebocytes may release substances that convert fibrinogen into fibrin, thus trapping blood cells. Spastic contraction of muscles at the place of a lost appendage may produce the same effect as clotting.

Gas exchange. Crustaceans use gills, general body surface, or some modification of such structures for breathing. Most gills are associated with the appendages (Fig. 14-6). In some cases, the exites of the basal segment have the functional aspects of respiratory surfaces, being flat or branched. Among the malacostracans, with few exceptions, gills are present on one or more pairs of thoracic appendages. Many different patterns of gill arrangement are found. They may be vascular sacs or lamellae or more complicated structures in which the lamellae are arranged with a central axis containing blood vessels. Such plans may be highly foliaceous or filamentous. In the decapods, gills are arranged in three series near the bases of the thoracic limbs and are enclosed in a pair of branchial chambers covered over by the carapace (Fig. 14-6, *D* and *E*). Water currents are facilitated by the beating of appendages, some of which are specialized for this purpose.

Terrestrial crustaceans may have the vascularized branchial chambers enlarged to serve as lungs. Isopods may have tufts of branching

509

tubules filled with air, similar to tracheae. Some small crustaceans have no gills but breathe through the body surface or through restricted regions of the body surface.

Excretion. The paired excretory organs of crustaceans lie at the base of the antennae or maxillae. Early in development there may be two pairs, one called the antennal and the other the maxillary glands. Usually only one or the other of the two pairs persists, but both may be found in the adults of some forms. The structure of both glands is similar, and they may represent survivors of a segmentally arranged pattern of nephridia. They are similar to the coxal glands of chelicerates. Each type is made up of a thin-walled end sac and an excretory tubule (Fig. 14-7). Some modifications may be found in this arrangement, such as labyrinth and a bladder. The end sac is a vestigial part of the coelom, and the tubule represents a persistent coelomoduct. In malacostracans the antennal glands are more typical and open externally on the underside of the bases of the antennae. The maxillary glands, when present, open at the bases of the second maxillae.

The waste product of crustaceans is mostly ammonia with some urea and a little uric acid, but much of this waste may diffuse through the gills. Uric acid is sometimes stored in special cells (nephrocytes); it may form a part of the white chromatophores; or it may be deposited in the integument and cast off with the molt. The excretory products may be secreted by active transport, or they may be separated from the blood by filtration. Experimental evidence is divided on this point (N. S. R. Maluf, J. A. Riegel, and L. B. Kirschner). Crustacean urine is mostly

FIG. 14-7

Crustacean excretory systems. **A,** In pagurid hermit crabs, bladder ramifies throughout much of thorax with single or paired extensions the length of abdomen. **B,** Maxillary gland in larva of branchiopod *Estheria.* **C** and **D,** Antennal gland of decapod *Astacus.* **C,** Parts of gland in natural relation. **D,** Parts of gland unraveled. **E,** Section through antennal gland of *Mysis.* (**B** and **E** modified from Grobben.)

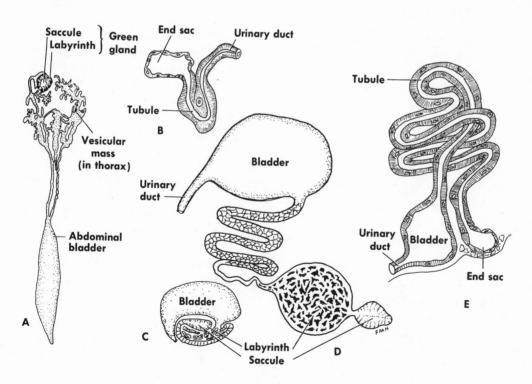

isotonic with the blood but may be hypotonic. *Artemia*, a saltwater form (and some others), produces a hypertonic urine. Water resorption may occur after filtration into the end sac in some cases, which could account for a more concentrated urine. The gills may be responsible for regulating the ionic balance of salts in the crustacean body by either expelling or absorbing salt. *Artemia* is also provided with special salt glands in its gills for the excretion of excess salt.

Nervous system. The nervous system is typical of arthropods, consisting of a large supraesophageal ganglion, or brain, connected by circumesophageal commissures with a double ventral nerve cord of segmentally arranged ganglia (Fig. 14-8). The primitive arrangement of the nervous system is best seen in the branchiopods in which the ventral nerve cord has a ladderlike pattern of 2 widely separated chains of ganglia (Fig. 14-8, *A*). The paired ganglia, one in each chain, are connected by double transverse commissures. This primitive arrangement of 2 chains is more or less coalesced in most groups.

In general the nervous systems of the different groups show a marked tendency toward concentration and fusion of the ganglia (Fig. 14-8, *B* and *C*). The chain ganglia may be drawn into a single mass representing the ventral chains, as in the true crabs. The brain also has many variations. In higher crustaceans ganglionic masses, which in lower forms are found on the esophageal commissures, are incorporated in the brain. More centers are found in the brain of decapods than in that of other crustacean groups. Neurosecretory cells in the brain of crustaceans produce hormones that pass down axons to the bloodstream to control chromatophores and trigger thoracic glandular cells for molting.

Each pair of ganglia in the ventral nerve cord usually gives off three pairs of lateral nerves for the innervation of the various organs and tissues. The posterior pair of each group of three pairs is mostly motor and runs to the flexor muscles of the body wall. The other two pairs of nerves are larger and are mixed motor and sensory in function. They innervate the appendages and the segments. The brain supplies pairs of nerves to the antennules, antennae, compound eyes, and ventral nerve cord (circumesophageal commissures). Other nerves may also run from the brain to various parts of the body, depending on the group.

Many crustaceans have giant fibers in the

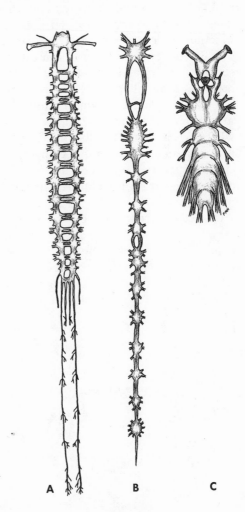

FIG. 14-8

Three types of crustacean nervous systems. **A,** *Branchinecta* (subclass Branchiopoda). **B,** *Astacus* (subclass Malacostraca, order Decapoda). **C,** *Argulus* (subclass Branchiura).

central nervous system that are adapted for rapid transmission of impulses and resultant quick reactions such as those in the escape mechanism. Decapods especially are provided with these fibers, which may reach a diameter of 200 microns. There are usually 2 dorsomedial giant fibers and 2 dorsolateral ones (Fig. 14-9). Each of the medial fibers has a large nerve cell in the anterior brain, but the axon it gives off crosses to the opposite side of the posterior brain. These giant fibers extend the entire length of the ventral cord and make synaptic junction with motor nerves from the ganglia. The dorsolateral giant fibers are made up of a series of neurons, each of which has its cell body in a ganglion of the nerve cord. Thus impulses for the dorsolateral giant nerves can arise from any of the ventral ganglia of the cord, whereas impulses for the dorsomedial giant fibers originate in the brain. A single impulse in either of the two types of giant fibers will produce muscular contractions for the rapid escape mechanism.

Crustacean reflexes are numerous, such as claw closing, autotomy, escape, feeding, sex behavior, etc. Some of these reflexes will occur even in the complete absence of the brain, which may exert a regulative control but is not necessary for the initiation of the process.

Sense organs. The sensory system of crustaceans is among the most extensive in the arthropods and has been much studied during the past two or three decades. In general the various senses are grouped into visual receptors, chemoreceptors, proprioceptors, tactile receptors, statocysts, and myochordotonal receptors. Some of these are special sense organs that are primarily specialized for a definite sensory impression, and others have a more general function. In any integrative reaction there is, of course, an overlapping and functional correlation of the different groups.

The eyes are of two types, the unpaired median, or nauplius, eye and the paired compound eyes. The median eye is characteristic of the common nauplius larval form and may be the only eye, as in the copepods. It may exist along with the compound eyes, or it may be vestigial or disappear altogether, as in most of the malacostracans. The median eye (Fig. 14-10) usually consists of 3 pigmentary cups, or ocelli. Each pigmentary cup contains columnar retinal cells, or photoreceptors, which are connected by nerves to the brain. Lenslike structures may or may not be present. The functions of these eyes may vary with the groups in which they are found. Some may function in movement perception, orientation by polarized sky-

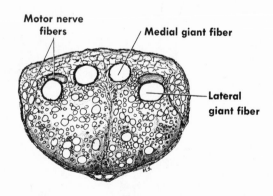

FIG. 14-9

Cross section of nerve cord of crustacean showing giant fibers. (From stained slide.)

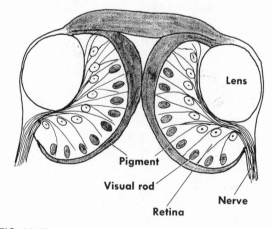

FIG. 14-10

Section through median eye (diagrammatic) of *Cypris*, an ostracod.

light, and visual recognition; but present knowledge does not justify definite conclusions about such matters. Some may be concerned only with distinguishing between light and dark for orientation.

A pair of compound eyes is characteristic of most crustaceans. They are similar to those of insects. Usually they are separated, each being placed on a movable peduncle or eyestalk, but they may be fused into a single organ in such forms as cladocerans and amphipods. Compound eyes may also be sessile or without a peduncle. In blind crustaceans of caves or other regions the peduncles may persist and become modified into spines for defense.

The compound eye (Fig. 14-11) is made up of a varying number of elongated, cone-shaped bodies, called ommatidia (Fig. 15-11). They are covered by a transparent membrane of the cuticle to form the cornea, which is divided into facets corresponding to the underlying ommatidia. The facets may be hexagonal or

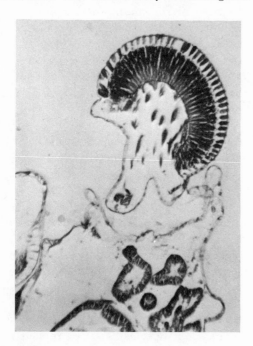

FIG. 14-11

Cross section through compound eye of fairy shrimp *Eubranchipus* (order Anostraca).

square in shape. The cornea is secreted by 2 corneagenous cells (modified epidermal cells) in each ommatidium. Lying behind the cornea is the long, cylindric, crystalline cone, functioning as a second lens and formed of 4 cone cells. Below this is a core of crystalline translucent material called the rhabdome, surrounded by 7 or 8 photosensitive cells, the retinular cells, which are stimulated by the light rays scattered and diverted by the rhabdome (Fig. 14-11). Retinular cells give off axons, or nerve fibers, which issue from the ommatidium and run to an optic ganglion that may be located in the eyestalk. The rhabdome itself may have a photoreceptive role and become the center of the first photochemical response. In turn, it would stimulate the retinular cells. The rhabdome may contain the light-absorbing pigment that has the absorption characteristics of rhodopsin in the rods of vertebrate eyes. The pigment appears to be most sensitive to light rays in the wavelength of green or yellow-green.

Each ommatidium is surrounded by a distal collar of black or brown pigmented cells around the crystalline cone and by a proximal collar of retinal pigment around the retinular cells at the base of the ommatidium. The manipulation of these two pigment collars makes possible two different forms of vision. In daylight the 2 collars can migrate centrally and distally and merge to form a collar completely surrounding each ommatidium. In such a condition each ommatidium will pick up those light rays that come in parallel to its long axis. Such an image is a mosaic (appositional) image; each unit sees only what is directly opposite it. For suffused light at night the collars are separated, leaving the middle of the ommatidium free of pigment. Thus each ommatidium can be stimulated by light rays entering from adjacent ommatidia (superposition image) as well as through the cornea, and thus the intensity of the light stimulation at night can be stepped up to compensate for the fewer light rays present at such times. Night vision, however, is far less precise than day vision.

Compound eyes are adapted for detecting slight movements. Any shift of light rays would be detected by a few or many ommatidia, according to the degree of movement. Several ommatidia could also be stimulated by the same beam of light by overlapping stimulation. The outer convex surface of the corneal facets ensures a fairly large field of vision so that in some stalked eyes the corneal surface can cover an arc of 180 degrees. Compound eyes show many grades of structural and functional components in a group as vast as the crustaceans. Some have only day- or night-adapted eyes. Superposition eyes are characteristic of deep-sea crustaceans and those that are strictly nocturnal in their habits. Visual acuity is to some extent correlated with the number of ommatidia in the compound eye. Some small terrestrial forms may have as few as 24 ommatidia in a compound eye; the large lobsters may have 12,000 to 15,000. In no case can it be stated that the visual acuity of a crustacean is comparable to that found in higher forms such as mammals, for the number of ommatidia never equals the number of rods and cones of the vertebrate eye. Investigations reveal that a certain amount of color discrimination is present in some crustaceans, especially in the malacostracans, in which chromatophore mediation occurs through the eye. Shades of yellow, red, and blue, for example, can be distinguished by *Crangon*.

One of the most common types of sense organs in crustaceans is the hairs or setae on the surface of the exoskeleton, which are variously modified for special functions. Some hairs are for touch reception, detection of water currents, or orientation. Such receptors are especially common on the appendages that are in most direct contact with the environment. Since the exoskeleton is a hard, insensitive armor that shields the underlying sensitive tissue from external stimuli, hairs and such receptors must communicate with the interior by channels similar to the prototrichs found in lizards and snakes, and for the same reason.

FIG. 14-12

Head end of lobster showing brain and nerve connection to statocyst in basal segment of right antennule. Below, section through statocyst (diagrammatic), with rows of sensory hairs cushioning statolith.

Tactile hairs or setae have chitinous shafts that articulate by ampullae to the exoskeleton.

Statocysts or ectodermal sacs are found at the base of the antennules, telson, and other parts in most decapods. They are open or closed sacs; their interiors are lined with sensory hairs and contain a statolith or statoliths, which may be grains of sand or secreted substances (Fig. 14-12). They are concerned with the maintenance of equilibrium.

Muscle or tendon receptor organs (or proprioceptors) have been described in all advanced crustaceans and may be found in varying degrees in the lower members. Each organ is made up of a modified muscle cell, a pair of which is found on both sides of the dorsal musculature of each abdominal segment, as well as in the extensor muscles of the seventh and eighth thoracic segments. Each proprioceptive organ has dendritic (sensory) connections with a multipolar neuron. The dendritic terminals are stretched either by the flexion of the abdominal muscles or by contraction of the receptor muscles that are supplied by motor fibers. There is also a differential rate of contraction in the 2 members of a pair of receptor muscles, 1 fast and the other slow. Much is

FIG. 14-13

Tip of antennule (diagrammatic) of brachyuran crab with aesthetascs, or long sensory hairs, each connected to cluster of sense cells. Nerve fibers from each cluster make up "olfactory nerve" component of antennular nerve. (After Hanström; modified from Waterman.)

yet to be learned about the complicated nature of the muscle receptor organs.

Whether or not crustaceans can hear, or have definite auditory organs, has never been resolved. In some way the statocysts seem to be involved in reactions to sound. Delicate hairs on certain appendages are apparently sensitive to vibrations of high and low intensity.

A special type of sensory receptor, called a myochordotonal organ, occurs in the lumen of the meropodite, or fourth segment of the walking leg. It consists of a thin membrane containing sensory cell bodies of bipolar neurons. The organ probably has a proprioceptive function.

A number of chemoreceptors are present in such places as the antennules, antennae, and other appendages around the mouth. These include aesthetascs (Fig. 14-13) (sensory hairs or plates), which occur in rows on the outer flagellum of the antennules and are innervated by the dendrites of certain bipolar neurons. Other possible chemoreceptors are represented by canals, pores, etc. Little is known about thermoreception in crustaceans. The two kinds of antennae are supposed to be sensitive to thermal stimuli. Thermal responses vary greatly in the different groups, especially to temperature gradients.

Reproduction

Most crustaceans are dioecious, but in the Cirripedia, a few parasitic isopods, and scattered instances in other groups they are hermaphroditic. Parthenogenesis occurs in some branchiopods, ostracods, and a few others. There may be alternation of generations in lower crustaceans. The gonads of crustaceans are usually elongated, paired organs in the dorsal region of the thorax or abdomen or sometimes in both. The gonoducts (sperm ducts or oviducts) are chiefly tubules that open either on a sternite or at the base of a pair of trunk appendages. There are some variations in the location of the opening from one group to another.

Sexual dimorphism may be pronounced. Either sex may be larger than the other. The males are often provided with clasping organs modified from almost any of the appendages (thoracic limbs, antennae, mouthparts, etc.). In the decapods the first and second pairs of abdominal appendages are used by the males for transmitting sperm. A penis is present in some crustaceans. In copulation (Fig. 14-14) the male may clasp the female and deposit sperm or spermatophores in or near the female seminal receptacles. The seminal receptacles may be located near the openings of the oviducts or in ectodermal, invaginated pouches of an adjacent segment.

Development

The eggs are usually carried on female appendages until hatched. They may also be contained in special brood chambers located on the female body, or they may be placed in a saclike structure, from which the eggs are expelled. In brood chambers a nutrient may be secreted for the young until they are released. Some shrimp attach the eggs to aquatic vegetation.

The nauplius larva is characteristic of many crustaceans (Figs. 14-1 and 14-15). It has an oval, unsegmented body and three pairs of ap-

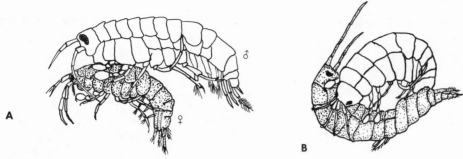

FIG. 14-14

Mating behavior in *Gammarus,* a freshwater amphipod. **A,** Precopulation. **B,** Copulation. (Modified from Kinne.)

FIG. 14-15

A, Nauplius larva of *Cyclops.* **B,** Metanauplius larva of *Cyclops* (order Cyclopoida). **C,** Zoea stage of *Crangon* (order Decapoda). (**A** and **B** modified from Claus; **C** modified from Sars.)

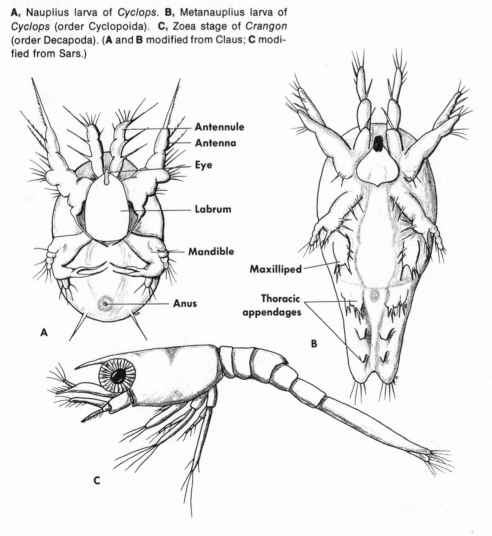

pendages (antennules, antennae, and mandibles). Of these appendages, the antennules are uniramous and the other two pairs are biramous. These appendages carry on the functions of sensory reception, feeding, and locomotion. A single median eye is found on the front of the head and represents a simple compound eye. The nauplius may be found only in the egg of the higher forms and may be absent altogether in some crustaceans. Life histories may be simple or complex. The young of some crustaceans resemble the parents in general structure as soon as hatched. After the nauplius stage in some groups the postlarva becomes a characteristic larva, such as the cypris larva of barnacles, the phyllosoma larva of spiny lobsters, the zoea and megalops of crabs, and the mysis of lobsters.

In the Malacostraca the nauplius larva is found only in some primitive decapods and a few others. In its further development the body of the nauplius larva elongates and, by successive molts, becomes segmented and acquires additional appendages (Fig. 14-15, *B*). New segments are added from a formative zone in front of the telson. A carapace is formed, and paired eyes appear at the sides of the head but not on stalks until later. This scheme of development appears to be the primitive one and is characteristic of the Branchiopoda and Copepoda. This primitive plan, however, is modified in most of the other crustaceans. Early stages are often passed through within the egg so that the hatched larva is more advanced than the nauplius. More advances are made at a single molt, and the regular order of succession of segments and appendages may be disturbed. This results, especially in the Malacostraca, in a larval form known as the zoea (Fig. 14-15, *C*), the most typical form of which is found in the Brachyura (true crabs). In the true crabs the posterior 5 or 6 segments are delayed in development, whereas the abdominal segments (often with appendages) are fully formed. After additional molts the adult features, except for size and sexual maturity are attained. The various groups of crustaceans show many modifications of this basic plan of development. The nauplius and zoea stages may be entirely suppressed, the young hatching as a postlarva. In the crayfish the young are juveniles, and larval development is suppressed.

Molting (ecdysis)

Because of its tough exoskeleton, a crustacean must shed its outer cuticle to grow. This process is under hormonal control and involves shedding not only the external cuticle but also the ectodermal linings of the foregut and hindgut, the tracheae, external parts of the eye reproductive ducts, etc. The process of molting which refers to all the processes involved in the shedding of the exoskeleton, may cease as soon as the crustacean attains sexual maturity; in others it may continue throughout life but at a decreasing tempo. Several steps are involved in this rather drastic procedure. There is first the preparatory stage (proecdysis), in which a new epicuticle is secreted beneath the old. During this time there is some resorption or dissolution of the old cuticle and the accumulation and storage of limy salts such as calcium in the hepatopancreas or in the gastroliths for use in the new endocuticle. The new cuticle is secreted to some extent while most of the old endocuticle is being digested by the enzymes of the molting fluid and is finished after ecdysis. Calcium especially is withdrawn from those strategic places where the old skeleton must be broken. The actual molting takes place by a swelling process in which blood is forced into certain parts of the body, water is absorbed from the gut, and excess air is taken in. This causes the old cuticle to split usually along the mid-dorsal line, and the animal emerges. Because of its helpless condition in the postmolt period, the animal seeks a safe shelter for the process. The periods between molts are called the intermolts. Much of the intermolt, however, is taken up with either the terminal process of the last molt or the beginning of the next molt.

The molt control mechanism involves mainly

the interaction of a molting hormone of the Y-organs or glands, located ventral and lateral to the eye socket, and a molt-inhibiting hormone of the X-organs, a group of neurosecretory cells of the eyestalk. The molting hormone initiates and integrates the metabolism of ecdysis, growth of tissue, and increase of size, whereas the molt-inhibiting hormone controls the molting gland activities by the degree of its own activity.

Regeneration and autotomy

Crustaceans have great power to regenerate lost parts, although many molts may be required before a new part attains the size of the lost part. There may also be some irregularities in the replaced parts as compared with the originals.

Autotomy is the voluntary parting with an injured appendage by muscular contraction. The limb is broken at a fracture plane especially provided. In many crustaceans this plane runs through the basipodite of the limb. At the fracture plane the integument is folded in to form a double-walled diaphragm, with a small opening for the nerves and blood vessels. When the limb is broken off at this point, the proximal wall of the diaphragm prevents loss of blood and promotes the clotting of blood for closing the ruptured opening.

Subclass Cephalocarida

Cephalocarids were first discovered in Long Island Sound in 1954. This small, shrimplike form is transparent and about one tenth of an inch long. The head is horseshoe shaped and covered by a head shield or carapace. The head bears two pairs of short antennae and a median (naupliar) eye on the ventral side. There are two pairs of maxillae. The trunk is made up of 19 segments, of which the first 9 bear paired appendages. The similar appendages are biramous (endopodite and exopodite), with a pseudoepipodite or plate as a type of third division. The terminal end (telson) of the abdomen bears 2 long spinelike structures. The ventral nerve

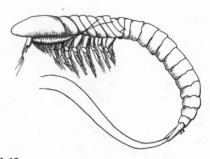

FIG. 14-16

Hutchinsoniella (subclass Cephalocarida), the most primitive crustacean. (Modified from Waterman.)

cord is composed of 2 separate cords connected by commissures. Cephalocarids are hermaphroditic and carry 2 egg sacs, each with an egg, suspended from the abdomen.

These tiny crustaceans are of primary interest because of their primitive characters, which place them among the most primitive crustaceans yet known. They share many characteristics of the ancient, extinct trilobites. In both groups all appendages except the first are used for locomotion and obtaining food. No appendage is specialized for a particular function. Both trilobites and cephalocarids have a seven-jointed endopodite, not found in other crustaceans. In cephalocarids a head limb (the second maxillae) is similar to the trunk appendages. This small group seems to represent an early stage in the evolution of crustaceans because its limb pattern is intermediate between the primitive arthropod limb plan and the pattern found in other crustaceans.

The original species found in Long Island Sound has been named *Hutchinsoniella macracantha* (Fig. 14-16). A similar species has been found in San Francisco Bay.

Subclass Branchiopoda

Most branchiopods live in fresh water, and some of them are the characteristic forms to be found in temporary pools during the spring and early summer. They are mostly absent from running water. One genus of them, the brine shrimp (*Artemia*), lives in salt lakes. Within the sub-

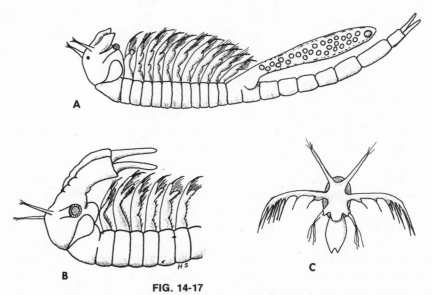

FIG. 14-17

Branchinecta, a fairy shrimp common in western regions of United States (order Anostraca). **A,** Adult female with brood sac. **B,** Anterior end of male showing enlarged antennae adapted as clasping organs. **C,** Nauplius larva. (Modified from Sars.)

class the various groups show many diverse characters (some of them primitive). The number of trunk segments ranges from 5 to 44 and may vary in the same species. The appendages are mostly restricted to the anterior part of the trunk. They are mostly foliaceous, being of a flat, leaflike form. A carapace is common to most members. It is a dorsal shield or in the form of a bivalved shell. These crustaceans have compound eyes and reduced mouthparts. The coxae of the appendages are provided with a flattened epipodite, which serves as a gill, so that most appendages are fitted for both respiration and locomotion. Most branchiopods are detritus and filter feeders, and their trunk appendages are adapted for such feeding.

The subclass Branchiopoda is commonly divided into four orders: Anostraca, Notostraca, Conchostraca, and Cladocera. The first three orders are sometimes grouped together in the division Eubranchiopoda, with body distinctly segmented and many thoracic appendages; and the order Cladocera is classified in the subdivision Oligobranchiopoda, with no external body segmentation, body enclosed by a folded carapace, and only five or six pairs of trunk appendages. The members of the division Eubranchiopoda are often called phyllopods be-

cause of their flat, leaflike appendages. The term "Phyllopoda" is sometimes used in place of Branchiopoda for the entire subclass. Their occurrence is very sporadic. A species may be abundant for several years in particular pools and then disappear entirely. A pond rarely contains more than one genus at a time.

■ ORDER ANOSTRACA

The members of Anostraca include the fairy shrimps (Fig. 14-17) and brine shrimps. They vary in size from 5 to 180 mm. in length. There are about 175 species of them, of which less than 30 are known in America. Their body is elongated, consisting usually of 19 segments, the first 11 segments of which (sometimes more) bear appendages. The first legless segments have the penes or the egg sac. There is no carapace. The eyes are compound and stalked (Fig. 14-11).

Fairy shrimps often have color, although many are transparent. Most of their colors are the result of refraction (green, blue, and reddish trimmings) and are very evanescent.

Anostracans have small antennules, generally, but the antennae in the male are strong, two-jointed, clasping organs. Fairy shrimp swim upside down. Two flattened cercopods are present on the telson. Their legs are foliaceous and, with exites, form filter chambers, into which water is sucked by leg movements and filtered through setae. In this way planktonic forms are collected and moved forward to the mouth in a ventral groove bordered with setules between the legs. Food particles in the grooves are formed into mucous chains, which are directed into the mouth mostly by the first maxillae.

The digestive system consists of a short esophagus and a midgut that is modified into a stomach. A dorsal tubular heart with ostia extends most of the length of all the body segments and forces blood to the tissues. Blood from the hemocoel is brought back to the heart. Blood contains colorless amebocytes and blood pigments such as erythrocruorin or some form of hemoglobin. The chief excretory organs are the coiled maxillary glands, which empty at the base of the second maxillae. *Artemia*, the brine shrimp, produces a hypertonic urine, although other crustaceans have a hypotonic or isotonic urine. Respiration takes place through the body surface and by means of the lamellar epipodites of the appendages. The nervous system is ladderlike, with a pair of ganglia and a pair of transverse commissures per segment as a rule. In copulation the male uses his large clasping organs to clasp the dorsal abdomen of the female and inserts the paired eversible penes into the female gonopore. The female develops a conspicuous ovisac at the junction of the thorax and abdomen when the eggs are extended from the uterus. Since fairy shrimp live in temporary spring pools, the clutch of eggs often must withstand desiccation when the pools dry up. Eggs will hatch in some species without undergoing drying or freezing. The young hatch as nauplii. Some Anostraca have as many as 17 instars in their life history.

Some common genera of fairy shrimp are *Branchinecta* (Fig. 14-17), *Eubranchipus,* and *Chirocephalopsis. Branchinecta gigas* in Oregon is the largest fairy shrimp (5-6 inches long).

■ ORDER NOTOSTRACA

Notostracans are commonly known as tadpole shrimp. They range in size from 20 to 90 mm. in length. There are twelve to fifteen species of them divided between two genera. Six species have been found west of the Mississippi River in North America. The number of their segments varies from 25 to 44, and each of the first 11 segments bears a pair of appendages and may be called a thorax. Behind these the segments may each have as many as 10 or more pairs of legs, which serve for both locomotion and respiration. Some of the segments are legless, and the posterior appendages taper off to a small size. Altogether the pairs of legs number 35 to 71. The telson bears 2 caudal filaments. The dorsal shieldlike carapace is rounded and flattened dorsoventrally and covers much of the anterior part of the body. The paired eyes are sessile, and the second pair of antennae are reduced or absent. The females carry their ovisacs on the eleventh pair of appendages. Their food is detritus. It is collected and transferred into the ventral groove by special setae (endites) or gnathobases, which are located on the basal part of the appendages, and then carried to the mouth. Tadpole shrimp live in the mud, where they burrow. Their dorsal tubular heart is short. Sperm is discharged through short vasa deferentia, which open separately by genital pores and not through penes. Some species are parthenogenetic. In copulation the male clasps the female at right angles to his own body. The eggs are carried by the female in a receptacle made by the fan-shaped flabellum and branchia on the eleventh paired trunk appendages. The eggs hatch into a nauplius larval form. The more familiar genus is *Triops* (Fig. 14-18). The only other genus is *Lepidurus. Triops,* by stirring up silt, is very destructive to rice fields.

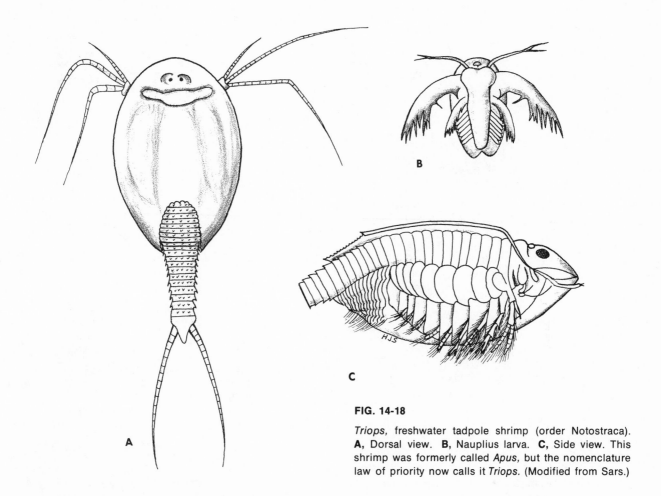

FIG. 14-18

Triops, freshwater tadpole shrimp (order Notostraca).
A, Dorsal view. **B,** Nauplius larva. **C,** Side view. This
shrimp was formerly called *Apus,* but the nomenclature
law of priority now calls it *Triops.* (Modified from Sars.)

■ ORDER CONCHOSTRACA

Clam shrimps, or claw shrimps, are mussel-
like crustaceans. About twenty-six species
are known from North America. The body is
enclosed in a bivalve shell that closes by a
transverse adductor muscle (Fig. 14-19). The
paired eyes are sessile. The trunk appendages
number 10 to 28 pairs. The telson is provided
with 2 clawlike cercopods. The legs are foliaceous, and there is a pair for each segment.
Filter chambers for the food are similar to those
in Anostraca. The second pair of antennae is
used for swimming as well as the trunk appendages. The clasping organs of the male are the
first two pairs of trunk appendages. In the
female elongated flabella of two pairs of posterior legs form a receptacle for holding a
mucous mass of developing eggs. Usually only
one generation is produced each year. The eggs
in some are brooded in a brood chamber beneath
the carapace.

Ecologically clam shrimps frequently occur in
muddy, alkaline waters. They crawl around on
the muddy substratum. They prefer warmer
waters for hatching and developing and are common later in the season than other phyllopods.

Of the many genera, the more common ones
are *Lynceus, Leptestheria* (Fig. 14-19, *A*), and
Eulimnadia (Fig. 14-19, *D*).

■ ORDER CLADOCERA

The small, familiar organisms of order Cladocera are from 0.15 to 3 mm. in length. They
number between 425 and 450 species. Clado-

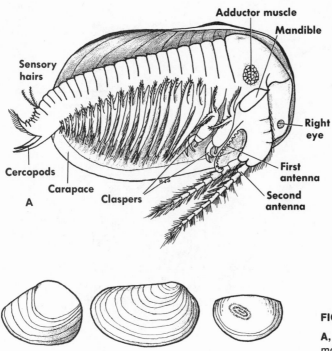

FIG. 14-19

A, *Leptestheria* (order Conchostraca), right valve removed. **B,** *Limnadia,* right valve. **C,** *Cyzicus,* right valve. **D,** *Eulimnadia,* male shell.

FIG. 14-20

Daphnia, the water flea, a common cladoceran. (From Encyclopaedia Britannica film, "Microscopic Life: The World of the Invisible.")

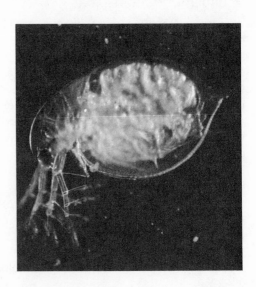

cerans are commonly known as water fleas and are closely related to the conchostracans, with which they are sometimes placed as suborders under the order Diplostraca. An unhinged valve shell, or carapace, encloses the trunk and trunk appendages but leaves the head and antennae free (Fig. 14-20). The first antennae are small, but the second are very large, biramous, and branched. The head projects ventrally as a beaklike rostrum, and the body is compact with no external segmentation. The head bears a sessile, median compound eye and a nauplius eye, or ocellus. The five or six pairs of trunk appendages are flattened, leaf shaped, and fringed with setae. Their movement creates currents of water for oxygen and food particles. The true abdomen is reduced, but there is a large postabdomen at the posterior end of the body. It is bent downward and forward and bears two terminal claws as well as two abdominal setae. In general the antennae are the chief organs of locomotion. Movement is jerky and by spurts.

The mouth is located near the junction of the head and body and has as mouthparts a labrum, a labium, a pair of stout mandibles, and a pair of pointed maxillae. The transparent body reveals most of the visceral organs. Food is collected and filtered by the setose appendages into the ventral median groove between the base of the legs. The mouth opens into an esophagus, which leads to the stomach and intestine. Digestive ceca may open into the anterior part of the intestine, which may be straight or convoluted. The anus is usually placed on the lower or dorsal part of the postabdomen.

The oval heart is on the dorsal side behind the head and has 2 ostia (Fig. 14-22, A). Blood

FIG. 14-21

Some internal structures of cladocerans. **A**, Circulation in heart and adjacent sinuses of *Simocephalus*. **B**, Central nervous system of *Daphnia*. (Modified from Herrick.)

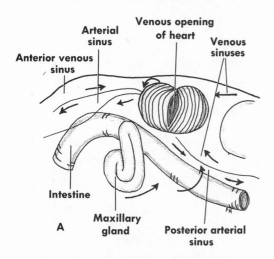

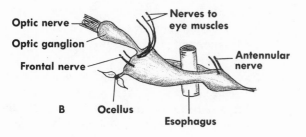

vessels are absent, and blood is circulated in the hemocoel among the many thin mesenteries (Fig. 14-21, A). The beat of the heart can easily be seen in the living animal. The rate of beat is influenced by temperature, time of day, and condition of the adult. The blood contains many colorless corpuscles. The blood is colorless in well-oxygenated water but pink in stagnant water. The respiratory pigment is erythrocruorin. Looped shell or maxillary glands function as excretory organs in the anterior part of the trunk. The nervous system is a double ganglionic chain, except that it may be concentrated in a few species. The brain lies dorsal to the esophagus (Fig. 14-21, B). In addition to the eyes, there are sensory abdominal setae and sensory hairs on the basal segment of the antennae.

Parthenogenesis is common among cladocerans. Several generations of female young are produced. After undergoing a single maturation in the ovary the eggs are released through the oviducts into the brood chamber. The number of eggs per clutch is usually between 10 and 20. After further development in the brood chamber the eggs hatch. Under certain conditions of temperature, scarcity of food, congestion, and waste accumulation, a generation arises with both males and females. Syngamic eggs with thick shells are then possible and are the winter eggs. The winter or dormant eggs give rise to females when they hatch. In some forms, such as *Daphnia* (Fig. 14-22), the large fertilized eggs are carried in special dorsal pouches (ephippia) that are cast off with the eggs when the animal molts. The ephippia serve as protective cases for the discharged eggs, protecting against drying and freezing, and can be dispersed on the mud, picked up by aquatic birds or by other agents. Some males also appear in the spring, but the factors controlling their appearance are still unknown. Populations of cladocerans may consist entirely of parthenogenic females, or they may consist of males, unfertilized sexual females, and fertilized sexual females. Population cycles are common in

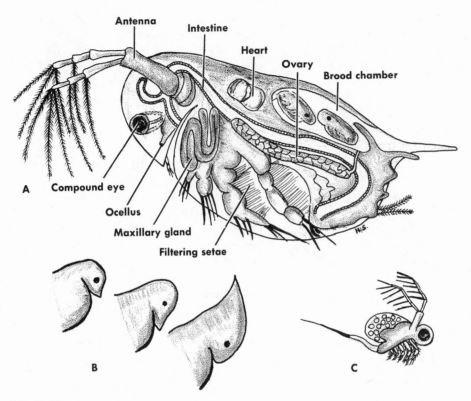

FIG. 14-22

A, Water flea *Daphnia pulex*, female (order Cladocera). **B,** Some seasonal head variations (cyclomorphosis) of *Daphnia*. Temperature seems to be controlling factor. **C,** *Polyphemus*, a predaceous cladoceran. Cyclomorphosis also occurs in rotifers, in which it may be correlated with their parthenogenetic development. (**A** and **B** modified from Pennak.)

cladocerans. Some may have one seasonal rise in numbers during the warmer months (monocyclic), and others may have two pronounced maxima in spring and fall (dicyclic). The mode of reproduction is influenced by environmental factors, according to experimental work.

Length of life is also highly variable under different environmental conditions. In general it is only 30 to 60 days but may be longer. The first molt produces the juvenile; three to five instars as a rule produce a sexually mature animal. Some cladocerans may undergo twenty-five instars during the adult stage.

Cyclomorphosis, or the changes in gross morphology occurring in successive seasonal variations, is striking in some cladocerans (Fig. 14-22, *B*). Females of *Daphnia pulex* and *Daphnia longispina* undergo a change from the round-headed condition, which they have in late fall to early spring, to a helmet-shaped one during the warm summer months. Later, in the fall, the head gradually reverts back to its original shape. Cyclomorphosis varies greatly with different kinds of aquatic habitats and is less pronounced in shallow water.

Although cladocerans are mostly freshwater forms, there are some marine forms. They are especially common in aquatic vegetation along pond, lake, and river margins. They may occur in both temporary and permanent pools. They have a wide temperature range, although some are restricted to warmer waters. They live on a great variety of foods such as protozoans, bacteria, algae, and organic detritus. The genera *Polyphemus* (Fig. 14-22, *C*) and *Leptodora* are predaceous and have appendages modified for seizing such prey as rotifers. *Anchistropus* is parasitic on hydra, whose favorite food is cladocerans such as *Daphnia*.

Subclass Ostracoda

The common name of the members of this subclass is seed shrimps, or mussel-shrimps. Ostracods have a worldwide distribution and comprise about 2,000 named species. They inhabit both fresh water and the sea. One or two species are terrestrial. Most are free living, but some such as *Entocythere* and *Sphaeromicola* live as commensals on the gills of crayfish and on amphipods and isopods. *Cypridopsis* preys on certain aquatic snails. Ostracods have some resemblance to conchostracans, for both have bivalved carapaces, but ostracods have fewer appendages. Ostracods vary in size from about 0.25 to 8 mm., although one marine species is more than 20 mm. long.

External morphology

The body and appendages are enclosed in a calcium-impregnated bivalve shell, which is oval shaped. Dorsally a noncalcareous hinge of cuticle holds the valves together. The shell has three layers: a waxy outer layer, a middle layer of organic substances and minerals, and an inner layer of chitin and protein. The shell may be pigmented (light and dark colors) and bear spines, tubercles, and setae. Adductor muscles close the valves. When the animal is active, the valves gape open and the locomotor appendages protrude. Ostracods lack concentric lines of growth on the valves.

The body has lost all traces of segmentation, and the trunk is much reduced. The head makes up about one half of the body. A single or double eye is usually present. The head region bears four pairs of appendages: antennules, antennae, mandibles and maxillae (Fig. 14-23). Both the first and second pairs of antennae are well developed. The first pair (antennules) have clawlike bristles for digging. The antennae are

adapted for locomotion and (in the male) for clasping the female during copulation. The mandibles are strongly toothed and bear 3-segmented palps. Each maxilla consists of a branchial plate and 4 basal processes, the outmost of which is palplike. The thoracic region usually has three pairs of appendages, but there may be fewer among the different species. They are commonly referred to as walking legs, but some are not used for that purpose. They may or may not be similar. The first thoracic legs may be modified in the males of some to form prehensile structures for grasping the female during copulation; the second thoracic legs have tapering claws; and the third thoracic legs are dorsally bent and specialized for keeping the inside of the shell cavity free from debris. The remainder of the trunk (sometimes referred to as the abdomen) has 2 unsegmented caudal or furcal rami, which are variable and may be mere spines.

Most ostracods live near or on the bottom substratum, often in the ooze. They burrow for food by means of beating movements, especially by the two pairs of antennae that are specialized for locomotion. Well-developed muscles are attached to both pairs of antennae as well as to the other appendages and to the mouthparts. The antennae may be modified in other ways for climbing vegetation or for swimming. Some ostracods can bounce or hop around. Some species are found on sandy bottoms and others on soft mud.

Internal morphology

The digestive system consists of the mouth, the foregut variously modified, the midgut, and the hindgut. The food consists of bacteria, algae, organic detritus, and dead animals. In general they are considered scavengers. Beating movements of the mandibular palps, the maxillae, and the branchial plates of the first legs create currents of water between the valves. Fine particles are strained out by the setae on the appendages and directed to the mouth. Inedible materials are mostly removed by the respiratory

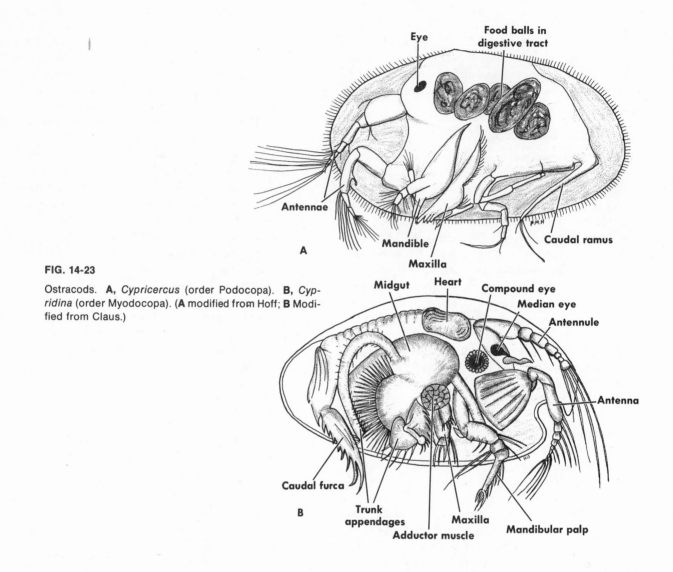

FIG. 14-23

Ostracods. **A,** *Cypricercus* (order Podocopa). **B,** *Cypridina* (order Myodocopa). (**A** modified from Hoff; **B** Modified from Claus.)

plates of the maxillae. Mucous strings of food may be employed by some species to convey food into the mouth. Large particles are broken up by the mandibles. The foregut may be modified into a short esophagus and a region where in some (order Podocopa) a type of gastric mill is found. The midgut is fitted for absorption, and in some a digestive gland on each side empties into it. Following the intestine are the hindgut and the anus, which lie at the base of the caudal rami.

Ostracods have both antennal and maxillary

glands, presumably excretory in function. A pair of large shell glands of unknown function opens near the base of the second antennae.

Most ostracods have no special respiratory organs such as gills. Respiration takes place through the inner epithelium of the shell lamellae, branchial plates of the appendages, and antennae setae. Heart and blood vessels are restricted to the order Myodocopa, in which the heart has two ostia and is provided with anterior and ventrolateral arteries.

The central nervous system is made up of a

brain (supraesophageal ganglion), a pair of circumesophageal connectives, a subesophageal ganglion, and a ventral chain of fused ganglia. In some species all the ganglia are fused in one mass. Nerves include the optic; those to the two pairs of antennae, the mandibles, and the maxillae; and those to the appendages. The sense organs include a nauplius eye (Fig. 14-10) and sessile compound eyes in the order Myodocopa. Other sense organs are auditory structures at the base of the antennules, an olfactory organ on one of the segments of the antennae, and setae or sensory hairs on various appendages.

Parthenogenesis is found among some members, but in general they are dioecious. Sex ratios may vary greatly in some species, with the males in a minority. The female system consists of a pair of tubular ovaries in the posteroventral region of the shell, paired oviducts with seminal receptacles, and genital openings between the bases of the third thoracic legs and caudal rami. Male organs include the testes (each of 4 coiled tubes placed between the valve lamellae), 2 sperm ducts (vasa deferentia), a pair of ejaculatory ducts, and a pair of penes just in front of the caudal rami. The ejaculatory ducts may not open through the penes in a few. In copulation the male becomes attached to the female shell dorsally and inserts the penes between the valves into female genital pores. The sperm are unusually long for a crustacean. Fertilization occurs in the seminal receptacle, and the eggs are brooded in the dorsal part of the shell cavity in some (*Darwinula*). In most forms the eggs are dropped into the water where they may be attached to vegetation. Eggs are viable for long periods of time under harsh conditions. Normally the eggs hatch within a few days to weeks after they are shed. The egg hatches into a bivalved nauplius. Sexual maturity is not attained until the ninth or last instar.

■ ORDER MYODOCOPA

The members of order Myodocopa are worldwide in distribution and are entirely marine.

They include the two families Cypridinidae and Conchoeciidae. The shell is provided with a notch through which the second antennae are extended when the shell is tightly closed. They have a heart, blood vessels, and compound eyes. No digestive glands empty into the intestine. The exopodites of the antennae, with well-developed muscles, are fitted for locomotion; the endopodites in the male are adapted for grasping. There are seven pairs of appendages and the caudal rami. A frontal organ on the head is also present. The subfamily Asteropinae are the only ostracods with true gills.

Cypridina (Fig. 14-23, *B*), *Pyrocypris,* and a few others are luminescent. The bluish light they produce is from secretions of a gland in the labrum. The light is under reflex control. The genus *Gigantocypris*, a deep-sea form, is the largest ostracod (10 mm. long).

■ ORDER CLADOCOPA

The members of order Cladocopa have no permanent aperture in their shell for the antennae. The shell can close tightly all around. They are marine and consist of one family, Polycopidae. Their biramous antennae are much flattened for locomotion. They have no heart or eyes. The body terminates in a process bearing clawlike spines. There are no trunk appendages.

The common genus *Polycope* has a smooth, circular shell. This is the smallest order of ostracods; they have been collected mainly in Atlantic water.

■ ORDER PODOCOPA

Order Podocopa includes mostly freshwater forms, but there are marine representatives also. It is a large order and is made up of four families based on the thoracic legs and caudal rami. The families are Cypridae, Darwinulidae, Cytheridae, and Bairdiidae. The second antennae are uniramous, and two pairs of trunk appendages are present. The valves of the shell have no aperture and are usually flattened on the ventral side. In the family Cypridae (Fig. 14-23, *A*) all the thoracic legs are different.

The first pair of thoracic legs is modified for mastication, and the endopodites form a prehensile palp in the male. The third pair of legs is adapted for keeping the shell clean. This family has marked seasonal population cycles and has both marine and freshwater species. The family Darwinulidae also has the first pair of thoracic legs modified for mastication, but the second and third pair are adapted for crawling. This family lacks caudal rami, and the female bears living young. One genus, *Darwinula*, with two species is known. One of the two species lives in fresh water. In the family Cytheridae the three similar pairs of thoracic legs are adapted for crawling, and there is no ejaculatory tube. Most members of this family are marine. In the family Bairdiidae only marine forms are found. The valves of their shells are unequal in size, the left being the larger. Their antennae are not adapted for swimming, but three similar pairs of thoracic legs are all adapted for locomotion. No ejaculatory ducts are present. Two common genera are *Bairdia* and *Bythocypris*. The fossil record of the Podocopa is better known than that of most other groups of ostracods, dating back to the Ordovician period.

■ ORDER PLATYCOPA

Order Platycopa consists of a single family of ostracods, the Cytherellidae. The biramous second antennae are large and have flattened segments. Neither pair of the antennae is used for locomotion. Heart and visual organs are absent. The three pairs of thoracic legs are non-leglike. All platycopids are marine. The only genus is *Cytherella*.

Subclass Mystacocarida

The small crustaceans of subclass Mystacocarida share with the Cephalocarida the distinction of being among the most primitive of living crustaceans. They were first discovered off the coast of New England in 1939. A different species has since been found along the coasts of Spain, France, and Italy, as well as another species off the coast of Chile. They are less than 5 mm. long and live in the intertidal sand. The body is elongated and wormlike. Head and thorax are separate. The paired maxillipeds are on a separate segment that is not fused with the head. The thorax proper has 4 segments, each with a pair of platelike appendages. The 6 abdominal segments have no appendages but bear a pair of caudal rami. A nauplius eye is present. The labrum is very large, and the mouth appendages bear setae that may be used in filter feeding. Sexes are separate, and the genital pore is on the thorax.

All three species belong to the genus *Dero-*

FIG. 14-24

Derocheilocaris (subclass Mystacocarida). This group shares with cephalocarids distinction of being among most primitive of crustaceans. (Redrawn from Pennak: McGraw-Hill encyclopedia of science and technology, vol. 8, McGraw-Hill Book Co.)

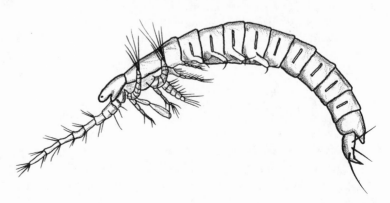

Their abundance and wide distribution have made them an important food item for fishes and other animals. Most of them are less than 2 mm. long, but some are as small as 1 mm. or less, whereas others may reach a length of 3 to 4 mm. Parasitic forms may be even larger. Most species are grayish and transparent, but freshwater forms in high altitudes may be orange, red, or other colors. There may also be seasonal variations in colors.

External morphology

Copepods have a similar basic pattern, but there are many variations among them. In general the body is cylindric and definitely segmented. It is composed of head, thorax, and abdomen. The head is composed of 1 segment, rounded anteriorly, and sometimes bears a rostrum. At its posterior end the head is fused with the first (and sometimes second) thoracic segment. The head and the fused segments of the thorax are often referred to as the cephalothorax, forming as they do a more or less fused compact structure. Compound eyes are absent, but there is a single, median (naupliar) eye. The true head region consists of five pairs of appendages: the antennules, the antennae, the mandibles, the first maxillae, and the second maxillae (Fig. 14-26). The first thoracic segment, which is fused to the head, bears a pair of maxillipeds. The two pairs of antennae are adapted for swimming, balancing, or grasping. The other anterior appendages are for feeding. Counting both fused and unfused segments, the thorax is usually composed of 6 segments. Each of the unfused thoracic segments bears a pair of swimming legs.

The abdomen consists of 1 to 5 segments and bears no appendages except a pair of caudal rami. In some species the seventh or genital segment has a pair of vestigial swimming legs. A common way of dividing the copepod body is to separate it into (1) an anterior metasome, consisting of the head, the cephalothorax, and the first two pairs of unfused thoracic segments and (2) a posterior urosome, composed of the

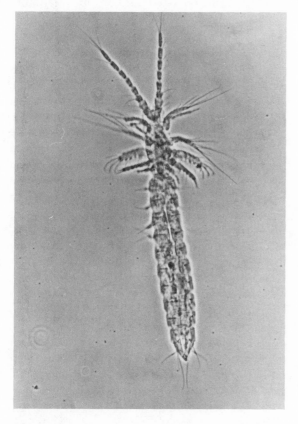

FIG. 14-25

Derocheilocaris typicus, ventral view (subclass Mystacocarida). Two other species of this subclass have been found since *D. typicus* was discovered in psammon habitats in 1942. (Courtesy R. P. Higgins.)

cheilocaris (Fig. 14-24). A photograph of one of the species is also shown in Fig. 14-25.

Subclass Copepoda

The copepods are strictly aquatic forms that are found in all kinds of water except that which is very salty. They represent the largest group of all the small crustaceans, the so-called Entomostraca. There may be as many as 6,000 named species. Most are free living, but some are parasitic or commensal on other animals.

last thoracic segments and the abdomen. There are many variations of fusion conditions among the different groups of copepods.

Basically the head and thoracic appendages are biramous, but there are many modifications of this plan. The antennules are uniramous and may contain as many as 25 segments. The antennae are usually biramous, much smaller than the antennules, and may have 9 segments in the endopodite. In some the exopodite is missing. The biramous mandibles may have 4 segments in both branches. The two pairs of maxillae and the maxillipeds may be biramous or uniramous. The biramous swimming legs (usually with three segments in each branch) are generally similar, but the last two pairs may be highly modified. They bear many setae. The sixth pair is lacking in the female and may be vestigial in the male. Locomotion in copepods is mostly by jerky movements produced by the action of the swimming legs. While the animal is swimming, the antennules are carried alongside the body. Some make use of the two pairs of antennae in swimming. A few swim upside down.

Internal morphology

Copepods have a variety of feeding habits, depending on whether they are planktonic or benthonic in their distribution. Some have their mouthparts adapted for filter feeding and others for seizing and biting. In filter feeders the antennae, mandibular palps, and the first antennae are provided with setae, which by rapid vibrations create a current of water alongside the body. This swirl of water is partially directed into a median filter chamber by the maxillipeds. From this current, particles of food are filtered, collected, and passed to the mouth by the action of the setae and endites of the maxillae and maxillipeds. The members of the order Cyclopoida have the mouthparts adapted for capturing their prey of unicellular plant and animal organisms. Some copepods are filter feeders as well as predaceous and have their mouthparts modified accordingly.

In transparent copepods the digestive system can be seen clearly because the food outlines its contour. The system is a more or less straight tube with few regional specializations. The anus is placed on the last abdominal segment be-

FIG. 14-26

Female *Diaptomus*, a calanoid copepod. (Modified from Herrick.)

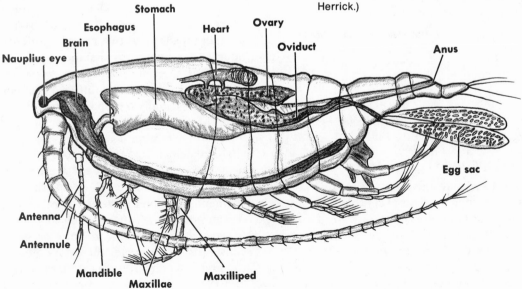

tween the bases of the caudal rami. A simple heart (Fig. 14-26) is present only in the order Calanoida (*Diaptomus*) and some parasitic forms; in others the blood is moved through the hemocoel by muscular actions of the body, appendages, and digestive tract. No gills are present, and gas exchange occurs through the body surface. Maxillary glands provide the means of excretion. The nervous system is mostly concentrated into a mass by coalescence of the ganglia. Distinct ventral ganglia are restricted to a few groups. Besides the nauplius eye, which shows up as a clear median pigmented spot near the anterior end, tactile sensory organs are scattered over the body, especially on the appendages. Light organs, which make use of luminescent secretion of glands in the integument, are found in some species and may be used for sex recognition.

Reproduction

The female reproductive system consists of an ovary (single or paired) and 2 glandular oviducts with diverticula. The genital opening is on the ventral surface of the first abdominal segment (Fig. 14-26). Near it are the separate openings of a pair of seminal receptacles. There may be connections between the seminal receptacles and the oviducts. The male system usually comprises a single testis and paired or single sperm ducts. The male genital opening is also on the first abdominal segment. The antennae, and sometimes the fifth pair of thoracic appendages, are modified for clasping the female's abdomen in copulation. Sperm are transferred to the female genital segment in spermatophores, which are stored in the seminal receptacles. Usually only one mating is necessary for the fertilization of all subsequent eggs laid by the female. Fertilized eggs are extruded into 2 ovisacs secreted by the oviducts. Some have only one sac and others have no sac at all. Each ovisac may contain 5 to 40 eggs. The female may shed the entire ovisac or shed the eggs singly into the water. The number of generations varies with the species. Some fresh-

water forms produce a generation at least every 2 weeks; others have three or four broods a year, and even fewer. In many the largest number of eggs is produced in the spring. In the familiar *Cyclops* the incubation period in the ovisac is from 12 hours to 5 days, and as soon as the newly hatched ones are shed, a new brood is produced in new ovisacs. Only one or two species are known to be parthenogenetic.

Development and life history

The copepod egg hatches into a nauplius larva, with its characteristic three pairs of appendages (antennules, antennae, and mandibles). There may be as many as 6 naupliar instars, but some have fewer, especially in the parasitic forms. In each successive instar new appendages are added until there are seven or eight pairs. When the fifth and sixth nauplius stage molts, the instar resembles the adult and is called a copepodid. In six copepodid instars the adult finally emerges. In parasitic forms there may be fewer copepodid instars. The time for complete development may be only a few days, or it may last several months.

Most of the parasitic crustaceans are found among the copepods. Most groups of animals are affected by parasitic copepods. Some orders are exclusively parasitic. Parasitism has produced striking structural modifications in many of their species. Certain aspects of parasitic forms will be described under the appropriate orders.

Habitats and ecology

Most copepods live in the sea. They form an important part of plankton populations. Most of them live in the upper 200 to 300 meters of water. However, there are others that are abyssal or bathypelagic. Some are bottom dwellers and many are found clinging to aquatic vegetation. Some species are adapted for living in the sand. Diurnal migrations are common among those that live near the surface waters.

Populations of copepods may reach tremendous numbers. Up to 600,000 or 700,000 have been found in each cubic meter of fresh water. They play important roles in converting plant into animal substance. They make up as much as 70% of the animal populations of plankton and serve as food for innumerable animals such as fishes and baleen whales.

Temperature may play an important role in their worldwide aquatic distribution. Some are definitely cold-water forms, and others prefer southern waters. Some are known to form cysts under harsh environments. Such encystments usually occur in some of the copepodid stages rather than in adult stages.

■ ORDER CALANOIDA

Most calanoids are planktonic forms and probably represent the most abundant and economically important of all marine animals. They occupy a unique position as the base of the animal food chain. The number of their species may be in excess of 1,200, which are divided into 120 genera. They are found in all levels of the ocean from the surface to abyssal depths. Many of them live in fresh water. They generally migrate toward the surface at night and return to deeper levels at day. Many of them bear very long antennae and plumose setae, which aid them in flotation. These free-living copepods have the metasome-urosome articulation between the fifth and sixth thoracic segments. Many of them are brightly colored, and some have brilliant displays of bioluminescence. They are mostly filter feeders and have certain mouth appendages specialized as passive filters. They possess a saccular or tubular heart, pro-

FIG. 14-27

Some copepods. **A,** *Calanus,* a planktonic form (order Calanoida). **B,** *Harpacticus* (order Harpacticoida) is found in marine, brackish, and fresh water. **C,** *Lernaea* female taken from gills of flounder (order Lernaeopodoida). **D,** *Lepeophtheirus* young attaches by filament to gills of fish (order Caligoida). **E,** *Doropygus* female commensal or parasitic in ascidians (order Notodelphyoida). **F,** *Monstrilla,* from marine plankton (order Monstrilloida). (Modified from: **A,** Miner; **B,** Lang; **C** and **D,** Scott; **F,** Davis.)

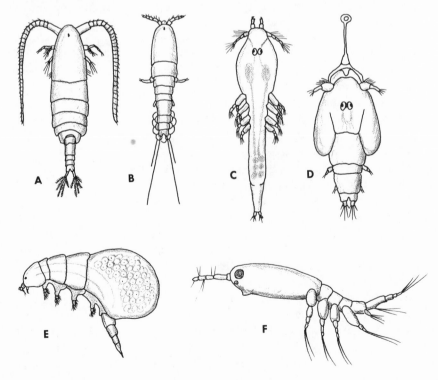

vided with a single, unbranched aorta. The sperm duct is single. Usually the antennae are not specialized for grasping the female, and the spermatophores are transferred to the female by the male thoracic appendages.

Common genera are *Calanus* (Fig. 14-27, *A*), *Diaptomus* (Fig. 14-26), *Metridia,* and *Calocalanus.*

■ ORDER CYCLOPOIDA

The members of order Cyclopoida are found in both salt and fresh water. Some are planktonic and others are benthonic. The order includes a number of parasitic forms as well as free-living forms. Some are intermediate hosts for human parasites. The body in general is oval shaped, with the abdominal part rather sharply set off from the anterior region. There are usually 10 segments in the male and 9 in the female, in which 2 segments are fused at the genital region. The metasome-urosome articulation is placed between the fourth and fifth thoracic segments. Two caudal rami with unequal setae are present. The antennae are uniramous, but most appendages are biramous. The antennules of the male are adapted for grasping the female. No heart is present, and the colorless blood is circulated by general body movements. The anus is found on the dorsal side of the last abdominal segment. The large ovisacs are paired. A dorsal brain is connected by circumesophageal connectives to a ventral chain of ganglia. Cyclopoids are not filter feeders; food is seized and eaten by the biting mouthparts.

Cyclops, of which there are many species, is the most common genus.

■ ORDER HARPACTICOIDA

Order Harpacticoida have a worldwide distribution, both in fresh water and in the sea. They are mostly free living and benthonic but are found at all levels. Some species are pelagic, commensal, or parasitic. There are about 1,400 species. They vary in length from 0.3 to 4 mm. In general the first thoracic segment is fused to the cephalothorax, and the last thoracic segment is a part of the abdomen. The male biramous antennules are specialized for grasping. The maxilliped and first part of the swimming legs may also be prehensile. The metasome-urosome articulation is between the fourth and fifth thoracic segments. The ventral ovisac may be single or double. The abdomen is unusually wide for copepods. The nervous system consists of a brain, subesophageal ganglion, circumesophageal ring, and a ventral chain of ganglia. Frontal organs are present in some. An aesthete, or long sensory process, may be present on the antennules and other appendages. The 2 oviducts open into a seminal receptacle. The species *Elaphoidella bideris* is parthenogenetic. Two other well-known genera are *Harpacticus* (Fig. 14-27, *B*), found in marine, brackish, and fresh water, and *Epactophanes,* a freshwater form often found in damp moss and on sandy beaches.

■ ORDER MONSTRILLOIDA

The small order Monstrilloida has about 35 species, most of which are rarely encountered. The adults lack antennae and mouthparts. In their life history the nauplius is free swimming, but other immature stages are parasitic on marine invertebrates, especially polychaetes. After emerging from the host the adults become free swimming in plankton populations. The digestive tract is vestigial, and food is absorbed directly from the host. Dorsolateral eyes may be present as well as the single median eye.

Monstrilla, a marine form (Fig. 14-27, *F*), is the best known genus.

■ ORDER NOTODELPHYOIDA

Notodelphyoids, like monstrilloids, are parasitic or commensal copepods. There are about 300 known species, all marine. Few can be called parasitic in a strict sense, but they live in the cavity or pharynx of tunicates, whose food they share. The metasome-urosome articulation is located between the fourth and fifth thoracic segments in males and between the first and second abdominal segments in females. Their size is similar to that of planktonic copepods. Because of their commensal habits, notodelphyoids are only slightly modified from free-living copepods. The exopodites of the thoracic limbs often bear claws for attachment to the host. The trunk and head appendages are usually small. Eggs are normally incubated in a large brood pouch in the dorsal region of the thorax. Larval development includes several naupliar instars. Infective stages are usually early copepodid stages.

Among the many genera are *Ascidicola, Notodelphys,* and *Doropygus* (Fig. 14-27, *E*).

■ ORDER CALIGOIDA

Caligoida are chiefly ectoparasites on freshwater and marine fishes. They live on the host's blood by becoming attached to the gills, fins, integument, and other parts of the body. They usually have a distinct carapace and fused segments. The antennae and maxillipeds are prehensile, and the sucking mouth is provided with pointed mandibles for piercing the host's skin. Swimming legs are much modified, with loss of setae, reduction of joints, and fewer numbers. There are about 400 species. The metasome-urosome articulation is between the third and fourth thoracic segments. They vary in size from a few millimeters to the largest copepod *Penella,* which may reach a length of several inches. There are many morphologic variations in the order. Many of these modifications are clinging adaptations. Copulation may occur in the last larval stage, with the male dying soon after.

Penella has been found in great numbers on whales as well as on other forms. *Caligus* on the skin and gills of fishes and *Lepeophtheirus* (Fig. 14-27, *D*) on the skin of freshwater fishes are other common genera.

■ ORDER LERNAEOPODOIDA

Lernaeopodoida are commonly known as fish maggots. They are ectoparasitic on freshwater

and marine fishes. The body has little or no segmentation, especially in the female, which appears wormlike. Males are very small but are distinctly segmented. Thoracic appendages may be reduced or absent. The second maxillae are strongly modified for attachment after they are buried in the host's tissue. Postembryonic development is reduced to two or three stages. The free-swimming larvae emerge from the ovisac in the first copepodid stage. Some are provided with a preformed, coiled frontal organ, which can be everted and embedded in the host's tissue. This tube finally degenerates when the larva emerges into the vermiform stage.

These ectoparasites feed on the blood or body fluids of the host. They number about 300 species.

Salmincola, Lernaea (Fig. 14-27, *C*), and *Brachiella* are common genera.

Subclass Branchiura

Branchiura are known as fish lice, of which there are fewer than 100 species, all included in the order Arguloida. Their morphology shows fewer variations than those found in most other groups of small crustaceans. The body is made up of a flattened, disk-shaped cephalothorax and a small, unsegmented, bilobed abdomen. They have a pair of antennules, a pair of antennae, a pair of mandibles, two pairs of maxillae, and four pairs of swimming legs. The appendages follow the longitudinal axis on the underside of the body and are covered by the thin cephalothorax. This adaptation fits them for their role of ectoparasites against the skin of a fish host. They have compound eyes. They are adapted for attachment by having hooks and suction cups on the underside of the body. The second to fourth pairs of swimming legs, which have a reduced number of segments, are used for copulation. There are usually no larval stages or nauplius, and development into a juvenile is direct. Sexual maturity requires about seven molts. They feed on blood and other tissue fluids by making rasping wounds with their mandibles. *Argulus* (Fig. 14-28) often

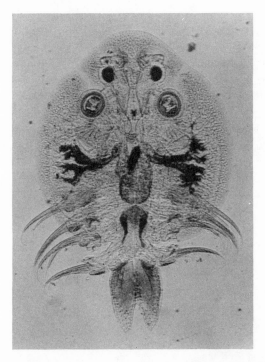

FIG. 14-28

Argulus (ventral view), a common ectoparasite on fish (subclass Branchiura). Adult fish lice range from 5 to 25 mm. in length. The 2 large suckerlike bodies are modifications of the first maxillae, and the 2 smaller dark spots are eyes.

leaves its host to attach its eggs to some object.

Argulus is the most common genus.

Subclass Cirripedia

The cirripedes in the adult stage are a group of attached crustaceans. They consist of the barnacles, acorn shells, and some related forms of different habits, including some parasitic forms of highly modified structure. They are exclusively marine. There are more than 800 species divided among five orders. Barnacles make up more than half of all the cirripedes.

Since there are few common characteristics that apply to all the orders, each will be considered separately. Their relationships are indicated by their larval development.

■ ORDER THORACICA

Barnacles can be divided into the stalked and sessile types. Stalked forms have a long peduncle, which comprises the preoral part of the animal. Most Thoracica are free living and some are commensal, but it is doubtful that any of them are truly parasitic. Barnacles will attach themselves to almost any object of the sea, dead or alive, to which they can fasten themselves (Figs. 14-29 and 14-30). Their fouling of ships and pilings is known in all seas.

The body of the stalked barnacle is enclosed in a soft mantle (capitulum), which is commonly protected by calcareous plates (Fig. 14-31). The capitulum includes all the body except the preoral part. The calcareous plates usually include the anterior keellike carina, 2 lateral terga, and 2 posterior scuta. By means of adductor muscles the margins of the terga and scuta can be pulled close together for protection or opened for extrusion of the appendages or cirri. In some barnacles there is a tendency for the number of calcareous plates to be reduced. In *Conchoderma*, often found on ships, nearly all the plates are greatly reduced or lost.

In sessile barnacles the body is enclosed within an immovable wall formed by the overlapping of 4 to 8 plates. The movable operculum of 2 to 4 plates is provided with muscles for opening and closing (Figs. 14-32 and 14-33). Sessile barnacles usually have wide attachment bases, which are calcareous or membranous. This corresponds to the preoral region of stalked barnacles and bears cement glands. In the sessile forms the plates are modified and adapted for different purposes from those in the stalked forms. The lateral plates become the mural plates in sessile barnacles. The mural plates in partial or complete fusion with the carina and rostrum form a complete vertical wall around the body of the animal. The other stalked barnacle plates, terga and scuta, cover the top of the animal. Acorn barnacles (Fig. 14-32) have a conical shell and are cemented directly to the rocks. Their orifice is opened or closed by 4 movable plates. The largest barnacle

FIG. 14-29

Large cluster of *Lepas* (gooseneck barnacles, suborder Lepadomorpha) attached to fishing-net glass float. Collected near Vancouver, British Columbia, in August, 1958. (From Hickman: Integrated principles of zoology, The C. V. Mosby Co.)

FIG. 14-30

Closeup of few of *Lepas* shown in Fig. 14-29. Note cirri extending between valves (order Thoracica, subclass Cirripedia).

is perhaps *Balanus psittacus*, which grows to a length of 8 to 10 inches with a diameter of 3 inches. There are many variations in the arrangement of the plates in sessile barnacles. Plates may be reduced, or they may be fused together in a continuous or compact wall. The terga and scuta may be lost altogether, or the wall surrounding the animal may consist of 1 tergum and scutum, with the other tergum and

FIG. 14-31

Mitella, a stalked barnacle (order Thoracica, subclass Cirripedia) showing long peduncle and plates covering capitulum. Compare plate arrangement with that of *Lepas.*

scutum making up the lid. Some forms are completely naked.

The arrangement of the body is such that the long axis of the mantle is at right angles to that of the body. This organization makes it possible for the cirri to be directed toward the mantle

FIG. 14-32

Group of acorn barnacles *Balanus* (order Thoracica, subclass Cirripedia).

FIG. 14-33

Balanus amphitrite. **A,** Enclosed by valves. **B,** Valves partially removed to expose soft parts. (Drawn largely from life.)

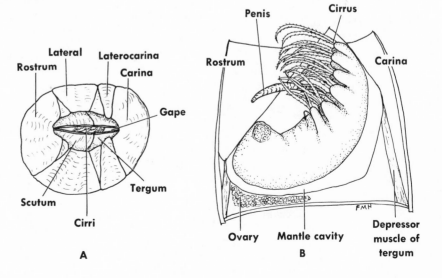

opening. The body is composed chiefly of a head and a thoracic region; the abdomen is lacking or represented only by the caudal rami or furca. The remains of the antennules are usually found in the cement glands, which serve for the attachment of the larva (Fig. 14-35).

This cement is one of the strongest adhesives known to man and has led to intensive investigations of its properties by the National Institute of Dental Research. The cement of the common acorn barnacle (*Balanus*) may be an adequate adhesive material in filling teeth. The cement is insoluble in most solvents, and chemically it is a polysaccharide, not a protein.

The antennae are lacking in the adult but are present in the larva. The mouthparts consist of a labrum, paired mandibles bearing palps, maxillulae, and maxillae (Fig. 14-34). There are usually six pairs of long, curled, biramous thoracic appendages or cirri. These are fringed with hairs and can be protruded from the slitlike opening of the mantle to sweep in small particles of animal or plant food. The 6 cirri on each side form a type of net, and the 2 nets come together to enclose the food, which is scraped off and passed to the mouth by the maxillulae.

The digestive system includes a foregut, an enlarged midgut with glands, a hindgut,

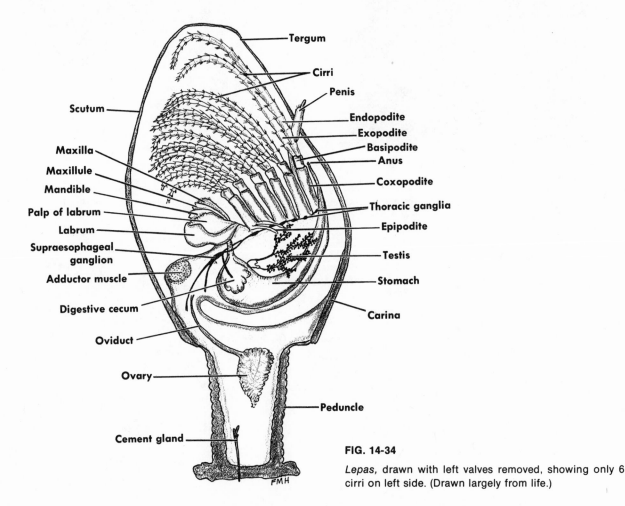

FIG. 14-34

Lepas, drawn with left valves removed, showing only 6 cirri on left side. (Drawn largely from life.)

and an anus that opens near the base of the cirri. Paired maxillary glands (excretory) open on the maxillae. Respiration occurs chiefly through the body surface and the mantle; circulation takes place through the lacunar spaces without a heart. The brain is located in front of the midgut, and the ganglia of the nerve chain are more or less fused into a few masses (Fig. 14-34).

Most barnacles are hermaphroditic. The reproductive system (Figs. 14-33 and 14-34) consists of ovaries that lie in the mantle wall (sessile) or in the peduncle (stalked), with paired oviducts opening on the first thoracic cirri, and of testes with paired sperm ducts enlarged into seminal vesicles that open into the penis. In copulation the penis can be projected out of the body and sperm deposited near the oviduct openings of an adjacent barnacle. Cross-fertilization is thus the rule, but self-fertilization may also occur. Marked sexual dimorphism is characteristic of many of the stalked barnacles. In some the males are small, dwarflike forms attached inside the mantle cavity of larger, normal individuals. These may

be miniature complemental males attached within hermaphroditic individuals, for example, *Scalpellum* and *Ibla*. Fertilized eggs are usually brooded in an ovisac or a thin lamella. The egg normally hatches into a free-swimming nauplius (less frequently into the cypris stage), with frontal horns on the carapace. After a number of molts the last nauplius instar gives rise to a cypris larva, in which the body is enclosed in a bivalve carapace with compound eyes and six pairs of thoracic appendages (Fig. 14-35, *A*). The number of larvae reaching the cypris stage is only a fraction of those released as nauplius larvae. The cypris larva settles down and becomes attached to a substratum by means of the cement glands of the antennules (Fig. 14-35, *B*). Many changes now occur that involve the infolding of the mantle and the formation of the 5 primary valves—a pair

FIG. 14-35

Metamorphosis of *Lepas*. **A,** Cypris stage. **B,** Attached larva. **C,** Young larva with developing valves still surrounded by cypris shell. (Modified from Korschelt and Heider.)

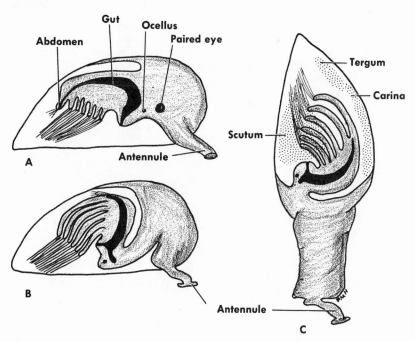

each of scuta and terga, and a carina—(Fig. 14-35, *C*). At the next molt the mantle fold opens, and the barnacle assumes the adult position. J. D. Costlow, Jr., and C. G. Bookhout (1953) have studied molting and growth in *Balanus improvisus* and have found that shell growth is a continuous process and has no correlation with molting. The barnacle is attached at its preoral surface, with the ventral surface at right angles to the attached surface and its posterior side directed upward.

The order Thoracica is commonly divided into the suborder Lepadomorpha (pedunculate barnacles), suborder Verrucomorpha (reduced pedunculate barnacles), and suborder Balanomorpha (sessile barnacles).

■ ORDER ACROTHORACICA

Acrothoracica are naked, boring barnacles that are permanent residents of holes in the shells of mollusks, corals, and other barnacles.

They have no calcareous plates and usually only four pairs of cirri. The posterior cirri are widely separated from the first pair. An abdomen is lacking or else is represented by the caudal furca. The ovary is present in the thick part of the mantle or disk, which is often the organ of attachment. Sexes are usually separate, with the males minute and usually attached within females. The size of most species rarely exceeds 10 mm. Only about a dozen species are known, of which *Trypetesa lampas* has a wide distribution.

■ ORDER ASCOTHORACICA

Ascothoracica are mainly parasites of coelenterates or echinoderms. They are naked barnacles but have a bivalve or saccular mantle. They are not attached at the preoral region as are most barnacles. Their mouthparts are mostly modified for piercing and sucking. The antennules, when present, are usually adapted

FIG. 14-36

Laura gerardiae, an ascothoracid barnacle attached to piece of coral *(Gerardia).* **A,** Minute animal is enclosed in mantle sac in which are papillae that penetrate into host tissues. Diverticula of digestive tract and ovaries extend through mantle. Circled portion is enlarged in **B. B,** Body of animal exposed and greatly enlarged. (After Lacaze-Duthiers; modified from Lankester.)

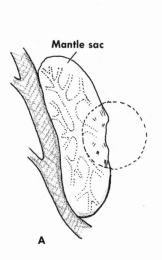

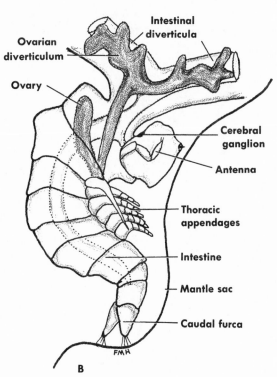

for clinging. The thoracic appendages, 6 or less, are reduced, and an abdomen may be present. There are no cement glands. Both the digestive tract and the ovaries have extensive diverticula, which are distributed in the mantle. Sexes are separate, with the males much smaller than the females.

There are some twenty-five species, the most representative genera of which are *Baccalaureus*, *Dendrogaster*, *Ascothorax*, and *Laura* (Fig. 14-36).

■ ORDER APODA

The members of the parasitic order Apoda are of doubtful taxonomic status. They have no mantle or appendages but are segmented and possess an abdomen. The mouth is suctorial. Antennules and cement glands are present. The digestive system is incomplete. They are hermaphroditic, but their development is not known. So far, the order is based (by C. Darwin) on one specimen found as a parasite in a stalked barnacle.

■ ORDER RHIZOCEPHALA

Probably the best known of the parasitic barnacles are the rhizocephalans because of their worldwide distribution and their habit of using other crustaceans, especially the Decapoda, as hosts. They are naked, but a thin bag-shaped mantle encloses the visceral mass, which consists chiefly of gonads. This saclike structure is fastened to the abdomen of the host by a short stalk. They have no appendages, segmentation, sense organs, or digestive system. A rootlike system, modified from the peduncle and composed of fine threadlike structures, penetrates to all parts of the crustacean host and sucks out its body fluids. The fertilized eggs develop in a brood sac, from which the larvae escape into the sea by an opening. The free-swimming nauplius develops into a cypris larva which, if it attaches to a crab, becomes a kentrogon larva. This larva injects into the host's intestine a cell mass that develops into a tumor and root system. In a few weeks the endoparasite emerges as a small sac. Those cypris larvae that settle on immature rhizocephalans become larval males. These males inject into the brood pouch a cell mass that later differentiates into sperm. The saccular animal becomes sexually mature some weeks later. This sex pattern, first discovered in *Peltogasterella*, may be general in rhizoceph-

FIG. 14-37

Sacculina, a barnacle parasitic on crabs (order Rhizocephala). *Sacculina* infects several species of crabs in which the cypris stage attaches itself to the base of a seta, transforms into a sac, develops a dartlike structure, and penetrates the thin cuticle. (From Hickman: Integrated principles of zoology, The C. V. Mosby Co.)

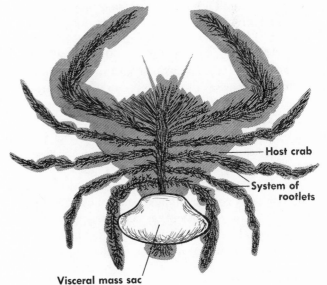

Host crab

System of rootlets

Visceral mass sac

alans, which are often called hermaphroditic because of the presence of sperm repositories and eggs in the same sac.

Parasitism by members of this order produces a characteristic parasitic castration in the infected host. Some of these striking changes involve secondary sex changes, such as the broadening of the abdomen of a parasitized male so that it resembles that of a normal female, and in parasitized immature female crabs an adult type of abdomen appears precociously. However, the growth of the swimmerets is retarded. Other changes may be the complete atrophy of the gonads in the host. Such changes may be explained by metabolic disturbances or hormonal imbalance. The parasite produces its effect by digesting away the androgenic gland, which regulates sex in crustaceans.

Sacculina (Fig. 14-37), a common parasite of crabs, is a classic example. Other genera are *Briarsaccus,* the largest member of the order (¾ inch in diameter), and *Loxothylacus,* which parasitizes the blue crab (*Callinectes*).

Subclass Malacostraca

The subclass Malacostraca is the largest taxon of the crustaceans. It includes the shrimps, crabs, lobsters, sow bugs, beach hoppers, wood lice, and a host of others and takes in the largest crustaceans as well as many smaller types. The group is worldwide in its distribution, embracing upward of 18,000 to 20,000 species. These species form a natural taxonomic group because of certain common characteristics, but there is an enormous diversity among the members within the group. The trunk is usually composed of 8 thoracic segments and 6 abdominal ones, each of which bears a pair of appendages. A carapace may be present, vestigial, or absent.

The malacostracans are usually divided into two arbitrary series: the Leptostraca, with 7 abdominal segments and an adductor muscle connecting the two halves of the shell; and the Eumalacostraca, with 6 abdominal segments and lacking an adductor muscle.

The common primitive type of crustacean (caridoid facies) from which the various members have specialized has a shrimplike body, a carapace enveloping the body, stalked and movable eyes, biramous antennules, antennae, thoracic limbs with swimming exopodites and respiratory epipodites, and one or more pairs of thoracic limbs forming maxillipeds as part of the mouthparts. In addition, the ventrally flexed abdomen is provided with five pairs of biramous swimming appendages (pleopods), a tail fan formed of the flat uropods (sixth abdominal appendages), and the telson. Some present groups of crustaceans that approach this primitive plan are the Mysidacea, Anaspidacea, and Eucarida.

Malacostracans have the female genital duct located on the sixth thoracic segment and that of the male on the eighth segment. Males often have the first two pairs of pleopods modified as copulatory organs.

Series Leptostraca
■ ORDER NEBALIACEA

Nebaliacea is the only order of the series Leptostraca and represents a marine group of fifteen to twenty species and varieties. Its members are relics of a much larger group that were common in the Paleozoic era. They average 4 to 12 mm. in length and live mostly as bottom dwellers in shallow water. The thorax and part of the abdomen are enclosed by a carapace with an adductor muscle in the head region joining the 2 valves (Fig. 14-38, *A*). A small, ringed rostral plate covers the head. The 8 thoracic segments are free, with clearly indicated suture lines. The thoracic appendages are similar and foliaceous. In some members the whole limb is rather flattened; in others the epipodite is small, whereas the endopodite and exopodite are long and slender. The thoracic limbs are usually adapted for creating water currents, from which food is collected into a ventral groove and then conveyed to the mouth (Fig. 14-38, *B* and *C*). The abdomen has 7 segments, the last of which is without appendages. The telson is provided with 2 articulated

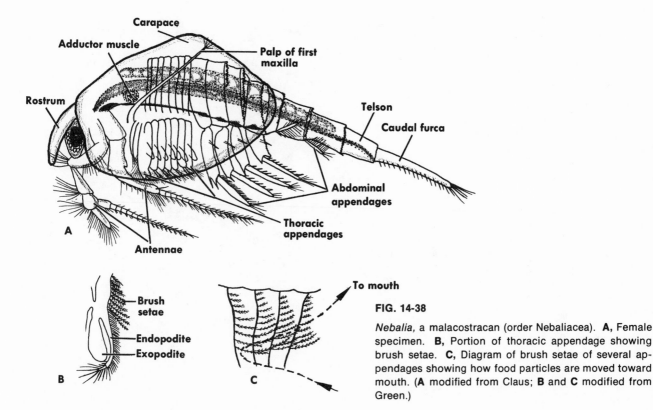

Carapace

Adductor muscle

Palp of first
maxilla

Rostrum

Telson

Caudal furca

Abdominal
appendages

Thoracic
appendages

A

Antennae

Brush
setae

Endopodite
Exopodite

B

To mouth

C

FIG. 14-38

Nebalia, a malacostracan (order Nebaliacea). **A,** Female specimen. **B,** Portion of thoracic appendage showing brush setae. **C,** Diagram of brush setae of several appendages showing how food particles are moved toward mouth. (**A** modified from Claus; **B** and **C** modified from Green.)

furcal rami. The first 4 biramous abdominal appendages are used for swimming; the last 2 uniramous appendages are reduced.

Gills are formed from the flattened thoracic epipodites. The breeding female has a fan of plumose setae that are arranged to form a basketlike brood chamber. The eggs hatch into a postlarval stage.

Most members of the order live in shallow, muddy bottoms or among stones and aquatic vegetation. Some, such as *Nebaliella,* have eyes, antennules, and other head parts adapted for burrowing. *Nebaliopsis* lives at great depths and feeds mostly on the eggs of various animals. Its midgut is provided with a large diverticulum for holding reserves of food.

Series Eumalacostraca
SUPERORDER SYNCARIDA

Superorder Syncarida includes two orders, the Anaspidacea and the Bathynellacea. Syncarids vary in length from 0.5 mm. to 5 cm. They are found mostly in caves, springs, and mountain lakes where they swim and crawl over the bottoms or on aquatic vegetation. As a group they are characterized by lack of a carapace, calcareous plates, and chelae. There are about thirty species in the superorder, and all are freshwater forms.

■ ORDER ANASPIDACEA

The primitive Anaspidacea are represented by only a few species, which are found in the fresh waters of Tasmania and South Australia. Anaspidaceans have a slender body that is almost uniformly segmented. Both their antennules and antennae are long. The biramous thoracic appendages are used mainly for swimming. The coxopodite of each thoracic appendage bears a pair of lamellar gills. The abdominal appendages, or pleopods, have fringed exopodites, but the endopodites are reduced or absent except in the male, in which the first two pairs are used as copulatory organs. Each

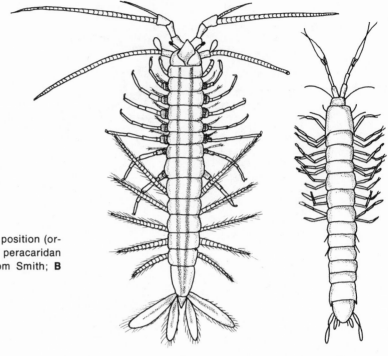

FIG. 14-39

A, *Anaspides,* a syncarid in natural walking position (order Anaspidacea). **B,** Male *Monodella,* a peracaridan (order Thermosbaenacea). (**A** modified from Smith; **B** modified from Karaman.)

mandible bears a palp and is fitted for macerating the food. The stalked compound eyes usually project from the dorsal wall above the antennae, but in some the eyes are sessile. The principal food is organic detritus, which covers the rocks of their habitat, and also dead insect larvae and worms. Some of these crustaceans have the second maxillae adapted for filtering. Currents of water are directed into a median channel to the mouth by the beating of the thoracic exopodites and maxillae. A heart and excretory maxillary glands are present. On the underside of the broad, flattened telson is the anus.

Anaspides (Fig. 14-39, *A*) is the best known genus and has been thoroughly studied. It is found in the lakes of Tasmania at high altitudes and may reach a length of 2 inches. It is considered the most generalized of modern malacostracans. *Paranaspides* and *Koonunga* are other genera from Tasmania and South Australia.

◼ ORDER BATHYNELLACEA

Bathynellacea are found in the subterranean waters of central Europe. They are less than 2 mm. long and are eyeless. The antennules and antennae bear clusters of setae. The thorax and abdomen are similar in structure, and each of the thoracic appendages has two epipodites that function as gills. Some of the abdominal appendages are reduced or absent. A typical telson is wanting, but the abdomen bears a pair of caudal furcae and a pair of elongated, jointed uropods. The eggs hatch directly into juvenile forms.

Bathynella and *Thermobathynella* are recognized genera.

SUPERORDER PERACARIDA

Superorder Peracarida is made up of six or seven orders, which include aquatic sow bugs, opossum shrimps, and some others. In these orders the young undergo development in a ventral brood pouch (marsupium), which is formed by oostegites or flat plates that extend medially from the coxopodites of the thoracic legs. The roof of the brood pouch is formed by the thoracic sternites. They also differ (except order Thermosbaenacea) from other malacostracans by having a 3-segmented protopodite in the antenna and a marked flexion of the thoracic legs between the fifth and sixth segments. The first of the thoracic segments is fused with the head.

■ ORDER THERMOSBAENACEA

The order Thermosbaenacea are small crustaceans of 1 to 4 mm. in length. This group is characterized by a dorsally located brood pouch formed by the carapace. The maxillipeds may have either an exopodite or an endopodite (sometimes both), and a respiratory epipodite. The body is mostly cylindric, with no division between the thorax and abdomen. There are no eyes, and the head and first thoracic segment are fused. The antennules are biramous and the antennae consist of exopodites only. There are five pairs of biramous thoracic appendages that lack epipodites. The last two thoracic segments lack appendages (except in the genus *Monodella*) (Fig. 14-39, *B*). Two pairs of appendages occur on the abdomen. The uropods are biramous with a one-jointed endopodite and a two-jointed exopodite.

The members of this order have been found in the transitional zone between fresh water and marine water in Italy and Yugoslavia.

The group is represented by two genera, *Thermosbaena* and *Monodella*. Only four or five species are found in the two genera.

■ ORDER SPELAEOGRIPHACEA

Order Spelaeographacea is represented by a single species found in a freshwater lake in Africa. It is a blind, shrimplike form about 6 to 9 mm. long. It has a short carapace united to the first thoracic segment. The thorax consists of 8 segments, the first one of which is fused to the head. The abdomen is made up of 6 segments. The antenna has a long flagellum. The rostrum is triangular, with a small ocular

scale on each side. The first pair of thoracic appendages forms the maxillipeds, each of which has a cup-shaped respiratory organ at its base. The other seven pairs of thoracic appendages are adapted for walking. Three pairs of oval gills are attached to the exopodites of the fifth through the seventh thoracic segments. Of the abdominal appendages, the first four pairs are biramous swimming paddles, the fifth pair is vestigial, and the sixth pair, with the telson, forms a tail fan. The incubation of the few large eggs occurs in the ventral brood pouch characteristic of the superorder. The only known species is *Spelaeogriphus lepidops*.

■ ORDER MYSIDACEA

The order Mysidacea is made up of about 550 species, and its members are commonly known as opossum shrimps, so called because of the ventral marsupium at the base of the last two pairs of thoracic legs in the female. Although they are mostly marine, some mysidaceans are found in fresh water. The body is composed of the typical malacostracan number of 19 segments, 5 of which form the head, 8 the thorax, and 6 the abdomen, with a tail fan of uropods and telson. Each segment has a pair of biramous appendages. A carapace covers the thorax but is not united with the last 4 thoracic segments. The eyes are stalked and movable. In cave-dwelling forms, eyes may be vestigial. The first pair of thoracic appendages (and sometimes the second pair) is modified as maxillipeds. The other thoracic appendages are filamentous, usually with setae, and are employed mostly for swimming. The abdominal appendages (pleopods) are generally all similar but may be modified or reduced in some species. Several mysidaceans (especially primitive forms) use pleopods for swimming, whereas others use the thoracic appendages for this purpose. Walking is usually done with the thoracic endopods. Mysids range from 3 to 85 mm. in length; some lophogastrids may reach as much as 200 mm.

Raptorial feeding is characteristic of the

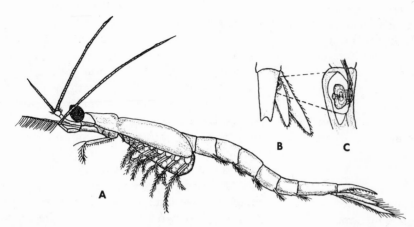

FIG. 14-40

Mysis, a peracaridan (order Mysidacea). **A,** Female with ventral marsupium. **B,** Telson and right uropod of *Mysis* showing statocyst near base of endopodite. **C,** Side view of statocyst of same showing statolith and sensory nerve. (**A** modified from Pennak; **B** and **C** modified from Sars.)

members of the suborder Lophogastrida. Living or dead animals are captured by the subchelate thoracic endopods and mandibular palps and are passed to the mandibles, where they are masticated. In the other suborder, Mysida, most of the members are filter feeders. Rotatory movements by the thoracic exopods, aided by the beating of the maxillipeds, produce currents that flow forward along the ventral food groove. The food is filtered by maxillary setae and passed into the mouth. Such filter feeders derive much of their food from the organic detritus of the sea floor. Lophogastrids have foliaceous thoracic gills and cause currents of water to flow backward over the gills by beating the lamellar epipod on each of the first thoracic appendages. The current of water flows through the branchial chamber lying between the thorax and carapace on each side and is directed over the gills from the second to the eighth thoracic appendages. In Mysida, which have no gills, the current of water is directed forward by the beating of the thoracic exopods, and the exchange of gases is made through the thin-walled, highly vascular lining of the carapace.

Sense organs include the eyes, usually with movable stalks, and statocysts with statoliths in the uropoda endopod (Fig. 14-40, *B* and *C*).

Developing eggs are carried in the marsupium 1 to 3 months (Fig. 14-40, *A*) and hatch directly into juveniles. The members of this order continue their growth just after each molt throughout life.

Mysids are active forms that dart among rocks and other shelters in their native habitats. During the day they feed mainly in the bottom waters, but at night they migrate to the surface. They may also migrate shoreward at night and away from the shore during the day.

Examples of mysids are *Neomysis* and *Mysis* (Fig. 14-40), North American freshwater forms. *Gnathophausia* is a deep-sea lophogastrid.

■ **ORDER CUMACEA**

Cumaceans occur throughout the world in all the seas, including the Caspian Sea. They vary in size from 1 to 35 mm. in length. There are 600 to 650 species recorded. All are characterized by an inflated carapace and a slender abdomen. The carapace, with a pair of anterolateral extensions (false rostrum), is fused to the first 3 or 4 thoracic segments and overhangs the sides to form branchial chambers. The eyes are sessile and are often fused into a single median eye, or they may be absent. There are eight pairs of thoracic appendages, three pairs of which are modified as maxillipeds. The epipodite of each first maxilliped bears

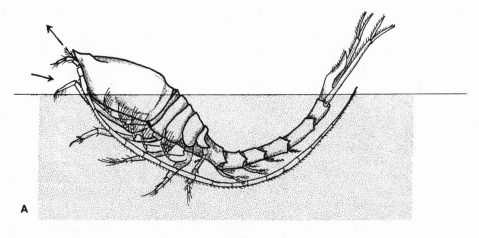

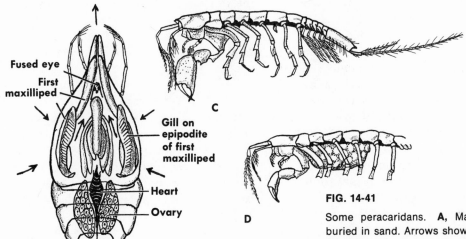

FIG. 14-41

Some peracaridans. **A,** Male *Diastylis,* a cumacean, buried in sand. Arrows show direction of water currents for feeding and respiration. Note long antennae, vestigial in female. **B,** Female *Diastylis,* dorsal view, showing direction of water currents created by epipodites of first maxilliped. **C,** Male *Apseudes* (order Tanaidacea). **D,** Portion of female *Apseudes* with marsupium. (**B** to **D** modified from Sars.)

filamentous gills (Fig. 14-6, *B*), which are enclosed in a branchial chamber formed by the epipodites and the carapace. Water is drawn over the mouthparts by the action of the epipodites (Fig. 14-41, *B*), passed upward through the branchial chamber, and then forced anteriorly by siphonlike structures on the first maxilliped on each side. The antennae are very long in the male (Fig. 14-41, *A*) but vestigial in the female. Some of the thoracic appendages are used mainly for swimming and burrowing. Pleopods are absent in the female, but one to five pairs occur in the male.

Many cumaceans are filter feeders. Food particles are filtered from the water current that is passed over the mouthparts by setae on the second maxillae. Some species scrape organic detritus from the sand by rotating the sand in cuplike modifications of the first and second maxillipeds and then pass the selected food to the mouth by the exhalant water current.

In their natural habitat, cumaceans live partially buried in the sand and mud of littoral regions (Fig. 14-41, *A*). Some live at considerable depths. The more common genera of Cumacea include *Cumopsis*, *Leptocuma*, *Diastylis*, and *Cyclaspis*.

■ ORDER TANAIDACEA

Most tanaidaceans are marine and have a wide distribution. They are benthonic and are found at various depths. They number about 350 species. Their size rarely exceeds 2 mm., and most are smaller. The body is elongated and is dorsoventrally flattened to some extent (Fig. 14-41, *C* and *D*). The first two pairs of thoracic segments are fused to the head to form a carapace that encloses a respiratory chamber on each side. The left mandible may bear a small movable process (lacinia mobilis), which may be absent on the right mandible. When eyes are present, they are usually located on immovable stalks. The thorax has eight pairs of appendages. The first pair is maxillipeds and the second pair, gnathopods, is large and chelate. The other six pairs of thoracic appendages are similar and are used for swimming except the third pair, which in some is specialized for burrowing. Exopodites are present only in the second and third pair. The pleopods of the abdomen may be reduced or absent, especially in the female.

The digestive system consists of a ventral mouth, a stomach provided with filtering and masticatory apparatus, a midgut, and a terminal anus. A feeding current produced by the epipodites passes over the mouthparts, carrying food particles filtered by the second maxillae. Some raptorial feeding also occurs in the group. Two pairs of hepatopancreas glands usually open into the midgut. Excretion is by a pair of maxillary glands. The nervous system is made up of a brain, a subesophageal ganglion, and a ventral chain of ganglia. Sense organs are represented by special sensory organs (aesthetascs) on the antennules of the male and by tactile sensory hairs or spines on both sexes. The exchange of gases occurs mainly in the inner surface of the carapace when the water

current produced by beating of the thoracic appendages passes anteriorly through the branchial chamber. The gonads are paired in each sex. The oviducts open at the base of the fourth pair of pereiopods; the vasa deferentia, with a common seminal vesicle, open on the last thoracic segment. Sexual dimorphism is most marked in the antennules of the two sexes, but there are other differences. Some forms are hermaphroditic. The eggs are incubated in a brood chamber formed by the flat plates (oostegites) of some of the thoracic appendages, and they hatch as postlarvae. Several broods are produced in succession by the female. Four larval instars may be necessary to produce the adult stage.

Apseudes (Fig. 14-41, *C* and *D*), *Sphyrapus*, *Tanais*, and *Heterotanais* are representative genera.

■ ORDER ISOPODA

Isopoda, a large order of 4,000 species, includes the sow bugs, pill bugs, wood lice, and others. They are found in the soil (terrestrial), in fresh water, and in the sea. Most of them are marine. Some are parasitic, but they are chiefly free living. They range in length from 1 mm. to 20 cm. (Figs. 14-42 and 14-43).

They have a dorsoventrally flattened body, which may be distinctly segmented or made up of segments more or less fused, especially in the abdominal region. The head region is made up of the true head fused to the first thoracic segment. The head is somewhat flattened; and the similar thoracic segments, as well as the abdominal segments, are expanded laterally to form projections (terga) over the basal parts of the appendages. The uniramous antennules are short or vestigial; the antennae with flagella are usually large, but may lack the exopodite. The compound eyes are sessile but may be vestigial in cave forms. There is no carapace. The thorax and abdomen are about the same width. Each of the 7 distinct thoracic segments bears a pair of pereiopods. The first pair of appendages is modified to form the maxillipeds, and

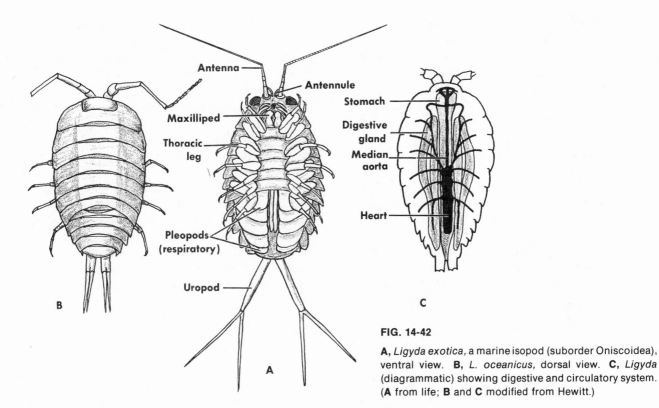

FIG. 14-42

A, *Ligyda exotica,* a marine isopod (suborder Oniscoidea), ventral view. **B,** *L. oceanicus,* dorsal view. **C,** *Ligyda* (diagrammatic) showing digestive and circulatory system. (**A** from life; **B** and **C** modified from Hewitt.)

the others are mostly for crawling. The thoracic appendages are uniramous, with the expanded coxal plate usually fused to the sternum. The first five pairs of abdominal appendages are foliaceous and modified for respiration. The last pair of appendages is the uropods, which may be styliform or fan shaped. The last abdominal segment is fused with the telson to form a pleotelson.

Isopods live on a variety of food, both living and dead. Many are scavengers and will eat both green and decaying vegetation. They do not employ filter feeding. The mouthparts are arranged into a ventrally projecting mound that is covered by the labrum and maxillipeds. The mandibles are strong and toothed. Feeding may be assisted by the anterior legs. The digestive system (Fig. 14-42, *C*) contains a short esophagus, a stomach with a gastric mill, and an intestine. One to three pairs of ceca are attached at the junction of the stomach and the intestine. Sym-

biotic bacteria help digest cellulose in such forms as *Porcellio.*

The large heart is found mostly in the posterior part of the thorax (Fig. 14-42, *C*). The heart discharges blood into the hemocoel by a large number of arteries. Blood is returned from the hemocoel to the pericardial chamber around the heart. Maxillary glands for excretion occur at the base of the second maxillae.

Respiration takes place mainly through the pleopods, the endopodites and exopodites of which are modified into lamellae or gills (Fig. 14-42, *A*). Some exchange may take place through the body surface. In some isopods the gills may be protected by a type of operculum formed by certain pleopods or by elongated, flexed uropods. The ventral surface of the abdomen may be covered by 2 rows of overlapping opercula.

The nervous system shows various arrangements in the different isopods. In general it consists of a supraesophageal ganglion (brain) with circumesophageal connectives, subesophageal ganglia, a double ventral cord with segmental ganglia in the thorax, and a group of fused ganglia in the abdomen. In some there is a tendency for the ganglia to be concentrated and fused anteriorly.

The males have their testes in the thorax and dorsal to the intestine. The sperm ducts open on the genital segment or seventh thoracic segment at the tips of 2 (or 1) projections. The 2 ovaries are long, baglike, and located dorsally. The oviducts have 2 openings on the ventral surface of the fifth thoracic segment. Each oviduct may also be provided with a seminal receptacle. The endopodites of the first two pairs of pleopods may be modified in the male

for transferring sperm into the gonopores during copulation. Fertilization of eggs takes place in the oviducts. Eggs are brooded in a brood pouch that is formed prior to egg deposition by the oostegites. The young hatch as juveniles, lacking only the seventh pair of thoracic appendages.

Freshwater isopods live mostly under rocks and debris in streams and ponds. They are rarely found on the open water. In the sea they have been found from intertidal zones to abyssal depths. Terrestrial forms live in humid regions; their cuticle is not adapted for living in dry, arid conditions, although some families can endure a drier climate than others can. One familiar isopod is the sow bug, which rolls up into a ball when disturbed. The marine form

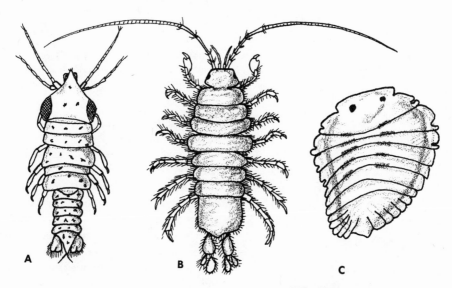

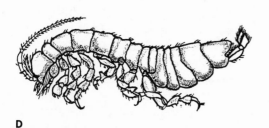

FIG. 14-43

Some isopods of different suborders. **A,** *Gnathia* (suborder Gnathiidea); segmented larva with sucking mouthparts; ectoparasitic on fishes (about 4 mm.). **B,** *Asellus* (suborder Asellota), a common freshwater form (up to 20 mm.). **C,** *Bopyrus* (suborder Epicaridea), female from gill chamber of *Palaemon* (8 mm.). Specimens taken from opposite side of host would be bent in opposite direction. **D,** *Phreatoicus* (suborder Phreatoicidea), female. (**A** and **B** modified from Smith; **C** redrawn from Green: Biology of Crustacea, H. F. and G. Witherby, Ltd.; **D** modified from Chilton.)

Limnoria can damage piers by eating the wood. Some isopods such as the suborder Gnathiidea, while still in the larval stage, are parasitic on the skin of fishes. *Cymnothoa* and related forms have piercing mouthparts for living on fishes. Members of the suborder Epicaridea are bloodsuckers on crustaceans.

The order Isopoda is commonly divided into seven suborders as follows:

Suborder Oniscoidea. Usually with distinct segmentation, vestigial antennules, and expanded coxal plates. Represented by terrestrial sow bugs, pill bugs, and wood lice. *Oniscus, Porcellio, Armadillidium, Ligyda* (Fig. 14-42).

Suborder Asellota. All abdominal segments fused with large telson; coxae never expanded to form plates; pleopods used as gills; commonly freshwater but also some marine types. *Asellus* (Fig. 14-43, *B*).

Suborder Valvifera. Valvelike uropods inflexed as opercula to cover pleopods; some fusion of abdominal segments; both marine and freshwater. *Idotea, Arcturus.*

Suborder Gnathiidea. Mouthparts adapted for sucking (females) or pinching (males); lateral uropods; ectoparasitic at some stage on marine fishes. *Gnathia* (Fig. 14-43, *A*).

Suborder Flabellifera. Flattened body; uropods expanded laterally with telson to form tail fan; mostly marine, but some are freshwater; free living, predaceous, and parasitic genera. *Cirolana, Limnoria, Cymnothoa, Bathynomus.*

Suborder Epicaridea (Bopyroidea). Suctorial mouthparts and piercing mandibles; parasites on marine crustaceans; body may be greatly modified. *Bopyrus* (Fig. 14-43, *C*), *Ione, Liriopsis.*

Suborder Phreatoicidea. Laterally compressed bodies; styliform uropods; genus *Phreatoicus* (Fig. 14-43, *D*) found in fresh water in New Zealand and South Africa.

■ ORDER AMPHIPODA

Next to the Isopoda, order Amphipoda is the largest in the superorder Peracarida. They number more than 3,800 species, about 50 of which are American freshwater forms. Altogether more than 600 species are found in fresh water in various parts of the world; the others are marine. Some are semiterrestrial. Four suborders are recognized in the group: Gammaridea, Hyperiidea, Caprellidea,

and Ingolfiellidea. Most of them are of modest size (3 to 12 mm. in length), but some may attain a length of more than 120 mm.

In general the order is characterized by having no carapace, sessile or unstalked eyes, first (and sometimes second) thoracic segment fused to the head, and a body flattened laterally. Thoracic exopodites are absent, and gills are located on the thorax. The first pair of thoracic appendages forms maxillipeds with fused coxae but lacks epipodites. Seven pairs of pereiopods are elongated, with coxae modified as plates. The second and third thoracic appendages are prehensile (gnathopods) (Fig. 14-44, *A* and *C*). Both antennules and antennae (without exopodites) are usually well represented. The abdomen has its appendages divided into an anterior group of three pairs of biramous swimmerets and a posterior group of three pairs of uropod-like appendages. The telson may or may not be lobed. There are many morphologic variations in such an extensive order. Some of these differences will be indicated in a summary of the suborders.

Some amphipods are pelagic and most of them can swim. However, the great majority are bottom dwellers. The thoracic appendages are used for crawling and walking, being assisted by the pushing of the uropods, which are directed backward. Grasping claws are used by some in climbing among vegetation. Many amphipods move on one side of their body, which undergoes a great deal of flexing and squirming movements in both swimming and crawling. Many species are active burrowers; some can construct mud tubes similar to those of certain crayfish or make unattached tubes of sand-shell fragments (*Siphonectes*) or hollow plant stems (*Cyrtophium*). Some tubicolous species even carry their tubes along with them (*Cerapus*).

Amphipods eat a great variety of foods of both plant and animal nature. Organic detritus is a common food of many filter feeders. In general they are scavengers. A few such as *Lafystius* are parasitic. Food is commonly

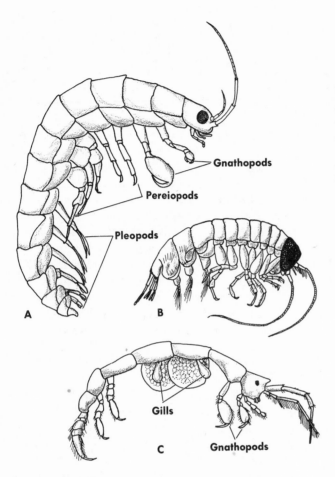

FIG. 14-44
Group of amphipods. **A,** *Orchestia,* the common sand flea, less than ¹/₂ inch long, found under damp seaweed at high-water mark along Atlantic coast (suborder Gammaridea). **B,** *Hyperia,* male, about ³/₅ inch long, lives in pits on underside of large medusae (suborder Hyperiidea.) **C,** *Caprella geometrica,* common in eelgrass of southern New England coast (suborder Caprellidea). (**A** from life; **B** modified from Davis; **C** modified from Miner.)

held by the prehensile gnathopods and other thoracic appendages and chewed by the mandibles. Filter feeders often use the antennae as filters in trapping the food, which is then scraped off the gnathopods. The digestive system consists of a short esophagus, a stomach with a gastric mill, an intestine, and an anus below the telson. Two or three pairs of ceca are associated with the midgut.

The tubular heart with ostia has an anterior and posterior artery (Fig. 14-5, *B*). Excretory organs are the antennal glands that open at the base of the second antennae. The coxal gills of the thoracic appendages are the chief organs of respiration (Fig. 14-6, *A*). These gills vary in structure and number among the various suborders and species, but they usually number two to six pairs. Water currents over the gills are created by the beating of the pleopods. The general body surface may play an important role in the respiration of terrestrial forms. Terrestrial amphipods must live in moist or humid conditions for their gills to function properly.

The nervous system consists of a brain, circumesophageal connectives, and 2 ventral nerve cords connected here and there by segmented ganglia. Some species are eyeless, but most have sessile compound eyes. There are many different arrangements of the eyes. Both antennules and antennae have olfactory and tactile properties. In some there are antennal statocysts. Most amphipods are negatively phototropic and are nocturnal feeders.

The sexes are separate, and the paired gonads

in each sex are elongated, tubular structures lying ventral to the heart. The male genital openings are on papillae located on the sternum of the last thoracic segment; the female oviducts open at the base of the fifth coxal plates. The female discharges her eggs into a ventral brood pouch formed by setose lamellae of the medial bases of the legs. In copulation the male transfers his sperm with his uropods to the brood pouch of the female. Development is direct and the young are juveniles. Marine forms may produce more than one brood a year, but freshwater species usually have only one. In 1954 the androgenic glands, which produce hormones for control of sex in crustaceans were first found in or near the testes in *Orchestia* (beach flea).

The major characteristics of the four suborders are summarized as follows:

Suborder Gammaridea. Distinct head, thorax, and abdomen; maxilliped with palp; large coxal plates on thoracic appendages; rather primitive ventral nerve cord with eight pairs of thoracic ganglia and four pairs of abdominal ganglia. Largest suborder of amphipods; includes all freshwater amphipods. *Gammarus, Hyalella, Orchestia* (Fig. 14-44, *A*), *Calliopius, Siphonectes*.

Suborder Hyperiidea. Distinct thoracic segments; maxillipeds without palps; large eyes; elongated appendages; mostly pelagic and exclusively marine. Double eyes in some (upper and lower portions), fused eyes in others; lateral arteries present in heart. *Hyperia* (Fig. 14-44, *B*).

Suborder Caprellidea. Vestigial abdomen; head fused to second thoracic segment; slender body; appendages reduced or absent. Often grotesque compared with members of other suborders. *Caprella* (Fig. 14-44, *C*) is a familiar genus found living on seaweeds and other marine vegetation.

Suborder Ingolfiellidea. Distinct thoracic segments and well-developed abdomen; vestigial abdominal appendages (except fourth and fifth pairs); eyeless (in some); four species. *Ingolfiella*.

SUPERORDER EUCARIDA

The important superorder Eucarida includes the lobsters, crayfish, shrimp, and crabs. The eucaridans include most of the larger and more specialized malacostracans. The diagnostic features of the group are the large carapace, which is fused to all of the thoracic segments; the movable, stalked, compound eyes; the lack of a brood chamber; and some form of metamorphosis in their development. There are two orders, the Euphausiacea and the Decapoda. Less than a hundred species belong to the euphausiaceans, but the decapods have between 8,000 and 9,000 species and are the largest order of crustaceans.

■ ORDER EUPHAUSIACEA

The euphausiaceans are a small group of shrimplike crustaceans that are chiefly pelagic. They have many primitive characteristics and in general retain the basic caridoid facies. They are about 1 to 2 inches in length. Many are surface forms and are important planktonic items of food (krill) for certain species of whales. Some, however, are found at great depths. Many undergo extensive diurnal, vertical migrations. They are worldwide in distribution and are strictly marine.

The carapace does not completely enclose the sides of the body, and the gills (of one series) are exposed. The maxillae have tiny exopodites. None of the thoracic limbs are modified as maxillipeds, but they may be specialized for filter feeding. The pleopods bear long setae and are used in swimming. The elongated telson has a pair of movable spines. The compound eyes are bilobed in some species for seeing upward and downward.

Most euphausiaceans are luminescent. They have on the abdomen photophores that give off a blue-green light (Fig. 14-45, *A* and *B*). The photophores are specialized organs with lens and reflector as well as light-producing cells (Fig. 14-45, *E*). The body is bright red; swarms of euphausiaceans sometimes so redden the water that seamen call it "tomato soup."

Euphausiaceans are mostly filter feeders and have their first 6 thoracic appendages modified as a filtering apparatus. Each of the endopodites of the appendages bears a fringe of setae, which form a net to capture food

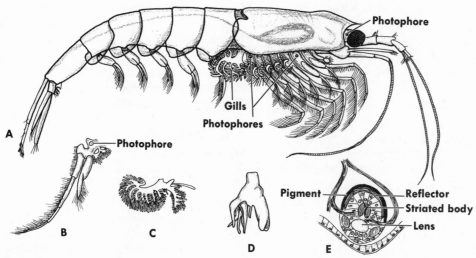

FIG. 14-45

Euphausia pellucida (order Euphausiacea). **A,** Adult female. Paired photophores are located on ocular stalk and thoracic appendages; unpaired photophores on abdomen between bases of first four pairs of pleopods. **B** and **C,** Second and eighth thoracic appendages of euphausiacean with attached gills. **D,** Endopodite of first male pleopod with copulatory apparatus unfolded. **E,** Section through a thoracic photophore. Striated body is apparently source of light. Photophores on ocular peduncle have no lens. (**A** modified from Sars; **B** to **E** modified from Calman.)

from the water current passing through it as the animal swims forward. A few have the third pair of thoracic appendages chelated for capturing prey and are predaceous. The digestive system is of the typical crustacean pattern.

Respiration is by means of foliose gills at the bases of the second to eighth thoracic appendages (Fig. 14-45, *A* to *C*). A heart with three pairs of ostia is present, and the blood has hemocyanin as its respiratory pigment.

The male transfers his sperm in spermatophores to the spermatheca of the female, where the eggs are fertilized. The fertilized eggs may be released into the water, or they may be retained on the undersurface of the female's body. There are many larval stages—the nauplius, metanauplius, calyptopis, furcilia, and postlarval (cyrtopia) stages—through which the larva passes in succession. Each of these, except the nauplius, has a number of substages.

Some of the genera are *Euphausia* (Fig. 14-45), *Meganyctiphanes, Nyctiphanes,* and *Bentheuphausia* (a deep-sea form).

■ ORDER DECAPODA

The largest order of the crustaceans includes the shrimps, crayfish, lobsters, and crabs. Most of them are marine; but the crayfish, certain prawns and shrimps, a few species of Anomura, and a few crabs have invaded fresh water and moist terrestrial habitats. These nonmarine members have fewer species compared with the more than 8,000 marine species. They vary in size from some commensal crabs, which are less than $\frac{1}{2}$ inch in length, to the giant Japanese crab, which may have a span of more than 12 feet between the tips of the chelipeds.

Although this extensive order is diverse, there are some common features that apply to the whole group. In general the diagnostic characteristics of the decapods are as follows: the first three pairs of thoracic appendages are modified as maxillipeds; the other five pairs of thoracic appendages are modified as legs; the first pair of legs are usually chelate and heavy (chelipeds); usually there is more than one

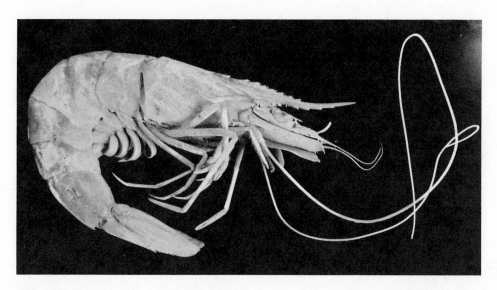

FIG. 14-46

Penaeus, an edible swimming shrimp (section Penaeidea, suborder Natantia).

FIG. 14-47

Crangon (Alpheus) heterochaelis, a snapping or pistol shrimp common among oyster reefs along coast from Virginia to Florida (section Caridea, suborder Natantia).

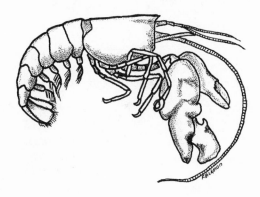

series of gills; the head and thorax are fused to form a cephalothorax; and the carapace overhangs the sides to form a branchial chamber on each side. The decapods may be divided into the long-tailed members (shrimps, lobsters, crayfish), and the short-tailed members (crabs). The long-tailed forms have a cylindric or laterally compressed carapace, usually with a rostrum. In the short-tailed crabs the carapace is broadened and flattened dorsally, usually without a rostrum. The short abdomen or tail in these forms is flexed under the thorax. This division is purely arbitrary and is not based on a natural relationship of members.

The order Decapoda is usually classified as follows:

Suborder Natantia. Body laterally compressed and well-developed abdomen; cephalothorax with keel-shaped, serrated rostrum; legs mostly slender; shrimps and prawns.

Section Penaeidea. Body with pleura of first abdominal segment overlapping those of second; first three pairs of legs usually chelate. *Penaeus* (Fig. 14-46), *Lucifer.*

Section Caridea. Body with pleura of second ab-

dominal segment overlapping those of the first; third pair of legs not chelate. *Crangon* (Fig. 14-47), *Heterocarpus.*

Section Stenopodidea. Body with pleura of second abdominal segment not overlapping those of first; third thoracic legs very stout, long, and chelate. *Stenopus.*

Suborder Reptantia. Body dorsoventrally flattened, or cylindric; cephalothorax with rostrum depressed or absent; abdomen well developed or reduced; legs usually heavy and lacking exopodites. Lobsters, crayfish, crabs.

Section Macrua. Body with well-developed abdomen and tail fan; carapace longer than broad and with

A

B

C

FIG. 14-48

A, Common crayfish *Cambarus.* **B,** *Cambarus,* female, "in berry," with eggs attached to swimmerets. **C,** Lobster *Homarus* (section Macrura, suborder Reptantia).

FIG. 14-49

Hermit crab *Clibanarius* in snail shell (section Anomura). See also Fig. 14-50.

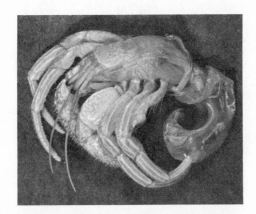

FIG. 14-50

Clibanarius, a pagurid, or hermit crab, removed from safety of its snail shell. Note vulnerable soft abdomen, which it never exposes except briefly when moving into larger home (section Anomura).

FIG. 14-51

Emerita talpoida, the mole crab, follows tidal waves up and down open ocean beach, burrowing quickly backward into sand as each wave recedes, leaving feathery mouthparts exposed to filter food from water (an anomuran).

FIG. 14-52

Callinectes sapidus, the blue crab, an important seafood product (section Brachyura). Note last pair of legs flattened into paddles for swimming.

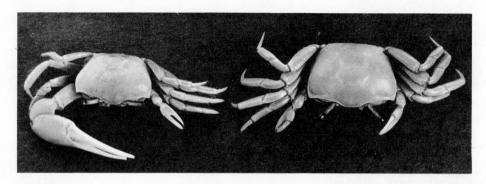

FIG. 14-53

Uca minax, male and female. Male (left) has one very large chela that he moves horizontally in front of him, giving rise to common name fiddler crab (section Brachyura, family Ocypodidae).

FIG. 14-54

Neopanopeus texana, a mud crab common along East coast (section Brachyura, family Xanthidae). (Drawn from life.)

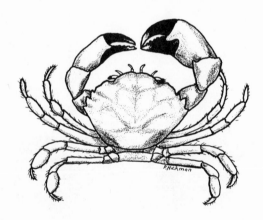

rostrum. Lobsters (Fig. 14-48, *C*) and mud shrimps. *Callianidea, Callianassa.*

Section Anomura. Body with abdomen usually bent ventrally without lateral plates; third pair of legs not chelate. Hermit crabs (Figs. 14-49 and 14-50) and sand crabs (Fig. 14-51). *Polyonyx* (commensal in *Chaetopterus*).

Section Brachyura. Body with abdomen reduced, without a tail fan, and bent under the thorax; cephalothorax very broad and short; rostrum absent. True crabs. *Callinectes* (Fig. 14-52), *Uca* (Fig. 14-53), *Neopanopeus* (Fig. 14-54).

In general, the Natantia are adapted for swimming. They use the pleopods, which are large and setaceous, for this function. On the other hand, the Reptantia are suited more for crawling, although many of them can swim. When they swim, they do so without their pleopods, which are not adapted for swimming. Many of the Reptantia use their legs for swimming, for example, certain crabs. Others use the strong, muscular abdomen for darting backward (crayfish, lobsters).

External morphology

The decapods have nineteen pairs of appendages (Fig. 14-2) and a pair of segmented eyestalks. The eyestalks of the compound eyes consist of 2 to 3 elongated segments. The eyes may be set in a notch of the carapace, or they may be covered by thin margins of the carapace. The antennules have a 3-segmented peduncle and usually 2 flagella. The antennae consist of 5 or fewer segments and a single flagellum. They may be very long (Fig. 14-46). They may also have a scalelike exopod. Their coxopodite is a small operculum over the opening of the excretory (antennary) gland.

The mouthparts number six pairs. In the mandibles the incisor process (biting edge) is toothless, and the molar process is usually a moundlike elevation behind the biting edge. There may be modifications of this arrangement. A 3-segmented palp, but no lacinia mobilis, is usually present on each mandible. The first maxillae (maxillulae), the second maxillae, and the first pair of maxillipeds are mostly leaflike and subdivided into lobes. Each maxilla usually bears a paddle-shaped scaphognathite ("bailer"), which fits the exhalant end of the branchial chamber. The second maxillipeds have broad, flattened segments whereas the third maxillipeds may be elongated or may form an operculum over the other mouthparts. The mouthparts have a plate or epistome above them and are enclosed laterally by the carapace. In crabs the buccal frame is more or less square and is formed by the fusion of the epistome with the carapace. This buccal frame is covered by the two plates, or operculum, of the third maxillipeds.

The five posterior pairs of thoracic appendages are the walking legs, although some of them may be modified. In the long-tailed decapods one or more pairs may be chelate; in the short-tailed decapods generally only the first pair is so modified. The legs may be modified into swimming paddles in swimming crabs (Fig. 14-52). The pereiopods in shrimp usually consist of 7 segments. In other decapods the second and third segments are commonly fused. In some, one or two pairs of pereiopods may be reduced or absent. In most long-tailed decapods, the first five pairs of pleopods (abdominal) are modified into swimming organs. The first one or two pairs in the male, however, may be modified into copulatory organs (gonopods); in the female they are reduced or absent. In the short-tailed decapods, the pleopods are used for sperm transfer in the male and for the attachment of eggs in the female. The pleopods may be reduced in the female. The last pair of abdominal appendages (uropods) in the long-tailed decapods form, with the median telson, a tail fan (Fig. 14-48, C). In most short-tailed species such as crabs the uropods are absent. However, in the hermit crab they are present and are twisted into a spiral for holding the crab in the snail shell where it normally lives (Fig. 14-50).

Internal morphology

The digestive system is divided into the foregut (stomodeum), the midgut, or mesenteron, and the hindgut (proctodeum). The foregut and hindgut are lined with chitin. In most decapods the foregut is modified into a short esophagus and a dilated stomach. The anterior, or cardiac, part of the stomach bears movably articulated ossicles of calcium, the gastric mill, which is controlled by a complex system of muscles (Fig. 14-4). The posterior, or pyloric, part has hairy or setaceous ridges that form a straining apparatus. A hepatopancreas, or liver, empties by ducts into the midgut. The anus is located at the ventral base of the telson.

The circulatory system (Fig. 14-5, C) consists of a short polygonal heart with paired ostia under the posterior carapace, a median anterior artery, paired antennal arteries, paired hepatic arteries, a median posterior artery, and a sternal artery to various organs. It also has a venous sinus system that carries blood to the gills and back to the pericardial sinus around the heart. Modifications of this pattern are found, especially in terrestrial forms. The respiratory pigment of the blood is hemocyanin, which is dissolved in the plasma.

Excretion is carried on chiefly by the antennal, or green, glands (Fig. 14-7, C and D), since the maxillary glands of the larva are transitory. The antennal glands are mostly compact and more complex than those in the other crustaceans. Each consists (in sequence) of an end sac, or saccule, which may be partitioned into cavities; a labyrinth of a spongy mass of canals; a nephridial, or excretory, tubule; and a bladder with a duct to the exterior. The bladder may be a simple vesicle (crayfish) or lobated (shrimp and some crabs). In some cases the bladder sends diverticula throughout a great part of the body. These may unite by lobes to form a single vesicle above the stomach or may anastomose to form complex networks, as in hermit crabs (Fig. 14-7, A). Nitrogen excretion (ammonia, urea, uric acid) occurs largely by diffusion through the gills. Uric acid may be collected by migratory cells called nephrocytes, or it may be deposited in the integument and cast off in molting. Crustacean urine is mostly isosmotic with the blood except in certain freshwater crayfish, in which the urine is hypoosmotic. Other freshwater crustaceans produce urine that is isosmotic with the blood. *Artemia*, the saltwater shrimp, may produce a hyperosmotic urine. Regulation of ionic concentration may be maintained by the gills or by the excretory tubule through secretion and reabsorption.

With one or two exceptions, all decapods have gills connected to the lateral walls of some or all of the 8 thoracic segments and to the basal segments of the thoracic appendages (Fig. 14-6, *D* and *E*). On each side of the thoracic segments there is typically a set of 4 gills. Each set consists of a pleurobranch attached to the lateral wall of the thoracic segment; 2 arthrobranchs fastened to the articular membrane between the first segment of the appendage and the body wall; and 1 podobranch attached to the coxa of the appendage. There are thus four series of gills, with omissions and modifications, running on each side of the thoracic segments of all decapods. No decapod has all the possible sets of this arrangement (8 sets, or 32 gills, on each side). The number of gills varies greatly with different groups. The lobster, for instance, has 20 gills on each side; the freshwater crayfish (*Cambarus*) has 17 on each side. The greatest number is 24 on each side (the primitive *Benthesicymus*). There may be as few as 3 gills on a side (*Pinnotheres*).

The gills are enclosed in a gill chamber covered by the lateral part (branchiostegite) of the carapace on each side of the thorax. The gill chamber has restricted access to the outside. The paddlelike scaphegnathite, or bailer, of the second maxilla on each side moves back and forth in a channel under the carapace and draws forward currents of water to bathe the gills so that gaseous exchange may take place. The water is expelled anteriorly in front of the head.

The direction of the water current may be reversed in burrowing forms, which may also extend the exhalant channel anteriorly to form a tubelike structure with the antennal flagella, antennal scales, and third maxillipeds. In some burrowing forms such tubelike siphons are used as inhalant siphons. Water may enter the gill chamber along the lower margin, which bears setae to prevent access of silt and mud. In the Brachyura, or true crabs, this arrangement is modified by having the free margins of the branchiostegite fitting snugly at the bases of the legs, leaving an opening for the ingress of water around the base of the chelipeds. The epipodites of the third maxillipeds form opercular valves for the inhalant opening. The course of the respiratory water current in brachyurans is through the inhalant opening, then posteriorly in the hypobranchial chamber, next dorsally over the gill lamellae, and finally anteriorly through paired openings in the upper lateral corners of the buccal frame. The fringed epipodites of the maxillipeds are used by crabs for cleaning the gills. The bases of the chelipeds and the coxa of the legs also have setae for filtering the incurrent water.

The basic plan of a gill structure involves an axis with numerous lateral branches. In the axis of the gill there is an afferent and an efferent branchial sinus. Blood passes from the afferent channel to the gill filaments and back to the efferent channel. According to the arrangement of the lateral branches, three types of gills are recognized: (1) the trichobranchiate, with filamentous branches that are arranged in series around the axis; (2) the phyllobranchiate, or lamellar, with flattened platelike branches arranged in two series along the axis; and (3) the dendrobranchiate, with extensive and often complex biserially arranged main branches that are subdivided by secondary branches. The last type is characteristic of the primitive Penaeidea, but the first two types have a wide distribution among the decapods. The Caridea and most of the Brachyura, for instance, have phyllobranchiae, whereas most of the Macrura have the trichobranchiae.

Terrestrial or semiterrestrial decapods have a different arrangement of gas exchange surfaces from that of typical gills, which are unsuitable for air breathing. Whether in air or water, the respiratory surface must be kept moist. Land decapods must have large moist respiratory surfaces that afford little evaporation of water. Terrestrial forms depend less on gill breathing or else have gills modified by sclerotized ridges that hold the filaments apart and prevent them from collapsing. Some, such as *Uca*, carry water in the gill chambers and circulate air through the water. In this form there are special respiratory openings between the third and fourth legs through which the air enters. In many land decapods the walls of the gill chamber have been transformed into vascular tufts, villi, and vascularizations so that the chamber functions as a type of lung. Along with these changes the gill covers may also have highly vascularized lacunar systems (*Ocypode*). The coconut crab *Birgus*, which is one of the best adapted of all terrestrial crabs, has an especially well-developed respiratory epithelium, with extensive blood lacunae and folds. Although oxygen consumption is greater in terrestrial than in marine decapods, the terrestrial forms can still survive when their gills are removed.

The Decapoda as a group show many variations of the nervous system, especially in the disposition of the ventral chain of ganglia. The maximum number of unfused ganglia posterior to the esophageal ganglion is 5 thoracic and 6 abdominal ganglia. This pattern is found in lobsters and crayfish (Fig. 14-8, *B*). In the true crabs (brachyurans) the ventral ganglia, both thoracic and abdominal, are fused together in one rounded mass in the ventral thorax. From this concentrated mass of ganglia, nerves radiate out to various parts of the body. Between these two extreme types there are varying degrees of coalescence of the ganglia among the different groups. Decapods have a system of visceral nerves that involves a gastric plexus formed by the anastomosis of a posterior cerebral nerve and a pair from the esophageal commissures. A cardiac ganglion as well as nerves from the central system regulates the heart action.

The Decapoda are perhaps the best equipped with sense organs of all crustaceans. The paired compound eyes are found in all except those species that live deep in the sea and in caves. The facets of the cornea are square or hexagonal in outline. The nauplius eye may persist as a vestigial organ in lower decapods. The stalk or peduncle of the compound eye consists of 2 or 3 segments, one of which may be lengthened to produce a very long stalk. In macrurans and some others the carapace above the eye is notched, and in the snapping shrimps the eye is covered with a thin margin of the carapace.

A pair of statocysts, or balancing organs, occurs on the proximal segment of the antennules (Fig. 14-12). They are saclike cavities lined with sensory hairs and open to the outside by large or small slitlike openings. In a few species the statocyst appears to be closed off from the exterior. A mass of sand grains held together by secretions from the sac wall acts as a statolith in each statocyst. Sensory hairs or setae from the walls of the sac are in contact with the statolith. At each molt the chitinous lining of the statocyst with the statolith is cast off. The lining and the grains of sand must be replaced at each molt. Gravitational pull on the statolith produces stimuli through the sensory setae and receptor cells that are attached to a branch of the antennular nerve. Various positions of the body alter the gravitational pull on the statoliths, producing different stimuli to the brain and leading to orientation of the body. Many investigations have been made to explain the complex movements and reflexes of the statocyst functions.

Olfactory setae are present on the outer flagellum of the antennules (Fig. 14-13), and tactile hairs are numerous on the body and appendages. Noise or stridulation involving the rubbing of one part of the body with another is produced in some (e.g., *Ocypode*). The snapping shrimp (Fig. 14-47) produces a loud noise by using a movable finger with a large process that fits into a socket on the immovable finger.

When the movable finger is locked or cocked by a special mechanism, the finger can be released abruptly by the contraction of a closer muscle and produces a sharp noise when it is snapped shut. This adaptation may be employed as a warning and to stun possible prey.

According to E. N. Harvey, at least seventeen genera of shrimp luminesce in some form or other. Luminescent organs are chiefly photophores, which are found on various parts of the body. Some shrimp are able to eject a luminous fluid from mandibular glands.

Most decapods are dioecious, but a few shrimps are protandric hermaphrodites. The paired testes lie mostly in the thorax but may extend into the abdomen. In hermit crabs they may lie wholly in the abdomen, and both may lie on the left side. Each testis is usually connected to the other by a diverticulum, or they may be fused together as a single organ. Each

of the paired vasa deferentia usually opens on the basal segment, or coxa, of the last pair of thoracic legs, or they may open on the last thoracic sternum. Parts of the sperm ducts are modified with glands for spermatophore formation. Just before it opens to the outside, each duct becomes a muscular ejaculatory duct. A single tubular penis is characteristic of some hermit crabs, and a pair of penes (modified first pair of pleopods) are found in true crabs. The ovaries are similar to the testes in shape and position. The oviducts usually open on the coxa of the third thoracic pereiopods except in the true crabs (brachyurans), in which

FIG. 14-55

Metamorphosis of *Penaeus* (suborder Natantia) showing different stages. **A,** Dorsal view. **B** and **C,** Ventral view. **D,** Side view. (**A** to **C** modified from Müller; **D** modified from Claus.)

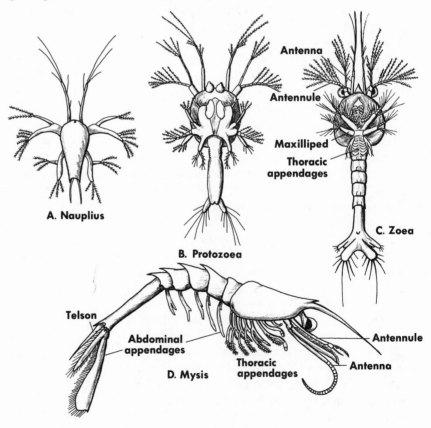

they open on the sternum. In the brachyurans each oviduct is modified as a seminal receptacle and vagina for the reception of the male penes. In many decapods a single median seminal receptacle, with a separate opening to the outside, may be formed by processes of the thoracic sternites.

Sexual dimorphism may be found in some decapods, such as the fiddler crab *Uca* (Fig. 14-53), in which one of the male claws is much larger than the other. Many species have distinct courtship rituals. Copulation of the true crabs involves the insertion of the first pair of male pleopods (with penes) into the female vagina and seminal receptacles. The sperm are discharged as spermatophores, which may break up into sperm as they are driven through the grooves of the first pleopods by the second pleopods. In crayfish and lobsters the tips of the first pleopods are inserted into the seminal receptacle, and sperm is transmitted along grooves in the pleopods. Other copulatory behavior patterns are known among decapods. In those decapods with seminal receptacles, where sperm are stored, fertilization occurs when the eggs are shed.

Some shrimp shed eggs directly into the water, but most decapods carry the eggs attached by a cement to the pleopods (Fig. 14-48, *B*). In some shrimp all primitive larval stages are represented (Fig. 14-55). A typical nauplius (with three pairs of appendages) is the first stage, followed by the metanauplius with four pairs of appendages. The protozoea, or third stage, with seven pairs of appendages has a carapace, unsegmented abdomen, and paired rudimentary eyes. In the fourth stage (zoea) there is further development of the eyes and carapace, and all thoracic appendages are present. In the last stage (mysis) the thoracic appendages replace the antennae as swimming organs, and an abdomen develops similar to that of the adult form (Figs. 14-55, *D*, and 14-56). There are many differences in the hatching stage among the decapods. In most of them the nauplius stage is completed in the egg, and the animal hatches as a zoea. The mysis stage may be replaced by postlarval forms. The postlarval stage in crabs is called megalops, with a large, unflexed abdomen and a full complement of pleopods for swimming (Fig. 14-57). The larval stages may be reduced or completely absent in freshwater decapods. Crayfish and freshwater crabs hatch as juveniles. Some terrestrial crabs return to the sea where the eggs pass through the larval stages of marine forms.

Molting is of course characteristic of all decapods. Ecdysis is absolutely necessary for an increase in body size. The young molt several times during their development; the adults less often. The process of molting is similar in all of them and has already been described. This period is a critical one, and the animal must hide away to avoid enemies.

Regeneration among the different members

FIG. 14-56

Mysis stage of lobster.

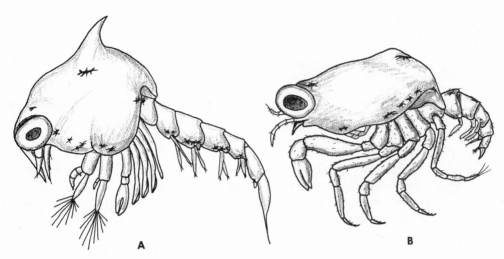

FIG. 14-57

Developmental stages of *Sesarma cinereum.* **A,** Fourth zoea. **B,** Megalops. (Modified from Costlow, Bookhout, and Monroe, 1960.)

varies to some extent but is greater in young animals than in adults. Lost limbs and some other structures are usually replaced. Sometimes the regenerated structure is different from the original (heteromorphosis). Crayfishes and crabs especially have the power of autotomy or self-amputation, the fracture occurring at a definite breaking plane, such as the basal segment of the endopodite of a cheliped. Special muscles bring about the breaking, and a valve or other adaptations prevent excess loss of blood.

Ecology and physiology of Decapoda

Such an enormous group of crustaceans as decapods has a great range of ecologic relationships. Their economic importance need not be mentioned here to any great extent. It will suffice to state that members of this group (shrimp, lobsters, crabs, etc.) are extensively used as food by people of all parts of the world. All of them serve as important food chains for other animal groups. Their ability to adapt to a wide range of ecologic conditions explains their wide distribution. Many striking ecologic behavior patterns are found among them. One

of these is the vertical migration of aquatic forms, in which there is a diurnal movement to considerable depths during the day and a return toward the surface at nighttime. This migration, characteristic of many groups of crustaceans, is mostly under the control of light conditions. Many prefer dim light intensities. The distance of migration varies with the group. Some larger forms may descend to a depth of several hundred meters; smaller species restrict the vertical movement to shorter distances.

Color changes by chromatophore action are also common in crustaceans. A great variety of pigments is found in their chromatophores, sometimes several in the same chromatophore (polychromatic). Both melanin and carotene derivatives are represented, such as black, brown, red, blue, and yellow pigments. The typical chromatophore is a stellate-shaped cell of many noncontractile processes. A blanching, or lightening, effect is produced when the pigment granules become concentrated toward the center of the cell; a darkening effect occurs when the pigment granules flow into and are dispersed through the processes. Many color patterns are possible, especially in certain shrimp that have trichromatic chromatophores of yellow, red, and blue pigments.

Much investigation has shown that the control of chromatophore behavior is directed by hormones from a neurosecretory system in the eyestalk or nearby. The sinus gland, which appears to be the center of hormone release, is found between the 2 basal optic ganglia. This sinus gland receives secretions through nerve fibers from secretory cells of the eyestalk and central nervous system. The sinus gland may not secrete hormones itself but may serve as a reservoir or releasing point of hormones. The hormones are produced by secretory cells in the nervous system and secretory masses in the eyestalk, such as the X-organ in the medulla terminalis (M.T.X. organ). Investigative methods involve the removal of the eyestalk, injections of eyestalk extracts, and then notation of the results in each case. Retinal pigment migrations for light and dark adaptations are also under the control of eyestalk hormones, although the mechanism is not fully known.

The process of molting involves control by hormones. A molt-inhibiting hormone apparently is found in the sinus gland and may be produced by the brain. On the other hand, a molt-accelerating hormone occurs in the eyestalk but not in the sinus gland. Hormones produced by the X-organ and sinus gland regulate the preparation of proecdysis, accelerate proecdysis, and control the amount of water taken up. The Y-organ, which is located in the head, may produce a hormone that stimulates the onset of proecdysis and also acts as the medium through which the molt-accelerating and molt-inhibiting hormones perform their effects. Cessation of molting may be due to degeneration of the Y-organ or an excess of molt-inhibiting hormone.

As stated previously, in 1954 the discovery of the androgenic gland was made in the amphipod *Orchestia*. It has since been found in many malacostracans. This gland is located between the muscles of the coxopodite of the last walking leg and near the end of the vasa deferentia except in the Isopoda, in which it is found within the testis. When the gland is removed, male characteristics are lost; the testes are converted into ovarian tissue. When the gland is trans-planted into a female, the ovaries become testes.

Physiologic clock mechanisms of pigment movement have been extensively studied in the fiddler crab *(Uca)* by F. A. Brown, Jr., and others. Normally the body color of *Uca* is dark during the day and light at night. This rhythm persists even when it is kept in continual darkness. Some phase shifting can be induced by low temperatures so that the rhythm can be stabilized at a new phase. Periodicities of color changes may also be related to the normal fluctuations of such environmental factors as ocean tides. Within their burrows during high tide, the usual nocturnal paleness of color is characteristic, reflecting their minimum metabolic rate. This produces a dispersion of black color through the daylight hours in accordance with the lunar day of 24.8 hours, which is different from the solar day of 24 hours. Two hypotheses have been advanced to account for the mechanism of persistent rhythms. One theory stresses the idea that the organism possesses, independently of its physical environment, periods of oscillation that match all the natural geophysical frequencies and that have been genetically determined by natural selection. An alternative theory holds that the organismic periods depend on a continuous response of the animal to geophysical rhythms that produce corresponding metabolic cycles in the organism. The adaptive nature of the cyclic mechanisms such as feeding, migration, and breeding enable the animal to best meet environmental changes. Whatever their causes and changes, cycles are built around a 24-hour period.

SUPERORDER HOPLOCARIDA

The only order of the hoplocaridans is the Stomatopoda, or mantis shrimps (Fig. 14-58). The order is made up of one family (Squillidae), which contains about 200 species. They represent, with the Decapoda, the largest of the crustaceans. They range in size from about 20 mm. to more than 30 cm. in length. They have elongated bodies that are dorsoventrally flat-

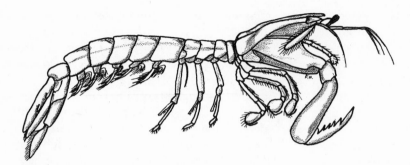

FIG. 14-58

Squilla, a mantis shrimp (order Stomatopoda).

tened. Part of the cephalic and thoracic segments are fused and covered by the dorsal carapace. The cephalic segments that carry the antennules and eyes are not covered by the carapace. The third and fourth thoracic segments are reduced, whereas the last four thoracic segments are unfused and not covered by the carapace. The carapace at its anterior end bears a flattened plate, the rostrum. The abdomen is broader than the thorax and is made up of 6 distinct segments. The tail fan consists of a median telson and 2 large uropods. The stalked eyes are large and movable. The antennules are much larger than the antennae, and each is made up of 3 segments and 3 flagella. The antennae each have 2 segments, a single flagellum, and a large fringed scale. The mouthparts consist of a pair of strong mandibles (sometimes with palps), a pair of small flattened maxillulae, and a pair of large flattened maxillulae. The first pair of the eight pairs of thoracic appendages is slender and hairy for cleaning, whereas the second pair is very large and adapted as a raptorial claw, with the distal segment folding back into a deep groove in the penultimate segment. By means of this claw the mantis shrimps catch their prey. The third to fifth pairs of thoracic appendages are provided with oval chelae and help bring the food to the mouth. The last three pairs of appendages are slender, without chelae, and are used as walking legs. They are the only biramous thoracic appendages, since the others lack an exopodite. In the males a long slender penis is modified from the base of each of the last pair of thoracic legs.

Each of the first 5 abdominal appendages bears a pair of well-developed biramous pleopods. The exopod of each of the pleopods bears a filamentous gill. The appendages of the last abdominal segment are the uropods.

The digestive system consists of a cardiac stomach provided with a pair of lateral ossicles for trituration, a pyloric stomach with filtering bristles, an intestine enveloped by a large digestive gland, and an anus that opens on the telson. The circulatory system is made up of an elongated, tubular heart extending through most of the body, a single anterior artery, a pair of anterolateral arteries, and a system of sinuses for returning the blood. Excretion occurs through a pair of maxillary organs. The nervous system consists of many fused ganglia in the anterior region, followed by a ventral chain of ganglia. The male reproductive system is made up of a pair of convoluted tubules in the abdomen and a pair of sperm ducts, each of which opens into the penis already mentioned. The female system contains a pair of ovaries located in the thorax and abdomen and a pair of oviducts that open on the middle of the sixth thoracic sternite.

When the eggs are discharged, the female carries the egg mass between the anterior thoracic legs. The larvae differ from the adults and are pelagic. They undergo many molts before becoming adults.

Stomatopods are marine crustaceans and are sometimes found in brackish water in tropic and subtropic regions. Most species live in littoral zones, but a few have been found at considerable depths. Most adult members are burrowing forms, usually in mud or sand. *Gonodactylus* has been found in coral reefs. Stomatopods are carnivorous and live on other crustaceans, mollusks, and fishes, for the catching of which their large raptorial claws are well suited. Some of them can make a snapping noise with their raptorial claws.

Besides *Gonodactylus*, other common genera are *Squilla* (Fig. 14-58), *Pseudosquilla, Coronida,* and *Odontodactylus.*

SOME RECENT INVESTIGATIONS ON LAND ADAPTATIONS OF CRUSTACEANS

With the exception of insects, perhaps no group of arthropods have been subjected to more research in recent years than have the crustaceans. This interest has been due to a number of factors. Although the Crustacea represent the only large class of arthropods that are far more common in the sea than on land, a considerable number of them have become adapted (at least partially) to a semiterrestrial or terrestrial existence. Many have undergone evolutionary changes toward a terrestrial existence, but relatively few have become completely terrestrial in their adaptations. Some of the Isopoda have adjusted to a life on land and are able to complete a life cycle on land. Even among the land isopods, all of them lose water unless the air is saturated, but some can survive better than others in drier air. As a group, crustaceans are not well equipped for life on land. On the other hand, many possess a hard skeleton and mostly internal fertilization, which may be useful preadaptations for land existence.

Land invasions among crustaceans are fairly recent geologically speaking. Isopods are considered among the first strictly land invaders. Crustaceans that have invaded land have under-

gone few changes in their morphology and physiology because of their land existence. Successful invasion of land was chiefly made by those that already had preadaptations for land existence.

Much investigation has been done to find out the problems crustaceans have had to face (and experimental evidence of their potentialities) in adjusting to land existence. A noteworthy symposium along this line was held in New York (1967), in which many specialists presented papers on various terrestrial adaptations of crustaceans. To understand the morphologic and physiologic adaptations of terrestrial crustaceans, it has been necessary to understand the differences between the characteristics of aquatic and terrestrial environments. This has been the crux of the present symposium.

In this symposium, research workers have presented experimental evidence to show how certain crustaceans have solved the problems required for successful invasion of land and the potentials possessed by others for terrestrial adaptation. Morphologic, physiologic, and behaviorial adaptations vary in different groups of crustaceans in their general adaptation to a land existence. Three ordinal groups of crustaceans have attained some success in living on land—the Isopoda, the Amphipoda, and the Decapoda. The most successful of these are the isopods, of which several suborders and families are terrestrial. Only one family (of several species) of Amphipoda have become truly terrestrial. The Decapoda have shown the least potentiality for terrestrial living. Phylogeny of the various groups may have played an essential role in terrestrial adaptation. The Amphipoda and the Isopoda are considered very ancient groups and may have had time to make some adjustments to a land existence. Crayfish of the decapods apparently have had time to become independent of the chloride ion and to develop a nonmetamorphic life cycle. They also have suitable adaptations of behavior for a land existence. Except for the freshwater crabs, the decapods have primitive, complicated larval metamorphosis instead of the direct development of the egg and the juvenile form at hatching, as found in the isopods and the amphipods. These two latter groups can live and breed entirely on land; the terrestrial decapods must return for spawning to either fresh water or the sea. Yet in certain respects, terrestrial species of decapods have the important terrestrial adaptations of resistance to desiccation and the facility to mobilize ions and water. To offset the loss of water by transpiration in higher temperatures, terrestrial decapods often have the capacity to take up water from a moist substratum and store water in the pericardial sacs. This adaptation is quite useful in molting when water serves to stretch the new soft exoskeleton before it hardens.

To ascertain how crustaceans have adapted to land, investigators have studied and experimentally tried to modify the various morphologic, physiologic, and behavioral characteristics that may have a bearing on the evolution of their terrestrial adaptation. These include the factors of lethal and regulative temperatures, spawning conditions, tolerance to varying salinities, environmental factors on crustacean larvae, osmoregulation, water balance, gaseous exchange, blood pigments, comparison of metabolic factors in marine and terrestrial crustaceans, nutritional and mineral metabolism, behavioral adaptations, the role of rhythmic systems, tidal cycles, migration route to land, and others.

Of the three groups that have become adapted to land conditions, some have emphasized behavioral modifications (Amphipoda) and others have stressed morphologic and physiologic adaptations (Isopoda and Decapoda). The variety of habitats to which terrestrial crustaceans have become adapted require different adaptations; the study of one species may be restricted in many cases to the conditions of a particular habitat, and generalizations of that habitat may not apply to other species.

Much of the information gleaned from this symposium may be useful to an understanding of land adaptations in crustaceans, but the thoughtful student must be impressed by the scanty knowledge and restricted study of some of the many factors that are involved in the basic adaptations of crustaceans to a land existence. In this symposium many investigators are apologetic about the shortcomings and lack of knowledge regarding many details of their research. This indicates that much more investigation is required before definite conclusions can be made. The modern study of biologic phenomena now stresses ecologic reactions, and behavior patterns loom as one of the key factors for an understanding of life in all its aspects, but this area must be investigated chiefly in future research.

REFERENCES

Barnard, J. L., R. J. Minzies, and M. C. Bacescu. 1962. Abyssal Crustacea. New York, Columbia University Press. Some of the results obtained from the deep-sea explorations of the Columbia University vessel Vema. Many new genera and species are described.

Barnwell, F. H. 1966. Daily and tidal patterns of activity in individual fiddler crabs (genus *Uca*) from the Woods Hole Region. Biol. Bull. **130**:1-17.

Bliss, D. E., and L. H. Mantel (editors). 1968. Terrestrial adaptation in Crustacea. Amer. Zool. **8**:307-700. Papers contributed at a symposium.

Brown, F. A., Jr. 1961. Physiological rhythms. In Waterman, T. H. (editor). The physiology of crustaceans, vol. 2. New York, Academic Press, Inc. The author presents evidence that the periodicity of organisms depends on a continued response of the organism to geophysical rhythms but with the possibility that there is also an endogenous component involved in the mechanism.

Buchsbaum, R., and L. J. Milne. 1960. The lower animals: living invertebrates of the world. Garden City, N. Y., Doubleday & Co., Inc. Several excellent photographs of the crustaceans with concise text descriptions.

Bullough, W. S. 1950. Practical invertebrate anatomy. London, The Macmillan Co. This work includes many examples of crustaceans.

Calman, W. T. 1909. Crustacea. In Lankester, E. R. (editor). A treatise on zoology, part VII, fasc. 3. London, A. & C. Black, Ltd. A classic account that has proved useful to all students of crustaceans.

Carlisle, D. B., and F. Knowles. 1959. Endocrine control in crustaceans. New York, Cambridge University Press. A good summary of what is known about this widely investigated subject.

Coker, R. E. 1939. The problem of cyclomorphosis in *Daphnia*. Quart. Rev. Biol. **14**:137-148. An account of the seasonal changes in the gross morphology of *Daphnia* and some explanatory theories.

Copeland, E. 1966. Salt transport organelle in *Artemia salenis* (brine shrimp). Science **151**:470-471.

Costlow, J. D., Jr., and C. G. Bookhout. 1953. Moulting and growth in *Balanus improvisus*. Biol. Bull. **105**:420-433. This paper discusses some of the factors that inhibit and promote molting.

Costlow, J. D., Jr. 1956. Shell development in *Balanus improvisus*. J. Morphol. **99**:359-416.

Edmondson, W. T. (editor). 1959. Ward and Whipple's freshwater biology, ed. 2. New York, John Wiley & Sons, Inc. Many keys and figures of North American freshwater crustaceans are included.

Green, J. 1961. A biology of Crustacea. London, H. F. & G. Witherby, Ltd. A good concise introduction to the crustaceans, with emphasis on their evolution, physiology, and behavior.

Herrick, C. L. 1884. Crustacea of Minnesota. Minneapolis, Johnson, Smith & Harrison.

Hessler, R. R. 1964. The Cephalocarida: comparative skeleton-musculature, vol. 16. Memoirs of the Connecticut Academy of Arts and Sciences.

Hartford, Academy of Arts and Sciences. A study of the structures of the most recently found and most primitive group of crustaceans.

Hoar, W. S. 1966. General and comparative physiology. Englewood Cliffs, N. J., Prentice-Hall, Inc.

Kaestner, A. 1959. Lehrbuch der speziellen Zoologie, part 1, Stuttgart, Gustav Fischer.

Lockwood, A. P. M. 1962. Osmoregulation in Crustacea. Biol. Rev. **37**:257-305.

MacGinitie, G. E., and N. MacGinitie. 1968. Natural history of marine animals, ed. 2. New York, McGraw-Hill Book Co. A well-known account of Pacific coast animals, including many crustaceans.

Manton, S. M. 1952. The evolution of arthropodan locomotory mechanisms, part 2. J. Linnean Soc. London (Zoology) **42**:93-117. The significance of locomotory devices in arthropods.

Manton, S. M. 1960. Concerning head development in the arthropods. Biol. Rev. **35**:265-282.

Passano, L. M. 1961. The regulation of crustacean metamorphosis. Amer. Zool. **1**:89-95. Metamorphic adaptive changes of crustaceans are mainly for dispersive purposes instead of for feeding as in insects.

Pennak, R. W., and D. J. Zinn. 1943. Mystacocarida, a new order of Crustacea from intertidal beaches in Massachusetts and Connecticut. Smithsonian Misc. Coll. **103**:1-11. Only one genus, *Derocheilocardis*, has been found in this new order.

Pennak, R. W. 1953. Fresh-water invertebrates of the United States. New York, The Ronald Press Co. Keys and figures for identification.

Prosser, C. L., and F. A. Brown, Jr. 1961. Comparative animal physiology, ed. 2. Philadelphia, W. B. Saunders Co. Many physiologic aspects of crustaceans are scattered through this important treatise.

Sanders, H. L. 1957. The Cephalocarida and crustacean phylogeny. Syst. Zool. **6**:112-128.

Schmitt, W. L. 1965. Crustaceans. Ann Arbor, University of Michigan Press.

Snodgrass, R. E. 1952. A textbook of arthropod anatomy. Ithaca, N. Y., Cornell University Press. A fine account of crustacean morphology is given in part 6 of this work.

Southward, A. J. 1955. Feeding of barnacles. Nature **175**: 1124-1125.

Tiegs, O. W., and S. M. Manton. 1958. The evolution of the Arthropoda. Biol. Rev. **33**:255-337.

Tombes, A. S. 1970. An introduction to invertebrate endocrinology. New York, Academic Press, Inc.

Van Name, W. G. 1936. The American land and fresh-water isopod crustaceans. Amer. Mus. Bull. **71**:1-535. A taxonomic treatment of wood lice and freshwater Isopoda.

Waterman, T. H. (editor). 1960-1961. The physiology of Crustacea, vols. 1 and 2. New York, Academic Press, Inc. One of the most important treatises on the functional aspects of crustaceans.

Wulff, V. J. 1956. Physiology of the compound eye. Physiol. Rev. **36**:145-163.

CHAPTER
15

Phylum Arthropoda

THE MYRIAPODS AND INSECTS

GENERAL CHARACTERISTICS

1. The arachnids and the insects are primarily terrestrial groups, with their structures and functions adapted for the medium of air. Both these groups have become adapted to arid and humid regions in contrast to most other land invertebrates (snails, nematodes, leeches, crustaceans, etc.), which are fitted mainly for special humid conditions. Only a few arachnids and insects have secondarily returned to an aquatic medium, usually in fresh water. This is all the more remarkable because the sea may be considered the basic home for most phyla.

2. Along with the insects are the myriapods, which are wholly absent from the sea. The myriapods are relatively a small group that do not compare ecologically or in numbers to the insects. With the insects they constitute a natural group of terrestrial arthropods, fitted in many ways for a successful invasion of the land and altogether independent of the arachnid line. Evidences of relationships between insects and myriapods may be based on anatomic similarities, which include the head formation, locomotor mechanisms, absence of digestive glands, presence of tracheae and malpighian tubules, and plan of circulatory system. Myriapods also have a great diversity of form. Typically their head may be divided into a sensory procephalon with one pair of antennae, and a gnathocephalon of mouthparts, consisting of the mandibles and 2 maxillary segments. The term Myriapoda is a common term

for several classes of animals, of which Diplopoda and Chilopoda are the chief classes. The Diplopoda, or millipedes, have diplosomites, each bearing two pairs of legs. Chilopoda, or centipedes, have long legs and a capacity for swift locomotion in line with their carnivorous habits.

3. Insects have a rather stereotyped morphologic and physiologic pattern in contrast to the wide diversity of crustaceans. However, this restrictive pattern has not prevented insects from evolving into more species than those of all other animals combined. Most of their success can be attributed to the presence of the arthropodian exoskeleton of epicuticle and procuticle—an exoskeleton that can resist deformation and can form the basis of lever systems and morphologic patterns of wide adaptability. By being more or less metamerically segmented, this regional segmentation has led to specialization, division of labor, and increased complexity of reaction and performance. With a firm exoskeleton and jointed appendages, insects have developed a muscular body wall of separate muscles (which are also striated) so that responsive contractions can be precise and localized. Many other factors have contributed to their success, such as the well-developed tracheal system for breathing, the wide diversity of food habits (which tend to lessen competition), the development of an excretory system of malpighian tubules with emphasis on water conservation, the develop-

ment (in some) of the highest social organizations known among animals, their adaptation to harsh climatic conditions, their abilities to protect themselves from predators, their highly evolved chemical and other senses for making adjustments to the environment, the power of flight so well developed in the majority of insects, and many others.

4. The insects with the other tracheate animals (Myriapoda and Onychophora) are the only groups in which the respiratory gases are carried to and from all body cells by a special system of gas-filled tubes (tracheae). In other animals, gas (CO_2 and O_2) may be carried partially in physical solution or, more commonly, by the body fluids in the form of chemical compounds or chemical complexes. Transport of gases by this method of diffusion (tracheae) is suited only to small organisms. However, tracheae in insects have been responsible for a high degree of useful specialization and activity.

It is agreed by most zoologists that there is a close relationship between the myriapods and the insects so that there is a logical reason for considering them together. In arthropod evolution it is commonly supposed that there were two main lines: (1) the chelicerates, in which the first pair of leglike appendages were transformed into chelicerae, and (2) the mandibulates—crustaceans, myriapods and insects—in which the first pair of appendages were either changed into an extra pair of antennae or were lost, and their second pair changed into mandibles. The crustaceans, although mandibulates, probably do not fit in with the myriapods and insects because crustacean mandibles are not homologous with those of insects and for other reasons. That the crustaceans and trilobites are closely related there seems to be no doubt. Although the myriapod-insect stem does not fit in with the trilobite-arachnid-crustacean stem in many particulars, all arthropods have so many structural and functional homologies of metamerism, molting, limbs, etc. that they all must be placed together in the same great phylum—Arthropoda. Statements pro and con

can be made about their monophyletic or diphyletic phylogeny, but this phylum is so extensive that many adaptive radiation stems and ecologic diversities could exist and each could lead to a vast group, at the same time retaining certain basic characteristics that wield them into a single phylum.

The terrestrial mandibulates—myriapods and insects—have followed two main evolutionary lines. In myriapods the body is divided into two tagmata—head and trunk; in insects the body has three well-marked tagmata—head, thorax, and abdomen. Evidences of relationship between these two groups are the uniform structure of the head, the similar distribution of appendages, the same method of development of the mandibles, the presence of tracheae for breathing, and the absence of a digestive gland, as well as other characteristics the two groups have in common.

Four groups or classes are considered under the term Myriapoda—the Chilopoda (centipedes), the Diplopoda (millipedes), the Pauropoda (pauropods), and the Symphyla (symphylans). All these have a head and an elongated trunk that has many leg-bearing segments. Each of these four classes has its own peculiar structures and habits of life, and each may be thought of as a natural group. Therefore, there is a tendency to consider each as a separate class of arthropods and to drop the term myriapod altogether, although it is a convenient designation when all such animals are lumped together.

■ CLASS CHILOPODA

The chilopods are commonly known as centipedes. They have worldwide distribution and are found in both temperate and tropic climates. They number several thousand species. Like all myriapods, they are ground dwellers, living in soil, rotten logs, and under stones. They are usually restricted to habitats with a high environmental moisture. Unlike the Diplopoda, they are exclusively carnivorous and are very active, agile animals.

Chilopods are slender, elongated, segmented,

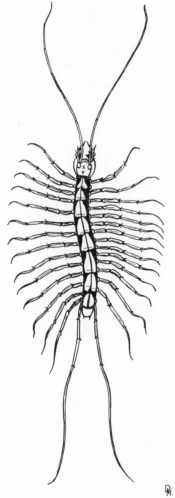

and flattened dorsoventrally (Figs. 15-1 and 15-2). They have an elastic cuticle and large trunk sclerites which do not articulate, but some rigidity is afforded by the large dorsal sclerites. The body in different species varies from 15 to 180 segments. The head has a pair of long, unbranched antennae with a varying number of segments, a pair of mandibles, two pairs of maxillae, and (in some) eyes that may be simple ocelli or modified, compound eyes. Many are eyeless. The first pair of flattened maxillae is located underneath the mandibles and forms a type of labium, which is not comparable to the labium of insects. Second maxillae are long, leglike, and project beyond the first maxillae. The first pair of trunk appendages (maxillipeds) is modified into large, raptorial pincers (prehensors), which are provided with poison glands for injecting venom through the terminal claw into the prey. With the exception of the last two or three segments, each of the other trunk segments bears a pair of small, seven-segmented, walking legs. The arrangement of the sternal and tergal plates varies with the different orders. Most centipedes are adapted for running, which they can do at a rapid rate. They have a high-geared type of gait

FIG. 15-1

Scutigera (Cermatia) forceps, the common house centipede (diagrammatic) (order Scutigeromorpha). (From Hickman: Integrated principles of zoology, The C. V. Mosby Co.)

FIG. 15-2

Scolopendra, a large centipede of the tropics and the American Southwest. All have poisoned claws, and some species may be dangerous to man (order Scolopendromorpha). (From Hickman: Integrated principles of zoology, The C. V. Mosby Co.)

and lack extensor muscles, but they have effective leg thrusts by hydrostatic pressure. Most of the legs are off the ground in rapid movement. Some have unusually long legs, with increasing leg lengths from anterior to posterior regions of the trunk to prevent the legs from stumbling over each other. The head, body, and limbs are covered with chitin, which is flexible where movement of the parts is necessary, especially at the joints of the appendages. The body is very deformable and can be extended, shortened, and flattened by the hydrostatic pressure of the hemocoel. Because of this, centipedes can insert themselves into crevices denied to most animals.

Centipedes live on slugs, insect larvae, other arthropods, and even larger prey (small snakes, mice) in certain cases. After capture, the prey is usually killed by the poison claws. The food is broken into pieces and masticated by the mandible. Although the bite of most centipedes may be painful, it is usually harmless to man. Bites by *Scolopendra*, however, may be dangerous to life, especially if the victim is a small child.

The digestive system is a straight tube, with one or two pairs of salivary glands in the region of the esophagus and a long hindgut, into which open a pair of malpighian tubules for excretion. The anus is located on the last trunk, or anal, segment. The vascular system consists of a tubular heart that runs through most of the body and possesses a pair of ostia and lateral arteries in each segment. Anteriorly the heart ends in a cephalic artery and a pair of arteries that join to form a supraneural vessel. Blood is returned to the heart through hemocoelic spaces.

Breathing takes place through air tubes, or trachea, that open by spiracles in the pleural region on each side of the body, except in the Scutigeromorpha, in which the openings are located in a dorsomedian position along the middle line of the body. Some centipedes have a pair of spiracles in each segment except for the first and a few of the last trunk segments. The tracheal tubes within the body usually branch and anastomose to the various tissues and possess spiral thickenings to prevent collapsing.

The male reproductive system contains a varying number of testes (1 to 24) that lie above the midgut and a single pair of sperm ducts that pass around the hindgut and open by a median genital opening, the penis, on the genital segment. The female system consists of a single tubular organ above the gut and an oviduct that also opens on the ventral surface of the genital segment. Small gonopods may be found on the genital segment, but they probably play no part in the copulatory process. Many centipedes brood their eggs, after depositing them on a substratum, by curling their bodies around them. Some simply lay their eggs, after a short retention on certain appendages, in the soil without further care. The egg contains much yolk, which serves as food for the developing young. Some of the young are hatched with the full complement of segments (epimorphic), whereas others have fewer segments than the adult (anamorphic).

RÉSUMÉ OF ORDERS

The class Chilopoda may be divided into two subclasses, the Notostigmophora and the Pleurostigmophora. The subclass Notostigmophora contains the single order Scutigeromorpha. The subclass Pleurostigmophora has lateral spiracles along each side of the body. Classification varies but usually includes three chief orders: Scolopendromorpha, Geophilomorpha, and Lithobiomorpha.

■ ORDER SCUTIGEROMORPHA

The members of this order are characterized by both primitive and specialized features. They have a single row of dorsal spiracles, long legs and antennae, compound-type eyes, 15 leg-bearing segments, and 8 terga. Their development is anamorphic, with only seven pairs of legs at hatching. The most familiar example is the common house centipede *Scutigera* (Fig. 15-1).

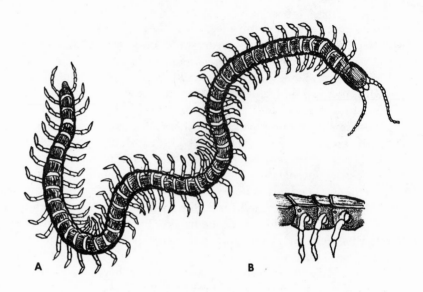

■ ORDER SCOLOPENDROMORPHA

This order of centipedes is one of the most abundant, especially in the tropics. It also includes the largest centipedes, some of which are almost a foot long. Some have eyes, whereas others are blind. They have 21 or 23 leg-bearing segments. Their tergal plates are similar in size, and their development is epimorphic. Genera include the well-known *Scolopendra* (Fig. 15-2) and *Otocryptops*.

■ ORDER GEOPHILOMORPHA

Geophilomorpha are the most specialized of all centipedes. They are adapted for burrowing in the soil and have long, dorso-ventrally flattened bodies and short legs (Fig. 15-3, *A*). They burrow very much as an earthworm does, by extending and contracting the trunk with their powerful muscles and hydrostatic force. Their short legs may act as setae do for anchorage during propulsion. The body wall is elastic, and the dorsal and ventral sclerites slide over each other with the aid of the intercalary sclerites. They can squeeze through narrow crevices with their pliable bodies. All are blind. They have a pair of lateral spiracles on all leg-bearing segments except the first and last. Their mouthparts show numerous variations.

FIG. 15-3

A, Geophilomorph centipede. **B,** Side view of 3 segments of *Lithobius,* a lithobiomorph centipede, showing variable length of tergal plates and arrangement of appendages and spiracles. (Modified from Snodgrass.)

They have from 33 to more than 180 leg-bearing segments. Their development is epimorphic. Common genera are *Geophilus,* the type genus, and *Strigamia.*

■ ORDER LITHOBIOMORPHA

These centipedes are characterized by having dissimilar tergal plates; usually a short tergum alternates with a long one. The antennae may have many segments, although they are usually short and thick. Lithobiomorphs possess 15 leg-bearing segments. Most of them have eyes. Their head and body are flattened, and their legs are short. Their development is anamorphic. The most representative genus is *Lithobius* (Fig. 15-3, *B*).

■ CLASS DIPLOPODA

The name Diplopoda refers to the most distinguishing feature of millipedes—the presence usually of two pairs of legs on each segment (Figs. 15-4 and 15-5). The double nature of the segment is indicated also by the presence of two pairs of ganglia in the ventral nerve cord and two pairs of heart ostia for each segment. The ontogenetic development indicates that the double nature of the segment has been derived

from the fusion of 2 segments that were originally single. Another common name for diplopods is thousand-legged worms. They have worldwide distribution.

In contrast to the carnivorous centipedes, the millipedes are mostly herbivorous or saprophytic. They vary in size from 2 mm. to more than 8 inches in length. The larger members are mostly tropic. More than 8,000 species have been described, but the number may eventually far exceed this number.

The body of the millipede is chiefly cylindric, although some may be dorsoventrally flattened (Fig. 15-5). The chitin of the body wall in most millipedes has calcareous deposits so that the walls are hard and brittle. There is no differentiation of the body into thorax and abdomen.

FIG. 15-4

Common millipede. Note rhythmic action of legs. Note also that most of the limb tips are on the surface and that certain clusters of limbs are drawn up ready for propulsive strokes. (From Encyclopaedia Britannica film, "Animals—Ways They Move.")

FIG. 15-5

Diplopoda. **A,** Polydesmoid millipede. **B,** Transverse section of fifth segment of polydesmoid millipede, anterior view. **C,** Juliform millipede *Arctobolus,* anterior and posterior ends. **D,** Ventral portion of middle segment of juliform millipede. **E,** Gonopods and sternum of seventh segment, anterior view, of *Arctobolus.* (Modified from Snodgrass.)

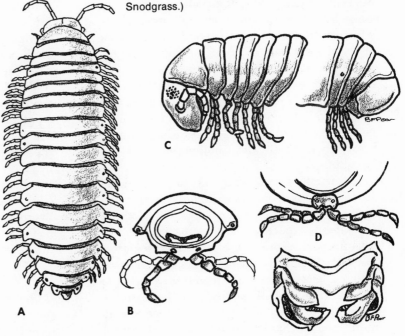

The trunk is composed of a variable number of similar segments (from 20 to more than 100). A cross section of 1 segment (Fig. 15-5, *B*) reveals that it is covered by a convex dorsal tergum, which may extend laterally as a tergal lobe; 2 small pleural plates extending ventrolaterally; and 2 small sternal plates on the ventral side. These segmented plates may be more or less fused.

The head of a diplopod is convex above and flattened beneath. It bears a pair of short antennae, usually of 7 segments. The antennae are located behind the anterior margin of the epistome and labrum and can be folded back flat on the body when burrowing. The mouth appendages are the large mandibles, whose convex sides cover the sides of the head, and the lower lip, or gnathochilarium. The mandibles are segmented, and each consists of a large basal part, with a movable gnathal lobe articulated at the basal part. The gnathal lobe bears teeth. The nature of the gnathochilarium is debatable. It may represent the first maxillae, in which case the second maxillae in diplopods are absent. Some diplopods are blind, but when eyes are present, they are simple ocelli, which may exist in small aggregates. The roof of the preoral food cavity is the epipharyngeal surface that extends posteriorly from the labrum and epistome; its floor is the gnathochilarium. The posterior floor of the preoral cavity is made up of a group of suboral lobes known as the hypopharynx. Most of the preoral cavity is filled by the gnathal lobes of the mandibles.

The first four trunk segments differ from the other segments or diplosegments, and some authorities call this region a thorax. Of these 4 segments, the first (collum) is legless, and the others have one pair of appendages apiece. Each of the diplosegments bears two pairs of walking legs arising close together against the midline (Fig. 15-5, *B* and *D*). The legs are six or seven jointed and in large millipedes may number more than 200 pairs. The last segment bears no legs and may represent the telson.

Millipedes walk with a slow, graceful movement. They do not perform the wriggling movement of the centipedes.

Millipedes are burrowers and require powerful propulsion as they push their way through soil and decaying vegetation, ingesting some of it on the way. Their movement is low geared, and the force they use in burrowing is exerted by the legs. The backward pushing stroke of the legs is relatively slow and is activated in waves along the body; the number of legs in a single wave is proportional to the force required for burrowing. This makes possible a strong but slow movement. The legs are very numerous and many legs may be included in each metachronal wave. It is thus seen that longer (duration) propulsive strokes are needed for power (millipedes) and shorter ones for speed (centipedes). The millipedes are geared like bulldozers, and the centipedes like racing cars (W. D. Russell-Hunter, 1969). The diplosegment (the fusion together in pairs) arrangement, the rigidity produced by ball-and-socket joints between the body rings, and the smooth fused cylindric segments facilitate the propulsive force as millipedes burrow (Fig. 15-4).

Some millipedes have special adaptations such as tergal lobes or lateral carinae for squeezing through crevices. When disturbed, millipedes roll themselves into a spiral for protection. Many of them also are provided with a pair of stink glands on each of most segments. The openings of these glands are located on the sides of the tergal plates or on the tergal lobes (Fig. 15-5, *A* and *C*). The secretion, which is yellow or red in color, has a pungent, offensive odor and contains among other constituents hydrocyanic acid and iodine. The larger millipedes can squirt this fluid a considerable distance from their bodies.

Internal morphology

The diplopod's food of plant and dead animal matter is moistened and masticated by the mandibles. Some Colobognatha have adaptive suctorial mouthpart arrangements or a beak for sucking juices out of plants. The digestive system is usually a straight tube, with the mouth and salivary gland behind the preoral cavity, a narrow esophagus, a long distended midgut, and a narrow hindgut into which open one or two pairs of excretory malpighian tubules. The anus is in the last segment and is provided with 2 anal valves. The foregut is lined with cuticle, and the midgut contains a cuticular tube (peritrophic membrane). Little is known about their digestion.

Respiration takes place through a profuse but largely unbranched tracheal system. The two pairs of spiracles in each diplosegment are located just anterior to the coxae of the two pairs of walking legs. The spiracles open into pits that are hollow, cuticular ingrowths of the body wall (apodemes) where leg muscles are attached. Each pit is protected by branching processes that prevent solid particles from entering with the air. The three pairs of anterior single segments may or may not be supplied with spiracles.

A heart runs along the length of the trunk above the alimentary canal; its anterior end is a short aorta. In the diplosegments the heart is provided with two pairs of ostia but in the anterior single segments, with only one pair of ostia.

The nervous system consists of a small, bilobed brain in the head above the esophagus, around which are commissures to the ventral nerve cord that runs the length of the trunk. From the brain, nerves pass to the antennae and to the eyes (when present). In each diplosegment are two pairs of ganglia, from each of which issues a pair of lateral nerves. Sense organs (in some) include eyes or simple ocelli, which vary greatly in number; tactile and chemoreceptors on the antennae; and a pair of sensory pits, or organs of Tömösvary, near the base of the antennae. The sensory pits may be covered by cuticle and may serve as hearing organs. The eyes show a variety of forms among different millipedes. In general the eye is cup shaped, with photoreceptor cells convergent toward its axis. Each eye usually has a thick convex lens; but in the Julidae, where there are several contiguous eyes in each group, there may be a common flat corneal covering, the inner surface of which projects a thick lens into each retinal cup. Most millipedes, with or without eyes, are negatively phototropic.

The male reproductive system consists of paired tubular testes, joined together and lying between the midgut and ventral nerve cord, and a sperm duct from each testis. The sperm ducts may open through a pair of penes near the coxae of the second pair of legs, a median penis, or a groove behind the coxae. Modified appendages, gonopods (Fig. 15-5, E), may be represented by one or both pairs of appendages of the seventh segment. They transfer sperm from the genital openings on the third segment of the male to the seminal receptacles of the female. In some millipedes only the first pair of appendages on the seventh segment is called gonopods. The structure of gonopods varies greatly among millipedes and has taxonomic value. The female system contains a single, long, tubular ovary between the midgut and nerve cord; an oviduct; and a uterus that divides into 2 vulvae. The vulvae open on the ventral surface of the third, or genital, segment near the coxae. Each vulva communicates with a seminal receptacle that receives the male sperm. In oniscomorph millipedes, gonopods are absent, and the male transfers his sperm in the form of spermatophores with his mouthparts. Fertilization is internal and probably takes place at the time of egg laying. The eggs, which vary in size and number, may be laid in a cluster in the soil, scattered singly, or placed in a specially constructed dome-shaped, mud or excreta nest formed by the female. In some cases the female may brood the cluster of eggs or lie coiled around the egg nest. The development of the eggs is anamorphic, and when the young hatch, they have only the first three pairs of legs and a few segments. Additional appendages and segments are added at each molt until the adult complement is reached.

CLASSIFICATION

Millipedes are classified chiefly according to the structure of the male gonopods, which are more or less characteristic for each species. Other useful taxonomic criteria are the types of segments, appendages, and mouthparts. Most species are restricted in their distribution and remain fairly close to their centers of origin. Few genera are common to more than one continent.

A brief classification of the subclasses and orders and the more common genera of diplopods is given on p. 460.

■ CLASS PAUROPODA

Of the four classes of myriapods, the pauropods are the least well known. They are minute, soft-bodied forms that rarely exceed 2 mm. in length. Their normal habitat is in moist soil, leaf litter, under bark and debris, or among decaying vegetation. They are often overlooked in places where they may exist in considerable numbers. Not more than 60 to 70 species have been described so far. They apparently have a wide distribution but are not found in deserts or colder zones.

The body of a pauropod consists of a conical head and a trunk of 12 segments. Each of 9 trunk segments bears a pair of legs. The legless segments are the first (collum), the eleventh, and the twelfth (pygidium). A 10-legged pauropod, *Decapauropus,* has been described. The entire dorsum of the trunk is covered by only 6 tergal plates, 1 plate for every 2 segments (Fig. 15-6, *A*). This tendency to form diplotergites does not extend to the formation of diplosegments or to the ventral region. Each of the tergal plates except the first bears a pair of long, lateral setae, which are tactile in function. The colum, or first segment, is small dorsally and expanded ventrally. Each side of the head has an oval area that covers a fluid-filled chamber and gives the appearance of an eye, but it is thought to be a sensory organ of unknown function. The head bears a pair of large, branched antennae. Each antenna

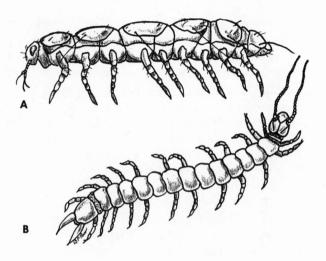

FIG. 15-6

A, *Pauropus silvaticus,* adult (class Pauropoda). **B,** *Scutigerella immaculata,* adult (class Symphyla). (**A** modified from Tiegs; **B** modified from Snodgrass.)

consists of a 4-segmented basal stalk and 2 branches. The dorsal branch terminates in a single flagellum; the ventral branch bears 2 short flagella, with a club-shaped sensory appendage between them. Each of the paired mandibles is a single piece bearing teeth and deeply buried in the head. The first pair of maxillae may be represented by the lower lips and may be comparable with the gnathochilarium of the diplopods. The pauropod legs are 6 segmented with a variable pretarsus consisting of a median claw, ventral lobe, and 1 or 2 accessory claws.

The ovaries are found between the gut and nerve cord; the oviducts discharge through a single median depression on the genital, or third, trunk segment. A sperm receptacle is connected to the genital depression. The adult testes lie above the gut, and the sperm ducts open through paired penes near the coxae of the legs. Eyes, tracheae, spiracles, and circulatory systems are lacking in the entire group. In their development, pauropods are anamorphic, similar to that of diplopods. They are considered to be more primitive in their organization than the diplopods to which they appear to be closely related. This relationship is most evident in the similarity of their head organization, position of gonads, and opening of the genital ducts on the third trunk segment. Representative genera are *Pauropus* (Fig. 15-6, *A*), and *Allopauropus*.

■ CLASS SYMPHYLA

The symphylans are small myriapods with centipedelike bodies (Fig. 15-6, *B*). They live in about the same kinds of habitat as do the pauropods, i.e., leafy mold, humus, and various kinds of debris. They often become greenhouse pests. Most of them are between 2 and 10 mm. long. Some 100 species are known. Although they resemble diplopods, some of their characteristics are shared with the lower insects. For this and other reasons, many authorities consider them to be possible ancestors of the insects, but this theory is questioned by others, especially by S. M. Manton, because the mandibular mechanism of the Symphyla must have evolved along a different line from that of the insects.

Their bodies are soft and slender and consist of 12 leg-bearing segments and a larger number of tergal plates. Some of the tergal plates are duplicate terga of certain segments (3, 5, 7). A thirteenth, or preanal, segment bears a pair of unjoined appendages that function as spinnerets and have silk-forming glands connected to them. The last segment is the small anal segment, which carries a pair of sensory hairs (trichobothria) and the anus.

The legs are much alike with 6 segments; the first pair, however, has only 5 segments and is usually smaller than the others. The terminal pretarsus of the legs is provided with an accessory lateral claw. Each leg arises from a moundlike structure, the limb base, or coxa. Near the base of the leg and attached to the body wall is a short stylus, or bristlelike process, and an eversible vesicle, the functions of which are not known. These vesicles and styli are absent on the first body segment.

The head resembles that of other myriapods, with the mouthparts lying beneath its surface. The epistome and bilobed labrum project forward, and the head may be flattened on top. The head bears on its dorsal surface a pair of long, many-segmented antennae that are simple and unbranched. Each mandible bears a movable gnathal lobe. The flattened gnathal lobes lie horizontally in the preoral cavity of the head; each is articulated to the basal plates of the mandibles. The mandibles are manipulated by a complex musculature. The first maxillae are long appendages lying against the underside of the head, whereas the second pair is fused to form the labium, or lower lip. The labium is a broad, oval, bilobed plate with 6 papilla-bearing lobes.

The tracheal system is restricted to the anterior region of the body. It consists of a single pair of spiracles arising in the head and tracheal tubes that ramify through the first 3 trunk segments. Although eyeless, symphylans have a pair of well-developed organs of Tömösvary near the bases of the antennae.

As in many other myriapods, the genital openings are located on the ventral side of the third trunk segment. Eggs are laid in small clusters and attached by stalks. The development is anamorphic, with the first instar having six or seven pairs of legs. The symphylans molt throughout their life, and some, such as *Scutigerella immaculata*, live several years.

The most familiar genera are the *Scutigerella* (Fig. 15-6, *B*), which include the garden centipede of the greenhouse, and *Hanseniella*. Both of these genera have been extensively studied, although many details of their life history are still lacking.

■ CLASS INSECTA (HEXAPODA)

It is not within the scope of this textbook to treat in detail the most extensive group of organisms within the animal kingdom. No group has had a more extensive coverage in texts and monographs than has this arthropodan class. Nevertheless, it would be a serious omission to ignore a group that has reached such a high peak of invertebrate evolution. The plan of the present work is to fit the insects into the morphologic patterns characteristic of the arthropod group.

The present discussion will first emphasize the general morphologic features of insects as a group and will then show how each of the principal orders has modified these features in its adaptive and ecologic relations.

An insect may be defined as a member of the Arthropoda, in which the body is usually divided into head, thorax, and abdomen (or three tagmata); with never more than three pairs of jointed legs carried by the thorax; and often with one or two pairs of wings on the second and third segments of the thorax. In addition, it has only one pair of antennae and usually both simple and compound eyes. The second maxillae are fused into a labium, or lower lip, and the mouthparts are variously modified and adapted for different feeding practices. Breathing occurs by means of tracheae. The genital openings are located toward the end of the abdomen, and the female is usually provided with an ovipositor. A technical definition of insects restricts the number of postcephalic somites to 14.

Most insects are strictly terrestrial and are found in almost every kind of habitat. Some are adapted for an aquatic existence, either for all or part of their life history. Only a few species are found in marine waters. There are more species in this class than in all the rest of the animal kingdom. The number of insect species already described has been estimated to be at least 800,000, with many millions yet to be described. In insect populations the number of individuals reaches fantastic proportions in even a small area. The variety of insects is also markedly increased by the occurrence of two or three different forms in most species, such as the metamorphic stages, larva and pupa, in addition to the adult. The genus *Phylloxera*, for example, has as many as 21 different forms. Since each form of a species usually has a different food habit, a great variety of habitats is occupied.

The economic importance of insects need not be mentioned here except to say that they pose some of the great problems confronting man. As a group, insects embrace within their food habits almost every kind of organic material and thus become the greatest pests to man's food supplies. On the credit side there are many beneficial insects, such as those that are essential in cross-fertilization of plants, those that are predaceous and parasitic and are used for biologic control of other insects, and those that are used as food by countless birds and other animals.

Insects show a great range in size. Some are of almost microscopic size. Certain beetles are only 0.25 mm. long. Most fall within the range of 5 to 40 mm. in length. Some tropical moths may have a wingspread of about 30 cm. Fossil records indicate that some early insects were much larger.

Insect evolution is a never-ceasing marvel to students of evolution. The fossil record indicates that they are not so old as many other groups. The earliest known fossil insects date about 300 million years ago to the upper Carboniferous age, but insects may have been in existence long before this. Many of these early insects were closely related to the dragonflies and cockroaches of the present time (Fig. 15-7). Early insects were more primitive than any other known insects and may have been close to the ancestors of the winged forms. Amber deposits of the Tertiary period of around 50 million years ago indicate that insects had already undergone an amazing evolutionary development. Many of the genera found in those deposits are still around. Living genera of bees, however, are practically nonexistent in amber deposits. Evidence also indicates that the earliest insects were wingless and similar to the present order Thysanura; they may have arisen from the ancestral stock of the Symphyla. The development of wings was the second stage in the evolution of insects and must have occurred at least 50 million years before the reptiles and birds had wings (F. M. Carpenter, 1953).

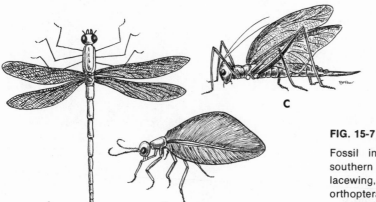

FIG. 15-7

Fossil insects. **A,** *Tarsophlebia*, Jurassic dragonfly, southern Germany. (×1.3.) **B,** *Mesopsychopsis*, Jurassic lacewing, Germany. (×1.) **C,** *Metoedischia*, Permian orthopteran, Russia. (×1.) (Redrawn from Moore, Lalicker, and Fischer: Invertebrate fossils, McGraw-Hill Book Co.)

The rapid evolution of insects in comparison with other major groups may have been due to their short generations, reproductive rate, gene fluctuations in relation to environment, behavior patterns, locomotion, etc. Their actual mutation rate may not be greater than that of other germ plasm lines. Another factor of primary importance in their rapid evolution is their small size, which enable them to fit into a great diversity of small ecologic niches that may have been denied to larger forms. Evolution involves not only alteration in the animal but also the availability of a suitable environment into which it can fit. On the other hand, their chitinous exoskeleton has been a factor in restricting the evolution of arthropods because of ecdysis. The fossil record further indicates that the number of insect groups now living is only a small percentage of those that existed in the past, although there have been some progressive changes in their morphology and physiology.

External morphology

Adult insects differ from other groups of the Mandibulata in the fusion of certain segments and the division of the body into three marked parts. The head, which is sharply defined from the other two divisions (thorax and abdomen), is made up of the fusion of 6 segments. The thorax consists of 3 segments that form a com-

pact, rigid unit, with an extensive musculature for the manipulation of the legs and muscles. The abdomen, which is made up basically of 11 segments, contains most of the visceral organs. In some cases the terminal abdominal segments are reduced or vestigial.

The integument of insects is basically the same as that of other arthropods. It consists of a cuticle produced by a single layer of epithelial cells (hypodermis). The basic constituent of the cuticle is chitin of a complex chemical nature, chiefly a nitrogenous polysaccharide, with other substances to give color and rigidity. A variety of cuticular outgrowths such as hairs, spines, and scales are attached to the integument. To grow, the old cuticle must be shed (ecdysis), as in other arthropods.

Head. The head varies greatly in the different groups of insects. It is made up of a procephalon (of 1 or 2 segments), to which the antennae, compound eyes, and ocelli are attached, and the gnathocephalon (of 4 segments) bearing the mouthparts. Head sutures rarely coincide with the sutures between the original segments so that it is difficult to make out the exact number of segments in the head. The first segment of the gnathocephalon is reduced and bears no appendages in insects but bears the chelicerae in arachnids. The other 3 segments of the gnathocephalon bear the mandibles and the two pairs of maxillae. The procephalon may correspond to the prostomium of annelids and the antennae to the appendages of the polychaete prostomium.

The antennae are of several forms and are usually made up of many segments. They may be slender and tapering, clubbed, filamentous, feathered, or serrated; each type may have developed with respect to specific environments (Fig. 15-8). A typical antenna may consist of three divisions: a basal segment, or scape; a small segment, or pedicel; and a long, many-segmented flagellum. Antennae usually have chemical and mechanical receptors for olfactory, auditory, and tactile stimuli (Fig. 15-9). The pedicel, or second antennal segment, often bears Johnston's sense organ, which is sensitive to vibrations. Most insects do not have intrinsic muscles beyond the pedicel of their antennae. A few insects have no antennae, and in others they are reduced to 1 segment.

Insect eyes are of two types, simple and compound. Simple eyes (ocelli) may be numerous in insect larvae, but there are never more than 3 in the adult. They are often situated in the angles between the compound eyes and the rims of the antennal sockets. Each simple eye consists of a cuticular disk, a lens, a vitreous body, and a group of retinal cells that are connected to nerve fibers running to the brain. Pigment

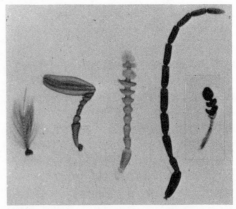

FIG. 15-8

Some types of insect antennae. From left to right: plumose (mosquito), laminate (May beetle), serrate (click beetle), moniliform (tenebrionid beetle), and clavate (water scavenger beetle). Antennae of insects have sensory functions and act as tactile organs, as organs of smell, and, even in some cases, as organs of hearing. The most delicate chemoreceptors occur on the antennae, and some are capable of receiving stimuli from distant sources (as in sex attractants) as well as by direct contact.

FIG. 15-9

Finer details of insect antennae. **A,** Longitudinal section of two terminal segments of antennae of worker bee. Numerous openings of sensory (olfactory) nature are seen in cuticle. (×83.) **B,** Sense organs (sensilla) on antennae of grasshopper *Romalea*. In this type of olfactory organ, sensory bristle ends in pit overlaid by thin membrane. (×1125.) (From Andrew: Textbook of comparative histology, Oxford University Press, Inc.)

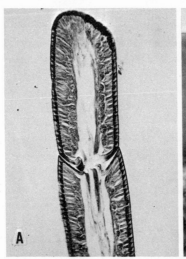
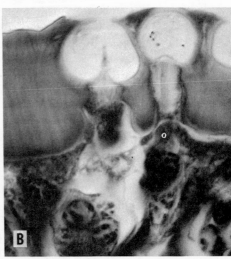

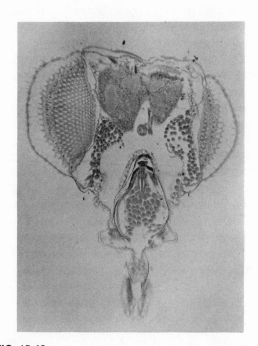

FIG. 15-10

Section through eyes of insect. The two large laterally placed masses showing many facets are the compound eyes.

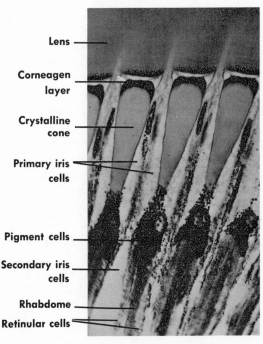

FIG. 15-11

Several ommatidia of compound eye of grasshopper *Romalea.* (×400.) Insect compound eyes have a range of wavelengths of light rays of about 2,500 to 7,000 Å as compared with those of the human, 3,800 to 7,600 Å. Insects thus have a wider range than man in the ultraviolet, but insects differ in their spectral sensitivity. (From Andrew: Textbook of comparative histology, Oxford University Press, Inc.)

may be absent in the ocelli of most insects. Ocelli probably do not form images.

The insect compound eye (Fig. 15-10) has the same basic structure as that of the crustacean compound eye already described, although certain details are different. The compound eye (paired) is made up of numerous ommatidia, each of which is a pencil-shaped structure that abuts on a facet of the cuticle (Fig. 15-11). Since each ommatidium lies at a slightly different angle from its neighbors, it points in a slightly different direction from that of its neighbors. The compound eyes may be composed of from only a few to several thousand ommatidia (more than 20,000 in dragonflies). Usually the compound eyes are so large that they comprise a great part of the head. Compound eyes produce a composite erect image made up of a mosaic of images from each om-

matidium. They are very well adapted for picking up motion and for precise information about distance. In some insects they are also sensitive to different colors. In bees the compound eyes can analyze polarized light and use skylight as a compass.

The mouth consists of four closely united parts: the labrum (upper lip), the mandibles, the maxillae, and the labium (lower lip) (Fig. 15-12). All these except the labrum are homologues of paired appendages found in other arthropods. The mandibles are heavy, triangular structures that are often provided with toothlike projections. They are commonly used for crushing food; however, in some insects they are modified into long, sharp stylets that, with other mouth structures, are adapted for sucking sap or blood. Each maxilla usually consists of a coxa with endites and a jointed palpus, but they may be modified to form piercing organs or sucking tubes (proboscis of butterflies and moths) (Fig. 15-13). The typical form of the labium, or lower lip, consists of an immovable basal part and a distal part, which is movably articulated with the basal part and bears a palp. The basal part in some insects is known as the postmentum, and the distal part is known as the prementum. The labium is really formed by the fusion of the second pair of maxillae, as in many other arthropods. The labrum, or upper lip, is a flat plate hinged to the head anterior to the mandibles. The labrum also is variously modified and, in the suctorial type of mouth, may serve as a sheath for the other mouthparts. The hypopharynx is a tonguelike lobe found behind the labrum. It is in the preoral cavity (cibarium) and furnishes saliva. The preoral surface of the labrum is highly sensitive and is called the epipharynx. In most insects the jaws project below the head (hypognathus); in primitive forms the jaws may be directed forward (prognathus).

The typical head is divided into three regions (Fig. 15-14): the vertex, which is the top of the head between the eyes; the frons, which is the region below the antennae; and the clypeus, which is below the frons, between the bases of the mandibles, and above the labrum. There are many modifications of this plan.

Thorax. The thorax is usually connected to the head by a membranous neck, or cervical

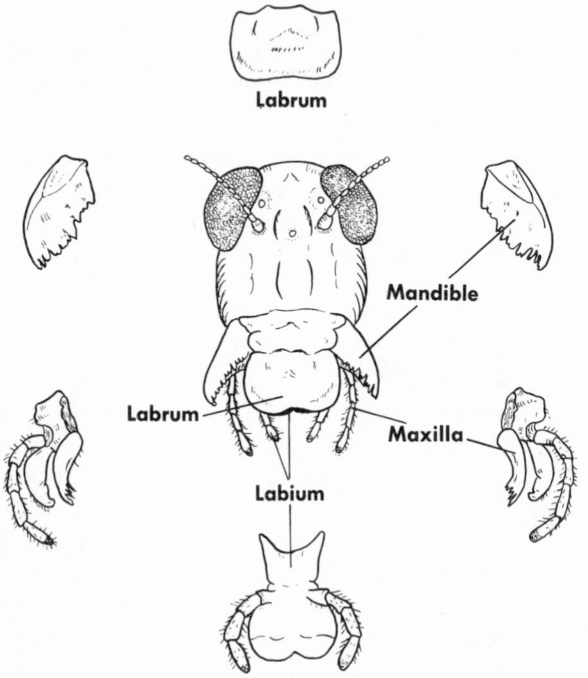

FIG. 15-12

Mouthparts of grasshopper (diagrammatic). (From Hickman: Integrated principles of zoology, The C. V. Mosby Co.)

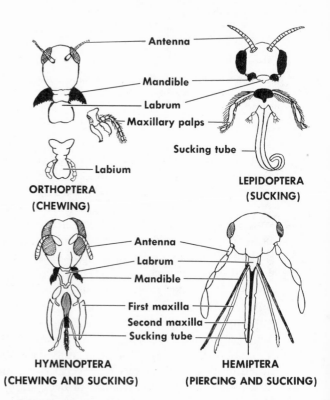

FIG. 15-13

Adaptations of insect mouthparts for various types of feeding. (From Hickman and Hickman: Laboratory studies in integrated zoology, The C. V. Mosby Co.)

sclerites, and is the center for locomotion in the insect body. The thorax is composed of 3 closely knit but recognizable segments. Each segment is made up of a dorsal tergum, lateral pleuron, and ventral sternum; each of these may consist of a number of sclerites except the sternum, which usually has only a single sclerite. The thoracic terga are called the notum, which is usually divided into two principal sclerites, the anterior alinotum and the posterior postnotum. The notum of the prothorax is a large saddlelike plate between the head and base of the wings in some insects. Because of its functions, the thorax forms a stoutly built, compact box with thick walls. The first segment, or prothorax, is usually the smallest of the 3 segments and bears the first pair of legs. Each of the next 2 segments, the mesothorax and metathorax, bears a pair of legs, and in the Pterygota, a pair of wings also (Fig. 15-14). In the Pterygota the walls of the pleural regions of these 2 segments are strengthened by thicker layers of chitin and by invaginations to form struts. The legs are walking legs and usually consist of 6 segments. Beginning at the body these

segments, in order, are called coxa, trochanter, femur, tibia, tarsus, and pretarsus (Fig. 15-14). The coxa forms a ball-and-socket joint with the subcoxa of the body wall; the trochanter is small (divided into two parts in dragonflies); the femur is the largest of all; the tibia is slender; the tarsus is divided into 5 or fewer subsegments; and the pretarsus bears a pair of lateral claws, between which may be a fleshy lobe, the pulvillus. Each true leg segment has its own musculature. In different insects these legs may be modified for special functions. The front legs may be adapted for prehension or digging; the hind legs (Fig. 15-15) may be modified for jumping or swimming; or legs may be adapted for clinging, etc.

In their development the wings arise as lateral outgrowths of the body; they are not modified appendages. Since they are double, they consist of upper and lower membranes. They are adult structures, and the only immature insects that have them are the mayflies. Wings are really flat integumental folds strengthened by cuticular thickenings that form branching, riblike structures called veins. The

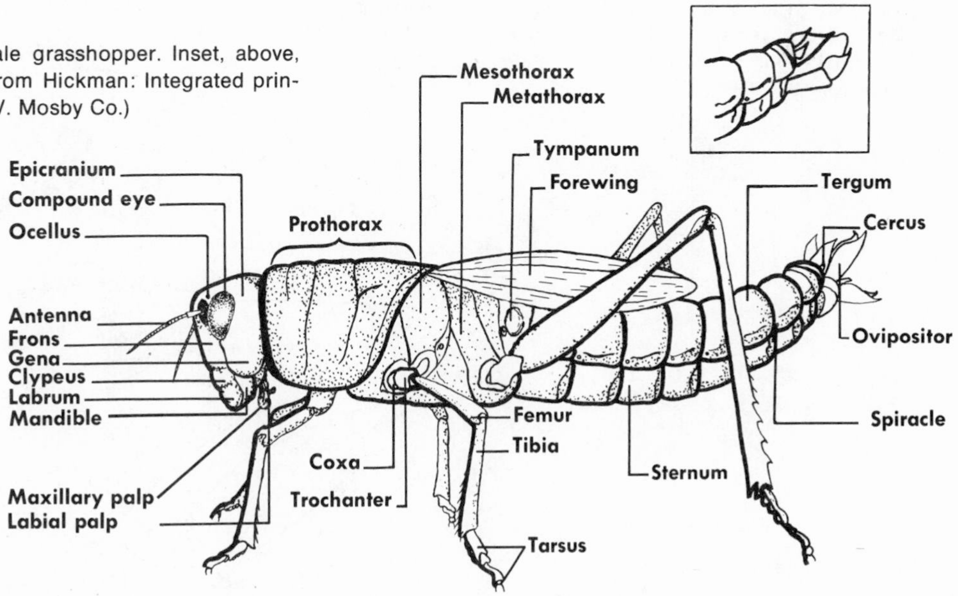

FIG. 15-14

External features of female grasshopper. Inset, above, posterior end of male. (From Hickman: Integrated principles of zoology, The C. V. Mosby Co.)

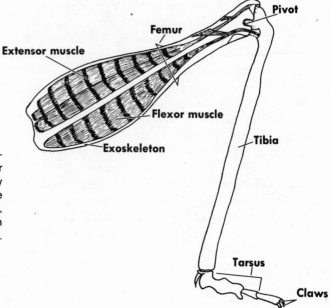

FIG. 15-15

Hind leg and femur muscles (diagrammatic) of grass-hopper. Muscles operating leg lie within hollow cylinder of exoskeleton, attached to internal wall from which they manipulate segments of limb on principle of lever. Note pivot joint and insertions of extensor and flexor muscles, which act reciprocally to extend and flex tibia. (From Hickman: Integrated principles of zoology, The C. V. Mosby Co.)

FIG. 15-16

Generalized wing venation of an insect showing the chief veins (capital letters) and cross veins (lower case letters). Comstock-Needham terminology scheme has been followed. Although wing venation varies in different insects, venational homologies may apply to most insects and have taxonomic value in certain species. Insect wings arise as double layers of epidermis. Between these layers are the tracheae around which cuticle thickens, and thus wings are strengthened. Abbreviations indicate principal veins and cross veins as follows: **C**, costa; **Sc**, subcosta; **R**, radius; **Rs**, radial sector; **M**, media; **Cu**, cubitus; **A**, anal. Cross veins: **h**, humeral; **c**, costal; **r**, radial; **s**, sectorial; **r-m**, radiomedial; **m-cu**, mediocubital; **cu-a**, cubitoanal.

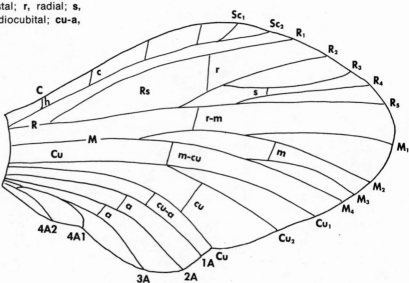

pattern of venation is distinctive for taxa as low as families (Fig. 15-16). Blood tracheae, and nerves pass through the wings. The main branches of the hollow veins are connected by cross veins. Wings have usually been the best fossilized parts of the insect body.

The mechanism of wing movement will be discussed in a section on locomotion in insects.

Abdomen. The abdomen is made up basically of 11 segments, but the eleventh is usually much reduced and may be represented only by appendages. Some insects may have fewer segments. The exoskeleton of the abdomen consists of a dorsal tergum and a ventral sternum connected by a membranous pleural region, which rarely contains sclerotized areas. In many immature insects, adult Apterygota, and male Odonata, segments 1 to 7 bear appendages. In the immature forms these abdominal appendages may be gills (mayfly nymphs), lateral filaments (larvae of aquatic Neuroptera), or prolegs (Lepidoptera larvae). In the adult Apterygota the abdominal appendages are usually styli. In the male Odonata they consist of a copulatory structure on the ventral side of the second abdominal segment. In many insects a pair of cerci extend back from the posterior end of the abdomen and are sensory organs derived from a pair of appendages. On the genital segments (8, 9, and also 10 in males) are the specialized appendages or external genitalia for copulation. These parts are absent from the primitively wingless insects. The stings of bees and wasps are modified ovipositors and are withdrawn when not in use. A pair of spiracles, or openings of the tracheal system, is usually found on each of the first 8 segments, as well as on the mesothorax and metathorax.

Internal morphology and physiology

The internal anatomy of insects has much in common with that of crustaceans. Most systems in the two groups are similar, with minor differences here and there. A basic unity of internal structures runs throughout the extensive group; it is the diversity of external features

that has formed most of the criteria for their classification.

Muscular system. The active Insecta group is well supplied with muscles that connect one part of the exoskeleton with another. The locomotor system is similar to that of the vertebrate tetrapods. In the tetrapods the stiffness of limb segments is due to bone; in arthropods the stiffness of each segment is due to the rigid cuticle that forms a hollow cylinder enclosing the muscles. In both tetrapods and insects, joint action is controlled by pairs of muscles acting reciprocally, and the limbs act as levers (Fig. 15-15). Whereas the muscles of the tetrapod limb are fastened to the outside of the bone and act over the joint, the muscles of the insect limb are attached to the inside wall of the cuticle and act over the joints from within. In soft bodies such as the caterpillar, movement is produced by a combined action of the muscular wall and the hydraulics of body fluids, similar to that of annelids. Skeletal muscles that attach to the body wall usually fasten onto the cuticle by fine tonofibrillae, which are reformed at each molt. Insect muscles are segmentally arranged in the abdomen but not in the head and thorax. Muscles are mostly striated and very strong for their size. Their innervation is different from that of tetrapods. Only a few nerve fibers run to each insect muscle. The muscle is usually supplied by a fast axon, which responds only when impulses in excess of 15 per second pass over it and produce relatively slow contractions. Insect axons usually supply several end plates to each muscle fiber and may allow local contractions of the fiber.

In walking, the insect uses an alternate tripod method, whereby the anterior and posterior limbs of one side and the middle limb of the other are used for support while bringing the other 3 limbs forward.

Digestive system. The basic plan of the alimentary canal consists of a foregut, midgut, and hindgut. The foregut (stomodeum) and the hindgut (proctodeum) are ectodermal in origin and are lined with cuticle but have a

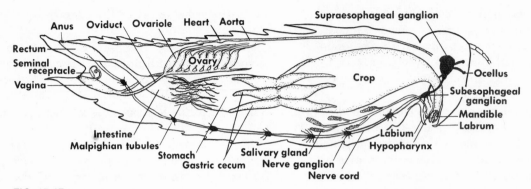

FIG. 15-17

Internal structure (diagrammatic) of female grasshopper. (From Hickman: Integrated principles of zoology, The C. V. Mosby Co.)

muscular coat. The midgut is endodermal in origin and lacks a cuticular lining. The alimentary canal may be straight or convoluted but usually has certain common, specialized regions. Valves and sphincters regulate the passage of food from one region to another. The first part is the preoral cavity, or cibarium, which is followed by the pharynx (which may have a pumping device) with a pair of salivary glands, and then by a slender esophagus. The esophagus is commonly provided with a dilated portion, the crop (Fig. 15-17), that often has teeth for grinding food. The last part of the foregut is the proventriculus, or gizzard. The midgut is the stomach, into which 8 double gastric pouches pour their digestive juices. Into the posterior end of the midgut or the anterior end of the hindgut 2 to 150 long malpighian tubules empty and discharge certain wastes. Urates are discharged with the feces. The hindgut consists of a narrow intestine and a wider rectum, which opens by an anus on the eleventh segment. Most digestion takes place in the crop and the midgut. Absorption of food occurs largely in the midgut, whereas most of the water is absorbed from the hindgut.

Since insects use many different kinds of organic foods, there are many different feeding and digestive mechanisms. Insects are adapted to all types of diets that include a great variety of living, dead, and decomposing animals, plants, plant and animal products, blood, and plant juices. Modifications of the mouthparts and digestive system are associated with the kind of diet and with the method by which food is obtained.

There are two general types of mouthparts—those adapted for chewing (mandibulate) and those adapted for sucking (haustellate). Each of these types is subjected to many variations in different insects. The chewing mouthparts probably represent the primitive plan. In this plan the mandibles, which are heavy and capable of cutting, tearing, etc., move sideways so that the insect is able to bite off and chew its food. This method occurs in the primitive Apterygota, Orthoptera, Odonata, Coleoptera, Hymenoptera, Isoptera, Neuroptera, and some others. It also includes the larval stages of many insects. The most generalized condition of chewing mouthparts is found in locusts and crickets, in which the mandibles are heavily sclerotized. A few mandibulate insects (e.g., bees) have modified the labium and maxillae into a tonguelike structure for sucking liquid food (chewing-sucking mouthparts). The larvae of predaceous diving beetles have such a device for sucking the body fluids of their prey.

The sucking, or haustellate, mouthparts are elongated or stylelike. This type is for piercing and sucking, but there are many adaptations for the same feeding habits, as the method may have evolved independently in the different groups of insects. This method, which is highly modified in different orders, include the thrips, Hemiptera, Homoptera, Diptera, fleas, sucking lice and Lepidoptera. In the Hemiptera the beak may be long and segmented and may originate at the front or rear of the head. The labium forms a segmented sheath and encloses 4 piercing stylets, the 2 mandibles, and the 2 maxillae. The labium does not pierce but folds back as the stylets pierce the tissue. The proboscis of the Lepidoptera is long and coiled and is formed of the 2 galeae, or outer lobes of the maxillae. The mandibles and hypopharynx are entirely lacking in this type, which is adapted for sucking nectar from flowers. When used, the proboscis is uncoiled by blood pressure; when unused, it recoils by its elasticity. In some of the Diptera (horseflies and mosquitoes) there are 6 stylets for piercing: labrum, mandibles, maxillae, and hypopharynx. The labium usually serves as a sheath for the stylets. The stylets are very slender and needlelike in mosquitoes, but broader and knifelike in horseflies. The labium in horseflies may be used to convey blood from a wounded tissue to the food channel between the hypopharynx and the grooved labrum. Houseflies (nonbiting) simply lap up liquid fluid or food liquefied by salivary secretions with their specialized labium, provided with lobes bearing transverse grooves.

Most insects take in food through the mouth, although some parasitic larvae can absorb food through the surfaces of their body. In a generalized form (e.g., cricket) the mouth appendages are the epipharynx (a median lobe on the ventral surface of the labrum) and the hypopharynx; these appendages surround the preoral cavity. The preoral cavity opens into the buccal cavity of the foregut (stomodeum). The small buccal cavity has a pair of salivary (labial) glands opening into it by a common duct. The secretions of the salivary glands vary. In many insects they contain certain enzymes (amylase, or invertase); in others the secretions lubricate the food for swallowing; in mosqui-

toes they contain an anticoagulant; and in the larvae of Lepidoptera they are modified to form silk glands for the cocoon.

Many insects eject digestive enzymes on food before food is ingested (flesh fly larvae and aphids). This extraintestinal digestion may also occur in the prey of ant lions and the larvae of predaceous diving beetles.

The short pharynx, which connects the buccal cavity to the esophagus, is mostly a conveyer of food and ends at a constriction where the stomodeum passes through the nerve ring. From the esophagus the food passes into the posterior part of the foregut, which often serves as a storage place for food and for the partial digestion of the food. A crop may also be formed as a diverticulum from the esophagus. Here salivary, or labial, glands (which open into the buccal cavity) secrete saliva to moisten the food and provide some digestive enzymes for preliminary digestion. In those insects that live on liquid food (cicadas and aphids), a part of the foregut may be modified into a sucking pump (cibarium) that is manipulated by longitudinal muscles in the wall of the foregut. When these muscles contract, the cavity of the foregut is enlarged, making a partial vacuum into which food is forced by the outside air. The proventriculus or gizzard at the posterior end of the foregut may have the cuticle lining formed into teeth for breaking up the food.

The midgut (ventriculus) with its thin muscular coat is lined with epithelial cells, but in many there may be a peritrophic membrane (of chitin and protein) lining, which is secreted by a collarlike extension of the foregut inside the midgut. This membrane is permeable and permits the exchange of both digestive enzymes and the digestive products ready for absorption. This lining protects and supports the delicate midgut wall from abrasion by food particles, and in some species it is secreted continuously to serve as a conveyer belt for the food down the gut. The final fate of the membrane is that it may be destroyed in the hindgut, or it may enclose the fecal pellets. Insects that feed on fluid food usually do not have a peritrophic membrane; those that feed on solid food usually have one. The midgut is the principal site of enzyme production, digestion, and absorption. The type of enzymes is associated with the diet of the insect. Some enzymes are very specific, such as those of the clothes moth that digests hair and keratin. Omnivorous insects produce a general group of enzymes (lipase, carbohydrase, and proteolytic). Blood-sucking insects emphasize the proteolytic enzymes. The midgut is often equipped with ceca, which are outpocketings of the gut. These may house bacteria that aid in digestion, or they may serve as storage sites. Not many insects produce enzymes that digest cellulose, but some are able to use cellulose as food because of microorganisms (bacteria and flagellate protozoans) present in their gut. These organisms can digest the cellulose, and the products are used by the insect host. Some insects (those feeding on juices) can extract a large percentage of the water in food (usually in the crop) before water comes in contact with enzymes.

Insects have above the same nutritional requirements as man, including the essential amino acids, B vitamins, sterols, nucleic acid derivatives, and minerals. Some essential accessory items of food are furnished by the symbiotic organisms. Most female mosquitoes must feed on blood to lay eggs. Experiments show that female mosquitoes cannot produce eggs if fed on a diet of sugar alone. Water requirements vary. Some get a great deal of water with their food (leaf-feeders) and must have considerable water loss, as do aquatic insect larvae. They are mostly concerned with salt retention. Those that feed on dry food must conserve their water and excrete excess salt. However, the water and salt content of the blood is about the same for all insects.

The hindgut (proctodeum) may absorb some food products, but water absorption and fecal formation appear to be its principal function, especially in the rectum, which has tall epithelial cells for this purpose. Most Homoptera, which feed chiefly on plant juices, have a peculiar filter chamber that serves to extract excess water. Although this filter chamber varies with different species, ordinarily it consists of certain parts of the midgut and hindgut, which are bound close together with connective tissue, so that most of the water bypasses the rest of the fluid food and is taken to the hindgut, whereas the rest is digested and absorbed in the midgut.

Excretory system. The chief excretory organs in insects are the malpighian tubules and rectal glands. The malpighian tubules are adaptive devices for life on dry land and are not associated in any way with nephridia and coelomoducts so characteristic of excretion in many animals. The malpighian tubules may be absent in some insects. They are not found in Collembola and are much reduced in other orders of insects. Their chief function is to remove nitrogenous and other wastes, as well as regulate the water and salt balance in the body fluids. These wastes are extracted from the blood by the tubules, which are attached to the gut at the junction of the midgut and hindgut (Fig. 15-17), and are passed into the hindgut, from which they are passed out of the anus along with the fecal waste of digestion. In contrast to aquatic forms that excrete nitrogen mostly in the form of ammonia, insects excrete nitrogenous waste chiefly in the form of uric acid. The toxic effects of ammonia are not present when it is diluted with water, which aquatic animals have in abundance. Besides, ammonia excretion requires a great deal of water, whereas uric acid requires little or no water to flush it out of the body, since uric acid is crystalline and relatively insoluble. In this way, water can be conserved in a land animal such as an insect.

Malpighian tubules have about the same histologic structure as that of the midgut but are ectodermal, not endodermal derivatives. They have muscular coats for agitating the tubules so that the latter can absorb materials from the blood. They are lined with cuboidal epithelial cells and have connective tissue in their walls. Movements of the tubular contents are facilitated by muscular movements and by the water passing in at the distal end of the tubule and out at the proximal end into the gut. The number of tubules varies from a few to more than 200.

Not all the waste of the insect body is removed by the malpighian tubules. Some salts are stored in the cuticle, to be removed at molting, and some salts are excreted through the walls of the gut. Uric acid may also be stored in the fat bodies. Waste is stored in the molted skin of the Collembola, which continue their molting during the adult life.

In some insects malpighian tubules procure waste materials and water throughout their length and discharge it into the gut; the water is reabsorbed in the rectum. In other insects, however, the tubule functions in a different way at its proximal and distal regions (Fig. 15-18). The distal part becomes filled with fluid urine, and the proximal part collects crystals of uric acid after water there is reabsorbed. The uric acid is sent to the rectum to be discharged. Water reabsorbed in the rectum (and also in the proximal tubule) returns to the blood and can enter the distal tubule with a new load of waste materials, thus producing an effective cycling process of water in waste disposal.

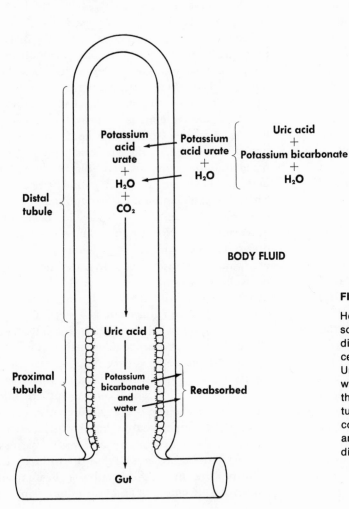

FIG. 15-18

How the malpighian tubule functions in the excretion of some insects. The tubule consists of two histologically distinct portions; a proximal part with brush-border lining cells and a distal part with a smooth epithelial surface. Uric acid combines with potassium bicarbonate ions and water in the body fluid under alkaline conditions to form the soluble potassium urate, which diffuses into the distal tubule. In the proximal tubule the pH shifts to an acid condition, and soluble potassium bicarbonate and water are reabsorbed, leaving uric acid crystals, which are discharged into the gut. (Adapted from Wigglesworth.)

Circulatory system. The circulatory system of an insect is an open one, for the insect body is a hemocoel that is divided into many sinuses by diaphragms. Sheetlike muscle bands in pairs make up a dorsal diaphragm which incompletely separates the region around the heart (pericardial cavity) from the main body cavity. The viscera lie below this dorsal diaphragm in the perivisceral sinus. Below the perivisceral sinus and separated from it by a ventral diaphragm is the ventral sinus where the ventral nerve cord is found.

The heart of insects is tubular and usually extends through the first 9 abdominal segments (Fig. 15-17). The heart has paired lateral openings (ostia), one pair per heart chamber, through which blood enters the heart. Ostial valves prevent backflow of blood in the chambers during the systole of the heart. The chambered heart occurs in only a few insects, for most insect hearts are continuous tubes. The anterior aorta, which is a continuation of the heart, is about the only real blood vessel present in insects. Blood flows forward in this aorta and is discharged into the head region. The blood flows backward in the perivisceral and ventral sinuses and then flows upward in the abdominal region to enter the pericardial sinus. From this sinus the blood enters the heart through the ostia. A thoracic blood vessel or "heart" occurs in rapid-flying insects and draws blood through the wings before discharging into the aorta.

Accessory pulsatile organs (booster hearts) in the thorax aid in moving the blood into the wing, and similar boosters near the base of the antennae aid the circulation through the antennae and legs. Booster hearts vary in location in different insects. Blood from the ventral sinus enters the legs and returns into the perivisceral sinus. Blood usually passes into the anterior veins of the wings and back through the posterior veins. Wing circulation is very necessary to prevent collapsing of the tracheae and in the spreading of the wings when the young insect emerges from the pupa. Blood flow is also aided by various body movements, especially by abdominal contractions during gas exchange and by the thoracic muscles during flight.

The rate of the heartbeat varies from about 12 to 160 beats per minute, but this rate depends to a certain extent on activity. Along the wall of the heart are small triangular muscles (alary) whose contractions may be involved in filling the heart and blood. Both myogenic hearts (the more primitive) and neurogenic hearts are found in insects. Little blood pressure is developed in the general flow of blood through the body, but localized pressures may aid in uncoiling the proboscis of butterflies and help in molting and hatching. Many aspects of blood pressure in insects are little understood at present.

Because of their tracheal system, insects carry little oxygen in their blood, which contains many times as much amino acids as compared with the blood of most vertebrates. Some oxygen is carried in the blood of midge larvae in which hemoglobin is present. The blood in insects, however, transports food, waste (especially uric acid), and hormones the same as blood of other animals. The blood consists of plasma and many types of cells (hemocytes). They average about 50,000 per cubic millimeter of blood or hemolymph—a figure far below that of vertebrates. Not much is known about the different functions of the hemocytes. Some are phagocytic for protecting the body from bacterial invasion, others store fat and carbohydrates, and still others can agglutinate to stop hemorrhage from wounds.

In connection with blood, the water balance is of the utmost importance to an insect. However, the maintenance of water and salt involves many tissues. In insects organic molecules (such as the amino acids) tend to replace inorganic ions in the osmotic regulation of the body fluids. The impermeable body wall prevents water loss by evaporation; some water loss may also be reduced by absorption of water by the rectum. The malpighian tubules are mainly concerned in water balance in aquatic insects and in those insects that take in much water with their food. These tubules can also get rid of excess salt when the water becomes too saline.

Nervous system. Insects have a typical arthropod nervous system, consisting of a dorsal brain of 3 fused ganglia and a pair of closely connected nerve cords (Fig. 15-17). Typically there is a pair of ganglia for each body segment, but this plan is often modified. The 2 nerve cords separate in the head and pass around the esophagus to form circumesophageal connectives to the brain. Autonomic fibers (both sensory and motor) pass from certain ganglia to the heart, gut, and other organs. Nerve cells are typical neurons with cell bodies and axons. The brain serves for general coordination, but it is not absolutely necessary for vital processes; a decapitated insect can still live, walk, and perform many other bodily processes. This is due perhaps to lack of inhibition from the higher centers. In the brain neurosecretory cells secrete a hormone that activates other hormones to control molting and the stages of metamorphosis (Fig. 15-19).

The brain consists of a protocerebrum associated with the compound eyes and ocelli; a deutocerebrum with nerves to the antennae; and a tritocerebrum innervating the labrum and foregut (Fig. 15-19). Connecting the brain with the subesophageal ganglion are the circumesophageal connectives. The brain has resemblance to that of the Annelida, but the insect brain has more complexity, as indicated by experiments. Its size varies in different insects but is usually larger in those with complex behavior. Ratio of brain volume to total body volume is much greater, for instance, in a honeybee than in most beetles. Although insects depend on central conduction and segmental reflexes just as annelids do, insects can adapt themselves more quickly to experimental interference. The ganglia of the nerve cord are large and can carry out complex movements in walking and other func-

FIG. 15-19

Portion of the central nervous system of an insect showing location of endocrine organs in head and prothorax. (Adapted from several sources.)

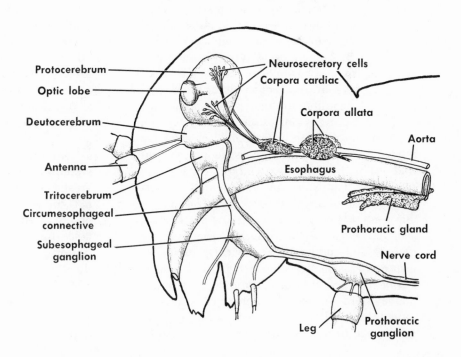

tional activities. For instance, it has been shown by experiments that the front legs of a praying mantis will still carry out the characteristic grasping movements if its ganglia and segment are completely isolated from the other parts of the body. The same can be observed in mating. The last ganglion of this animal has a copulatory center that is inhibited by the subesophageal ganglion. When the subesophageal ganglion is removed (together with its inhibitory center), copulation is carried out even though the female may be destroying the anterior end of the male at the same time (K. D. Roeder, 1935).

Of the three major divisions of the brain, the protocerebrum is generally composed of large protocerebral lobes and an intercerebral region. It contains many important centers, such as the pons cerebralis (an association center), the corpus centrale (which receives fibers from other parts of the brain), and the pedunculate bodies (which vary with the complexity of behavior patterns), but its most complex center is connected to the compound eyes. The deutocerebrum is a relay center for the antennae, with which it is connected by a commissural tract. Although small, the tritocerebrum is associated with the nerves to the labrum and is connected through the circumesophageal connectives with the subesophageal ganglion. Its 2 lobes are separated by the esophagus and are connected by a commissure under the esophagus.

Although there is typically a ganglion for each segment in the central nerve cord, insects have tended to modify this plan by fusion so that there may be fewer ganglia than there are segments. The subesophageal ganglion (Fig. 15-19) is composed of three pairs of fused ganglia and connects with nerves the somites associated with the mouthparts, the salivary glands, and some of the cervical muscles. It also contains neurosecretory cells that secrete hormones. The ventral nerve cord is typically double and is made up of a chain of segmental ganglia. Both the thoracic and abdominal ganglia in the nerve cord are often fused, such as

those of adult Diptera, in which the 3 thoracic ganglia are all fused into 1. On the other hand, *Chironomus* has separate ganglia for each segment in the thorax and abdomen. Giant fibers occur in the ventral nerve cord of some (cockroaches), where they are used for rapid transmission; they may have evolved independently, for they are restricted to only a few insects. In the general makeup of the ganglion, motor fibers are restricted to the dorsal side and sensory to the ventral side. There are also other regional distributions of neurons. The autonomic nervous system may be represented by the stomodeal nervous system, which has its center in the frontal ganglion of the brain and may involve functions of the visceral organs.

Sense organs. The eyes of insects were described in the section on the head. That insects have an amazing sensory system is indicated by a great array of evidence. The variety of stimuli to which they can respond is still not entirely known in spite of much investigation. Insects are sensitive to waves of light and sound, chemical stimuli of taste and smell either directly or indirectly, temperature and humidity, color stimuli, ultraviolet light, ultrasonic sound waves, gravity, etc.

The hard cuticle of insects is usually not conducive to the formation of sense organs, but insects above all other arthropods have overcome this handicap and produced sense organs of the greatest variety and sensitivity. Because all insects have had to face the same basic problems of sensory structure, they have followed a common plan in developing their sensory systems. As already mentioned, the exact mechanism by which a nerve impulse is aroused is not known, yet many theories have been proposed as possible explanations. Some theories hold that substances may directly affect the sensory cells of the receptor, whereas other theories suggest that stimulating substances may react with something in the receptor to produce agents that are involved in the actual stimulation of an impulse. Some organs that may appear very much alike in insects may have different functions. Some scents, for instance, can be detected by one sex and not by the other, as illustrated by the sensitivity of the male moth, but not of the female, for a sex attractant.

Stimuli may trigger certain activities and inhibit others. Stimuli may come from an outside environment or they may come from within due to change of movement or position. Sense organs in insects may be entirely foreign to our own.

The wide range of sensory information that insects have and the organs involved may be indicated as follows:

1. Tactile organs react through hair organs of hair sensilla (Fig. 15-9, B) The impulses may be generated by some deflection of the hair. Such a sensillum is usually placed in a socket covered by a membranous, flexible cuticle. Such organs are especially numerous on the wings and antennae.

2. Campaniform sense organs have a cuticle in the shape of a dome with a sensory nerve ending. When the dome or plate is altered by tension changes in the exoskeleton by vibrations of wind or water currents, the sense organs are stimulated. These organs may also be used for proprioreception.

3. Sense organs responsive to gravity and pressure are found in most insects, especially in aquatic species. Joints in insects have sensitive tactile hairs that register information about position; pressure and gravity may be detected by campaniform sensilla.

4. Scolopophorous organs are made up of sensilla, each of which consists of a cap cell and a cell that encloses the tip of a sensory cell. The sensory ending has a sensory tip (scolops), which contains a central axis that is attached to the cuticle above. These organs have a wide distribution on the insect's body, and some are known to detect vibration. They may be involved in the stimuli associated with acceleration, such as those involved in flight.

5. Tympanal receptors are the most specialized of the auditory sense organs and are especially characteristic of crickets, grasshoppers, and cicadas. These receptors are types of scolopophorous organs whose sensory cells are attached to tympanic membranes. Beneath the tympanum is an air sac that picks up vibrations to which the attached receptor is sensitive. Some sound is detected by many insects with hair sensilla. Mosquitoes have hair on the antennae, one kind of which is called Johnston's organ in the second antennal segment. Tympanic organs may be located in different places in different groups, but are usually located on the first abdominal segment in cicadas and some grasshoppers. Airborne sound can be localized by sound receptors and are very sensitive to amplitude modulation.

An insect is unaffected by differences in the frequency of sound as long as the sound can be detected. Nor does it detect differences in the pitch of a sound in the higher frequencies. The rhythmic features of an insect's song are important aspects of a sound as far as another insect is concerned. Noctuid moths are capable of detecting the high-pitched echolocating sounds of bats and can take evasive action. Some aquatic insects (back swimmers, whirligig beetles) are able to avoid obstacles by receptors on their antennae. Sound plays many roles in insect behavior, for they are alert to environmental conditions. Sound also serves for insect communication.

6. Besides the compound eyes, insects have at least three other types of photoreceptors: dermal receptors, dorsal ocelli, and lateral ocelli. Dermal receptors are supposed to be rare in insects, but several examples of sensitivity to illumination of their bodies are known (V. B. Wigglesworth, 1950). Tenebrionid larvae (mealworms) have been found sensitive to light all over the body, but not all light reactions in insects are due to dermal receptors, as no single receptor system at the cellular level can explain all types of dermal light reactions. The exact nature of the structure of dermal light receptors is largely unknown, but they appear to be in all major phyla.

Dorsal ocelli are organs that react to differences in light intensity and do not form perceptible images, for the light is focused below the retina. Lateral ocelli, however, are capable of forming images on the light-sensitive regions of the retinal cells. Each ocellus is made up of a single lens over a light-sensitive retina which has only a few sensory cells and is not adapted to perceive form. Ocelli may aid in flight movement, since they are often present in winged forms and absent in those without wings, but there are exceptions. Much experimentation has been done on the sensitivity to light of the compound eye. Of course, the large compound eye contains far fewer ommatidia than the number of rods and cones in the vertebrate eye, but from the flower-visiting habits of bees, they can discriminate form to some extent. Honeybees are blind to red, but can distinguish colors in the region of the greens and blues. Most bees are insensitive to waves of light longer than 6,500 angstroms, but some can see ultraviolet light as short as 2,500 angstroms.

At present, there are many difficulties in analyzing sensory receptors in insects in both compound eyes and ocelli, but such experiments have shown a considerable range in visual acuity among the different species.

7. Chemoreception in insects covers a wide range of sensory impressions, and many of the types of sense organs already described involve chemoreception in some form. The chief difference between a chemoreceptor hair and one that is a mechanoreceptor is that the former is rigid, not set in a socket, and its sense cells extend into the axis core of the hair. Both senses of taste and smell are commonly associated with chemoreceptors because they are stimulated by molecules that react with them chemically. Most sensitive chemoreceptors are found in the antennae of insects. Experimentation has shown that the removal of the antennae raises the level of substances needed to cause response by the insect. Other chemoreceptors are common in the maxillar palps. As yet, few generalizations can be made about the specialization of chemoreceptors, but the evidence is strong that it exists, such as that of humidity, which is very important in the selection of the proper environmental conditions for living, sites for egg deposition, food, and other factors.

In general it may be said that insects are living examples of the sensitivity of the organism and the way its behavior patterns have fitted it to react meaningfully and successfully to the environment. Perhaps this feature, more than any other, has been responsible for the amazing diversity of the group.

V. G. Dethier (1971), a leading investigator of chemosensory organs of insects, has recently given a summary of some experimental work on the gustatory sense of flies and the olfactory sense of caterpillars. Chemoreceptors of insects are primary neurons and lack synaptic junctions until their axons enter the central nervous system, compared with the complexity of the vertebrate plan of many interneurons forming a maze of multiple innervations and connections. Compared with that of a vertebrate, an insect's sense organs consist of few receptors. A single hair, for instance, may represent the total receptor field of taste discrimination. Dethier and others have shown that each of the 4 common receptors for taste in insects is sensitive to more than one compound. When pure compounds of a substance are mixed, different interactions occur, and discrimination between different mixtures can be detected, as indicated by electrophysiologic responses (generation of action potentials on a record). Although insects have few sensory units as compared with higher forms, it has been shown that single units (receptors) have wide performances and thus relatively few units can supply the central nervous system with much information. It is seen that some compromise takes place between the total generality and total specificity of receptor units. There is no special receptor for each kind of environmental information, nor is there an omnireceptor for all kinds of information.

Gas exchange. Most insects use one of the most efficient gas exchange systems known — the system of tracheae. Surface respiratory exchange in small insects with large surface area can take place directly through a permeable cuticle, such as in the wingless Collembola that have no trachea. When exchange takes place at the body surface, the blood must be responsible for the distribution of oxygen. With the exception of *Chironomus* larvae (midge flies), which have hemoglobin in the blood plasma, insect blood has a low capacity for carrying oxygen. Some respiratory exchange occurs in the body surface of most insect larvae and to some extent in large adult insects. If small insects live in moist soil conditions, they can breathe by absorption of oxygen through the moist body surface. Since most insects have neither oxygen-carrying blood nor an elaborate circulatory system, they have developed the method of transporting oxygen by having the respiratory pockets in the form of special tubes

extending into all parts of the body — the tracheal system (Figs. 15-20 and 15-21).

The tracheal system is an invagination of the cuticle and is represented by an extensive network of branching and anastomosing tubes that convey air directly to the tissues. This system communicates with the exterior by means of openings called spiracles. There are usually two pairs of these spiracles in the thoracic region and eight pairs in the abdomen, a member of each pair on each lateral surface. There are many variations of this arrangement, which can be considered the most primitive in those that have tracheae. There are two main types of spiracles: those with an open tracheal system, in which the spiracles are merely openings into the integument without devices for closing them, and those spiracles that lie in pits, or atria, which are usually provided with filters of fine hair and devices for opening and closing them. The first type is characteristic of some of the Apterygota. Most insects have the second type, with the filtering and closing mechanisms for keeping out dust and other foreign particles (Fig. 15-21, *A*). In all those insects without tracheal systems (apneustic), all the spiracles are closed so that exchange

FIG. 15-20

Large tracheae of grasshopper *Romalea* showing spiral thickenings of walls. (×142.) (From Andrew: Textbook of comparative histology, Oxford University Press, Inc.)

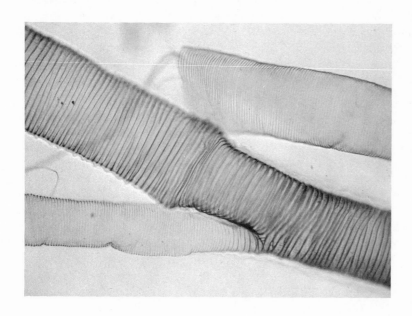

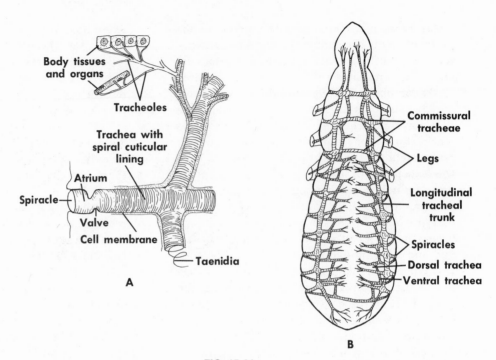

FIG. 15-21

A, Relationship of spiracle, taenidia, tracheae, and tracheoles (diagrammatic). **B,** Generalized arrangement of insect tracheal system (diagrammatic). Air sacs not shown. (Adapted from several sources.)

of gases must take place through the body surface, or through outgrowths of it that are called gills (endoparasites and most aquatic larval insects).

The opening and closing of the spiracles must be well-regulated to meet the needs of the insect at any particular moment. This regulation is done in various ways. The respiratory center in the brain of some is stimulated by increases in the tension of carbon dioxide, such as an increase in metabolic activity, and this tension is carried by the arterial blood to the center. In the locust and some others the respiratory movements are under the control of a ventilation center in the metathoracic ganglion. This control may be exerted by direct innervation or by neurosecretion. In some insects the spiracles are provided with receptors that are specifically sensitive to carbon dioxide. Spiracles of the opening and closing types are usually provided with muscles for the operation of the mechanism of regulation.

The general plan of the tracheal system varies in different species and orders, but there is usually a pair or more of longitudinal trunks with cross-connections with each other (Fig. 15-21, *B*). These longitudinal trunks may have been formed by the anastomoses between the main tracheae of adjacent segments. Longitudinal trunks may be found in various locations and formed essentially in the same way. Connections between the trunks originate from anastomoses of corresponding branches on the right and left sides of the trunk. The main branches of the tracheae lead away from the spiracles and break up into finer and finer tubules. The lining of the tubules is a spiral ribbon or thickened rings of the cuticle (taenidia) to prevent the tubes from collapsing (Figs. 15-20 and 15-21, *A*). The tracheae widen in various places, forming internal air sacs. These sacs may be found where muscle movements will fill and empty them. Air sacs may have no taenidia and are very sensitive to ventilation pressure. Oxygen is not absorbed from the tracheae but from small dead-end tubes, the tracheoles, at the ends of the finest branches. These tracheoles are formed in special end cells that persist after the tracheoles are formed. The tracheoles are more or less filled with fluid and are not more than 2-cell widths away from tissue cells. These tracheoles may branch in a network over the cells, or they may penetrate the large cells and muscle fibers. The cuticle in the tracheoles is not shed during molting, as is that of the lining of the tracheae.

The methods of ventilation have been studied in many insects. The grasshopper, for instance, will relax the abdominal muscles for some 0.25 of a second, and for about the last 0.2 second of this time will open the anterior 4 spiracles for air. Just before the abdominal muscles contract, these spiracles close. The contraction of the abdominal muscles compresses the air sacs and drives the air into the smaller and deeper air sacs, from which air passes into the tracheoles by diffusion. After contracting the abdominal muscles for periods up to 1.5 seconds, the last half of this period is used to open the other spiracles for the release of air. In this way, ventilation pressure gradients are formed for the compression of the air sacs and the delivery of the oxygen to the tracheoles.

Many insects live as adults, larvae, and nymphs in the water and get their oxygen in one or the other of two ways—from oxygen dissolved in the water or from atmospheric air. In small soft-bodied aquatic forms gaseous exchange is mostly by diffusion through the body wall, especially during the early periods of life. But tracheae in some form are generally present in aquatic immature forms. In nymphs and larvae many have special adaptations for gaseous exchange in water. Some of these are in the form of tracheal gills, which may be located on various parts of the body. For instance, mayfly nymphs have leaflike structures on the sides of the first 7 abdominal segments. Those in dragonfly nymphs have folds (sup-

plied with tracheae) in the rectum, and water is alternately passed in and out over the folds through the anus. Damselfly nymphs have the gills in the form of 3 leaflike structures at the end of the abdomen. In some nymphs (stone flies) the gills are fingerlike and occur around the bases of the legs. In all these immature forms there may be extensive exchange of gases through the body surface, as well as through the tracheal gills. Adult aquatic insects usually get their oxygen either through spiracles at the water surface or from an air bubble or film of air trapped somewhere on the surface of the body, such as under the wings. In some insects (riffle beetles) a region of very fine hair is used in holding the bubble and an exchange between this layer of water and the gases dissolved in water may be established on a permanent basis of constant renewal. Some aquatic forms (e.g., water scorpions) have a breathing tube at the end of the body, which can be extended to the surface of the water. In this connection the rat-tailed maggot (*Eristalis*) is provided with a telescoped breathing tube that can be extended at will to the surface for considerable distances.

Reproductive system. Insects are almost universally bisexual. Some special forms of reproduction, such as parthenogenesis (aphids), polyembryony—in which several larvae arise from the same egg (Chalcididae)—and paedogenesis (*Miastor*), are very exceptional. The male organs (Fig. 15-22) include the paired testes, each of which empties into a vas deferens with a seminal vesicle; an ejaculatory tube formed by the union of the vasa deferentia; and accessory glands for secreting fluid to assist in the transfer of sperm. The ejaculatory tube opens on the subgenital plate between the ninth and tenth abdominal sterna. The female system (Fig. 15-22) consists of paired ovaries, each ovary being made up of chains of ova (ovarioles); a pair of oviducts that unite to form a vagina; and the genital opening (vulva) on the ventral side of the eighth segment. Special glands and a spermatheca for

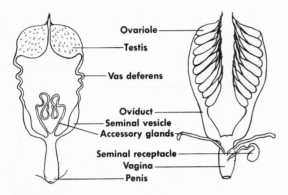

FIG. 15-22

Reproductive organs of grasshopper. Male on left; female on right. (From Hickman: Integrated principles of zoology, The C. V. Mosby Co.)

FIG. 15-23

Early development of egg of honeybee *Apis*. **A,** Centrolecithal egg. **B,** First division of nucleus. **C,** Eight-cell stage is really syncytium. **D,** Blastula. (**A** to **D** longitudinal section.) **E,** Gastrula. Ventral thickening will produce embryo; dorsal blastoderm becomes outer serosa layer; end and side folds will form amnion around embryo. **F,** Median ventral plate gives rise to mesoderm and endoderm; lateral plates give rise to ectoderm. (**E** and **F** transverse section.)

the reception of sperm empty into the oviduct. Fertilization of the egg occurs as the completed egg in the oviduct passes the opening of the spermatheca, where sperm from the male have been stored. Sperm are sometimes transferred to the female in the form of spermatophores. There are many variations of this plan of reproduction in the different groups of insects. An opening (micropyle) in the eggshell makes possible the entrance of the sperm. Most insects are oviparous, but viviparity is found in some flies, such as the tsetse fly *(Glossina)*.

The eggs of insects belong to the centrolecithal type (Fig. 15-23). Such eggs are yolky and are provided with vitelline membranes and strong chorionic shells. By incomplete cleavage involving mitosis of only the nuclei, the egg becomes a syncytium of yolky cytoplasm containing the cleavage nuclei. These nuclei migrate to the peripheral cytoplasm, where they form a cellular blastoderm. A ventral thickening of the blastoderm forms a germ band that will produce the embryo, with the lateral and dorsal cells forming the serosa membrane. End and side folds of the germ band become the amnion that encloses the embryo in an amniotic cavity. From the ventral line along the germ band,

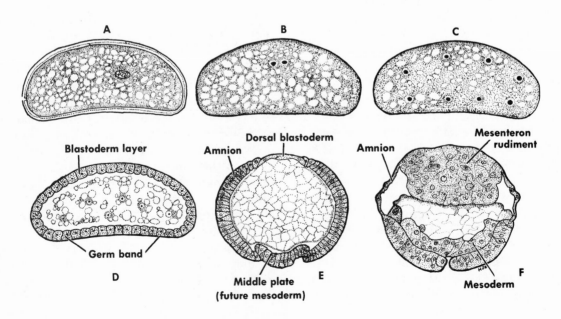

certain cells pass by invagination and overgrowth to form a mesoderm above the germ band. The part left outside forms the ectoderm. By cross furrows the mesoderm layer gives rise to a series of segments that become the head and appendages. The endoderm lies against the yolk and will form the midgut. The foregut (stomodeum) will start as a pit at the anterior end, and the hindgut (proctodeum) will arise in a similar manner from the posterior end. The origin of other organs and systems has already been mentioned.

In the cytoplasm of cockroach eggs (where they were first found) are bacteria-like particles known as Blochmann bodies. During embryonic development they can be traced back to somatic cells (mycetocytes). Blochmann's bodies may have descended from free-living microorganisms that have become modified to form a symbiotic relationship in the insects where they are found, which include, besides cockroach eggs, many of the Homoptera and the weevils (Coleoptera). The particles seem to be involved in oxidative metabolism and are transmitted through the maternal line. They stain very much as bacteria do with Giemsa stains, and they also have other resemblances to bacteria. However, they cannot be cultivated in bacteriologic media. Some authorities classify them with other cell particles as plasmids. Since they contain DNA, they may in their endosymbiosis aid in the synthesis of enzymes (U. N. Lanham, 1968).

Locomotion in insects

The ability to move is one of the basic characteristics of all animals, and this capacity is well developed in insects. Although insects are typically adapted for terrestrial existence, many can live in water, as already described. Insects represent one of the few groups that have the power of flight, and this capacity has been one of the greatest features of their success. They thus have met in some way or other the power to move on land, in water, and in the air, and in all these media they have been quite successful. Perhaps no aspect of insect life has received more attention by investigators in recent years than that of locomotion, especially the ability to fly, which in insects involves many ways and problems not found among the other few groups that have the power of flight.

Certain requirements are basic for all forms of locomotion in animals. There must be sources of energy and there must be tissues that are specially adapted for movements. Although it is true that nearly every part of the animal's body must be integrated and coordinated for movements, certain parts carry the effective burden of movement and are emphasized in all investigations of this type of activity.

The first requirement is naturally that of energy sources, which are demanded in all processes of life. The ultimate source of energy is food. The demands made on most animals for energy in such activities as locomotion are perhaps greater than those in any other process of living. In most animals, glycogen and fat are the principal sources of energy, and it is so in insects. Insects undergo metabolic processes similar to those of other animals. Some of their reactions may not involve direct usage of oxygen, but an efficient source of oxygen supply, which is given by the tracheal system, is an absolute necessity. Like other animals, insects require the machinery of metabolism—the use of organic phosphates; the formation of ATP, or adenosine triphosphate; the reactions of ATP with muscle proteins; and the different metabolic pathways with the end products of carbon dioxide and water.

The fuel consumption of insect muscle (per gram of muscle), especially in flying, is enormous and may be on the order of 1,000 calories per gram of muscle per hour. This is many times higher than the fuel consumption of vertebrate muscle tissue.

Of the tissues most directly involved in locomotion, muscles stand supreme. In arthropods the muscles are all striated, even those of the viscerae. Although both smooth and striated muscles are found in all metazoans, for controlled, rapid movements, striated muscles are by far the better; their presence to such an extent in arthropods, especially insects, indicates a significant aspect of insects' general life activities. In most insects, muscles are very strong so that in these animals, muscles are known that can lift many times the weight of the animal and can enable many of them to jump prodigious distances compared with their weights. The general power of a muscle varies with the size of its cross section. The mass of the animal body varies with the cube of the linear dimensions so that as the body decreases, the muscles within it are more powerful. In insects (which is true for all active animals) the muscle tissue makes up a large proportion of the whole body weight. Only the abdomen is free of muscle to any extent, as it contains most of the other organ systems.

We shall consider three common forms of locomotion among insects: swimming, walking, and flight.

Many aquatic insects are good swimmers, especially the aquatic beetles. In swimming, they make use of their middle and hind legs; the front legs are mostly for seizing and tearing food. Their propulsive legs are flattened with a fringe of hair and furnish most of the thrust. In whirligig beetles the swimming legs are short and flattened, whereas the front legs are elongated and slender. In predaceous diving beetles the paired legs on a segment move together and the paired legs of each of the middle and hind legs alternate; in water scavenger beetles the legs of a pair move alternately. Leg strokes are from 3 to 10 per second in some diving beetles and up to 60 in whirligig beetles. The hairs on the legs are so manipulated that they give the power stroke a great resistance and far less resistance in the backstroke. Other aquatic insects have about the same method of swimming.

The most common pattern of leg movements in walking involves the use of a tripod method by which the first and third left legs and middle right leg act together as one tripod and the other 3 legs as the other tripod. In this way, insects typically keep 3 of their 6 legs on the ground in going about and can make abrupt stops while remaining stable throughout. Other types of leg movements may occur in which 4 or 5 legs are on the ground at the same time. For any method each leg involved must be so coordinated that it undergoes a sequence of activities of alternation of flexion and extension for propulsion. When the insect increases its speed, it does not alter its gait, but simply decreases the time of the thrust and recovery strokes so that the stride length is increased. In other words, they perform the same movements but at a swifter pace.

Stepping movements are controlled by ganglia and may be triggered through impulses that come through the ventral nerve cord from various segments. It is thought that the coordination of the various legs is brought about by intersegmental reflexes in the nerve cord. It appears that all insects have two nerves for each muscle fiber: one for quick twitches, as in jumping and rapid movements, and the other (due to a chain of impulses) for relatively slow movements, as in walking and the maintenance of posture. Collembola (springtails) have a special structure, the furcula, for jumping around. Grasshoppers in jumping elevate their bodies with the front and middle legs and then suddenly extend their hind legs. Muscles involved in the spring are located in the hind femora, which extend the tibiae.

Flight in insects has enabled them to spread quickly to new environments and new sources of food, to search for mates, to be protected from enemies, etc. As far as is known, insects were the first animals to achieve flight, as their record of flight goes back about 350 million years ago (early Devonian times). To escape predators, they may have learned to glide from high to low elevations. The first gliders were no doubt strong jumpers so that flattened projections or planes from the sides greatly aided them in leaping. Each wing arises at the margin of the tergum and pleuron on a process of the lateral exoskeleton, which serves as a pivotal point for the wing. To provide the upward and downward movements of the wing, an alternate depression and elevation of the notum, or tergum, occurs where the wing is hinged with a pleuron support. These movements are caused by dorsoventral tergosternal muscles that depress the notum and elevate the wings and by longitudinal intertergal muscles that elevate the notum and cause the downstroke of the wings (Fig. 15-24). These two types of muscles are the indirect flight muscles that in

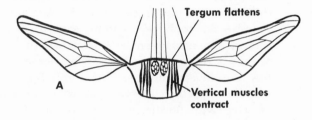

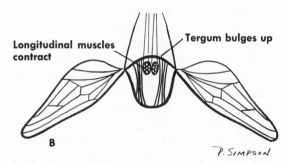

FIG. 15-24

Flight mechanism of insect (diagrammatic). Indirect flight muscles have no direct connection with wings but change shape of body wall to which wings are attached. **A,** When vertical muscles between tergum and sternum contract, tergum flattens and causes wings to rise. **B,** When longitudinal muscles contract, tergum bulges upward and causes wings to be lowered. Direct muscles attach to the wing bases, running from the sternal and pleural regions of the thorax to be inserted on sclerites at the wing bases. These muscles provide forward thrust and lift as well as rotation of the wings. See text description of flight. (From Hickman: Integrated principles of zoology, The C. V. Mosby Co.)

higher forms of insects consists of two fiber groups running at right angles to each other and attached to the elastic cuticular plates of the thorax. In this way the thoracic deformation produced by the longitudinal muscles is reversed by the contraction of the vertical muscles. During flight or wing movement these muscles contract alternately and relax alternately. These two groups of flight muscles range from tibial leg muscles in slow-flapping primitive insects (grasshoppers, dragonflies) to fibrillar muscles that fill most of the thorax in the higher insects. Fibrillar muscles are often the pinkish-yellow color because of the great number of respiratory enzymes. They also contain many mitochondria.

The first winged insects had two pairs of wings that operated independently in flying, and such a primitive condition is the case today with many groups of insects. However, most modern insects have tended to reduce the number of flight planes to one. This is well seen in beetles that have changed the front wings into protective sheaths for the hind wings that perform the actual flight movements, although the outstretched wing sheaths in flight may play a part in the aerodynamics of flight by providing fixed lift-generating surfaces. In bees and wasps the smaller hind wings are hooked to the front pair and thus both front and hind wings move as a single pair in flight. Other insects have the 2 wings on each side overlapped at their bases. In the Diptera, the only winged order with just a single pair of wings, the hind wings have been converted into small balancing organs, the halteres, which have the same frequency in vibration as the wings. The halteres move through an arc of 180 degrees, and any change in the direction of the insect produces a gyroscopic effect that can be detected by campaniform sensilla at the base of the halteres.

A mere up-and-down movement of the wing produced by the indirect flight muscles does not result in flight so that there are changes in the wing angle at the top and bottom of each wing stroke. These angular changes are necessary to obtain a resultant force of lift and forward thrust. In their upward and downward sweeps, the wing deflection assumes a figure-eight motion. The wing moves in a continuous sweep. The main propulsion takes place during the downward stroke, with the anterior edge of the wing being lower and the posterior edge higher. During the upward stroke the anterior edge of the wing is higher than the posterior one. Tilting of the wing aids in spilling air at the trailing edge, and this tendency is intensified by the figure-eight movement. This pattern of movement varies among the different kinds of insects, but in general it creates a zone of increased pressure behind and below the insect and a zone of decreased pressure in front of and above it. The insect moves along this gradient of pressure. Those insects that have narrow, fast-moving wings with a pronounced figure-eight component are the fastest flyers. Those with little tilt and weak tendencies of a figure-eight motion are much slower in flight.

The muscles involved in flight are of two types—indirect and direct. The indirect muscles, already described, produce the upward and downward movements of the wings by changing the shape of the thorax. Direct muscles are attached to the wing bases and run from the sternal and pleural regions of the thorax to be inserted on the sclerites at the base of the wing. In the dragonfly, for instance, power is applied to the wings through the direct muscles, each of the 4 wings having elevator and depressor muscles. Here the two pairs of the wings beat out of phase with each other. In this way the posterior pair of wings meets the backward flow of air before it is disturbed by the forward pair.

In general, it may be stated that the movements during active flight involve upward and downward movements, forward and backward movements, changes in the wing shape by twisting and folding, and movements of the wing on its longitudinal axis. An insect wing movement in forward flight resembles an airplane propeller, for it draws air from above and in front and drives it backward, thus producing low pressure above and in front and a high pressure behind. This results in a lifting and forward motion. In hovering, the wing movements of an insect are oscillatory and not rotary, the action of the oscillation being horizontal.

In insect flight there is often a high frequency at which the wings may beat. Large butterflies have only 4 or 5 beats per second. Most grasshoppers and dragonflies have wingbeats of 15 to 20 per second. Some bees may have frequencies as high as 300 beats per second, and mosquitoes have even higher frequencies (600 to 1,000 per second). A distinct nerve impulse for each beat cannot exceed 50 per second and may even be less; when the frequency is greater than 40 to 50 beats per second, the muscle contracts several times for each stimulus. High-frequency wingbeats are associated with fibrillar muscle, which has the peculiarity of contracting several times with each motor impulse. Each impulse generates a tension at which the muscle oscillates and is independent of motor nerve impulse frequency. On this account many insects have an asynchronous flight mechanism and can maintain a high frequency of wingbeat until the oscillations fade away. The orders of insects that have the asynchronous type are the Coleoptera, Hymenoptera, Diptera, and certain Hemiptera. The insects of other orders have flight muscles that contract and relax in direct response to nervous impulses.

Insect behavior

The behavior of insects has always fascinated the field naturalist, as well as the laboratory investigator. Although the behavior of insects may be thought to consist of responses to stimuli, much of their behavior seems to involve more than this. Much of our information about their behavior has been derived by removing parts of their nervous system and observing how the insect responds without the missing parts.

Experiments of this nature have been common for many years and some of the investigators have become famous from their experiments and observations. Yet with all the exploring that has been done, there remains a lot more information that has not been discovered about this most common group of animals in the animal kingdom. The significance of their abundance and their adaptability to nearly every kind of ecologic niche indicate that their behavior patterns are well integrated to habitats where they are found. Many of the acts of insects are highly stereotyped, but much of their behavior involves complex activities in which there are series of different acts. It is difficult to make generalizations about insect behavior that will apply to all, for their activities are as varied as their body forms. One can only point out a few clues here and there regarding the type of behavior one is likely to find.

First of all, the nature of insect life is such that there is little opportunity for them to acquire behavior patterns based on emulation of experienced members. In most insects there is little or no contact between immature stages and adults. Each stage with its own specialized way of existence may last only a short time. However complex behavior patterns may be, such as those of orientation, foraging, feeding, mating, and reproduction, they are as innate and inevitable as anything carefully planned can be. All their acts have such adaptive perfection and elaborate sequence of stages that one is led to believe that these acts are guided by foresight and intelligence. But in spite of such stereotyped behavior, learning and memory of a rather high order is possible with some of them.

Insects are small and have a small central nervous system with far fewer components than that of vertebrates. They must rely on built-in or instinctive patterns of behavior that produce accurate response to a restricted number of environmental stimuli. Their stereotyped behavior is more predictable than that of many other groups of animals.

Although there is no reason for an investigator to attribute purpose to their acts, no one can deny that their behavior reactions do have definite goals and are woven into their hereditary patterns. Insects must do things a certain way to survive. Although faced with the same problems of perpetuation of young for the next generation, the various species of solitary wasps studied by the Peckhams many years ago reveal a great variety of individual traits towards effecting the same purpose. These solitary wasps dig a burrow, find and paralyze a caterpillar or other prey, put it in the bottom of the burrow, attach their eggs to it, and close the entrance (some by using a pebble as a hammer).

Flexibility of behavior patterns do occur in the same species. As the Peckhams have shown, solitary wasps may modify the performance of a certain act in a sequence of separate activities. They found that the wasp does not always sting a caterpillar in the same part of the body when she is preparing a supply of food for the young larvae. Some wasps oversting their prey and kill it, thus defeating their purpose of keeping it alive for the young to feed on. In those that capture spiders for their young before digging their burrows, the paralyzed spider may be hung on a plant out of the way of red ants if they are around, or the paralyzed spider may be left on the ground if no such danger threatens until the wasp is ready to deposit it in the burrow.

Although investigations performed many years ago on the behavior of insects were more often right than wrong, more sophisticated methods are now employed in studying behavior patterns. Analysis of insect behavior is difficult, and much more work must be done before significant generalizations can be made about its nature. Animals often perform quite differently in their natural surroundings and in the laboratory. A honeybee, for instance, may panic and do random movements under laboratory conditions, and valid conclusions about its natural reactions are impossible or inaccurate. However, the same insect in nature behaves in a very competent way. The honeybee knows its own terrain intimately and remembers the utmost details about it. The field naturalist therefore has a definite role to perform in many cases of insect behavior. Simple orientation problems, for instance, involving a fixed response to a single environmental factor may be observed correctly in the laboratory. Some insects will always keep away from light sources and shifting the light may produce a corresponding shift in orientation, for the sensory response is simple and adaptive. But the complex, light-compass orientation of bees and ants pose many problems. These insects may navigate long distances with amazing precision on their foraging expeditions. Many appear to have built-in time sense that permits them to compensate for sun displacements and other geophysical phenomena. If bees are fed a rich supply of their food at a certain time of day, they will return promptly to the same spot 24 hours later even though they have no solar cues to guide them. If the beehive is moved to a different region, bees will get their compass bearings and search out the food at the correct distance at the proper time of day.

Many experiments are performed today by alteration or removal of the brain or other parts of the nervous system to determine details about insect organization and function. Physiologic research on the relation of neural function to behavior patterns attempts to subdivide complex organs into simpler components for analysis. In this way neurons or groups of neurons are carefully analyzed in isolation and their functions inferred from alterations in the behavior of the animal.

Such competent investigators as K. D. Roeder, V. B. Wigglesworth, J. W. S. Pringle, L. E. Chadwick, E. O. Wilson, J. Ten Cate, and a host of others have given fresh insights into the many problems of insect behavior, and the results of their work deserve the attention of students interested in this fascinating problem.

Social life in insects

Insects have some of the most highly social organization in the animal kingdom. The term colony is commonly applied to the complex societies they form. This organization among insects has arisen independently several times in unrelated groups of insects. Ordinarily, a society of social insects does not accept members from other colonies of the same species. The best examples of social organization among the insects are the termites (Isoptera) and the social wasps, bees, and ants (Hymenoptera). Typically, social insects are differentiated into castes, which are specialized in structure, function, and behavior. Although there are variations in the organization, principal castes include the reproductives (chiefly king and queen) and the sterile members (chiefly workers and soldiers). In termites and some higher ants, the sterile castes include workers and soldiers of both sexes, but among the social wasps, bees, and other ants, the workers are the only sterile members (nonreproductive females).

Social insects may have arisen from subsocial individuals that had shown some communal life and division of labor. The members of a colony are subordinated to the life of the community and are bound together by chemical and physiologic mechanisms rather than structure. Termites seemed to have come from certain ancestral cockroaches; social wasps belong to the one family Vespidae; ants form the family Formicidae, which may have come from solitary wasps; and social bees have evolved from solitary bees.

All social insect colonies have some resemblance to an organism in that they have the structure, heredity, integration, and unit organization of an organic entity. Such a biologic unit must be self-sustaining and must acquire its energy from its environment. It must also have the properties of protection against others, maintain its ecologic position, and perpetuate its kind. The organization of a colony of social insects separates these fundamental adaptations by division of labor among its members because of greater efficiency associated with specialization of function. This separation of function requires coordination, for the group must perform as a biologic unit.

Caste determination is due to a number of factors, such as extrinsic factors, nutritional factors, and genetic factors. Among the Hymenoptera (bees, wasps, and ants) the factors of nutrition and genetics form the basis of differentiation. Males develop from unfertilized eggs and are haploid. Queens, workers, and soldiers develop from fertilized eggs and are females (diploid). Differences between worker and queen are mainly differences in their quality and quantity of food. Queens are fed royal jelly for a few days and are then fed beebread (honey and pollen). The workers and drones, or those larvae destined to be such are fed entirely on beebread. Castes in termites are due to extrinsic factors rather than to genetic ones. Eectohormonal inhibition by which reproductives and soldiers give off secretions (ectohormones) containing inhibitory substances are passed on to the nymphs. By mutual feeding (trophallaxis) this substance prevents the nymphs from developing into like forms (soldiers and reproductives). Certain additional members of the caste may differentiate when trophallaxis is lacking or fails to reach undifferentiated nymphs.

Colony formation is the result of many factors, such as temperature, overcrowded conditions, moisture, or other factors. In the social wasps and bumblebees, the impregnated queen hibernates over the winter and starts a new colony in the spring. The first brood are all workers, whose duties are foraging, nest building, and care of young. It is not until late summer that males and young queens appear. Only the young fertilized queen survives the winter. Honeybees propagate colonies by swarming. Each swarm consists of the old queen and many workers and produces a new colony. C. G. Butler (1961) found that the queen has over her body a so-called queen substance (secreted by her mandibular glands) that inhibits workers from becoming queens when they share this substance. Swarming may come about in an overcrowded colony because this substance may not be distributed properly to all workers.

Insects have many ways of communicating with each other. Many produce sounds (some of which are not heard by man), which play an important role in communicating with each other. Stridulatory structures involve almost any part of the body. Body secretions (pheromones) that pass to the outside of the body may cause specific reactions in others of the same species (E. O. Wilson, 1965). Pheromones include sex attractants, the queen substance, and other types whose influences are useful to the colony. Attractants are now used in the detection and control of insect pests by baiting traps with attractants and destroying the insects with potent insecticides. Since attractants may be highly specific, emphasis is also placed on their isolation and chemical identification (M. Beroza, 1971). The language of the bees (von Frisch, 1950) represents one of the most revealing methods of communicating known among insects. It has been studied by other workers, and some additions have been made to von Frisch's original concept (A. M. Wenner, 1964; H. Esch, 1967). It may be summarized as follows: A worker bee or scout that discovers a good source of nectar returns to the hive and communicates precisely just where it is located, the direction and distance to the source, the scent of the flower sources, and the richness of the find. These are all communicated by differential dances: the amount of nectar and pollen collected; the source of the food with reference to the sun's position, as indicated by the dance toward and away from the sun; the distance of the food, as indicated by the rapidity of the tail-wagging; sound vibrations of the wings, as indicated by the average number of vibrations in proportion to the distance from the hive; and the inclinations to the right or left, as indicative of the angle of the sun to the right and to the left of the food source.

Ants' nests often harbor other arthropods (myrmecophils) that are not ants. Such guests include beetles, mites, flies, bees, and others, which live in a peculiar guest-host relationship with the host ants. The secret of success of the guests in living with such fastidious and aggressive insects as ants is thought to be due to

Table 11. Some common larval forms in the metamorphosis of invertebrates*

Porifera—amphiblastula, parenchymula
Coelenterata—planula, actinula, scyphistoma, strobila
Ctenophora—Cydippid larva
Platyhelminthes—Müller's larva, miracidium, sporocyst, redia, cercaria, cysticercus
Rhynchocoela—Pilidium, Desor's larva
Annelida—trochophore
Echiurida—trochophore
Sipunculida—trochophore
Phoronida—actinotroch
Ectoprocta—cyphonautes larva
Entoprocta—trochophore-like
Acanthocephala—acanthor
Mollusca—trochophore, veliger, glochidium
Crustacea—nauplius, zoea
Insecta—nymph, caterpillar, pupa, grub, etc.
Echinodermata—dipleurula, bipinnaria, echinopluteus, auricularia, etc.
Hemichordata—tornaria
Ascidiacea—the tadpole larva

*A larval form is unlike its parents and must undergo a metamorphosis before assuming its adult characteristics. Those early embryos that look like their parents except in size are called juveniles.

the ability to communicate in a language similar to that of the ants. Although this relationship between the host and guests varies with different species, some beetles (*Atemeles*, for instance) actually live in the brood chambers of the ants, where beetle larvae may eat the small ant larvae with impunity, as well as some of their fellow larvae and thus restrict their own competitive populations. Communication may involve visual, chemical, and mechanical cues. Secretions similar to pheromones are given off by both larvae and adults of the beetle and serve to appease aggressive tendencies of the ants. Such a guest-host relationship may have evolved by natural selection of adaptive changes in the guest (B. Hölldobler, 1971).

Metamorphosis and growth

After hatching, insects change not only in size but, often to a striking extent, in form also (as well as many other animals as shown in Table 11) (Figs. 15-25 and 15-26). Insects attain their maximum size by successive molts (ecdysis). The number of molts varies with the species, some having far more than others. The period between any two molts is called the instar, or growth period. In the life history of most insects there is a succession of instars followed finally by the adult instar. Metamorphosis refers to the abrupt changes in form from one distinctive stage to another in the life history. There are various degrees of metamorphosis from those that have none at all to those in which the adult has no resemblance to the larval stages. Classification of insects is often based on the type of metamorphosis they have. The following represents such a classification:

Ametabola. This group includes orders that have no metamorphosis, such as Collembola, Protura, and Thysanura. Stages in development: egg, growth stages, and adult.

Metabola. This artificial taxon refers to insects that have some distinct changes in form and size from larva to adult. The group is subdivided into three distinct types:

Paurometabola. These orders of insects have simple, gradual metamorphosis in which immature forms (nymphs) resemble adults except in size. They are represented by such orders as Orthoptera (Fig. 15-27), Isoptera, and Hemiptera. Early instars in some are called larvae if succeed-

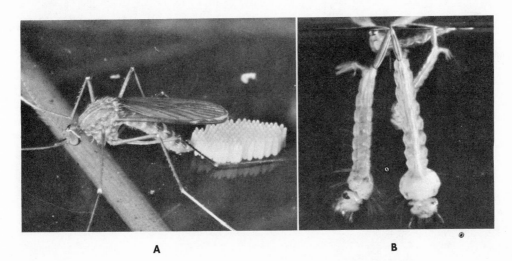

A   B

FIG. 15-25

A, Mosquito *Culex* lays her eggs in small packets or rafts on surface of standing or slowly moving water. **B,** Mosquito larvae are familiar wrigglers of ponds and ditches. To breathe they hang head down, with respiratory tubes projecting through surface film of water. Motion of vibratile tufts of fine hairs on head brings constant supply of food. (From Encyclopaedia Britannica film, "Mosquito.")

FIG. 15-26

Development of honeybee (holometabolous metamorphosis). **A,** After hatching, tiny larvae are fed for several days by young worker bees; then more food is placed in cells and cells are capped. **B,** Larvae feed and grow another day or two and pupate for 13 days. Then young bees gnaw away cell cap and emerge to take over their first duties within hive. (From Encyclopaedia Britannica film, "The Honey Bee.")

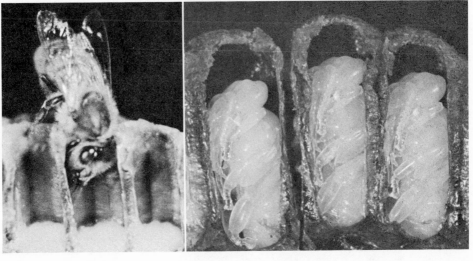

A   B

FIG. 15-27

Praying mantid illustrates paurometabola, or gradual metamorphosis. Mantid eggs are laid in fall in large masses, covered with quickly drying, tough mucus and attached to branches and plant stems. **A,** In spring dozens of young mantids emerge from case, each a pale, tiny copy of the adult. **B,** After several moltings to develop its wings, adult mantid reaches length of about 2½ inches. Beneficial insect that lives on other insects. (**A** and **B** from Hickman: Integrated principles of zoology, The C. V. Mosby Co.; **B** courtesy Joseph W. Bamberger.)

A

B

A

B

C

FIG. 15-28

Ecdysis in cicada. **A,** Having outgrown its cuticle, cicada emerges from cuticle through longitudinal dorsal split and leaves it empty, still attached to leaf. **B,** Last molt; wings are now developed. **C,** Adult cicada. (From Encyclopaedia Britannica film "Insect Life Cycle, The Periodical Cicada.")

ing instars are somewhat different. Stages of development: eggs, series of nymphs, and adult.

Hemimetabola. Insects belonging to this group have incomplete metamorphosis represented by series of immature aquatic forms called naiads. Although naiads do have gills, they are still similar to adults except for size, body proportions, and lack of wings. Hemimetabolous orders are represented by Odonata, Ephemeroptera (Ephemerida), and Plecoptera. Stages of development: egg, series of naiads, and adult.

Holometabola. These insects have complete or indirect metamorphosis that includes resting stage (pupa) between larva and adult (Fig. 15-26). Larva has a series of instars different from pupa and adult. Majority of insects belong to this group. Stages of development: egg, larva, pupa, and adult (imago).

Ecdysis and metamorphosis are under the control of hormones. In every molt a new skin with changes is laid down underneath the old (Fig. 15-28). The old skin ruptures as the insect takes in water or air and increases in size. The new skin is at first wrinkled and allows the body to expand before hardening. A sequence of ecdyses eventually brings about the adult form. In holometabolous insects the hormone-dependent process is controlled chiefly by ecdysone, a hormone produced in the prothoracic gland, which is activated by a brain hormone from the neurosecretory cells of the brain (H. A. Schneiderman and L. I. Gilbert, 1964). Another hormone, the juvenile hormone from the corpora allata glands, prevents the metamorphosis of the insect larva into the pupa under the influence of ecdysone. As long as the juvenile hormone is active, each molt simply results in a larger larva; but when this hormone declines, the ecdysone is free to bring about the change from larva to pupa and then to the adult.

Metamorphosis has evolved along with the evolution of wings in insects. In many insects, wings are restricted to the reproductive stages, during which distribution of the species can be of greatest advantage. In time, however, the evolution of this plan has resulted in fitting the wingless larva and winged adult into differ-

ent ecologic niches and different food habits. In this way the insect could restrict its energy for growth and development to the larval and pupal stages and its energy for locomotion and distribution to the adult stage.

RÉSUMÉ OF INSECT ORDERS

Insects may be classified in various ways, depending on what basic plan is emphasized. Such classification could be made with reference to the presence or absence of wings, presence or absence of metamorphosis, types of metamorphosis, types of mouthparts, etc. A natural classification would emphasize the fossil record, comparative structures, and relationships, but sufficient details are not always available to make such a classification satisfactory to all students of entomology.

The following brief summary of the major orders gives the student a general idea of the diversity of structure and function in the largest group of invertebrate animals. (There is a considerable divergence of opinion about the number of orders.)

■ CLASS INSECTA (HEXAPODA)—INSECTS

Body with head, thorax, and abdomen distinct; single pair of antennae; mouthparts modified for chewing, piercing and sucking, or lapping; thorax of 3 segments, each with a pair of jointed legs, and last 2 segments with one or two pairs of wings, or none; abdomen typically of 11 segments without ambulatory appendages; respiratory system of branched tracheae; excretory system of malpighian tubules; nervous system of brain and ventral double nerve cord; eyes of both simple and compound types; sexes separate, with genital openings near end of abdomen; development with or without true metamorphosis; chiefly terrestrial or freshwater.

Subclass Apterygota (Ametabola)

Primitively wingless insects; true metamorphosis absent; abdomen with ventral appendages other than sexual appendages and cerci.

■ ORDER PROTURA

Minute size (less than 1.5 mm.); body without eyes or antennae; abdomen of 12 segments; each of first three with one pair of small appendages; have 9 abdominal segments on hatching, with an extra segment

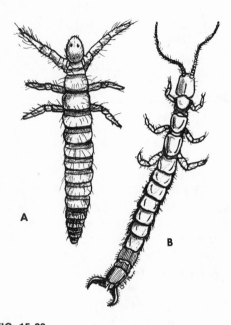

FIG. 15-29

A, *Acerentulus* (order Protura). **B,** *Heterojapyx* (order Diplura) has caudal forceps.

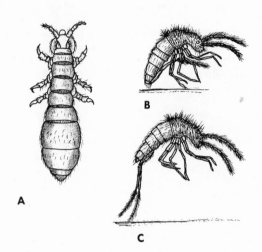

FIG. 15-30

Springtails (order Collembola). **A,** *Anurida.* **B** and **C,** *Orchesella* in resting and leaping positions.

added to abdomen at each of first three molts; mouthparts of long, slender stylets for piercing; some with spiracles and tracheae; found in moist and decaying vegetation; about sixty to ninety species have been described. *Acerentulus* (Fig. 15-29, *A*) is typical genus.

■ ORDER COLLEMBOLA (SPRINGTAILS)

Body up to 5 mm. long; primitive wingless insects; antennae of four joints; mouthparts chewing or sucking; simple eyes; abdomen of 6 segments with springing organ (furcula) on fourth abdominal segment (Fig. 15-30); furcula released by catch (hamula) on third abdominal segment; ventral tube (collophore) produces sticky secretion as adhesive organ on first abdominal segment; leg appendages lack tarsi but have pretarsus with claw; respiration by tracheal system or through integument; no true metamorphosis; live in humid leaf mold; often found on snowbanks in spring of year; some 2,000 species; earliest known insect fossil (*Rhyniella*) belongs to this order. *Entomobrya* and *Achorutes* are common genera.

■ ORDER DIPLURA (JAPYGIDS)

Body mostly 8 to 10 mm. in length, but some larger; no eyes; antennae many segmented; abdomen of

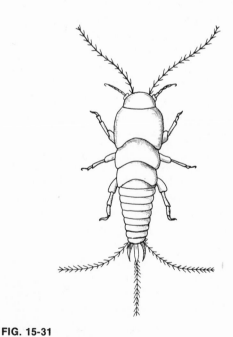

FIG. 15-31

Silverfish *Lepisma* (order Thysanura), often found in homes.

11 segments, with 2 slender cerci or with pincers (Fig. 15-29, *B*); mandibles and maxillae enclosed in head pouches; thorax of 3 segments with three pairs of legs; respiration by tracheal system; live in rotting logs and damp humus; about 400 species. Common genera are *Campodea* and *Japyx*.

■ ORDER THYSANURA (BRISTLETAILS)

Body up to 30 mm. in length with soft integument, sometimes scaly; mouthparts mandibulate with mandibles and maxillae fully exposed; antennae and maxillae long; eyes may be compound, ocelli, both, or absent; abdomen of 11 segments with 2 or 3 many-segmented cerci; live under stones and leaves and around human habitations; about 150 species; silverfish (*Lepisma*) (Fig. 15-31) eats starch of books, and firebrat (*Thermobia*) is often found in warm places in homes.

[NOTE: Since mouthparts of first three orders of Ametabola are mostly hidden in the head, they are referred to as Entotropha; order Thysanura has visible mouthparts (Ectotropha).]

Subclass Pterygota (Metabola)

Wings usually present, but some with vestigial wings or none at all; wingless forms probably arose from winged ancestors; abdominal styli lacking; larva hatched with full complement of adult segments; abdominal appendages restricted to genitalia and cerci; metamorphosis incomplete or complete; includes most insects.

Division Hemimetabola, including Paurometabola (Exopterygota)

Young stages are nymphs with compound eyes; wings develop externally; metamorphosis gradual or incomplete.

■ ORDER ORTHOPTERA (GRASSHOPPERS, CRICKETS, ETC.)

Body with two pairs of wings, first thick and leathery and second membranous and folded beneath first; wings may be absent; mouthparts for chewing; nymph resembles adult; abdomen usually with cerci and ovipositor; well-developed sound-producing and sound-perceiving organs; sound may be produced by rubbing inner faces of femora against outer wings, or by rubbing wings together; eggs frequently laid in clusters (oothecae); mostly live on plants; an extensive order (25,000 species) of several families. Principal families and genera are Locustidae or Acrididae, true grasshoppers (*Romalea*); Gryllidae, true crickets (*Gryllus*); Phasmidae, walking sticks (*Anisomorpha*); Blattidae, cockroaches (*Periplaneta*); Mantidae, praying mantids (*Stagmomantis*); Tetrigidae, grouse locusts (*Tetrix*); Tettigoniidae, long-horned grasshoppers (*Microcentrum*); Gryllotalpidae, mole crickets (*Gryllotalpa*) (Fig. 15-32).

■ ORDER DERMAPTERA (EARWIGS)

Body 60 to 200 mm. in length; chewing mouthparts; tough, shiny bodies with slender antennae; forewings short and leathery, hindwings large and membranous, with radial veins, and folded under forewings at rest; some without wings; abdomen with cerci modified into strong, movable forceps; mostly herbivorous, eating both living and dead vegetation, but may be carnivorous in tropics; about 1,100 species; mostly nocturnal, terrestrial forms, living under stones, refuse, and bark of trees; European earwig *Forficula auricularia* (Fig. 15-33), introduced in this

FIG. 15-32

Mole cricket *Gryllotalpa* (order Orthoptera) showing forelegs adapted for digging.

country, has become a garden and household pest; *Labia* is a common native genus about 6 to 8 mm. long.

■ ORDER PLECOPTERA (STONE FLIES)

Body soft with chewing mouthparts; size 15 to 50 mm. in length; primitive in structure; except for wings and tracheal gills, aquatic immature forms resemble adults; long antennae with setae; membranous wings held close to back when resting, with hind wings pleated and hidden; tarsi three jointed, and abdomen usually with 2 many-segmented cerci; naiad usually with filamentous gill behind each leg, but gills may be attached to sides of thorax and abdomen or at tip of abdomen; naiad life may extend from 1 to 3 years; undergoes many molts before assuming adult condition; adult life relatively short; naiads live under stones in clean, swift streams, and adults found on stream banks around vegetation; eggs laid in large clutches in or near water; in some places, stone flies emerge each month of year; mostly herbivorous, but some are carnivorous; about 1,500 species. *Perla* (Fig. 15-34) and *Pteronarcys* are common genera.

■ ORDER ISOPTERA (TERMITES)

Body soft with abdomen joined broadly to thorax, biting mouthparts; social insects of many castes and divisions of labor; castes include wingless, sterile workers and soldiers (Fig. 15-35, *B* and *C*), sexual males and females (Fig. 15-35, *A*), with similar membranous wings that are carried flat on back at rest and are shed at breakage suture after dispersal

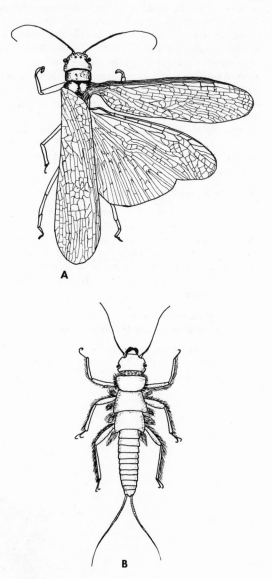

A

FIG. 15-33

A, Earwig *Forficula* (order Dermaptera). **B,** Forcepslike cerci of earwig.

FIG. 15-34

A, Adult stone fly *Perla* (order Plecoptera), with one pair of wings outspread. **B,** Naiad of stone fly.

flight; after reaching favorable place for new colony, males and females mate; since female lacks sperm receptacle, male fertilizes her from time to time; queen may lay 30,000 eggs each day; some queens, by feeding on special diet, may be 4 or 5 inches long and may live many years; in addition to castes already mentioned, there are many intermediate castes among the termites, such as supplementary reproductives, including fertile soldiers, several types of sterile male and female workers, and male and female soldiers; also a nurse caste; castes are far more numerous in termites than in other social insects such as ants and bees; variations in castes include no working caste in primitive species and no soldier castes in others; food mostly cellulose from both living and dead vegetation; cellulose digested in many termites by symbiotic, intestinal protozoan flagellates, which supply necessary enzymes; to supplement their cellulose diet, termites eat their own dead and fecal material for protein; young may become infected with necessary symbiotic protozoans by eating dead termites or by licking anus of adults; live in some form of closed nest, which may be located in tunnels within soil or in wood; in dry tropical regions they build

huge mound nests up to 25 feet high (termitaria) with soil and dry fecal material forming hard surfaces; within termitaria, an orderly arrangement of cells and galleries fitted for special purposes is found; usually different castes cannot be distinguished until after third molt; regulation and determination of caste differentiation appear to be under influence of ectohormones that inhibit nymphal development of individuals of same sex or caste, thus tending to keep caste numbers within bounds; this process of passing inhibiting substance to nymphs through mutual feeding is called trophallaxis; some undifferentiated numphs may not come under influence of trophallaxis and become additional members of caste; initial trend of development is toward reproductives, but if inhibited, soldiers or workers result; cause great economic damage to wooden structures wherever found, especially in tropics, but also are of benefit in building up soil by aeration and releasing nutrients; more than 2,100 species in six families have been described in all regions except the Arctic and Antarctic. Familiar and important genera are *Reticulitermes*, *Kalotermes*, and *Macrotermes*.

■ ORDER EMBIOPTERA (EMBIIDS)

Body soft and straight sided, varying in length from 3 to 25 mm.; chewing mouthparts; females wingless, but males usually have subequal, membranous wings that are folded when not in use; legs short with 3-segmented tarsi; forelegs adapted for spinning with spinning organs; cerci short and two jointed; spinning organs located in swollen basal segment of tarsi; construct silk-lined galleries in grass or under objects; live on decayed vegetation; about 1,000 species, chiefly tropic in distribution. Common genera are *Embia* (Fig. 15-36), *Pararhagadochir*, and *Gynembia*.

■ ORDER ODONATA (DRAGONFLIES, DAMSELFLIES)

Body large and slender, often brightly colored; compound eyes large (20,000 to 30,000 ommatidia) and antennae vestigial; 3 ocelli; chewing mouthparts; paired, transparent, membranous wings equal in size (up to 7 inches in wingspread), with a complex network of veins; each wing has a notch (nodus) midway along its leading edge and a distal, pigmented stigma; wings usually held in horizontal position while at rest (dragonflies); thorax divisible into prothorax and fused meso-metathorax; legs crowded forward for capturing prey in flight and for perching, but not for walking; tarsi three jointed; cerci small; male abdomen at end has claspers for grasping female head or prothorax with female behind male in tandem fashion while flying; copulation usually occurs at rest with

FIG. 15-35

Some castes of termites (order Isoptera). **A,** Winged sexual adult (left wings not shown). **B,** Worker. **C,** Soldier.

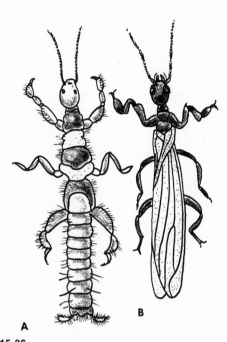

FIG. 15-36

A, Female embiid. **B,** Male embiid (order Embioptera).

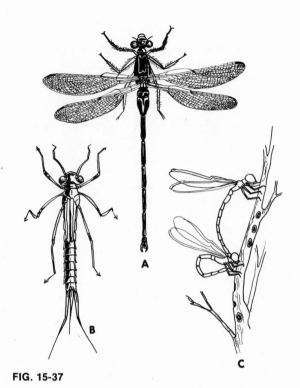

FIG. 15-37

A, Damselfly *Archilestes* (order Odonata). **B,** Damselfly naiad. **C,** Female damselfly ovipositing her eggs in plant stem, while male holds her in position.

female looping her abdomen forward to male accessory genital organs to receive sperm; eggs are dropped in water or inserted in plants (Fig. 15-37, *C*); immature forms, naiads or nymphs, are aquatic (Figs. 15-37, *B*, and 15-38), with hinged prehensile labium that can be thrust out to capture prey; naiads breathe with flattened, terminal gills or with rectal gills; development into adult requires many molts and up to 5 years; transformation usually above water on vegetation; newly emerged adults (tenerals) are weak; adults are mostly active on clear, sunny days, and males establish territories; some species migrate long distances; highly predaceous, capturing almost any flying insects and even recently hatched fishes; serve as intermediate hosts for amphibian and avian trematodes; mass migrations occur frequently over long distances; order divided into suborders (sometimes ranked as orders) Anisoptera (dragonflies) and Zygoptera (damselflies); anisopterans have thick thorax and large wings held laterally when resting; zygopterans are slender, with transparent wings held vertically over back when resting. Nymphs have 3 external, leaflike caudal gills; about 5,000 species with worldwide distribution; dragonflies more numerous than damselflies. *Gomphus, Anax,* and *Libellula* are common genera of dragonflies;

FIG. 15-38

Dragonfly naiad (order Odonata).

Lestes, Agrion, and *Argia* are common American damselflies.

■ ORDER EPHEMEROPTERA (EPHEMERIDA) (MAYFLIES)

Body soft, grayish, and up to 25 mm. long; mouthparts of adults vestigial; most primitive winged order of insects; may have derived from ancestral Thysanura; prothorax and metathorax reduced in size and may be consolidated with the enlarged mesothorax; wings membranous and held vertically above back at rest, with forewings much larger; wing expanse up to 75 mm.; setiform antennae and small to large eyes; 1- to 5-segmented tarsi; slender legs with forelegs of male specialized for grasping female; very long, segmented cerci, 2 or 3 in number (Fig. 15-39); eggs are dropped in water in packets or fixed to surface of rocks in currents; naiad stage lasts from a few weeks to 2 or 3 years; naiads have paired tracheal gills on back of each of first 7 abdominal segments; mouthparts for chewing, usually with segmented maxillary and labial palpi; feed on plant material, although a few are predators; legs adapted for burrowing or clinging; unique subimago stage follows naiad when flies emerge at surface of water (sometimes underneath surface) and shed their skins; subimago has functional wings, but does not feed, and is immature sexually; subimago within minutes or hours molts to adult, or imago, stage, only instance of an insect molting in its winged form; adult emerges with transparent wings and vestigial mouthparts; adults undergo rhythmic dancing and mating flight, oviposit, and die without feeding; shortest adult life among insects; more than 1,500 species, 500 of which are found in the United States. *Ephemera, Hexagenia,* and *Potamanthus* are common genera.

■ ORDER MALLOPHAGA (BITING LICE)

Body flat and up to 5 mm. long; parasitic on birds and mammals, mostly among feathers of birds; body with long, smooth hairs; biting mouthparts with well-developed mandibles; antennae and eyes reduced; wingless; thorax short or fused; legs short, with or without tiny claws; ovipositor very much reduced; eggs are laid singly and attached to hair or feathers of host; young similar to adult, and all stages on host; feed mostly on feathers, hair or skin scales; most harm to host from irritation when lice population is excessive; about 2,800 species. *Menopon* (Fig. 15-40, *A*), *Goniodes,* and *Trichodectes* are familiar genera.

■ ORDER ANOPLURA (SUCKING LICE)

Body depressed, flat, and up to 6 mm. long; head small; mouthparts (retractile) of 3 slender stylets adapted for puncturing and sucking; antennae 3 or 5 segmented; thorax fused; eyes reduced or absent; tarsi one jointed and enlarged on one or more pairs for grasping a hair; ovipositor reduced; eggs (nits) attached to hairs by glue from ovipositor; young escape from a lid at free end of egg; young resemble adult; exclusively ectoparasitic on mammals; feed on blood of host and transmit typhus fever, trench fever,

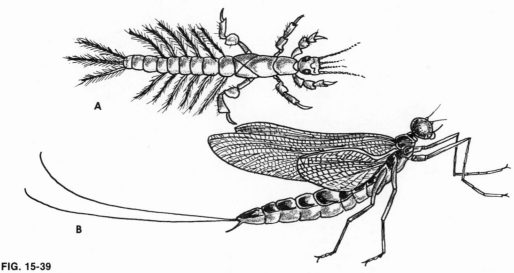

FIG. 15-39

Mayfly (order Ephemeroptera). **A,** Naiad. **B,** Adult.

relapsing fever, and others; about 250 species. *Pediculus humanus* (head and body louse) (Fig. 15-41) and *Phthirius pubis* (crab louse) are common ectoparasites of man.

■ ORDER PSOCOPTERA (CORRODENTIA) (BOOK LICE, BARK LICE)

Body up to 10 mm. long; chewing mouthparts, maxillae with lacinia; antennae long; wings present or absent; tarsi 2 or 3 segmented; cerci absent; head and eyes large; gradual metamorphosis; some live under bark or on foliage of trees, others in old books and museums; live on molds, glue, cereals, dead insects, and organic debris; about 1,250 species, 150 of which occur in the United States. Common genera are *Psocus*, *Psocathropos*, *Trogium*, and *Atropos* (Fig. 15-40, *B*).

■ ORDER THYSANOPTERA (THRIPS)

Body from 0.6 to 14 mm. in length and slender; piercing cone-shaped mouthparts; wingless, or with two pairs of similar, narrow, membranous wings with few or no veins; wings fringed with long hairs; tarsi with one or two joints and terminating with bladder-like organ on tip; antennae with 6 or 9 segments; parthenogenesis common, and males unknown in some species; one to ten generations each year; many species give birth to living young; suck out juices of plants; very destructive to fruit and as vectors of plant diseases; some have wide host range and feed on many kinds of plants; some are carnivorous bloodsuckers; two suborders are recognized, Terebrantia and Tubulifera; members of Terebrantia have saw-like ovipositor for inserting eggs in epidermis of plants; Tubulifera lack this ovipositor and lay eggs on surface; about 3,500 species; common genera are *Hercothrips*, *Heliothrips*, and *Thrips* (Fig. 15-40, *D*).

■ ORDER ZORAPTERA

Body up to 2.5 mm. long; a rare order related to the termites or book lice; discovered in 1913; wings, when present, may be shed; antennae 9 segmented; tarsi 2 segmented; cerci short; chewing mouthparts;

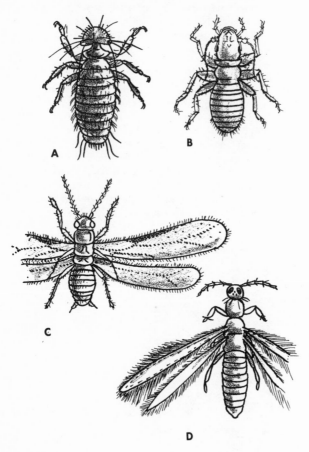

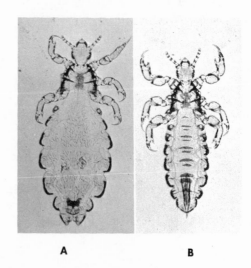

FIG. 15-40

A, Common chicken louse *Menopon* (order Mallophaga). **B,** Wingless book louse *Atropos* (order Psocoptera). **C,** Winged female *Zorotypus* (order Zoraptera). **D,** *Thrips* (order Thysanoptera).

FIG. 15-41

Head and body louse *Pediculus humanus* (Anoplura). **A,** Female. **B,** Male.

often found in small colonies near termite nests; may have caste social system; may occur in old sawdust piles; about twenty-one species known, all in genus *Zorotypus* (Fig. 15-40, *C*).

■ ORDER HEMIPTERA (TRUE BUGS)

Major order of insects; size range about 2 to 100 mm. in length; wings present or absent; forewings with thick basal area, hindwings membranous and fold under forewings; mouthparts elongated into a

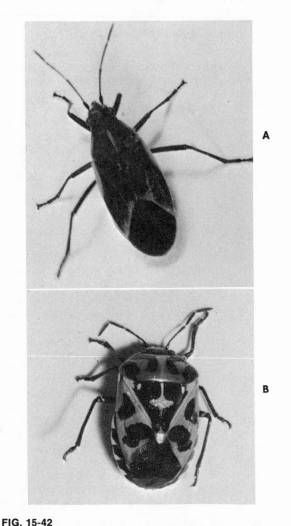

FIG. 15-42

A, Box elder bug *Leptocoris,* family Coreidae (order Hemiptera). **B,** Harlequin cabbage bug *Murgantia,* family Pentatomidae (order Hemiptera).

piercing-sucking mechanism with a sheathlike labium; stylets are grooved and slide up and down during feeding, mandibles adapted for holding substance that is being penetrated; antennae of 4 or 5 segments; compound eyes, and often with 2 ocelli; pronotum of thorax large; triangular mesothoracic scutellum; metathorax partly fused with first abdominal segment; front legs in some predaceous forms may be chelate, whereas middle and hind legs may be adapted for swimming in some; tarsi usually 3 segmented; 2 claws are commonly present; abdomen of 10 to 11 segments, some of which may be hidden; males usually have ninth segment developed as genital capsule, with which a pair of claspers may be associated; female usually has ovipositor arising from eighth and ninth segments; spiracles (open or closed) mostly dorsolateral; midgut of stinkbugs and squash bugs has specific symbiotic bacteria; about 25,000 species of worldwide distribution except Antarctica; live on sap or animal body fluids; many of great economic importance; about forty to fifty families represented, the most familiar of which are Corixidae (water boatmen), Nepidae (water scorpions), Belostomatidae (giant water bugs), Gerridae (water striders), Cimicidae (bedbugs), Coreidae (squash bugs) (Fig. 15-42, *A*), Reduviidae (assassin bugs), Lygaeidae (chinch bugs), Pentatomidae (stinkbugs) (Fig. 15-42, *B*), Miridae (plant bugs), and Tingidae (lace bugs).

■ ORDER HOMOPTERA (CICADAS, APHIDS, SCALE INSECTS, TREEHOPPERS, LEAFHOPPERS)

Often included as a suborder under Hemiptera; large number of diverse species; body mostly small, but cicadas reach length of several centimeters; 4 membranous wings in winged form, and all wings are of same structure; wings roofed over body at rest; mouthparts of piercing-sucking type, with beak placed well back on ventral side of head; beak of two pairs of stylets (maxillae and mandibles) ensheathed by labium; male scale insects have only 2 wings; digestive system may form filter chamber for food, certain unneeded parts of which are shunted around the digestive part of intestine; periodical cicada occurs in many broods in the United States; for different broods in group, nymphs grow 13 to 17 years, feeding on plant juices from tree roots; final instar of cicada emerges from ground and, on tree limb or trunk, molts to form adult; female cicada deposits eggs in woody twigs with sharp ovipositor; male has vibrating organ on ventral part of metathorax with which it sings; number of instars varies greatly with different species; many destructive to plants and some serve as vectors of disease to plants; parthenogenesis common in aphids; more than 30,000 species have been

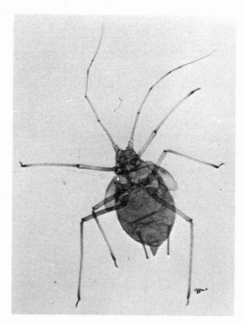

FIG. 15-43
Aphid (order Homoptera).

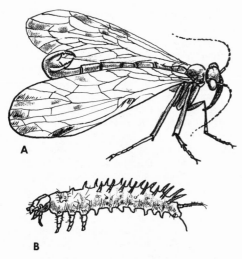

FIG. 15-44

A, Scorpion fly *Panorpa* (order Mecoptera), male. Note recurved abdomen with scorpionlike clasping organs. **B,** Larva of scorpion fly.

named, but actual number is far in excess of this; two suborders, Auchenorhyncha and Sternorrhyncha, are recognized; between thirty and forty different families, the common ones of which are Cicadidae (cicadas or locusts) (Fig. 15-28), Cercopidae (spittle bugs and froghoppers), Membracidae (treehoppers), Cicadellidae (leafhoppers), Aphidae (true aphids) (Fig. 15-43), and Coccidae (scale insects and mealybugs).

Division Holometabola (Endopterygota)

Young stages (larvae) without compound eyes; wings develop internally; complete metamorphosis.

■ ORDER MECOPTERA (SCORPION FLIES)

Body small to medium sized; head prolonged into beak, with chewing mouthparts; head held vertically; antennae long; eyes large; 4 long, slender, membranous wings with many veins; wings at rest held as roof over back; wings absent in some; end of male abdomen (clasping organ) may be recurved over back, scorpionlike; larvae resemble caterpillars; pupae with free appendages; both larvae and adults carnivorous; mostly in moist woodlands; about 250 species. Common genera are *Panorpa* (Fig. 15-44) and *Bittacus.*

■ ORDER TRICHOPTERA (CADDIS FLIES)

Body size 3 to 25 mm.; soft bodied; feeble mouthparts suitable for lapping liquids; antennae long; two pairs of wings, well veined and hairy, folded tentlike over body (Fig. 25-41, *A*); body hairy, may be in form of scales; adults live up to several months and crawl into fresh water, where they lay their eggs under stones or debris; eggs hatch into aquatic larvae, which construct movable cases of leaves, sand, and gravel bound together by silk secreted by larval glands (Fig. 15-45, *B*); cases often funnel shaped to capture food; some do not build cases but live in crevices or other retreats; most larvae are omnivorous, living on algae and small organisms carried to them by water currents; some are entirely predaceous and live on other insect larvae; free-living larvae spin cocoons in which they pupate; case-building larvae pupate within their anchored cases; pupae emerge as adults on vegetation or other objects; about 7,000 species distributed among some twenty-six families; cosmopolitan. Common genera are *Hydropsyche, Helicopsyche, Rhyacophila,* and *Goera.*

■ ORDER NEUROPTERA (DOBSONFLIES, LACEWINGS, ANT LIONS)

Body small to large and soft; chewing mouthparts; all wings are about equal, membranous, with many cross veins; wings roofed over abdomen at rest; long

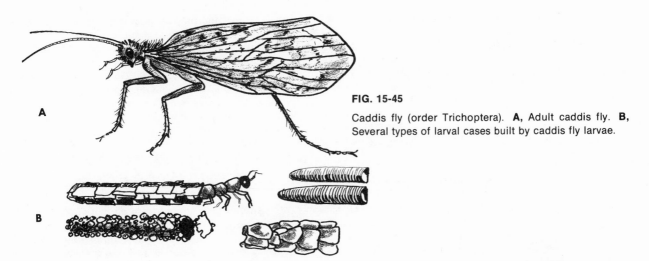

FIG. 15-45

Caddis fly (order Trichoptera). **A,** Adult caddis fly. **B,** Several types of larval cases built by caddis fly larvae.

antennae; tarsi have five segments; cerci absent; larvae carnivorous, with mouthparts modified for piercing and sucking; suborder Megaloptera (dobsonflies—Fig. 15-46 and alderflies) has cross veins rather than branching veins near anterior margin of wings; aquatic larvae usually with tracheal gills and biting mouthparts; hellgrammites, larvae of dobsonfly *Corydalus cornutus* (Fig. 15-47), are prized as bait by fishermen; suborder Planipennia (lacewings and ant lions) has wing veins branching at margin; larvae mostly terrestrial (aphis lions and ant lions) and with sucking mouthparts; eggs of lacewings on stalks; larvae of ant lion (Myrmeleontidae) dig circular pits in sand for trapping ants as prey; larvae of ant lions feed on aphids; pupation occurs in silken cocoon from which adult escapes through hole; about 4,000 species; most primitive insects to have complete metamorphosis. Common genera are *Corydalus* (Fig. 15-46), *Chrysopa,* and *Myrmeleon.*

■ ORDER LEPIDOPTERA (MOTHS, SKIPPERS, BUTTERFLIES)

Body with wings spread from 2 to 300 mm.; mandibles usually lacking; adults with hairs or flattened setae (scales) on body, legs, and wings; varicolored; most with two pairs of membranous wings of equal size; antennae long, of many segments; coiled proboscis formed by maxillary galeae; proboscis extended by blood pressure, coiled by diagonal muscles; fluid is sucked up by muscular pump formed by the pharynx and buccal cavity; large compound eyes, but ocelli often absent; mesothorax large and may overlap the metathorax; venation of wings primitive with few cross veins; the larvae or caterpillars may be naked or adorned with setae, spines, hair, and tubercles;

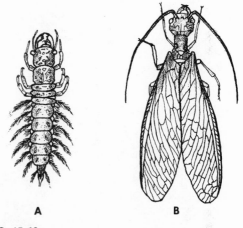

A   **B**

FIG. 15-46

Dobsonfly *Corydalus* (order Neuroptera). **A,** Naiad (hellgrammite). **B,** Adult.

FIG. 15-47

Hellgrammite larva of dobsonfly *Corydalus* (order Neuroptera), used by fishermen as bait.

larvae have three pairs of thoracic legs and a variable number of pairs of forelegs or false legs; larvae usually have six pairs of ocelli but no compound eyes; silk for spinning their cocoons is produced by the labial glands of long simple tubes that unite to form a spinneret opening at the front of the labium; no silk cocoon in butterflies, but chrysalis is suspended by a silken loop. Classification of such an extensive order naturally varies, but the 100,000 to 200,000 species are grouped by many authorities into two suborders of unequal size, Homoneura (Jugatae) and Heteroneura (Frenatae); suborder Homoneura has forewings and hindwings similar in shape and venation, with mandibulate mouthparts and vestigial mandibles; includes ghost moths and a few others; suborder Heteroneura has hindwing smaller than forewing, of different shape, and with fewer veins; sucking mouthparts; of the more than 150 families, the following heteroneurans deserve notice: Aegeriidae (clearwing moths), Cossidae (carpenter moths), Tineidae (clothes moths), Tortricidae (leaf rollers), Psychidae (bagworms), Eucleidae (slug caterpillar moths), Geometridae (measuring worm moths), Bombycidae (silkworm moths), Saturniidae (giant

FIG. 15-48

Swallow-tailed butterfly *Papilio* (order Lepidoptera). (From Encyclopaedia Britannica film, "Introducing Insects.")

silkworm moths), Sphingidae (hawk moths), Noctuidae (owlet moths), Arctiidae (tiger moths), Hesperiidae (skippers), Papilionidae (swallow-tailed butterflies) (Fig. 15-48), Pieridae (white and sulphur butterflies), Lycaenidae (blue, copper, hairstreak butterflies), Nymphalidae (4-footed butterflies); the two great groups of Lepidoptera are the moths (by far the greater number) and the butterflies; moths are mostly nocturnal with featherlike or threadlike antennae without knobs; butterflies are diurnal with threadlike antennae bearing knobs at the end.

■ ORDER DIPTERA (TRUE FLIES, MOSQUITOES, GNATS, MIDGES)

Adults range in size from minute to those that have a wingspread of 100 mm.; only one pair of wings (forewings) with few veins; hindwings represented by a pair of balancers, or halteres; some wingless; third largest order; mouthparts for lapping or piercing, often in form of proboscis; labium serves as a guide for the piercing mandibles and maxillae, as well as for the sucking tube of the elongated epipharynx and hypopharynx (many variations); body divisions of head, thorax, and abdomen distinct; antennae usually of 2 basal segments and a flagellum with one or more segments; compound eyes large, and usually 3 ocelli; mesothorax large and united to prothorax and metathorax; legs with claws mostly alike, although they may be modified in some for raptorial or other functions; beneath claws a membranous pad, or pulvillus, is usually found; abdomen commonly shows 4 to 9 visible segments; setae and bristles over the head and body are common and are valuable taxonomic characters; metamorphosis complete with egg, larva (maggot), pupa, and imago; eggs may be placed in almost any situation; many, such as mosquitoes, have aquatic larvae with head and chewing mouthparts; other larvae live in mud, damp wood, moist earth, blood, fecal material, and organic refuse, larval skin often serves as cocoon for pupa, or in some pupae developing adult may be free; economic importance of group very great; some are pollinators of flowers; many destroy other arthropods directly or by laying eggs in their bodies when developing larvae use host as food; diseases transmitted to man by dipteran vectors are malaria, yellow fever, dengue, filariasis, equine encephalitis, kala-azar, sleeping sickness, tularemia, and many virus diseases; flies can spread from infected persons such germs as typhoid fever, dysentery, cholera, etc.; domestic animals much affected by screwworms and botflies that produce direct damage on tissues and hides; 85,000 species; about 140 families divided between two suborders: Orthorrhapha, in which the adult emerges from the pupa by a straight, longitudinal, dorsal slit; and Cyclorrhapha, in which the adult escapes from its puparium (its transformed last larva skin) by pushing off the anterior end of it. Some of the principal families of Orthorrhapha are Tipulidae (crane flies); Psychodidae (mothflies), Culicidae (mosquitoes) (Fig. 15-25), Simuliidae (blackflies), Tabanidae (deerflies and horseflies), Stratiomyiidae (soldier flies), Asilidae (robber flies), Chironomidae (midges), Muscidae (houseflies); some families of Cyclorrhapha are Drosophilidae (pomace flies), Sarcophagidae (flesh flies), Tachinidae (tachinid flies), Oestridae (botflies), Calliphoridae (blowflies), Hypodermatidae (warble flies), and Conopidae (wasp flies).

■ ORDER SIPHONAPTERA (FLEAS)

Body laterally compressed, compact, and around 3 to 4 mm. in length; no wings; ectoparasites on mammals and some birds; legs modified for jumping; mouthparts for piercing and sucking; body armed with spines and setae; coxae enlarged, and tarsi five jointed; antennae short and concealed in grooves; simple eyes only, or none; eggs laid on host pelt or on places where host lies; segmented larvae are legless and eyeless, and feed on host feces, host's blood obtained from adult flea feces, organic debris, and vegetable matter; larva spins a cocoon where molting takes place to form the pupal stage; may remain as nonfeeding pupa up to a year, or much shorter, before emerging as adult; adults live exclusively on blood of host; usually live only part of the time on host and spend part of the time in debris of substratum; transmit various diseases, especially bubonic plague, and act as intermediate hosts for cat and dog tapeworms; general pests of irritation on all hosts; about 1,000 species; common species are *Pulex irritans* (man, rat) (Fig. 15-49), *Ctenocephalides canis* (dog), *C. felis* (cat), and *Xenopsylla cheopis* (rat).

■ ORDER COLEOPTERA (BEETLES AND WEEVILS)

Body minute to large size with heavy cuticle; chewing mouthparts with some modifications; mouthparts include labrum, mandibles, maxillae, and labium; forewings modified into horny veinless elytra, and form tight, straight line down back; hindwings membranous with few veins and folded under forewings at rest; elytra usually extend to abdomen, but may be truncate; some wingless; abdomen lacks first sternum; usually seven or eight pairs of abdominal spiracles; antennae usually 11 segmented; prothorax is free with large notum; mesothorax and metathorax are fused; head usually with a pair of compound eyes and 3 small ocelli; number of tarsal segments varies in different groups, but 5 is a common number; tarsal formulas are useful in taxonomy; larvae grublike with thoracic legs usually present; change from larva to pupa may occur in cocoon but usually in wood cell or debris; mature adult emerges from the pupal skin; many variations in their complete metamorphosis; ecologic diversity surpasses that of all other groups; occupy all animal habitats, except marine; most beetles are plant feeders and will eat almost any form of organic substance; some are parasitic on other insects or animals; others are predators; some are beneficial as scavengers; both larvae and adults of the lower taxa often differ greatly in their niche requirements so that an enormous range of ecologic relationships is set up; many species have established symbiotic associations with ant and termite colonies; many such beetle inquilines give off exudates that are highly prized by ants and termites; about 275,000 species, or 40% of all insects; two to four suborders and some 160 families are recognized; suborder Adephaga have filiform antennae and hindwings with one or two cross veins; suborder Polyphaga have diversified antennae and hindwings lacking cross veins; among the chief families are Cicindelidae (tiger beetles) (Fig. 15-50), Carabidae (ground beetles), Dytiscidae (predaceous diving beetles), Gyrinidae (whirligig beetles), Staphylinidae (rove beetles), Silphidae (carrion beetles), Lucanidae (stag beetles), Tenebrionidae (darkling beetles), Hydrophilidae (water scavengers), Coccinellidae (ladybird beetles), Lampyridae (glowworms and

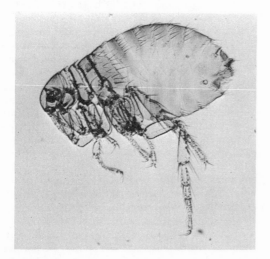

FIG. 15-49

Common flea *Pulex irritans* (order Siphonaptera), female, ectoparasitic on man and other mammals.

FIG. 15-50

Tiger beetle *Cicindela* (order Coleoptera). Note sensory hairs on legs. (Courtesy Encyclopaedia Britannica Films, Inc.)

fireflies), Elateridae (click beetles), Dermestidae (skin beetles), Anobiidae (powder-post beetles), Psephenidae (water penny beetles), Cerambycidae (long-horned beetles), Chrysomelidae (leaf beetles), and Circulionidae (weevils or snout beetles).

■ ORDER STREPSIPTERA (TWISTED-WINGED INSECTS)

Size minute; mouthparts for chewing, or vestigial; male hindwings fan shaped, but forewings club shaped (pseudohalteres); female wingless and legless, endoparasitic in other insects; female, as primary larva, bores into host and becomes an egg-producing sac with loss of eyes, antennae, and legs; head and thorax of female fused into a cephalothorax; females and larvae remain parasitic in bees, wasps, and homopterous bugs; absorb nutrition from host; the order is often considered a highly specialized parasitic branch of the Coleoptera, and may be placed as a family under that group; hypermetamorphosis and life history unique in the far-flung patterns of insect life; 200 to 300 species in six families. Two common genera are *Xenos* and *Stylops*.

■ ORDER HYMENOPTERA (ANTS, BEES, WASPS)

Body length from 0.5 to 50 mm. and more or less hairy; most species with 4 membranous wings, the forewings larger, with reduced venation; hindwings locked to forewings by hooklets; wings absent in some, especially the workers of ants, females of some solitary wasps, and many parasitic forms; mouthparts for chewing or chewing-lapping; compound eyes moderate to very large; usually 3 ocelli; antennae vary from short (about 12 segments) to long with many segments; legs usually slender, but modified in some for carrying pollen; abdomen constricted at posterior end of first segment, which is fused into hind end of thorax; females have ovipositor modified for piercing, sawing, or stinging; ovipositor very long in ichneumon flies; some have caterpillar-like larvae, but many larvae are legless and maggotlike with well-developed head; pupae usually have free appendage and silk cocoons; most species are dioecious, but parthenogenesis of several types is common; sexual dimorphism, with males usually smaller than females, is common; most species are solitary, but specialization for social life has reached one of its highest peaks in this order; behavior patterns and complicated adaptations for rearing young and general life habits are in line with their high social organizations; employment of sign language for communication among honeybees represents a high level of intelligent behavior; nearly all members of the order are terrestrial, living in or near the earth's surface; most adults feed on plant nectar or honeydew; some are parasitic; larvae may be herbivorous, but many live as parasites on hosts provided for them by the females; several hymenopterans have economic importance both injurious and beneficial to man's interests and animal economy; more than 120,000 species have been described and distributed

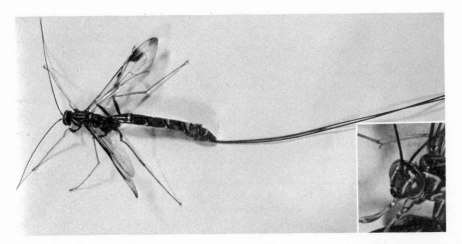

FIG. 15-51

Ichneumon fly (family Ichneumonidae, order Hymenoptera). Inset, head of ichneumon fly. Some species of this wasp use the long ovipositor to drill a hole in wood and lay their eggs on or near certain larvae (of beetles and other wood-boring forms) which are parasitized and destroyed by the larvae of the wasp. The drilling of a hole in hard wood by the hairlike ovipositor represents one of the most ingenious methods known among insects.

FIG. 15-52

Bald-faced hornet *Vespula* (family Vespidae, order Hymenoptera.)

among many families; two suborders, Symphyta and Apocrita, are commonly recognized; the Symphyta have abdomen broadly joined to the thorax, larvae provided with legs, and include more primitive members; Apocrita have abdomen more or less constricted at the base, legless larvae, and include majority of Hymenoptera; some of important families of Hymenoptera are Tenthredinidae (sawflies), Ichneumonidae (ichneumon flies) (Fig. 15-51), Chalcididae (chalcid wasps), Cynipidae (gall wasps), Scelionidae (scelionid wasps), Formicidae (ants), Vespidae (hornets, yellow jackets) (Fig. 15-52), Sphecidae (mud-dauber wasps), Megachilidae (leaf-cutting bees), and Apidae (honeybees, bumblebees).

REFERENCES

Beermann, W., and U. Clever. 1964. Chromosome puffs. Sci. Amer. **210**:50-58 (April).

Beroza, M. 1971. Insect sex attractants. Amer. Sci. **59**:320-325.

Burror, D. J., and D. M. De Long. 1971. An introduction to the study of insects, ed. 3, New York, Holt, Rinehart & Winston, Inc. The best up-to-date treatise on insects.

Bronn, H. G. (editor). 1926-1934. Klassen und Ordnungen des Tierreichs. Leipzig, Akademische Verlagsgesellschaft. Volume 5 of this great work is devoted to the Chilopoda, Diplopoda, Symphyla, and Pauropoda.

Butler, C. G. 1961. The efficiency of a honeybee community. Endeavour **20**:5-10 (January).

Carpenter, F. M. 1953. The geological history and evolution of insects. Amer. Sci. **41**:256-270.

Carter, W. 1963. Insects in relation to plant diseases. New York, John Wiley & Sons, Inc. A comprehensive account of insect vectors and the plant viruses they transmit.

Cavill, G. W. K., and P. L. Robertson. 1965. Ant venoms, attractants, and repellents. Science **149**:1337-1345.

Cleveland, L. R. 1928. Symbiosis between termites and their intestinal Protozoa. Biol. Bull. **54**:231-237.

Cleveland, L. R. 1949. Hormone-induced sexual cycles of flagellates. I. Gametogenesis, fertilization, and meiosis in *Trichonympha*. J. Morphol. **85**:197-296.

Corbet, P. S. 1963. A biology of dragonflies. New York, Quadrangle Books, Inc.

Davey, K. G. 1965. Reproduction in the insects. San Francisco, W. H. Freeman & Co., Publishers. A summary of the different methods insects employ.

Dethier, V. G. 1955. The physiology and the histology of the contact chemoreceptors of the blowfly. Quart. Rev. Biol. 30:348-371.

Dethier, V. G. 1971. A surfeit of stimuli: a paucity of receptors. Amer. Sci. 59:706-715.

Eisner, T. E. 1966. Beetle's spray discourages predators. Natur. History 75:42-47 (February).

Evans, H. E. 1963. Predatory wasps. Sci. Amer. 208:144-154 (April).

Frisch, K. von. 1950. Bees: their vision, chemical senses, and language. Ithaca, N. Y., Cornell University Press.

Frost, S. W. 1965. Insects and pollinia. Ecology 46:556-558. Pollinia are traplike pollen masses in such plants as milkweeds and orchids, which insects frequently carry away attached to their appendages.

Gilmour, D. 1965. The metabolism of insects. San Francisco, W. H. Freeman & Co., Publishers.

Grassé, P. P. (editor). 1949. Traité de zoologie, vol. 9. Paris, Masson et Cie. A comprehensive and authoritative account.

Harker, J. 1960. Endocrine and nervous factors in insect circadian rhythm. Cold Spring Harbor Symposium. Quant. Biol. 25:279-287.

Hölldobler, B. 1971. Communication between ants and their guests. Sci. Amer. 224:86-93 (March).

Hoyle, G. 1953. Slow and fast nerve fibers in locusts. Nature 172:165.

Jacobson, M., and M. Beroza. 1964. Insect attractants. Sci. Amer. 211:20-27 (August).

Johnson, C. G. 1963. The aerial migrations of insects. Sci. Amer. 209:132-138 (December).

Kennedy, J. S. (editor). 1961. Insect polymorphism. London, Royal Entomological Society. Polymorphism results from the simultaneous occurrence of several genetic factors or gene arrangements in a population that produces discontinuous phenotypic effects. Its ecologic significance is not clear in all cases.

Kukenthal, W., and T. Krumbach (editors). Handbuch der Zoologie. 1926, vol. 4, Progoneata, Chilopoda; 1929, vol. 52, Myriapoda, Geophilomorpha; 1930, vol. 54, Chilopoda, Scolopendromorpha; 1937-1940, vols. 68-70, Diplopoda, Polydesmoidea. Berlin, Walter de Gruyter & Co. A very extensive account of these groups.

Lanham, U. N. 1968. The Blochmann bodies: hereditary intracellular symbionts of insects. Biol. Rev. 43:269-286.

Luscher, M. 1961. Air conditioned termite nests. Sci. Amer. 205:138-145 July.

McKittrick, F. A. 1964. Evolutionary studies of cockroaches; Memoir 389. Ithaca, N. Y., Cornell University Agricultural Experiment Station.

Manton, S. M. 1954. The evolution of arthropodan locomotory mechanisms. IV. The structure, habits and evolution of the Diplopoda, J. Linnean Soc. (London) (Zoology) 42: 299-368.

Nur, U. 1970. Evolutionary rates of models and mimics in Batesian mimicry. Amer. Nat. 104:477-486.

Peckham, G. W., and E. G. Peckham. 1905. Wasps social and solitary. New York, Houghton, Mifflin Co.

Roeder, K. D. 1935. An experimental analysis of the sexual behavior of the praying mantis. Biol. Bull. 69:203-220.

Roeder, K. D. 1965. Moths and ultrasound. Sci. Amer. 212: 94-102 (April).

Russell-Hunter, W. D. 1969. A biology of the higher invertebrates. New York, The Macmillan Co., Publishers.

Schneiderman, H. A., and L. I. Gilbert. 1964. Control of growth and development in insects. Science 143:325-333.

Smith, R. F. (editor). 1965. Annual review of entomology, vol. 10. Palo Alto, Calif., Annual Reviews, Inc. This annual review is useful in the evaluation of current work in the field of insects.

Swan, L. A. 1964. Beneficial insects. New York, Harper & Row, Publishers. The useful side of a group that has been the chief target of the pesticide war.

Tiegs, O. W. 1947. The development and affinities of the Pauropoda, based on a study of Pauropus silvaticus. Quart. J. Microscop. Sci. 88:165-267, 275-336.

Walker, E. M. 1953. The Odonata of Canada and Alaska, vol. 1. Toronto, University of Toronto Press. A systematic study of the dragonflies.

Wenner, A. M. 1964. Sound communication in honeybees. Sci. Amer. 210:116-124 (April).

Wigglesworth, V. B. 1965. Principles of insect physiology, ed. 4. Cambridge, Cambridge University Press.

Williams, C. M. 1946. Physiology of insect diapause; the role of the brain in the production and termination of pupal dormancy in the giant silkworm Platysamia cecropia. Biol. Bull. 90:234-243.

Wilson, E. O. 1965. Chemical communication in the social insects. Science 149:1069-1071.

Part V □ The Deuterostome animals

Phylum Echinodermata

GENERAL CHARACTERISTICS

1. Echinoderms are the only major invertebrate phylum of deuterostomes. Other invertebrate deuterostomes are the Chaetognatha, Pogonophora, Hemichordata, and the protochordate group of the Chordata. Echinoderms are enterocoelous coelomates.

2. Of their many unique characters, perhaps the most outstanding is their radial symmetry, which seems to have been acquired secondarily from a bilateral ancestor. Echinoderms are the only animals that begin life with bilateral symmetry and then metamorphose into radially symmetric forms. They are predominately pentamerous, that is, the body is arranged in 5 parts around a central axis, as is seen in the 5-armed sea stars and brittle stars and in the 5-petaled pattern on the surface of the sand dollars.

3. Echinoderms are unsegmented, have no head and no cephalization, and have an oral-aboral axis rather than an anteroposterior axis.

4. They possess an endoskeleton of calcareous ossicles, with spines, and covered with epidermis that is usually ciliated.

5. There is an extensive coelom subdivided into distinct systems, such as the perivisceral coelom, the perihemal coelom, the aboral sinus system, and the water-vascular system. Three types of fluid are usually present—the perivisceral fluid, the blood of the hemal system, and the fluid of the water-vascular system. Coelomocytes may pass freely from one division to the other. The coelomic fluid is the main transport system.

6. The unique water-vascular system provides a hydraulic system of tube feet (podia) used in locomotion, respiration, and sensory reception. The podia are typically arranged in areas called ambulacra, which alternate with areas lacking podia and called interambulacra.

7. There is a complete digestive system except in ophiuroids, which lack an anus. Excretory organs are absent. Respiratory adaptations are different in each group and include such devices as dermal branchiae, tube feet, respiratory trees, and bursae.

8. The nervous system is diffuse and uncentralized and consists basically of three systems of nerve rings and radiating nerve nets. The oral (ectoneural) system is stressed in all forms except the crinoids and gives rise to sensory and motor nerve fibers.

9. Echinoderms are mostly dioecious, with simple gonads and gonoducts. Fertilization is usually external. Development involves holoblastic, radial cleavage, free-swimming ciliated larval stages, and metamorphosis to a radially symmetric form.

PHYLOGENY AND ADAPTIVE RADIATION

1. Although the fossil record of the echinoderms extends back to the lower Cambrian period, it throws no light on the nature of the first ancestor. Along with other deuterostomes, echinoderms may have sprung independently from coelenterates, but no clear relationship between them and any other phylum has been established. The resemblance between the larva of the sea cucumbers and asteroids and that of

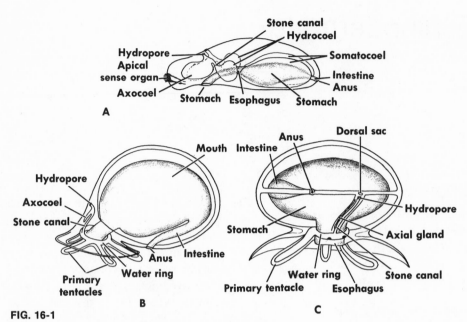

FIG. 16-1

Theories of echinoderm phylogeny. **A,** Dipleurula ancestor. **B,** Early bilateral pentactula ancestor. **C,** Later pentactula ancestor after assumption of radial condition. (**A** modified from Bather; **B** and **C** after Bury, from Hyman: The invertebrates; Echinodermata, McGraw-Hill Book Co., Inc.)

the hemichordates, once considered evidence of relationship between echinoderms and chordates, is now questioned in the light of convergent larval evolution.

Two hypothetical ancestors have been proposed. The dipleurula ancestor was a bilateral, humpbacked, vermiform animal with three pairs of coelomic compartments (Fig. 16-1, *A*). The pentactula (more widely accepted today) had, in addition to the three pairs of coelomic sacs, a circlet of 5 tentacles around the mouth, which were extensions of the coelomic sacs and which gave rise to the water-vascular system (Fig. 16-1, *B*). The pentactula evolved into a sessile attached form and in time underwent a gradual shift in body structures to assume the radial symmetry so compatible with sessile life (Fig. 16-1, *C*). Some of them later detached

and assumed a free-living existence but retained the radial symmetry.

Within the phylum, authorities do not agree on relationships between the classes. Larval development among the classes does not always agree with the paleontologic record. Crinoids, the most primitive of the existing echinoderms, cannot be related to any of the extinct noncrinoid groups of Pelmatozoa, nor can the origin of the other groups be traced directly to the crinoids. The holothuroids' primitive characteristics and their lengthened shape, which causes them to lie on one side and thus assume a more bilateral symmetry, may be considered primitive characters inherited from a bilateral ancestor but also may have been acquired secondarily. That the asteroids (sea stars) and ophiuroids (brittle stars) are closely related is shown by their fossil record, the similarity of their larvae, and the general pattern of their skeletons.

2. Regarding their adaptive radiation, there seems to be little controversy about the potentialities of the water-vascular system and of the skeletal ossicles in guiding the evolution of the echinoderms into the many species found at present. They are primarily bottom-dwelling

forms, and their adaptations have largely conformed to the requirements of such an existence. Some of the fossil forms had no radial symmetry, and others had a blend of bilateral and radial symmetry; whatever their ancestor may have been like, the typical echinoderm form, as we know it, is an outcome of adaptations to a sessile, benthonic existence, and its basic morphologic pattern has restricted evolutionary divergence within the limits of such a pattern.

• • •

The echinoderms include such familiar marine forms as sea stars, sea urchins, brittle stars, sea cucumbers, and sea lilies. The name means "spiny-skinned," which refers to the projecting spines or tubercles that give a rough, warty appearance to the surface of most echinoderms. Their morphologic organization is unique in the animal kingdom, and their structural peculiarities tend to separate them from all other phyla. However, they have certain basic structures that give them high ranking in the evolutionary pattern of the animal kingdom. Of all invertebrate groups, they are placed the closest to the chordates and represent the most advanced of the invertebrate phyla within the Deuterostomia, to which the chordates belong.

Except for a few planktonic forms, echinoderms are mostly bottom dwellers. They are strictly marine, and none is adapted to fresh water. Only a few are found in brackish water. In the ocean they are found in almost every conceivable habitat—in all depths and at all latitudes. Many are found in muddy bottoms (holothurians and echinoids), obtaining their organic food from the mud passed through their digestive tracts. Other echinoderms are found in sandy or rocky bottoms. Still others (asteroids and ophiuroids) are commonly found in littoral zones and also in fairly deep waters. In general, echinoderms are sluggish animals. The most rapid among them are the ophiuroids. Echinoderms range in size from less than 1 inch to 2

or 3 feet in diameter. Fossil forms were often much larger.

No other invertebrates surpass the echinoderms in the complexity of their skeleton, which has been responsible for an enormous range in their variations. The outstanding basic characteristic of the phylum, from the earliest fossils to present-day species, is their crystalline, calcareous endoskeleton. Each element of the skeleton consists of calcium carbonate laid down in the form of a fine honeycomb of crystalline calcite. All skeletal parts of ossicles, spines, and plates have this lattice arrangement of calcite crystals, each being a crystallographic unit. Interspersed through this lattice arrangement is the soft organic material that is replaced by calcite during fossil formation. Ossicles or plates may articulate with each other, or they may be fused to form a rigid test.

The fossil record of echinoderms is impressive, for it extends back to the early Cambrian period. During this time many structural patterns evolved, some of which were quite different from those of existing groups. Most echinoderm fossils that have been recovered were laid down in shallow sea deposits. There seems to be no doubt, however, that echinoderms were also laid down in deep-sea deposits that are unknown at present. Modern echinoderms have been dredged from great depths in recent times. From the Ordovician period, fossils of all the major classes of echinoderms except holothurians have been found. The earliest fossils were the stem-bearing echinoderms (Fig. 16-2, C and D), such as crinoids, most classes of which did not survive beyond the Mesozoic era. Fossils of the free-moving echinoderms are more common in later geologic formations. Echinoderms are well adapted for preservation as fossils because of their calcareous hard parts.

About 5,700 species of present-day echinoderms are recognized. Of the fossil echinoderms, more than 5,000 crinoid fossil species have been found, in comparison with 600 to

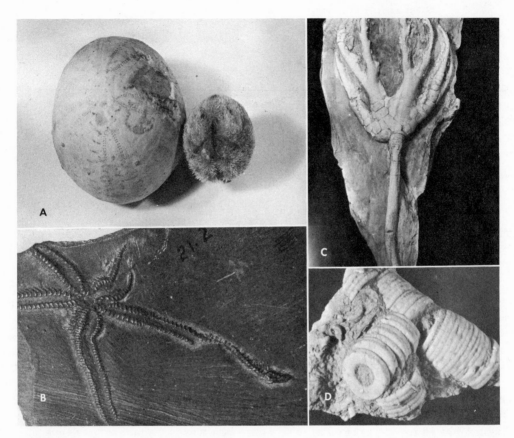

FIG. 16-2

Some fossil echinoderms. **A,** Heart urchins (spatangoids), past and present. Left, fossil type (from Florida) known from Eocene to Recent times; right, *Moira atropus,* present-day heart urchin that lives on our eastern shores. **B,** Devonian asteroid *Roemeraster,* Germany. **C,** Mississippian crinoid *Onychocrinus,* Indiana. **D,** Fossil crinoid stems.

650 living species of that class. The eleutherozoans are more numerous and varied today than those represented by the fossil record. Of the free-moving forms, the echinoids make up more fossil species than all the other classes combined because of the rigid construction of their skeletons.

Many higher echinoderm taxa, including classes, are wholly extinct; others are represented by only a few modern types. For instance, the class Somasteroidea is wholly extinct except for one living species, *Platasterias latiradiata,* found in the Pacific Ocean near Mexico a few years ago. These extinct echinoderms are included, along with existing forms in many classifications of echinoderms but have been omitted in the classification below.

CLASSIFICATION OF EXISTING ECHINODERMS

Phylum Echinodermata. Body of pentamerous radial symmetry, enterocoelous coelomate; bilateral symmetry as larvae; body usually of 5 ambulacra that bear tube feet (podia), alternating with interambulacra around an oral-aboral axis; calcareous endoskeleton of separate plates usually bearing spines; jawed pedicellariae (in some); coelomic canals forming a water-vascular system, with projections to the exterior (tube feet or podia); definite head and brain absent; dioecious; external fertilization; entirely marine.

Subphylum Eleutherozoa. Stalkless and free-living; mouth usually on lower surface.

Class Somasteroidea. Primitive arms; bilobed ampullae; mostly fossil sea stars; single living species, *Platasterias*.

Class Asteroidea. Star-shaped or pentagonal; 5 or more arms merged gradually to central disk; endoskeleton flexible; short spines and pedicellariae. Sea stars.

Order Phanerozonia. *Astropecten, Luidia, Dermasterias, Ctenodiscus.*

Order Spinulosa. *Patiria, Henricia, Echinaster, Pteraster, Asterina, Solaster.*

Order Forcipulata. *Asterias, Pisaster, Leptasterias, Heliaster.*

Class Ophiuroidea. Five slender arms, sharply set off from central disk; no suckers or pedicellariae. Brittle stars and basket stars.

Order Ophiurae. *Ophiothrix, Ophiura, Ophioderma, Ophiolepis.*

Order Euryalae. *Gorgonocephalus, Asteronyx.*

Class Echinoidea. Body globular, heart- or disk-shaped; endoskeleton usually a rigid test of closely fitted plates; covered with movable spines; 3-jawed pedicellariae. Sea urchins, heart urchins, sand dollars.

Subclass Regularia. Globular form and mostly circular, sometimes oval; pentamerous symmetry, with 2 rows of ambulacral plates; lantern present; mouth and anus at oral and aboral poles.

Order Lepidocentroida. Mostly extinct. *Claveriosoma, Asthenosoma.*

Order Cidaroidea. Mostly extinct. *Cidaris, Eucidaris, Notocidaris.*

Order Diadematoida. Sea urchins. *Arbacia, Lytechinus, Echinus, Strongylocentrotus, Echinometra, Diadema, Salenia.*

Subclass Irregularia. Mostly flattened oval to circular test; irregular test with anus shifted out of apical center and mouth centered or eccentric; lantern present or absent.

Order Holectypoida. Mostly extinct. *Echinoneus.*

Order Cassiduloida. Mostly extinct. *Cassidulus, Apatopygus.*

Order Clypeastroida. Sand dollars, cake urchins. *Clypeaster, Echinarachnius, Dendraster, Mellita.*

Order Spatangoida. Heart urchins. *Moira, Echinocardium, Hemiaster, Lovenia, Meoma.*

Class Holothuroidea. Body elongated in oral-aboral axis; body wall thin to leathery; endoskeleton of microscopic ossicles embedded in body wall; no arms, spines, or pedicellariae. Sea cucumbers.

Order Elasipoda. *Pelagothuria, Peniagone.*

Order Aspidochirota. *Stichopus, Holothuria, Parastichopus, Actinopyga.*

Order Dendrochirota. *Thyone, Cucumaria, Thyonepsolus.*

Order Molpadonia. *Molpadia, Caudina.*

Order Apoda. *Synapta, Leptosynapta, Chiridota.*

Subphylum Pelmatozoa. Body in cup- or calyx-shaped skeleton; mouth and anus on upper surface; stemmed or free-living; all but one class extinct.

Class Crinoidea. Stalked or stalkless; arms mostly branched; sessile or free-swimming. Sea lilies and feather stars.

Order Articulata. *Antedon, Neometra, Cenocrinus.*

EMBRYOLOGY

Echinoderms have the radial and indeterminate cleavage characteristic of the deuterostomes. Eggs are mainly homolecithal and are fertilized in the sea water or in brood pouches. In radial cleavage the characteristic pattern is the formation of 4 equal blastomeres by two vertical cleavage planes (Fig. 16-16). The third cleavage plane, horizontal and equatorial, divides the 4 blastomeres into 8 equal blastomeres. The fourth cleavage plane is vertical and divides the embryo into 16 cells, of which 8 are equal-sized mesomeres, 4 are small micromeres at the vegetal pole, and 4 are large macromeres. At the fifth cleavage, the blastomeres are divided about equally, forming 32 cells. The succeeding cleavages are irregular, and the blastomeres do not all divide at the same time. As the blastomeres continue to divide, the blastula stage, a hollow sphere with a single layer of cells, occurs. By invagination at the center of the vegetal region at gastrula stage develops, with a narrow archenteron, or primitive gut. The blastopore, or a region near it, becomes the anus; and the mouth is formed by the breaking through of a stomodeum, a characteristic of the deuterostomes. Most cells of the archenteron are endodermal, but some mesodermal cells may also be found among them. The coelom is formed by outgrowth pockets, or vesicles, of the archenteron that separate and spread through the blastocoel. The cells of the coelomic walls are mesoderm.

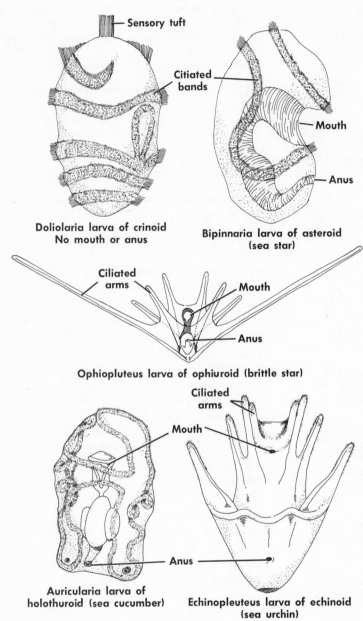

Doliolaria larva of crinoid
No mouth or anus

Bipinnaria larva of asteroid
(sea star)

Ophiopluteus larva of ophiuroid (brittle star)

Auricularia larva of
holothuroid (sea cucumber)

Echinopleuteus larva of echinoid
(sea urchin)

FIG. 16-3

Larval types of the different echinoderm classes.

FIG. 16-4

Bipinnaria larva. (Photograph by Roman Vishniac, courtesy Encyclopaedia Britannica Films, Inc.)

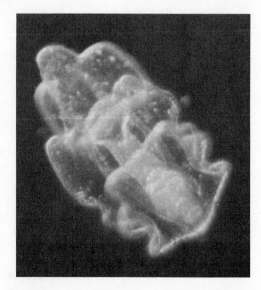

Such a method of mesoderm formation is described as enterocoelous, in contrast to the schizocoelous method of protostomes. The process of coelomic pouch formation varies among the different echinoderms, but eventually it gives rise to the three pairs of coelomic cavities that are called the axocoel (which opens dorsally by a hydropore), hydrocoel, and somatocoel. The mesenteries of the gut are formed by the meeting of the somatocoels above and below the gut. The free-swimming larva is bilateral and varies with the different classes. Among the eleutherozoans there are two well-marked larval types — the pluteus group, with long arms and easel-shaped forms (ophiuroids and echinoids), and the auricularia group, with barrel-shaped forms and lobe bands (holothurians and asteroids) (Fig. 16-3). A bipinnaria stage (Fig. 16-4) often follows the auricularia stage in asteroids. The ciliary bands, so characteristic of echinoderm larvae, have come from modifications and specializations of the cilia that are found on each of the blastomeres from the blastula stage on. These ciliary bands are for locomotion. The doliolaria larva of crinoids has some of the characteristics of the auricularia larva, with its shape and ciliary bands, but it never feeds; it obtains all its nourishment during its brief swimming period from enteric yolk cells. After a planktonic existence as a free-swimming form, the larva is transformed into an adult by a metamorphosis that varies with the different classes.

Subphylum Eleutherozoa
■ CLASS ASTEROIDEA

The familiar sea stars are star-shaped echinoderms composed of a central disk that merges gradually with the arms, of which there are typically 5 in number (but there may be more) (Fig. 16-5). They are also the only eleutherozoans that have open ambulacural grooves. These grooves are on the oral surface and are provided with 2 to 4 rows of locomotor podia (tube feet), with suckers and radial water

FIG. 16-5

Some West coast asteroids. **A,** *Dermasterias.* **B,** *Pycnopodia.* **C,** *Pisaster.* (From Hickman: Integrated principles of zoology, The C. V. Mosby Co.)

canals on the outer side of the ambulacral plates.

Sea stars are usually brightly colored with carotenoid pigments that fade in most preservatives. They are found along the coasts of all the oceans and live in all types of bottoms — muddy, sandy, and rocky. A few are found on sea algae. Several species occur at depths

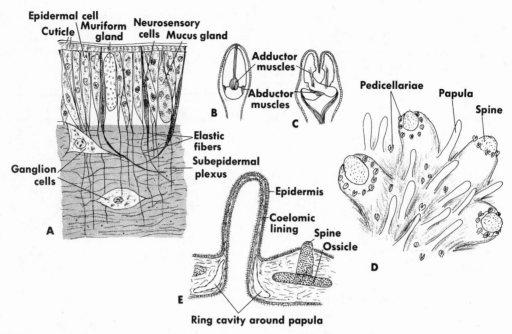

FIG. 16-6

A, Section through cuticle, epidermis, and underlying radial nerve of *Asterias*. **B,** Straight forceps type of pedicellariae. **C,** Crossed scissors type of pedicellariae. **D,** Small portion of aboral surface of live *Asterias* showing spines, papulae, and pedicellariae. **E,** Section through papula. (Semidiagrammatic.)

of 3 miles or more. They are most abundant and varied in tropic regions, especially coral reefs.

GENERAL MORPHOLOGIC FEATURES
External anatomy

Sea stars have a flattened, flexible body that is generally pentamerous in arrangement. Ciliated columnar epithelium with a thin cuticle covers the entire body (Fig. 16-6, *A*), and the body surface is rough because of projecting tubercles, spines, and ridges (Fig. 16-7). The mouth is located on the underside, or oral surface, in the center of the oral disk (Fig. 16-7, *B*). The upper, or aboral, surface bears the very small anus in the center of its disk (Fig. 16-7, *A*). At one side, between 2 of the arms, is the madreporite (Fig. 16-7, *A*), a small but conspicuous circular, grooved plate. Below each

arm and lining the ambulacral groove are the two longitudinal series of ambulacral plates or ossicles, which meet in the midline (Fig. 16-8, *E*). They form a roof over the groove (Fig. 16-8, *A* and *G*) to protect the radial water vessel, the radial nerve, and podia. Along the margins of the grooves on each side are movable spines that can form a protective barrier over the grooves. Pincerlike pedicellariae of microscopic size are found on the skin or at the base of the spines (Fig. 16-6, *B*, *C*, and *D*). These are protective in function. Six or seven major types are recognized and are useful in classification. Sea stars also have mucous gland cells in the epidermis (Fig. 16-6, *A*) that produce a gelatinous secretion for protection and cleansing. At the arm tip is the terminal tentacle (Fig. 16-7, *A*), with a red pigment spot where there are a number of ocelli. Most spines of the aboral surface are projections of the skeletal plates, but those on the oral side may articulate with their plates by a socket arrangement. Dermal gills or papulae, small delicate projections of the coelom, are scattered between the ossicles and function in respiration (Fig. 16-6, *D* and *E*).

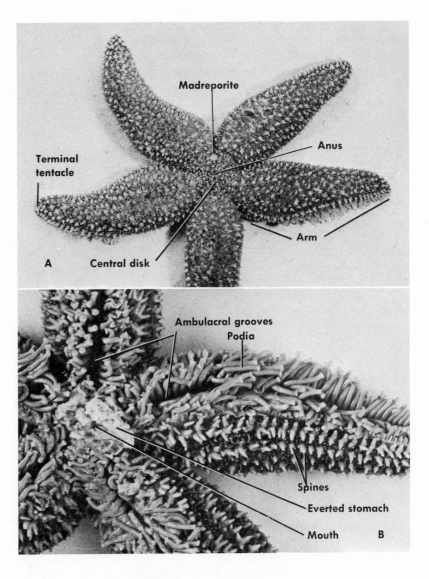

FIG. 16-7

A, Aboral view of sea star *Asterias.* **B,** Oral view of disk and arm of *Asterias.* (From Hickman: Integrated principles of zoology, The C. V. Mosby Co.)

Sea stars have a considerable size range. Some are not more than 1 cm. in greatest diameter; and others are almost a yard in diameter. Those species that have more than 5 rays tend to vary in the number of their rays.

Internal anatomy

The endoskeleton (Fig. 16-8) consists of ossicles that have the characteristic crystalline calcite structure described in an earlier section. It is made up of a deep supporting part of cal-

careous plates bound together by connective tissue in the dermis of the body wall and a superficial part of spines and tubercles that project from or rest on the ossicles of the deeper part. The superficial part bears a thin covering of epidermis and dermis that is often worn off in exposed places. The calcareous ossicles, which make up the major part of the skeleton, may overlap, or they may be farther apart. They are of various shapes, and when bound together, form a closely knit or reticulated skeleton, with scattered spaces for the emergence of papulae (Fig. 16-8, *B* and *D*).

The body wall contains on its coelomic side a thin muscular layer of outer circular and inner longitudinal muscle fibers for moving the arms. The ambulacral system is provided with several muscles for manipulating its ossicles.

The water-vascular system follows the general echinoderm pattern, with certain modifications. The system is derived from parts of the coelomic sacs, and its canals are lined with flagellated epithelium and filled with fluid. In sea stars the system is chiefly concerned with

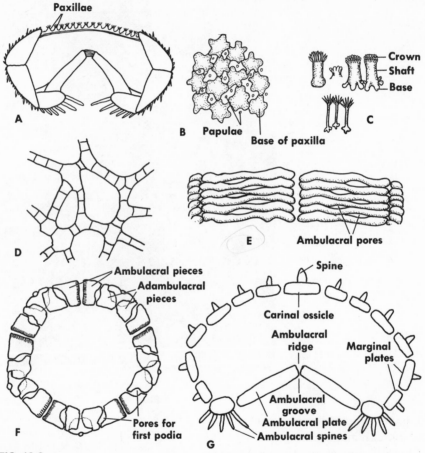

FIG. 16-8

Endoskeletal structure of asteroids (schematic). **A,** Section through arm of *Astropecten* (order Phanerozonia) showing paxillae, or modified spines, found in some stars. **B,** Portion of endoskeleton from inside showing basal plates of paxillae. **C,** Several types of paxillae, each consisting of basal plate supporting rod crowned with minute spines. **D,** Reticular arrangement of aboral endoskeleton of *Asterias* (order Forcipulata). **E,** Ambulacral ossicles of *Asterias* showing pores for podia, staggered to form 2 rows of pores on each side of groove. **F,** Peristomial ring of ossicles of *Asterias*. **G,** Skeleton in schematic section through arm of forcipulate star. (**B, C,** and **G** modified from Hyman; **F** modified from Chadwick, 1923.)

locomotion. The canals are connected with sea water outside by means of a conspicuous madreporite located in an interradial position on the aboral side (Fig. 16-7, A). It is a round calcareous sieve plate, with ciliated pores located at the bottom of furrows (Fig. 16-9). Each pore leads into a pore canal, of which there may be more than 200. The pore canals unite to form collecting canals that open into an ampulla beneath the madreporite. There may be 1 to 5 madreporites in *Linckia* and some others.

From the ampulla a vertical stone canal (Figs. 16-9, 16-10, and 16-11) passes in an oral direction and opens into the water ring, or circular canal, which is located on the inner side of the peristomial ring of ossicles around the mouth region. The inner side of the water ring gives rise interradially to five pairs of small pouches called Tiedemann's bodies. One member of a pair may be missing at the place where the stone canal enters the ring, making a total of 9 Tiedemann's bodies (Fig. 16-10). In some asteroids the ring canal also gives off interradially on its inner side 1 or more elongated polian vesicles suspended in the coelom. The functions of the Tiedemann's bodies, as well as that of the polian vesicles (absent in the common *Asterias*), are chiefly unknown.

From the outer side of the water ring a radial water canal extends the length of each arm and ends in the terminal tentacle (Figs. 16-10 and 16-11). This radial water canal runs on the oral side of the ambulacral ridge of ossicles that forms the center of the ambulacral groove (Fig. 16-12). From each side of the radial canal, lateral canals branch off and pass between the ambulacral ossicles on each side of the groove. Each lateral canal has a valve and runs to a podium provided with an expanded bulb (the ampulla) at its inner end in the coelomic cavity. Externally the podium forms a tubular projection of the body wall into the ambulacral groove.

FIG. 16-9

A, External view of madreporite plate of *Asterias*. Furrows are perforated with ciliated pores. **B,** Vertical section of madreporite and aboral end of stone canal. **C,** Horizontal section through stone canal, showing relative positions of stone canal and axial organ in axial sinus. (Modified from Chadwick, 1923.)

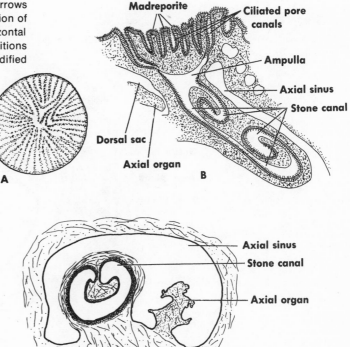

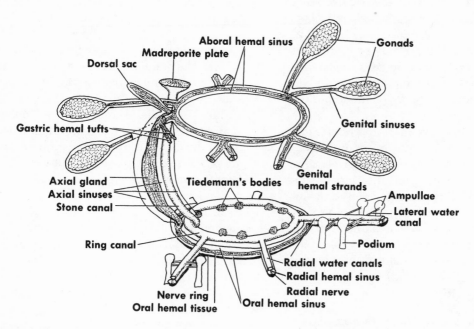

FIG. 16-10

Water-vascular and hemal systems of *Asterias* (diagrammatic).

FIG. 16-11

Vertical section through disk and part of 1 arm of sea star (diagrammatic).

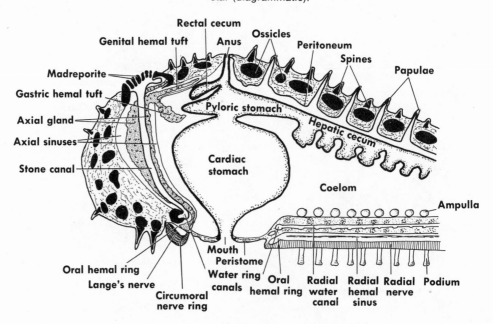

The outer end of the podium usually bears a sucker for attachment. The ampulla is a muscular sac consisting of peritoneum on both its inside and outside surfaces, between which are circular and longitudinal muscle fibers and connective tissue. The podium is covered with flagellated epithelium on the outside and peritoneum on the inside. Between these layers are connective tissue and longitudinal muscle fibers. The number of podia varies in the different species. Those with a large number have the rows staggered so that there may be 4 rows instead of 2 along the ambulacral groove (Fig. 16-8, E). This difference is brought about by the alternate length of the lateral canals, some long and some short. Those species with only 2 rows of tube feet have lateral canals of the same length. The ring and radial canals have few muscles, but consist of a lining of flagellated coelomic epithelium, with connective tissue in the wall.

The water-vascular system operates on the principle of a hydraulic system. Since each

podium is a closed cylinder, when the valve in the lateral canal closes, the contraction of the bulblike ampulla at its inner end forces fluid into the podium and extends it as a flexible process that can be manipulated in any direction by the longitudinal muscles in its wall. These podia retain constant lumina by their sheaths of connective tissue made up of circular and longitudinal fibers so that increase in pressure from the ampulla involves elongation of the podium without wasteful lateral bulging of the podial wall. When the tip of the podium touches a substratum, the withdrawal of the center of the terminal sucker lessens the pressure in the tip and causes the sucker to adhere. When the longitudinal muscles contract, fluid is forced back into the ampulla, and the animal is drawn forward by the shortening of the podia. Fluid pressure for the operation of the tube feet may be a function of all parts of the water-vascular system, such as the madreporite, stone canal, ring canal, and radial canals. Any slight fluid loss in the system is compensated by sea water intake through the madreporite. The podia can perform independently or in a coordinated manner. It is thought that

FIG. 16-12

Cross section through arm of young sea star (diagrammatic).

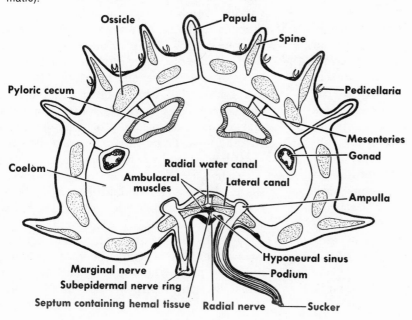

FIG. 16-13

Oral view of large sun star, attached by podia to side of aquarium.

coordination of the podia in movement is brought about chiefly by the presence of a nerve center at the junction of each radial nerve with the nerve ring, but an explanation of all the varied movements of the podia is still vague. Locomotion, especially on a soft substratum, makes little use of the suckers, the podia performing a stepping process that involves swinging of the podia followed by contraction that moves the animal forward. The coordinated action of the podia and their suckers is quite powerful and can draw the animal up a vertical surface without much effort (Fig. 16-13). When inverted, the sea star can right itself by twisting one or more arms until some of the suckers are attached to a substratum as an anchor and then slowly rolling over. Some asteroids do not have suckers but use only stepping motions, with little or no gripping of the substratum.

Sea stars are carnivorous and feed on a variety of animals, such as crustaceans, mollusks, other echinoderms, fishes, etc. Deep-sea forms feed on the organic detritus of mud by a filter-feeding method. Many have a decided preference for shellfish. *Asterias* is highly destructive to oyster beds.

The digestive system consists of a mouth, stomach, digestive glands, intestine, and anus and extends between the oral and aboral sides of the disk (Fig. 16-11). The mouth is in the middle of the peristomial membrane on the oral side and leads by a short esophagus into a large saclike stomach that is divided by a constriction into a lower cardiac stomach and a smaller, aboral pyloric stomach.

The cardiac stomach has pouched walls and is connected by paired mesenteries or gastric ligaments to the body wall. The star-shaped pyloric stomach lies above the cardiac stomach and is connected by ducts to a pair of large pyloric ceca in each arm (Figs. 16-11 and 16-14). These ceca are digestive glands that are suspended in the coelom by dorsal mesenteries. Each cecum is made up of a longitudinal duct with many lateral ducts, into which secretory lobules open. The secretory lobules are lined with tall flagellated epithelium containing granular, storage, and mucous cells. Food reserves are found in the storage cells and enzymes in the granular, or secretory, cells. The longitudinal duct along each cecum may enter the pyloric stomach separately or it may join with the other duct of the same arm before doing so. From the pyloric stomach a short intestine, provided with a varying number of rectal or intestinal ceca, ascends to the anus. An anus (and sometimes an intestine) is absent in many sea stars, and the feces are discharged through the mouth.

Histologically the digestive tract is similar to the body wall, as it is lined with flagellated epithelium beneath which is a layer of muscle and connective tissue. Coelomic epithelium covers the coelomic side of the digestive tract.

Asteroids feed by two general methods. Those with short rays and suckerless podia swallow the prey whole (Phanerozonia) and digest it in the stomach, casting out the indigestible part through the mouth. *Asterias* and other sea stars with long flexible arms and podia with suckers feed by everting the cardiac stomach upon the prey, which is held enclosed by the arms. By contracting the body wall muscles, they increase the pressure of the coelomic

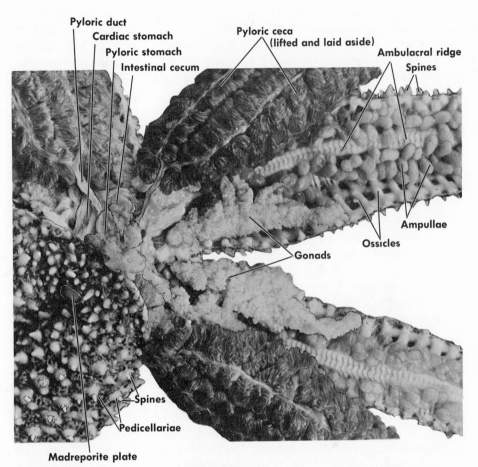

Pyloric duct
Cardiac stomach
Pyloric stomach
Intestinal cecum
Pyloric ceca (lifted and laid aside)
Ambulacral ridge
Spines
Ampullae
Ossicles
Gonads
Spines
Pedicellariae
Madreporite plate

FIG. 16-14

Aboral view of *Asterias* showing portions of three arms dissected to expose internal structures. Parts of two undissected arms at left. (From Hickman: Integrated principles of zoology, The C. V. Mosby Co.)

fluid, causing the cardiac stomach to be everted through the mouth. The gastric ligaments prevent the cardiac stomach from everting too far. The stomach engulfs the prey and pours on it powerful digestive juice, with enzymes from the pyloric ceca or stomach wall. This results in the digestion of the prey into a type of broth, which is then sucked up into the body. Retraction of the stomach occurs when the stomach wall muscles contract and the body wall relaxes.

Since many sea stars feed on bivalves (clams and oysters), the problem of opening the valves of the prey is involved. During feeding, the sea star humps itself over the bivalve and grasps it with its podia, while holding onto the substratum with the distal ends of its arms. By manipulating itself the sea star brings the gape

of the bivalve close to its mouth. Opening the valves was long thought to be brought about by the pull of the podia upon the valves with a force sufficient to overcome the adductor muscles of the bivalve. It has now been firmly established that minute gapes in the imperfectly closed valves offer an opening sufficient for the entrance of the sea star's stomach (A. L. Burnett and A. Christensen), although the pull of the sea star itself probably produces small gapes by the pull of the podia (which may act in relays) (M. Lavoie, 1956). The force of the pull may

643

reach 5,000 grams in some species. It appears that the stomach of a sea star can squeeze through an opening as small as 0.1 mm. The sea star does not pull continuously with its podia on the valves but only intermittently. If the valves close on the stomach, it is not injured.

Digestion in sea stars is mostly extracellular, although some intracellular digestion may occur in the pyloric ceca. Such enzymes as protease, amylase, and lipase are found in the digestive juice produced by the ceca. Because of the nature of their food, only a small amount of feces is produced. In those forms that feed by mucus-ciliary methods (e.g., *Porania*), ciliary currents and mucus entangle microscopic food and carry it to the mouth. J. M. Anderson (1960) has shown that the pyloric stomach of such asteroids is provided with five pairs of Tiedemann's pouches that act as flagellary pumping organs for sucking food from the stomach into the pyloric ceca.

The coelomic cavity of asteroids is very large and encloses the digestive system and reproductive organs. In addition, there are some other coelomic cavities such as the axial sinus, genital sinus and marginal sinus (Figs. 16-10 and 16-11). Coelomic spaces are lined with flagellated cuboidal epithelium. The coelom is filled with coelomic fluid, which has the composition of sea water, plus certain coelomocytes that are amebocytes with phagocytic properties. Echinoderms have little or no ability to regulate their internal osmotic pressure. They cannot regulate the intake of water, but they do have some ability to retain ions in their cells. However, to do so, they must expand a great deal of energy. Coelomocytes originate from the coelomic epithelium and not from Tiedemann's bodies, as formerly supposed. The cells of Tiedemann's bodies are specialized for absorption and phagocytosis and may remove extraneous dissolved materials (J. C. Ferguson, 1966).

The hemal, or blood, system (Fig. 16-10), which is not so pronounced in asteroids as in some other echinoderms, is made up mostly of fine channels, or lacunae, enclosed in coelomic spaces (often called perihemal sinuses). The principal hemal vessels are the oral hemal ring around the peristome; the radial hemal sinus (hyponeural canal) that runs into each arm from this ring along the midline of the ambulacral groove; and the axial sinus, which contains the stone canal and axial gland and arises from the oral hemal ring, passing aborally into the genital sinus and ampulla of the stone canal. The axial gland, a brownish, spongy body of unknown function, may be the remnant of a heart (Figs. 16-10 and 16-11). Coelomocytes are present in the fluid of the hemal system. The two main types of coelomocytes are filiform amebocytes and bladder amebocytes. The hemal system may be useful in distributing digested products, although its general functions are unknown.

Respiration in asteroids is effected by the papulae, or skin gills (Fig. 16-6, *D* and *E*), and by the tube feet; but general body surface may be used, especially in those species that lack papulae. The papulae are evaginations of the body wall that pass between the ossicles; their histologic structure is similar to that of the body wall, with flagellated epidermis on the outside and coelomic flagellated epithelium on the inside. Respiratory mechanisms differ in each class of echinoderms and some have no respiratory structures.

Asteroids have no excretory system. Coelomocytes may be responsible for getting rid of unwanted materials, and the pyloric ceca apparently are also used for the same purpose. Ammonia and urea appear to be the chief nitrogenous waste. The loaded coelomocytes leave through the papulae or podia; soluble wastes leave at the body surface.

The nervous system, like that of all echinoderms, is a diffuse one, having a strong connection with the epidermis. It is commonly described as consisting of three interrelated systems. The chief one of these systems is the oral, or ectoneural, system just beneath the epidermis. It consists of a nerve ring, radial nerves (Figs. 16-10 to 16-12), and a subepidermal plexus. The nervous center is the pentagonal circumoral nerve ring near the periphery of the peristomial membrane. This nerve ring supplies nerves to the esophagus and to the peristomial membrane. At each radius it gives off a radial nerve that runs as a V-shaped mass along the bottom of the ambulacral groove in each arm to the terminal tentacle. The radial nerve, which has both bipolar and multipolar ganglion cells, sends nerve fibers to the podia and ampullae, and the radial nerves are connected to the subepidermal plexuses that innervate all the body wall appendages.

The second or entoneural nervous system consists mainly of a marginal cord (a thickened part of the subepidermal plexus) at the outer margin of the ambulacral groove along the length of the arm on each side. These marginal cords give

off a pair of lateral motor nerves to the muscles of each pair of ambulacral ossicles. The nerves continue to the coelomic lining, beneath which they form a plexus that innervates the muscle layers of the wall.

The third or hyponeural system is a plate of nerve tissue (Lange's nerve) (Fig. 16-11) in the lateral oral wall of the hyponeural sinus beneath the coelomic epithelium. The system innervates the transverse muscles between the ambulacral ossicles in the roof of the hyponeural sinus. Nerves from the same system pass to the peristomial region, where they form 5 interradial thickenings in the floor of the ring sinus, aboral to the main nerve ring.

The coordinated action of the podia depends on the nerve ring and radial nerves. If a radial nerve of 1 arm is cut, the coordination of the podia in that arm is destroyed, and the podia

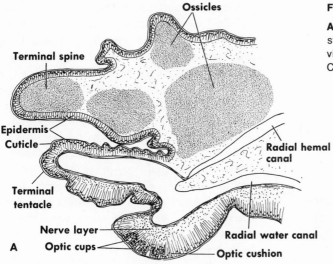
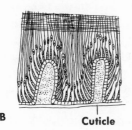

FIG. 16-15

A, Longitudinal and sagittal section through tip of sea star ray, showing optic cushion and cups. **B,** Enlarged view of vertical section of 2 optic cups. (Modified from Chadwick, 1923.)

645

do not work in unison with those of the other arms. A nervous center seems to exist at the junction of each radial nerve with the nerve ring; and when the sea star is progressing, the nerve center of the base of the leading arm has dominance over the nerve centers of the other arms and causes all their podia to move in harmony together.

The sensory system has few specialized sense organs. Sea stars cannot see, but they can detect differences in light intensities. An eyespot is located at the base of the oral surface of each terminal tentacle of the arm (Fig. 16-15). It consists of many pigment-cup ocelli that together form an optic cushion. Each cup, or ocellus, is made up of epidermal cells with red pigment granules and is covered externally by a cuticle, beneath which in some species is a lens formed from the epidermis. The retinal cells, or photoreceptors with bulbous enlargements, project into the cavity of the cup. The number of ocelli in an optic cushion increases with age, but usually ranges between 50 and 200.

In addition to the eyespots, there are numerous neurosensory cells found everywhere in the epidermis. They are long, slender cells with distal threadlike processes that run to the cuticle and proximal fibers which join the subepidermal plexus (Fig. 16-6, A). They may serve for reception of contact and chemical stimuli and are especially abundant in the podial suckers, at the bases of the spines, and in the terminal tentacles. Some variations of the sense organs are found among the different species.

Asteroids are chiefly dioecious, although hermaphrodites may occasionally be found in those that normally have separate sexes. In this case the individual may be male at first and female later. Usually there is a pair of gonads for each arm, but there may be more in some species. Each gonad (Fig. 16-14) is a feathery cluster of tubules, the size of which varies with the seasons. At the spawning season they are very large; when inactive, they are only small tufts.

Each gonad is enclosed in a genital coelomic sac, which is derived from the genital coelomic sinus and is attached at its proximal end near the junction of the arm and disk (Fig. 16-10). Each of the gonads has a small gonopore that usually opens aborally between the bases of the arms on the central disk, or in those with ventrally located gonads (e.g., *Asterina*) it opens orally. In the breeding season of some species the testes are pale in color, and the ovaries are pink or orange; but in general the sexes have no external sexual dimorphism.

A form of copulation may occur in a few species in which the male lies on top of the female, his arms alternating with hers. Fertilization usually occurs in the sea, however, where the eggs and sperm are shed. Stimulation of gamete shedding appears to be brought about by several factors during the breeding season, which usually occurs once a year. Egg sea water will cause the male to shed his sperm, and sperm sea water will activate the female. It has been shown that the maturation and shedding of sea star eggs are stimulated by an identified secretion from neurosecretory cells in the radial nerves (A. B. Chaet and R. A. McConnaughy, 1959).

Brooding methods are found in some species. The female retains the fertilized egg mass in an oral concavity formed by the arching of her disk and the bending of the arms. Aboral brood chambers formed by spines also occur. Most asteroids, however, simply shed their gametes into the water and give no further attention to the fertilized egg. Sea stars usually become sexually mature in about a year, but they are known to grow for 3 or 4 years.

DEVELOPMENT

Asteroids have two types of development, indirect and direct. The indirect type has a free-swimming larval stage, and the eggs are homolecithal with little yolk. The direct type has large, yolky eggs, and the free-swimming stage is omitted. This type usually broods its eggs in some manner.

The development of the indirect type of egg

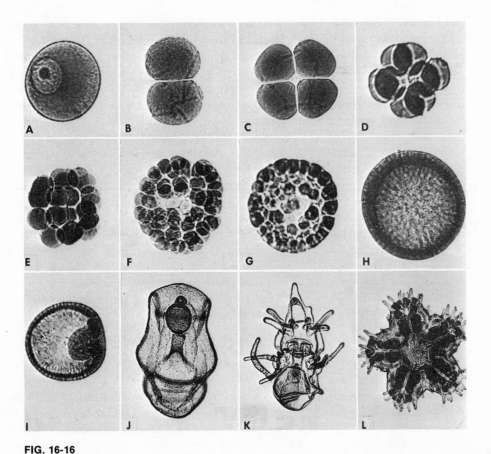

FIG. 16-16

Development of sea star *(Asterias).* **A,** Unfertilized egg. **B** to **F,** Early cleavage stages. **G** and **H,** Blastula stages. **I,** Gastrula. **J,** Bipinnaria larva. **K,** Brachiolaria larva. **L,** Young sea star. (Courtesy Carolina Biological Supply Co., Burlington, N. C.)

(Fig. 16-16) may be described as follows: cleavage occurs soon after fertilization and is equal and indeterminate; by the second day a spherical, ciliated blastula is formed and becomes free swimming; by invagination an elongated gastrula is formed, the blastopore of which later becomes the anus, as is characteristic of deuterostomes. While forming, the archenteron gives off some mesenchyme into the inner blind end of the blastocoel. The archenteron gives rise to 2 outpocketings, or lateral pouches, that bud off, elongate, and lie on each side of the archenteron. Later, the two pouches connect anteriorly to form a U-shaped coelom. From the left pouch a narrow evagination forms a hydropore at the dorsal surface. The mouth originates from an ectodermal inpocketing (stomodeum), and the digestive system differentiates into the esophagus, stomach, and intestine, which are derived from the archenteron. The posterior ends of the lateral pouches are eventually pinched off and become the right and left somatocoels. In *Asterina,* the common European sea star, the anterior portion of the archenteron is cut off as a single coelomic sac to develop into a coelom, and the posterior archenteron becomes the digestive tract.

When the larva begins to swim, its body is covered with cilia; but later the cilia are restricted to an exterior band of many loops that are used in locomotion. Still later, projections or arms arise on each side of the body. The loco-motor bands extend along the arms, which often become long. At this stage the larva is called bipinnaria (Figs. 16-4 and 16-16, *J*). The bilaterally symmetric larva swims in a clock-wise rotation, with its anterior end foremost. It begins to feed, primarily on diatoms, as soon as its digestive tract is completed.

After a few weeks of free swimming, 3 addi-tional lobes (brachiolar arms) appear at the anterior end, and the larva is termed a "brachio-laria." The new arms differ from those of the bipinnaria in having a prolongation of the coelom in them and being tipped with adhesive cells. The arms, together with a suckerlike glandular region found between the bases of the 3 arms, form an attachment device. When the brachiolaria settles to the bottom, it attaches itself by the sucker at the anterior end, which becomes a type of stalk.

Metamorphosis now occurs. The anterior end degenerates and is absorbed, and the posterior end becomes enlarged. Five lobes appear on the right side of this posterior end, which becomes the aboral surface of the future sea star, and the left side becomes the oral surface. Arms are formed as extensions of the body. From the hy-drocoel the water-vascular system develops, and two pairs of outgrowths from the hydrocoel in each lobe become the first tube feet. Skeletal elements begin to be formed as soon as the first lobes or arms appear. Most parts of the larval digestive system degenerate and are formed anew in the metamorphosis to the sea star condition. Most of the coelom arises from the somatocoel already described. Within 2 months after development starts and sufficient podia are formed for holding to a substratum, the young sea star (Fig. 16-16, *L*), not more than 1 mm. in diameter, is detached and begins a free existence. In this development the bi-lateral symmetry has gradually been replaced in most particulars by radial symmetry.

The larval stages of sea stars may form a considerable part of planktonic populations, especially in tropic and temperate seas. In the Arctic and Antarctic seas, however, where echinoderms tend to brood their young and have direct development, they form only a small part of the plankton. The same appears true of those that live in great depths of the ocean, where most asteroids have large eggs and brood their young.

Regeneration of lost parts is common among asteroids. Some species are known to reproduce asexually by transverse division (*Allostichaster*). Almost any part of the arm can be regenerated. Some species can regenerate the whole animal from a severed arm (*Linckia*). In *Asterias* a complete regeneration of the animal will take place if part of the disk is attached to 1 arm, es-pecially if the madreporite is included; but regeneration is a slow process, and many months may be required for complete replace-ment of parts.

■ CLASS OPHIUROIDEA

The members of class Ophiuroidea are com-monly called brittle stars or sometimes basket stars and serpent stars. They are readily dis-tinguished from the members of class Asteroidea by the arms, which are usually sharply demar-cated from the central disk (Fig. 16-17). There are also many other differences between the two classes. Ophiuroids have no ambulacral groove, their podia are small and without ampullae, and their arms are composed of jointed seg-ments produced by the fusion of ambulacral ossicles. Ophiuroids do not use their podia for locomotion, but move by whiplike lashing move-ments of their arms, a process that resembles the vigorous action of serpent tails. They are found in all oceans at nearly all depths, from tidewater down. Many are shallow-water forms and hide in the algae or the mud with their arms protruding. They also frequently attach them-selves to other animals by means of their twin-ing arms. Because they are negative to light and positive to contact with a solid (stereotropism), they conceal themselves during the day and are active mainly at night.

There are about 1,900 living species of ophiu-

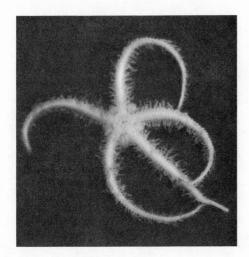

FIG. 16-17

Oral view of *Ophiothrix angulata,* a brittle star (Ophiuroidea).

roids belonging to more than 200 genera. Two orders, Ophiurae and Euryalae, are recognized. Members of Ophiurae can move their arms (which do not branch) only sidewise in a horizontal plane. Those of Euryalae have flexible arms (often branched), which can coil upward or downward in a vertical plane. Their fossil record dates back to the Devonian period. It shows that many Paleozoic families had open ambulacral grooves.

GENERAL MORPHOLOGIC FEATURES
External anatomy

Most ophiuroids are 5 armed, but a few species are regularly 6 or 7 armed. A 5-armed species may have an extra arm as an abnormality. The general appearance of all ophiuroids is very similar. Their small flattened disk has a rounded, pentagonal, or scalloped contour (Fig. 16-17) and is usually not more than 1 inch in diameter (much smaller in many species). The basket stars (Gorgonocephalidae) may have disks of 3 or 4 inches in diameter and an arm spread of 20 inches. The aboral surface of the disk may be smooth, granular, or covered with small spines. In many cases it may have embedded plates. A pair of large radial shields is found at the base of each arm. There are no openings on the aboral surface, for an anus is lacking.

The arms may be smooth or provided with spines. Their interior is filled with a series of articulating ossicles. They have a jointed appearance because of four longitudinal series of calcareous plates, or shields, which may be concealed partly by the skin. Each arm joint corresponds to an internal skeletal ossicle and is covered aborally by an aboral shield, orally by an oral shield, and on each side by a lateral shield. Modifications of this plan may occur in which the aboral and oral shields are reduced, with an enlargement of the lateral shields. The lateral arm shields correspond to the adambulacral ossicles of asteroids. Each lateral shield usually has about 3 to 15 spines in a vertical row. Many variations in size, shape, and arrangement of these spines occur in the different species. In some species there are poisonous spines, and in others the spines are transformed into hooks (*Ophiothrix*). Ambulacral grooves are totally lacking on the oral side.

There is a pair of small podia or tentacles on the oral surface of each joint. One member of each pair of podia emerges on each side between the lateral and oral arm shields. The podia in ophiuroids are mostly sensory in function, although they may play a minor role in locomotion by gripping objects. It is thus seen that the arm of ophiuroids consists of a series of jointed segments, each of which is made up of a "vertebra" that articulates with the others by means of ball-and-socket joints. In basket stars this articulation allows movement in all planes, but in serpent stars it permits only lateral mobility. The vertebrae are connected and manipulated by four sets of longitudinal muscles that are especially strong on the oral side of the arm, where they are inserted in muscle pits in the ossicle on the side toward the disk. Each vertebra is the result of the fusion of 2 ossicles lying side by

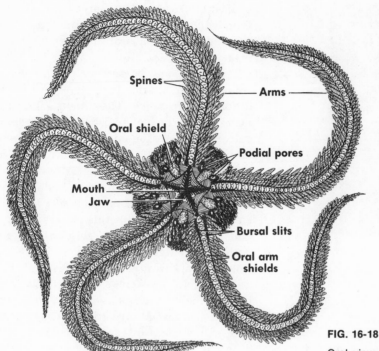

Spines

Arms

Oral shield

Podial pores

Mouth

Jaw

Bursal slits

Oral arm
shields

FIG. 16-18

Oral view of brittle star (Ophiuroidea). (From Hickman:
Integrated principles of zoology, The C. V. Mosby Co.)

side. On the aboral side of each vertebra is a
dorsal groove, and on the oral side is an ambulacral channel that houses the radial water
vessel, nerve, and blood vessel.

The skeleton of the disk is made up of the
mouth frame, the genital plates between the
genital pouches and arms, the buccal shields,
and certain other plates or scales. In the center
of the oral disk is the star-shaped mouth that
is surrounded by five interradial triangular jaws
made up of plates (Fig. 16-18). Each jaw consists of several plates, the most important of
which are the half-jaws and the oral or buccal
shield. The oral ends of the half-jaws are usually the exposed triangular part of the jaw. The
oral shield covers most of the oral surface of
the jaws. The jaws have teeth along their margin and also at their tip, where the teeth extend
aborally in a vertical row. The jaw apparatus is
manipulated by two concentric series of muscles — the external interradial and the oral inter-

nal interradial muscles. One of the buccal
shields is commonly modified to form a madreporite, which has an oral location in ophiuroids.
On the oral side the arms, with all their parts
except their aboral shields, are embedded in
the oral disk and extend inward to the angles
of the mouth.

The body surface of ophiuroids is covered
with a cuticle, but cilia or flagella are restricted
chiefly to the bursal slits and the surface of
the disk. The syncytial epidermis is continuous
with the dermis, which has embedded in it the
endoskeletal shields, along with a connective
tissue matrix and scattered cells. Some ophiuroids are highly colored with pigments of carotenoid, melanin, riboflavin, and xanthophyll.
Pigment spots sensitive to light are found on
the aboral surface of the arm ossicles in some
species. Melanin is a screen against radiation.
Others have luminescent pigments, especially
riboflavin. These pigments are found in the

dermis. There are no pedicellariae or papulae in the skin. External appendages are restricted to spines and podia. The larger spines are movable, like those of asteroids, and are mounted on tubercles.

Ophiuroids can move rapidly when disturbed. As already mentioned, the arms are well provided with longitudinal muscles that link the vertebrae together. Movement is produced either by trailing or by pushing ahead one of the arms while the others, by rowing movements against the substratum, thrust the body forward in a series of jerks. Movements show many variations. During locomotion the ophiuroid holds its disk above the substratum. Some move by grasping objects with their arms and pulling themselves along. A few apparently can swim by trailing 1 arm behind and performing stroking movements with the others (*Ophiacantha*). A sand-burrowing ophiuroid (*Amphiura*) can burrow directly downward by digging with its podia and remain covered in sand except for its arm tips.

Internal anatomy

The ophiuroid water-vascular system is typical of the echinoderm group, with certain modifications (Fig. 16-19). The madreporite is usually single, although *Ophiactis* has 5. The madre-

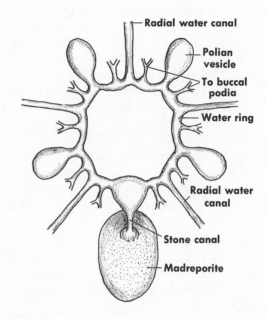

FIG. 16-19

Water-vascular ring of ophiuroid (schematic).

FIG. 16-20

Cross section of arm of ophiuroid (schematic).

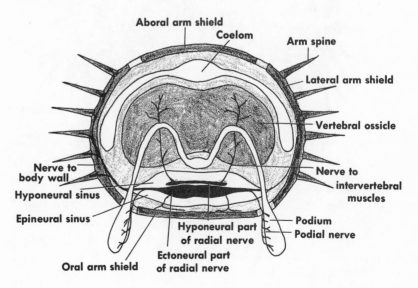

porite (a buccal shield pore), which is found on the oral surface, usually has 1 pore canal, but may have more in some species. The stone canal ascends to the water ring, which occupies a groove on the aboral surface of the jaw apparatus. There are no Tiedemann's bodies, but usually 4 polian vesicles are borne by the water ring. From the water ring a radial canal descends toward the oral side, enters the arm base, and runs along the arm, aboral to the hyponeural canal, to the terminal tentacle. In each vertebral ossicle the radial canal gives off a pair of lateral or podial canals, which may pass directly or by a loop to the podia (Fig. 16-20). Ampullae are absent, but a valve is present where the podial canal enters the podium. The

water-vascular system is lined throughout with flagellated coelomic epithelium.

The alimentary canal of ophiuroids is the simplest among the echinoderms. It consists of only a mouth and a large saclike stomach. There is no intestine or anus. The stomach, usually with 10 marginal pouches, is restricted to the disk (Fig. 16-21), except in the Ophiocanopidae, in which it sends its blind ceca into the arms. The mouth, which is a circular opening in the center of the 5-angled peristomial membrane, leads to a very short esophagus that may be provided with a smaller circular cavity on the inner side of a large one. Muscles from these cavities are involved in the opening and closing of the mouth. The esophagus opens into the sacciform stomach, which fills most of the interior of the disk. The peripheral pouches of the stomach are closely applied to the inner aboral surface of the disk, where they are attached by mesenteries. The bursae are placed between the pouches. The stomach wall consists

FIG. 16-21

Ophiuroid with aboral disk cut away to show principal internal structures. Only bases of arms shown. (From Hickman: Integrated principles of zoology, The C. V. Mosby Co.)

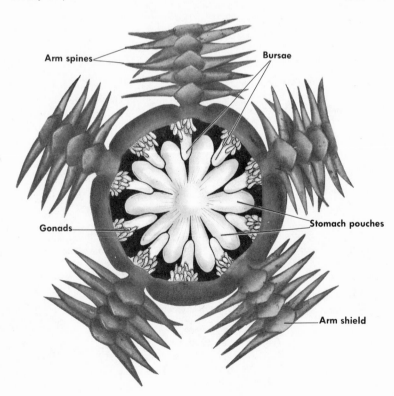

of an internal lining of flagellated epithelium, a nervous stratum, a layer of connective tissue, circular muscle fibers, and an outer flagellated coelomic covering. Gland cells seem to be absent.

Ophiuroids live on bottom detritus and small organisms. They usually eat whatever they can find, although some are selective in their food habits. Certain predatory ophiuroids are known to live on coral polyps. When they feed, small food particles are passed to the mouth by the podia and by ciliary action, and large particles are carried toward the mouth by rhythmic sweeping of the arms.

The ophiuroid coelom is restricted to space in the disk around the organs, a narrow arm space between the vertebral ossicles and the arm shields, and certain minor coelomic cavities (hyponeural radial sinuses, axial sinus, and genital sinuses) (Figs. 16-20 and 16-22). The coelomic fluid contains coelomocytes and certain ameboid cells. The axial complex is similar to that of the asteroids, except that it is turned orally because of the location of the madreporite. The axial sinus runs from the hyponeural ring sinus to the ampulla beneath the madreporite. Most of this sinus is filled with the axial gland and stone canal. The axial gland develops from the coelomic wall (the axocoel), which encloses the stone canal. The axial gland is considered a hemal center that connects the oral and aboral hemal rings, but its real function is largely a mystery.

Most of the hemal channels in ophiuroids are enclosed in coelomic channels. The oral hemal ring lies inside the hyponeural ring canal and gives off a radial hemal channel into each arm at the aboral side of the radial canal. The radial hemal channel sends hemal channels to the podia. The aboral hemal ring just mentioned sends channels to the bursae and gonads and receives gastric branches from the stomach. The hemal channels, or blood lacunae, have walls of nucleated membranes or a thin layer of connective tissue. The hemal fluid contains amebocytes of three or four types, but no bladder or filiform amebocytes. Some ophiuroids have coelomocytes containing hemoglobin.

Ophiuroids have special organs for respiration called bursae (Fig. 16-21). These are saclike invaginations of the oral wall of the disk near the arm bases. There are 10 of these bursae, each located between 2 stomach pouches. Each bursa opens to the outside by an elongated slit close to the arm base on the oral surface of the disk. In *Ophioderma* there are 2 openings to each bursa because the middle part of the slit has fused. The bursae are lined with flagellated epithelium, and on the outside they are covered with peritoneum. Calcareous deposits may be found in the dermis of the bursal wall. Water

FIG. 16-22

Longitudinal section of ophiuroid (schematic) through interradius of disk and portion of 1 arm.

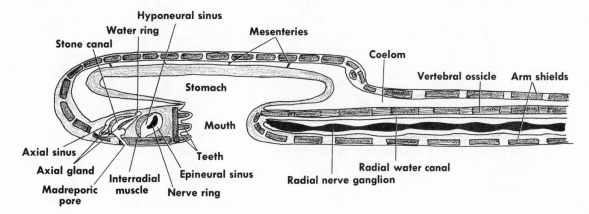

currents for gaseous exchange are constantly entering and leaving the bursae through the slits, a process faciliated by the beating of the flagellae. A pumping or suction action produced by the aboral disk wall may aid in the movement of the water in some species.

The nervous system of ophiuroids is similar to that of asteroids, consisting chiefly of an outer ectoneural system (sensory and motor) and an inner hyponeural system (motor) separated by only a thin membrane (Fig. 16-20). The two systems are actually superimposed; the nerve ring and radial nerves represent each system. The nerve ring is found in a groove near the aboral surface of the jaw apparatus between the outer water ring and the inner hemal ring. This nerve ring supplies nerves to the esophagus, to the buccal or first podia in the arm, and to the jaw apparatus and its teeth. From the nerve ring a radial nerve runs through each arm beneath the radial hemal channels and along the oral faces of the vertebral ossicles (Fig. 16-22). A ganglionic enlargement at the level of each vertebral ossicle is present in the radial nerve. A pair of motor nerves (from the hyponeural system) is given off at each ganglion in the arm to the intervertebral muscles. Between the ganglionic enlargements a podial nerve and a pair of lateral sensory nerves are given off and pass to the podia and body wall. A ring ganglion, formed by the podial nerve around the base of the podium, supplies the podium with a nerve. The chief coordinating center of the nervous system appears to be the nerve ring.

Most ophiuroids are dioecious, although some are known to be hermaphroditic (Amphipholis). Usually the sexes are very much alike, but in some species the female carries the dwarf male on her disk. The saclike gonads are confined to the disk, being attached to the coelomic side of the bursae (Fig. 16-21). Each gonad opens indirectly to the exterior by way of the bursa to which it is attached. When the gonads are ripe, the gametes are discharged into the bursae through a rupture or temporary opening and

are carried out by the water currents. The number and arrangement of gonads vary. There may be one or more gonads per bursa. In some species they may be paired in the basal arm joints and discharged through the body wall. Fertilization occurs chiefly in the water, where development takes place. Brooding of the young in the bursae (where few or many embryos are present) is a common method of Arctic and Antarctic ophiuroids. The embryo of the viviparous Amphipholis has a nutritional attachment to the bursal wall. Most ophiuroids do not become sexually mature until they are 3 or 4 years old. Their life span has been estimated at 5 years, although some may live much longer.

Regeneration of lost parts is common in the class. Autotomy, or shedding of the arms, is also practiced. In some species the entire disk and at least 1 arm must be present for complete regeneration of the whole animal. Some species regularly reproduce asexually by transverse division, especially in the 6-armed Ophiactis, so that such serpent stars may have one set of 3 large arms and one of 3 small arms.

DEVELOPMENT

Those ophiuroids with small eggs that are shed into the water pass through an indirect type of development (considered the most primitive), which involves a free-swimming larval form called an ophiopluteus (Fig. 16-3). Those with large eggs that are brooded have a direct type of development and omit the free-swimming stage. In indirect development the egg undergoes holoblastic and equal cleavage. A free-swimming and flagellated blastula stage is formed within 24 hours. The gastrula stage occurs (by embolic invagination) within 2 or 3 days. The coelomic sac is formed either as a separate sac from the tip of the archenteron or as a pouch on each side of the archenteron. The pluteus larva swims about during metamorphosis. Many aspects of development are similar to those in asteroids, although there are modifications of the basic echinoderm pattern here and there. There is no attachment stage during the development, and at the end of the metamorphosis the small brittle star sinks to the bottom to begin its adult existence.

■ CLASS ECHINOIDEA

Class Echinoidea are the sea urchins, sand dollars, and heart urchins. They are pentamerous echinoderms without arms, but with a compact body enclosed in an endoskeletal test or shell that is formed from fused plates bearing movable spines. Their general shape may be oval, globose, or a similar design, with the body flattened along the oral-aboral axis. Although they have no arms, their radii bear five double rows of podia that are arranged as meridians between the oral and aboral poles. Some echinoids show a tendency toward secondary bilateral symmetry.

A   B

FIG. 16-23

Regular sea urchins. **A,** Oral view of *Arbacia punctulata,* an East coast urchin. **B,** Aboral view of *Strongylocentrotus franciscanus,* the giant red urchin of our Pacific shores.

FIG. 16-24

A, Group of regular sea urchins showing wide variety of spines. **B,** Cidaroid sea urchin (oral view) with large, paddle-shaped primary spines. (From Encyclopaedia Britannica film, "Echinoderms [Sea Stars and Their Relatives].")

A   B

There are about 850 existing species divided among 225 genera. They are classified into two subclasses. The Regularia (Endocyclica), which are radial and have a centrally located aboral anus, are more or less spherical or globular and include the sea urchins (Figs. 16-23 and 16-24). The Irregularia (Exocyclica) are bilateral, with a marginal anus, and may be oval or circular and flattened. They are represented by the heart urchins (Figs. 16-2, A, and 16-29), sand dollars (Fig. 16-30), and cake urchins (Fig. 16-31). The class as a whole has left a very impressive fossil record (Figs. 16-2, A, and 16-30, A).

Echinoids vary in size from a few millimeters to around 20 to 30 cm. in diameter. Although many species are darkish or dull in color, some have bright colors of red, green, orange, and purple. The pigment echinochrome (naphthaquinone), found in no other echinoderm and responsible for the red and purple colors of certain echinoids, is present both in the spines and the test.

Echinoids are found in all seas, but are absent from the deepest ocean depths. They are present from intertidal zones to depths of several thousand meters, but are most common in the littoral zone. Members of the subclass Irregularia are partial to sandy substrates, whereas the Regularia prefer rocky bottoms. The faunas of echinoids have regions of local distribution that vary greatly in the different seas. Some species (e.g., *Echinoneus* and *Echinocardium*) have a wide distribution, whereas others are very restricted. The littoral echinoids are most common in the Indo-Pacific region. Other regions with rich echinoderm faunas of interest to the new world are the West Indies–Florida area, the Panama area or eastern tropic Pacific region, the Gulf of California, and the Puget Sound region.

GENERAL MORPHOLOGIC FEATURES
External anatomy

As the names of the two subclasses Regularia (Endocyclica) and Irregularia (Exocyclica) suggest, echinoids are commonly divided into regular and irregular groups.

Subclass Regularia

Sea urchins (Figs. 16-23 and 16-24) are characterized by a body that is globose or spherical in shape and an armature of many movable spines. Their oral and aboral surfaces are differentiated; they move on their oral surface, which tends to be flattened as a consequence, whereas their aboral surface is somewhat dome shaped. The sea urchin's body is thus radially arranged around a polar axis, with the oral pole bearing the mouth and peristome region and the aboral pole bearing the anus and periproct region. The body surface between the poles is divided into 5 ambulacral regions with rows of podia and 5 interambulacral regions without podia. These two types of regions alternate with each other as they run in a meridional direction between the 2 poles.

The mouth in the center of the oral surface is surrounded by a peristomial membrane, which is thickened along the mouth edge into a lip (Fig. 16-25). The membrane contains small embedded ossicles for support. Five pairs of modified podia (buccal podia) form a circlet near the mouth, and five pairs of little bushy gills are found at the edge of the peristome. The peristome usually bears on its surface small spines and pedicellariae. At the aboral pole is the periproct, which is a circular membrane also bearing embedded endoskeletal plates with small spines and pedicellariae (Fig. 16-26). The periproct contains the anus, usually near its center, but sometimes eccentric in position.

The spines are arranged in rows in both the ambulacral and interambulacral areas. The spines may be all the same size, but usually they are shortest at the poles and longest around the sides. Often the large and small spines are intermingled over the entire body. The length and size of spines vary greatly with different species (Figs. 16-23 and 16-24).

Each spine has a concave socket that rotates over a tubercle on the test, thus forming a ball-

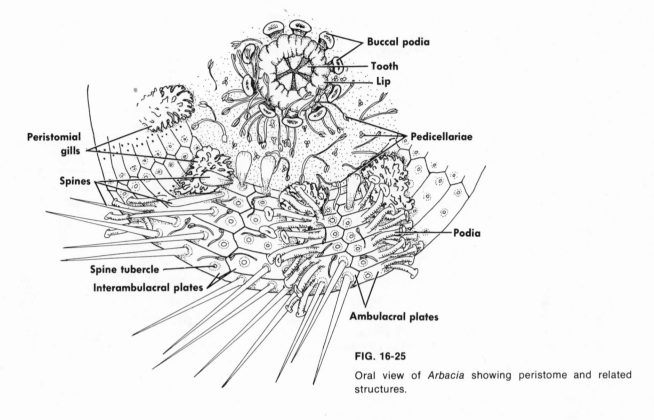

Buccal podia

Tooth

Lip

Pedicellariae

Peristomial gills

Spines

Podia

Spine tubercle

Interambulacral plates

Ambulacral plates

FIG. 16-25

Oral view of *Arbacia* showing peristome and related structures.

FIG. 16-26

Aboral view of *Arbacia* showing periproct and related structures.

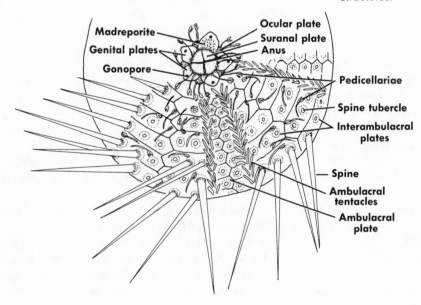

Madreporite

Genital plates

Gonopore

Ocular plate

Suranal plate

Anus

Pedicellariae

Spine tubercle

Interambulacral plates

Spine

Ambulacral tentacles

Ambulacral plate

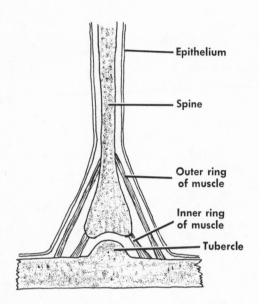

FIG. 16-27

A movable spine from a sea urchin showing relation of spine to its tubercle and musculature involved. (Modified from Russell-Hunter.)

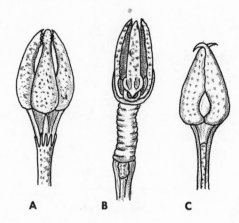

FIG. 16-28

Types of sea urchin pedicellariae. **A,** Globiferous *(Eucidaris).* **B,** Tridentate *(Arbacia).* **C,** Bidentate (sand dollar).

and-socket joint (Fig. 16-25). The spine bears a milled ring at its base, to which muscles are attached. A circle of longitudinal muscles, which are attached at one end to the milled ring of the spine and at the other end to the marginal depression of the test around the tubercle, can point the spine in any direction. An inner set of muscles (the "catch" muscles) hold the spine in an erect position (Fig. 16-27).

Spines are sometimes classified into primary, which are long for defense and walking; and secondary, which are short and may be used to protect the muscles around the primary spines. In the more familiar echinoids (*Echinus, Strongylocentrotus,* Fig. 16-23, *B*) the distinction between the two types is not evident. Spines are found in a variety of forms as well as sizes. Some are straight and circular in outline; others are flattened or triangular. Still others are curved. Poison spines, especially the secondary, occur in the genus *Asthenosoma* and are harmful to man.

The spines are composed of calcite, which

is arranged in an inner axial zone and outer cortex. The entire spine behaves optically as a single crystal unit.

The pedicellariae are microscopic defense organs found only in the asteroids and echinoids (Fig. 16-25). They consist of a head, usually composed of 3 jaws (sometimes 2, 4, or 5), set with flexible necks upon stalks. The stalk contains an internal skeletal rod that reaches to the flexible neck or sometimes to the head. An internal calcareous valve supports and determines the shape of each jaw. The jaws articulate against each other and are operated by muscles, which close and open them.

The function of pedicellariae is to discourage intruders and to catch small prey. They may also be used for cleaning the body surface. Some contain poison glands (globiferous, Fig. 16-28, *A*). Surrounding the outer side of each jaw are 1 or 2 poison glands that open by ducts just below the terminal tooth of that jaw. The toxin paralyzes small prey. Other types of pedicellariae are the tridentate (Fig. 16-28, *B*), tri-

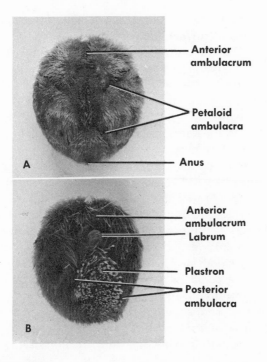

Anterior
ambulacrum

Petaloid
ambulacra

Anus

A

Anterior
ambulacrum
Labrum

Plastron
Posterior
ambulacra

B

FIG. 16-29

Moira atropos, the heart urchin (subclass Irregularia).
A, Aboral view. **B,** Oral view. Note large spoon-tipped
spines on plastron used for burrowing. Long lateral
spines are used to throw sand sidewise when burrowing.
Spines in ambulacra are short. Minute spines (clavules),
located in tract on aboral side, create water currents and
help clean sand off test.

phyllous, and ophiocephalous. Of these, the
tridentate is the largest and most common.
Its head has 3 elongated jaws that usually
meet only distally.

On the ambulacral areas of most echinoids
are hard, solid bodies called sphaeridia. These
may be organs of equilibrium. They are located
on various regions of the body, such as the
center of the ambulacra, on the oral side, or
even on the aboral surface when numerous. In
Arbacia there is 1 sphaeridium on each ambulacrum located near the edge of the peristome.

The podia in sea urchins (Fig. 16-25) are
arranged in five double rows on the ambulacra
and extend from the peristome to the periproct.
In some sea urchins (e.g., Cidaroidea) the podia

are even found on the peristome as far as the
edge of the mouth. In these the special buccal
podia are absent. Most podia are of the locomotor type, just as they are in asteroids. They
are usually long and slender and are provided
with terminal suckers. Both the suckers and
the stalk are supported by calcareous spicules, which are C-shaped in the stalk region.
The podia (tentacles) of the aboral side may
lack terminal suckers (Fig. 16-26) and may
have sensory functions.

The gills are thin-walled outgrowths that
are attached to the edge of the peristome (Fig.
16-25). They are restricted to the sea urchins,
in which they are found in all members except
one group (Cidaroidea). The gills open into the
coelomic cavity around Aristotle's lantern.

Sea urchins use both the podia and spines in
their locomotion. They can move in any direction with any ambulacral area forward. When
placed on their aboral surface, they right themselves by attaching suitable podia to the substratum and contracting them. Righting may be
assisted by using their long spines. Some
can burrow by means of the Aristotle's lantern and by manipulating the spines around
the circumference.

Subclass Irregularia

The irregular echinoids include the heart
urchins and sand dollars. The heart urchins
are commonly known as spatangoids (Figs.
16-2, *A,* and 16-29), and the sand dollars as
clypeastroids (Figs. 16-30 and 16-31). They
differ in many important respects from the
regular sea urchins.

In the heart urchins the periproct and anus
lie outside the apical system of plates, being
displaced posteriorly along one interambulacrum or interradius, although the apical plates
remain at the aboral pole. The mouth and peristome, along with the center of the oral surface,
have moved anteriorly. The body is somewhat
flattened and oval or rounded in shape. The
circumference of the test (the ambitus), viewed
orally or aborally, tends to be circular or some-

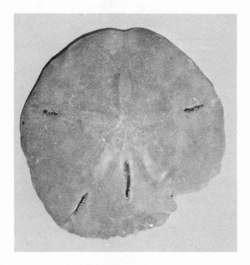

FIG. 16-30

Aboral view of test of sand dollar *Mellita quinquiesper-forata,* a clypeastroid common to our East coast. Five slits or lunules (one broken) are basis for species name. Common West Indian sand dollar *M. sexiesperforata* has 6 lunules. Lunules may be useful in burrowing (Irregularia).

FIG. 16-31

Cake urchins, past and present (subclass Irregularia). **A,** Fossil clypeastroid. Such fossil types date from Upper Cretaceous period to Recent times. **B,** Test of modern clypeastroid, the common West Indian sea biscuit.

what angled. The oral surface is flat and the aboral surface hemispheric or convex.

The aboral ambulacral area is petal shaped, i.e., like petals radiating from a center, and is called petaloid. The oral ambulacral area is leaflike in arrangement and is known as a phyllode.

In the heart urchins the podia are not adapted for locomotion, which is done by the spines. Some of the aboral podia (petaloid region) are leaflike and branched, have no skeletal support, and are probably respiratory in function. Other types are found in this group, varying from slender tapering forms to those that have stellate disks. The oral ambulacral regions (phyllodes) of heart urchins bear special penicillate podia in place of the buccal podia of sea urchins. These podia are covered with erect club-shaped projections that are supported by an internal skeletal rod. They are chemoreceptive in function and also assist in food getting.

Spines in heart urchins are curved and held more or less parallel to the body. Spines vary in shape and size with the different regions of the body. Very minute spines called clavules, which are ciliated and flattened at the ends, are found in special tracts (fascioles). They function in maintaining water currents and in removing sand grains from the test. Other

A   B

types of spines are specialized for burrowing (Fig. 16-29).

The sphaeridia in heart urchins are usually placed in shallow depressions on the plates of the phyllodes and along certain posterior ambulacra. Pedicellariae are usually of two types with only 2 jaws (Fig. 16-28, C) and scattered among the spines. No poisonous types of pedicellariae are found in this group.

The second group of irregular echinoids, called the clypeastroids, is made up of the sand dollars and cake urchins (Figs. 16-30 and 16-31). They usually have a much-flattened body in the oral-aboral axis and a circular or oval ambitus. The aboral surface is frequently arched dorsally, which produces a cross-section profile of a convex aboral and a flattened oral side. The ambulacra are petaloid on the aboral side. Radiating grooves from the peristome to the ambitus occur on the oral surface and are used for transporting food to the mouth. Phyllodes, which are characteristic of the spatangoids, are absent in the clypeastroids. The mouth is midorally located, and the periproct and anus are usually also on the oral side in the posterior interambulacrum. The aboral and oral surfaces (except the ambulacral grooves containing tube feet) are densely covered with tiny spines, which give the animal a hairy appearance. Many sand dollars bear elongated slit-like holes (lunules) that perforate the test in a symmetric arrangement around the marginal area. They arise as marginal indentations and later are enclosed by the growth of the test. There are 2 or more of them wherever they occur. The keyhole sand dollar (Mellita) (Fig. 16-30) is a familiar example of a sand dollar in which there are 5 or 6 lunules. No real function has been assigned to the lunule, although some authorities think it may aid in burrowing. Several sand dollars also have indentations along the posterior margin (Rotula). Some sand dollars do not exceed 10 mm. in diameter, whereas others reach a diameter of 1 foot (Sperosoma).

In sand dollars there are chiefly two kinds of podia: (1) the small suckered podia that are found all over the test in both ambulacral and interambulacral regions and help the spines in locomotion and (2) the large simple or lobulated branchial podia found in the petaloids and used for respiration.

The irregular echinoids are especially well adapted for burrowing. They move only with their anterior end forward. Movements are restricted to the use of spines because their podia are specialized for a variety of other functions (sensory, respiratory, food getting, etc.) The spatangoids burrow by erecting themselves on their oral spines and tossing the sand sidewise by means of curved lateral spines. Most of the time the heart urchins remain buried in the sand, with part of the aboral surface exposed to the outside. Water currents through the burrow are kept moving by the large spines on the aboral surface. Sand dollars, which lie buried just under the surface of the sand, in some instances pile up sand into mounds with their podia and then push themselves into the mound with their spines. Other methods of burrowing are also employed.

Internal anatomy

The body wall of echinoids has about the same organization as that of asteroids (Fig. 16-32). It consists of an external epidermis of one-layered epithelium covered with cilia, a dermis of connective tissue that includes the endoskeleton, and a flagellated coelomic lining. A cuticle may be present. Gland cells are found in the epidermis of some echinoids (Diadema). Pigment in the form of chromatophores or granules is found in the epidermis and dermis. Melanin is especially common in echinoids. A body wall musculature is absent. The external appendages have about the same histologic structure as the body wall, except that cilia tend to disappear from the spines with age and from the suckered terminals of podia. A layer of the ectoneural nervous system is present at the base of the epidermis. Poisonous pedicellariae occur in sand dollars.

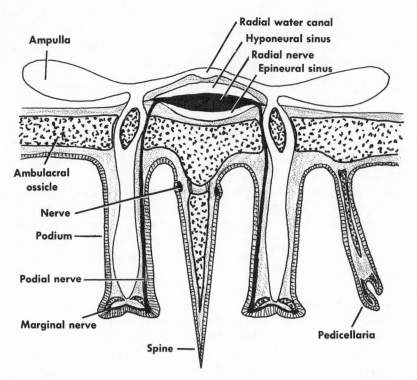

Ampulla

Radial water canal
Hyponeural sinus
Radial nerve
Epineural sinus

Ambulacral
ossicle

Nerve

Podium

Podial nerve

Marginal nerve

Spine

Pedicellaria

FIG. 16-32

Cross section of radius of echinoid (diagrammatic).

The endoskeleton of echinoids is made up of closely fitting calcareous plates or ossicles so that the whole forms a solid, immovable test. Only the peristome and periproct plates are free and flexible. The skeletal plates (which can be seen to the best advantage in the cleaned test) are arranged in rows from the oral to the aboral pole (Figs. 16-25 and 16-26). These plates are organized into 5 ambulacral areas of 2 plate rows alternating with 5 interambulacral areas of 2 plate rows. This arrangement produces 20 rows of plates, 10 of which are ambulacral and 10 interambulacral. The ambulacral areas are usually not so wide as the interambulacral areas, and they bear holes through which the podia pass. In all other echinoderms the podia pass between the plates, not through them.

Both ambulacral and interambulacral plates

bear numerous tubercles on which the spines articulate in life. In the sea urchin test there is usually one large, primary tubercle on each plate. These primary tubercles occur in meridional rows, with the smaller tubercles scattered among them. In the irregular echinoids (heart urchins and sand dollars) there is no such arrangement of large tubercles, but the plates bear many small tubercles, to which the tiny spines are attached.

The apical system of plates in regular echinoids surrounds the periproct. This system consists of 5 large genital plates, 1 of which serves as the madreporite, and 5 small ocular plates (Fig. 16-26). The genital plates, each with a gonopore, are in line with the interambulacra, and the ocular in line with the ambulacra. Each of the ocular plates has a pore for the emergence of the modified papillate podium (sensory). In the irregular echinoids the periproct lies outside the apical system, which usually has only 4 plates arranged close together with the madreporite shifted posteriorly, although there are many different arrangements of the apical system.

The water-vascular system is similar to that of other echinoderms (Fig. 16-34). It consists of a madreporite (a modified genital plate), a stone canal with an axial gland, a water ring with 5 polian vesicles above Aristotle's lantern, and 5 radial canals extending from the water ring and running under the 5 ambulacral areas of the test. These radial canals ascend to the inner surface of the test, where each gives off a terminal tentacle through a pore in an apical ambulacral plate. Each radial canal gives off, alternately to each side, lateral canals (with valves) for the ampulla of the podia (Fig. 16-32). From each ampulla 2 canals pass through the ambulacral plate (by a pore pair) and unite into a single canal on the outer surface of the test to form the lumen of the podium. In young stages and in some of the podia of irregular echinoids there is only a single canal connecting the podium to the ampulla. In older stages of some regular echinoids the ambulacral plates may fuse into compound plates with more than one pore pair. The arrangement of pore pairs has taxonomic value.

The spongy axial gland (Fig. 16-34) is not enclosed in the axial sinus, but is hollow and lined with coelomic epithelium. It represents a coelomic cavity of the right axocoel. It is connected to the hemal ring, but little is known about its function except that it may be excretory.

The mouth is located in the center of the peristome on the oral surface of the test in regular echinoids. In the irregular forms (spatangoids) it is shifted anteriorly and may lie opposite the interambulacrum, which contains the anus. The mouth is provided with a lip and passes into a small buccal cavity, where the teeth of the masticatory apparatus or Aristotle's lantern project. From the buccal cavity the pharynx or esophagus runs vertically through the lantern to the intestine. In all regular echinoids and in some of the irregular members the lantern of Aristotle, a unique chewing structure, is present (Fig. 16-34).

The lantern is an intricate complex of calcareous plates and muscles that is fitted for seizing and chewing food (Fig. 16-33). It consists of 5 jaws in the form of calcareous plates (pyramids), each with a pointed tooth that projects into the mouth region. The pyramids are fanglike structures, composed of 2 half-pyramids fused by a median suture. They are arranged radially in an interambulacral position. Each pyramid is connected by muscles to the adjacent pyramid on each side.

On the inner side along the midline of each pyramid is a calcareous band, the upper end of which is curled into a dental sac. This is the tooth band, the lower end of which is composed of hard calcareous material forming the tooth, which projects into the buccal cavity. When the teeth are worn away by usage, continuous growth of the tooth band replaces them.

At the aboral end of each pyramid is a crossbar (the epiphysis). The crossbars are sometimes joined to form a ring around the edge of the lantern. In the ambulacral positions 5 radial rotulas interlock with the epiphyses, and above these rodlike pieces are 5 arched rods, each composed of 2 ossicles (the compasses) and branched at the ends. An inturned flange of the test around the peristome forms the perignathic girdle to which the protractor muscles, which push the teeth through the mouth, are attached. Other muscles (retractors) pull the lantern back and open the teeth. Altogether, some 60 muscles operate the different parts of Aristotle's lantern. The variations in the structure of the lantern afford a basis for classification. Of the irregular echinoids, the sand dollars have a modified Aristotle's lantern, and the heart urchins have none at all.

A cecum may be present at the junction of the esophagus and stomach. The intestine is fastened in place by mesenteries; first it makes an almost complete circuit around the inside of the test in a counterclockwise direction (seen aborally) and then another partial circuit in the opposite direction but aborally to the first (Fig. 16-34). The first circuit may be considered

FIG. 16-33

Aboral-lateral view of dissected Aristotle's lantern of *Arbacia*.

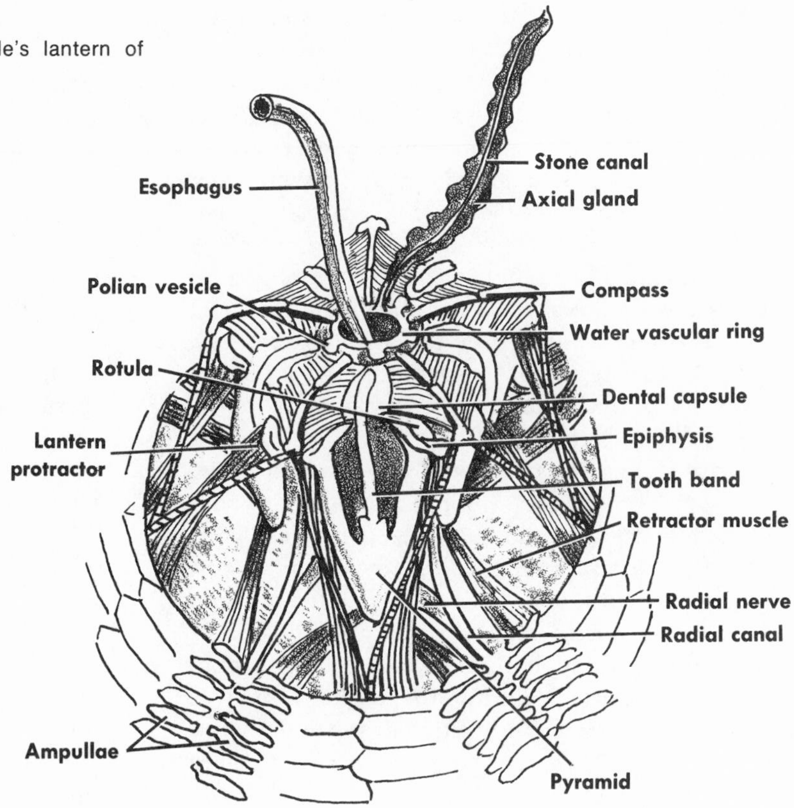

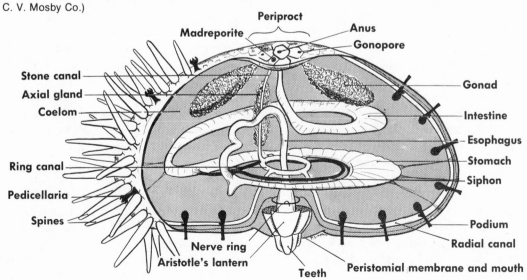

Esophagus

Stone canal

Axial gland

Polian vesicle

Compass

Water vascular ring

Rotula

Dental capsule

Epiphysis

Lantern protractor

Tooth band

Retractor muscle

Radial nerve

Radial canal

Ampullae

Pyramid

FIG. 16-34

Chief internal structures of echinoid (diagrammatic). (From Hickman: Integrated principles of zoology, The C. V. Mosby Co.)

Periproct

Madreporite

Anus

Gonopore

Stone canal

Axial gland

Coelom

Gonad

Intestine

Esophagus

Stomach

Siphon

Ring canal

Pedicellaria

Spines

Podium

Radial canal

Nerve ring

Aristotle's lantern

Teeth

Peristomial membrane and mouth

the small intestine and the second the large intestine. The large intestine joins the rectum, which empties through the anus located within the periproct.

In many echinoids a slender tube called the siphon (Fig. 16-34) runs along the intestine from the beginning of the small intestine to near the large intestine. This siphon enables the water removed from the food to bypass that part of the intestine where food is concentrated for digestion.

Sea urchins, or regular echinoids, will eat almost any kind of organic material. Some are mainly carnivorous and others herbivorous. Most are scavengers when the necessity arises. The irregular echinoids (heart urchins and sand dollars) feed mostly on organic particles in the sand where they lie buried.

In the sand dollar, sand and the organic material it contains are passed over the aboral surface by club-shaped spines and deposited in the sand, either through the lunules or at the rear end of the test (I. Goodbody, 1960). The finer food particles drop down between the spines and are carried to the oral surface. On the oral side special ciliated tracts convey the food to the ciliated ambulacral grooves and thence to the mouth. The modified Aristotle's lantern in sand dollars cannot be extended through the mouth. Their intestine makes but a single loop around the test, and the anus is located at the posterior ventral end.

The spatangoid, or heart urchin, makes use of its pencillate podia on the phyllodes in its feeding behavior. While it is buried in the sand, these extensible podia are thrust out through the exposed aboral surface to explore the sand surface and pick up food particles with their adhesive ends. Then the podia contract and pass the food to the spines, which send it to the mouth. In the heart urchin the intestine makes two circuits, but the anus lies at the posterior ventral end.

Histologically the digestive system of echinoids is lined with ciliated, tall epithelial cells. Muscle and connective tissue are found in the wall, which is covered on the outside with flagellated epithelium. Glands or glandular crypts are found in the lining epithelium of the first two fifths of the stomach (N. D. Holland and J. A. Lauritus, 1968) but are absent from the intestine. Digestion and absorption probably occur in the intestine.

The enzymes of sea urchins are similar to those of the mammal and have about the same chemical structure. Proteases are synthesized by the pancreas but have an extra fragment that prevents their activity until they are discharged in the intestine.

The coelomic cavity is spacious in regular echinoids, less so in the irregular echinoids. In addition to the major coelom, there are some minor coelomic cavities, such as the peripharyngeal cavity that encloses Aristotle's lantern. This cavity in some echinoids has 5 elongated sacs (Stewart's organs) that are expansion chambers for the fluid in the peripharyngeal cavity. The coelom is filled with a fluid that has about the same composition as sea water, plus a large number of coelomocytes of many types. Some of them are phagocytic with large pseudopodia; others are nonphagocytic with short pseudopodia. The number of coelomocytes per unit volume of fluid varies greatly. Some have as many as 4,000 to 7,000 per cubic millimeter of fluid.

The hemal system consists of a hemal ring around the esophagus on the aboral side of Aristotle's lantern, radial sinuses beneath the radial canals of the water-vascular system (Fig. 16-32), the ventral marginal and the dorsal marginal sinuses on the ventral and dorsal sides of the intestine, respectively, and small vessels that supply the organs or tissues directly. The sinuses do not have a definite lining, as they are made up of connective tissue and a covering of peritoneum. Little is known about circulation in echinoids.

Some echinoids have five pairs of gills attached to the peristome (Fig. 16-25). The gills are outpockets of the body wall and communicate with the peripharyngeal coelomic cavity.

On the aboral part of the lantern a special muscle (the elevator of the compasses) is attached to the compasses already described; when this muscle contracts, the volume of the peripharyngeal cavity is increased, and fluid is drawn out of the gills. Other muscles, the depressors of the compasses, pull the compasses down and force fluid into the gills. This pumping occurs only when the echinoid needs oxygen. Respiratory movements are controlled by the nerve ring, which is stimulated by an increase in carbon dioxide content of the fluid. Peristomial gills are lacking in the irregular echinoids, which use the modified podia of the petaloids for breathing.

The chief excretory products appear to be ammonia and urea. Coelomocytes ingest waste substances and carry them to the gills or body wall for elimination or deposit. The axial gland may also serve as an excretory organ, for coelomocytes laden with granules are numerous in it.

The nervous system has about the same pattern as does the water-vascular system. The ectoneural is the chief nervous system. It consists of the circumoral ring that surrounds the pharynx within the lantern (Fig. 16-34); the 5 radial nerves that arise from the nerve ring, pass between the pyramids of the lantern, and follow the underside of the test in the midline of the ambulacra, just beneath the radial canals of the water-vascular system (Fig. 16-32); and nerves given off by the radial nerves that penetrate the test to the podia, body wall (where an extensive subepidermal plexus is formed), spines, and pedicellariae. Each of the podial nerves ends in a nerve network in the terminal sucker (Fig. 16-32), and the subepidermal plexus supplies the spines and pedicellariae with a network of nerves. A hyponeural nervous system of 5 plaques of nervous tissue on the aboral surface of the nerve ring sends nerves to the muscles of the lantern. Some regular echinoids may also have an entoneural nervous system in the form of a ring in the inner surface around the periproct, which gives off nerves to the gonads.

Sense organs are represented in the echinoids by tactile and taste organs in the spines, podia, and pedicellariae. The sphaeridia have functions of equilibrium. Some echinoids have photosensitive eyespots in the aboral ectoderm, most of which are negatively phototropic.

The echinoids are dioecious, although hermaphrodites may occur as anomalies. Except in the case of brooding females, there is no sexual dimorphism. The regular echinoids have 5 gonads suspended by mesenteries along the interambulacra on the inner side of the test (Fig. 16-34). In the irregular echinoids the gonads may be reduced to 3 or 4 because the gonads of certain interambulacra are destroyed by the migration of the periproct. The gonads are quite large when ripe and may extend from the aboral center to the lantern. Each has a short gonoduct that opens through a gonopore in a plate of the apical system of the aboral end. Histologically, each hollow gonad is covered with coelomic peritoneum and lined with germinal epithelium, with muscle and connective tissue between.

Both sperm and eggs are shed into the sea, and fertilization is external. Some echinoids

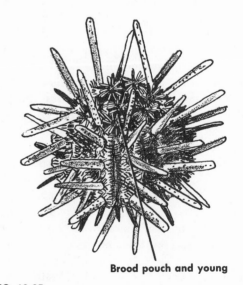

Brood pouch and young

FIG. 16-35

Cidaroid sea urchin with brood pouch full of young.

of both subclasses brood their young, especially in the Antarctic (Fig. 16-35). The brood pouch may be a depression on the petaloids, around the periproct, or enclosed in spines on the peristome.

DEVELOPMENT

The eggs, mostly small except in brooding species, have a holoblastic, equal type of cleavage. After the 8-cell stage the 4 cells at the vegetal pole divide unequally into 4 small micromeres and 4 large macromeres. Further division of the micromeres and macromeres results in a coeloblastula of about 1,000 cells that are mostly equal in size, except for the micromeres at the vegetal pole. The blastula is flagellated and rotates within the fertilization membrane before the embryo is freed. The stage of the swimming blastula is reached about 12 hours after fertilization.

At the flattened vegetal pole the micromeres proliferate, sending mesenchymal cells into the interior. The invagination is embolic, and the resulting archenteron proliferates and produces secondary mesenchyme at its tip. At the inner end of the archenteron a double coelomic sac is formed and develops into the coelom. The conical gastrula develops into a characteristic echinopluteus larva with 4 to 6 long arms and an easel-shaped form (Fig. 16-3). There are many varieties of the echinopluteus among the different species of echinoids. The echinopluteus is small and planktonic. Several months may elapse before the larva attains full development. The metamorphosis of the echinopluteus into a young urchin is rapid, requiring only an hour or so.

After metamorphosis, development continues. At metamorphosis only five pairs of podia are present, but additional pairs arise at the oral side, forcing the first ones aborally. At first the apical plates cover the entire aboral surface, but as the ambulacral and interambulacral plates form, the apical plates cover only a small surface. In echinoids that brood their young, a free larval stage is omitted, and metamorphosis may occur in 2 or 3 days.

■ CLASS HOLOTHUROIDEA

Holothurians are commonly known as sea cucumbers because of their resemblance to the vegetable. In some ways they are the most aberrant group of echinoderms. They are elongated along an oral-aboral axis. Their shape forces them to lie on their side instead of on their oral surface, which leads to a differentiation of dorsal and ventral surfaces and a secondary bilateral symmetry. Their skeleton is reduced to microscopic ossicles embedded in the body wall. They have no arms and no pedicellariae. Ambulacral grooves are lacking, but ambulacral and interambulacral areas run from the anterior to the posterior end. Tube feet may or may not be present. The buccal podia are modified into a ring of 5 or more tentacles around the mouth.

The holothurians have worldwide distribution in all seas and at all depths. They have benthonic habits and are common in intertidal zones, where they are found on mucky bottoms, usually concealed from sight. Certain areas have much larger faunas than others, such as the Indo–West Pacific area, Australian coasts, Hawaiian Islands, coasts of the Mediterranean Sea, and many others. Along our Pacific coast the various species of *Stichopus* (Fig. 16-36, *B*), *Parastichopus*, and *Cucumaria* are among the most familiar. Along the Atlantic coast, common genera are *Cucumaria* and *Thyone* (Fig. 16-36, *A*).

Some 1,100 species and 170 genera of holothurians are recognized at present. They vary in size from 3 cm. body length to 1.5 meters, with a diameter of several inches. Most common North American species are about 6 to 10 inches long. The largest forms are tropic. They display a wide range of color. Many species are dull brown, black, or greenish. Some are red, yellow, violet, and other shades. Striped patterns are common.

Because of the nature of their rather diffuse skeleton, holothurians have not left an impressive fossil record. Impressions of their bodies have been found in the Cambrian Burgess Shale of British Columbia, and their skeletal spicules

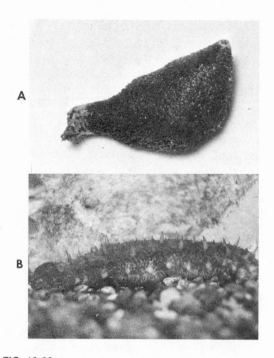

FIG. 16-36

Holothuroidea. **A,** *Thyone briareus,* a common Atlantic coast sea cucumber. **B,** *Stichopus californicus,* the California cucumber.

and ossicles have been found in limestones and other rocks dating from the Mississippian period.

GENERAL MORPHOLOGIC FEATURES
External anatomy

The body varies in shape from globose to vermiform and cylindric. Since the side on which holothurians lie is usually flattened, it is called a sole and bears the locomotor podia. This gives the animal a secondary bilateral symmetry. On the dorsal surface the podia are usually represented by warts and papillae. The creeping sole consists of 3 ambulacral areas (the trivium) and 2 interambulacral areas; the dorsal surface has 2 ambulacral areas (the bivium) and 3 interambulacral areas. The podia of the sole may be arranged in 3 longitudinal rows, corresponding to the 3 ambulacral areas (*Stichopus*), or they may be scattered all over

the sole (*Holothuria*). In the common *Thyone* and some others, podia are present on both sole and dorsal regions of the body although the suckers are best developed on the ventral podia. There is a tendency in many holothurians for the number of podia on the sole to be reduced and their radial distribution lessened. Podia are entirely absent in Apoda.

Sea cucumbers are able to move about but mostly at a slow rate. Those with differentiated creeping soles can perhaps move best of all, using their podia very much as asteroids do. Some use muscular waves of contraction, much like those of a caterpillar. They can right themselves when placed on their dorsal sides by twisting the podia of the oral side around and attaching them to a substratum. Apodous holothurians can move by using their tentacles as holdfasts. Burrowing, which is common among the order Apoda, is accomplished by pushing into the sand with alternate contractions of their circular and longitudinal muscles, using their tentacles to shove the sand aside. Animals without tentacles cannot burrow. *Thyone* buries the middle part of its body, leaving both ends above the surface. Swimming is accomplished in young *Leptosynapta* by curving into a U shape and thrashing the ends of the body together in a type of scissors kick. The deep-sea apodous form *Pelagothuria* has a sail or web formed by the papillae or modified podia for locomotion.

The body surface in most sea cucumbers is leathery and somewhat slimy, with microscopic endoskeletal ossicles embedded in the dermis (Fig. 16-37). In the family Psolidae the skin plates are large and overlap (imbricated), forming a type of armor. The microscopic ossicles assume many shapes and sizes and are useful in taxonomic studies (Fig. 16-38). The ossicles are supposed to represent the persistent embryonic state of the skeleton (neoteny). Ossicles are present in the tentacles, the papillate podia, the adhesive disks of the locomotor podia, and even in the internal structures.

The mouth may be terminal or displaced

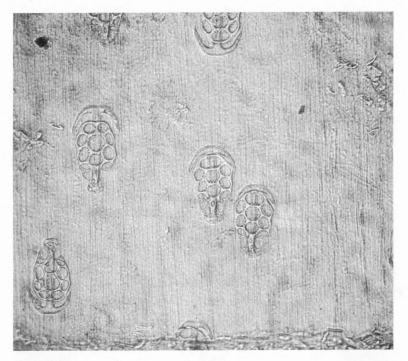

FIG. 16-37

Surface view of whole mount of skin of sea cucumber *Synapta* showing microscopic endoskeletal ossicles scattered through dermis, Compare with Fig. 16-38, **E.** (From Andrew: Textbook of comparative histology, The Oxford University Press, Inc.)

FIG. 16-38

Some types of ossicles in sea cucumbers. **A,** Radial piece *(Holothuria).* **B,** Anchor type *(Molpadia).* **C,** Table type *(Holothuria).* **D,** Biscuit type *(Holothuria).* **E,** Section through body wall of *Synapta* with anchors in place. Compare **E** with Fig. 16-37. (**B** after Danielsson and Koren; **E** after Woodland; both modified from Hyman.)

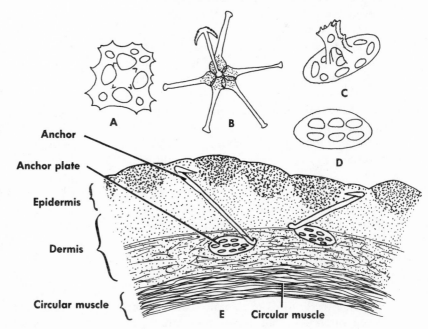

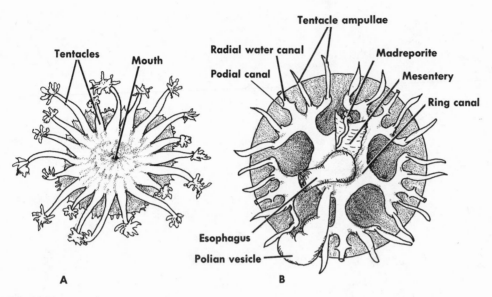

FIG. 16-39

A, Anterior or oral view of *Holothuria*. **B,** Water-vascular system (diagrammatic) of *Holothuria*. (Modified from Bather.)

dorsally or ventrally. It is circular in outline and surrounded by a buccal membrane (Fig. 16-39, *A*). Around it in a single circlet are 10 to 30 modified buccal podia, known as tentacles. These podia contain extensions of the water-vascular system from the radial canals. They may or may not be of the same size. Both mouth and tentacles can be pulled into the body by retractor muscles, and the body wall made to close over them. In the order Dendrochirota the tentacles are much branched (dendritic) as in *Thyone* and *Cucumaria*. Sometimes they are attached to a short central stalk, which has horizontal branches (order Aspidochirota), or they may be fingerlike (order Molpadonia). The anus is terminal in the orders Molpadonia and Apoda but is displaced in the other orders to the dorsal or ventral side. The anus may be encircled by small papillae or calcareous plates.

Internal anatomy

The body wall varies in thickness among the different orders. Its surface has no cilia, and a thin cuticle covers the underlying epidermis. The epidermis consists of a single layer of tall epithelial cells that tend to merge with the dermis underneath. The epidermis also has sensory and gland cells, which are common in the Apoda.

The dermis makes up the greater part of the body wall and consists of connective tissue that encloses the calcareous ossicles (Fig. 16-38, *E*). The pigment, which gives color to the animal, may be in the form of chromatophores in the dermis or as free granules in both epidermis and dermis. A nerve plexus is contained in the dermis, and a subepidermal plexus may occur under the epidermis. Under the dermis is a layer of circular muscles that may form a complete cylinder, or the layer may be interrupted by five single or double bands of longitudinal muscles that run along the ambulacral areas. The muscle fibers are of the smooth type. The inner surface of the body wall is lined with ciliated epithelium.

The water-vascular system is similar to that of other echinoderms in its basic plan (Fig. 16-39, *B*). It consists of a water ring around the proximal pharynx, 5 radial canals running the length of the body wall between the longitudinal and transverse muscles, small lateral canals to

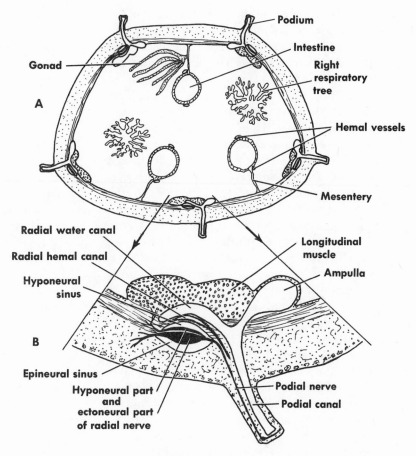

Gonad

Podium

Intestine

Right respiratory tree

Hemal vessels

A

Mesentery

Radial water canal

Radial hemal canal

Hyponeural sinus

Longitudinal muscle

Ampulla

B

Epineural sinus

Hyponeural part and ectoneural part of radial nerve

Podial nerve

Podial canal

FIG. 16-40

A, Section of holothurian (schematic). **B,** Enlarged section of 1 of radii shown in **A**. (Modified from Lang.)

the podia, and a stone canal to an internal madreporite.

The water ring around the base of the pharynx bears one or more elongated or rounded polian vesicles that hang down in the coelomic cavity and open by a narrow neck into the ring canal (Fig. 16-39, B). As many as 12 or more polian vesicles are found in the two common orders Dendrochirota and Aspidochirota, although some orders have the original condition of only 1. The vesicle's wall is similar to that of the water ring. Polian vesicles may serve as expansion chambers for the maintenance of pressure in the water-vascular system. The stone canal, with calcified wall, arises from the water ring or ring canal and passes to the madreporite. There may be more than 1 stone canal and more than 1 madreporite. The madre-

porites are internal (except in the order Elasipoda); and coelomic fluid, rather than sea water, enters the water-vascular system through them.

Of the 5 radial canals, 3 lie in the trivium, or ventral side, and 2 in the bivium, or dorsal side (Fig. 16-40, A). After leaving the water ring, the radial canals pass upward through the outer wall of the aquapharyngeal bulb to the inner side of the calcareous ring, where the canals give off branches to the tentacles. Then proceeding through notches or holes in the radial plates of the calcareous ring, the radial canals run posteriorly along the inner surface of the ambulacral areas. Here the radial canals supply lateral or podial canals (with valves) to the podia and their ampullae (Fig. 16-40, *B*). Locomotor podia are projections of the body wall found chiefly on the sole. They may or may not have suckers. Their corresponding ampullae are sacs projecting into the coelom. The papillate podia on the dorsal surface are also provided with lateral canals from the radial canals, but these have nothing to do with locomotion. The radial canals terminate in the last podia or end podia of the ambulacral areas. The water-vascular system is poorly developed in the orders Molpadonia and Apoda.

The digestive system of holothurians consists typically of mouth, pharynx, esophagus, stomach, intestine, cloaca, and anus (Fig. 16-41). One or more of these divisions may be absent. The mouth, already described, leads into the pharynx, which is supported by mesenteric strands in a coelomic cavity, called the aquapharyngeal bulb. The beginning of the pharynx is encircled by a calcareous ring usually made up of 5 radial and 5 interradial plates bound together by connective tissue. This ring serves as an insertion point for the longitudinal muscle bands and retractor muscles of the tentacles and as a support for the pharynx and water ring. The pharynx may lead into a slender esophagus, which opens into a short, muscular stomach, or, as in *Cucumaria*, it may pass directly to the intestine. The holothurian intestine may be three or four times the length of the body and is looped in a definite arrangement in the coelom. The last part of the intestine is sometimes called the large intestine. The terminal parts of the large intestine are called the cloaca in those forms that have respiratory trees. The cloaca is attached to the surrounding wall by radiating strands of connective tissue and muscle. The digestive system is supported by a mesentery that is continuous or else one that is divided into three portions.

The wall of the digestive tract is made up of a ciliated or nonciliated epithelial lining, an inner layer of connective tissue, a circular and longitudinal muscle layer, an outer thin layer of connective tissue, and a ciliated peritoneal covering. Gland cells may occur in the pharynx and stomach. Coelomocytes assist in digestion by carrying intestinal enzymes into the lumen and by distributing products of digestion.

Many holothurians, especially the dendrochirotes, are plankton feeders, making use of tentacles with mucous secretions for catching small organisms. The tentacles deliver their food catch by being thrust into the mouth and pharynx, where the food is wiped off. Other holothurians feed on deposits that are passed into their mouths by their tentacles or are swallowed as the animals burrow in the deposits of the substratum. The organic part of the ingested material is digested, and the indigestible part is voided through the anus. A great deal of substrate can pass through an animal in a short period of time.

Some members of the order Aspidochirota, such as *Holothuria* and *Actinopyga,* have small white or reddish tubules, called the organs (tubules) of Cuvier, attached usually at the base of the respiratory tree. When irritated, the animal shoots out of its anus these tubules (threads) by contractions of the body. These threads wrap around an intruder and render it helpless. When the sea cucumber breaks away, the threads are left behind and new ones are regenerated. These threads may also be used to trap small organisms, as well as to repel predators.

The hemal system is well developed in holothurians (Fig. 16-41). It is made up of a hemal ring around the pharynx, radial hemal sinuses that parallel the water ring and radial canals of the water-vascular system, and a dorsal and ventral sinus along the small intestine and part of the large intestine. The intestinal sinuses give off a large number of small channels to the intestine. The hemal system is best developed in its relation to the intestine. From the ventral sinus, running along the descending intestine, a transverse connection crosses over to the ventral sinus running along the ascending intestine. The dorsal sinus is connected to the

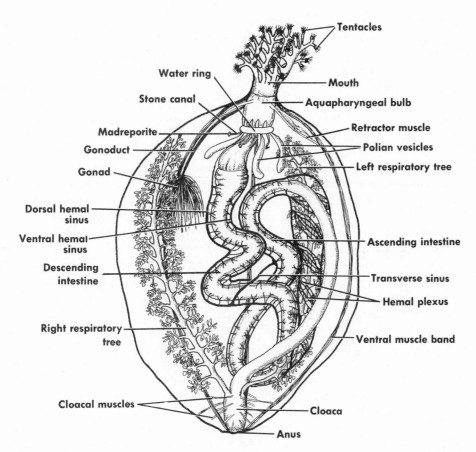

Tentacles

Water ring

Stone canal

Mouth

Aquapharyngeal bulb

Madreporite

Retractor muscle

Gonoduct

Polian vesicles

Gonad

Left respiratory tree

Dorsal hemal
sinus

Ventral hemal
sinus

Ascending intestine

Descending
intestine

Transverse sinus

Hemal plexus

Right respiratory
tree

Ventral muscle band

Cloacal muscles

Cloaca

Anus

FIG. 16-41

Internal structure of common sea cucumber *Thyone*
(schematic).

ascending small intestine by a network of
lacunae called the rete mirabile. A collecting
vessel joins this network with the ascending
small intestine by numerous branches. Part
of the rete network runs to the terminal portion
of the left respiratory tree. There are variations in the distribution of the hemal system in
the various species.

The hemal vessels do not have definite linings, but their walls contain coelomic epithelium, connective tissue, and muscle. The hemal
fluid is about the same as that of coelomic
cavities and contains coelomocytes of many
types that are formed in the walls of the hemal

channels (C. L. Prosser and C. Judson), not in
the polian vesicles as formerly supposed.

Respiration occurs in the skin, tube feet, and
the respiratory trees. The respiratory trees (Fig.
16-41) represent a unique breathing device
of tubules and are found in all orders except
Apoda and Elasipoda. They are evaginations of
the digestive tract at the anterior part of the
cloaca near the entrance of the large intestine.
Each respiratory tree consists of a main stem
from which branches with terminal vesicles
spring. There are 2, left and right, and they
occupy the space between the intestinal loops
in the coelomic cavity. The 2 trees may enter
the cloaca by a common stem or separately.
Water is drawn into and out of the trees by
rhythmic contractions of the cloaca and tree
tubules. When the cloaca dilates, water flows

in and fills it. The anal sphincters then close the anus, and the contraction of the cloacal wall forces water into the respiratory trees. Water is forced out by relaxation of the cloaca, contraction of the tubules, and opening of the anus. Several pumping cycles are required to fill the trees, but only one vigorous expulsion is necessary to expel all the water.

Experimental evidence indicates that oxygen in the inspired water diffuses through the terminal vesicles into the coelomic fluid and thence into the hemal system. Oxygen requirements of sea cucumbers are very low.

Excretion takes place by means of the coelomocytes and the respiratory trees. Most waste is in the form of ammonia compounds, although information about excretion is scanty. Urea and purine nitrogen have been identified in some. Nitrogenous waste in crystalline form is picked up and conveyed by coelomocytes to the respiratory tree, gonadal tubules, or to any hollow organ with an external opening.

Although the apodans lack respiratory trees, they possess ciliated funnels called urns that are found on various parts of the body, especially on the mesenteries. The urns are lined with ciliated columnar epithelium and are covered with coelomic epithelium. They may be clustered together to form trees of urns. Their cavities contain coelomocytes, which have accumulated waste in them. Coelomocytes later carry this waste to the body wall, or they let it fall into the coelom to form masses called brown bodies.

The nervous system of holothurians follows the pattern of the water-vascular system, and is found between the canals and the body wall. The system consists of a nerve ring, radial nerves, and their numerous branches. The circumoral nerve ring is a flattened bandlike structure in the buccal membrane close to the base of the tentacles and usually within the calcareous ring. It supplies a nerve to each of the tentacles and also nerves to the buccal membrane and pharynx. Five radial nerves pass from the nerve ring, one to each of the ambulacral areas (Fig. 16-40, *B*). On leaving the ring, each radial nerve passes outward through a notch or hole in the plate of the calcareous ring. The radial nerves are located in the innermost dermal layer of the body as they pass through the ambulacral areas external to the radial water vessel. On the outer side of each radial nerve is the epineural sinus (noncoelomate), and on the inner side is the hyponeural sinus (coelomate) (Fig. 16-40, *B*). The radial nerves are ganglionated and divided into a thicker ectoneural part and a thinner hyponeural part. The ectoneural division furnishes ganglionated nerves to the podia and to a body wall plexus that supplies the sense organs. The hyponeural division (motor) innervates the body-wall muscles. The hyponeural division is found only on the middle course of the radial nerve, not at the ends. No aboral nervous system is found in holothurians. The nerve ring apparently does not have a dominant role in nervous integration.

Sense organs are restricted mainly to sensory cells in the epidermis. The whole body surface is sensitive to light, and in some synaptids (order Apoda) there are eyespots at the base of the tentacles. Like the eyespots of sea stars, these can detect shadow movements but cannot form images. The surface of synaptids also has warty elevations containing sensory buds surrounded by glands. These sense organs send nerves to ganglia of the dermal nerve plexus.

Statocysts have been demonstrated in a number of holothurians. These may be located on the radial nerves near their exit from the calcareous ring and are hollow spheres with lithocytes. Each statocyst sends a nerve to the adjacent radial nerve. Most statocysts are found around the nerve ring, but some are distributed along the ventrolateral radial nerves. All holothurians are negatively phototropic, and burrowing species are positively geotropic.

The power of autotomy is one of the most striking characteristics of the holothurians. This phenomenon is best shown in a process called eviceration, in which certain visceral organs are completely eliminated through a rupture in the cloacal region or in the body wall. Mention has already been made of the tubules of Cuvier by which these sticky threads are cast out in defense against predators or other invaders. Evisceration is a more drastic process. Under conditions of crowding, high temperatures, excessive irritation, foul water, and injuries to the body, the sea cucumber ruptures itself by a forceful contraction of the body (the coelom of which is filled with fluid) and, depending on the species, emits one or both respiratory trees, the digestive tract (except the ends), and the gonads. Sometimes (e.g., *Thyone*) the anterior end is ruptured and the anterior part of the viscera is shed. Such eviscerated individuals usually recover under normal conditions and undergo regeneration of the missing parts. The course of regeneration varies, but in many species studied the cloaca seems to be the center of regeneration. If the digestive tract has been shed in evisceration, the stubs of the esophagus and cloaca send outgrowths along the remnants of the mesentery, which meet to form a new tract in some 25 days. During the period of regeneration the animal does not feed. In *Stichopus*, regeneration of a new digestive tract is made from the cells along the torn edge of the mesentery. Although evisceration has been observed mostly in sea cucumbers under laboratory conditions, there is some evidence that it may occur in nature when the temperature is unusually high and the oxygen content of the water is low.

Usually each half of a transversely bisected animal forms a complete animal; in smaller sections only the section with a cloaca does so.

Most holothurians are dioecious, and sexual dimorphism is not evident except in those females that brood their young. Hermaphroditic forms are sometimes found, especially in the order Apoda; but a hermaphroditic individual is not male and female at the same time; it is usually protandrous.

The reproductive system comprises a gonad and its duct (Fig. 16-41). The gonad is located in the anterior part of the coelom beneath a mid-dorsal interambulacrum. It consists usually of a tuft or cluster of simple or branched tubules united basally. In some species such as *Stichopus* there may be 2 tufts, 1 on each side of the mesentery. Some authorities consider each tuft to be a separate gonad. From the common hollow base into which the tubules enter, a gonoduct passes in the mesentery to the gonopore, which may be mounted on a genital papilla. The gonopore is found either within or near the tentacles. The gonoduct sometimes divides so that more than 1 gonopore is present. The gonadal tubules are lined with germinal epithelium and covered with peritoneum. The gonoduct is lined with ciliated epithelium and covered with connective tissue.

Most species spawn during spring and summer. Fertilization is external in sea water. Dendrochirote and apodous species brood their young, especially in cold temperatures. Such brooding forms produce large yolky eggs. Brood pouches are commonly found on the sole or dorsal body areas, where there may be depressions. The invaginated brood pouches may actually be internal in position. Spawning females catch the eggs released from the gonopore with their tentacles and transfer them, sometimes with the aid of special podia, to the brood depressions. In a few cases (e.g., *Synaptula*), eggs develop in the coelom, where they are discharged from the gonads and fertilized in some unknown manner. The young are discharged from the female through an anal rupture.

Asexual reproduction occurs in a few species such as *Cucumaria planci* by transverse fission. Asexual reproduction seems to be restricted to young individuals.

DEVELOPMENT

Holothurians follow the deuterostome pattern of equal, holoblastic, and radial cleavage. Except in species that brood their young, in which development is direct and the young hatch as juveniles, development occurs in the sea, where the embryo becomes a planktonic member.

The usual pattern of a coeloblastula with embolic gastrulation is followed. The gastrula becomes flagellated, escapes from the fertilization membrane, and develops into a free-swimming member. Mesenchyme is usually given off from the tip of the archenteron and fills the blastocoel with entomesoderm. The coelom is formed from the distal part of the archenteron, and the anlage of the digestive tract from the proximal part. The blastopore becomes the anus.

The coelomic sac separates from the archenteron and constricts, making an anterior hydrocoel and axocoel and a posterior somatocoel. The somatocoel later divides into the left and right somatocoels. Only the left hydrocoel and axocoel, which will form the definitive stone canal, are present. In the middle of the ventral surface a stomodeum forms and connects with the digestive tract.

About the third day of development the embryo becomes the free-swimming auricularia larva (Fig. 16-3). The transparent larva is provided with a flagellated band that makes a number of loops and curves similar to that of the bipinnaria of asteroids. The larva is usually about 1 mm. in length, but it may be longer. The auricularia is soon changed into a barrel-shaped larva (doliolaria), which has 5 hooplike rings that have arisen by the partial degeneration and reconstruction of the flagellated band of the auricularia. In some cases the embryo bypasses the auricularia stage altogether and becomes the doliolaria.

In the doliolaria the mouth, which is in the middle of the ventral surface of the auricularia, now has shifted to the anterior pole, and the anus has shifted to the posterior pole. In its metamorphosis to an adult the young sea cucumber has 5 tentacles and 1 or 2 functional podia and is sometimes referred to as a pentactula. Additional podia and tentacles appear later.

Subphylum Pelmatozoa
CLASS CRINOIDEA

Crinoids are stalked or unstalked and represent the most primitive of all clases of echinoderms. They have a flower-shaped body, which is placed at the tip of an attached stalk (sea lilies) or is free as an adult (feather stars). Most living crinoids belong to the stemless feather stars (Fig. 16-43). Crinoids are the sole

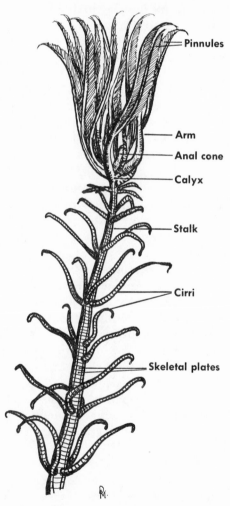

FIG. 16-42

General structure of stalked crinoid, or sea lily. (From Hickman: Integrated principles of zoology, The C. V. Mosby Co.)

surviving members of the Pelmatozoa, which flourished during the Paleozoic era (Fig. 16-2, C and D). The oldest crinoids date from the lower Cambrian, but the origin of this group must lie somewhere in the Precambrian period.

Crinoids are distinguished from other echinoderms by their shape and by the structure of their skeleton (Fig. 16-42). The body is globular or disk shaped and is enclosed in an armor of symmetrically arranged calcareous plates. It bears a number of outspread appendages called arms that are usually subdivided into forks and many branches. The pinnulated arms and stem (when present) consist of many segments formed of calcareous ossicles, so joined that they permit a considerable degree of flexibility.

Although the crinoids gave rise to no other groups of echinoderms, the members of the subphylum Pelmatozoa to which they belong have laid down the basic plan of the modern echinoderms as we know them. The evidence from all disciplines indicates that the following features can be traced back to the pelmatozoan ancestor: the distinctive echinoderm characteristics of radial symmetry superimposed on a bilateral symmetry, attachment usually at some stage of the life history, the coil of the gut, the development of the hydrocoel and the water-vascular system, the development of a free-swimming, ciliated bilateral larva, the pentamerous body of ambulacral and interambulacral plates, and the system of canals growing out of coelomic cavities.

The pelmatozoans were one of the earliest groups to exploit fully the sessile mode of existence. By having numerous arms provided with food grooves and sensory structures, they were able to explore and to gather food from all directions. A heavy armor of plates ensured them of maximum protection against the dangers to which sessile forms are exposed. The arrangement of internal organs such as nerves and blood vessels was influenced by the above characteristics. Crinoids perhaps developed many of these tendencies to the fullest

extent possible. By adaptive radiation the basic pelmatozoan pattern has been modified to form the various groups of modern echinoderms. Zoologists have been able to explain how some, but not all, of these changes could have occurred.

About 625 living species of crinoids are known, of which approximately 80 are stalked. This is a modest number compared with the more than 5,000 fossil species known to paleontologists. The heyday of the crinoids was the remote geologic past, and modern species represent only a remnant of their former greatness. The trend has been toward an unattached existence.

Crinoids live in all seas except the Baltic and Black. None is found in fresh water. Feather stars are usually more common in shallow water and occur most abundantly in shallow tropic lagoons. More stalked forms (sea lilies) are found in deep water than stemless species (feather stars). Some sea lilies, for instance, have been dredged from depths as great as 5,000 meters. Many crinoids are also found in very cold water in Arctic and Antarctic seas. Their distribution is spotty, however, for they are scarce in certain regions of both the Atlantic and Pacific oceans. The genus Antedon (Fig. 16-43) has a wide range of distribution from cold water seas to warm. Feather stars prefer rocky bottoms, but sea lilies, especially those with long cirri, are often dredged from muddy sea floors. The same species may exhibit great variations in size range under different environmental conditions, being much larger in cold than in warm waters. Crinoids are often found in dense aggregations in a favorable location because of the feeble powers of dispersal of their free-swimming larvae.

Crinoids are often infested with large numbers of other animals. Some of these may establish symbiotic relationships of commensalism and parasitism. The curious polychaete myzostomes are ectoparasites on crinoids. These parasites live by inserting their pharynx into the food channels of the crinoids and sucking up

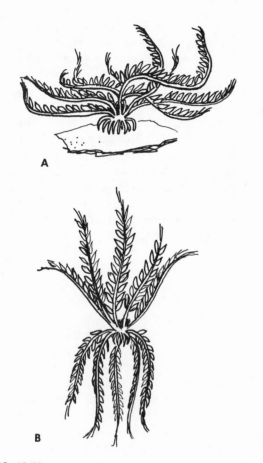

FIG. 16-43

Feather star *Antedon* (suborder Comatulida), adult free-swimming stage, about natural size. **A,** At rest. **B,** Position of arms in swimming. (Modified from Chadwick, 1907.)

the food. Carnivorous snails (*Stylina, Sabinella*) are able to bore with their proboscis through the theca and suck out the soft parts. Many crustaceans are harbored in some way by crinoids. Hydroids frequently attach themselves to a crinoid stem as a substratum. Usually sea lilies, which are chiefly found in deeper waters, are bothered less by other animals than are feather stars.

Crinoids range in size from 3 cm. across the outspread arms to more than 1 meter across. Most free-swimming crinoids are about 1 foot across the arms. In present stalked crinoids

the average stalk rarely exceeds 2 feet, but in extinct species it was much longer (50 to 70 feet).

Feather stars may be highly colored because of conjugated carotenoids. They may be green, purple, yellow, or red. *Antedon bifida* owes its red color to echinochrome. The arms are frequently banded with different colors.

The class Crinoidea includes three orders or subclasses, which are wholly extinct, and one order or subclass of living crinoids. The three extinct groups are Inadunata, Flexibilia, and Camerata. All modern crinoids belong to the order (subclass) Articulata. Fossil crinoids had the general structural features of extant species, although differences in many details of these general features do exist. The calyx of fossil species tended to constitute more of the crown, and the tegmen included more plates and was less leathery than in modern crinoids. Stems were usually much longer in the fossils, and stemmed species were relatively more common. Many fossil crinoids had an asymmetric arrangement of plates, in contrast to the regular pentamerous organization of modern forms.

The Articulata, or existing crinoids, are comprised of the stalked orders or suborders (Isocrinida, Millericrinida, Cyrtocrinida) and the unstalked suborder (Comatulida). As already stated, the suborder Comatulida includes the vast majority of living crinoids.

GENERAL MORPHOLOGIC FEATURES
External anatomy

The crinoid has a pentamerous body proper, called the crown, and an attachment stalk. The stalk is characteristic of sessile sea lilies, but is vestigial or has disappeared in feather stars. The latter group is attached when young, but breaks off from its stalk when mature and swims away. The stalk has a cylindric or polygonal contour and its internal skeletal ossicles give it a jointed appearance. It is often fringed with small, jointed cirri arranged in whorls around it (Fig. 16-42). Although the stem is lost in feather stars, some of the circlets of cirri

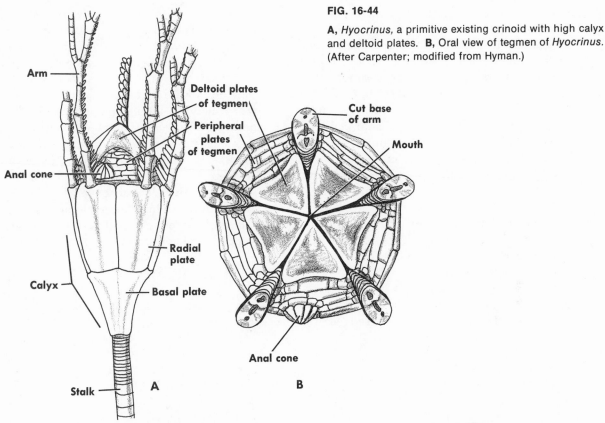

FIG. 16-44

A, *Hyocrinus,* a primitive existing crinoid with high calyx and deltoid plates. **B,** Oral view of tegmen of *Hyocrinus.* (After Carpenter; modified from Hyman.)

persist at the aboral end of the crown (Fig. 16-43). The cirri, the terminal joint of which is claw shaped, are used for grasping a substratum when needed (feather stars) and when the stalk becomes dislodged (sea lilies). The stalk bears at its terminal end rootlike extensions or disklike enlargements for attachment.

The crown consists of the skeletal calyx or aboral body wall enclosing the viscera, the membranous tegmen or oral body wall, and a set of arms (brachia) that originate at the periphery between the calyx and the tegmen (Fig. 16-44). The calyx (theca) is composed of two circlets, each of 5 plates. The plates of the lower circlet are called basals and those of the upper circlet radials. Such a calyx is said to be monocyclic. When a third circlet of infrabasal plates is present below or aboral to the basal plates, the calyx is called dicyclic. The

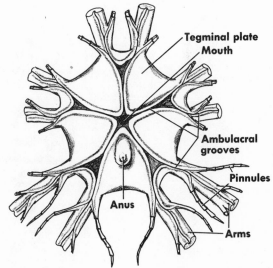

FIG. 16-45

Oral view of tegmen of *Antedon,* arms removed. (Modified from MacBride.)

radials make up the skeletal elements at the beginnings of the arms.

The mouth is usually located near the center of the tegmen (Figs. 16-44 and 16-45). From the mouth 5 ambulacral grooves run to the arms, thus dividing the tegmen into 5 ambulacral and 5 interambulacral areas. The anus, usually located at the tip of the anal cone, is present in one of the interambulacra.

Each of the 5 arms, the primitive number, is usually subdivided by 1 or more forks, forming at least 10 arms, and sometimes many more (up to 200 in some comatulids). The arms vary in length and shape. Some are long and slender; others are short and wide. They are usually shorter in cold water than in warm. The average length of an arm is 4 or 5 inches, but it may be shorter or longer. The arms may be of different lengths in the same animal. They are composed of brachial ossicles, which may be arranged in a single series (uniserial) or in a double series (biserial). They carry lateral appendages called pinnules. Pinnules have the same structure as the arms; and although their main function is to capture food and bear reproductive organs, they are differentiated into types with specialized functions (tactile and protective). The oral ambulacral grooves on the oral surface of the tegmen also extend along the arms and pinnules so that each arm and pinnule has a ciliated groove along its center (Figs. 16-45 and 16-46). A groove may be lacking in some of the specialized pinnules. These ambulacral grooves bear raised edges that are scalloped into projections (lappets), which alternate on the two margins

FIG. 16-46

Transverse section of crinoid arm. (After Bather; redrawn from Lankester: A treatise on zoology, Adam & Charles Black, Ltd.)

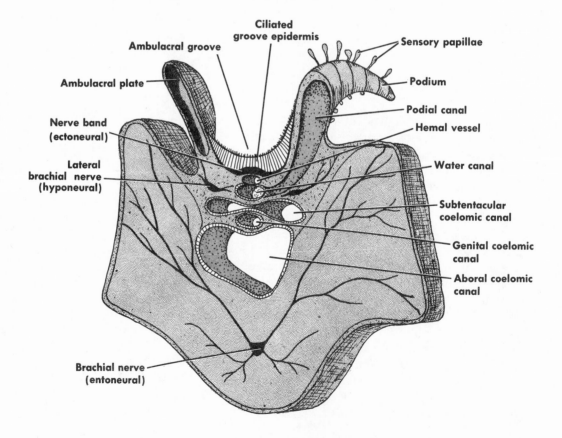

of a groove. On the inner side of the base of each lappet is a group of 3 podia. These lappets and podia are also found on the pinnules as well as on the arms. Small spheric bodies (saccules) are also located on the outer side of the lappets along the ambulacral grooves. Their function is not known, but pigment often collects in them in preserved specimens. Along the surface of the tegmen the ambulacral grooves bear reduced lappets with a single podium, not in groups of 3.

The free-swimming comatulids, when inactive, are usually found attached by claws on certain cirri to some object. When detached, they can swim by raising and lowering a set of arms alternately with others (Fig. 16-43). When the arms are raised, the pinnules of those arms are flexed against the arm axis; when lowered, the pinnules are fully extended. Comatulids can also crawl by catching hold of objects with the adhesive ends of the pinnules and pulling themselves along by contracting their arms. All have the oral surface uppermost when they swim or crawl. The well-known *Antedon* is a very active comatulid and is capable of diverse movements. The sessile sea lilies can swish their arms and pinnules and bend their stems for adjustment to the varying conditions of water currents.

Internal anatomy

The endoskeleton composes most of the body wall, and its arrangement is of great importance in taxonomy. The endoskeletal ossicles are of the fenestrated calcite type characteristic of echinoderms. They are made up mostly of calcium carbonate and silicon dioxide. The ossicles are bound together by connective tissue ligaments. The stalk is made up of a single row of superimposed disk-shaped ossicles called columnals, which may be like or unlike throughout the stem. The end surfaces of the columnals are adapted for articulation with each other and some flexibility is possible. The stem lacks muscles, but the columnals are bound together by elastic ligaments. The cirri, which are carried by the nodes of the stem, fit into articulating sockets, and consist of ossicles called cirrals held together by elastic ligaments. The cirri likewise have no muscles. The arms and pinnules also have a similar solid structure of superimposed ossicles held together by ligaments, but unlike the stem and cirri, they are provided with muscles.

The body surface may or may not have a cuticle. The poorly developed epidermis is mostly a syncytial epithelium that is not sharply delineated from the underlying mesenchyme and may be lacking altogether. The ambulacral grooves are lined with tall, ciliated cells that are supporting and sensory in function. The mesenchyme of the body wall contains a gelatinous matrix, together with stellate cells and fibers.

The muscular system consists chiefly of intersegmental muscle fibers that are restricted to the arms and pinnules. Flexor muscles enable the arms to bend inward toward the disk, and elastic ligaments enable them to be extended. Such an arrangement is also found in the pinnules. Muscle fibers appear to be a mixture of smooth and striated fibers.

The water-vascular system plan is similar to that of other echinoderms, consisting of a ring canal, radial vessels, and podia (Figs. 16-46 and 16-47). The ambulacral grooves have already been described. Underneath each groove lies a canal of the water-vascular system. The water ring canal gives off a radial water canal into each arm, and the radial water canals send branches alternately on the side to each pinnule. Also, in each arm and pinnule a group of 3 podia of the ambulacral groove receives alternately on each side a water canal that divides and sends a branch to the podium. The podia lack suckers and ampullae and are covered with cilia. They are well supplied with nerves, and their primary function is to serve as respiratory organs and, by producing food currents along the ambulacral grooves to the mouth, for feeding. A madreporite is lacking, but the water ring canal around the mouth gives off at each interambulacral area about 30 short stone

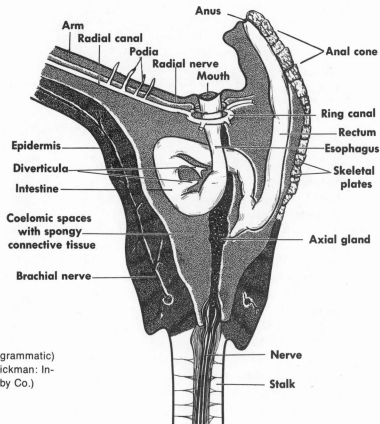

FIG. 16-47

Section through calyx of typical crinoid (diagrammatic) showing principal internal structure. (From Hickman: Integrated principles of zoology, The C. V. Mosby Co.)

canals that open into the coelom. Many ciliated funnellike canals pass from the outside through the wall of the tegmen to the coelomic cavity. These little funnels, which may number as many as 1,500, allow sea water to enter the coelom and maintain a proper fluid pressure in the body and in the water-vascular system.

The digestive system of crinoids consists of a mouth in or near the center of the tegmen; a short esophagus of tall ciliated cells provided with goblet cells; an intestine or midgut that first runs aborally and laterally, then makes a complete turn in a clockwise direction around the inner surface of the calyx before turning orally; and an anal cone with an anus through which the intestine or rectum opens (Fig. 16-47). A number of diverticula often branch off from the intestine on its inner side. The entire digestive system is lined with ciliated epithe-

lium, and a reduced circular muscle layer and connective tissue are found in its walls. Gland cells are abundant in certain regions of the digestive tract, but little is known about the digestive process itself.

Crinoids are suspension feeders and live chiefly on small plankton organisms and organic detritus. When feeding, the crinoid is fastened to the substratum by its cirri and extends its arms out horizontally, with the tips of the arms slightly curved upward. In this way a collecting bowl with the mouth in the center is formed. The pinnules have both large and small podia (in groups of 3) along each side of the food groove. The podia bear small papillae (external projections of epithelium), with small sensory processes at their distal ends. The papillae are provided with mucous glands, which shoot out mucous traps for small prey. The tube feet, or podia, then flick the food particles into the food groove, which carries the food to the mouth. A paralytic toxin for narcotizing small organisms may be secreted by the pinnules.

The coelom of crinoids is divided into spaces by the extensive strands of connective tissue (Fig. 16-47). Calcareous deposits are often found scattered through this connective tissue. The lining of the coelomic spaces, all of which communicate with each other, is a cuboidal epithelium. One of the coelomic spaces is the axial sinus that runs vertically through the center of the body and encircles the esophagus (Fig. 16-47). From the axial sinus a series of 3 single and one pair of coelomic canals, parallel to each other, extend into each arm. Some of these continue into the pinnules. At the aboral end of the axial sinus is a chambered organ composed of 5 radially arranged compartments and surrounded by nervous tissue. The 5 compartments end blindly at the oral end and become the canals of the cirri at the aboral end. The coelomic fluid and tissues contain many kinds of coelomocytes that wander freely in most places of the body.

The hemal system is mostly a network of sinuses, or lacunae. The lacunae do not have a definite lining, but are enclosed in connective tissue with a coelomic epithelial covering. Part of the hemal system is formed into plexuses around the esophagus and beneath the tegmen. Branches from these plexuses extend into the spongy organ, which lies along one side of the axial gland and which may form coelomocytes. A hemal space extends into each arm and carries sex cells (Fig. 16-46). The digestive tract is also supplied with hemal lacunae. The blood in the hemal system is not the same as coelomic fluid but contains a high protein content. It contains no distinctive cells except the ubiquitous coelomocytes. At least five types are found—ameboid, fusiform and red corpuscles, and others.

Respiration can take place through the body surface, especially that of the extensive arm and pinnule system. The podia also are thought to play an important role in gaseous exchange.

There are no kidneys or other excretory organs, but coelomocytes may collect waste and deposit it in connective tissue.

The nervous system is made up of three systems at different body levels: (1) The ectoneural (oral) system consists of a subepidermal band running just under the epidermis of the ambulacral groove in each arm and its pinnules, where nerves supply the epidermis, the inner surface of the podia, and other parts (Fig. 16-46). The bands from all the grooves converge into 5 bands in the tegmen and in the mouth region form a nerve sheath around the esophagus and intestine. (2) The hyponeural (deeper oral) system forms a pentagon lateral of the water ring of the water-vascular system (Fig. 16-47). From this pentagon, nerves are sent to the tegmen podia, anal canal, and other internal organs; and a pair of lateral brachial nerves is sent into each arm, one on each side (Fig. 16-46). These supply the musculature of the walls of the vessels in the water-vascular system and the outer surface of the podia. (3) The entoneural system (aboral) consists of a cup-shaped mass in the apex of the calyx and a series of nerves to the arms and the various aboral regions (Figs. 16-46 and 16-47). The chief center of this system, or the cup-shaped mass, sends nerves directly to the cirri (comatulids) or forms a nervous sheath in the coelomic canal that runs through the columnals of the stem of stalked crinoids; from this sheath, nerves are sent to the cirri at the nodes. From the same mass center of the aboral system 5 brachial nerves pass through a pentagonal nerve ring in the radial plates of the calyx and then through canals in the arms. In each arm the brachial nerve gives off a pair of nerves to the flexor muscles and adjacent epidermis, a pair of nerves to the aboral part of the arm, a pair of commissure connectives to the lateral brachial nerves of the hyponeural system, and a pair of nerves to each pinnule. The entoneural system is the dominant system in crinoids, whereas in other echinoderm classes the ectoneural (oral) system is the dominant one.

Sense organs are represented only by the

papillae of the podia (Fig. 16-46), by free nerve endings, and by sensory cells in the epidermis of the ambulacral regions. Crinoids so far as is known are negatively phototropic.

The sexes in crinoids are separate, but sexual dimorphism is lacking. The gonads are usually found in the genital pinnules as masses of sex cells, although they may also occur in the arms. The sex cells develop within the genital canal (a coelomic extension within the arm, Fig. 16-46), which gives off a branch to each genital pinnule where the gametes are deposited. When the sex cells are mature, the walls of the pinnules rupture and release the eggs at spawning. In some (e.g., *Antedon*) the eggs may adhere to the pinnules by adhesive secretions; but they usually fall into the sea, where they are fertilized. Development occurs in the membrane of the egg, and is followed by the stage in which a ciliated larva hatches and swims about before attachment. A few Antarctic species brood the young in saclike invaginations of the pinnule walls.

Regeneration of lost parts is rather well marked in crinoids. They can regenerate arms as well as pinnules and cirri. They can even regenerate four of the five pairs of arms that are lost simultaneously. The aboral nervous center is necessary for regeneration.

DEVELOPMENT

In development the egg undergoes a regular deuterostome type of holoblastic, radial cleavage, which produces a typical coeloblastula. Gastrulation occurs by a modified embolic-type method. After the completion of gastrulation and the formation of the archenteron, the blastopore closes; the archenteron becomes a closed, elongated sac in the blastocoel surrounded on three sides by entomesodermal mesenchyme. The archenteron sac now divides into an anterior vesicle (enterohydrocoel) and a posterior vesicle (somatocoel). The somatocoel divides into a right and left somatocoel. In the meantime the enterohydrocoel assumes a crescent shape and curls around the middle of the somatocoel at right angles to the long axis of the latter. The dorsal part of the crescent forms a process that becomes the left axocoel; the ventral part of the crescent forms a process that separates as the hydrocoel (the left hydrocoel of other echinoderms). The right axocoel and right hydrocoel never form. The remaining portion of the enterohydrocoel, together with its two ends (horns), fuses to become the enteric sac or future intestine. The embryo elongates in an anteroposterior axis, and at its anterior end a tuft of cilia develops. Around the embryo by ectodermal differentiation, 4 or 5 ciliated bands about equally spaced appear along the axis.

The larva hatches as a free-swimming doliolaria (Fig. 16-3) without a bipinnaria stage. The larva has no mouth but lives off the yolk of the egg during its few days of swimming existence. It then attaches itself by a small adhesive pit located near the apical tuft of cilia so that the anterior end becomes the attached aboral end and the posterior end becomes the oral end. The metamorphosis of the larva into a stalked sessile form involves a stalked larva known as the pentacrinoid stage (about 3 mm. long) which lasts for several months (*Antedon*). The stalk is produced by an elongation of the underpart (aboral) of the larva, which rests with its terminal plate on a substratum. Buds for the future arms appear on the upper or oral side of the calyx cup as 5 projections. Each of these splits into 2 forks later to form the 10 arms common among comatulids. The pinnules, which are really miniature arms, are produced by the alternate forking of the arms to left and right. After the formation of cirri the pentacrinoid larva breaks loose from its stalk at the junction of the centrodorsal plate (formerly the top stem joint) with the top columnal of the stem and then leads a free existence.

SOME INVESTIGATIONS ON ECHINODERMS OF CURRENT INTEREST

At the end of the preface of her classic monograph on echinoderms, Dr. Libbie H. Hyman stated that she saluted the echinoderms as "a noble group especially designed to puzzle the zoologist." Written almost a score of years ago, her statement holds true today, for although many investigations have been made and are being made, the group of echinoderms as a whole poses more unsolved problems than any other group in the animal kingdom. The very fact that they do represent an enigmatic group of animals has been a challenge to many research individuals so that more attention has been directed toward them in recent years.

The echinoderms are more advanced in many ways than the coelenterates, yet the former have many primitive characters. They are deuterostomes but are radially symmetric. Their sensory organs are of the simplest and throughout the phylum are poorly developed. Their nervous system is more or less diffused, consisting chiefly of three networks. No brain can really be assigned to any part of it, yet wonderful coordination of hundreds of tiny locomotor units (podia) is performed. Echinoderms transport their food and waste particles with coelomocytes—a primitive method. Their coelom is departmentalized but has an enterocoelous formation. The genital system is about as simple as it can be, with the sex products in most cases shed into the sea to develop. They make use of a hydraulic system in most of their movements, but unlike the annelids, they have developed a specialized system that is located within one compartment of the coelom. In some the interlocking system of ossicles makes growth possible only in a rectilinear pattern. In short, echinoderms are examples of forms that rank high in their overall advancement in the animal kingdom but retain or modify in odd ways primitive characters that make it difficult to elucidate their relationships or phylogeny with other groups. Yet this group of bizarre animals retains throughout a basic pattern of structural features that they may highly modify for particular purposes. Certain features are diagnostic of the whole phylum but are of functional importance in whatever way they may be modified.

Many of the problems of the echinoderms still unresolved are the exact functions of the axial organ, the Tiedemann bodies, the polian vesi-

cles, the lack of the ampullae in connection with the ophiuroid water-vascular system, the evolution of pentamerism, and many others.

Some investigators have been more successful at solving other functional aspects of echinoderm behavior. An interesting one is the shedding of gametes by a shedding substance in sea stars. In 1959 A. B. Chaet and R. A. McConnaughy were the first to find out that extracts prepared from the radial nerves and injected into the perivisceral coelom of ripe individuals resulted in the discharge of large numbers of gametes. This discovery served to stimulate many other investigators along this line. Chaet in 1964 proposed the hypothesis that this gamete-shedding substance was a neurohormone of a polypeptide nature, as it was digested by trypsin. In 1967 M. J. Imlay and A. B. Chaet in a more sophisticated investigation found this neurosecretory substance and studied it more in detail in the same sea star, *Asterias forbesi*. They were able to isolate the substance in three distinct regions of the radial nerve complex. They found no evidence that this neurosecretory substance migrated from some other part of the nervous system, as it does in vertebrates. These investigators were not able to find measurable quantities of the hormone in other tissues, although some was found in the ring nerve, which is a continuation around the mouth of the radial nerve. The shedding substance was found in the radial nerves of both male and female sea stars. Stains commonly used to detect neurohormone substance were employed and gave positive results. Chaet has some evidence that there is an inhibitory factor (shedhibin) and draws the conclusion that the normal control of gamete release is a function of the balance between shedding substance and shedhibin. Shedhibin apparently builds up during the development of the ovary. Shedding occurs in a ripe individual when shedhibin disappears. This shedding substance has been found in the radial nerves of many species.

Chaet, who has been one of the most active investigators in this line of research, has given some explanation of the way this substance may operate. He thinks it performs a twofold physiologic role. One role is involved in the contractions of the ovary a few minutes before the actual release of eggs. Calcium appears to be necessary for this contraction, as H. Kanatani (1964) also discovered. The other role is concerned with maturation activity in the intact egg, as shown by the germinal vesicle breakdown. Kanatani also believes the shedding substance may be responsible for dissolving the extracellular material holding the mass of eggs together.

H. Kanatani and M. Ohguri (1966) working on some Japanese species of sea stars, found that the testis responded more readily to the action of the shedding substance than did the ovary. The shedding substance appears to be identical in each sex. They also discovered some shedding substance in the tube feet and body wall. The active substance was found in the coelomic fluid only at the time of spawning. They suggested that the shedding substance may be distributed to the gonads by the nerves running there.

Another organ of echinoderm structure that has remained more or less imperfectly known is the axial organ. This organ is variously formed in the different echinoderm classes. It is the belief of N. Millott (1967) that the organ is imperfectly understood because its structure has been incorrectly described by investigators. Millott has studied the axial organ in echinoids and has indicated that recent investigation has emphasized three important features of the organ: the contractile vessel running through it; the confluence within the organ of the perivisceral coelom, the water-vascular system, and the hemal system; and the ability of some echinoids to survive without the organ. An axial sinus (a special coelomic cavity for the organ) is lacking in echinoids. Millott has described the glandular portion of the organ as a spongy mass fastened together by connective tissue and muscle and consisting of three kinds of cavities: the central cavity

(lumen) lined with peritoneum and spanned by the contractile vessel, a system of canaliculi communicating with the perivisceral coelom, and the ramifying lacunar system of spaces surrounding the canaliculi and continuous with the contractile vessel by means of irregular channels. R. A. Boolootian and J. L. Campbell (1964) studied the contractile vessel in *Strongylocentrotus purpuratus* and found that the vessel contracted rhythmically. They called this structure a primitive heart, but these rhythmic pulsations in echinoids appear to be restricted to *Strongylocentrotus* and *Arbacia punctulata*. N. Millott (1966) believes that the function of the axial organ is chiefly an integrated defensive response to injuries, based on his evidences of coelomocyte clots which seal damaged areas, and the ingestion of phagocytes of foreign substances, which contaminate the perivisceral fluid. Many others have worked on the same organ, but there is still no final agreement regarding the exact function of the axial organ.

Tiedemann's bodies represent another enigmatic structure in echinoderms. Present in all asteroids are 9 or 10 irregularly shaped bodies on the inner side of the ring canal. Each body consists of numerous radiating tubules within a stroma of connective tissue and muscle fibers, the whole surrounded by peritoneum. The electron microscope shows that the cuboidal epithelial cells lining the tubules each bear a flagellum in the midst of cytoplasmic ridges broken up into processes resembling microvilli. The cells include mitochondria, Golgi apparatus, and certain inclusions. Within the lumen of each tubule free cells may be collected. Many functions have been assigned to these bodies, such as coelomocyte formation, filtering devices, and enzyme formation. However, none of these theories is wholly satisfactory, and many zoologists regard the function of Tiedemann's bodies as unknown.

J. C. Ferguson (1966) studied Tiedemann's bodies in young specimens of *Asterias forbesi* by injecting sterile sea water labeled with H[3]-

thymidine into their rays. In this way, nuclei which have undergone replication become incorporated with the H³-thymidine and are radioactive. Examination of the Tiedemann bodies for radioactive cells revealed that some cells are produced by the bodies but not in excess of maintaining their own normal growth. There were no evidences that cells were produced sufficiently for normal replacement of coelomocytes throughout the body. Radioactive coelomocytes were found elsewhere in the body but too soon after injection to have originated in the Tiedemann bodies. Ferguson concludes that the bodies (and the organs associated with the hemal system) filtered the body fluids and removed foreign substances from the fluids, since these organs are in communication with all the major body spaces of echinoderms. Other recent investigations reveal several stages of desquamation of cells from the tubule lining in the formation of coelomocytes. This diversity of conclusions indicates that the exact function of Tiedemann's bodies is still largely unresolved.

Much concern also has been caused by the sea star *Acanthaster planci,* better known as the crown of thorns, that is causing destruction of the polyps of coral reefs, such as the Great Barrier Reef of Australia. This sea star feeds on the living coral polyps, and the resulting dead coral skeletons are subjected to erosion processes that destroy the coral reefs. It is not yet known just why this particular sea star has suddenly increased in numbers, although the problem is being extensively investigated. The practical control methods of using predators such as the large mollusk (*Charonia tritonis*) and the shrimp (*Hymenocera elegans*) may prove useful. How long it takes the reefs to regenerate is not known definitely, but some think regeneration is more rapid than formerly believed possible.

REFERENCES

Anderson, J. M. 1960. Histological studies on the digestive system of a starfish, *Henricia*, with notes on Tiedemann's pouches in starfishes. Biol. Bull. **119**:371-393. Tiedemann's pouches are used as pumping organs for drawing food from the stomach of the starfish into the pyloric ceca.

Anderson, J. M. 1965. Studies on visceral regeneration in sea stars. III. Regeneration of the cardiac stomach in *Asterias forbesi* (Desor). Biol. Bull. **129**:454-470. Found normal feeding possible within 15 days after operation.

Bather, F. A. The Echinoderma. In Lankester, E. R. (editor). 1900. A treatise on zoology, part 3, London, Adam & Charles Black, Ltd. A work of great importance on the basic structures of echinoderms.

Boolootian, R. A. 1963. The physiology of Echinodermata, Washington, D. C., vol. 3, pp. 113-138. XVI International Congress of Zoology. A symposium represented by the following participants: J. M. Anderson—Aspects of digestive physiology among echinoderms. R. A. Boolootian—Reproductive physiology of echinoderms. A. Farmanfarmaian—Respiration in echinoderms. D. Nichols—Comparative histology of the echinoderm water-vascular system. J. E. Smith—The functions and properties of the echinoderm nervous system. G. Vevers—Pigmentation of the echinoderms. M. Yoshida—Photosensitivity in echinoids.

Boolootian, R. A. 1962. The perivisceral elements of echinoderm body fluids. Amer. Zool. **2**:275-284.

Boolootian, R. A. (editor). 1966. Physiology of Echinodermata. New York, John Wiley & Sons, Inc.

Boolootian, R. A., and J. L. Campbell. 1964. A primitive heart in the echinoid *Strongylocentrotus purpuratus*. Science **145**:173-175. Found rhythmic pulsations in contractile vessel of axial organ.

Burnett, A. L. 1955. A demonstration of the efficacy of muscular force in the opening of clams by the starfish, *Asterias forbesi*. Biol. Bull. **109**:355-368. No narcotic is used, but the starfish takes advantage of a small gape produced by an imperfectly sealed edge or by its own muscular force to insert its cardiac stomach.

Buschbaum, R., and L. J. Milne. 1960. The lower animals; living invertebrates of the world. Garden City, Doubleday & Co., Inc.

Chaet, A. B., and R. A. McConnaughy. 1959. Physiologic activity of nerve extracts. Biol. Bull. **117**:407-408.

Clark, A. H. 1915-1950. A monograph of the existing crinoids, vol. 1, parts I to IV. U. S. Nat. Mus. Bull. **82**:1-700. This incomplete monograph represents the most outstanding work on crinoids.

Cockbain, A. E. 1966. Pentamerism in echinoderms and the calcite skeleton. Nature **212**:740-741. Pentamerism may have arisen as a unique solution to the problem of constructing a rigid test from a few calcite crystal plates.

Cuenot, L. Echinoderms. In Grassé, P. P. (editor). 1948. Traité de zoologie, vol. 11, Paris, Masson et Cie. An authoritative work on echinoderms.

Feder, H. M. 1955. On the methods used by the starfish, *Pisaster ochraceus*, in opening three types of bivalved mollusks. Ecology **36**:764-767.

Ferguson, J. C. 1966. Cell production in the Tiedemann bodies and haemal organs of the starfish, *Asterias forbesi*. Trans. Amer. Microscop. Soc. **85**:200-209.

Fitch, W. M., and E. Margoliash. 1967. Construction of phylogenetic trees. Science **155**:279-284.

Galtsoff, P. S., and V. L. Loosanoff. 1950. Natural history and methods of controlling the starfish (*Asterias forbesi Desor*). U. S. Bur. Fisheries Bull. **49**:75-132.

Goodbody, I. 1960. The feeding mechanism in the sand dollar, *Mellita sexiesperforata*. Biol. Bull. **119**:80-86. The author describes how small particles of food are carried by cilia and ciliated food tracts to the ambulacral grooves and mouth.

Harvey, E. B. 1956. The American Arbacia and other sea urchins. Princeton, Princeton University Press.

Holland, N. D., and J. A. Lauritus. 1968. The fine structure of the gastric exocrine cells of the purple sea urchin, *Strongylocentrotus purpuratus*. Trans. Amer. Microscop. Soc. **87**:201-209.

Horstadius, S. 1955. Reduction gradients in animalized and vegetalized sea urchin eggs. J. Exp. Zool. **129**:249-256.

Hyman, L. H. 1955. The invertebrates; Echinodermata, vol. 4. New York, McGraw-Hill Book Co., Inc. The most up-to-date treatise on this group.

Ikeda, H. 1941. Functions of the lunules of *Astriclypeus*. Annot. Zool. Japan **20**:1-24. This paper describes how the lunules function in burrowing by passing sand through them to the aboral surface.

Imlay, M. J., and A. B. Chaet. 1967. Microscopic observations of gamete-shedding substance in starfish radial nerves. Trans. Amer. Microscop. Soc. **86**:120-126.

Jennings, H. S. 1907. Behavior of the starfish *Asterias forreri* de Loriol. Univ. Calif. Pub. Zool. **4**:53-185. The author describes several behavior patterns, including the righting movement.

Kanatani, H. 1964. Spawning of starfish: action of gamete-shedding substance obtained from radial nerves. Science **146**:1177-1179.

Kanatani, H., and M. Ohguri. 1966. Mechanism of starfish spawning. I. Distribution of active substance responsible for maturation of oocytes and shedding of gametes. Biol. Bull. **131**:104-114. Active substance found only in coelomic fluid during spawning of sea star.

Lavoie, M. 1956. How sea stars open bivalves. Biol. Bull. **111**:114-122.

MacGinitie, G. E., and N. MacGinitie, 1968. Natural history of marine animals, ed. 2. New York, McGraw-Hill Book Co. The echinoderms included in this well-known work are mostly those of the Pacific coast.

Marshall, A. 1884. Nervous system of Antedon. Quart. J. Microscop. Sci. **24**:121-172. A classic paper on the nervous system of crinoids.

Millott, N. 1966. A possible function of the axial organ of echinoids. Nature **209**:594-596.

Millott, N. (editor). 1967. Echinoderm biology. (Symposium of the Zoological Society of London.) New York, Academic Press, Inc.

Nichols, D. 1960. The histology and activities of the tube-feet of *Antedon bifida*. Quart. J. Microscop. Sci. **101**:105-117.

Nichols, D. 1961. A comparative histological study of the tube feet of two regular echinoids. Quart. J. Microscop. Sci. **102**:157-180.

Pearse, A. S. 1908. Behavior of *Thyone*. Biol. Bull. **15**:25-48. A revealing study of a sea cucumber.

Ricketts, E. F., and J. Calvin. 1952. Between Pacific tides, ed. 3. (Revised by J. W. Hedgpeth.) Stanford, Calif., Stanford University Press. A unique book of seashore life. These authors with MacGinitie (see previous reference) are perhaps the most outstanding American students of marine life.

Stephens, G. C., J. F. Van Pilsum, and D. Taylor. 1965. Phylogeny and the distribution of creatine in invertebrates. Biol. Bull. **129**:573-581.

Swan, E. F. 1961. Seasonal evisceration in the sea cucumber *Parastichopus californicus*. Science **133**:1078-1079. Seasonal evisceration occurs chiefly in the month of October.

Wald, G. 1963. Phylogeny and ontogeny at the molecular level. Proceedings of the Fifth International Congress of Biochemistry **3**:12-51.

Yonge. C. M. 1958. Ecology and physiology of reef-building corals. In Buzzati-Traverso, A. A. (editor). Perspectives in marine biology. Berkeley, Calif., University of California Press.

CHAPTER 17

Minor deuterostome animals

PHYLA CHAETOGNATHA, HEMICHORDATA, AND POGONOPHORA

GENERAL CHARACTERISTICS

1. Phyla Chaetognatha, Hemichordata, and Pogonophora may be considered minor ones because they have relatively few species and do not rank among the larger, more dominant groups in the animal kingdom. Most of the animals making up the great majority of those that fill the earth's ecologic niches belong to a few phyla—arthropods, mollusks, chordates, etc. In a phyletic sense, however, some of the minor phyla have given a basic understanding of evolutionary diversity and make many contributions to the overall blueprint of the animal kingdom. Often these minor phyla deserve far more consideration than is commonly given them. Important clues to different animal patterns may thus be found in obscure places.

2. The chaetognaths differ from the other deuterostomes in many ways. Although they are coelomates, they lack a peritoneum as a coelomic lining. Their early development reveals their deuterostome position, for they have a modified enterocoelous development, radial cleavage, and gastrulation by invagination. They have chitin (a rare feature in deuterostomes) in their grasping spines. Their threefold coelomic regions may be homologous to the axocoel, hydrocoel, and somatocoel of higher forms (e.g., hemichordates). They may represent an independent branch of deuterostome evolution.

3. The Pogonophora are coelomate animals entirely without a digestive tract. The trunk has a threefold division (protosome, mesosome, and metasome) similar to that of the hemichordates. Other deuterostome characteristics are the enterocoelous body cavity, a pair of coelomoducts, and radial cleavage (although segmentation of the blastomeres is uneven). The unpaired coelom of the protosome extends branches into the tentacles, which may function in nutrition. They lack a real gastrula, with no formation of a blastocoel and blastopore. Their closest relatives appear to be the hemichordates.

4. The phylum Hemichordata are primitive deuterostomes that may be related to the Pogonophora. Some zoologists place the pogonophores as a class under the hemichordates. The coelom of the hemichordates is also tripartite (a protosome with a single protocoel, a mesosome with paired mesocoels, and a metasome with paired metacoels). Branchial gill slits connect the lumen of the pharynx with the exterior for the ejection of water. Some are known to produce a current of mucus on the body surface of the proboscis and carry particles of food to the gut. Hemichordates have a type of radial cleavage and a coeloblastula that gastrulate by epiboly, or the growth of cells downward to cover the cells of the vegetal region. A blastopore quickly closes over, and an anus is formed near the earlier blastopore. In isolecithal eggs of indirect

development, a tornaria larva (very much like the larva of sea stars) is formed. The class Pterobranchia are small colonial forms, and usually live in tubes. They have characteristics of the larger class Enteropneusta but are considered more primitive.

The evidence at present seems to indicate that these minor deuterostomes are closely related to the echinoderms and somewhat distantly to chordates.

PHYLOGENY AND ADAPTIVE RADIATION

1. The chaetognaths do not seem to fit into any relationship with other phyla except that they are deuterostomes. Their larvae, however, show no resemblance to those of echinoderms and hemichordates. They appear to have diverged quite early from the line of ancestry leading to the deuterostomes (L. H. Hyman), and at present they represent an isolated group that fits nowhere in the evolutionary scale (P. P. Grassé).

On the other hand, hemichordates show affinities to both echinoderms and chordates. The resemblance (in indirect development) between the tornaria larva of hemichordates and the bipinnaria larva of echinoderms is considered to be fundamental in indicating a close relationship between hemichordates and echinoderms in spite of the warning of Hyman that divergent specializations and resemblances may occur among larval forms, rendering exact relationship interpretation difficult. The more primitive class of the hemichordates, the Pterobranchia, with their lophophore and tentacular arms, have a sessile feeding mechanism.

In the recently evaluated interpretation of the development of the Pogonophora, the presence of the tripartite division of the body (protosome, mesosome, and metasome) found also in hemichordates may indicate a fundamental relationship between the two phyla. There is no larval stage in pogonophores, but they have developmental stages similar to those of the early stages of the enteropneusts.

2. The general morphology and physiology of the chaetognaths are adapted for a planktonic existence. *Spadella* is the only genus that has departed from a planktonic pattern of living. Their fins are excellent flotation devices, for their powers of locomotion are feeble and restricted. Evolutionary divergence occurs chiefly by fitting species for differences in temperature, light, and ocean depths.

Of the hemichordates, the primitive pterobranchs are sessile and lophorphorate ciliary feeders and have little evolutionary divergencies in their rather uniform habitats. The more active Enteropneusta have more species. Many of their species have a form of mucusciliary feeding by trapping small organism with the mucous secretion on their proboscis and directing their trapped food into the gut. Burrowing forms especially consume sand, from which organic material is extracted. Their evolutionary divergency probably occurs because of specialized adaptations for different habitats involving respiratory devices and the expulsion of excess water and inorganic substance.

The tentacles of porgonophores have shown the greatest evolutionary diversity, varying in number from 1 to 268 among the eighty or more species discovered to date. These animals not only capture food but also form digestive cavities with their tentacles, where digestion occurs, as Pogonophora have no digestive system.

■ PHYLUM CHAETOGNATHA

The common name for members of phylum Chaetognatha is arrowworms. They belong to the Deuterostomia along with the echinoderms, hemichordates, pogonophorans, and chordates, although they lack most of the characteristics found among the other deuterostome members. Chaetognaths are among the most common forms found in planktonic populations in all oceans and at all latitudes. They are also found at considerable depths in the seas. Only one genus (*Spadella*) is an exception to their planktonic existence. This particular genus is enthonic, living mostly in shallow water, where it clings to plants and rocks by means of adhesive papillae. The abundance of chaetognaths in plankton is attested by the fact that as many as 1,000 or more specimens have been found in a cubic meter of sea water. Many of the species show a marked stratification of distribution. Some are found only in the limit of light penetration, living in depths not exceeding 200 meters (epiplanktonic). Other species are present in the zone of 200 to 1,000 meters' depth (mesoplanktonic), and still others live in depths below 1,000 meters (bathyplanktonic). It seems that each species is restricted to a definite depth zone for its habitat. Temperature may be the critical factor in their stratification. Their horizontal distribution, however, varies with latitude because of the variation in water temperatures between cold and warm regions. Most chaetognaths seem to belong to the warmwater, epiplanktonic zone. No one species is found in all depths and at all latitudes. However, the genus *Eukrohnia* has been found from the Arctic to the Antarctic by way of the colder water at considerable depths in the tropic regions. Some are restricted to zones of the continental shelf (neritic species) and others to areas normally far from shore (oceanic species). The degree of salinity may also be a factor in their distribution because shore water is sometimes diluted by fresh water from continental sources.

Little is known about the fossil record of chaetognaths because of their soft bodies. In the Burgess shale deposits of the Middle Cambrian period of British Columbia, 3 fossil specimens of this group have been discovered (*Amiskwia*). They were about 20 mm. long and had some of the typical features of chaetognaths, although there were some differences.

Chaetognaths are parasitized by several groups of animals. Protozoan parasites are represented by parasitic amebas, flagellates, ciliates, and sporozoans. Certain stages (metacercariae) of trematodes, the definitive hosts

of which are fishes, have been found in the intestine and coelom of chaetognaths. Some adult trematodes are also harbored. Copepods may serve as intermediate hosts in the life history of these parasites. Other parasites found in arrowworms are nematodes and copepods.

CLASSIFICATION

Phylum Chaetognatha. Bilaterally symmetric; coelomate and enterocoelous mesoderm formation; body divided into head, trunk, and tail; body with lateral and caudal fins; head with eyes and chitinous teeth and jaws (in some); cerebral and ventral ganglia connected by a pair of nerves; body wall musculature of longitudinal muscles arranged in 4 bands; digestive track straight with muscular walls; no excretory, respiratory, or blood systems; hermaphroditic; self-fertilization; juvenile free-swimming larva; marine. Not divided into taxa higher than genera, about eight of which are recognized. The number of species does not exceed fifty. *Sagitta* is most common genus. Other genera are *Spadella, Pterosagitta, Bathyspadella, Krohnitta, Zahonya, Eukrohnia,* and *Heterokrohnia.*

GENERAL MORPHOLOGIC FEATURES
External anatomy and physiology

Chaetognaths range in size from 5 to over 100 mm. in length. Most species are between 12 and 25 mm. The body is fusiform or torpedo shaped (Fig. 17-1). It is divided into head, trunk, and tail. These divisions may correspond respectively to the protosome, mesosome, and metasome of the hemichordates. The anterior, ventral vestibule and mouth are almost surrounded by rows of lateral chitinous spines or hooks for catching prey. On the anterior end in *Sagitta*, the most common genus, are also 2 or 4 paired rows (anterior and posterior) of short chitinous teeth for holding the prey. Some species have only one row of teeth, and others none at all. The head and spines are covered by a hood (a fold of the body wall) when swimming and thus lessen the resistance of the head (Fig. 17-2). The head bears two dorsal eyes on the rear part. On the elongated trunk are one or two

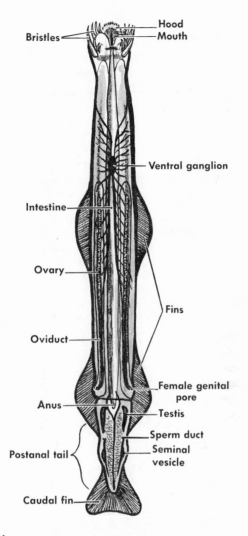

FIG. 17-1

Ventral view of arrowworm *(Sagitta)*. (From Hickman: Integrated principles of zoology, The C. V. Mosby Co.)

pairs of lateral, horizontal fins formed from the epidermal body wall. A large caudal, or tail, fin embraces the posterior end. The body is covered with a thin cuticle that overlies the single or stratified layer of epidermal cells. They are the only invertebrates with a stratified epidermis (L. H. Hyman, 1959).

Internal anatomy and physiology

Besides the epidermis and cuticle, the body wall (Fig. 17-3) contains 4 bands of longitudinal muscles (2 dorsolateral and 2 ventrolateral) and a basement membrane on which the epidermis rests and which also forms special plates in the head. The coelom is not lined with peritoneum but is divided into compartments represented by one coelomic space in the head, paired coelomic sacs in the trunk, and 1 or 2 (sometimes more) coelomic sacs in the tail region (Fig. 17-3, *A* and *B*). The coelomic spaces are separated by septa. The head musculature is complicated, being made up of special muscles for operating the hood, the teeth, and the spines.

Chaetognaths do not swim much but drift passively most of the time. The fins are merely for flotation and play no role in swimming. Swimming is accomplished mostly be movements of the caudal fin and the contraction of the longitudinal body muscles. This enables them to dart forward in spurts with rest periods

FIG. 17-2

Preserved specimen of arrowworm *(Sagitta),* with head mostly enclosed in hood (head at left). (From Hickman: Integrated principles of zoology, The C. V. Mosby Co.)

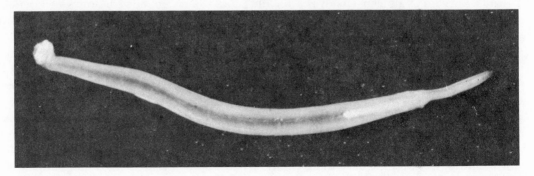

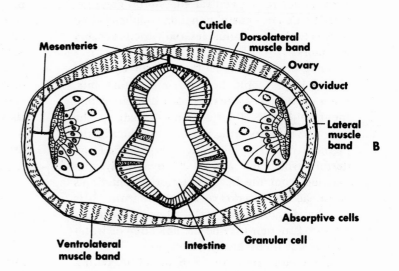

FIG. 17-3

A, Cross section of *Sagitta* through region of postanal tail showing 4 coelomic cavities. **B,** Cross section of *Sagitta* at level of ovaries. There are 2 coelomic spaces at this level. (Modified from Burfield, 1927.)

between. They cannot constrict their bodies because of the absence of circular muscles.

The simple digestive system consists of a cuticle-lined vestibule formed by the inturning of the center of the head, a mouth leading into a bulbous pharynx, a straight intestine with a pair of lateral diverticula at its anterior end, and an anus opening ventrally at the septum separating the trunk and tail (Fig. 17-1). The digestive tract contains muscle in its walls, especially in the pharynx, where both longitudinal and circular muscles occur. The intestine is lined with columnar epithelium containing both granular secretory cells and absorptive cells (Fig. 17-3, *B*). The digestive tract is supported by dorsal and ventral mesenteries formed by extensions of the basement membrane of the body wall.

Chaetognaths are mostly carnivorous, living on copepods, small worms, larva, eggs, crustaceans, and occasionally detrital particles. Prey is captured with the aid of the grasping spines or hooks and shoved into the mouth with the assistance of the small anterior and posterior teeth. Digestion is thought to be extracellular. The entire process is very rapid.

Specialized respiratory and excretory organs are absent and both respiration and excretion may occur through the body surface. The coelomic fluid serves as a circulatory medium. It is a colorless fluid with small granules and is kept in movement by ciliary action, although there is no coelomic epithelium or peritoneum to bear cilia.

The nervous system consists of ganglia, connecting commissures, and nerves. The chief part of the nervous system is a circumenteric ring surrounding the pharynx. This ring is composed of a large dorsal cerebral ganglion, which sends 2 frontal commissures to a pair of vestibular ganglia at the sides of the mouth. The vestibular ganglia, in turn, are connected to each other by a subpharyngeal commissure, thus completing the ring. The cerebral ganglion supplies nerves to the head region, such as the eyes, and from its sides sends 2 circumenteric connectives to the ventral subenteric ganglion (Fig. 17-1). The ventral ganglion supplies paired nerves to the trunk muscles and various sensory receptors.

Sense organs include the pair of eyes, the ciliary loop, tactile bristles, and ciliary pits. Each of the eyes consists of 5 combined pigment cell ocelli, 1 lateral and 4 median. The eyes are inverse, with retinal cells facing the pigment cup. The ciliary loop is modified epidermis with cilia and is an oval body that extends from the head posteriorly, sometimes as far as the anterior region of the trunk. It is usually demonstrated only when stained. Its sensory function may be the detection of water currents or their chemical nature.

All chaetognaths are hermaphroditic, with a pair of tubular ovaries in the trunk just anterior to the trunk-tail septum and a bandlike testis on each side of the tail, just behind the septum (Fig. 17-1). An oviduct is found along the lateral side of each ovary and ends posteriorly in its own seminal receptacle. The oviduct consists of an outer tube and an inner tube; only the inner tube opens into the seminal receptacle. Each seminal receptacle opens through a vagina and gonopore in front of the trunk-tail septum. A sperm duct passes posteriorly from each testis to a seminal vesicle in the lateral body wall; along its course the duct opens by a funnel into the tail coelom. Spermatogonia leave the sperm duct through these funnel openings and pass into the coelom, where spermatogenesis is completed. When mature, the sperm pass back into the sperm duct and go to the seminal vesicle, where they are formed

into spermatophores. There is no male gonopore, and the spermatophores are released by rupture of the seminal vesicle. The spermatophores adhere to the body surface or to the fins after their discharge. Sperm are then released and make their way to the female gonopore to enter the seminal receptacles. Fertilization is internal. Eggs mature in the ovaries at different times; as they mature, they pass through an improvised channel formed by 2 special cells to the oviduct nearby (Fig. 17-3, B). Fertilization occurs when the sperm migrate up the oviduct from the seminal receptacle to the eggs. Self-fertilization occurs in *Sagitta*, but in *Spadella*, fertilization is reciprocal by exchange of sperm. In *Sagitta*, fertilized eggs with a jelly coat are discharged singly into the sea, where they float as planktonic objects. In the genus *Krohnitta* the eggs are carried in packets attached to the back.

DEVELOPMENT

Chaetognaths have radial and equal cleavage and form a coeloblastula. Gastrulation occurs at about the 50-blastomere stage and is embolic. The coelom is enterocoelous and is formed from two pairs of coelomic sacs that arise from 2 backward folds of the anterior wall of the archenteron. Embryology is mostly direct, and the young larvae (1 mm. long) hatch as juveniles within about 48 hours. *Sagitta* is noted for the early formation of its germ cells, which can be detected during gastrulation as a group of 2 cells in the anterior wall of the archenteron. The adult structures all appear within a week after hatching, and there is no metamorphosis.

◼ PHYLUM HEMICHORDATA

The members of phylum Hemichordata were formerly considered to be a subphylum of the phylum Chordata because of their unmistakable affinities with the chordates. As such they were judged to be the lowest of the chordates. However, the general organization of hemichordates is different from that of chordates except that certain characteristics are common to each group. Among the common characteristics are the presence of gills and a short anterior structure considered to be a notochord. Many recent authorities, however, believe that the notochord in these animals is not a true notochord but a stomochord instead. Many hemichordate features are invertebrate in type.

The hemichordates are vermiform, enterocoelous coelomate animals that are strictly marine, living usually in shallow water. Some members are colonial. The body and coelom are divided into three successive regions—proboscis, collar, and trunk. The first division has a single coelomic cavity, and the other two each have paired coelomic sacs. The two groups that make up the phylum are superficially very dissimilar, although they possess certain basic characters in common. General descriptions given here apply more to the enteropneusts than to the pterobranchs.

Enteropneusts live mostly in burrows or under stones, usually in the intertidal zones. Some have been collected at depths of several hundred meters. Burrowing is accomplished by thrusting the proboscis forward into the muck or mud and then by contracting the longitudinal muscles, forming a peristaltic bulge that passes posteriorly. The bulge anchors the animal in preparation for another thrust. As the worm moves forward, grains of sand are carried backward by cilia to the collar, where a sand girdle is formed. By additions from the forward end, the girdle is pushed posteriorly over the trunk. Not all members of the group form burrows, however.

Hemichordate distribution is not well known. Their secretive habits and fragile bodies make their collection difficult. They usually favor warm and temperate waters. In general their distribution is worldwide in particular regions, although none has been reported from the Antarctic regions. Several species are found around the British Isles. The genus *Saccoglossus* (*Dolichoglossus*) is found on both Atlantic and Pacific coasts. Other genera and species also occur on the Atlantic coast.

The size ranges of the class Enteropneusta (acorn worms) is from 20 to 2,500 mm. in length, with a breadth of from 3 to 200 mm. The members of the class Pterobranchia are much smaller, not exceeding 14 mm., and most are under 5 mm. (not including the stalk).

About seventy species of enteropneusts and three small genera of pterobranchs are recognized.

A few fossil pterobranchs have been described from the Cretaceous period and from the Paleocene epoch, but the wholly extinct graptolites, which closely resemble pterobranchs, have left a very good record. Other hemichordates have left little or no record.

CLASSIFICATION

Phylum Hemichordata. Body vermiform or vaselike; coelomate and enterocoelous; notochord (stomochord) restricted to preoral region; body typically divided into three unequal regions (proboscis, collar, trunk); no tail; 3 coelomic spaces corresponding to body divisions; paired gill slits, many, few, or none; nerve tissue both dorsal and ventral, partly hollow; with or without a lophophore bearing tentaculated arms; free living or sessile; some colonial; dioecious; asexual (in some); marine.

Class Enteropneusta. Body vermiform, long, and contractile; many gill slits; straight intestine; tentaculated arms absent; dioecious, with saclike gonads in coelom along trunk; tornaria larva (in some). Three families are recognized: Harrimaniidae (*Saccoglossus*), Spengelidae (*Spengelia*). Ptychoderidae (*Balanoglossus*).

Class Pterobranchia. Body small, vaselike; aggregated or colonial in secreted tubes (coenecia); one genus, *Atubaria*, without coenecium; tentaculated arms; gut U-shaped; dioecious and asexual reproduction; with or without gill slits.

Order Rhabdopleurida. True colonies with zooids in organic continuity; each zooid in a secreted tube; 2 tentaculated arms; gill slits absent; 1 gonad. *Rhabdopleura.*

Order Cephalodiscida. Aggregated in common coenecium; zooids unconnected; four to nine pairs of tentaculated arms; one pair of gill slits; 2 gonads. Two genera are recognized: *Cephalodiscus, Atubaria.*

Class Planctosphaeroidea. A class known only by its transparent larva with highly branched ciliary band and complex internal structure. *Plancto-sphaera.*

■ CLASS ENTEROPNEUSTA
GENERAL MORPHOLOGIC FEATURES
External anatomy and physiology

Hemichordates have the tricoelomate structure of deuterostomes. The elongated, wormlike forms are commonly called acorn or tongue worms. Some do not exceed 20 mm. in length, although others may reach a length of 2,500 mm. (*Balanoglossus gigas*). Their cylindric, flaccid, mucus-covered body is divided into a tonguelike proboscis (protosome), a short cyclindric collar (mesosome), and an elongated trunk (metasome) (Fig. 17-4). The proboscis is connected to the collar by a narrow proboscis stalk. The short ringlike collar projects anteriorly (collarette) over the proboscis stalk and the posterior part of the proboscis and ventrally bears the mouth. The trunk, which makes up most of the body, bears along its length certain

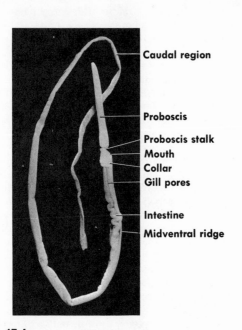

FIG. 17-4

Preserved specimen of tongue worm (*Saccoglossus* [*Dolichoglossus*], class Enteropneusta).

anatomic landmarks that correspond generally with various internal features. A middorsal and a midventral ridge indicate the location of median blood vessels and nerves (Figs. 17-4 and 17-5, *A*). Just back of the collar on each side of the middorsal ridge is the branchial region, which bears a longitudinal row of gill pores. The lateral region on each side of the anterior part of the trunk shows a genital ridge, or else 2 genital wings, underneath which are the gonads, whose gonopores to the exterior cannot be detected.

The postbranchial part of the trunk, which is relatively undifferentiated externally, may show the darkened outline of the intestine through the body wall, forming the hepatic region. In such a case the remaining part of the trunk is called the caudal region. In many genera these external divisions are not evident. The anus is found at the terminal end of the trunk.

The body is covered with ciliated columnar epithelium that bears many gland cells in certain areas, especially in the trunk. Plankton and other small organisms are picked up by the mucus (which may contain an amylase) of the proboscis and carried toward the mouth by ciliary currents in nonburrowing forms (Fig. 17-5).

Internal anatomy and physiology

In addition to the epidermis, the body wall contains a well-developed nervous layer in the base of the epidermis; a basement membrane that forms the proboscis skeleton, composed of a median plate in the proboscis stalk and 2 posterior horns in the roof of the buccal cavity; a muscular sheath of outer circular and inner longitudinal fibers in the proboscis and collar regions or mostly longitudinal fibers in the trunk; and a connective tissue and muscle fiber layer that fills most of the coelomic cavities. A peritoneum is lacking except in *Protoglossus*.

The coelom of hemichordates is divided into three segments: the proboscis, or protocoel, with a single cavity; the collar, or mesocoel,

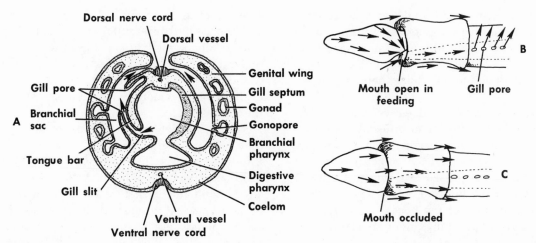

FIG. 17-5

A, Transverse section of pharynx (diagrammatic) of a hemichordate. **B,** Side view of anterior end of a hemichordate with mouth open, showing direction of currents created by cilia on proboscis and collar. Food particles from detritus or plankton are directed toward mouth; rejected particles are directed to outside of collar. **C,** When mouth is occluded, all particles are rejected and passed onto the collar. Nonburrowing and some burrowing hemichordates utilize this method of feeding. (**B** and **C** adapted from Russell-Hunter.)

FIG. 17-6

Diagram of longitudinal section of anterior part of *Saccoglossus (Dolichoglossus).* (From Hickman: Integrated principles of zoology, The C. V. Mosby Co.)

with 2 cavities; and the trunk, or metacoel, with 2 cavities. The protocoel is present in the posterior part of the proboscis and opens by a middorsal pore to the exterior (Fig. 17-6), the paired cavities of the mesocoel open by a pair of canals and pores on each side of the middorsal line; the paired metacoel cavities, separated by the mesenteries of the gut (Fig. 17-5, *A*), have no external openings. The three types of cavities are separated from each other by a transverse septum or by other tissue.

The muscular system, already described, consists of smooth muscle fibers of coelomic origin, which course through the coelomic spaces and the inner surface of the epidermis. The muscular arrangement varies with the different re-

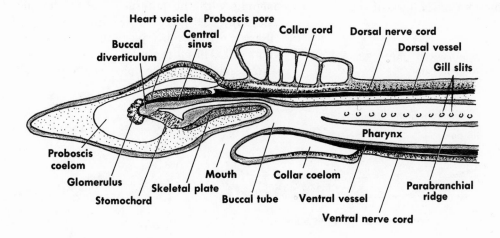

gions of the body, especially in the proboscis and collar where the fibers may crisscross in complicated patterns, often filling up much of the coelomic spaces. The trunk musculature is composed of longitudinal fibers beneath the epidermis.

Digestive system. The digestive tract is a straight tube with little or no musculature in its walls. In the ventral region of the collar between the collarette and the proboscis is the large mouth, which passes to a buccal cavity in the center of the internal collar (Fig. 17-6). From the dorsal part of the buccal cavity projects anteriorly a buccal diverticulum, formerly thought to be a notochord but now often called a stomochord. It is considered to be merely an extension of the gut. The buccal cavity opens posteriorly into the pharynx, in the dorsal half of which on each side are the U-shaped openings that connect with the gill pouches (Fig. 17-6). The ventral half of the pharynx is part of the digestive tract (Figs. 17-5, A and 17-6). These two regions of the pharynx may be separated by lateral constrictions. Following the pharynx in the digestive tract is the esophagus, which may be differentiated in certain regions. In some members of the families Harrimaniidae and Spengelidae, the esophagus communicates with the exterior by canals and pores. The remaining part of the digestive system is the intestine, the posterior part of which is the rectum that opens to the outside by the anus. The anterior part, or darkish hepatic

FIG. 17-7

Balanoglossus (class Enteropneusta). (After Spengel; redrawn from Cambridge Natural History.)

division of the intestine, has in some cases dorsal saccular differentiations (Fig. 17-7) and can usually be distinguished from the posterior region by its dark brown or greenish inclusions. Glands for the production of enzymes are found in the hepatic region, but little is known about their digestion.

Feeding mechanisms of Enteropneusta. One of the most interesting physiologic aspects of the Enteropneusta is their method of feeding. In addition to the habit of the burrowing forms that engulf sand and extract organic substance from it as it passes through their enteron, many enteropneusts are suspension feeders with a unique method of obtaining their food. Many investigators in recent years have studied this method, which is also found to some extent in burrowing forms. The investigations are well summarized by E. J. W. Barrington (1965).

According to investigations, the principal feeding mechanism is the collection of food particles by the mucus on ciliary tracts of the proboscis and partly on the anterior part of the collar (Fig. 17-5, B and C). Food (plankton and other small organisms) is carried backward and ventrally to the preoral ciliary organ (groovelike body on the posterior face of the proboscis). Here the food threads are collected together in mucous ropes that pass to the mouth. Large particles or unwanted substances involve a rejection mechanism that halts the regular mucus feeding method by closing the mouth opening with the anterior edge of the collar. These rejected particles are then carried to the ciliary paths running posteriorly on the collar and trunk. The ciliary preoral organ may test the water and food as they pass to the mouth and

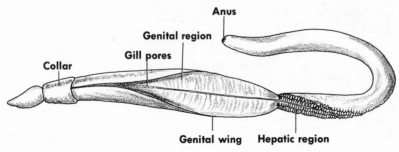

may form a part of the selectivity apparatus. Investigations show that the ciliated surface of the trunk and the posterior part of the collar do not play a role in the feeding process but only in the rejection mechanism. By micropinocytosis some ingestions of food materials may take place in the skin as the food particles pass along the ciliary tracts of the proboscis. In those species in which the pharynx is subdivided, the gill slits are food strainers. They pass the water to the outside but retain food and sand and send them to the ventral gutter of the pharynx.

The principal feeding method described may be evidence that the Enteropneusta evolved from pterobranch-like ancestors, as the mucous trapping is similar to that of the food-trapping of the ciliated tentacles of the Pterobranchia.

Circulatory and excretory systems. The circulatory system includes a middorsal vessel, in which the blood is carried anteriorly; a midventral vessel; and a system of sinus channels (Fig. 17-5, A). Near the anterior level of the collar, the dorsal vessel, which runs in the dorsal mesentery of the digestive tract, expands into a venous sinus. The venous sinus then passes anteriorly into the central sinus, located below the heart vesicle, or pericardium, which has muscle fibers in its lower wall. The contraction of this lower wall drives blood through the central sinus into a special plexus organ, the glomerulus. The buccal diverticulum, central sinus, heart vesicle, and glomerulus together form a projection complex into the proboscis coelom. This complex is covered with peritoneum, the fingerlike evaginations of which contain blood sinuses and produce the organ known as the glomerulus. The sinuses in the glomerular evaginations contain blood from the central sinus. The glomerulus may free the blood from waste; hence it is called an excretory organ. From the glomerulus, vessels or arteries pass backward on each side to the ventral longitudinal vessel, which runs posteriorly beneath the digestive tract almost to the anus. Blood from the longitudinal vessel passes through an extensive plexus of sinuses to the gut and body wall. From this plexus the blood goes to the dorsal vessel, thus completing the circuit.

Branchial system. The branchial apparatus is supposed to be concerned with gaseous exchange in breathing. The dorsal branchial region of the pharynx is placed just above the ventral digestive region and bears a longitudinal series of gill slits that vary in number from a few to a hundred or more on each side of the body (Fig. 17-6). This condition is called pharyngotremy, or perforations in the walls of the pharynx. New pairs are added as the animal grows older. The gill slits, or pharyngeal openings, do not open directly to the outside, but into branchial sacs (Figs. 17-5, A, and 17-8). The gill slits in the inner pharyngeal wall are U-shaped, produced by the downgrowth of tongue bars from the pharyngeal wall. Each slit opens into a separate branchial sac, which opens to the exterior by a dorsolateral gill pore that lies with the other gill pores in a longitudinal furrow along the branchial region of the body. The septa, or pharyngeal walls between clefts, and the tongue bars are ciliated and provided with a plexus of blood sinuses from the ventral longitudinal vessel. The cilia produce water currents through the mouth and through the gill slits. These water currents are important in pulling mucous strands of food into the mouth, but the gill apparatus does not capture food particles, which may have been its primitive function. Respiratory exchange of gases may occur between the blood sinuses of the septa and tongue bars and the water currents passing through the gills.

Nervous and sensory system. The nervous system consists of a net of nerve fibers in the base of the epidermis over the entire body and middorsal and midventral longitudinal cords formed from the thickening of the epidermal nervous layer (Fig. 17-6). The Hemichordata thus share with the Coelenterata and the Echinodermata a primitive nerve net or plexus. The 2 longitudinal nerve trunks are connected at

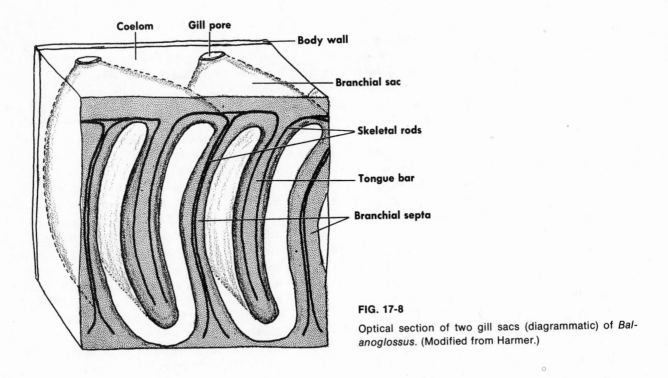

FIG. 17-8

Optical section of two gill sacs (diagrammatic) of *Balanoglossus*. (Modified from Harmer.)

the anterior end of the trunk by a circumenteric ring. The dorsal cord continues into the collar as the collar cord, and the ventral cord terminates at the collar. In some the collar cord is hollow and may open to the outside by neuropores; in other acorn worms it may contain scattered cavities. The collar cord seems to represent the chief nervous center of enteropneusts. Conduction in the trunk can take place through the epidermal plexus, as well as through the ventral and dorsal nerve cords. The collar cord is known to have giant nerve cells, the processes of which run posteriorly in the dorsal nerve trunk and to the ventral cord by way of the circumenteric ring.

Sense organs are represented by neurosensory cells throughout the epidermis. They are especially abundant in the proboscis. The preoral ciliary organ on the ventral side of the proboscis base may have chemoreceptor functions. W. N. Hess (1937) discovered photoreceptor cells in *Saccoglossus (Dolichoglossus) kowalevskii*.

Reproductive system. The sexes are separate

FIG. 17-9

Burrow of *Balanoglossus clavigerus*. U-shaped burrow is provided with a round opening for posterior end and a funnellike opening for anterior end. The burrow, formed by thrusts of proboscis, is lined with mucus, which cements the sand particles together. The anterior end of the burrow may have more than one branch, either of which may be used by the worm in its movements. (Modified after Stiasny, from Hyman.)

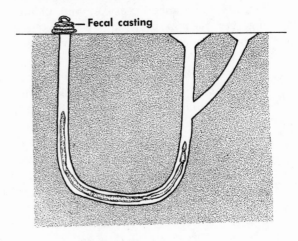

in enteropneusts, and there is no sexual dimorphism. The gonads are saclike and lie on each side of the trunk in 1 or several longitudinal rows in the anterior region (Fig. 17-5, A). They may lie behind the gill slits in some species. The region containing the gonads may bulge to the outside as wings or ridges. Each gonad has a necklike extension, which opens to the outside by a pore that may be located in the same groove as the gill pores.

The eggs are large (1 mm.) and yolky in the family Harrimaniidae but are small in the other two families. In burrowing species (Fig. 17-9), eggs are discharged from the burrows in coils that form masses of several thousand. Shortly thereafter, sperm are emitted from the burrows of the males. The masses of fertilized eggs are broken up by tidal action, and the embryos are scattered in the sea.

DEVELOPMENT

Indirect development involves a pelagic tornaria larva (Fig. 17-10), but in direct develop-

FIG. 17-10

Tornaria larva. This larva, which has remarkable resemblance to bipinnaria larva of sea stars, was first described by Johannes Müller, but E. Metschnikoff was first to point out its true systematic position in 1869.

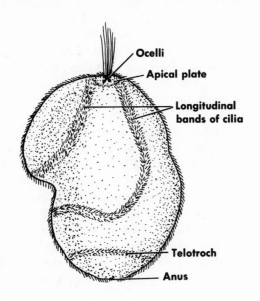

Ocelli

Apical plate

Longitudinal
bands of cilia

Telotroch

Anus

ment there is no tornaria. The cleavage is holoblastic, equal, and of the radial type, producing a coeloblastula, which by embolic invagination forms the archenteron. Where the blastopore closes marks the location of the future posterior end of the animal. The animal elongates in the anteroposterior axis, forms cilia, rotates in the egg membrane for a time, and then hatches to begin a planktonic existence. Direct development usually has a free-swimming larva, as does indirect development; and up to the larval stage the major features are about the same in both types of development. In indirect development the larva gradually assumes the tornaria characteristics by forming sinuous ciliary bands and losing cilia elsewhere (Fig. 17-10). At a certain stage the tornaria resembles the bipinnaria larva of asteroids. Tornariae range in size from less than 1 mm. to 4 or 5 mm. in length. After a planktonic existence of several days or weeks the tornaria undergoes several changes, sinks to the bottom, and assumes an adult existence. In direct development the tornaria larva is bypassed, and the ciliated larva develops directly into the adult condition. Tornariae are found in *Balanoglossus, Ptychodera,* and some others. The common American species *Saccoglossus (Dolichoglossus) kowalevskii* has direct development without a tornaria.

■ CLASS PTEROBRANCHIA
GENERAL MORPHOLOGIC FEATURES

These small members of the hemichordates have many basic resemblances to the enteropneusts. Pterobranchs may or may not have gill slits. They inhabit both shallow and deep water, chiefly the latter. Most species occur in the southern hemisphere. Of the three genera, two live in aggregations or colonies in secreted tubes called coenecia, which are attached to various kinds of bottoms—rocky, muddy, sandy, or gravelly. They may also be attached to other sessile animals, such as sponges and bryozoans. Many species have been collected in the Antarctic and sub-Antarctic regions. They are small

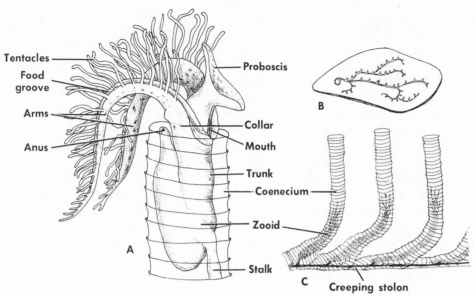

Tentacles
Food
groove
Arms
Anus
Proboscis
Collar
Mouth
Trunk
Coenecium
Zooid
Stalk
Creeping stolon
A
B
C

FIG. 17-11

A, *Rhabdopleura,* a pterobranch hemichordate in its tube.
B, A small colony of *Rhabdopleura* on a bit of shell. **C,**
Portion of colony with zooids retreated into tube. (**A** after
Delage; **B** and **C** adapted from Hyman, after Schepotieff.)

animals, usually within the range of 1 to 5 mm.
(not including the stalk). The members of *Cephalodiscus* are free living, but many individuals
live together in coenecia that may be joined together. In *Rhabdopleura* (Fig. 17-11) the individuals form true colonies, with the members
in organic continuity, although each is enclosed
within a tube. Individuals have extensible body
stalks, which attach the animal to its tube
and which also bears buds. The individuals of a
colony are called zooids. The body is divided
into three regions of protosome (proboscis),
mesosome (collar), and metasome (trunk). The
corresponding body cavities of these divisions
are protocoel, mesocoel, and metacoel (Fig.
17-12). The body is vase shaped, and the gut is
a loop with the anus opening on the dorsal side
of the collar (Fig. 17-12). The proboscis is
shield shaped and is often called the buccal
shield. It is tilted toward the ventral side and
partially hides the crescent-shaped mouth.
On its dorsal side the collar bears the ten-

taculated arms (lophophore). In *Cephalodiscus*
the collar bears five to nine pairs of arms; in
Rhabdopleura a single pair. Each arm is composed of a single stem (containing a groove)
with many side tentacles that increase in number with age. Food in the form of small organisms is caught by the adhesive secretions of the
tentacles and arms and is conveyed by ciliary
grooves to the mouth. *Cephalodiscus* draws into its mouth a stream of water that is discharged
through its gill slits.

Cephalodiscus has one pair of gill slits;
Rhabdopleura has none. The gill apparatus
contains no tongue bars, and the pharyngeal
region continues as the esophagus without
marked demarcation. A slender diverticulum
of the alimentary canal at the base of the
proboscis is called a stomochord (Fig. 17-12).

The sexes are separate, but asexual reproduction from the stalk buds is common. A ciliated
larva is found in *Cephalodiscus*. In *Rhabdopleura* new individuals are formed by budding from a creeping stolon; each young zooid
buds off a new stolon from its stalk, thus resulting in colony formation (Fig. 17-11, *C*).

Atubaria differs from the other two genera
in having no coenecia, no bud formation, and

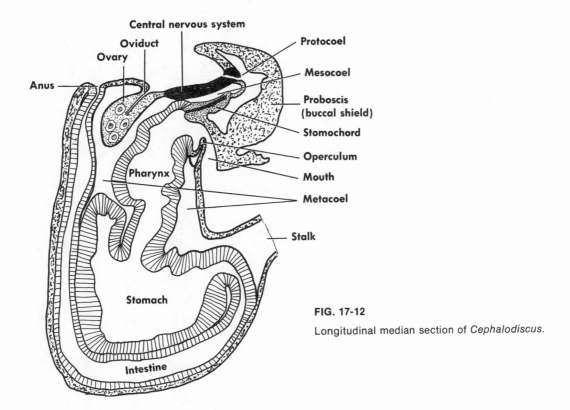

FIG. 17-12

Longitudinal median section of *Cephalodiscus*.

in being solitary. Its structure is similar to that of *Cephalodiscus*. It uses its stalk to anchor itself to hydroids or other objects.

■ PHYLUM POGONOPHORA (BRACHIATA)

The latest phylum in the animal kingdom to be appraised is the Pogonophora, or beard worms, which were first discovered in 1914 in the deep sea off the coast of Indonesia. Although a number of French zoologists studied the animals in the years following their discovery, it remained for the Russian investigator A. V. Ivanov to establish them (1955) as a separate phylum, which he called Brachiata.

Since they were first found, their geographic range of discovery has been greatly extended; it is now believed that they inhabit all seas. The Bering Sea, the Sea of Okhotsk, and the seas of the Malay Archipelago seem to have the richest fauna and the greatest number of species at present. But new species are continually being discovered elsewhere, such as in the waters around New Zealand, Norway, and Africa. Recently they have been found in great abundance off the coast of California.

Pogonophores are marine forms, usually but not exclusively found at abyssal depths of the sea. As new species are discovered, many have been found in relatively shallow water (22 meters). Some species are able to live in both shallow and abyssal waters. The greatest depth at which they have been found is about 10,000 meters, but most have been discovered in depths of from 150 to 1,500 meters. They are sessile animals. They live in closely fitting cylindric tubes that are much longer than the animal (Fig. 17-13) and are made of an organic complex of chitin and protein. The tubes may be uniformly smooth or made up of rings alternat-

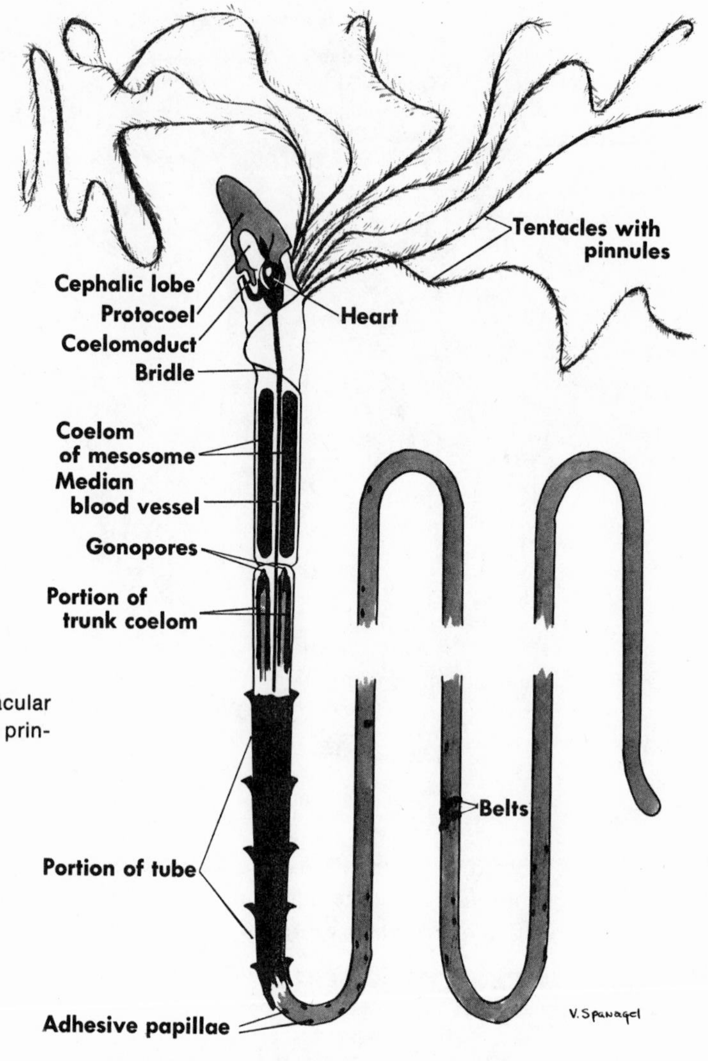
Tentacles with pinnules
Cephalic lobe
Protocoel
Coelomoduct
Heart
Bridle
Coelom of mesosome
Median blood vessel
Gonopores
Portion of trunk coelom
Belts
Portion of tube
Adhesive papillae
V. Spanagel

FIG. 17-13

Pogonophore with tentacles separated from tentacular crown (diagrammatic). (From Hickman: Integrated principles of zoology, The C. V. Mosby Co.)

ing with elastic interspaces. The diameter of the tube varies between 0.1 and 2.8 mm. and its length up to 150 cm., depending on the species. Part of the tube usually projects above the surface of the seabed, and the remaining part is deeply embedded in the muddy ooze and sediment. However, the anterior part of the tube usually has thin walls and is so flimsy that it collapses and lies flat on the surface.

Pogonophores may be found in great abundance wherever they occur. Since they are filter feeders, their presence in any locality depends on the abundance of food particles and bacterial fauna suspended in the water, plus a steady bottom current. In land-locked seas and deep trenches the pogonophore fauna seems to be more diversified than in the open ocean.

CLASSIFICATION*

Order Athecanephria. Protosome separate from mesosome; pericardial sac close to dorsal part of heart;

*Ivanov (1963) describes 71 species of the 80 or more that have been discovered.

tentacles one or many, and always separate; coelomoducts divergent with lateral external pores; postannular portion of metasome without transverse ventral rows of adhesive papillae; spermatophores spindle shaped; two families; five genera. The chief genera are *Oligobrachia* and *Siboglinum*.

Order Thecanephria. Protosome may not be externally separate from mesosome; coelomoducts convergent with median pores; tentacles may be fused together; postannular region of metasome with transverse ventral rows of adhesive papillae; spermatophores flattened or leaf shaped; three families; nine genera. The chief genera are *Heptabrachia*, *Diplobrachia*, *Lamellisabella*.

GENERAL MORPHOLOGIC FEATURES
External anatomy and physiology

Pogonophores are cylindric worms with long, slender bodies. They vary in length and diameter, with the length usually 100 to 600 times the breadth (Fig. 17-13). One species (*Siboglinum minutum* is only 0.1 mm. in breadth, whereas some have a breadth of 2.5 mm. The length varies from 4 or 5 to 45 cm. The elongated body is composed of three sections: protosome, mesosome, and metasome. The two anterior sections (protosome and mesosome) may be more or less fused into an anterior region that is separated from the metasome or long trunk region by a diaphragm. The trunk is subdivided into a preannular section that extends to the girdles, or belts (annuli), and a postannular region posterior to the girdles. The 2 or 3 belts, or girdles, are elevated regions on muscular ridges and bear toothed platelets. The girdles may or may not encircle the body and help to anchor it in the tube. Dorsal and ventral sides are not evident, but the dorsal side is considered by some authorities to be the one on which the cephalic lobe with the brain is located and the ventral side the one to which the tentacles are attached.

FIG. 17-14

Cross section of tentacular crown of pogonophore *Lamellisabella*. Since pogonophores have no digestive system, the tentacles may so cluster together to form an intertentacular cavity where digestion may occur. (Modified from Ivanov.)

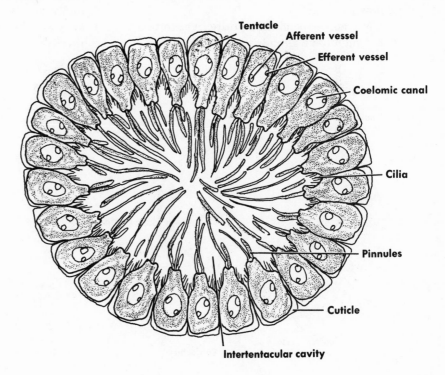

Tentacle

Afferent vessel

Efferent vessel

Coelomic canal

Cilia

Pinnules

Cuticle

Intertentacular cavity

The protosome is the first division of the body and carries the dorsal cephalic lobe and the ventral tentacles. In some species it may be divided from the mesosome by an annular groove. The ciliated tentacles are extensions of the body and coelomic cavities and are attached to the front end of the protosome, where they are usually arranged in the form of a crown. They number from 1 to 268. They may be fused to form a cylinder (Fig. 17-14) or even twisted to form a spiral cavity in which digestion of particulate matter may occur. Each tentacle receives a single nerve from the brain. The tentacle contains a coelom (lined with peritoneum) in which afferent and efferent blood vessels run. Each tentacle bears on its inner surface a fringe of filiform pinnules arranged in rows. These may be lacking in some species. Capillaries from the tentacular blood vessels occur in the pinnules. Adhesive papillae, or oval protuberances with cuticular plaques, are found on ridges of the metasome, especially in the preannular regions. They help the animal cling to the inside of the tube as it moves up and down.

Pogonophores spend their entire lives in their tubes, which have already been described.

Internal anatomy and physiology

The body wall is made up of a cuticle, an epidermis of columnar epithelium with glands, and circular and longitudinal muscles (Fig. 17-15). The coelom (protocoel) in the protosome is single and opens to the surface by a pair of ciliated coelomoducts (Fig. 17-13). The coelomic canals of the tentacles originate from the protocoel. The coelomic sacs in the mesosome and metasome (mesocoel and metacoel) are paired. The mesocoel, unlike that of hemichordates, has no coelomoducts, but the metacoel has a pair, just as the protocoel does. A peritoneal membrane lines the coelomic cavities.

There is no digestive or respiratory system. Respiration is probably a function of the tentacular apparatus, where an extensive surface is exposed to water by the tentacles and their pinnules. Nutrition is carried on by the tentacular apparatus. Pogonophores have external and extracellular digestion. Digestive enzymes, secreted by the epidermis of the tentacles or glands at the base of the pinnules, act on the food collected by the tentacular ciliary bands and digest the food within the intertentacular cavity. Absorption of the food takes place through the walls of the tentacles.

FIG. 17-15

Cross section in mesosomal region (diagrammatic) of pogonophore *Siboglinum,* showing body wall and internal structure. (Modified from Ivanov.)

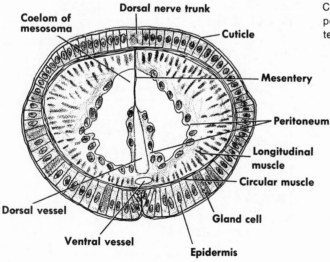

Coelom of mesosoma · Dorsal nerve trunk · Cuticle · Mesentery · Peritoneum · Longitudinal muscle · Circular muscle · Gland cell · Epidermis · Ventral vessel · Dorsal vessel

Excretion takes place through the coelomoducts of the protocoel (Fig. 17-13). Certain cells in the mesosome and metasome may store excretory products.

Pogonophores have a closed vascular system, which is well developed. A dorsal and a ventral blood vessel (Fig. 17-15), running the entire length of the body, are connected at the anterior end by blood vessels of the tentacular system and at the posterior end by transverse commissural vessels. A ventral heart, differentiated from the ventral vessels, is found in the mesosome. Blood flows forward in the ventral vessel and backward in the dorsal vessel.

The nervous system is located within the epidermis (Fig. 17-15). In the cephalic lobe of the protosome is a ganglionic mass (brain), from which arises a dorsal nerve trunk, with nerves to the tentacles. Neurochords, or giant nerve fibers, are found in the dorsal nerve trunk. Little is known about the innervation of the internal organs.

Pogonophores are dioecious. The gonads are located in the metacoel. The male system consists of a pair of testes and a pair of sperm ducts. The sperm ducts are long and ciliated and open ventrally at the mesosome-metasome septum. The testes are large and, with their ducts, fill most of the posterior half of the metasome. The mature sperm are packed in spermatophores to which filaments are attached that may be used in the transfer of the spermatophore

FIG. 17-16

Spermatophore of *Lamellisabella*, with long attached filament. (Modified from Ivanov.)

(Fig. 17-16) to the female. The female system consists of a pair of long ovaries, which lie in the anterior end of the metacoel, and a pair of oviducts. The oviducts are actually a pair of U-shaped coelomoducts that have a funnel-like opening to the outside.

DEVELOPMENT

Eggs are laid by the female in the anterior part of the tube in which she lives, where fertilization takes place and the entire development occurs. A clutch may consist of 10 to 45 eggs laid in a row in front of the animal. Eggs are rich in yolk and, after fertilization, are surrounded by a fertilization membrane.

The irregular nature of cleavage and gastrulation has already been noted. The first cleavage plane runs obliquely across the egg and divides it into 2 equal blastomeres. At the second cleavage both blastomeres divide unequally, with the 2 larger ones lying at the anterior and posterior poles and the two smaller ones between. Many subsequent cleavages are unknown at present; but there is a distribution of the larger blastomeres to the hind end of the embryo, which early shows a bilateral symmetry. Although there is no trace of an enteric cavity (except an endodermal solid cord that does not develop a gut), the coelomic cavities arise as a pair of sacs from a central cell mass. The coelom is thus considered enterocoelous in its formation. The coelomic pouches arise in very much the same manner as those of the hemichordates. The complete embryo is provided with a ciliary band on the protosome, and part of the mesosome bears a ciliated tract. On the metasome are also some ciliated bands and bundles of short bristles. When removed from the maternal tube, the larva can swim around, but normally it remains in the tube and is not free swimming; its locomotor devices are simply a remnant of ancestral conditions.

When fully formed, the young pogonophore leaves the tube and begins to make a tube of its own. Little is known about its subsequent differentiation of structures.

REFERENCES

Beauchamp, P. de. Chaetognatha. In Grassé, P. P. (editor). 1960. Traité de zoologie, vol. 5. Paris, Masson et Cie. The most recent description of the group.

Bieri, R. 1959. The distribution of planktonic Chaetognatha in the Pacific and their relationship to the water masses. Limnol. Oceanogr. **4**:1-28.

Bullock, T. H. 1940. Functional organization of the nervous system of Enteropneusta. Biol. Bull. **79**:91-113.

Burfield, S. 1927. Liverpool Marine Biological Commission, Memoir 28, Proceedings and Transactions of the Liverpool Biological Society **28**, pp. 1-104. An excellent monograph.

Dahlgren, U. 1917. Production of light by Enteropneusta. J. Franklin Inst. **183**:735-754. Luminescence is restricted to the family Ptychoderidae. Photogenic cells are modified epidermal cells of the goblet type. These glands discharge the slime that produces the luminescence.

Dawydoff, C. Stomocordes. In Grassé, P. P. (editor). 1948. Traité de zoologie, vol. 11. Paris, Masson et Cie. This is one of the outstanding treatises on the Hemichordata, although the author has proposed a different name for the group.

De Beer, G. 1955. The Pogonophora. Nature (London) **176**:888. An early appraisal of the group.

Hartman, O. 1954. Pogonophora Johansson, 1938. Syst. Zool. **3**:183-185.

Hess, W. N. 1936. Reaction to light in *Ptychodera*. Washington, D. C. Carnegie Institution of Washington. Papers from Tortugas Laboratories **31**:77-86.

Hess, W. N. 1937. Nervous system of *Dolichoglossus kowalevskii*. J. Comp. Neurol. **68**:161-171.

Hyman, L. H. 1959. The invertebrates; smaller coelomate groups, vol. 5. New York, McGraw-Hill Book Co., Inc. The best account in English.

Ivanov, A. V. 1963. Pogonophora. (Translated and edited by D. B. Carlisle.) New York, Consultants Bureau. This treatise by the most active investigator of the phylum gives a fine evaluation of its present status. Seventy-one of the eighty known species are described in it. Illustrated, with up-to-date bibliography.

Kuhl, W. Chaetognatha. In Bronn, H. G. (editor). 1938. Klassen und Ordnungen des Tierreichs, vol. 4, part 4. Leipzig, Akademische Verlagsgesellschaft. The most comprehensive treatment of the phylum.

Manton, S. M. 1958. Embryology of Pogonophora and classification of animals. Nature (London) **181**:748-751.

Morgan, T. H. 1894. The development of *Balanoglossus*. J. Morphol. **9**:1-86. A classic description by the famous geneticist.

Shipley, A. E. *Sagitta*. In Harmer, S. F., and A. E. Shipley (editors). 1896. Cambridge natural history, vol. 2. London, The Macmillan Co. A brief account.

Southward, A. J., and E. C. Southward. 1963. Biology of some Pogonophora. Washington, D. C. Proceedings of the Sixteenth International Congress of Zoology **1**:96. In this and other accounts these investigators have discussed the ecology and distribution of this newly described phylum. Pogonophores are found chiefly in muddy sand at depths of from 150 to 1,500 meters and are very spotty in their distribution because of the absence of mobile stages in their life cycle.

Swedmark, B. 1964. The interstitial fauna of marine sand. Biol. Rev. **39**:1-42.

Van der Horst, C. J. Hemichordata. In Bronn, H. G. (editor). 1927-1939. Klassen und Ordnungen des Tierreichs, vol. 4, part 4. Leipzig, Akademische Verlagsgesellschaft. The most authoritative and comprehensive account of the Hemichordata.

The invertebrate chordates
(the protochordates)

CHARACTERISTICS

1. Two subphyla of chordates, the urochordates, or tunicates, and the cephalochordates, are really invertebrates because they lack a vertebral column, although they are built on the chordate ground plan to some extent. They are also called protochordates because they possess (at some time in their life history) the chief diagnostic characteristics of chordates—the notochord, dorsal hollow nerve cord, and pharyngeal clefts. (The third subphylum of the chordates is the vertebrates, composed of backboned animals, which make up 95% of the phylum, or about 43,000 species.)

2. Of the three classes of urochordates—the Ascidiacea, Larvacea, and Thaliacea—the Ascidiacea, or sea squirts, are by far the most common and the most interesting from an evolutionary point of view. The adult sea squirt has little resemblance to a chordate form, for in its sessile existence it really has only one basic chordate characteristic, the pharyngeal cleft modified in connection with its mucusciliary filter feeding. The ascidian adult may be basically a primitive type representing direct descendants of hemichordate-like, filter-feeding ancestors, and its adult condition is not due to sessile degeneration.

3. The ascidian tadpole is a brief free-swimming larval stage in some ascidians, and serves as an archetype of the chief diagnostic characteristics of the chordate plan—notochord, dorsal nerve cord, pharyngeal gill slits, and muscular postanal tail. The tadpole quickly attaches itself and undergoes a metamorphosis of rotation of body organs and a reduction of the tail by a shrinkage of the epidermis, which draws the notochord and muscles into the body, where metamorphosis is completed.

4. Both urochordates and cephalochordates belong to the Deuterostomia. Cleavage is bilateral and mosaic (an unusual condition for a deuterostome) in urochordates. Gastrulation occurs by epiboly and invagination, and the large archenteron obliterates the blastocoel.

5. Cephalochordates represent a simplicity that may be due to the retention of larval characteristics. They possess a well-developed notochord, a typical hollow dorsal nerve cord, and many gill slits.

6. In its development the cephalochordate amphioxus is almost a classic case of simplicity and regularity. The cleavage is total, bilateral, and radial. Gastrulation occurs by invagination, and the anus originates from the blastopore. The enterocoelous coelom is formed by the budding off of three pairs of coelomic sacs from the anterior end of the archenteron. At first the larva is asymmetric but later acquires bilateral symmetry.

PHYLOGENY AND ADAPTIVE RADIATION

1. The attached ascidian is thought to represent the primitive ancestral condition and the brief ascidian tadpole stage, a neotenous form from which vertebrates finally evolved. The

probable neotenous origin of class Larvacea may indicate how vertebrates could have arisen. Tunicates may have developed from the chordate line at an early time, for they have no coelom and no segmentation. On the other hand, cephalochordates have a restricted coelom that may have a tripartite division embryonically. They also have a series of myotome units that make up the bulk of the tissues.

2. The ascidians as adults are mostly sessile, filter-feeding animals that remove plankton from the currents of water which pass through the pharynx. They have undergone considerable evolutionary divergence because they are adapted to a great variety of marine habitats. Solitary ascidians are usually large and require more space in their habitats, being attached to rocks, piles, and similar places where they are often exposed to violent wave action. Colonial, or compound, forms are best adapted to flat surfaces where water is fairly calm and clean. Colonial species are usually small, although the whole colony may be large, with the members sharing in a common cloaca having a single aperture. Some ascidians may also be grouped together to form a social group. The classes Thaliacea and Larvacea are chiefly adapted to a planktonic existence in the upper surfaces of the sea.

The cephalochordates live in sand with the head protruding in their sedentary feeding position, although they do swim around in short spurts. The small number of species show varying amounts of asymmetry, especially with regard to the arrangement of the gill slits.

The phylum Chordata includes the tunicates (urochordates), the cephalochordates, and the vertebrates. They possess the three most distinguishing characteristics of the phylum — a notochord, a gelatinous supporting rod above the gut; a dorsal hollow nerve cord; and paired pharyngeal gill slits on the sides of the pharynx. All these characteristics are found at some stage in the life history of a chordate animal, although they may be restricted to an embryonic stage. The urochordates and cephalochordates are small groups of marine animals whose chief interest to zoologists is the light that they throw on the possible origin of the phylum. They lack a backbone or vertebral column; they are therefore invertebrates and lie within the scope of this work.

Subphylum Urochordata (Tunicata)

Urochordates, or tunicates, differ from other chordates in the following ways: the notochord is chiefly restricted to the tail and posterior region of the body of the larval form and disappears during metamorphosis; no typical chordate metamerism is present; the typical coelom of chordates is reduced, and the main cavity is a kind of hemocoel characteristic of arthropods; an atrium developed from dorsal lateral invaginations of the epidermis is found around the pharynx in all ascidian and thaliacean tunicates but is absent in all vertebrates. A secreted tunic of living cells that contain supporting structures of calcareous deposits and cellulose forms an encasement for the body, and although they are deuterostomes, they have determinate instead of indeterminate cleavage.

Tunicates are strictly marine animals of cosmopolitan distribution throughout all seas from the intertidal zones to the abyssal depths. The most common and primitive group of tunicates is the sea squirts. Some tunicate species have a wide distribution through all latitudes; others have a restricted, local distribution.

About 2,000 species of tunicates are known to zoologists. They are usually classified in accordance with the position of their reproductive glands and the nature of their gill apparatus. Three classes are recognized—Ascidiacea, Thaliacea, and Larvacea (Copelata).

The class Ascidiacea is by far the largest and best known. Ascidians are sedentary animals with typical life histories and metamorphosis. The other two classes, Thaliacea and Larvacea (100 species), are pelagic or free-swimming groups and are specialized for a planktonic existence. Although the three classes have differences from each other, many characteristics of both morphology and physiology are shared by all of them.

Tunicates are of great interest in the evolutionary interpretation of the chordates. Formerly the adult sessile, or attached, form of tunicates was considered to be degenerate and the tadpole stage a recapitulation of a primitive condition. It is now thought that the attached ascidians are the primitive ancestral condition and that the chordate-vertebrate group arose from ocean-bottom forms. A plausible theory, widely accepted at present, is that vertebrates arose from tunicate tadpoles that became neotenous (retention of larval characteristics) by mutation and eliminated the sessile stage. These tadpoles, by swimming up continental streams to exploit the rich organic detritus, developed structures such as muscles, nerves, segmentation, and sense organs, which are largely lacking in the primitive sessile ancestors. The class Larvacea of existing tunicates may well be considered neotenous, with later stages omitted in its evolution. The early divergence of the urochordates from the chordate line of descent may explain the lack of segmentation in that group.

■ CLASS ASCIDIACEA

A common name for the ascidians is sea squirts. They are common in marine littoral waters, where they may be found attached to rocks, piers, shells, seaweed, and other objects. A few species live at considerable depths in the water. Common genera of ascidians along our coasts are *Ciona*, *Styela*, *Molgula*, and *Botryllus*. Most ascidians are attached, but *Polycarpa* and some others are free living.

External anatomy and physiology

The body of ascidians is cylindric or globose in shape and attached at one end by a stalk to a substratum (Fig. 18-1). It is enclosed by a test or tunic, which is lined by a membranous mantle with muscle fibers and blood vessels. The tunic contains tunicin (cellulose) and is secreted by the mantle. Cellulose is found in only a few other metazoans, such as sessile hemichordates and (recently) in the skin of man. At the end opposite the one attached to the substratum are 2 external openings that can be extended as siphons, the top incurrent siphon and the side excurrent siphon. The incurrent, or oral, siphon leads into the pharynx (branchial) region, and the excurrent, or atrial, siphon (dorsal side of

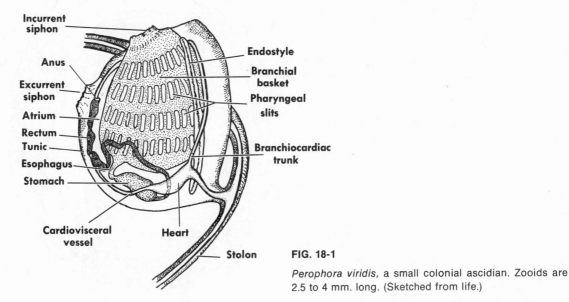

FIG. 18-1

Perophora viridis, a small colonial ascidian. Zooids are 2.5 to 4 mm. long. (Sketched from life.)

animal) leads away from the atrium. Water currents are kept moving in and out of the 2 siphons by the beating of cilia on the gill slits of the pharynx. The cilia of the different cells around the gill slits are known to beat slightly out of phase with each other to produce metachronal waves. The striated muscle of the body wall muscles can produce contraction of the body, causing the familiar squirts from the siphons.

Internal anatomy and physiology

Surrounding the pharynx (except where the pharynx is attached to the midventral body wall) is the atrium, which opens through the excurrent, or atrial, siphon (Figs. 18-1 and 18-4, *B*). The atrium is lined with ectodermal epithelium, which also lines the atrial siphon, and is continuous with the epidermis on the outside. The atrium contains strands of tissue or muscle to prevent overexpansion, and the siphons are provided with sphincters to regulate

intake and outflow of water currents. Below the atrial cavity is the visceral or abdominal cavity, which contains the digestive tract and other visceral organs.

The pharyngeal chamber is often barrel shaped and occupies a large part of the animal (Figs. 18-1 and 18-2). It is lined with endoderm and covered by ectoderm, with a mesenchymal framework between. Its entrance is guarded by a ring of tentacles for selecting food particles. The walls of the pharynx are pierced by small slits, through which water passes from the pharyngeal region into the atrium and thence through the atrial siphon to the outside. On the ventral side of the pharynx and extending through its whole length is a groove, the endostyle, the lateral walls of which are

FIG. 18-2

Cross section (diagrammatic) through pharyngeal region of tunicate. Arrows indicate direction of food-water currents.

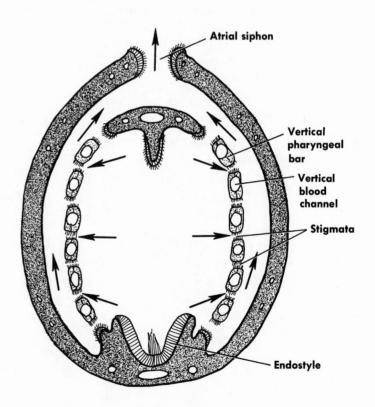

Atrial siphon

Vertical pharyngeal bar

Vertical blood channel

Stigmata

Endostyle

ciliated and the bottom of which contains mucus-secreting cells and long flagella (Figs. 18-1 and 18-2). From a pit at its posterior end the endostyle sends ciliated peripharyngeal grooves or ridges anteriorly, where the ridges form a band of glandular tissue just behind the anterior ring of tentacles. This band is a glandular tissue and secretes mucus. A dorsal membrane (dorsal lamina) runs posteriorly to the esophagus along the dorsal side. This membrane bears languets, or fingerlike projections, involved in the feeding mechanism.

The pharyngeal cavity between the dorsal membrane and the ventral endostyle bears the specialized gill slits called stigmata. The stigmata are curved or spiral shaped and represent subdivided gill slits (Fig. 18-3). They are arranged in horizontal rows that are separated by horizontal and vertical bars, thus producing a grid pattern for greater surface area.

Ascidians are mucus-ciliary feeders and live on the plankton organisms brought into the pharynx by water currents. The feeding mechanism involves the coordinating action of the two components—the water currents and the mucous entanglement of the food particles. An enormous amount of water for the size of the animal is strained for food. The principal sites of mucous formation are the endostyle and the anterior band of glandular tissue. Strands of mucus secreted by the endostyle (Fig. 8-1 and 8-2) are carried upward or anteriorly on each side by cilia on the pharyngeal wall and by the ciliated peripharyngeal grooves. As the strands of mucus cross the pharynx, the food particles carried by the water currents through the stigmata are trapped on the mucous cords and passed to the dorsal midline of the pharynx. Here the food is rolled into cords by the ciliated dorsal membrane, with the aid of the languets and tentacles. The mucous strands, now laden with trapped food, are passed posteriorly in the direction of the esophageal opening on the dorsal side of the pharynx, toward which the dorsal membrane is directed.

The digestive system of sea squirts varies but consists of a ciliated tube or narrow esophagus, enlarged stomach with glandular outgrowths (liver), and a more or less looped intestine (Figs. 18-1 and 18-4). In certain ascidians (*Ciona*) a tube arises as a double evagination at the base of the pharynx; the evaginations fuse together and then run as a simple tube parallel to the intestine, which it may surround. It is called an epicardium. In *Molgula* the epicardium transforms into a renal organ alongside the heart and stores excretory uric acid crystals. The epicardium may also play a role in budding. The anus opens into the cloacal region of the atrium. Digestion is extracellular. A pyloric gland composed of a network of vesicles occurs on the outer wall of the intestine and opens into the stomach. It may have excretory functions.

Some waste substances are excreted at the surface of the tunic, and some are carried by specialized cells (nephrocytes) to certain regions (intestine and gonads), where they accumulate to form vesicles or excretory storage organs.

The blood-vascular system is open, and capillaries are absent (Fig. 18-1). The tubular heart, with pacemakers at each end, alternates the direction of blood flow, resulting in a periodic reversal of blood flow. The reversal is controlled

FIG. 18-3

Spiral stigmata in *Corella*. Greater surface is possible by this arrangement. In most tunicates, stigmata are straight or curved. (Modified from Herdman.)

FIG. 18-4

A, Colony of *Clavelina*. All members of colony are connected by common blood system, and buds on stolons are vascular outgrowths of body and epicardium. **B,** *Clavelina* showing thorax and long abdominal region. (**A** modified from Herdman; **B** modified from Brien.)

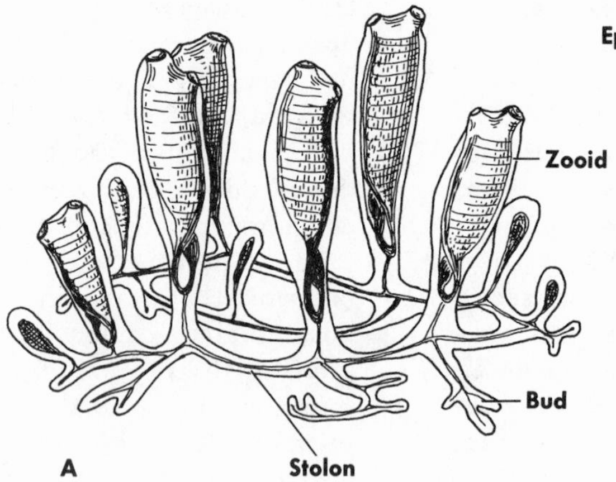

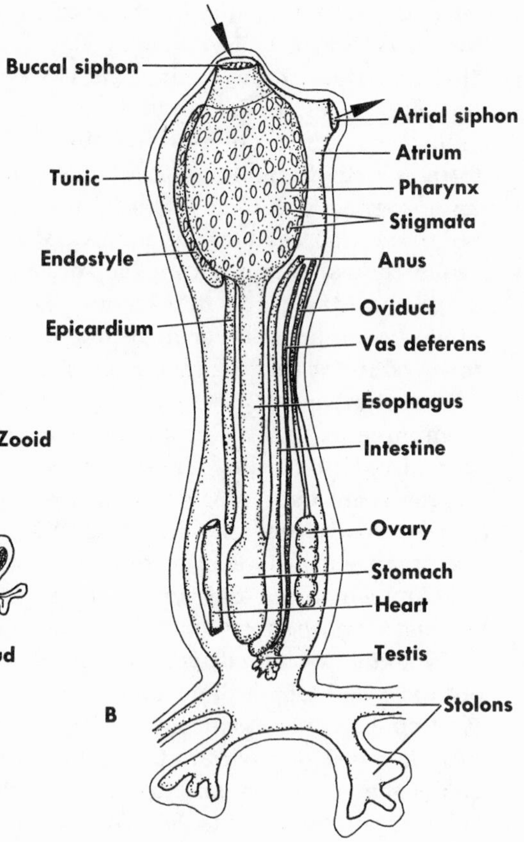

by 2 pacemakers. The heart is a specialized part of the pericardium and is located near the stomach. Each end of the heart opens into a large sinus channel. Blood is carried by the subendostylar channel to the endostyle region and to the gill slits where respiratory exchange occurs. From the pharyngeal region the blood is carried by a channel beneath the dorsal membrane back to the digestive tract and other organs, and from there it is carried to the dorsal end of the heart by a dorsal abdominal sinus. This pattern of the blood channels is similar in some ways to that of fishes.

No respiratory pigment is found in the blood plasma, the osmotic pressure of which is slightly higher than sea water. The plasma contains blood cells of several types. In some ascidians, oxides of vanadium and free sulfuric acid are found in the plasma and in certain cells. The

function of these two substances is unknown, although the sulfuric acid tends to keep the vanadium in a reduced state.

The nervous system consists of a cerebral ganglion (brain), located in the body wall between the 2 siphons, and a variable number of nerves that originate in the ganglion and run to the siphons, gill slits, and visceral organs. A neural gland lying near the brain communicates by a duct with the dorsal corner of the gill chamber. This structure is thought by some investigators to be homologous with part of the vertebrate pituitary gland.

Sensory cells of tactile and chemoreceptive functions are abundant around the siphons (both inside and out), around the buccal tentacles, and in the atrium. These are probably used in sampling the water.

Tunicates are nearly all hermaphroditic,

although the male gonads may develop before the ovaries. Usually there is a single testis and a single ovary, which are located within or posterior to the intestinal loop (Fig. 18-4, *B*). Both gonads are attached to the atrial wall, with the testis below the ovary. The ovary is a saccular structure, with germinal areas and follicles for the developing eggs. Mature eggs are carried by a single oviduct, which runs along the intestine and opens into the cloacal region in front of the anus. A sperm duct leads from the testis alongside the oviduct and opens into the cloaca. *Molgula, Styela,* and some others have paired gonads of each sex, with one set consisting of a male and female gonad located in the right body wall of the branchial region and the other set within the digestive loop. In such cases the gonoducts are also paired. Fertilization may be internal or external, with some forms viviparous and others oviparous. Some species are solitary (*Molgula, Styela,* and others), and some are colonial or compound (*Botryllus, Clavelina*). Solitary ascidians are generally large, but colonial individuals are usually small, although the colony may be large. In some colonies the individuals are separate entities but are united by stolons that may be long and trailing; in others the stolons are short, and the individuals form compact groups (Fig. 18-4, *A*). In some colonial fo.... the basal parts of the individuals are joined together (*Clavelina*), and in the highly specialized or compound forms many individuals share the same test (*Botryllus,* Fig. 18-5). In some of these compound ascidians each member has its own inhalant siphon, but the excurrent opening is a common cloacal chamber with 1 aperture in the center of the colony. Many other arrangements occur in compound ascidians. In the so-called social ascidians a peculiar type of colony is formed by the grouping together of two entirely different genera such as *Clavelina* and *Stolonica.*

Tunicates, with few exceptions, have a high capacity for regeneration of missing parts. A related phenomenon is budding, or asexual

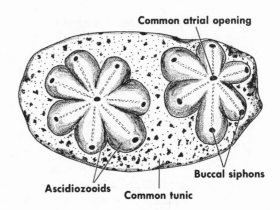

FIG. 18-5

Colony of *Botryllus* with common excurrent siphon and separate incurrent siphons. (Modified from Milne-Edwards.)

reproduction, which has been much studied, especially by N. J. Berrill, in recent years. Budding is chiefly a separation of a fragment or part of an individual and the subsequent growth and maturation of the fragment. Buds may detach and form independent individuals, or they may stay together and produce a colony. The nature of budding in tunicates is highly variable and complex. Usually it involves some form of constriction by the epidermis; the tissues are divided into separate masses, the nature of which depends on the part separated. In some a bud or blastozooid first appears on the stolon as a growth with food reserves. When the stolon and bud are separated from the parent, or the parent degenerates, the bud develops (*Clavelina,* Fig. 18-4, *A*). In others, after the thorax degenerates, the abdomen breaks up into buds. Nearly all colonies are formed by budding in some way, although not all budding results in colonies. It may be simply a method for increasing the number of independent individuals, nor is it restricted to the adult condition, for it often occurs in the larval stage.

Development

The embryologic development of tunicates has largely revealed the true status of the group, which was formerly placed among the Mollusca. The eggs of the solitary species tend to be smaller than those of the colonial forms, which have more yolk. Fertilization in most solitary species occurs in sea water; in colonial species internal fertilization is the rule. In *Botrylloides* the eggs are fertilized in the ovary, and development to the tadpole stage occurs there or in the atrium. The tadpole finally escapes through the body wall. Most colonial forms brood their young in the atrium.

There are many variations in the development of the different species of tunicates. In the common genus *Styela* the egg has a yellow-pigmented, a gray yolky, and a clear yolk-free area. The cleavage is holoblastic and radial but determinate (contrary to most other deuterostomes, in which it is indeterminate). In the 16-cell stage the blastomeres are arranged in two tiers of 8 cells, each with the yellow-pig-

mented region located in 4 of the vegetal cells. A flattened coeloblastula appears in about the 40-cell stage. The future ectoderm of the animal side is made up of thick cells, whereas the future endoderm of the vegetal side is flat. Gastrulation is epibolic and produced by an invagination of the vegetal pole cells, followed by an overgrowth of the prospective ectoderm. Within the archenteron the yellow-pigmented cells make up the mesoderm, which is not formed by an enterocoelous method, and the other vegetal cells produce the endoderm. The blastopore, which later closes, marks the

FIG. 18-6

A, Sagittal section of anterior end of free-swimming ascidian tadpole *(Ciona).* **B,** Four stages in metamorphosis of simple ascidian tadpole from attachment of tadpole to sessile adult stage. See text for description of process. (Modified from Kowalevsky and others.)

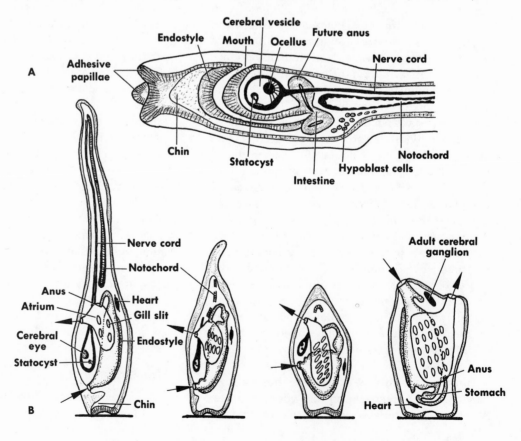

posterior end of the embryo. The dorsal layer of ectoderm anterior to the blastopore gives rise to the neural tube, which becomes the brain and spinal cord along the middorsal line. Beneath the neural tube the mesoderm forms the notochord, which is the forerunner of the vertebral column in vertebrates. The notochord enlarges to become the future axis of the tail, and the larva is hatched.

During later development, the larva elongates and becomes an appendicularia larva, commonly called the ascidian tadpole (Fig. 18-6, A). At this stage the tadpole has a nonfunctional digestive tract, two pairs of gill slits, a cerebral eye, statocysts, and 3 anterior adhesive papillae by which it can fasten itself to a substratum. The mouth, which later becomes the inhalant siphon, is located anteriorly but may not be opened during the larval stage. The mouth passes into the pharynx, which is followed by a looped digestive tract. A ventral endostyle occurs along the pharynx, and the atrium is represented by a small pocket. A tunic secreted by the ectoderm covers the entire larva and in the dorsal and ventral tail region forms a fin.

Metamorphosis begins after a short free-swimming but nonfeeding existence when the larva sinks to the bottom and attaches itself to a substratum by the adhesive papillae (Fig. 18-6, B). It is now known that the tail is not absorbed by phagocytosis, but a shrinking of the epidermis pulls the notochord and muscles into the body where structures are recognized. A rapid growth of the chin region near the adhesive papillae and the mouth results in the mouth being carried upward until it is opposite the point of attachment. Internal organs undergo a similar rotation of 90 to 180 degrees. A new nervous system replaces the old larval one. A differentiation of the intestine forms a stomach and liver. The circulatory system, which was started during the larval stage, now forms a heart. The number of gill slits increases, and the gonads are laid down. The final stages involve a functional use of the siphons. Free-swimming larvae are absent in tunicates that

have suitable substrata for attachment without the necessity of moving around to find one.

■ CLASS THALIACEA

The tunicates of class Thaliacea are pelagic and form part of the plankton population of the ocean, especially in tropic and semitropic waters. They are characterized by absence of a tail, inhalant and exhalant siphons opening at opposite ends of the body, a degenerate nervous system, and gill clefts not divided by external longitudinal bars. In many the body is barrel or lemon shaped and encircled by rings of muscle strands. These circular muscle bands are complete in *Doliolum;* incomplete in *Salpa.* Contractions of these muscles drive water through the exhalant (atrial) opening and aid in locomotion by a type of jet propulsion. In *Pyrosoma* the gill clefts are numerous and tall dorsoventrally; in *Salpa* there are only 2, each occupying most of a side of the pharynx. Many, including *Pyrosoma,* have brilliant luminescence.

Asexual budding occurs in all thaliaceans. The sexually reproducing stages are derived as blastozooid buds from the individual (oozooid) formed from the egg. In the colonial *Pyrosoma* there is no larval stage. Each member produces a meroblastic egg, which develops in the parent atrium into an oozooid embryo. The oozooid gives rise to 4 buds along a stolon, which are liberated by the rupture of the parent atrium. When the oozooid degenerates, the 4 buds on the stolon form a new colony by secondary budding. In the solitary *Salpa* the single egg in the atrium develops by a placental attachment, and the oozooid breaks free from the parent to form a chain of buds that later separates to form the sexual individual. In *Doliolum* (also solitary) there is a tailed larval stage. Three eggs produced by the parent are shed into the water where each develops into a tailed larva. The larva metamorphoses into an oozooid bearing a stolon with buds. These prebuds form several generations of blastozooids, which are attached to the parent oozooid and which form individuals of three kinds: gastrozooids, for nutrition;

phorozooids, for carrying the third type; and gonozooids. The gonozooids are later liberated and become the sexual-reproducing individuals.

■ CLASS LARVACEA (COPELATA)

The small transparent tunicates of class Larvacea are also pelagic and planktonic. They represent sexually mature forms that have retained the larval organization in many respects. Their body has a resemblance to the ascidian tadpole described in an earlier section. They have a permanent tail and a much simplified internal structure. Their test, or "house," does not contain tunicin and encloses or is loosely attached to the body. From time to time the animal disengages itself from the house and secretes a new one. The house, which is formed of gelatinous material secreted by the epidermis, is provided with an incurrent and an excurrent orifice. The incurrent orifice has a water-intake filter in the form of a screen of fine fibers. A food-concentrating apparatus, which strains out plankton from the water current, is also present in the test and serves as a second filter. The food from this second screening is carried to the mouth by the water current, which is created by the beating of the tail.

The body is a small compact structure, with the tail attached to the ventral side. The mouth is located at the anterior end of the body, which bears 2 simple gill clefts opening directly to the outside. The pharynx is also provided with a food filter, and the intestine opens to the exterior ventrally or to one side.

The nervous system is composed of a brain (a compact ganglion) and a nerve cord. There is no eye, but a statocyst lies near the brain. The Larvacea are hermaphroditic, and the gonads are located in the posterior end of the body. A notochord and a caudal ganglion are found in the tail.

Oikopleura is one of the most familiar genera in the Larvacea, of which there are about sixty species.

Subphylum Cephalochordata

The members of this subphylum are commonly called lancelets. Another name for them is amphioxus. Lancelets are small fishlike animals that rarely exceed 2½ inches in length, although there is one species in southern China that reaches 6 inches.

The adults are found in coarse sand in shore water of the subtemperate and tropic oceans of the world. They burrow rapidly in the sand, anterior end first, by means of the vibrating action of the body and assume a resting condition with the anterior end exposed to the water. They swim mostly at night, by lateral movements of their body.

Only two genera and twenty-nine species are recognized in the Cephalochordata.

Asymmetron. Gonads present on right side only; asymmetric metapleural folds, the right one con-

FIG. 18-7

Mature specimen of *Branchiostoma,* the cephalochordate commonly known as amphioxus. Gonads appear as row of block-shaped structures beneath myotomes. (From Hickman: Integrated principles of zoology, The C. V. Mosby Co.)

tinuous with the ventral fin; may or may not have a caudal appendage.

Branchiostoma. Gonads paired; both metapleural folds similar and ending behind the atriopore; no caudal appendage.

Branchiostoma virginiae (Fig. 18-7) is a common species along the southern Atlantic coast (United States); *Branchiostoma californiense* is common along the Pacific coast south of the San Diego region.

GENERAL CHARACTERISTICS

The lancelet is laterally compressed and tapered at each end, with the postanal tail that is characteristic of all true chordates. A low median dorsal fin occurs along the body, passes around the tail as the tail fin, and continues forward to the atriopore as the ventral fin. A pair of metapleural folds extends along the anterior two thirds of the ventral surface (Fig. 18-8). Both atriopore and anus open ventrally. The buccal cavity (oral hood) on the ventral anterior end is fringed with sensory cirri. These can spread out to open the buccal cavity or bend inward to form a protective grating over the cavity. The body is covered by a single layer of epidermis protected by a thin cuticle or layer of mucus. Beneath the epidermis is the dermis of connective tissue. There are no glands or

chromatophores in the integument, although some pigment is present in the epidermis near the anterior end.

Internally amphioxus has a notochord extending the length of the body above the alimentary canal (Fig. 18-8). It is a solid flexible cord of gelatinous material and vacuolated cells filled with fluid and surrounded by a connective tissue sheath (Fig. 18-10, *A*). Other supporting structures are the cartilage-like reinforcement of the oral hood, gill bars, fin rays, and cirri. Longitudinal muscles are arranged in chevronlike segments called myotomes (Fig. 18-7). Contraction of these muscles produces the lateral movements for burrowing and swimming, as the muscles on each side are out of phase with those of the other side. Propulsion is thus produced by a laterally flexing tail reinforced by its internal skeleton, the notochord. Transverse muscles in the floor of the atrial cavity between the metapleural folds serve to compress the atrial cavity to discharge water.

Above the notochord lies the dorsal nerve cord, with a small central cavity (neurocoel). It is dilated anteriorly to form the median cerebral vesicle. An eyespot of black pigment may be present, and usually two pairs of so-called cranial nerves are given off from the anterior region of the cord. Pairs of ventral motor nerves

FIG. 18-8

Structure of young amphioxus *(Branchiostoma).* (From Hickman: Integrated principles of zoology, The C. V. Mosby Co.)

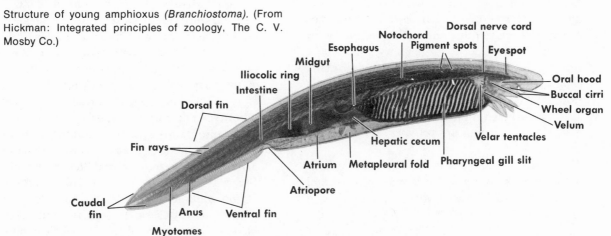

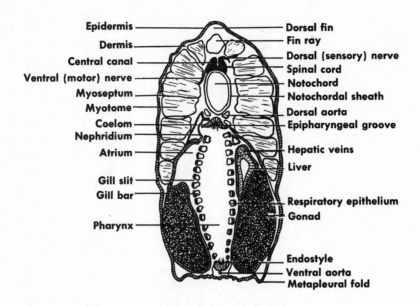

Epidermis — Dorsal fin
Dermis — Fin ray
Central canal — Dorsal (sensory) nerve
Ventral (motor) nerve — Spinal cord
Myoseptum — Notochord
Myotome — Notochordal sheath
Coelom — Dorsal aorta
Nephridium — Epipharyngeal groove
Atrium — Hepatic veins
Liver
Gill slit
Gill bar — Respiratory epithelium
Gonad
Pharynx — Endostyle
Ventral aorta
Metapleural fold

FIG. 18-9

Cross section of amphioxus through pharynx and gonads. (From Hickman and Hickman: Laboratory studies in integrated zoology, The C. V. Mosby Co.)

connect the cord with the myotomes, and pairs of dorsal sensory nerves run to the skin and gut.

The straight digestive system starts with the oral hood. At the posterior end of the hood is a membranous velum with sensory tentacles, which guard the mouth and strain out large particles from the water current. The mouth itself is a circular opening in the velum. On the lateral walls of the oral hood, ciliary patches direct the water current and food particles toward the mouth. The large pharynx has more than a hundred pairs of diagonal gill slits separated by bars containing skeletal rods (Fig. 18-8). The edges and inner surfaces of the gill bars bear cilia for drawing water from the mouth and driving it outward through the atrium.

Feeding in lancelets occurs by a filter-feeding or mucus-ciliary method. When the animal is at rest in its burrow with its anterior end protruding and mouth open, the cilia on the inner surface and sides of the numerous gill bars draw the feeding current of water, laden with minute organisms, into the mouth and pharynx. As the water passes through the pharynx and between the gill bars into the atrium, the endostyle (Fig. 18-9) secretes a stream of mucus, which is swept up the side walls of the pharynx in the form of longitudinal strings by the frontal cilia of the gill bars. The food in the water currents is strained off by the gill bars, trapped in the mucous strings on the walls of the pharynx, and rolled upward along ciliated tracts into a hyperbranchial groove that concentrates the mucous strings of food into a continuous food strand which passes into the straight intestine. Near the end of the intestine is a ciliated iliocolic ring where the cord of food and mucus is rotated (Fig. 18-8), thus assisting the action of the enzymes on the food. Digestion is partly extracellular and partly intracellular. Absorption takes place in the intestine.

The circulatory system of amphioxis is similar in its basic plan to that of primitive chordates. There is no heart, and capillaries are absent. Blood is driven by the muscular contraction of the walls of the larger vessels. The blood is aerated as it passes through the gill bars. The blood of cephalochordates is colorless and has

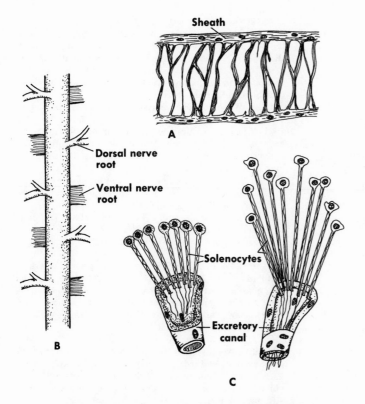

FIG. 18-10

A, Section of notochord of *Branchiostoma* showing vacuolated cell structure. **B,** Portion of nerve cord of *Branchiostoma* showing arrangement of dorsal and ventral nerve roots. **C,** Comparison of nephridia (in part) of marine polychaete (left) and amphioxus (right) showing solenocytes and their flagella. (**A** modified from Herdman; **C** modified from Goodrich.)

no respiratory pigment. Some investigators doubt the oxygenation of blood in the gills (although there are lamellae on the bars of the gill clefts), but they believe that this process occurs in the lacunae of the matapleural folds.

The excretory system consists of many protonephridia (100 or more) in the coelom, which empty into gill slits. The protonephridium bears along its curved side several tufts of solenocytes, or specialized flame cells (Fig. 18-9, *C*). A solenocyte is an elongated tubular cell with a long flagellum extending down its interior from a protoplasmic mass. By diffusion through the solenocyte wall, waste and excess fluids are excreted from the coelom into the atrium. The solenocytes are in close contact with the blood vessels. Elsewhere in the animal kingdom, solenocytes are found only in some polychaetes and a few others (Fig. 18-10, *C*).

The small coelom shows the tripartite division in the pouches of the archenteron during development but not in the adult condition.

The sexes are separate in amphioxus and there is no sexual dimorphism. There are usually 25 to 33 gonads on each side (Fig. 18-7). The mature gametes are shed by bursting through the wall into the atrium and thence to the exterior. Fertilization is external in the sea water. Breeding usually occurs in early spring.

The eggs of amphioxus have so little yolk, or else it is so uniformly distributed in them (isolecithal), that they represent a classic example of regular cleavage and gastrula formation. In general, amphioxus has the deuterostome pattern of embryology, involving radial cleavage and enterocoelous formation of mesoderm.

EVOLUTIONARY STATUS OF AMPHIOXUS

Although amphioxus shows many basic features common to all chordate animals, it can hardly be called a generalized ancestral type because of such unique characters as the many gill slits, the presence of solenocytes, the atrium, and some others. The animal apparently arose from sessile ancestors that had a tadpole stage. Some of the characters have been retained from an early sessile form, such as the employment of the large pharynx with its numerous gills for a food-catching mechanism, its lack of a brain, and its behavior of remaining stationary in a burrow for a large part of its existence. It may be considered a divergent line from the early ancestral line leading to the vertebrates.

REFERENCES

Barrington, E. J. W. 1937. The digestive system of *Amphioxus*. Phil. Trans. Roy. Soc. (London) **228**:269-312.

Barrington, E. J. W. 1965. The biology of Hemichordata and Protochordata. San Francisco, W. H. Freeman & Co., Publishers.

Berrill, N. J. 1936. The evolution and classification of ascidians. Phil. Trans. Roy. Soc. (London) **226**:43-70.

Berrill, N. J. 1955. The origin of vertebrates. New York, Oxford University Press, Inc. This author stresses the ascidian tadpole theory.

Bigelow, H. B., and I. P. Farfante. Lancelets, cyclostomes, and sharks. In Tee-Van, J., and others (editor). 1948. Fishes of the western North Atlantic, part 1. New Haven, Conn., Yale University Press, Sears Foundation for Marine Research.

Brien, P., and P. Drach. Procordes. In Grassé, P. P. (editor).

1948. Traité de zoologie, vol. 11. Paris, Masson et Cie. A comprehensive treatise on the lower chordates. One of the best accounts of this group.

Bullough, W. S. 1950. Practical invertebrate anatomy. London, The Macmillan Co. This author gives practical dissection methods for members of all three subclasses of tunicates and for amphioxus.

Garstang, W. 1928. The morphology of Tunicata. Quart. J. Microscop. Sci. **72**:51-187.

Hatschek, B. 1882. Studien ueber der Entwicklung des *Amphioxus*. Arbeiten aus dem zoologischen Institute der Universitat Wien. **4**:1-88. A classic work by the great German investigator.

Herdman, W. A. 1904. Ascidians and *Amphioxus*. In Harmer, S. F., and A. E. Shipley (editors). Cambridge natural history, vol. 7. London, The Macmillan Co.

Huus, J., H. Lohmann, and I. F. W. Ihle. 1933. Tunicata. In Kukenthal, W., and T. Krumbach (editors). Handbuch der Zoologie, vol. 5, part 2. Berlin, Walter de Gruyter & Co.

Van Name, W. G. 1945. The North and South American ascidians. Amer. Mus. Natur. Hist. Bull. **84**:1-476. The best American treatise on the group. Keys to species are given.

Willey, A. 1894. Amphioxus and the ancestry of the vertebrates. New York, Columbia University Press. A classic study of the famous animal.

Young, J. Z. 1962. The life of vertebrates, ed. 2. New York, Oxford University Press, Inc. The author believes that amphioxus has retained the ciliary feeding behavior of the early chordate ancestors and that it has undergone little change since the Silurian time. A good description of its structural features.

Glossary

acetabulum True sucker in trematodes, usually of circular shape and raised rim.

aciculum (pl. acicula) Special strong seta for stiffening parapodium in polychaetes.

acontium Fine thread originating as a septal filament at the edge of a mesentery in the anthozoan gastrovascular cavity.

actinotroch (actinotrocha) Modified trochophore larva in the phoronids.

aerobic Requiring oxygen in respiration.

aesthetascs Slim, minute sensory processes on the antennae of certain crustaceans.

aesthetes Small light-sensitive structures on the calcareous plates of certain Amphineura.

afferent Leading toward a given position.

agamete Single germ cell that develops into an individual.

aliform Wing shaped.

allometry Difference between mean growth rate of body and that of its organs.

ambulacra Radially arranged series of plates that bear the rows of tube feet, or podia, in echinoderms.

ameboid movement Process of putting forth pseudopodia, as in an ameba or white blood corpuscle.

amictic Pertaining to the diploid egg of rotifers or the females that produce such eggs.

amphiblastula Larval stage of certain marine sponges in which the animal pole has flagellated cells and the vegetal pole unflagellated cells.

amphid Anterior chemoreceptor sense organ in certain nematodes.

amplexus Clasping process of copulation in certain animals.

ampulla Bulblike internal end of a tube foot in echinoderms.

anabiosis State of suspended animation, especially in the tardigrades.

anaerobic Oxygen-independent type of respiration.

anamorphic Having fewer segments in young stages than in the adult.

anastomosis Union of two or more hollow organs, such as blood vessels.

androgenic gland Sex-regulating gland in the amphipod *Orchestia* and some others.

anisogametes Unlike gametes in fertilization.

annulus Ringlike part, often mistaken for a true segment in leeches.

antennae Long, sensory, paired appendages on the head of many arthropods.

antennule One of the first pair of antennae on the head of crustaceans; usually smaller than the second pair of antennae.

antrum Chamber at the end of the male or female system in flatworms; opens into the gonopore.

apodeme Internal projection in an arthropod skeleton; usually for the attachment of muscles.

apopyle Pore between radial canals or flagellated chambers and the spongocoel in sponges.

archenteron Primitive gut of a metazoan embryo, formed by the process of gastrulation.

areoles Irregular, rounded, and thickened cuticular areas in some Nematomorpha.

athecate Absence of a hydrotheca around hydranths in certain hydrozoans.

atrium Large anterior chamber containing the pharynx in cephalochordates and tunicates; the peribranchial chamber.

autotomy Self-amputation of a body structure by reflex action.

autotrophic nutrition Ability to use simple inorganic substances for synthesis of more complex compounds, as in chlorophyll-bearing organisms.

avicularium Differentiated individual, or zooid, in a colony of ectoprocts; of the general shape of a bird's head and beak.

axoneme Fibril extending from the flagellum to its basal granule.

axopodium Pseudopodium containing an axial rod and a cytoplasmic envelope, as in Heliozoida.

benthonic Referring to the bottom of lakes and seas.

biramus Having two branches.

blastomere Early cleavage cell of an embryo before the gastrula stage.

blastopore In embryology, the mouthlike opening of a gastrula.

blastula In embryology, an early stage of a mass (usually hollow) of cells just before the gastrula stage.

blepharoplast (kinetosome) Intracellular granule of the flagellum from which the axoneme extends; may be fused with the basal granule.

bothrium One of 2 or 4 shallow sucking grooves on the scolex of certain tapeworms and characterized by weak muscularity and lack of inner muscle layer; variable in shape and form as a holdfast organ.

botryoidal tissue Loosely arranged mesenchyme tissue in certain leeches.

brachial Referring to the forelimb or arm.

branchial Referring to gills.

branchiostegite Expanded lateral part of the carapace, serving as a gill covering or gill chamber in certain crustaceans.

buccal cavity Mouth cavity; the region just inside the mouth opening.

bursa Saclike cavity or pouch forming part of the female reproductive system. The pouchlike bodies with genitorespiratory functions in ophiuroids.

byssus Anchorage fibers produced by many marine bivalves.

calotte Anterior cap of cells in certain Mesozoa.

calyx Visceral body of certain invertebrates; the cup-shaped central part of the body.

capitate Referring to short tentacles with the nematocysts packed in terminal knobs.

capitulum Part of a barnacle enclosed in the mantle. The swollen end of a hair or antenna.

capsule Cluster of ganglia in annelids, sometimes called the follicle. An investing sheath of connective tissue.

captacula Tentacles with suckerlike tips on the head of certain scaphopods for sensory functions and food getting.

carcinology Study of crustaceans.

cephalization Tendency to concentrate in the head region.

cephalon Head region.

ceras (pl. cerata) Club-shaped projection containing tubules of the digestive glands and nematocysts on the back of certain nudibranchs.

cercaria In the life history of trematodes, the tadpole-shaped larva that breaks out of the snail or other intermediate host.

cercopod One of a pair of long filamentous projections of the posterior end of certain crustaceans.

cervical groove Transverse groove indicating the general boundary between the head and thoracic areas of the carapace of certain crustaceans.

chela Pincer claw on an arthropod appendage.

chelicerae First pair of appendages in arachnids; for seizing and crushing prey.

chilarium One of the pair of appendages at the posterior end of the cephalothorax in certain arachnids.

chitin Horny substance of a nitrogenous polysaccharide forming the hard part of the outer integument in arthropods, mollusks, and other invertebrates.

chloragogue Special spongy tissue that envelops the intestine of certain annelids; probably aids in excretion.

chlorocruorin Green, iron-bearing respiratory pigment in certain polychaetes.

cingulum Outer, ciliated, girdlelike band, or zone, on the corona of a rotifer.

cirrus Any slender appendage of various forms, such as the featherlike arm of a crinoid, the filamentous appendage of a barnacle, etc.; temporary projection formed by the turning inside out of the ejaculatory duct and serving as a male copulatory organ in many invertebrate phyla.

clitellum Saddlelike glandular structure on some midbody segments of certain annelids; forms mucus for cocoon formation and copulation.

clypeus Median anterior plate on an insect's head.

coeloblastula Simple blastula with cavity (blastocoel).

coelomocyte Corpuscle or amebocyte in the coelomic or pseudocoelomic fluid of invertebrates.

coelomoduct Duct that carries gametes or excretory products from the coelom to the exterior.

coenenchyme Common tissue connecting adjacent polyps in alcyonarian coelenterates.

coenosarc Axial living part of certain hydrozoan coelenterates; often surrounded by a perisarc.

coenosteum Calcareous skeleton of colonial corals.

collenchyma Undifferentiated mesenchyme with loosely scattered cells.

colloblasts Adhesive cells found on the tentacles of ctenophores.

collum Collar or neck region.

commensalism Symbiotic relationship between two different species whereby one derives an advantage and the other derives no advantage or disadvantage.

conchiolin Albuminoid organic matrix of a mollusk shell.

coracidium Early larval stage of certain tapeworms.

corona Ciliated zone surrounding the anterior end of a rotifer, often called the wheel organ.

coxal gland Excretory gland at the base of the leg in some arthropods.

coxopodite Basal segment of a pereiopod in crustaceans.

craspedote Pertaining to a coelenterate medusa with a velum.

crystalline style Cylindric proteinaceous rod in a cecum of the stomach or intestine of certain bivalves and gastropods; releases enzymes as it rotates.

cyclomorphosis Gradual changes in gross morphology during successive seasonal generations, as in Cladocera.

cypris larva Free-swimming stage developed from the nauplius in barnacles.

cysticercoid Larval stage in certain tapeworms; resembles the cysticercus but has a tail.

cytopharynx Funnel-shaped food canal leading from the cytostome into the cell of ciliates.

cytopyge Area in the body wall of ciliates that serves as an anus for the discharge of undigested materials.

cytosome Cell body inside the plasma membrane.

cytostome Mouth of certain holozoic protozoans.

dactylozooid Type of polyp specialized for a tactile role in hydrozoan coelenterate colonies.

Desor larva Modified pilidium larva that remains inside the egg membrane and lacks certain pilidium characteristics.

determinate cleavage Fate of blastomeres is fixed early in development; mosaic cleavage in which all the blastomeres are required for complete development.

detritus Any fine particulate debris of organic or inorganic origin.

dextral Right-handed; in gastropods, when facing the aperture, the coils turn to the right; having the aperture to the right of the shell axis.

diapause Period of suspended development or growth in certain insects and other invertebrates during stages of their life cycle.

diaxial Movable at right angles to the long axis of the body.

dioecious Pertaining to an organism in which the male and female reproductive systems are in different individuals.

dipleurula Hypothetical ancestral form of echinoderms; a general term for various types of ciliated echinoderm larvae.

doublura In horseshoe crabs, the ventral surface of the marginal reflected carapace of the prosoma.

ecdysis Periodic molting or shedding of the exoskeleton to permit increase in size of arthropods, nematodes, and others.

ectolecithal Pertaining to zygotes with peripheral yolk granules.

elytra Pair of anterior heavy wings of beetles.

embolus Anterior end of the palp in male spiders through which sperm is injected into the female.

endite Basal median lobe of many arachnid and crustacean appendages.

endolecithal Pertaining to zygotes with enclosed yolk granules.

endopodite Median branch of a biramus appendage.

endosome Intranuclear body, other than chromatin granules, within the nuclear membrane.

endosternite Sclerotized process on the inner surface of the cephalothoracic exoskelton in certain arthropods for muscle attachment.

endostyle Ciliated ventral groove in the pharynx of tunicates, cephalochordates, and some others; collects food particles and passes them on.

enterocoelous formation of coelom Type of mesoderm formation that involves the pouchlike outfolding of the archenteron.

enteron Part of the digestive tract derived from the endoderm; sometimes used as a general term for the alimentary canal.

ephippium Posterodorsal egg pouch in certain Cladocera.

ephyra Immature, free-swimming jellyfish from asexual division of the strobila.

epiboly Growth of ectodermal cells downward to cover the vegetal cells during gastrulation.

epicuticle Outermost, hardened waxy layer in an arthropod cuticle.

epigastric furrow Groove separating the region of the book lungs from the posterior part of the abdomen.

epigynum Flap covering the genital pore in female spiders.

epimeron Posterior lateral part of a pleuron in certain arthropods.

epimorphic Having a full complement of segments in the young.

epiphysis Small peripheral bar in Aristotle's lantern.

epipodite Respiratory process attached at the base of the trunk appendage in certain arthropods.

epistome Flap that covers the mouth of ectoprocts.

epitoky Seasonal modification in posterior part of marine polychaete; posterior part becomes swollen with gonads and eggs or sperm.

erythrocruorin Invertebrate respiratory pigment of several varieties found in annelids, echinoderms, mollusks, and some others.

eutely Constant number of cells in all adult members of a single species.

exite Lobe on the outer margin of a foliaceous appendage in certain crustaceans.

exopodite Lateral branch of biramous appendage.

filopodium Pseudopodium in the form of filamentous projections that may branch but do not anastomose, as in *Amoeba radiosa*.

flame bulb Hollow bulb-shaped structure of 1 or more cells provided with one or many cilia (flagella) and forming part of the protonephridial system in many invertebrates; excretory and osmoregulatory in function.

foliaceous appendage Flat, leaflike subdivided appendage in certain crustaceans.

frustule Nonciliated planulalike bud that develops into a polyp in some hydrozoans.

funiculus Fibrous strand connecting the lower end of the digestive tract to the body wall in Ectoprocta.

furca Any forked process.

gamont Gametocyte that develops into sporoblasts in sporozoan life histories.

gastrolith Calcareous body in the cardiac stomach wall of certain crustaceans.

gastrula In embryology, an invaginated blastula; in some cases it is hollow and sometimes a solid sphere of cells.

gemmule Cystlike body of freshwater sponges able to withstand winter conditions and to germinate new sponge when conditions are favorable.

giant fiber Large nerve fiber that can transmit impulses faster than normal fibers; refers especially to those in the nerve cord of annelids, arthropods, and squids; a neurochord in the nerve trunk.

glabella Median part of the head of a trilobite and some others.

glochidium Bivalve larval stage in freshwater mussels.

gnathobase Basal process on the appendages of certain arachnids; usually for crushing or handling food.

gnathochilarium Appendage of united second maxillae, or the lower lip, of millipedes.

gnathosoma Anterior part of the body of ticks and mites, chiefly the mouth and mouthparts.

gonangium Asexual reproductive polyp in certain coelenterates.

gonophore Asexual bud of a polyp or polypoid colony in certain hydroids; produces medusae or gametes.

gonotheca Vaselike sheath around the blastostyle of certain coelenterates.

gonozooid Polyp specialized for reproduction in certain hydrozoans.

Gymnolaemata A group of ectoprocts with circular lophophore.

hectocotylus Modified arm in certain cephalopod mollusks for use as a copulatory organ.

hematochrome Bright red pigment occurring in some plantlike flagellates; allied to chlorophyll.

hemocoel Blood cavity consisting of spaces between tissues; found in arthropods and some others with reduced coeloms.

heterogamy Condition in which two unlike gametes unite to form a zygote.

heterotrophic nutrition Type of nutrition that depends on complex organic food materials originating in other plants and animals.

holoblastic cleavage Complete and nearly equal cleavage of cells in early embryology.

holophytic nutrition Type of nutrition by which green plants are able to make food from simple inorganic substances.

holozoic nutrition Animal-like nutrition that involves ingestion and metabolic utilization of solid food particles or living prey.

hydatid Fluid-filled cyst that produces scolices and daughter cysts in the life history of some tapeworms.

hydrotheca Vaselike sheath around the hydranth, or terminal portion, of certain hydroid polyps.

hypertonic Condition in which a concentration of dissolved materials gains solvent through a differentially permeable membrane from a solution with a lower solute concentration.

hypodermic impregnation Injection or passage of sperm, often in packets (spermatophores), directly through the epidermis.

hypodermis Epidermis that secretes an overlying cuticle.

hypostoma Median anterior mouthpart in certain arachnids; in Hemiptera, the lower part of the anterior surface of the head.

hypotonic Condition in which a concentration of dissolved materials loses solvent through a differentially permeable membrane to a solution with a concentration of greater dissolved materials.

idiosoma Posterior part of ticks and mites, making up most of the body in these forms.

indeterminate cleavage Type of cleavage in which the early blastomeres do not have a predetermined fate in forming tissues and organs; if separated, early blastomeres will each give rise to a complete larva.

infauna Animals that burrow in the bottom of the ocean or fresh water.

infraciliature System of granules and fibrils lying beneath the pellicle and underlying the cilia in ciliates.

infusorigen Hermaphroditic cell cluster in the rhombogen of Mesozoa.

inquiline Organism living in the nest or burrow of another animal; may or may not be a commensal.

insemination Introduction of sperm into the female.

integument Outer covering, or skin and its derivatives, of an animal.

introvert In sipunculids, the anterior part, or proboscis, which may be withdrawn into the trunk.

isogamy Condition in which the 2 gametes of a union are alike morphologically.

Johnston's organ Auditory organ on the second segment of each antenna in certain insects.

kinetosome Small body or basal granule underlying the cilium or flagellum and capable of self-reproduction; may be involved in controlling the motion of the flagellum or cilium.

kinety Row of basal granules, or kinetosomes, with their associated fibrils (kinetodesmas).

labium Lower lip of the mouth in insects; sometimes bilobed.

labrum Fleshy upper lip of the mouth in insects.

lacinia mobilis Movable process near the mandibular teeth in certain crustaceans and insects.

lacuna Space in the tissues, serving as part of the circulatory system.

larva Early feeding stage in the life cycle of an animal; unlike the adult.

Laurer's canal In some trematodes, a narrow canal running from the oviduct to the dorsal body surface where it may open; may function in copulation.

lemnisci Pair of bodies in the neck or pseudocoel of Acanthocephala, serving as reservoirs for the lacunar fluid when the proboscis is retracted.

littoral Shallow part of lakes and seas along the coast; the sea floor from the shore to the edge of the continental shelf.

lobopodium Blunt, fingerlike pseudopodium, as in *Amoeba proteus*.

lophophore Anterior ridge bearing ciliated tentacles in Ectoprocta, Brachiopoda, and Phoronida.

lorica Rigid case or exoskeleton of certain protozoans and rotifers.

macromere One of the larger cells resulting from unequal cleavage during embryology.

macronucleus Polyploid nucleus in ciliates produced by repeated divisions of chromosomes; the larger vegetative nucleus of the two types of nuclei in ciliates.

macroplacoid Small sclerotized plate embedded in the pharyngeal muscles of a tardigrade.

madreporite Sievelike porous plate (usually on the surface) that enables the fluid to pass in and out of the water-vascular system in echinoderms.

malpighian tubule Excretory, blind tubular glands opening into the anterior part of the hindgut of many terrestrial arthropods.

mantle Extension of the body wall in mollusks and brachiopods; involved in secreting the shell and many other functions.

marsupium External brood pouch.

mastax Muscular portion of a rotifer pharynx containing the trophi.

mastigonemes Branching fibrils extending from the flagella in some flagellates.

maxilla Head appendage (one or two pairs) modified in various ways for food handling in many arthropods.

maxillary glands Two small glands posterior to the mouth in certain crustaceans; probably excretory in function.

medusa Free-swimming stage or form in the life history of many coelenterates; jellyfish.

megalops Larval stage just before the adult stage of marine crabs.

Mehlis' gland Cluster of unicellular glands surrounding the ootype in some trematodes and cestodes; formerly called the shell gland.

membranelle Compound ciliary organelle made up of several fused transverse rows of cilia, usually occurring in a buccal cavity leading to the cytostome.

merozoite Cell resulting from multiple fission, or schizogony, of a schizont in certain Sporozoa.

mesenteron Midgut.

mesoglea Amorphous jellylike material between the epidermis and gastrodermis in sponges and coelenterates; when it has cells and fibers, it is considered a third germ layer (mesoderm).

mesopsammon The interstitial fauna or microfauna in the interstices of sediments.

metacercaria Larval stage in trematodes in which the larva loses its tail and becomes encysted on an intermediate host.

metagenesis Alternation of generations; the succession of asexual and sexual individuals in a life history.

metamerism Linear repetition of body parts (metameres, or segments) at regular intervals along the anteroposterior axis.

metanephridium Type of nephridium with open inner ends; the inner end called the nephrostome and the external opening the nephridiopore; common in annelids, sipunculids, mollusks, and others.

metasoma Posterior part of the body in arachnids.

metastoma Labium.

micromere One of the smaller cells resulting from

unequal cleavage during embryology; contrast with macromere.

micropyle Small pore in the chorion of an insect egg through which sperm enter.

microthriches Fingerlike projections of the outer surface of the tegument in certain cestodes, which may interdigitate with the microvilli of the host's intestine.

mictic Pertains to the haploid eggs of rotifers or the female that produces such eggs.

miracidium Ciliated free-swimming larva that hatches from a trematode egg; the first larva in the life cycle of trematodes.

mosaic Type of development in which each blastomere is fated to become a predetermined part of the whole embryo after fertilization.

motorium Ciliate organelle with its connecting fibers, serving as a motor-coordinating center.

myochordotonal organ Small proprioceptor in the third segment of each leg of decapod crustaceans.

myoneme Contractile filament in the cytoplasm of protozoans.

mysis One of the first three larval stages of certain decapods; has thirteen pairs of appendages.

myxopodium Condition of anastomosing rhizopods in certain ameboid protozoans.

nacreous layer Innermost lustrous layer of a mollusk shell.

nauplius Free-swimming larval form typically of three pairs of appendages; characteristic of many crustaceans.

nauplius eye Unpaired median eye of crustaceans.

nectochaeta Type of free-swimming larva in some polychaetes.

nectophore Swimming, bell-shaped medusa of a siphonophore colony.

nematogen Individual dicyemid mesozoan found in young cephalopods; gives rise to vermiform larvae.

nephridiopore External opening of a nephridium.

nephridium Tubular organ for excretion and osmoregulation in annelids, mollusks, arthropods, and other invertebrates.

nephromixium Combination of coelomoduct and nephridium into a compound organ with both genital and excretory functions.

nephrostome Ciliated internal opening of a nephridium as in a metanephridium.

neritic Pertaining to that part of the continental shelf between the low-tide line and depths of about 600 feet, including both bottom and water.

neuron Nerve cell, including cell body, dendrites, and axons.

neurosecretory A neuron that produces hormones.

notochord Longitudinal elastic rod lying between the nervous system and the digestive tract in the embryos of all chordates and in the adults of some.

nuchal organ Glandular pit on the head of certain polychaetes; may have sensory functions.

ocellus Complex photoreceptor with a lenslike body enclosed in pigment.

olynthus A vase-shaped early stage in the development of calcareous sponges.

ommatidium One of the slender visual units of the compound eye, made up of a complex of structures.

oncosphere Six-hooked larva that hatches from the eggs of certain tapeworms.

oostegites Flat plates extending medially from the thoracic appendage of certain crustaceans (Peracarida).

ootype Thickened portion of the oviduct surrounded by Mehlis' glands in trematodes and cestodes; formerly considered the place where the shell is formed.

operculum Flaplike covering of many types, such as the gill cover of crayfish, plates of a barnacle shell, horny covering of the aperture in gastropods, etc.

opisthaptor Sucker at the posterior end of certain trematodes (Monogenea).

opisthosoma Abdomen, or posterior part of body, in arachnids.

opsiblastic Thick-shelled gastrotrich egg for withstanding harsh environmental conditions.

organelle Specialized part of a cell having a specific function.

organism Any living individual whether plant or animal.

osphradium Sensory area in or near the siphon for sampling incoming water in certain mollusks.

ossicles Calcareous plates or rods making up the echinoderm endoskeleton.

ovicell Greatly modified zooecium serving as a brood pouch for the early embryo in certain marine Bryozoa.

oviger One of a pair of accessory legs in male pycnogonids for carrying eggs.

oviparous Producing eggs that hatch outside the mother's body.

ovisac Egg receptacle.

ovoviviparous Reproductive pattern in which eggs develop within the mother's body without nutritive aid from the mother.

palmella Stage in the life cycle of many flagellates involving the aggregation of numerous individuals, simulating a colony.

palp (palpus) Projection, often sensory in function, on the head or near the mouth of some invertebrates; often attached to a head appendage.

palpon Tactile zooid making up part of the siphonophore colony.

papilla Any nipple-shaped projection, large or small.

papulae Dermal branchiae, or skin gills, in echinoderms.

parabasal body Staining body of various shapes that lies near the base of the flagellum and is often attached to the blepharoplast by a rhizoplast in some free-living and parasitic flagellates.

paragnath Sclerotized jawlike teeth in certain polychaetes. One of 2 lobes forming the lower lip in crustaceans.

paramylum Polysaccharide serving as a reserve food supply in certain phytoflagellates.

parapodium Fleshy, lobelike projection on each side of each somite in polychaetes, used for locomotion.

paraxial Movable in the anteroposterior plane of the body axis.

parenchyma Type of mesenchyme with closely packed vacuolated cells filling in spaces between body organs; the chief functional cells or tissues of an organ in contrast to the supporting tissues (stroma).

parenchymula Stereogastrula.

parietal Pertaining to the walls of a cavity.

parthenogenesis Development of an organism from an unfertilized egg.

pathogenic Producing a diseased condition.

paxillae Modified spines, resembling mushrooms, in certain sea stars.

pectines Pair of comblike sensory organs on the ventral surface of the abdomen in scorpions.

pedal gland Gland in the foot of rotifers.

pedicellariae Minute pincerlike structures around the spines and dermal branchiae, or dermal gills, of certain echinoderms for keeping the surface of the body free from debris.

pedicle Stalk, or stalklike support, such as that of a brachiopod.

pedipalp One of the second appendages of arachnids.

peduncle Any stalklike structure supporting another structure.

pelagic Pertaining to the open waters of the sea and lakes away from the shore.

pellicle Thin, outer envelope that covers many protozoans and lies in contact with the plasma membrane.

penial spicules Sclerotized copulatory hooks in the cloaca of most male nematodes; used to spread open the female gonopore.

pentamerous Arranged in 5 or multiples of 5.

pereiopod One of the paired appendages on most thoracic segments of many crustaceans and modified for seizing and for locomotion.

periostracum Outside chitinoid layer of many mollusk shells.

peristalsis Rhythmic involuntary muscular contraction along a hollow organ.

peristome Area around the cytostome; usually the specialized region around the buccal cavity.

peronium One of the thick epidermal tracts from the base of the tentacles to the margin of the exumbrella in certain medusae.

pharetrone Type of sponge in which the spicules are united into a network.

phasmid Precaudal sensory organ or gland in certain nematodes.

photophore Small light-producing region in luminescent animals.

phreatic Refers to ground water or the upper parts of subsoil water.

Phylactolaemata A group of ectoprocts with horseshoe-shaped lophophore.

phyllosoma Free-swimming larval stage of spiny lobsters.

phylogeny History or evolution of a group of organisms.

pilidium Characteristic helmet-shaped larva of many marine nemertines.

planuloid larva In some coelenterates, an asexually produced larva with the general morphology of a planula; develops into a polyp.

plasmodium Multinucleated mass of naked protoplasm without definite size or shape and bounded by a plasma membrane.

plasmotomy Type of binary fission by which certain multinucleate protozoans divide into 2 or more multinucleate masses without concurrent mitosis.

pleopod Biramous appendage for swimming (swimmeret) in many crustaceans.

plerocercoid Larval stage that develops into the adult in the life cycle of certain tapeworms.

pleuron Lateral plate, forming part of a typical segment in arthropods.

plicate Having many small ridges.

pneumostome Pulmonary aperture in certain gastropods.

podia Tube feet in echinoderms.

polar capsule Capsular structure containing a coiled polar filament that may be discharged to form an anchoring device for the germinating disk; in spores of some sporozoans.

polian vesicles Blind fingerlike ceca projecting from the ring canal in certain echinoderms; function largely unknown, but may be the formation of expansion chambers for the water-vascular system in some.

polybostrichus Many-curled condition, as in the epitoky of some polychaetes.

polymorphism Presence of two or more morphologic forms together in the same species.

polyp Sessile stage, or form, in the life history of many coelenterates; in some coelenterates the life cycle is exclusively polypoid.

polyphyletic Pertaining to a group of organisms derived from more than one known evolutionary line.

polypide Soft parts of a single ectoproct individual.

prehensor In centipedes, one of the paired maxillipeds modified for grasping.

primitive Not specialized; the retention of characters pertaining to the early ancestral condition.

procercoid Elongated larval stage in certain tapeworms; the larva before the plerocercoid larva.

proctodeum Ectodermal invagination that becomes the anus or the cloacal aperture and hindgut.

proctostome Single opening of the gastrovascular cavity; serves as both mouth and anus.

procuticle Innermost layer of the arthropod cuticle; mostly a protein-chitin complex; also called endocuticle.

proglottid Segment of a tapeworm; contains a set of both male and female organs.

prohaptor Sucker at the anterior end of trematodes (Monogenea).

proloculum First chamber formed in the foraminiferan shell.

proprioceptor Receptors that give information to the brain regarding the movement and position of muscles.

prosoma Anterior part of the body, or cephalothorax, in certain arthropods.

prosopyle Small pore that connects the incurrent canal with the radial canal or the flagellated chambers in syconoid and leuconoid sponges.

prostomium Preoral dorsal half-segment at the anterior end of annelids.

protandry Condition in which sperm are produced first and then the eggs by the same gonad.

protaspis First larval instar of a trilobite.

protocephalon Primary head of a nauplius larva.

protoconch Larval shell of a gastropod.

protonephridium Invertebrate excretory unit of a flame bulb system.

prototroch Preoral ring of cilia in a trochophore larva.

psammolittoral Refers to the intertidal areas of sandy beaches, or the interstitial biota of such regions.

psammon The microfauna and microflora inhabiting the interstices between grains of sand of sandy beaches, or the psammolittoral biota.

pseudocoel Body cavity not lined with peritoneum; mesoderm may line the inner surface of the body wall, but is lacking over the enteron.

pustule system Number of connected vacuoles forming a reservoir, usually containing pink fluid in dinoflagellates.

pygidium Terminal body segment in many invertebrates; the anal segment in annelids.

pyrenoid Body associated with a chromatophore or chloroplast; may function in starch formation.

racquet organ Certain sense organs on the hind legs of solpugids.

radial cleavage Type of cleavage in which the axes of early cleavage spindles are either parallel or at right angles to the polar axis, so that the resulting blastomeres are above or below one another in tiers.

radiole Spine of sea urchins. One of the crown of pinnate tentacles in certain polychaetes.

radula Filelike tongue in most mollusks.

redia In the life history of many trematodes, a larval stage produced by a sporocyst.

renette Ventral excretory cell in certain nematodes.

retinula In the basal part of an arthropod compound eye, a group of pigmented cells, each of which is attached to a nerve fiber of the optic ganglion.

retrocerebral organ Anterior glandular organ of unknown function in rotifers.

retroperitoneal Not enclosed within the peritoneal cavity.

rhabdite Rodlike type of rhabdoid in turbellarians.

rhabdoids Minute rodlike structures in the integument of some turbellarians; may form mucus on disintegration; the rhabdite is one of three types of rhabdoids.

rhabdome Long, pigmented rodlike mass in the center of the retinula of an arthropod compound eye.

rheotactic Pertaining to sense organs that can detect water currents.

rhizopodium Pseudopodium with branching and anastomosing filaments, as in many foraminiferans.

rhombogen Individual dicyemid mesozoan found in mature cephalopods; gives rise to infusoriform larvae.

rhopalium Marginal sense organ of certain medusae; a tentaculocyst.

rhynchocoel Dorsal cavity that houses the proboscis in nemertines.

rhynchodeum Short anterior dorsal cavity in nemertines opening to the outside by the proboscis pore; the proboscis opens anteriorly into the rhynchodaeum.

rosette Shape resembling a rose because of its pleated condition; the posterior attachment organ in Cestodaria.

rostellum Projection on the anterior end of the scolex of tapeworms; may bear hooks.

rostrum Median pointed process at the anterior end of the cephalothorax in certain decapod crustaceans.

sagittocyst Pointed epidermal vesicle provided with an explosive rod for defense in certain acoel Turbellaria.

saprozoic Nutrition by absorbing simple organic materials and salts from the surrounding medium and synthesizing them into protoplasm.

scaphognathite Modified second maxilla of crayfish and lobsters for bailing currents of water over the gills.

schizocoelous formation of coelom Type of mesoderm formation that involves a splitting of mesodermal cords growing from the region of the blastopore.

schizont Intracellular stage in the life history of a sporozoan in which the trophozoite gives rise to merozoites by multiple fission.

sclerite Hard plate forming part of the exoskeleton of arthropods, especially in insects.

scleroprotein Fibrous protein found in connective tissue.

sclerotized Pertaining to specially hardened areas of the arthropod exoskeleton.

scolex Anterior end of a tapeworm; usually provided with holdfast structures.

scutellum Part of the dorsal thoracic exoskeleton in some insects.

scyphistoma Polyp stage of a scyphozoan jellyfish with tetramerous symmetry and a gastrovascular cavity of four divisions.

seminal vesicle Enlarged part of the male reproductive system where sperm are stored; sometimes a fluid is secreted here also.

sensillum Cell or group of cells associated with chitinous structures and functioning as a sense organ.

septum Membranous partition separating two cavities.

seta Type of hairlike projection of the exoskeleton, or integument, of arthropods and annelids.

setosal Bearing bristles.

setule Threadlike bristle.

sinistral Left-handed; in gastropods, when facing the aperture, the coils turn to the left; having the aperture to the left of the shell axis.

sinus Cavity, recess, or depression; an expanded blood vessel or blood cavity.

sinus gland Ductless gland in the eyestalk of certain crustaceans and associated with the X organ; regulates molting and controls the physiology of chromatophores.

siphon Specialized tube of the mantle of certain pelecypods and cephalopods, or in the test of tunicates; for the entrance and exit of water.

siphonoglyph Groovelike, ciliated lip of an anthozoan stomodeum.

solenia Gastrodermal tubes, or continuations of the gastrovascular walls of adjacent polyps, by which the polyps communicate with each other in the colonies of Alcyonaria.

solenocyte Type of flame cell, or bulb, with a single long flagellum.

specialized Adapted by structure or function for a particular mode of life in its evolutionary development as contrasted with its early ancestral condition.

spermaducal gland Small gland associated with the vas deferens of oligochaetes.

spinneret Conical posterior structure bearing openings to silk glands in spiders.

spiracle External opening to the tracheae or book lungs in arthropods.

spiral cleavage Type of cleavage in which the cleavage spindles are diagonal to the polar axis, resulting in successive alternating tiers of cells or a spiral arrangement.

spirocyst A type of nematocyst restricted to the taxon Zoantharia of the Anthozoa.

spongin Fibrous network skeleton of certain sponges, consisting of a sulfur-containing protein.

spore Cell encased in a resistant covering and capable of developing into an organism.

sporocyst Saclike larval stage in many trematodes; usually embedded in the viscera of an intermediate host. In certain protozoans, a sac in which spores are formed.

sporogony Process of spore formation resulting in the production of sporozoites.

sporozoite Last stage in the sexual cycle of Sporozoa.

statoblast In Bryozoa, a bud or germ for resisting winter conditions.

statocyst Sense organ of equilibration; usually consists of a vesicle with lithocytes and a movable statolith.

stercoral pocket Cecum connecting with the rectum in many arachnids and serving as storage place for feces.

stereogastrula Any type of solid gastrula; a parenchymula.

stigma Light-perceiving organelle in green flagellates and some colorless ones; usually contains reddish pigments. A spiracle, or breathing pore, in certain arthropods.

stigmata Breathing pores. Slits in the pharyngeal wall of tunicates.

stolon Horizontal stemlike attachment of certain invertebrates; gives rise to individuals at intervals.

stoma Mouthlike opening, or mouth cavity, in some invertebrates.

stomochord Anterior diverticulum of the buccal cavity in hemichordates; may be considered a type of notochord.

stomodeum In embryology, the ectoderm-lined mouth

and foregut of the enteron; in class Anthozoa of coelenterates and phylum Ctenophora, the ectodermal-lined pharynx.

streptoneury Visceral loop (especially the nervous system) in the form of a figure 8 in certain mollusks.

stridulate Producing sound by rubbing parts of the body together.

strobila Scyphistoma stage of scyphozoan jellyfish made up of a stack of potential jellyfish (ephyrae). The main body of a tapeworm, not including the anterior scolex; the chain of proglottids.

symbiosis Intimate association of two organisms of different species, which may be expressed as commensalism, mutualism, or parasitism.

syncytium Mass of protoplasm or a cell containing several nuclei not separated by cell boundaries.

syngamy Formation of a zygote by the union of male and female gametes; fertilization or conjugation.

synkaryon Fertilization nucleus formed by the union of two pronuclei.

tachyblastic Thin-shelled gastrotrich egg.

tagmosis Union of body segments into functional groups.

tapetum Iridescent pigmented area of the choroid coat and retina of the eye for reflecting the light rays.

tectin Substance similar to chitin (pseudochitin) in the shells of forams.

tegmen Covering of the oral disk of crinoids.

tegument External covering, formerly called the cuticle, in cestodes.

telopodite Segmented appendage of the trilobite trunk; also those segments of an insect limb distal to the coxa.

telotroch Preanal tuft of cilia in certain larval forms (e.g., a trochophore).

telson Posterior projection of the last body segment in some arthropods; also the terminal stinging segment of a scorpion.

tentaculocyst Small club-shaped sense organ along the margin of certain medusae; rhopalium.

tergum Dorsal surface of a body segment in arthropods; may consist of one or more sclerites.

theca Sheath such as the test, or armor, of a protozoan.

thecate Having a hydrotheca around the hydranth in coelenterates.

Tiedemann's bodies Small ceca of the ring canal in many echinoderms and supposed by some authorities to furnish coelomocytes.

torsion Twisting process in certain gastropods that alters the position of the viscera and other structures through 180 degrees from the primitive plan.

toxicyst A type of trichocyst that contains a toxin for paralyzing prey.

tracheae Air tube respiratory system in insects and other arthropods.

trichites Rodlike bodies forming an internal skeleton around the mouth and gullet of certain ciliates; similar to trichocysts, but are not discharged.

trichobothrium Sensory hair in certain arthropods.

trochal disk One of the ciliated disks of which the corona is composed in certain rotifers.

trochophore larva Type of larva common (with some modifications) to several invertebrate groups.

trochus Inner ciliated ring at the anterior end of a rotifer.

trophi Set of sclerotized jaws embedded in the mastax of rotifers.

trophozoite Growing, or vegetative, stage in the protozoan (especially in the Sporoza) life cycle.

urn Minute vase-shaped bodies that collect excretory particles in the coelom of certain Sipunculida and may be voided through the nephridia.

urosome Posterior or abdominal region of an arthropod.

velarium Subumbrella extension in certain scyphozoan medusae, analogous to a true velum.

veliger larva Free-swimming larva that develops from the trochophore in many marine mollusks; has the beginning of a shell, mantle, and foot.

velum Shelflike projection extending inward from the periphery of hydrozoan medusae.

vibraculum Type of modified zooid of whiplike shape that aids in keeping the bryozoan colony free from debris in certain marine species.

X organ Ductless gland closely associated with the sinus gland, the secretions of which prevent or delay molting.

Y organ Ductless gland in the head of crustaceans; apparently accelerates molting.

zoarium Colony of ectoprocts.

zoea One of the early larval stages of marine crabs.

zonite Body segment in Kinorhyncha and millipedes.

zooecium Secreted covering of the individual making up a colony of ectoprocts.

zooid Individual member of a colony of animals.

Index